# Electronic Techniques

Seventh Edition

# Electronic Techniques

## SHOP PRACTICES AND CONSTRUCTION

Robert S. Villanucci
Alexander W. Avtgis
William F. Megow

*Wentworth Institute of Technology*
*Boston, Massachusetts*

Upper Saddle River, New Jersey
Columbus, Ohio

**Library of Congress Cataloging-in-Publication Data**

Villanucci, Robert S.
    Electronic techniques : shop practices and construction / Robert
S. Villanucci, Alexander W. Avtgis, William F. Megow. — 7th ed.
      p. cm.
    ISBN 0–13–019566–9
    1. Electronic apparatus and appliances—Design and construction.
2. Electronic packaging.   3. Electronics.   I. Avtgis, Alexander W.
II. Megow, William F.   III. Title.
TK7836.V55               2002
621.381—dc21                                      00-050195
   CIP

Editor in Chief: Stephen Helba
Assistant Vice President and Publisher: Charles E. Stewart, Jr.
Assistant Editor: Delia K. Uherec
Production Editor: Tricia L. Rawnsley
Design Coordinator: Robin G. Chukes
Cover photo: FPG
Cover Designer: Becky Kulka
Production Manager: Matthew Ottenweller
Electronic Text Management: Karen L. Bretz
Illustrations: Carlisle Communications, Ltd.

This book was set in Times New Roman by Carlisle Communications, Ltd. and was printed and bound by Courier Kendallville, Inc.
The cover was printed by Phoenix Color Corp.

10 9 8 7 6 5 4 3 2 1
ISBN 0-13-019566-9

*This book is dedicated to
Annette, Gloria, and Helen*

# Preface

As has been the case for many years, the electronic packaging industry continues to undergo rapid changes and there appears to be no end in sight to its progress. The adoption of new device packages for high-density assemblies, in conjunction with more functionally complicated printed circuit board technologies, has continued to bring about even more sophisticated design, construction, and packaging techniques that require today's technician to possess special skills. Furthermore, today's packaging technicians are expected to be more familiar than ever with the general theory of operation behind the components and devices with which they are working. Thus, it is the purpose of this seventh edition to continue to present a practical and realistic approach for developing these new skills in planning, designing, and constructing electronic equipment.

Due to the wide acceptance of the first six editions of *Electronic Techniques,* much of the material has been retained. Several changes, mostly additions and updates, have been incorporated into the seventh edition and the *Instructor's Resource Manual,* both to strengthen and to add a new dimension to the material.

### What's New in the Seventh Edition?

1. Material written specifically in order to strengthen the projects in Unit 7. A new Chapter 27 on *Electronic Devices for Student Projects* has been developed to provide the packaging technician with the theory of component and device operation as well as the practical background knowledge necessary to more completely understand and troubleshoot the student projects contained in the last two chapters of the book. While broad in overall range and coverage, the theoretical and practical material presented in Chapter 27 is specific to the devices chosen for student projects that follow and is of sufficient detail so as to be useful. The new chapter provides the following material:

   (a) Theory of operation and useful information on *semiconductor diodes.* Both the *1N914 small signal* and *1N4001 rectifying diodes* are considered in detail. There is also a section that presents general information on common light-emitting diodes, or *LEDs.*

   (b) A section on both the operating theory and the usage of *digital displays* is included in this chapter on electronic devices. The *MAN6760E seven-segment display—*

selected for its high visibility factor in most ambient light conditions—is considered first. Based on LED technology, these displays are chosen when power consumption is not an issue. For applications sensitive to excessive power drain, that is, portable battery-powered devices, a popular *liquid-crystal display, or LCD,* is discussed.

(c) Because some of the projects in Unit 7 use discrete devices, a section on *semiconductor transistors* is provided. Both the *2N3904 NPN* and complement *2N3906 PNP* transistors are included. In this section, the general theory of operation is covered briefly, but applications are restricted to those BJT devices that are used as electronic switches.

(d) Another semiconductor switch included in chapter 27 is the *silicon-controlled rectifier.* Intended for DC applications only, the ubiquitous *C106B1 SCR* is discussed. Information on testing this important device provided.

(e) A very popular and useful integrated circuit, or *IC,* used extensively in the student projects of this book, is the *operational amplifier.* In addition to some general theory of operation regarding this important group of linear ICs, information on their use as both a voltage comparator and as an amplifier is provided. The *LM324, LM358,* and the *OP-07* op amps are used to detail the applications presented in this section.

(f) While not employed extensively in the student project section, some devices that can only be grouped as *digital logic* are used, and therefore included here. Elementary information to form a basic understanding of use for both the *74H07* hex buffer and the *CA4030* exclusive-OR gate is presented in a separate section of the chapter.

(g) Many of the major topics in packaging presented throughout the body of this book—and also in both of the student project chapters—revolve around popular power supply (AC-to-DC converter) circuits. Therefore, a section on *high-power voltage regulator ICs,* which covers both the theory of operation and some circuit design considerations, is presented here. The popular *7805* three-terminal fixed (5-V) regulator and the *LM317* three-terminal adjustable (1.2- to 26.6-V) regulator ICs are covered because these are the devices unique to student projects.

(h) Portable measurement equipment, operated typically from battery power, often requires a stable reference to facilitate a good design. To accomplish this task the circuit designer generally employs a *low-power reference IC,* either voltage or current, depending on the application. The *REF-02,* a precision, fixed 5.00-V stable reference; and the *REF200,* a precision 100-$\mu$A current source; are both discussed. Basic test procedures are presented.

(i) The *NE555 timer* is introduced and both its internal architecture and its external pin configuration considered. This IC, which for some has reached almost universal status, is presented in two applications: a *time-delay, power-on* circuit and a basic *square-wave oscillator.*

(j) Several of the student projects include *sensors* to measure physical parameters such as temperature and pressure. We have therefore included some basic practical information on the use of three sensors: (1) the *UUA41J1 thermistor*—a nonlinear device to measure temperature; (2) the *AD590 current transmitter*—a linear IC used also to measure temperature; and (3) the *SCX-15 ANC sensor*—a linear IC designed to measure absolute pressure.

(k) Information on the general fabrication and the theory of operation of *piezoelectric devices* such as those used in an audible buzzer is included. This technology, used in the microcontroller project of Chapter 29, is a popular solution in such common household products as smoke and gas detectors, as well as burglar alarms.

(l) Finally, we have included some technical information on a group of devices referred to as *integrated circuit converters.* The *7660 switched-capacitor inverter,* used to create a dual-supply from a single DC source (that is, create $-9$ V from $+9$ V), is addressed first. The *ICL7106 analog-to-digital converter* (ADC), a common IC building block in many handheld electronic products requiring a digital display (LCD), is discussed in some detail. Finally, information on the *CA3162E/CA3161E ADC* chip set, similar in functionality to the ICL7106 but chosen when power drain is not a consideration and an LED display is warranted, completes our discussion of student project devices.

2. A new *Section 13.10* has been added to complement the conventional single-sided, double-sided, and multilayer PCB fabrication process begun in Chapter 9 and completed in Chapter 14 of this book. Referred to in the industry as chemical or "wet" processing, the text material now includes information regarding the fabrication of prototype single- and double-sided PCBs using mechanical methods—a printed circuit board prototype fabrication system that employs only drilling, engraving, milling, and routing methods. Designed around a unique software-controlled XY milling/drilling machine specifically tailored to rapidly produce prototype PCBs, this technology is ideally suited for companies and universities that require making prototype PCBs more quickly than a conventional "wet" process and do not wish to incur the expense of establishing a conventional chemical process. Finally, this computer-driven XY milling machine technology can be expanded to produce prototype multilayer (MLB) PCBs. MLBs, however, require the addition of some chemical processing steps.

3. *Section 1.3* of *Chapter 1* has been updated to include the use of a newer version of *PSpice* to simulate the unregulated or "raw" power supply section of the switching-regulator packaging design project. This update reflects the changes in this popular simulation software.

4. New Appendices have been added to the seventh edition (a) to complement the new material presented in Chapter 13, Section 10 regarding fabrication of PCBs by mechanical means and (b) to provide a quick reference to manufacturers' and suppliers' Web sites.

## To the Educator

This text has been designed to fulfill the needs of a typical electronic shop course, manufacturing processes program, or comprehensive package design project. The material is directed toward educational institutions such as technical–vocational schools, technical institutes, and junior colleges, as well as industrial and military training programs.

The training of a skilled craftsperson requires not only clear and detailed explanations, but also visual and graphic aids. For this reason, over 500 figures and drawings have been included to illustrate fundamental techniques used in electronic design, construction, and packaging. In addition, many exercises have been included to help readers develop a thorough understanding of basic concepts.

Although the chapters are grouped in units of similar material, individual chapters may be studied independently. No prerequisite knowledge of electronic circuits is necessary to understand the package design and fabrication techniques presented, although a deeper insight may be realized if the technician has a fundamental background in electronic circuits and devices. Chapter 1 does, however, require some background in electronics for a more complete understanding of material presented therein. To this end, and to help the interested student gain more insight into the projects presented in Chapters 28 and 29, a new Chapter 27 on electronic devices has been added to this edition. See the information under Unit 7 below for more information on this topic.

**Unit 1:** In Chapter 1, the design of a 5-V switching regulator power supply with popular software tools is introduced and used as a teaching vehicle for all the remaining topics in this first unit. Chapter 2 discusses the general factors that must be considered in packaging any electronic system. Chapter 3 reintroduces the switching-regulator project and its packaging design. In Chapter 4, preliminary considerations of the package are converted to detailed engineering drawings and sketches.

**Unit 2:** The four chapters in this unit detail the information necessary to design both single- and double-sided printed circuit boards. Chapter 5 supplies pertinent information on available printed circuit board material to make an informed selection. Chapter 6 covers the logical design process necessary to generate a "sketch" of the component placement and conductor pattern routing for single-sided PCBs. Then, in Chapter 7, using a general purpose CAD drafting package, detailed information on keystrokes and procedures necessary to convert the "sketch" into a conductor pattern artwork is covered. AutoCAD, Release 12, was chosen to illustrate this chapter, but the presentation of PCB design concepts can be adapted to other manufacturers' products. This unit concludes with Chapter 8, which presents the design of both double-sided and four-layer multilayer PCB. In this chapter a software tool developed specifically to design PCB, Eagle 2.6, was selected to illustrate the use of a Schematic Module, Layout Editor, and Autorouter to develop the top and bottom conductor pattern artworks for a simple double-sided PCB. Concepts such as Electrical Rules Check, Design Rules Check (DRC), and RATSNEST are presented.

**Unit 3:** Chapters 9 through 15 provide detailed information on fabricating both single- and double-sided printed circuit boards. Chapter 9, which introduces the photographic process, includes information on diazo film to make positive phototools and covers the use of a densitometer. Chapter 10 supplies detailed information on the fabrication of single-sided printed circuit boards using the print-and-etch method. Chapters 11 through 13 cover the plated-through-hole process, the print-plate-and-etch technique for double-sided board fabrication, and the use of dry-film photo resist. Information on an entirely mechanical PCB prototype fabrication process is included at the end of Chapter 13. Chapter 14 provides basic design and fabrication procedures for multilayer boards (MLBs). Visual inspection and testing procedures for printed circuit boards are presented in Chapter 15, which also includes the use of a microprocessor-based bed-of-nails tester, a Caviderm, and microsectioning techniques.

**Unit 4:** Chapter 16 covers printed circuit board hardware and component assembly, and Chapter 17 includes information on PCB soldering. Topics range from basic hand soldering to industrial processes such as wave soldering. This unit concludes with Chapter 18, which details information on surface-mount technology, including component assembly and soldering. New equipment and techniques, specifically intended for surface-mount component assembly, are introduced and covered in some detail.

**Unit 5:** The layout, fabrication, forming, and finishing of sheet-metal chassis elements are covered in Chapters 19 through 24.

**Unit 6:** Chassis hardware assembly and wiring techniques are covered in Chapters 25 and 26, combining the various elements into a finished system.

**Unit 7:** This unit begins with Chapter 27, which covers the theory of operation for the all of the devices to be encountered in the next two project chapters. While not generally considered necessary information from a strictly packaging point of view, this practical information is useful if one wishes to gain more insight into the total problem of electronic packaging. Chapter 28 contains a selection of simple exercises designed to strengthen the material in each unit. Chapter 29 provides a broad selection of advanced packaging projects to further develop student interest and provide practice in developing the skills discussed in this book.

### Software

In keeping with the industrial trend of using more computer software packages to facilitate electronic designs, enhance quality, and improve productivity by reducing the packaging time cycle, five separate software packages have been presented in this edition. These software packages are: AutoCAD (general purpose drafting software), Eagle (PCB layout software), Switchers Made Simple (dc-to-dc converter design software), PSpice (circuit analysis software), and HP VEE (data acquisition and process control software). Each application package was chosen as a representative example to illustrate the solution to a particular problem presented. Many other products, however, are available from a very large group of suppliers, a group that is expanding rapidly. Please note that the selection was made without regard to price. The prices of the software illustrated in this book range from free (available from the manufacturer) to industrial grade products—large programs that are rather expensive. A list of the names and Web sites of the manufacturers or suppliers for each is provided in Appendix XXVI.

## A Word About Safety

Modern industrial printed circuit board fabrication processes require the use of hazardous chemicals, as described in this book. We cannot emphasize strongly enough that the handling, use, shipping, storage, and disposal of these chemicals must be performed under supervision by people trained in this area and who are familiar with local OSHA and EPA regulations. This is a learning text and, as such, describes in detail the chemical processes required. It, however, does not provide the necessary training or conditions required for the proper handling of these chemicals. We urge that the section on safety concerns and practices (following the Preface) be read carefully. This section is intended to help those interested in finding the necessary information on safety before attempting to work with any chemical or processes. Also the **CAUTIONS** in many chapter introductions should be consulted before attempting certain processes and procedures described throughout the text.

## Acknowledgments

We express our gratitude to Dr. Robert F. Coughlin and Professor Frederick F. Driscoll, our colleagues at Wentworth, for their many helpful comments and suggestions. We thank the staff at T-Tech, Inc., Atlanta, GA, who furnished photographs and technical support. We are also grateful to the many at Contact East, Inc., North Andover, MA, for their technical support. Our many thanks are extended to Kristen Marino for employing her formatting and image-setting skills in reproducing all of the new photos and artworks supplied electronically by manufacturers. Special thanks goes to the reviewers who provided helpful feedback: Leonard Leeper, Front Range Community College; Gilbert Ulibarri Jr., Salt Lake Community College; Garth Fisher, Walla Walla College; and Vincent Kasab, Erie Community College. Finally, we are grateful to our families for their continued support and patience throughout the preparation of the seventh edition of *Electronic Techniques: Shop Practices and Construction*.

*Robert S. Villanucci*
*Alexander W. Avtgis*
*William F. Megow*

# Safety Concerns and Information Sources

Techniques presented in this book involve the use of hand, machine, and power tools, electrical equipment, and printed circuit board processing systems, in addition to hazardous chemicals and materials. The authors, therefore, urge readers intending to perform any of the procedures described to first seek information regarding applicable safety practices. You must be fully aware of safety rules and regulations, precautions to take, and safety equipment to use to prevent putting your health at risk, sustaining personal injury, or causing injury to others working nearby. To help you become safety conscious, concerns for safety appear in the introductory sections of chapters that contain techniques where safety practices must be exercised.

There are many organizations from which to obtain the most current safety practices and procedures or from which approved safety equipment is in compliance with specific conditions, equipment, and materials being used. Three examples of such organizations are the American National Standards Institute (ANSI), the National Institute of Occupational Safety and Health (NIOSH), and the National Fire Protection Association (NFPA). ANSI is a privately funded organization that identifies and coordinates the development of industrial/public, national, consensus standards for safety. Many ANSI standards apply to the safe design and performance of equipment and safe practices and procedures. NIOSH is an agency that tests and certifies respiratory and air sampling devices and recommends exposure limits to the Occupational Safety and Health Administration (OSHA). OSHA is part of the U.S. Department of Labor and is the regulatory and enforcement agency for safety and health in U.S. industrial sectors. NFPA is an organization that promotes fire protection and prevention and safeguards against loss of life and property. NFPA publication 704M is the code for showing hazards of materials on familiar diamond-shaped container labels. The selection and use of equipment such as safety glasses or goggles, full-face shields, ear protectors, dust masks, respirators, and chemical-resistant gloves, clothing, and footwear is primarily based on the item being in compliance with and meeting of ANSI and NIOSH approval/certification.

Other helpful sources of technical safety information are distributors of safety equipment and materials. They will answer questions regarding the most appropriate equipment for your specific needs. Their catalogs also include ANSI and NIOSH certification and approval numbers.

In an environment in which hand, machine, or power tools are used, injuries may be sustained if you are unfamiliar with the correct use of the tool. You must develop respect for even the simplest cutting or forming tools. Very often tool manufacturers include safety information regarding correct operating procedures. It is essential that you put this information to use. For additional safety information, the Government Printing Office may be contacted for listings of available occupational safety literature.

In procedures where hazardous materials or chemicals are used, health risks are of great concern. Inhalation, skin absorption, and ingestion are the principal ways by which hazardous materials and chemicals enter the body. You must be aware of the safeguards required for your protection. Manufacturers are required to supply, on request, comprehensive information about their chemicals and materials that can pose health problems if improperly used. Mandatory requirements for providing Material Safety Data Sheets (MSDS), labeling chemicals, and making safety training available in the workplace are the result of the Occupational Safety and Health Act (OSH ACT), which established the Hazard Communication Rule (HCR). To better understand the type of information provided in an MSDS, a sample form that the manufacturer must prepare for you, on request, is shown in Appendix XII. To further assist you, chemical container labels give the identity of the chemical, who made it, why it is hazardous, what might happen if it is mishandled, and how to protect yourself. It is important that no chemical or material be used before the information provided in the MSDS and on the container label is read, understood, and applied.

When in doubt about any procedure or your safety in the workplace, consult the MSDS provided at the Right-To-Know (RTK) station in your facility (academia or industry). Also know who your designated safety officer is so that you can obtain assistance and safety training if required. Although safety rules and regulations are in place for your protection, it is also always an individual responsibility to be in compliance.

Addresses and phone numbers of several sources of safety information and equipment are provided in Appendix XIII.

Finally, it is important to apply ergonomics in the workplace. Ergonomics is the study of human characteristics for the most appropriate design of both the living and work environments. In simple terms, applying ergonomics is the science of adapting equipment, procedures, and surroundings to adjust the environment to fit people better. Discomforts such as neck and back problems and eye strain can be avoided, as well as other disorders, such as carpal tunnel syndrome, which is caused by performing repetitive motions with the hands over a long period of time. If the causes of discomfort are left unchecked, long-term health problems can develop and permanent damage may occur. Using ergonomics can make your job easier, more pleasant, and safer by adjusting the job to you, reducing physical and mental stress, and preventing illness and injury. As an example, refer to Chapter 7, Section 7.0, for applying ergonomics when using a computer.

# Contents

# 29 Advanced-Level Student Projects   577

# Unit

# ONE

# Circuit Design and Project Planning

# 1

# Designing a 5-Volt Switching-Regulator Power Supply

## LEARNING OBJECTIVES

*Upon completion of this chapter on the design of a 5-volt switching-regulator power supply, the student should be able to*

- Identify the three terminals of a standard household wall receptacle and explain its connection to the service entrance panel, and then to the power company lines.
- Understand the function of the transformer, bridge rectifier, and filter capacitor of an unregulated power supply.
- Use a circuit analysis software package called PSpice to model the unregulated supply, and obtain its performance characteristics.
- Know the differences between an unregulated and a regulated power supply's output.
- Compare and contrast the advantages and disadvantages of linear regulators and switched-mode dc-to-dc converters.
- Use a design software package called *Switchers Made Simple* from National Semiconductor first to choose the correct switching regulator IC and then to define the components of a buck-type converter.

## 1.0 INTRODUCTION

In this first chapter we begin the design of a high-quality, reliable piece of electronic equipment. The system chosen to illustrate electronic design and packaging principles is a switched-mode power supply. When completed, this power supply project will output a regulated 5 volts dc with a ripple voltage of less than 100 millivolts, and deliver a load current of up to 1 ampere.

This project has several desirable characteristics. Its relatively straightforward theory is easy to understand, it incorporates the most current integrated circuit technology available for switched-mode power supply design, and it requires the use of a printed circuit board (PCB) for effective operation. Further, computer-assisted techniques are available to both analyze the unregulated supply section and then design the switched-mode dc-to-dc portion of the project.

Before we begin to examine the numerous and common packaging problems involved in the preliminary planning stage of an electronic system, we will consider the electronic design, analysis, and "breadboarding" phases of the switching power supply circuit. Each phase of this design is considered in detail in the sections to follow.

*CAUTION:* When breadboarding and testing any electrical circuit, be sure that all wiring is completed and checked against the schematic and test lead connections are made before power is applied. Do not attempt to make any changes in the circuit while power is applied ("live"). Before applying power to the circuit or attempting to take any measurements, obtain appropriate instruction by having your work checked by the instructor.

Electrical shocks, whether large or small, can be fatal under certain circumstances. Also, the section on safety, outlined at the beginning of the book, should be reviewed for more specific information.

## 1.1 STANDARD HOUSEHOLD/INDUSTRIAL POWER

The most convenient source of electrical power is a standard household wall outlet, or receptacle, like that shown in Fig. 1.1. In the United States, the available voltage is usually a 115-volt ac sine wave at a power frequency of 60 hertz, and the receptacle is wired in accordance with the National Electric Code.

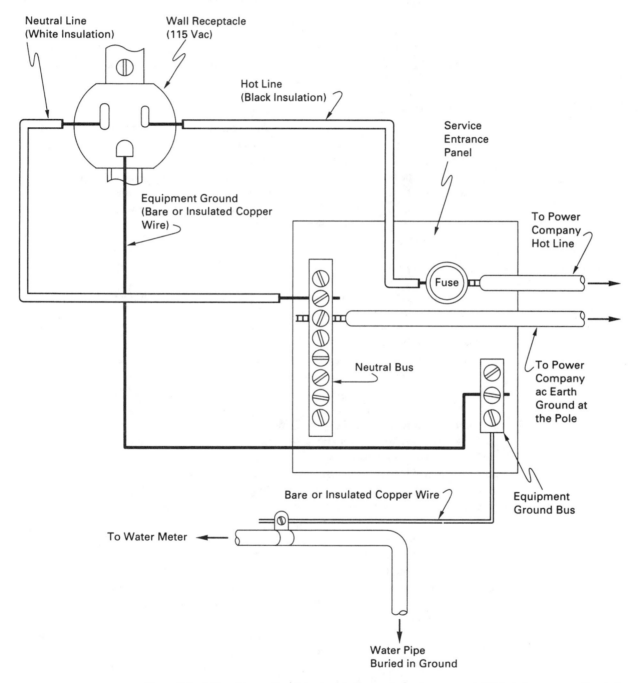

**Figure 1.1** 115-volt ac wall outlet wired to the service entrance panel. Wiring shows grounding techniques and standard connections to power company lines.

The wide slot of the receptacle is called the *neutral,* or *common, connection.* It is routed to the neutral bus at the service entrance panel with a white insulated wire. The neutral bus is then wired to the power company's ac earth ground at the power pole.

The narrow slot of the receptacle is the *hot connection* and is wired to a fuse in the service entrance panel with a black (or red, blue, etc.) insulated wire. This fused hot wire is then connected to the power company's 115-volt ac hot wire. Thus, the hot wire is at 115-volts ac with respect to the neutral wire. The hot line is fused at the service panel with either thermally activated fuses or circuit breakers. If American Wire Gauge (AWG) No. 14 wire is used at the receptacle, the line is fused for 15-A operation, and a 20-ampere fuse is used for AWG No. 12 wire.

Finally, the U-shaped slot of the receptacle is connected to earth ground. The connection at the wall receptacle is made with either a bare copper wire, green insulated wire, or using the metal shield or conduit pipe of the household service. At the service panel the connection to ground is made with a bare copper wire connected to a water pipe buried in the ground. The U-shaped slot of the system is known by many names: *safety, equipment, chassis, utility, third wire,* or *U-ground.*

In a good power distribution system, the U-shaped slot should be at zero volts with respect to earth or ground. However, it is not uncommon to measure a several-volt potential between the neutral wire and this third wire ground. This is due to return currents flowing through the wire resistance of the neutral line. *CAUTION:* It is both dangerous and often unnecessary to measure the voltage between the hot and neutral slots directly. Before attempting this measurement, obtain appropriate instruction from your instructor.

Since most modern electronic equipment runs on low-voltage dc, an ac-to-dc converter is needed to convert the 115-volt ac household power at the wall receptacle into a lower dc voltage to power the equipment. The first stage in this conversion process is usually an unregulated power supply.

## 1.2 THE UNREGULATED POWER SUPPLY

Figure 1.2 illustrates a typical unregulated supply. When plugged into a household receptacle this "raw" supply will convert the ac input voltage into a low-level dc output voltage. The main components in Fig. 1.2 are a step-down transformer ($T_1$), four rectifying diodes ($D_1$ to $D_4$), and a filter capacitor ($C_1$). While there are other support components (fuse, on/off switch, indicator lamp, etc.) needed to complete this subsystem, only the main components are discussed in detail here.

**Step-Down Transformer**   The step-down transformer consists of a primary winding connected to the household power and a secondary winding connected to the diode rectifier circuit. Since both windings are wound on a laminated steel core with insulated wire, the transformer provides electrical isolation between household power and the output of the unregulated supply. Furthermore, the transformer is manufactured with approximately 10 times more primary winding turns than secondary winding turns. With 115-volts ac applied to the primary, the secondary winding output is therefore reduced to about 12.6-volts ac (12.6-volts rms), hence the name *step-down transformer.*

The transformer chosen for this project is referred to as a filament transformer and has a nameplate rating of 12.6-volts ac at 2 amperes. By these specifications the manufacturer guarantees that the secondary winding output ($V_{sec}$) will not fall below the rated 12.6-volts ac when the secondary is delivering its rated load current of 2 amperes rms. To guarantee this output level and overcome winding resistance at full load, the transformer manufacturer winds less than the theoretical 10:1 turns ratio on the core. Therefore, at no load, when the secondary winding is delivering very low load currents, the secondary output will be above the rated 12.6-volts ac, dropping to its rated value when the rated load current of 2 amperes is drawn from the secondary winding.

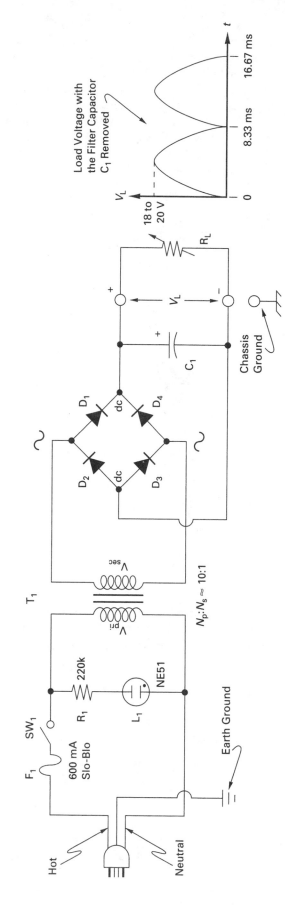

**Figure 1.2** An unregulated power supply, with four bridge rectifier diodes, outputs pulsating dc when filter is removed.

**Diode Bridge Rectifier**    The four diodes ($D_1$ to $D_4$) shown in Fig. 1.2 are wired in a configuration called a full-wave bridge rectifier. The transformer's secondary winding is connected to opposite corners of the diode bridge. These are referred to as the ac connections. Note the $\sim$ symbol in Fig. 1.2. The filter capacitor and load are connected to the remaining corners of the bridge, labeled the dc connections. Note also the dc symbol in Fig. 1.2. This circuit will convert the transformer's secondary ac voltage, $V_{sec}$, into a pulsating dc output as shown.

To understand how the diode bridge functions as a rectifier, assume that the filter capacitor is removed and replaced with just a load resistor ($R_L$). Then assume that the transformer secondary is going through its positive half-cycle. When the $V_{sec}$ exceeds two diode drops (about 1.2 volts), diodes $D_1$ and $D_3$ begin conducting current through the load. The output polarity is shown and the load voltage raises to about 1.2 volts less than the peak secondary voltage.

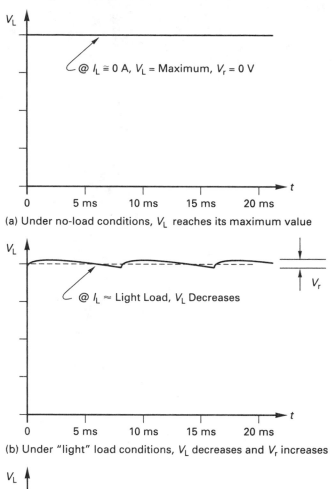

(a) Under no-load conditions, $V_L$ reaches its maximum value

(b) Under "light" load conditions, $V_L$ decreases and $V_r$ increases

**Figure 1.3**   An unregulated supply exhibits a decrease in dc load voltage as $I_L$ increases. Also ripple voltage increases as $I_L$ increases.

(c) Under full-load conditions, $V_L$ reaches its minimum value

During the negative half-cycle, the polarity of the transformer's secondary voltage reverses, diodes $D_2$ and $D_4$ begin conducting, and current is routed through the load in the *same direction* as $D_1$ and $D_3$ during the positive half-cycle. The result is a pulsating dc output voltage. Under light loads, when the load current ($I_L$) is low, the peak output voltage will be maximum, dropping in value as $I_L$ increases until, at a maximum load of 2 amperes, the peak output voltage will drop to about 18 to 20 volts.

**Filter Capacitor**   When a filter capacitor ($C_1$) is connected across the load, the output voltage more closely resembles the desired constant value voltage. During each half-cycle, when the secondary voltage is above the load voltage, the transformer furnishes current to the load and charges the filter capacitor. When the transformer's secondary voltage drops below the load voltage, the capacitor furnishes load current in an attempt to maintain the output voltage constant. The transformer is "disconnected" from the load by the rectifying diodes during this time.

Under no-load conditions, when $I_L$ is 0 amperes, the filter capacitor charges to its maximum level, and the output is shown in Fig. 1.3a as a constant dc voltage. When a load is connected, that is, when $I_L$ increases, the average or dc output voltage decreases due primarily to transformer-winding resistance losses. In addition, since the capacitor must furnish load current by discharging through the load, a small component of ripple voltage ($V_r$) will be present at the output (Fig. 1.3b). Under maximum-load conditions, when $I_L$ equals 2 amperes for the transformer specified, the dc output will decrease to its minimum value and the ripple voltage will increase to its maximum level (Fig. 1.3c).

With a basic understanding of how the unregulated section of our power supply works, we turn to a circuit analysis software package called PSpice. PSpice will be used to analyze the performance of the unregulated section of our power supply.

## 1.3 CIRCUIT ANALYSIS USING PSPICE

Before we can design the switching regulator portion of this project, it will be necessary to determine several parameters of the unregulated supply section. Specifically, we must determine the raw supply's dc output voltage at both no load ($I_L = 0$ A) and at full load ($I_L = 1$ A). Also useful will be the worst-case ripple voltage when the load current equals 1 A.

Until recently it would have been necessary to breadboard the components of Fig. 1.2 on the bench, connect an adjustable load resistor $R_L$ across the filter capacitor, and set it to draw the necessary maximum-load current. We would then measure the average output voltage across $R_L$ with a dc-reading voltmeter. Ripple voltage $V_r$ could be found with an ac-coupled oscilloscope connected at the output. The measurements would be repeated at no-load (or a very light load, i.e., about 1% of the maximum) condition.

Today, breadboarding a circuit and analyzing its characteristics can be done more conveniently using a personal computer and one of the several commercially available circuit analysis software packages. To illustrate this new method of "breadboarding," the authors have chosen to use the PSpice software to analyze the performance of the unregulated supply. PSpice® is a registered trademark of Cadence Corporation.

If PSpice is used, one begins an analysis of this type by first creating a circuit schematic. Upon program startup, a blank *schematic page* will be displayed to which parts and wiring can be added to create the desired circuit. Parts can be called by first clicking on the toolbar's *Draw* menu and then clicking *Get New Part*. A full list of parts is available to the user, grouped by common function and alphabetized. You select the needed part by first highlighting it with a left mouse click. Clicking *Place* locates the device symbol on the schematic page where it can be moved with the mouse. Once at the desired location a left mouse click anchors the first symbol on the schematic page. If duplicate parts are needed—as in the case of the four diodes used in Fig. 1.4—a second device symbol (or third or fourth) of the same part style will automatically be made visible. This new symbol can

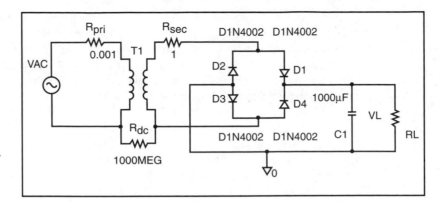

**Figure 1.4** The transient analysis of the unregulated supply is achieved by running a PSpice analysis.

be placed on the schematic page as before. A right mouse click will exit the *Place* routine. The PSpice part identifications for the components shown in Fig. 1.4 are listed below.

$V_{AC} \rightarrow$ VSIN

T1 $\rightarrow$ XFRM_LINEAR

D1 – D4 $\rightarrow$ D1N4002

All resistors $\rightarrow$ R

C1 $\rightarrow$ C

Ground symbol $\rightarrow$ AGND

Once all the parts are located and placed on the schematic page they are wired electrically by first selecting *Wire* from the toolbar's *Draw* menu. The *pencil* icon that appears is then used—very much like an actual pencil—to make the electrical connections.

## 1.4 PSPICE SIMULATION OF AN UNREGULATED SUPPLY

Figure 1.4 is a redrawn circuit schematic of the raw supply. Note that ac voltage source $V_{AC}$ is being used here to duplicate the wall receptacle output voltage—115 $V_{AC}$ @ 60 Hz. To configure this part properly, double-click the left mouse button on the $V_{AC}$ symbol to display its *dialog box*. Three attributes need to be set for this component. The first is the voltage amplitude. To set this parameter, first highlight the text *VAMPL=*, then move the cursor to the *Value* box and enter the desired value. In this case we want to set VAMPL equal to 170 V, the peak value of Edison. Clicking once on the *Save Attr* button assigns this value to $V_{AC}$. Follow this same procedure to set the frequency (*FREQ =*) to 60 Hz and the offset voltage (*VOFF =*) to O V. The updated part will appear on the schematic page when you click the *OK* button. Note that this voltage source drives the primary winding of the transformer $T_1$.

You can also change the *reference designator label* for any part—that is, $V_1$ to $V_{AC}$ or $R_1$ to $R_{pri}$—by first highlighting the *reference designator label* and then typing in the desired label in the *dialog box* that opens for each part.

The step-down transformer $T_1$ is being modeled as two inductors coupled together magnetically. The primary and secondary winding resistances of this part are also being shown, external to the device, as part of the transformer model. The primary and secondary winding inductances needed to set the attributes of this part, are in units of henrys, and are obtained by measurement. The values for the transformer selected by the authors are given below.

L1_VALUE = 4 H

L2_VALUE = 0.06 H

Primary winding resistance $R_{pri}$ is included since PSpice will not allow a voltage source like $V_{AC}$ to be connected directly across an inductor. The 1-milliohm (0.001 $\Omega$) value for $R_{pri}$

was chosen so that it would not have an adverse effect on circuit performance. Secondary winding resistance $R_{sec}$ is set to 1 $\Omega$, a typical measured value for a good quality filament transformer. Finally, this transformer's coefficient of coupling (*COUPLING*) is set to 1, the coupling value of a lossless transformer.

One requirement of PSpice is that each node must have a dc path to ground (zero node) created by the AGND symbol. This is necessary so that the analysis software can find a small-signal bias solution. Resistor $R_{dc}$ is added to the circuit schematic so that the primary winding has a path to ground, and satisfies this software's requirement. This component's value (attribute) is set to 1000 M$\Omega$ (recorded as 1000MEG in PSpice), resulting in negligible effect on circuit performance.

The bridge rectifier consists of four silicon-rectifying diodes represented in PSpice by four D1N4002 devices. Note that in order to neatly represent these components the orientation of the anode and cathode terminals for $D_1$ and $D_3$ must be opposite in orientation to diodes $D_2$ and $D_4$. This is readily accomplished when initially calling each part from the software library. With the part highlighted (displayed in red on the schematic), a counterclockwise (CCW) rotation of 90° can be made by depressing the computer's control key simultaneously with the R key—*Ctrl+R*. A 180° CCW rotation is accomplished by striking the Ctrl+R keys combination twice.

An electrolytic capacitor $C_1$, with a typical value of 1000 μF, is added for output filtering and completes the "raw" supply circuit. The analysis of this supply will be performed twice: once under no-load conditions, when $I_L = 0$ A, and then a second time under full-load conditions, when $I_L = 1$ A. The load resistor is place in parallel with the filter capacitor $C_1$. To ensure that the forward-biased bridge rectifier diodes are always conducting a current, the no-load condition is analyzed with a "very light load" by setting $R_L$ to 10,000 $\Omega$. $I_L$ should be only a few milliamps under this load condition. Resetting $R_L$ to 15 $\Omega$ and again analyzing the circuit, can test for full-load conditions.

Power supply circuit performance is accomplished by configuring the software to perform a *transient analysis*. To configure PSpice for this type of solution, click the *Analysis* button on the software's toolbar and then click *Setup. Enable* the transient function—by clicking the left mouse button in the appropriate box—and then open it by clicking *Transient.* The dialog box that appears allows you to set the total time for the analysis (*Final Time*) and the number of times this analysis will be performed (*Print Step*). For our problem we want to analyze the circuit's performance every 0.1 ms for a total of 34 ms or about two complete cycles of 60 Hz input. Therefore, *Final Time* = 34 ms and *Print Step* = 0.01 ms should be set in the appropriate locations and both the *Transient* and *Analysis Setup* dialog boxes should be closed to complete the transient analysis setup.

## 1.5 LOADING PSPICE ONTO THE COMPUTER

The authors used an IBM-compatible computer for all their PSpice development work. The first task is to load the analysis software onto the computer. To load the evaluation version of PSpice onto the hard drive (C:), a subdirectory called SPICE is first created. This subdirectory is created with the following keystrokes:

```
C:\>MD SPICE<CR>
```

Note that the underscored information is the keystrokes you must enter after the DOS prompt, and <CR> indicates either carriage return or enter.

The following is then keyed in to change to the newly created SPICE subdirectory:

```
C:\>CD SPICE<CR>
```

Next load the first 5.25-inch PSpice floppy disk into drive A:, and copy the files onto the C: drive by the following keystrokes:

```
C:\SPICE>COPY A:*.* C:\SPICE<CR>
```

*Designing a 5-Volt Switching-Regulator Power Supply*

Repeat this last step for each floppy disk supplied with the evaluation version of PSpice.

The circuit analysis program is now loaded onto the hard drive of your computer. In the next section we use this program to analyze the unregulated supply.

## 1.6 RUNNING A CIRCUIT ANALYSIS

The authors stored the PSpice schematic shown in Fig. 1.4 in circuit file Power_Supply.sch. To run an analysis on this circuit file, click *Analysis* and then *Simulate*, or simply depress the *F11* key. The software will perform the analysis, and then display a blank graph labeled *Probe*. PROBE is a graphics postprocessor that is available to display the results of a simulation. Since we want to display a trace of the output voltage, click first on the toolbar's *Trace* menu item and then click *Add*. An *Add Traces* box appears identifying a list of voltages and currents that can be graphed as a function of time. The voltages in this list are referenced to the 0 node and the currents are as indicated.

From Fig. 1.4 we note that the load voltage is across $R_L$. Select *V(RL:1)* by first highlighting it with the mouse pointer and then pressing the left mouse key. A hard-copy result is shown in [Fig. 1.5a]. Since the analysis was done at no-load conditions ($R_L = 10$ k$\Omega$),

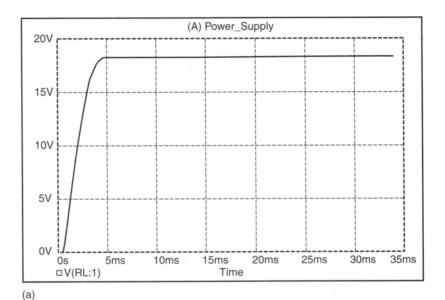

(a)

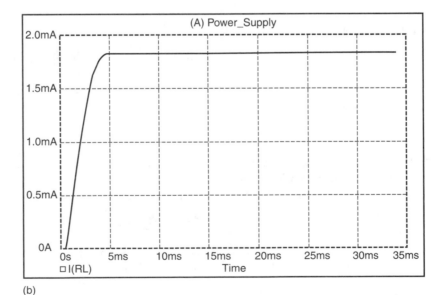

**Figure 1.5** PSpice hard-copy output for $R_L = 10$ k$\Omega$: (a) Graph of load voltage ($V_L$) vs. time ($t$); (b) Graph of load current ($I_L$) vs. time ($t$).

(b)

we learn that the raw supply will have a maximum average dc output voltage of about 19 V with essentially no ripple ($V_r = 0$ V) under these conditions.

Figure 1.5b shows the PROBE output for $I_L$ versus time ($t$) for the same no-load condition. This graph was generated using the steps outlined above except $I(RL)$ was selected for display. Under these no-load conditions, $I_L$ does not exceed about 2 mA.

To obtain the full-load graph, double-click the load resistor symbol (or value, i.e., 10 kΩ) and modify the component value to 15 Ω. Again using the analysis procedure detailed in this section, we generate the data shown in Fig. 1.6. Under full-load conditions, note that $V_L$, the maximum average dc load voltage, drops to about 15 V, and the peak-to-peak ripple voltage $V_r$ increases to a peak-to-peak value of about 5 to 6 V. Figure 1.6b further demonstrates that the supply is delivering an average dc load current of about 1 A. This is the unregulated power supply data that is needed to design the switching regulator circuit. Switching regulator design will be shown in Section 1.10.

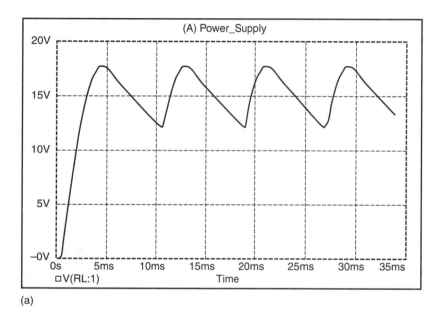

(a)

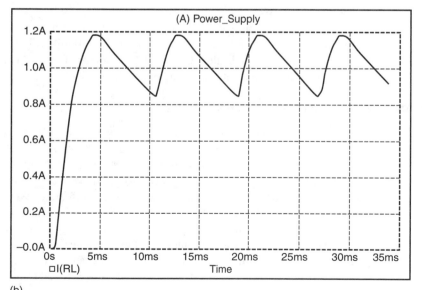

**Figure 1.6** PSpice hard-copy output for $R_L = 15$ Ω: (a) Graph of load voltage ($V_L$) vs. time ($t$); (b) Graph of load current ($I_L$) vs. time ($t$).

(b)

*Designing a 5-Volt Switching-Regulator Power Supply*     **11**

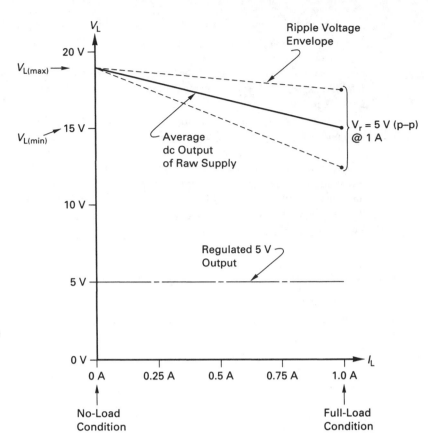

**Figure 1.7** The average dc output of an unregulated supply varies from a no-load value of 19 volts to a full-load value of 15 volts. As $I_L$ increases, $V_r$ also increases.

## 1.7 THE NEED FOR A REGULATOR

The data developed using PSpice to analyze the performance of the unregulated section of the project are plotted in [Fig. 1.7]. This plot, referred to as a *load regulation curve,* characterizes circuit performance. The solid line reveals that the dc output voltage varies drastically from 19 to 15 volts as load current increases from approximately 0 to 1 ampere. Further, the ripple voltage, shown as a pair of enveloping dashed lines, increases from 0 at no load to about 5 volts peak-to-peak under full-load conditions. The load regulation curve dramatically illustrates that load voltage is not constant, and that the ripple voltage, at full load, is excessive.

To obtain an output, constant at 5 volts dc, and reduce the ripple voltage of the raw supply to less than 2% of its dc output voltage (<100 millivolts), a regulator is needed. In the next section we look at the two available regulator types and compare and contrast the advantages and disadvantages of each.

## 1.8 LINEAR AND SWITCHING REGULATORS COMPARED

The output of a raw supply like the circuit shown in Fig. 1.2 can be regulated by choosing either a linear regulator or a switched-mode dc-to-dc converter. Three-terminal linear regulator ICs have been used widely since the early 1970s, gaining immediate acceptance and becoming a spectacular success as a low-cost solution to the problem of regulation. They are available with both positive and negative outputs with either fixed or adjustable voltages. Since all the engineering has been completed by the device manufacturer and packaged into a simple three-terminal case, they are essentially a "drop-in" solution, often requiring only the design of a heat sink and the addition of bypassing capacitors to become completely useful. While three-terminal linear regulators do not rival discrete designs, they do have good specifications for line and load regulation, output ripple, and transient response.

Linear regulators, however, have disadvantages that must be addressed if an informed regulator selection is to be made. First among these is the problem of efficiency. As shown in Fig. 1.8a, the pass transistor of the linear regulator appears in series with the load and is between the output of the raw supply and the load resistor. The pass transistor operates continuously in its linear region, always conducting load current. Under worst-case conditions the raw supply delivers 15 V dc and 1 ampere. Since the input power is $P_{in} = 15$ V (1 A) $= 15$ W and the output power is $P_{out} = 5$ V (1 A) $= 5$ W, the efficiency of the linear regulator is

$$\% \text{ Efficiency} = (P_{out}/P_{in})\ 100\% = (5\ \text{W}/15\ \text{W})\ 100\% = 33.3\%$$

In other words, the linear regulator package must waste

$$P_{reg} = (15\ \text{V} - 5\ \text{V})\ (1\ \text{A}) = 10\ \text{W}$$

This translates into heat, and a corresponding increase in packaging size (heat sink) and cost. In effect the linear regulator is wasting 66.6% of the input power to perform the task of regulation.

Another drawback is that linear regulators are step-down devices only, requiring a larger input voltage from the raw supply that it will output. Step-up is not possible in linear regulator technology.

While not quite a drop-in solution, switching regulator designs have advanced tremendously in the last few years. This is due primarily to the demand for high-efficiency regulators in battery-powered equipment, such as laptop computers and cellular telephones, to name just two. Wasting 66.6% of a battery's power on regulation is unacceptable in portable applications. Also switched-mode dc-to-dc converters or switching regulators lend

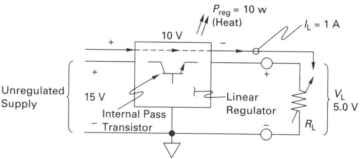

(a) The pass transistor of linear regulator is always on, conducting current, and dissipating power continuously

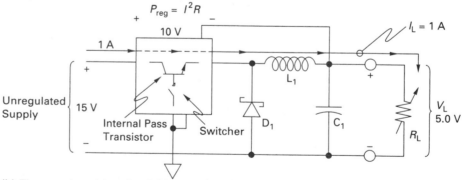

(b) The pass transistor of switching regulator is switched on and off, dissipating power only when it is on

**Figure 1.8** The efficiency of a switcher can approach 90%.

themselves to step-down (buck), step-up (boost), inverting (buck–boost), and dual-output (flyback) modes of operation.

However, the main reason for selecting a switching regulator is its inherent efficiency. As with the linear regulator, the pass transistor in a switcher is in series with the load (Fig. 1.8b). Again, under worst-case load conditions, when the raw supply is delivering 15 volts at 1 ampere, we evaluate the performance. The pass transistor in a switching regulator is never operated in its linear region, but switched on (into saturation) or off (into cutoff). Since the on resistance of a transistor is quite small (fractions of an ohm in large power applications), the power wasted in the pass element is low even at high currents ($P_{reg} = I^2R$). When the pass element is switched off, the current is low (approaching 0 A), and again the power dissipated in the regulator is low. Switching regulators can achieve efficiencies in excess of 90%, wasting less than 10% of the input power on regulation.

The major disadvantage of selecting a switching regulator is the larger values of output ripple voltage $V_r$. It is not uncommon for a switching regulator to have an output ripple component in excess of 2% of its regulated dc output voltage. Further, this ripple component is at the switching frequency of the IC and can vary from 40 kilohertz to 1 megahertz, depending on the device selected. High-frequency, high-energy spikes are difficult to remove from the output of a switcher.

What makes the selection of a switching regulator an attractive alternative today is that some manufacturers now supply their devices with a software design and analysis package that makes the task of selecting the correct IC switcher, and then designing the filter section, both simple and straightforward. Gone are the days when the design of these nonlinear circuits was best left to a small coterie of engineers who specialized in this area.

We introduce just such a design solution in the sections to follow.

## 1.9 NATIONAL SEMICONDUCTOR'S *SWITCHERS MADE SIMPLE* SOFTWARE

This section introduces a software package from National Semiconductor for the computer-aided design (CAD) of the Simple Switcher-based dc-to-dc converter. It was developed to aid the designer working with National Semiconductor's LM2575, LM2576, and LM2577 switching regulators. The software runs on an IBM-compatible computer with MS-DOS version 2.0 or higher and requires 512K of memory.

The first task is to load the software onto the computer. At the DOS prompt, switch to the A: drive by typing the following:

```
C:\>A:<CR>
```

Load the 5.25-inch floppy disk *Switchers Made Simple,* available free from the manufacturer, into drive A: and then type

```
A:\>SIMPLE<CR>
```

To get to the main menu from the displayed company log, press any key twice. The main menu is shown in Fig. 1.9. A review of the information presented shows that this software package supports four types of switching supplies: boost, flyback, buck, and buck–boost, which is a mode of operation that generates a negative output from a positive input voltage without the use of an isolation transformer.

Before using the software to design a switching regulator, you can view the help file by pressing <F1>. Key in the information within the <> brackets to access this information file. The supplemental information in this README.DOC file is quite extensive and useful. You use the <PgDn> and <PgUp> keys to scroll through the file and press <END> to exit back to the main menu.

Design boost(1), flyback(2), buck(3) or buckboost(4) converter? (1/2/3/4)

This program supports four types of power supplies.

Boost:
Used to step up the input voltage, e.g. Vin = 5V, Vout = 12V

Flyback:
Used for multiple output voltages, positive or negative, with the possibility of isolation. Both step up and step down are possible. High output voltages may be achieved. A transformer is required instead of an inductor.
E.g. Vin = 5V, Vout1 = 15V, Vout2 = –15V, or
Vin = 5V, Vout1 = 15V, Vout2 = 12V, or
Vin = 20V, Vout = 100V

Buck:
Used for stepping down a voltage, e.g. Vin = 10V, Vout = 5V

Buck-Boost:
Used for generating a negative voltage from a positive one without isolation, e.g. Vin = 5V, Vout = –5V

To view the helpfile, PRESS <F1> NOW.
To exit this program at any time, press <ESC>

**Figure 1.9** The main menu of *Switchers Made Simple* provides the user with the option of designing four different types of switching regulators.

## 1.10 DESIGNING A SWITCHING REGULATOR FOR A 5.0-VOLT, 1.0-AMPERE OPERATION

Again refer to the main menu shown in Fig. 1.9. Since our unregulated supply outputs a voltage larger than the output of the switcher, press <3> to select a buck mode design. You are then prompted for information, some related to the unregulated supply previously analyzed with PSpice, and some information about the ambient temperature extremes and output design requirements. The following data were entered for our design using the arrow keys to move down the list and <CR> to enter the data:

$$\text{Vinmin} = 15.00 \text{ V}$$
$$\text{Vinmax} = 19.00 \text{ V}$$
$$\text{Tamax} = 40.00 \text{ C}$$
$$\text{Tamin} = 10.00 \text{ C}$$
$$\text{Vout} = 5.00 \text{ V}$$
$$\text{Ilmax} = 1.00 \text{ A}$$
$$\text{Diode} = \text{Schottky}$$

The first two entries are from the PSpice analysis of the unregulated supply, and the ambient temperature extremes that follow are typical operating environment values. The regulator's output is set with the data entered for Vout and Ilmax. Finally, select Schottky for the catch diode.

Press <End> to load the information and answer yes <Y> to the question *Use standard inductor?* While a custom inductor could be wound on a magnetic core, we will choose a standard off-the-shelf part. At this point the software will select the appropriate regulator IC and design the input and output filter component values. Figure 1.10 is a display of the hard-copy output information. We will return to this information, but for now note that all the component values are presented and that the design predicts a ripple voltage less than 82 millivolts, well within the 2% of $V_{out}$ required for this design.

--------------------------------------------------------------------

### Circuit Parameters

```
            Vinmin:    15.00 V
            Vinmax:    19.00 V
             Tamax:    40.00 C
             Tamin:    10.00 C
              Vout:     5.00 V
             Ilmax:     1.00 A
             Diode: Schottky
```

### Misc Calculated Information

```
                Mode: Continuous
  Peak switch current:     1.15 A
               ESRmax:     0.13 Ohms
               ESRmin:    65.27 mOhms
              Vripple:    81.49 mV
        Crossover Freq:     8.25
         Phase margin:    31.89 Deg
         Junction Temp:    72.16 C
```

### Component List

```
Cout:330.00 uF
                       ESR:     0.10 Ohm
                      Vmax:    25.00 V
           ECE-B1EF331   : Panasonic
Cin:47.00 uF
                      Vmax:    27.00 V
L:330.00 uH
                      Imax:     3.45 A
               430-0635: AIE
               PE-53117: Pulse
                RL2447: Renco
D1:3.00 A
                      Vmax:    20.00 V
               1N5820: Motorola
              MBR320P: Motorola
U1:
         LM2575T-5: National Semiconductor
```

**Figure 1.10** The hard-copy output of our design with all values specified.

Next you are prompted to answer several questions. Each is listed below with the authors' answers.

Change any component value? <u>N<CR></u>

Perform thermal check? <u>Y<CR></u>

Will you use a K(TO-3), T(TO-220), M(SO24), N(16 L N)? <u>T<CR></u>

Note that by selecting a TO-220 plastic package, the internal junction temperature of the switching regulator will rise to only about 72.2°C and no heat sink will be required. The

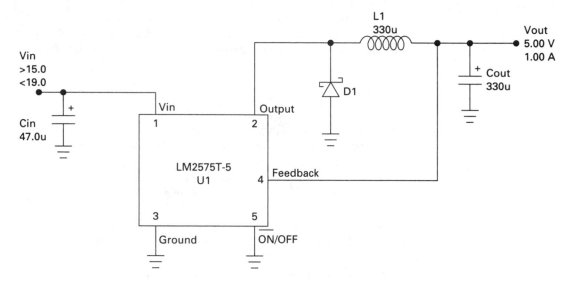

**Figure 1.11**  Hard-copy printout of the buck-type switching regulator designed by the *Switchers Made Simple* software.

software's ability to do a thermal check, prior to fabrication, greatly simplifies the overall design. Next the software prompts for the following information:

Will a heat sink be used? N<CR>

Modify this design? N<CR>

Would you like to save this design? N<CR>

We answered no to the last question because we will get a hard-copy output from the software.

After all questions are answered, pressing any key will instruct the software to draw the switching regulator with accompanying components identified. Figure 1.11 is the complete circuit schematic displayed for this switcher's design. If a printer is available, a hard-copy output can be obtained. After the schematic is displayed, the software prompts you for a parts list, available by pressing any key. The printout is shown in Fig. 1.10.

## 1.11 COMPONENTS FOR THE BUCK CONVERTER

Return to the hard-copy output of the *Switchers Made Simple* software shown in Fig. 1.10 and the circuit schematic shown in Fig. 1.11. The first subheading, *Circuit Parameters,* is a printout of both input and output specifications for the design. It is followed by the subheading *Misc calculated information,* where pertinent operating information about the design is tabulated. Of particular note is *calculated ripple voltage* (Vripple). At a specified 81.49 mV, this value is below the maximum 2% of $V_L$ required by the design. Note that $V_L = V_{out}$.

The final subheading, *Component List,* itemizes the switching regulator components. These five elements, which combine to create a switched-mode dc-to-dc converter, are the (1) input filter, Cin; (2) controller, U1; (3) inductor or choke, L; (4) output filter capacitor, Cout; and (5) a "catch" diode, D1. The selection of each item was made by the software and is considered in some detail here. General information on topics such as component size, construction material, and device configurations is given. The prime objective is to acquaint the reader with the kinds of components that are suitable for use in switching regulators along with those that are not acceptable.

**Input Filter (Cin)**   The input filter is not critical and a standard 47-microfarad aluminum electrolytic, with a WVDC rating of 35 volts, is chosen. During the package design phase, however, we have to ensure that the lead length of this device is kept as short as possible.

**Controller (U1)**   U1 is chosen to be National Semiconductor's LM2575T-5 switcher. Packaged in a TO-220 case, the switcher develops a fixed output of 5.0 volts dc. This low-power switching regulator employs a principle called pulse-width modulation (PWM). In this PWM IC, a constant-frequency square wave oscillator regulates output voltage by increasing switch on time when $V_{out}$ drops below a preset value, and decreases switch on time when $V_{out}$ goes above the same preset value. The output voltage $V_{out}$ is sensed by the feedback pin of the IC. Switching frequency is normally set at 52 kilohertz.

PWM controllers do have larger output ripple voltages than other types of controllers, and the PWM controllers draw more quiescent currents than other types, but they are far less critical to circuit performance and the coil's inductance value. One necessity of all switching regulators is that they be supplied with a low-inductance, low-impedance path to ground. This precludes the use of conventional wiring techniques and requires a PCB ground plane for effective operation.

**Inductor (L)**   The inductor or coil used in a switch-mode regulator functions primarily as an energy-conversion element. It stores energy in its powdered-iron (or Mo-Permalloy) core material, in the form of a magnetic field, during the charging phase and then delivers the stored energy to the load during the discharge phase. Selecting an inductor value often includes many considerations. First among them might be maximizing output current capacity while minimizing physical size to optimize a design. This type of optimization, however, often requires nonstandard coils, which is not our intention. We will adopt a more practical approach using only off-the-shelf components and accepting the value of inductor chosen by the software.

In general, larger values of inductance provide for a larger output capacity (due to higher core storage capacity) and lower ripple currents. Lower inductor values have the effect of reducing maximum output capacity and raising ripple current but are physically smaller. We chose the PE-53117 coil from Pulse Engineering, with a 330-microhenry inductance value, for our design.

**Capacitor (Cout)**   The main purpose of the output capacitor is to reduce the output ripple voltage. Both value and component type are important when selecting a filter capacitor for switch-mode regulators. Typically, capacitor values will run the gamut from 10 to 10,000 microfarads for this type of application, with the larger valued components achieving the lower ripple voltages. To understand the reason for this, note that the model of a typical filter capacitor includes capacitance (*C*), effective series inductance (ESI), and equivalent series resistance (ESR). In switching regulator designs, the value of capacitance *C* and ESR both contribute to total ripple, but ESR is often the most important factor. Lowest ripples are achieved with large values of *C* and low ESRs. Since the value of ESR is the major contributor of ripple in a switching regulator, every effort should be made to reduce it.

ESR is directly proportional to capacitor volume in cubic inches, not the value of capacitance in microfarads. All else being the same, the ESR of a larger value filter capacitor is usually smaller than that for a smaller value capacitor of the same physical construction. Standard aluminum capacitors have an ESR in the range of 0.05 to 0.9 ohms for capacitor values between 100 and 1000 microfarads. One way to reduce this relatively high value of ESR is to parallel capacitors. This has the dual effect of reducing resistance and increasing effective capacitance. The downside of this approach is increased pc board space. A better method is to choose one of the high-frequency or low-ESR parts with an effective series resistance of less than 0.1 ohm as specified by the software. Several manufacturers, such as Cornell Dubilier, Mallory, Panasonic, and Sprague, make capacitors for switching regulators. The Panasonic ECE-B1EF331, 330-microfarad capacitor was selected for this project.

When selecting a capacitor, avoid Tantalum dielectrics. With ESR values below 0.05 ohms they can cause instability and unwanted oscillations. Finally, the capacitors best suited for switched-mode applications are those that have extended-foil construction (for low ESR) and radial leads (for low ESI).

Since capacitor lead length contributes to total resistance (ESR), lead length must be kept as short as possible. Layout is important if minimum ripple is to be realized. Additionally, a 0.1-microfarad disk-type capacitor across the output filter can be very useful in reducing high-energy spikes generated during power transistor switching. Ceramic types of this value have an ESR value of about 0.1 ohm or lower.

The filter capacitor working voltage rating (WVDC) for $C_{out}$ should be a minimum of 25 volts.

**Diode (D1)**   In the buck converter of Fig. 1.11, the catch diode D1 provides the return path for output current when the switcher is off. Fast-recovery or Schottky-type diodes are often chosen as the catch diode in these designs. Using the ubiquitous 1N4001 as a catch diode will prove disastrous. Its very slow recovery time specification generates excessive internal heat that will quickly destroy the device. Schottky diodes are often a better choice for the catch diode in low-output voltage switch-mode regulators. These diodes are not pn junction devices, but use a silicon-to-metal technology to create rectification employing the Schottky barrier principle. The symbol for a Schottky diode is shown in Fig. 1.11.

Lightly doped n-type silicon material (cathode) is bonded to a "barrier" metal like aluminum, which acts as the p-type (anode) region. When a forward bias is applied to a Schottky diode, electrons in the n-type material gain sufficient energy to cross the metal-to-silicon interface, where they quickly lose energy and become free electrons in the metal. Forward current flow is due only to majority carriers (electrons). Since there is no minority carrier (holes) current flow, there is no stored charge and no direct reverse recovery time. When a Schottky diode is reverse biased, conduction ceases as quickly as 10 nanoseconds ($10 \times 10^{-9}$ seconds). Schottky power diodes have very low forward turn-on voltages. At low currents, $V_F$ can be as small as 0.25 volt but will rise to about 1 volt at its maximum current rating. The one drawback to these devices is their relatively low breakdown voltage, which never exceeds much more than 60 volts, prohibiting their use in high-voltage designs.

Another practical consideration worthy of note is diode lead length, which is especially true in high-power designs. The lead length between the diode, capacitor, and power switching transistor should be made as short as possible.

Finally, we selected Motorola's 1N5820 Schottky diode for this project.

## 1.12 FINAL POWER SUPPLY CIRCUIT SCHEMATIC

A complete circuit schematic showing the unregulated power supply section combined with the switch-mode dc-to-dc converter is shown in Fig. 1.12. All parts are identified on the circuit, as well as important packaging and wiring information. In the following two chapters of this unit, we look at the packaging problem, trade-offs, and decisions that are made to develop a high-quality, reliable prototype package.

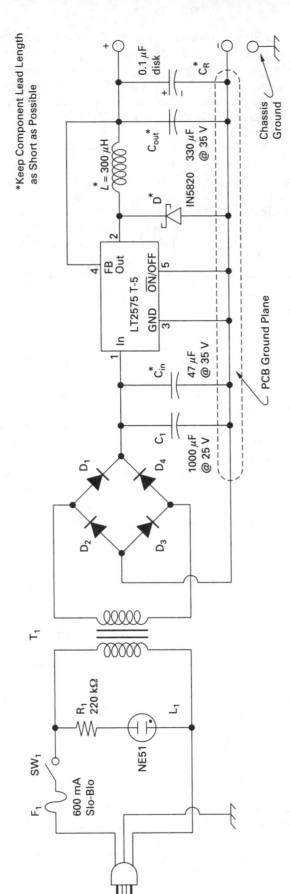

**Figure 1.12** A complete circuit schematic of the 5-volt at 1-ampere switching regulator power supply.

## EXERCISES

### A. Questions

1.1    List at least three common names used to refer to the U-ground terminal of a standard household receptacle.

1.2    List the main components of an unregulated (raw) power supply.

1.3    Explain the difference between no-load and full-load conditions as they relate to power supply load current.

1.4    In the transient analysis setup, modify the final time to display 3 complete cycles of a 60 Hz input.

1.5    Explain the steps necessary to load PSpice onto the hard drive (C:) of an IBM-compatible computer.

1.6    Discuss why linear regulators are less efficient devices than switched-mode dc-to-dc converters.

1.7    Identify and explain the main drawback in using a switching regulator over an equivalent linear regulator.

1.8    Explain pulse-width modulation (PWM) as it relates to switched-mode regulators.

1.9    Discuss the reason for selecting low-ESR-rated output filter capacitors for switching regulator applications.

1.10   Explain why a standard 1N4001 rectifying diode cannot be used as the catch diode in a switching regulator design.

### B. True or False

Circle *T* if the statement is true, or *F* if any part of the statement is false.

| | | | |
|---|---|---|---|
| 1.1 | In the United States, household wiring is performed in accordance with the National Electric Code. | T | F |
| 1.2 | The narrow slot of a household receptacle is connected to the neutral wire and fused at the service entrance panel. | T | F |
| 1.3 | A step-down transformer's secondary winding voltage is less than its primary winding voltage. | T | F |
| 1.4 | During maximum load current conditions, an unregulated power supply will exhibit maximum ripple voltage. | T | F |
| 1.5 | The load regulation curve of an unregulated supply dramatically illustrates the need for an output regulator. | T | F |
| 1.6 | Linear regulators can support both step-up and step-down designs. | T | F |
| 1.7 | Linear regulators are less efficient than switched-mode dc-to-dc converters. | T | F |
| 1.8 | The switching frequency of an LM2575T-5 switcher from National Semiconductor is approximately 52 hertz. | T | F |
| 1.9 | Ripple voltage is reduced by lowering the ESR rating of the output filter capacitor. | T | F |

### C. Multiple Choice

Circle the correct answer for each statement.

1.1    In the United States, household power is delivered at 115 volts ac at a frequency of (*50 Hz, 60 Hz*).

1.2    A diode bridge rectifier ($D_1$ to $D_4$) converts a transformer's secondary voltage output into a (*pulsating dc, filtered dc*) output.

1.3    A typical raw supply exhibits its worst-case ripple under (*no-load, full-load*) conditions.

1.4    The typical filter capacitor for a 1-ampere raw supply is (*1000 µF, 1000 pF*).

1.5    Battery-powered portable equipment usually demands a (*linear, switching*) regulator.

1.6    The inductor or coil used in a switched-mode regulator circuit functions primarily as a(n) (*energy storage, power dissipation*) device.

1.7    At low forward-current levels, Schottky power diodes turn on at typically (*0.25 V, 0.6 V*).

## D. Matching Columns

Match each item in Column A to the most appropriate item in Column B.

| COLUMN A | | COLUMN B | |
|---|---|---|---|
| 1. | Diode bridge | a. | Postprocessor |
| 2. | Schottky diode | b. | White |
| 3. | Unregulated | c. | Turns ratio |
| 4. | PSpice | d. | Switching regulator |
| 5. | Mo-Permalloy | e. | Filter |
| 6. | Neutral | f. | Catch diode |
| 7. | Probe | g. | Resistance |
| 8. | Dc-to-dc converter | h. | Raw |
| 9. | 5:1 | i. | Analysis software |
| 10. | ESR | j. | Isolation |
| 11. | Transformer | k. | Rectifier |
| 12. | Capacitor | l. | Henry |
| 13. | $L_{pri}$ | m. | Magnetic |

# 2 Planning and Designing Electronic Equipment

## LEARNING OBJECTIVES

*Upon completion of this chapter on the planning and designing of electronic equipment, the student should be able to*

- Understand the electrical criteria involved in packaging an electronic system.
- Select the appropriate type of chassis.
- Place components and hardware in their optimum positions in a package.
- Incorporate maintainability in package design.
- Incorporate human engineering in package design.

## 2.0 INTRODUCTION

A high-quality piece of electronic equipment can result only from thorough, careful planning and application of the fundamental mechanical and electronic design factors. Of these factors, *reliability* is one of the most important. A highly reliable piece of equipment will result if all the design factors that govern packaging are considered during the preliminary planning stages.

In this chapter we examine the numerous, common packaging problems involved in the preliminary planning stage of constructing electronic equipment. Consideration is given to *space* and *weight requirements, component* and *hardware selection, chassis materials* and *configurations, component* and *hardware positioning, maintenance,* and *human engineering.* Specific problems are related to the packaging of a *switching regulator power supply* that serves as an example packaging problem in the first unit of this book. Each step of the packaging phase is applied, in sequence, to this power supply, which illustrates some common packaging problems.

## 2.1 SPACE AND WEIGHT REQUIREMENTS

Once the need for a system has been determined, the mechanical and electronic design engineers will stipulate the many specifications for the completed unit to ensure that it will be compatible with its predicted environment. For example, packaging for *vehicular* installation often introduces complex dimensional and weight restrictions. If space is at a premium and size and weight must be kept to a minimum, the system must be packaged in subassemblies or adapted to a modular design that will fit into the various-shaped spaces that are characteristic of vehicles such as ships and air or space craft. *Stationary* equipment, such as that used in homes and laboratories, is normally subject to fewer severe environmental constraints and is designed around chassis, cabinets, racks, panels, and consoles that

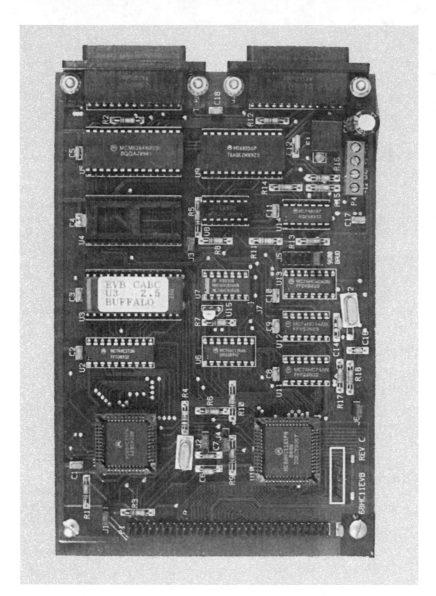

**Figure 2.1** Eight-bit microcontroller system developed as a double-sided pc board.

are commercially available. Modern stationary equipment makes extensive use of double-sided and multilayer printed circuit boards [Fig. 2.1].

## 2.2 COMPONENT AND HARDWARE SELECTION

Once the component values of the system have been designed, the circuits are tested in experimental or *breadboard* form. As noted in Chapter 1, breadboarding can effectively be accomplished with software analysis packages such as PSpice. It is generally the responsibility of the design engineer to determine the final selection of component values and ratings. Technicians must also be acquainted with the factors that govern these selections in order to enable them to choose the best available component for the specific applications. The final choice of components and hardware will be somewhat dependent on *cost* and *availability,* but the prime concern should be reliability. *Size, shape,* and the *finished appearance* of the unit also dictate part selection.

All electronic equipment should be as compact as possible, without reaching the point at which extreme component density prohibits assembly and service. Very often, the size of the unit is predetermined by its application, and the technician is compelled to work within established limits. The variation in size and mounting characteristics among commonly available components of the same electrical ratings makes it possible to satisfy even extreme dimensional restrictions.

The first criterion used in selecting components is, of course, that they meet the circuit's electrical demands with high reliability. Manufacturers' rated tolerance must be within design values. The closer the tolerance to the rated value of the component, the more it costs. Since cost is always a consideration, a technician should never choose components with closer tolerances than are necessary for the particular application. For example, if a resistor value is designed for 220 ohms with 10% tolerance, it certainly would be uneconomical to select a 1% resistor, since it may not appreciably improve circuit performance or reliability.

*Voltage, current, power levels, frequency,* and *continuous* or *intermittent operation* are the major circuit parameters that govern the selection of both the component value and type. In prototype construction the maximum electrical ratings (*stress levels*) of the components must be *derated* to ensure reliability. Derating provides a sizable safety factor so that components will operate well within their rated stress limits. In general, resistors are operated about 50% below their power stress levels. As an example, if a circuit resistor is required to dissipate 1/4 watt, the resistor actually chosen would have a 1/2-watt rating. On the other hand, a limit must be placed on overderating or compactness will be compromised. Using a 2-watt resistor where a 1/4-watt resistor will operate within derating values or an electrolytic capacitor with a working voltage of 100 volts where 10 volts would be more than adequate would sacrifice size, weight, and expense without improving circuit performance.

*Environmental factors* and *human engineering* also come into play in the determination of which component or style of hardware to use. To illustrate this, hermetically sealed components would be necessary for protection against extremely humid conditions. From the standpoint of safety, external accessibility, and appearance, an extractor post fuse holder would be selected over a fuse block [Fig. 2.2].

It is important that the technician realizes and appreciates that, in many cases, component and hardware selection is one of compromise. This compromise comes about through the combination of availability, cost, size, and tolerance. Regardless of the choice of components, in the final analysis the system must function within the established operating specifications. Otherwise, it is of little value, even though it may be well packaged.

Hardware items are sometimes a matter of personal choice. However, the size of such items must be realistic and fit the general scheme of the system. Miniature and subminiature hardware should not be selected when high component density is not required. In many instances, the selection of hardware such as switches, indicator lamps, connectors, jacks, and plugs is determined by those styles and types that are conventionally used in similar applications. If the unit being designed is to be used in conjunction with other units in an

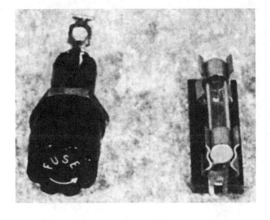

**Figure 2.2** Extractor post fuse holder and fuse block.

existing system, much of the hardware would be selected on the basis of mechanical and electrical compatibility and similar physical appearance. As an example, an existing system whose interconnections are terminated in *Cannon* connectors would require a special cable if a newly designed addition to the system used *Jones*-type connectors. Again, if a system makes use of *pc* (printed circuit) board connectors, additional boards used in the same system must be of a similar type [Fig. 2.3].

With the ever-widening use of integrated circuits, packaging has become more complex and a heavy reliance on printed circuitry has resulted. *Printed circuits* provide the only realistic solution to the problems associated with mounting transistors and the intricate wiring that is characteristic of integrated circuits. In the preliminary planning stage of construction, the technician is faced with the problem of using partial or total printed circuitry in a design. The solution to this problem often dictates the selection of the most suitable component and hardware types and styles. Although complex, printed circuits have reduced many packaging problems involving size, weight, and production uniformity. Since such a large percentage of components is designed for use with printed circuit boards, the technician will encounter little problem in finding components available. Certain designs mandate printed circuit boards for functionality. An example of such a design is the switching regulator power supply introduced in the previous chapter. The return (ground) path of all components in a switching design must be a large copper ground plane for effective operation. Short lead length and low inductance paths are also requirements for proper circuit functioning.

Some circuits require variable components, which are used only for initial adjustments for optimum circuit performance. Once these adjustments are made, ready access to the component is not necessary. These components are normally chassis-mounted or adapted to printed circuit design, as shown in Fig. 2.4.

**Figure 2.3** Modern printed circuit board connectors.

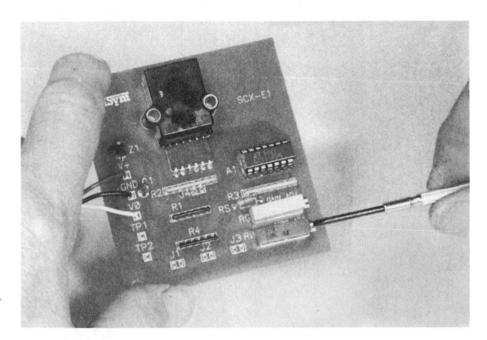

**Figure 2.4** Screw-adjustment potentiometer for printed circuit applications.

## 2.3 CHASSIS MATERIALS AND CONFIGURATIONS

Sound mechanical design is an essential element of electronic equipment packaging. Components must be provided with a sturdy mounting base that will hold them in a fixed position so that they may be electrically connected. The metal frame on which the components are normally mounted is called the *chassis*. The overall dimensions of the chassis are determined by the final component layout of the circuit. The chassis configuration is selected through an evaluation of the following factors: *circuit function, size, area restrictions, rigidity, component accessibility,* and *available manufacturing and fabricating facilities.*

The two basic types of chassis are the *box* and the *flat plate* configurations. Other chassis shapes are variations of these. Figure 2.5 illustrates several common commercially available chassis that are widely used in industrial and consumer product applications.

Factors that determine the style of chassis to use in a particular application are *weight of the heaviest components; use of controls and indicators,* such as switches, potentiometers, meters, and lamps; *circuit and operator protection; portability;* and *appearance.* If heavy items are to be mounted, a simple U-shaped chassis, as shown in Fig. 2.5a, may not provide enough strength, especially if aluminum is the chassis material. A more sturdy design, such as the box chassis in Fig. 2.5b, should be considered. Here, tabs on the long sides of the chassis are folded behind the short sides and allow the corners to be mechanically secured by sheet metal screws, rivets, or spot welds. Small self-contained circuits are packaged in chassis-enclosure combinations, as shown in Fig. 2.5c. Some circuit applications, particularly test equipment, require extensive use of meters, knobs, and indicator lamps, and in addition they must be portable. A reasonable selection for this application is the chassis and enclosure configurations shown in Fig. 2.5d or e, which provide a front panel for mounting controls and indicators, and, in some cases, handles for improved portability. Figure 2.6 illustrates a typical test equipment package.

Chassis are generally fabricated from sheet metal but may also be cast. The most common types of materials used for chassis fabrication are *aluminum, aluminum alloys, low-carbon cold- or hot-rolled steel,* and *molded plastics.* Special applications may call for chassis constructed of *copper, brass,* or *magnesium.* The choice of materials for chassis depends on *strength, weight, environmental conditions, finish, electrical and thermal*

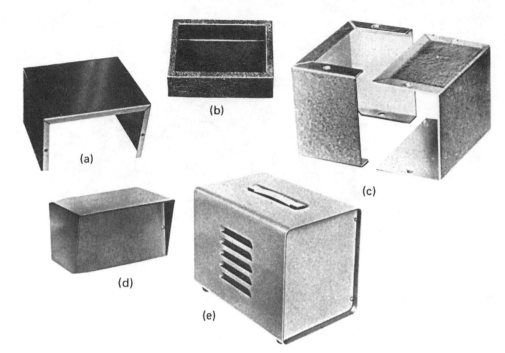

**Figure 2.5** Commercial chassis configurations: (a) U-shaped; (b) standard box; (c) small utility box; (d) front panel and enclosure; and (e) portable equipment case. *Courtesy of Bud Radio, Inc., Willoughby, OH.*

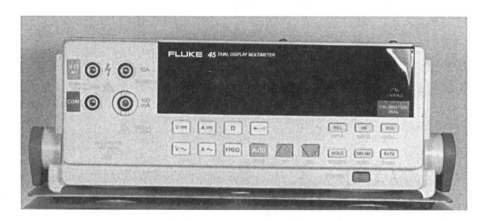

**Figure 2.6** The front panel of a digital multimeter package.

*requirements, cost,* and *special circuit demands* such as shielding. Aluminum chassis are the most readily available commercially in a variety of height, width, and depth dimensions with *gauges* (metal thicknesses) ranging from Nos. 14 to 20 in a wide selection of alloys and *tempers* (metal hardness). (See Appendix I for a table of metal gauges and their decimal equivalent thicknesses for aluminum and steel.) Aluminum used in chassis construction is light, strong, easy to work, and relatively inexpensive. It also requires no protective finish, since it does not normally corrode. Steel chassis need to be finished with a protective coating to prevent corrosion and to improve the appearance. Although steel chassis are more difficult to work with than aluminum and require additional care and finishing, they are more rugged than a comparable gauge in aluminum.

If circuit design characteristics include frequencies such as *VHF* (very high frequencies—30 to 300 megahertz) and *UHF* (ultrahigh frequencies—300 to 3000 megahertz), copper or copper-plated chassis are used because their conductivity is higher than that of aluminum or steel. This characteristic is especially desirable if the chassis is used as part of the electrical circuit.

Once the chassis configuration and material have been determined, the next step is to establish the size of the finished chassis. Factors that determine the size are the *number* and the *dimensions* of the components and the *component density* and *positioning*.

## 2.4 COMPONENT AND HARDWARE PLACEMENT

The packaging of an electronic system involves the positioning of parts and components to determine their optimum density and location, finished chassis size, and a balanced layout. Both electrical and mechanical factors are taken into consideration when packaging. Logical sequence dictates that mechanical criteria are applied first.

Large, heavy components such as transformers and filter chokes should be placed near the corners or edges of the chassis, where the greatest support exists. If several heavy components are being used, they should be evenly distributed over the entire chassis. Power and filament transformers, like the one used for the switching regulator power supply, are normally positioned to the rear of the chassis to avoid the ac line cord running near or through critical circuit areas. Fuses, cable connectors, and infrequently used binding posts and jacks for external antennas or remote speaker connections also should be positioned to the rear (Fig. 2.7a). Control knobs, switches, frequently used binding posts or jacks, meters, dials, and indicator lamps are generally located to the front of the chassis (Fig. 2.7b). These items are mounted not directly to the chassis but to a front panel or cabinet. Electrical connections between the panel-mounted and chassis-mounted components are then made directly or to intermediate terminal points. Panel-mounted hardware and components should be positioned to provide balance in their appearance and to ensure ease of operation, unobstructed visibility, and a neat and orderly wired package.

Whenever possible, speakers should be front-positioned so that the sound is directed toward the operator. At times, however, when tone quality is important, a larger speaker is required and it may be necessary, because of physical size, to position it elsewhere. Sacrificing position for tone quality is not an uncommon practice.

Electrical criteria often impose restrictions on component placement. Common circuit problems that must be considered in packaging to ensure proper circuit performance are *thermal radiation* from components such as resistors, switching regulator ICs, and power transistors; *ac hum* introduced into an audio circuit; and *distributed lead capacitance and inductance,* which can drastically affect audio circuits and rf (radio-frequency) circuits. These circuit problems are discussed in the following paragraphs.

Low-power transistors and integrated circuits are heat-sensitive devices and must be positioned a reasonable distance from any component that generates considerable heat. Components that give off a large amount of heat include power transistors and switching regulator ICs. In many applications, normal air convection does not provide for adequate heat dissipation and *heat sinks* are necessary. Heat sinks are metal configurations that, when brought into physical contact with the transistor case, absorb and dissipate a large portion of the generated heat. The chassis itself can be used as a heat sink, but power transistors are manufactured with the collector electrically connected to the case. If this creates an electrical problem when mounted directly to the chassis, an insulating washer is used. This washer electrically insulates the case from the chassis without causing severe adverse effects to the thermal conductivity between the case and the chassis. This mounting arrangement is shown in Fig. 2.8. Thermally conductive grease is also used on both sides of the insulating washer to increase thermal conductivity. If chassis sinking alone would not provide adequate heat dissipation, fin-type sinks like those shown in Fig. 2.9a are employed. If low-power devices require heat sinking, the type shown in Fig. 2.9b, which is not chassis-mounted, is often used. The proper size and configuration of sink required is predetermined during the circuit design stage. The problem of providing adequate space for the sink is left to the technician.

What is commonly referred to as *ac hum,* or 60-hertz hum, is a condition common to most electronic circuits employing power transformers. A 60-hertz signal, usually originating

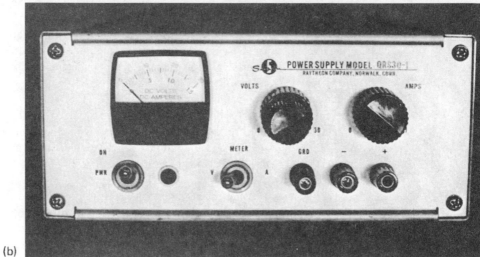

**Figure 2.7** External hardware and component placement: (a) typical rear-mounted hardware; (b) typical front-mounted components and hardware.

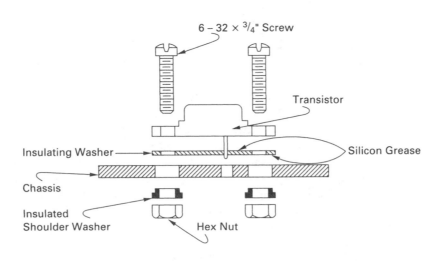

**Figure 2.8** Correct mounting arrangement of a power transistor to a metal chassis.

**(a)**

**Figure 2.9** Typical heat-sink configurations: (a) chassis-mountable heat sink for medium- and large-power devices; (b) heat-sink style for low-power devices. *Courtesy of Thermalloy, International Electronic Research Corporation and Wakefield Engineering Company, Inc.*

**(b)**

at the power transformer, can be induced into the input stages of the circuit. This induced signal is then amplified along with the desired signal, resulting in a steady hum. Shielding the signal lead or, by proper placement, making the input lead very short, will minimize this problem. It is advisable, however, to make it a common practice to locate power transformers, ac leads, or any source of electromagnetic radiation away from the signal input leads.

If circuitry employing radio frequencies is to be packaged, special attention must be given to the lead lengths required by the component and hardware positions. Because of distributed lead capacitance and inductance, which become a problem at these higher frequencies, wires used for electrical connections must be kept as short as possible. This is especially true at the input stages, as in the case of a receiver's local oscillator section, wherein the leads are extremely susceptible to noise pickup. Figure 2.10 shows an example of a compact rf package with the components positioned so as to keep the leads as short as possible.

If antenna positioning is involved in the circuit layout, it is important to locate the antenna input as close to the input circuit as possible to avoid stray pickup. In addition, if the

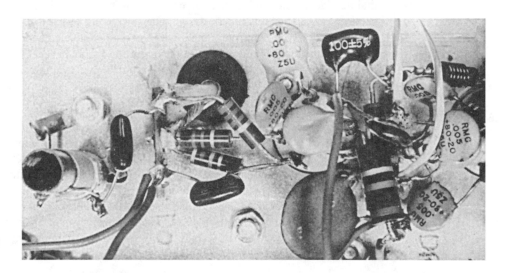

**Figure 2.10** Typical rf package design.

antenna is designed to be totally contained within the unit, it should be mounted on a non-metallic surface and kept as far away from metal elements as possible. This avoids the problem of picking up undesired frequencies. Component and hardware placement is by far the most involved and time-consuming phase of planning a package. Performance and reliability of the system will depend heavily on this phase of design.

## 2.5 MAINTENANCE

Maintainability is a measure of the ease, accuracy, and time required to restore a circuit or system to its normal mode of operation once a failure or abnormal function is detected. Maintenance covers two general procedures. First is the repair of failures and defects. Second is *preventive maintenance,* which deals with normal or routine checks and servicing of equipment. The object of preventive maintenance is to keep equipment operating trouble-free and at optimum performance. Certain precautions should be taken in the design of any package to minimize unnecessary maintenance. Whenever leads must pass through a hole in a metal chassis or bend around the edge of a bracket, either a grommet or clamp should be used to prevent abrasion of the lead insulation, which would cause a short circuit.

High ambient temperatures within enclosures can eventually cause components to overheat and fail. To ensure cool, trouble-free operation, approximately 6 cubic inches of space per watt of power to be dissipated is necessary if normal air convection is to be relied on for cooling a unit. Vents or holes must be provided, preferably at the top of the enclosure surrounding the circuit. If units are stacked in a rack, those units that dissipate the most heat should be located in the uppermost position to prevent the heating of components in other units. Whenever possible, liquid cooling and blowers should be avoided. Liquid cooling systems are subject to failure, and blowers could cause air circulation to cool certain components but overheat others. Air filters are often required when blowers are used, thus adding another maintenance dimension to the unit. Liquid cooling and blowers also add bulk and weight to the unit.

Enclosures, in addition to allowing circuits to function without overheating, must also be tight enough to isolate the components and wiring from dust or other contaminants that could accumulate in the circuit and cause failures. If large holes have to be formed in an enclosure for speakers, the entry of dust can be minimized by using a *speaker cloth.* For additional protection of the speaker cone, a grid or wire screen can be used with the cloth.

Well-designed enclosures minimize maintenance problems through the various types of circuit protection they afford. They can, however, present problems in terms of accessibility. Ease in maintenance is directly related to component accessibility. By locating access plates, doors, and openings near those components that are most likely to need maintenance, repair and service can be achieved easily, quickly, and accurately. Safety precautions should be built into the design when providing access to electronic circuits. Safety devices may have to be incorporated with the removable elements to avoid the possibility of electrical shock when working on the circuit through any access opening. Providing for maintenance in packaging design is a compromise involving three factors: *sufficient environmental protection, accessibility,* and *safety.*

## 2.6 HUMAN ENGINEERING (ERGONOMICS)

In electronic packaging design, human limitations associated with the use of electronic equipment must be considered. These human limitations involve such things as vision, arm reach, and hand manipulation, which must be considered in order to select the most suitable controls and visual displays. The ease and accuracy with which the unit is operated depends on the proper selection and placement of these components.

Meters should not be so small as to make the reading of major and minor scales difficult. The type of scale used should fit the function of the unit. For example, if only circuit conditions are to be monitored rather than the finite values of electrical parameters, the use of smaller meters having fewer scale divisions is desirable. If, on the other hand, accurate measurements are required, a larger meter with numerous minor scale divisions is preferable.

Glare must be avoided when continuous dial and scale readings are necessary. If reflected light in the unit's environment is excessive, glare shields should be employed. In addition, the panel should be finished in a color possessing low reflective characteristics (see Chapter 24).

Control knobs and switches should be mounted within easy and comfortable reach of the operator and should not be crowded. If they are, accurate individual adjustments will be difficult to make and other settings may be disturbed. Keeping with accepted conventions in mounting panel elements is very important. For example, toggle switches that are to be operated in a vertical manner should be on when the lever is in the upward position. Potentiometer adjustments in a clockwise direction should cause an increase in the magnitude of the electric quantity. In rack- or console-mounted equipment, all visual displays and indicators, such as meters and indicating lamps, should be mounted as close as possible to the eye level of the operator.

Stability is essential, especially if the unit is portable and relatively light. Some controls, especially rotary switches, require substantial torque to actuate and may cause the unit to become unsteady while in use. This problem can be reduced or overcome by either mechanically securing the unit to a mounting surface or attaching rubber pads to the bottom corners. The pads not only provide greater friction but also offer protection to the working surface, in addition to serving as electrical insulators for the unit.

The consideration of human limitations and capabilities, then, must be of concern to the technician designing a package. Many manufacturers employ human engineering consultants to render advice about the placement of control elements. A unit may function properly, but it will not be acceptable if it is difficult to control, read, or adjust.

In the next chapter, we use the basic concepts discussed here to plan the package design for the switching-regulator power supply project designed in Chapter 1.


## EXERCISES

### A. Questions

2.1    List at least five design factors that go into the planning of a piece of electronic equipment.

2.2    Of those factors listed in Question 2.1, which do you consider the most important, and why?

2.3    What electrical criteria impose restrictions on component placement?

2.4    What are the two basic types of chassis from which all other variations are derived?

2.5    What is the purpose of a *heat sink*?

2.6    What is meant by *human engineering* in package design? Why is it important?


### B. True or False

Circle *T* if the statement is true, or *F* if any part of the statement is false.

| | | |
|---|---|---|
| 2.1 | Stationary equipment design imposes fewer restrictions than does that for vehicular equipment. | T    F |
| 2.2 | A U-shaped chassis provides more strength than a box chassis. | T    F |
| 2.3 | Large and heavy components should be placed near the corners or edges of a chassis. | T    F |
| 2.4 | The most common material used to make a chassis is copper. | T    F |
| 2.5 | In designing high-frequency circuits, the lead lengths should be as short as possible. | T    F |

## C. Multiple Choice

Circle the correct answer for each statement.

2.1     One of the most important design factors is (*reliability*, *appearance*).

2.2     In circuit design involving ultrahigh frequencies, the chassis should be made of (*copper*, *plastic*).

2.3     Transistors and integrated circuits (*are*, *are not*) affected by heat.

2.4     In the design of rf packages, the antenna must be positioned (*close to*, *far from*) the signal input circuit.

## D. Matching

Match each item in Column A to the most appropriate item in Column B.

| COLUMN A | COLUMN B |
|---|---|
| 1. Design factor  e | a. Aluminum |
| 2. Packaging | b. Megahertz |
| 3. Circuit parameter | c. Support |
| 4. Derating | d. Shoulder washer |
| 5. Hermetically sealed | e. Reliability |
| 6. VHF  b | f. Humidity |
| 7. Heatsink  d | g. Component position |
| 8. Insulation | h. Limits |
| 9. Brown and Sharp | i. Voltage |
| 10. Transformers  i | j. Silicon grease |

# 3 Packaging Design

## LEARNING OBJECTIVES

*Upon completion of this chapter on the packaging design of a switching-regulator power supply, the student should be able to*

- Understand the function of each subcircuit of the switching-regulator power supply system.
- Select the appropriate chassis configuration to house the power supply system.
- Differentiate between *chassis ground* and the *high-quality ground* of the power supply.
- Understand the selection and positioning criteria for the components and hardware assigned to the front panel.
- Understand the requirements of critical components and the materials and techniques available to support them.
- Develop a preliminary hand sketch of the chassis and enclosure elements and the approximate placement of all the major components of the switching-regulator power supply project.

## 3.0 INTRODUCTION

A switching-regulator power supply, designed in Chapter 1 and employing many different components and devices, is used as a means of illustrating the practical applications of packaging techniques discussed in the first unit of this book. In this way, familiarity is gained with the application of the current design considerations, conventions, and practical procedures for sound electronic packaging.

Since a basic understanding of the electronic circuitry involved is essential to fully interpret and apply the necessary procedures for packaging, a comprehensive discussion of the switching regulator's circuit operation was given in Chapter 1. That information is revisited here with special attention paid to general packaging considerations, chassis design, and placement of critical components.

## 3.1 BLOCK DIAGRAM

A block diagram of the switching-regulator power supply designed in Chapter 1 of this unit is shown in Fig. 3.1. This type of diagram identifies the function and interrelationship of each of the system's subcircuits. Refer back to Fig. 1.12 for a detailed circuit schematic of this switching-regulator power supply project. The three main subcircuits of the switching-regulator power supply system are (1) the step-down transformer and associated input circuitry, (2) a bridge rectifier and filter to complete the unregulated or "raw" supply circuit, and (3) the dc-to-dc converter or switching regulator.

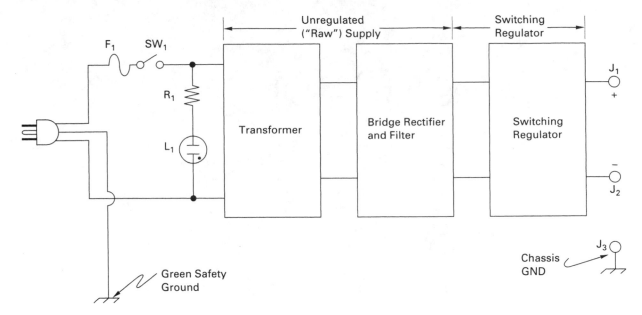

**Figure 3.1** Block diagram of the switching-regulator power supply.

The three main subcircuits act together as follows: The 115-volts ac household power is first reduced in amplitude by the step-down transformer and then converted to an approximate dc level, with considerable ripple, by the bridge rectifier and capacitive filter or "raw" supply. Then, the output of the raw supply is regulated to a precise 5 volts at the output of the switching-regulator circuit.

## 3.2 POWER SUPPLY PACKAGE DESIGN

Initially, a parts list will be prepared during the design phase of a project. The parts list for this switching regulator project is shown in Fig. 1.12 as part of the actual circuit schematic and in Fig. 1.10 as part of the hard copy from using the *Switchers Made Simple* software. The parts list identifies components by part number, and usually gives tolerances and electrical ratings for each item. However, hardware, chassis material selection, and finish are decisions often left to the package designer.

Cost and availability of parts often serve as secondary factors during the circuit design phase, but should be carefully considered when designing the final package. Forced substitutes with different dimensions or tolerances can cause considerable rework of a package, resulting in lost time and effort. It often falls to the package designer to consider alternative components and devices when developing a new design, but all options should be reviewed with the circuit designer for compatibility before any changes are made final.

Normally, a switching-regulator power supply has few space or weight restrictions. Therefore the first phase in the package design will be the selection of the chassis material. The type of aluminum that will be used for the chassis elements of this project will be *16-gauge No. H-1100*. This choice is made on the basis of the desirable characteristics of this material—easily worked, sturdy, lightweight, conductive, and readily available. The heaviest component, a transformer, weighs less than 22 ounces and is easily supported by this material. In addition, aluminum readily lends itself to the basic finishing techniques discussed in Unit 5, Chapter 24.

Two common styles of chassis and enclosures that lend themselves to the switching regulator are shown in Fig. 3.2. A box-type chassis to which is mounted a separate front panel is shown in Fig. 3.2a. In this configuration, all the parts not positioned on the front panel are placed on the box chassis element. A cover, bottom plate, and back panel are then required to complete this type of package. Although this package style provides good strength, it does have several undesirable features. First, unless a cover, bottom, and back plates are provided, there will be exposed wiring, which could result in a hazardous condition. Second, the height of the box chassis would have to be very low to accommodate the transformer and still maintain an aesthetically pleasing, low silhouette design. The three-element configuration shown in Fig. 3.2b overcomes these undesirable features and will be used to provide a stable platform onto which the switching-regulator project can be built.

The three-piece chassis and enclosure shown in Fig. 3.2b has a *subchassis* element fastened vertically to the *front-and-side* element. These elements, together with the three-sided *cover* shown, result in a completely enclosed package. All items not mounted on the front panel will be placed on the sides (same piece of sheet metal as the front) or on the subchassis element. The cover fastens to both the subchassis and the front-and-side panel element to form a rigid enclosure. Unlike the box-type chassis, removal of the single cover element will expose all the components for wiring.

A preliminary series of sketches of the component layout is undertaken next. These sketches are made simultaneously with the technique of positioning the components and hardware. Each of these subcircuits and accompanying components are discussed as they relate to package and chassis design.

### 3.2.1 The Front-and-Side Panel
Two primary factors need to be considered in the design of a front panel: *visibility* and *ease of use*. The determination as to which components

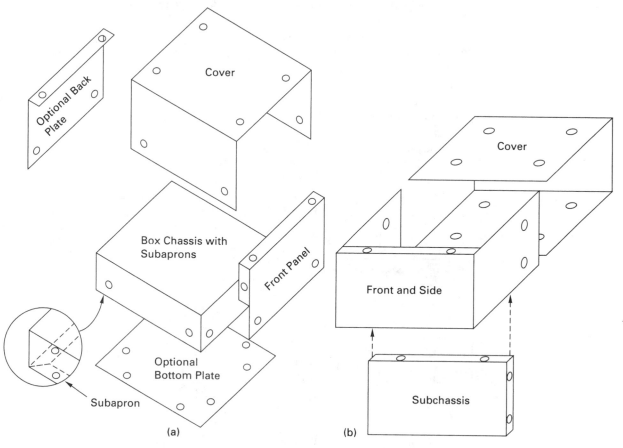

**Figure 3.2**   Custom chassis and enclosure designs. (a) Box chassis with several enclosure, panel, and plate elements. (b) Rigid three-piece chassis and enclosure configuration chosen for the switching-regulator power supply project.

will be mounted to the front panel is often dictated by those components that are used most often in the convenient operation of the equipment. A review of the block diagram of the project in Fig. 3.1 (and Fig. 1.12) suggests that the following components be considered for front-panel mounting:

1. $F_1$ (fuse holder and fuse)—*optional*
2. Line cord—*optional*
3. $J_1$, $J_2$, $J_3$, (+, −, and chassis ground terminals)
4. $SW_1$ (on/off switch)
5. $L_1$ (pilot lamp—power *on* indicator)

We begin the front-panel layout by first considering the optional items to determine if it is necessary for front-panel mounting. Of the components listed above, both the line cord and fuse with holder are classified as optional. It is common practice in laboratory equipment to provide easy access to the circuit's fuse by placing it on the front panel for ease of replacement. This is especially important when frequent circuit overloads are expected. However, overloads are not a problem in this project. The use of the LM2575-5 IC, with its internal protection circuitry, makes fuse replacement rare, occurring only under extreme fault conditions. For this reason, $F_1$ will be positioned on the right side of the front-and-side panel, directly forward of the line cord (refer ahead to Fig. 3.5). Note that it is good engineering safety practice to first fuse household power, available at the three-pronged receptacle, to protect all components within the system.

The line cord should be provided with a strain-relief bushing for safety, and a three-lug terminal strip should be mounted inside and directly above the line cord. The *hot* and *neutral* leads can thus be wired to the two outside insulated lugs, and the center lug, directly connected to the chassis, will be used to attach the green "safety" ground lead directly to the chassis. In this way the chassis will be properly grounded when the power supply is plugged into an ac receptacle. In addition, the chassis (conductive aluminum) will be used to make the electrical connection to jack $J_3$ on the front panel.

One front-panel design that meets the requirements of *visibility* and *ease of operation* is shown in Fig. 3.3. This is but one of several arrangements that could meet the desired criteria.

The three terminal jacks, $J_1$ (+), $J_2$ (−), and $J_3$ (chassis ground, GND), are positioned central and toward the bottom of the front panel so that leads extended from these terminals to any load will not interfere and obscure the remaining element of the front panel. Note that the power supply's high-quality ground or negative terminal, $J_2$, must be electrically insulated from the chassis (and chassis ground, GND) with a nylon shoulder washer.

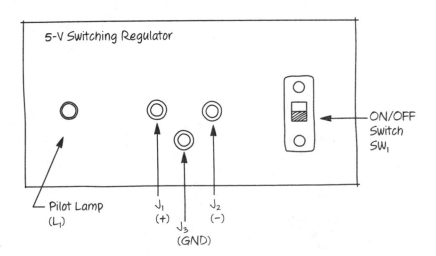

**Figure 3.3** Suggested front-panel design that provides easy access to jacks and on/off switch.

The on/off switch ($SW_1$) is mounted directly to the right of the terminal. It is given this prominent position because it is the most frequently used hardware in the project. It is common convention in the mounting of single-pole, single-throw (SPST) switches to have power applied to the circuit when the switch lever is *up* (or to the right in horizontal mounting) and the power off when the switch is in the *down* position (or to the left). This switch is associated with lamp $L_1$, which serves as a pilot light to indicate when power is on. Because the lamp requires no adjustment or access, it is positioned to the left on the front panel.

**3.2.2 The Subchassis Element**   Note the location of the three-pronged line cord and extractor-type fuse holder, both on the right side of the front-and-side panel (refer ahead to Fig. 3.5). This positioning dictates that the step-down transformer be placed on the right side of the subchassis to minimize internal wire length. With the on/off switch located on the right side of the front panel, a twisted pair of leads between the on/off switch, transformer primary, and indicator lamp will be required. The twisting of leads tends to cancel the generated magnetic field about the leads, thereby reducing 60-hertz hum.

The remaining components associated with both the bridge rectifier and filter as well as the switching regulator are fragile and require the rigidity of a printed circuit board. Figure 3.4 illustrates the major components and hardware placements on the subchassis element. Note that the PCB will be mounted in a rectangular chassis cutout with machine screws and that connections to the secondary of the transformer and access to the output jacks are made via turret terminals swaged to the PCB.

A double-sided PCB with conductor paths on the top side and a large *ground plane* of copper on the bottom side will provide the low-impedance, low-inductance paths necessary for good switching-regulator operation (refer to Unit 2, Chapter 8 for additional information on double-sided printed circuit board design). Also, a ground plane PCB design

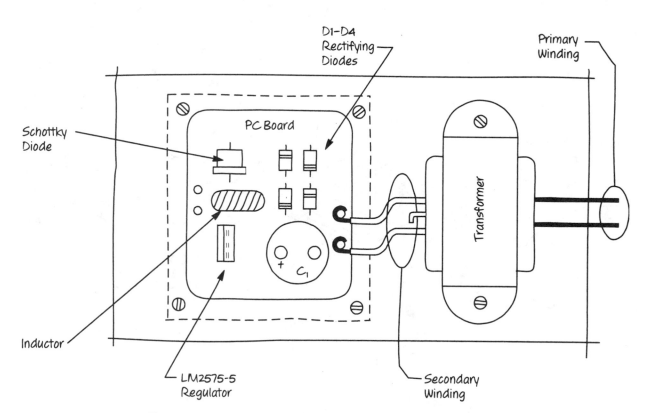

**Figure 3.4**   Major component and hardware placement for the subchassis element and the double-sided pc board.

facilitates short component lead lengths for the Schottky diode, inductor, and output capacitor of the dc-to-dc converter, a requirement dictated by this type of switching-regulator design.

### 3.2.3 The Cover Element
As discussed in Chapter 1, the LM2575-5 switching-regulator IC will dissipate some heat under high-load conditions. The cumulative heat developed by this component must be adequately removed to maintain proper circuit operation. Heat removal is accomplished by providing a series of ventilation holes in both the top and bottom sections of the cover element. These holes are shown in Fig. 3.5, an exploded view of the three chassis elements. In addition, rubber feet are mounted to the bottom surface of the cover to ensure that the bottom ventilation holes are not blocked. Thus, cool air will circulate up through the power supply, carrying the hot internal air out through the top ventilation holes.

## 3.3 DETAILED CONSTRUCTION DRAWINGS

Figure 3.5 shows the results of the major component positioning and mounting-space requirements. From this layout the optimum-size front-and-side panel, subchassis element, and cover can be designed and fabricated. To convert the final layout into useful detailed

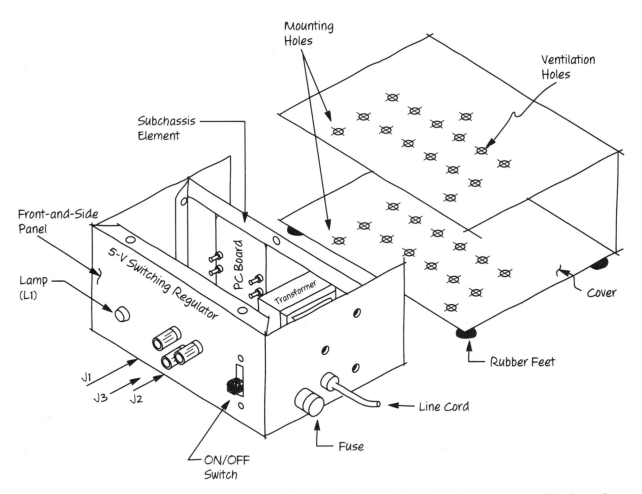

**Figure 3.5** Engineering sketch of all chassis elements including major components and hardware placement for the switching-regulator power supply.

drawings for actual construction of the unit, accurate measurements of components and their locations must be made. Procedures for determining accurate dimensional information and the application of this information in the preparation of detailed construction drawings are discussed in the next chapter.

## EXERCISES

### A. Questions

3.1 What are the significant differences between *block diagrams* and *circuit schematics*? *more detail.*

3.2 List the three main subcircuits of the switching regulator project and explain the function of each. *115 vac reduce in amplitude* ↓ *steplown → dc → filter*

3.3 What are the primary factors to consider in designing a front panel? *visibility and ease of use*

3.4 What is the difference between the *chassis ground* and the *high-quality ground* of a power supply?

3.5 Explain one method of providing adequate ventilation to an enclosed chassis.

### B. True or False

Circle *T* if the statement is true, or *F* if any part of the statement is false.

3.1 The position of individual components can be determined quickly from a block diagram of the system.   **(T)   F**

3.2 Block diagrams show the continuity of wiring connections between all components.   **T   (F)**

3.3 The purpose of a raw supply is to convert ac into an approximate dc voltage.   **(T)   F**

3.4 The purpose of the switching regulator is to convert the approximate dc voltage of the raw supply into a constant value.   **(T)   F**

3.5 The size and style of all components can always be determined from the parts list.   **(T)   F**

### C. Multiple Choice

Circle the correct answer for each statement.

3.1 The exact electrical connections for each component in an electric circuit is best shown by a (*block, schematic*) diagram.

3.2 The ripple of a raw supply is often (*low, high*).

3.3 The negative terminal of a power supply is referred to as (*chassis, high-quality*) ground.

3.4 Good design dictates that the fuse of a switching regulator power supply be mounted on the (*front, side or rear*) of the package.

3.5 Aluminum is selected for the chassis of our power supply system because it is a good (*conductor, insulator*).

### D. Matching

Match each item in Column A to the most appropriate item in Column B.

| COLUMN A | COLUMN B |
|---|---|
| 1. Fuse *e* | a. Chassis |
| 2. Twisted pair *b* | b. 60-hertz |
| 3. H-1100 *a* | c. Unregulated |
| 4. Ease of use *Front panel* | d. Safety |
| 5. Switching *f.* | e. LM2575-5 |
| 6. Raw supply *c* | f. Design element |

# 4 Preparing Detailed Drawings

## LEARNING OBJECTIVES

*Upon completion of this chapter on preparing detailed drawings, the student should be able to*

- Make accurate measurements using common precision instruments such as scales, calipers, and micrometers.
- Produce engineering sketches for the chassis elements.
- Know the basic line conventions.
- Know the four basic dimensioning techniques.
- Select between several methods of preparing good-quality drawings.
- Produce assembly drawings showing a variety of views.

## 4.0 INTRODUCTION

Once the preliminary plans for a package are completed and engineering sketches of the chassis configuration are prepared, *detailed drawings* must be developed to facilitate construction. These drawings entail the final positioning of all components and hardware. Sizes of these components and hardware items are first measured to obtain overall dimensions necessary to select and complete optimum orientation and position. Many of these measurements will also determine the size and location of mounting holes during the fabrication of the chassis. These detailed drawings must include all pertinent information with respect to component assembly and positioning and chassis fabrication and finish. It is important to mention here that someone other than the technician who designed the original package may be assigned the task of construction. For this reason, the detailed chassis drawings must be completely informative, accurate, and clear.

This chapter contains specific information on techniques of using measuring instruments such as scales, calipers, micrometers, and templates. Also included is relevant information on engineering sketches, various modern drafting methods, dimensioning, and procedural information for developing chassis layout drawings along with assembly and auxiliary views for detail work.

## 4.1 MEASUREMENTS WITH A SCALE

In the packaging design stage, as described in Chapter 3, approximate positioning of components and hardware was estimated by physically placing the components in the desired locations. From this preliminary method, component density can be observed. It is now necessary to take all dimensions and transfer this information first into an engineering sketch and then into detailed working drawings.

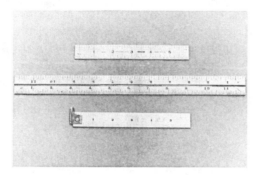

**Figure 4.1** Steel rules.

To obtain most of the required measurements, a simple *steel scale* is commonly employed. Steel scales are available in a variety of styles, including *steel tape rule, steel rule,* and *hook rule.* The 6- and 12-inch steel rules and the 6-inch hook rule are the types commonly used in electronic construction (Fig. 4.1). The major scale division on these rules is the inch, with half, quarter, eighth, sixteenth, thirty-second, and sixty-fourth subdivisions. Steel scales are also available with decimal graduations having major and minor divisions of 0.1 inch and 0.020 inch, respectively. This rule is desirable when the dimensioning format of the sheet metal drawings is in decimal form. The smaller graduation scales are used where close tolerance measurements are required.

Overall length and depth measurements for the chassis determined by the final component layout can be made by alternately positioning and reading the scale along the lines bounding the area previously established in the planning stage. The height of the unit is determined by reading the vertical measurement of the tallest component together with the desired enclosure clearance (see Fig. 4.2a).

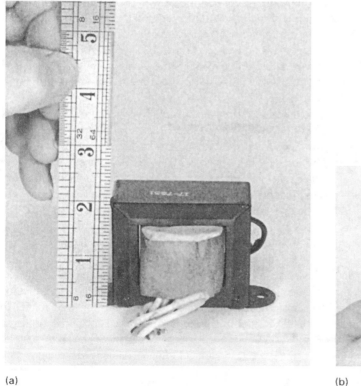

(a)

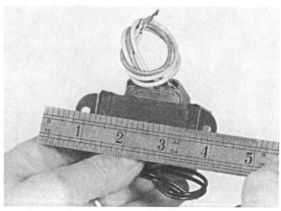

(b)

**Figure 4.2** Method of determining both (a) height and (b) on-center dimensioning.

*Preparing Detailed Drawings*

Component measurement for mounting purposes can also be made with a scale. As an example, a method for determining on-center dimensions for mounting holes is shown in Fig. 4.2b. Mounting information, however, is usually provided by the manufacturer in the form of a paper template.

For maximum accuracy when using the 6- or 12-inch steel rule, the scale should be positioned perpendicular to the surface being measured. In this way the graduation marks on the scale will be in direct contact with the work, which will eliminate the possibility of error due to the thickness of the rule if it is placed flat on the work. In addition, neither end of a rule should be used as the reference or starting edge. For accurate measurements, the work should be aligned with the first major inch graduation mark. This assures that scale-edge deformities will not cause errors.

The hook rule, shown in Fig. 4.1 (bottom), provides an accurate reference edge along the inside face of the hook. Thus, the zero graduation mark is perfectly aligned with the edge of the work, assuring more exact measurements.

## 4.2 MEASUREMENTS WITH CALIPERS

Although rules are adequate for most measurement applications, *calipers* are necessary if direct measurements are impossible. Calipers, shown in Fig. 4.3, are classified as *outside* or *inside,* depending on the shape and position of the legs as well as their specific use. The main parts of the caliper are the *legs, spring crown,* and *adjusting knob.*

The outside caliper is readily adaptable to measuring diameters of round stock, such as the shaft of a potentiometer or rotary switch. This caliper is held by the spring crown cupped in the palm of one hand with the thumb and forefinger free to apply a rotational force to the adjusting knob (Fig. 4.3a). Rotating the adjustment knob will close the legs of the caliper on the surface to be measured. Once the inside tips of the legs contact the surface to be measured, a back-and-forth motion of the caliper along the surface will indicate the correct measurement. The calipers are set properly if there is a *slight friction* as the inside tips of the legs move across the widest portion of the work. Practice is necessary to acquire the correct "feel" for this optimum setting. With the leg spacing set, the measurement can now be determined by placing the caliper along the edge of a scale and reading the distance between the inside of the legs. This procedure is illustrated in Fig. 4.3b.

Drill gauges can also be used as a quick means of measuring round bodies such as diodes, transistors, and integrated circuits. Figure 4.4 shows the measurement of the outside diameter of a transistor using a drill gauge.

If greater accuracy than can be obtained from the basic caliper is required, *dial calipers* are used. The 6-inch dial caliper is common in electronic construction. This precision instrument consists of *two pairs of jaws* (one for *inside* and one for *outside* measurements), a *dial indicator,* and a *thumb adjusting knob.* This instrument is capable of accuracies from 0.001 to 0.0001 inch (Fig. 4.5). Before using the dial caliper, a zero check should be made to determine if the dial indicator is properly calibrated with the jaws. First, the contacting surfaces of the outside jaws are wiped clean with a soft cloth to remove any foreign matter or dust. The thumb adjustment screw is then rotated clockwise to close the outside jaws. The dial indicator should read zero when the jaw surfaces make contact. If not, the rack and pinion must be adjusted. The two securing screws at both ends of the rack gear are loosened and the rack is slowly moved until the dial indicator reads exactly zero. The securing screws are tightened and the instrument is set for accurate measurements. Figure 4.6 illustrates the proper method of supporting the dial caliper when making measurements. With the adjusting knob rotated to close the outside jaws on the work, the correct amount of jaw pressure is obtained through a *slipping action* whereby continued rotation will result in the adjusting knob slipping with no further forward motion imparted to the jaws. This action eliminates damage to the caliper and also measurement error.

All settings should be checked for correct jaw alignment before measurements are read. To ensure accuracy, the jaws of the calipers are slightly rocked over the work (see Fig. 4.6). The minimum deflection of the dial pointer will result in the most accurate reading.

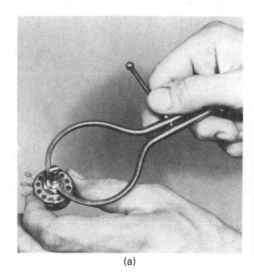 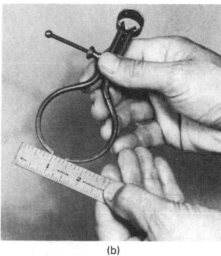

(a)                                    (b)

**Figure 4.3** Outside calipers used to measure round stock: (a) correct method of setting caliper tips; (b) distance between tips is determined with a steel rule.

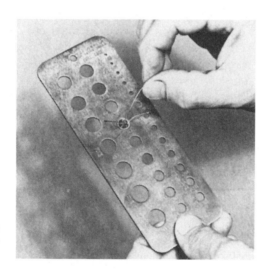

**Figure 4.4** Drill gauge used to measure the case diameter of a transistor.

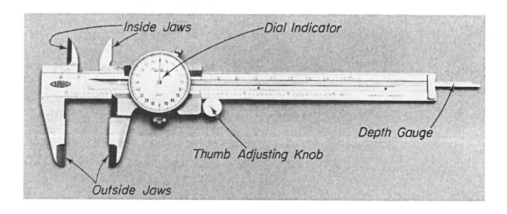

**Figure 4.5** Dial caliper.

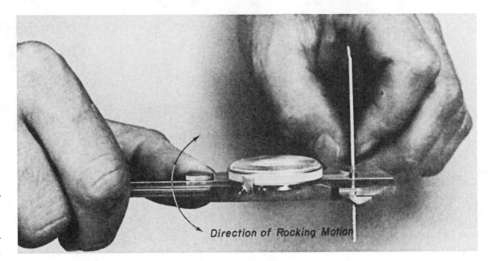

**Figure 4.6** Dial caliper used to obtain accurate measurements with rocking motion to determine minimum reading.

Direction of Rocking Motion

## 4.3 MEASUREMENTS WITH A MICROMETER

The micrometer, like the dial caliper, is capable of measuring to accuracies of 0.001 inch and estimates to 0.0001 inch. The 1-inch outside micrometer is typical, although larger-capacity micrometers are available in addition to special-purpose micrometers, such as screw-thread measuring instruments.

The 1-inch micrometer is used primarily to measure diameters and thicknesses. This micrometer, shown in Fig. 4.7, consists of a *frame, anvil, spindle, barrel, thimble,* and *ratchet stop.* The proper use of this micrometer is as follows. The little finger is hooked around the frame while the thumb and forefinger of the same hand are free to rotate the thimble (Fig. 4.8). Clockwise rotation of the thimble will close the spindle on the anvil, which is rigidly supported to the frame. Counterclockwise rotation of the thimble widens the distance between the spindle and the anvil. The work to be measured is held in the other hand and placed on the anvil. The thimble is rotated to close the spindle on the work. Once the spindle makes *light contact* with the work, the thumb and forefinger are moved to the ratchet stop. Clockwise rotation of the ratchet stop for two "clicks" will apply the correct amount of pressure to the work. Excessive pressure may damage the finely threaded spindle or mar the machined surfaces of the anvil and spindle, thus causing measurement errors. It is good practice to maintain continuity in measurement by duplicating the applied pressure. The two "clicks" of the ratchet stop will assure this duplication of pressure.

The spindle has a precision of 40 threads per inch. Thus, one complete rotation will move the spindle exactly $\frac{1}{40}$ or 0.025 inch. Each minor graduation mark on the barrel scale is therefore equal to 0.025 inch. Each major numbered division on the barrel scale represents 0.100 inch, requiring four complete turns to move from one major numbered division to the next. The thimble has 25 graduation lines. Each line represents $\frac{1}{25}$ of $\frac{1}{40}$ of an inch, or 0.001 inch. Therefore, as the thimble is rotated from one division to the next in a counterclockwise direction, the spindle moves 0.001 inch away from the anvil.

To obtain the spindle-to-anvil opening, three readings are necessary. First, the highest-*numbered* graduation shown on the barrel is noted. (Each numbered graduation represents 0.100 inch.) The second reading is the number of minor divisions following the major number on the barrel, each minor division representing 0.025 inch. The third reading is the lowest-numbered graduation line on the thimble that most closely aligns with the horizontal scale line on the barrel. This value is in one-thousandths (0.001) of an inch and is added to the sum of the preceding readings to obtain the final measurement.

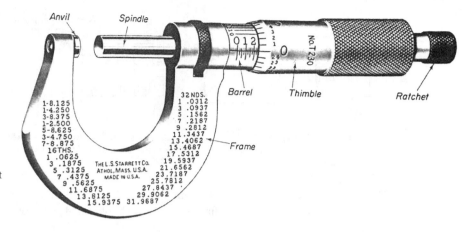

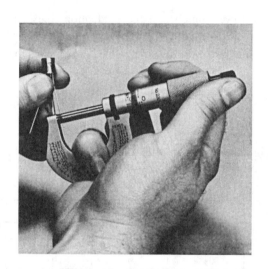

**Figure 4.7** One-inch micrometer with ratchet stop. *Courtesy of the L.S. Starrett Company, Athol, MA.*

**Figure 4.8** Correct positioning of the 1-inch micrometer.

Figure 4.9a represents a reading of 0.337 inch obtained as follows:

*Reading 1.* The largest visible number on the barrel is 3, or 0.300 inch.

*Reading 2.* The number of smaller graduations after the number 3 is 1, or $1 \times 0.025 = 0.025$ inch.

*Reading 3.* The lowest number of the graduation lines on the thimble that most closely aligns with the horizontal scale line on the barrel is 12, or 0.012 inch.

The sum of the three readings is 0.337 inch.

Some micrometers are equipped with a *vernier scale* to extend the measurement accuracy to one ten-thousandths (0.0001) of an inch. This vernier scale is immediately above the barrel scale and its graduation marks extend along the barrel to the thimble scale. The vernier and thimble scales both run horizontally. Each of the 10 graduation lines of the vernier scale, marked 0 to 9, represent $\frac{1}{10}$ of the value of the divisions on the thimble scale— that is, $\frac{1}{10}$ of $\frac{1}{1000}$, which equals $\frac{1}{10,000}$ (0.0001) inch.

To obtain a reading with a micrometer equipped with a vernier scale, the previously discussed three readings are first taken. The fourth reading is obtained by noting which graduation line of the vernier scale aligns with any line on the thimble scale. This reading,

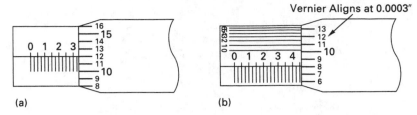

(a)　　　　　　　　　　(b)

**Figure 4.9** Illustrations of micrometer scale readings: (a) readings to the one-thousandth inch using the barrel and thimble scales; (b) readings to the ten-thousandth inch using barrel, thimble, and vernier scales.

in one ten-thousandth (0.0001) of an inch, is added to the sum of the other three readings. Figure 4.9b represents a reading of 0.4583 inch, obtained as follows:

*Reading 1.* The largest visible number on the barrel is 4, or 0.400 inch.

*Reading 2.* The number of minor graduations after the number 4 is 2, or 2 × 0.025 = 0.050 inch.

*Reading 3.* The lowest number of the graduation lines on the thimble that most closely aligns itself with the horizontal scale line on the barrel is 8, or 0.008 inch.

*Reading 4.* The number of the division on the vernier scale that aligns with a division on the thimble is 3, or 0.0003 inch.

The sum of the readings is 0.4583 inch.

## 4.4 ENGINEERING SKETCHES

A preliminary step in the preparation of the finished detailed working drawings is the development of *engineering sketches*. Figure 4.10 shows a sheet metal sketch of the switch-

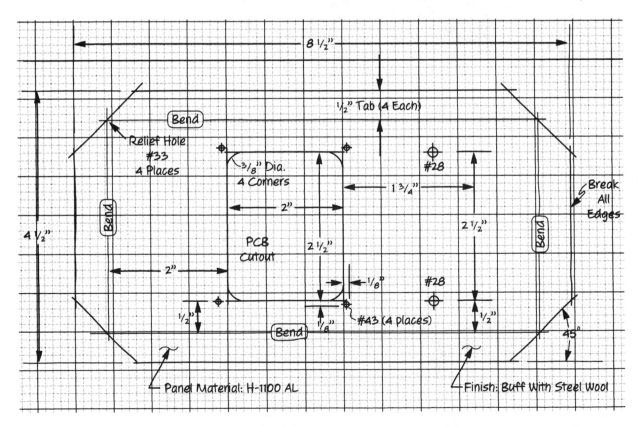

**Figure 4.10** Engineering sketch of power supply subchassis element.

ing regulator's subchassis element. Using quadruled paper results in a relatively clear picture without the need of special equipment. Note that the engineering sketch shown in Fig. 4.10 contains not only *dimensional information* but specifications for *finish, type of material,* and *hole positioning.* Also useful are manufacturer's specifications that provide hole-spacing information for mounting holes, pin sizes and configurations, case dimensions, and other pertinent information to aid in layout.

## 4.5 METHODS OF CREATING DETAILED WORKING DRAWINGS

To produce good-quality drawings, the technician must select from conventional drafting equipment or employ a computer-aided drafting (CAD) software package, such as AutoCAD, Release 12, described in Chapter 7 of Unit 2. While some drafting work is still being done by hand techniques, it seems apparent that the trend has shifted away from this method and toward computer-generated drawings. There are many reasons for this shift, chief among them being the proliferation of low-cost, user-friendly, general-purpose software tools and personal computers. Today's technician must become more skillful in the new computer technology, while still maintaining a passing acquaintance with the older hand-drawing techniques.

What follows is a brief overview of the typical drafting equipment necessary to generate high-quality drawings by hand methods and by CAD techniques.

**4.5.1 Hand-Drafting Equipment**   Tools common to the technician wanting to develop detailed drawings by hand-drafting techniques are (1) *drafting paper and pencils,* (2) *drafting boards,* (3) *T-squares,* (4) *triangles,* (5) *protractors,* (6) *engineering scales,* and (7) *drafting instruments and templates.*

***Drafting Paper and Pencils.*** Knowledge of the size and type of paper used in drawing is helpful. Commonly available drawing paper sizes are A ($8\frac{1}{2}$ by 11 inches), B (11 by 17 inches), C (17 by 22 inches), and D (22 by 34 inches). Larger sizes are available; K size (40 by 50 inches) is normally the largest used in engineering work. The paper size is, in many cases, selected to be compatible with the filing system in use. In all cases, the paper should be of sufficient size to show detail work without crowding, yet not so large as to have excessive unused areas or overly large borders. The preferred paper for electronic drafting is the *HP* (hot-pressed) *cream* or *buff-colored* type. This paper is sufficiently heavy and hard-surfaced to be used for high-grade, accurate detailed work. *CP* (cold-pressed) paper is less expensive and is used for general drawing or sketching. The surface of this type of paper will not hold up to erasing as well as the HP type. When blueprints are to be made from pencil drawings, a white, light-weight bond paper should be used. This type of paper allows blueprints to be produced directly without the need for tracing.

Pencil lead is available in a variety of grades from extremely hard (9H) to very soft (7B), with a medium grade of H or 2H selected for general-purpose technical drawings. Mechanical supports are preferred over wood.

***Drafting Boards.*** Constructed from straight-grained wood, drafting boards for electronic drawings should be flat, with true edges, and of a convenient size—18 by 24 inches being a common choice.

***T-Square.*** Although available in various lengths, the 24-inch T-square is suitable for the drafting board selected previously. Blades made of hardwoods such as maple, with clear celluloid edges, are preferred.

***Triangles.*** To facilitate drawing angles, the 45° and the 30 by 60° triangles are important items for hand-drafting. Most often, they are made of clear or transparent celluloid, with a 6- to 8-in.-long leg.

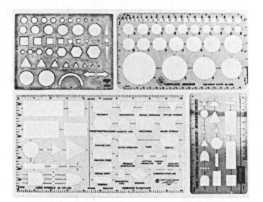

**Figure 4.11** Drafting and electronic layout templates.

***Protractors.*** When angles cannot be obtained with the common triangles, the protractor is used. For most applications, the clear or shaded celluloid protractor is satisfactory.

***Engineering Scales.*** Mechanical engineer's scales are the most useful in electronic drafting. These scales provide full-size, one-half, one-fourth, and one-eighth divisions that are satisfactory for most size reductions.

***Drafting Instruments and Templates.*** Drawing instrument sets contain *dividers, compasses,* and *inking pens,* and are necessary to produce quality, accurate plates. Occasionally, however, time is more important than precision. The use of templates, such as those shown in Fig. 4.11, can aid in producing good quality drawings with a considerable savings of both time and effort. Templates are used extensively as a quick means of reproducing electronic symbols and other shapes, particularly circles. They consist of various shaped cutouts in a piece of clear or shaded plastic. When these cutouts are outlined with a pencil or pen, the result is a uniform and neatly shaped electronic symbol or figure.

**4.5.2 Computer-Aided Drafting Software**   The basic building blocks for a typical computer-aided drafting system consist of a personal computer with color monitor, a keyboard and mouse as input peripherals, and a printer or plotter as the output peripheral. This hardware, together with the appropriate CAD software package, forms a complete drafting/design system. Two of the more popular software packages used in industry are AutoCAD and VersaCAD. Both of these products are very powerful and are designed for general-purpose drafting work, with a large selection of features and options.

AutoCAD, Release 12, was selected to illustrate this technology and is described in some detail in Chapter 7. Since it is a general-purpose drafting package, the primary operations of drawing lines, rectangles, and circles, as well as dimensioning, labeling, making modifications, and saving and printing files, are generic and apply equally to developing the chassis drawings described in this chapter as well as designing the PCB artworks of Chapter 7. Refer to Chapter 7 for a discussion of the basic methods and techniques employed in developing drawings using AutoCAD, Release 12.

## 4.6 BASIC LINE CONVENTIONS FOR DRAWINGS

Since the primary objective in engineering drafting is to transmit detailed information pictorially, a basic knowledge of the various types of lines used and their application is essential. The symbolic lines conventionally accepted are shown in Fig. 4.12. Whether by conventional drafting means or by CAD, the line width will vary depending on the type of symbol used. Thin lines are used for *dimension, center, extension, long-break,* and *section*

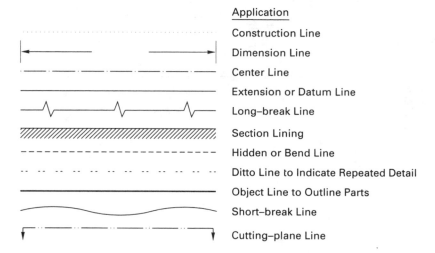

Application

Construction Line

Dimension Line

Center Line

Extension or Datum Line

Long–break Line

Section Lining

Hidden or Bend Line

Ditto Line to Indicate Repeated Detail

Object Line to Outline Parts

Short–break Line

Cutting–plane Line

**Figure 4.12** Basic line conventions used in engineering drafting.

symbols. Medium thickness is used for *hidden* and *ditto* lines. Thick lines are reserved for *object, break,* and *cutting-plane* lines.

## 4.7 DIMENSIONING FOR CHASSIS LAYOUTS

Several acceptable dimensioning techniques are used in chassis layouts, the choice of which depends on the configuration being dimensioned and the degree of accuracy required. These techniques are datum-line (or base-line), center-line, continuous, and tabulation dimensioning.

*Datum-line,* or *base-line, dimensioning* uses any two edges that are perpendicular to each other. These reference edges are called *datum lines.* Each dimension made uses one of the datum lines as a reference. This technique eliminates cumulative error, since each dimension is independent of all others. This method of dimensioning is often used for close-tolerance work, especially when mating parts are to be fabricated. Figure 4.13 illustrates base-line dimensioning.

In *pure center-line dimensioning,* the stock is first equally divided vertically and horizontally. The locations of all holes are made from these reference lines. No other lines are used as a reference. In electronic layout, datum-line dimensioning is preferable when locating holes, since edge reference facilitates this layout, as described in Chapter 20.

When holes are to be formed that match a particular pattern on a component, such as the mounting holes on a speaker, a variation of pure center-line dimensioning is preferable. Base-line or pure center-line dimensioning would create a larger tolerance in this application. For this reason, independent dimensioning of related holes or of hole patterns should be avoided. In this technique the center of the largest or major hole in the pattern is located from datum lines. All other hole centers are dimensioned using these major hole center lines as a reference. Figure 4.14 shows an example of center-line dimensioning.

As drawings become more complex, it is necessary to use both datum-line and center-line dimensioning, especially when the exclusive use of either one would adversely affect clarity. Figure 4.15 shows an example of combining both techniques on one drawing. Note that the use of both techniques reduces the number of dimension lines crossing. Although it is acceptable practice for dimension lines to cross, dimensions may become difficult to read if this practice is used extensively.

When tolerances are not critical, *continuous dimensioning* may be used. This method is also shown in Fig. 4.15, and is commonly used in combination with center-line dimensioning. This technique of dimensioning can cause large tolerances and should be used only when critical dimensions are not required.

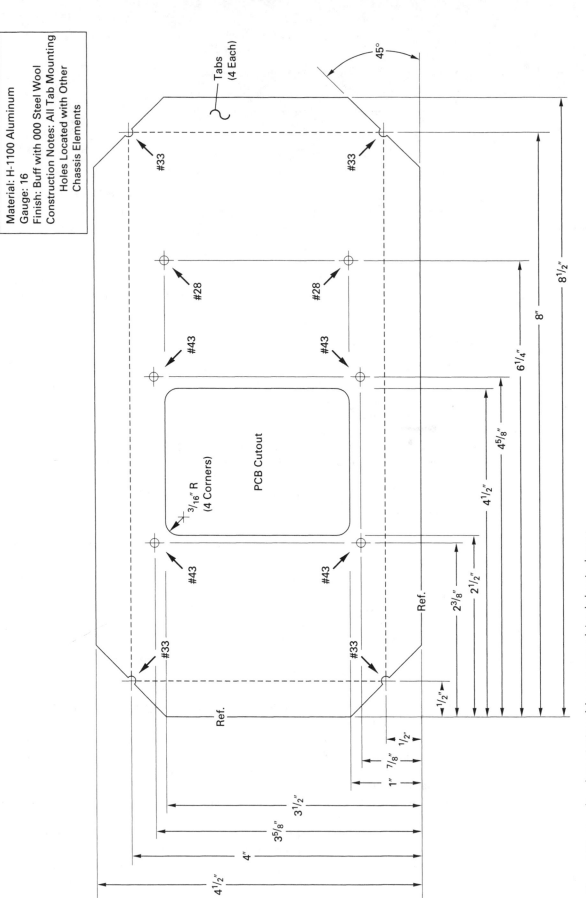

**Figure 4.13** Base-line dimensioning of the power supply's subchassis element.

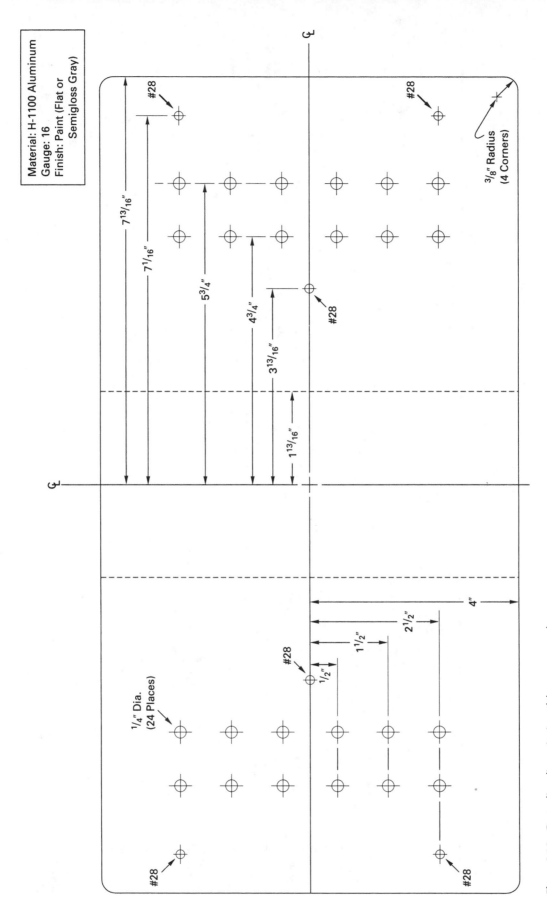

**Figure 4.14** Center-line dimensioning of the power supply cover.

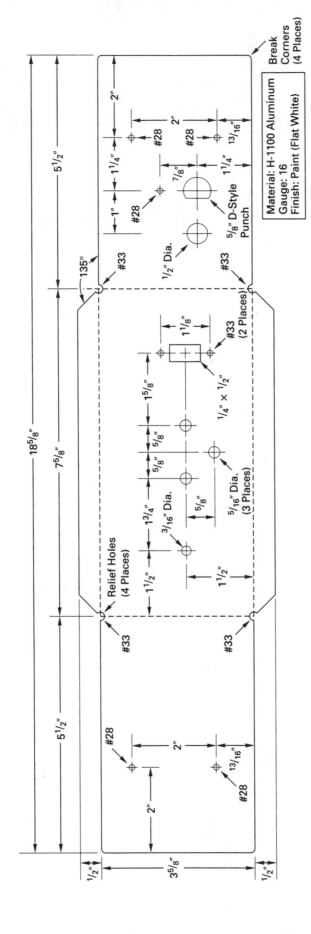

**Figure 4.15** Continuous dimensioning of the power supply front-and-side chassis element.

When dimensioning complex, high-density hole patterns, hole-center locations may be presented in *tabular* form, with no dimension lines shown on the drawing. Two datum lines are selected and identified with an appropriate designation, such as *X* and *Y,* for the horizontal and vertical datum lines, respectively. All hole centers are then identified literally, the same designation being used for common-size holes. A table identifying each hole, with its size and distance from each datum line, is then prepared and attached to the drawing.

When a high degree of dimensional tolerance is required, measurements should be taken in decimal form. Since this degree of accuracy is not normally required in prototype construction, fractional measurements are used throughout this section. With fractional dimensioning, accuracy to $\frac{1}{64}$ inch is easily obtainable. Maximum allowable variation above and below the nominal value by which the given dimension may differ from the drawing to the actual work appears with the dimension where applicable.

## 4.8 ASSEMBLY DRAWINGS

After a prototype has been completed, final drawings and photographs are made of all assemblies and subassemblies. These drawings may be of several variations: *two-dimensional top, bottom* and *side views; three-dimensional views; isometric projections;* and *exploded views.* The designated views must contain all pertinent information concerning the proper location and orientation of each component and mechanical part. All parts shown in assembly drawings should be identified and labeled consistent with the designations used on the schematic diagram. A common arrangement is a numbering system relating to the parts and accompanying the drawings.

Auxiliary views are frequently necessary, especially when complex or high-density packaging obscures some of the major sections of the package. An auxiliary view is a drawing of the obscured portion of the package after eliminating obstructions of the primary views. Figure 4.16 shows a partial assembly layout for a complex amplifier project, including an auxiliary view.

Photographs are particularly advantageous because of their high quality and three-dimensional perspective characteristics. In addition, nomenclature and reference information can be superimposed on photographs. Drawings, however, are absolutely indispensable for fabrication, assembly, and wiring because of the excessive amount of explanatory information necessary.

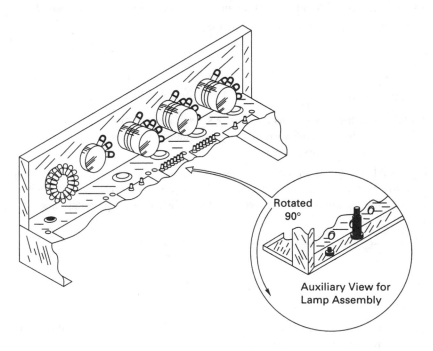

**Figure 4.16** Front panel assembly drawings for a complex amplifier project.

The next phase of development, after the package planning has been finished and the detailed working drawings for the sheet metal elements are completed, is to design printed circuit boards. In Unit 2, information is given on designing both single-sided and double-sided printed circuit boards, an important element in virtually all electronic packages.

Note that each unit in this book is a separate entity in itself. The switching-regulator power supply has been used as an instructional vehicle in this unit only, and is not used in the units to follow.

## EXERCISES

### A. Questions

4.1    How should a scale be placed when measuring the distance between mounting holes to obtain an accurate on-center dimension? *use inside the caliper*

4.2    Why is it not advisable to use the end of a rule as a reference when taking measurements? What reference should be used?

4.3    Describe two methods for obtaining the diameter of round components.  *outside    * inside — hole   dial caliper  NO hole*

4.4    How many steps are required to obtain the reading using a micrometer equipped to read to one ten-thousandth of an inch?

4.5    What information is shown on an engineering sketch? *dimensional*

4.6    How are the sizes of drafting paper designated? *A → K (size)*

4.7    What type of drawing paper is preferred for electronic drafting? *H*

4.8    Describe each of the following dimensioning techniques:

   a. Datum-line

   b. Center-line

   c. Continuous

   d. Tabulation

4.9    Which dimensioning technique results in large tolerances? *decimal form — continuous*

4.10   When a high degree of dimensional tolerance is required, in what form should measurements be taken?

4.11   When are auxiliary views on a drawing required?

### B. True or False

Circle *T* if the statement is true, or *F* if any part of the statement is false.

*0.001*

| | | |
|---|---|---|
| 4.1 | The thimble on a micrometer has 25 divisions, each representing 0.010 inch. | T **(F)** |
| 4.2 | Engineering sketches are easily produced using quadruled paper without the need of special drafting equipment. | **(T)** F |
| 4.3 | Pencil grades H to 2H are suitable for general-purpose work in technical drawings. | **(T)** F |
| 4.4 | As paper size increases (i.e., A, B, C, D), the height and width dimensions double. | T **(F)** |
| 4.5 | Datum-line dimensioning uses any two reference edges that are perpendicular to each other. | **(T)** F |
| 4.6 | When tolerances are critical, continuous dimensioning is used. *(not require)* | T **(F)** |

### C. Multiple Choice

Circle the correct answer for each statement.

4.1    For maximum accuracy in taking measurements with a scale, the work should be aligned with the (*end, first major graduation*) of the scale.

4.2      Micrometers are capable of measuring accurately to (*0.0001, 0.001*) inch.

4.3      Engineering sketches should be produced on (*plain, quadruled*) paper.

4.4      Another name for base-line dimensioning is (*datum, tabulation*).

4.5      It is (*acceptable, not acceptable*) to combine several dimensioning techniques to achieve clarity.

4.6      When dimensional tolerance is not critical, (*center-line, continuous, base-line*) dimensioning may be used.

## D.   Matching

Match each item in Column A to the most appropriate item in Column B.

| COLUMN A | COLUMN B |
|---|---|
| 1.  Engineering sketch | a.  Spindle, anvil |
| 2.  Micrometer | b.  Continuous |
| 3.  Outside | c.  $\frac{1}{10,000}$ inch |
| 4.  Subdivisions | d.  Isometric projections |
| 5.  Steel scale | e.  Caliper |
| 6.  Dimensioning | f.  Quadruled paper |
| 7.  Assembly drawing | g.  $\frac{1}{4}, \frac{1}{8}, \frac{1}{16}, \frac{1}{32}$ inch |
| 8.  Vernier | h.  Hook rule |

## E.  Problems

4.1      Read the measurements shown in Figs. 4.2 and 4.3b.

4.2      Read the micrometer settings shown in Fig. 4.7 and those in Fig. 4.17.

4.3      Read the vernier caliper settings shown in Fig. 4.18.

4.4      Measure the body and lead diameters and body lengths of $\frac{1}{4}$-watt through 10-watt resistors. Also determine the body and lead diameters of several electrolytic capacitors and the body thickness and lead spacing of various-sized disk capacitors. Tabulate all measurements for future reference.

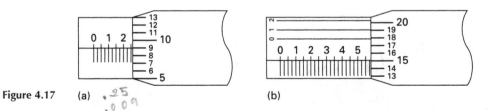

**Figure 4.17**   (a)          (b)

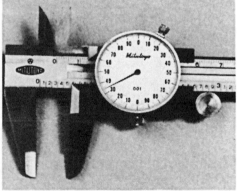

**Figure 4.18**

# Unit
# TWO

## Designing Printed Circuit Boards with CAD Techniques

# 5 Printed Circuit Board Materials

*Upon completion of this chapter on printed circuit board materials, the student should be able to*

- Know the classification of printed circuit boards.
- Be familiar with the characteristics of printed circuit cable.
- Be familiar with the various grades of rigid printed circuit laminates and their characteristics.
- Understand the properties of copper foil on rigid printed circuit boards.
- Select the appropriate grade of printed circuit laminate and copper foil thickness to satisfy the specifications for a variety of applications.

## 5.0 INTRODUCTION

With the advent of the transistor and the more recent heavy reliance on integrated circuits, electronic packaging and wiring problems have been compounded. No longer can the technician rely solely on hand-wiring methods; now designs must be oriented toward more extensive use of printed circuitry. Many modern electronic systems would be virtually impossible to package without incorporating printed circuits into their design.

Printed circuits are *metal foil conducting patterns,* usually copper, bonded to a *substrate* (insulating base material) for support. The metal pattern serves as the connecting medium for the electrical components that are assembled on the opposite side of the board. Component leads are fed into holes that are drilled or punched through the base material and foil. These leads are soldered to the conducting pattern to form the complete printed circuit.

The advantages of printed circuits over conventional wiring methods are that they (1) adapt easily to miniaturization and modular design, (2) provide uniformity in production, (3) cost less, (4) virtually eliminate wiring error, and (5) minimize assembly and inspection time. Printed circuit techniques lend themselves readily to mass production, resulting in a highly reliable package. Although the format of this text is focused on prototype construction, it must be assumed that eventually the package will be released for production. Consequently, the technician should design the package with an eye toward this end by selecting materials and configurations that will reduce construction problems.

To completely explain the sequential processes involved in printed circuit board design and fabrication, the next 13 chapters are devoted exclusively to these topics. This chapter discusses the initial step in printed circuit construction—the selection of the most appropriate type of printed circuit board material. Topics discussed are classification of printed circuit boards, insulating base materials, conductive foils, bonding, and tabulated information of board characteristics to aid the technician in selection.

## 5.1 CLASSIFICATION OF PRINTED CIRCUIT BOARDS

Printed circuits are designed as either *rigid* or *flexible* and are further classified into *single-sided, double-sided,* or *multilayer* conductors. Regardless of the type, all printed circuit boards are formed with an insulating base material on which a conducting foil is either chemically or mechanically bonded. When processed, the printed circuit board provides both an electrical wiring path and a mechanical support for the components.

Rigid printed circuit boards of the single-sided type are the most frequently used, finding extensive application in relatively uncomplicated circuitry. Figure 5.1a shows a rigid single-sided blank before processing. The completed conductor pattern ready for component mounting is shown in Fig. 5.1b.

Rigid printed circuit boards of the double-sided type consist of conducting foil bonded to *both* sides of the insulating base. This type of board is employed when circuit complexities make it difficult or impossible to develop practical wiring layouts on one side only. Double-sided boards also reduce the overall size of a comparable circuit constructed on a single-sided board, since components are mounted on both sides. Figure 5.2 shows an assembled double-sided board. Interconnections among the conductive patterns on both sides of the board are typically accomplished by plated-through holes, which is a manufacturing process discussed in Chapter 8. With the use of these interconnections, wires or component leads can be soldered between the foil patterns on opposite sides of the base materials.

Multilayer rigid printed circuit materials were developed to satisfy the needs of the more recent state of the art, which demands extremely complex wiring layouts in a relatively small space. Multilayer board material selection as well as fabrication are discussed in detail in Chapter 14.

*Flexible printed cables* consist of conductors laminated between layers of insulation and exposed for soldering only at the terminal points, as shown in Fig. 5.3. Flexible printed cables are used in a manner similar to ordinary cable but are extremely flexible. In addition, the flat configuration of these circuits make them desirable for system interconnections where a round cable may prove to be too cumbersome. *Teflon, polymide, polyester, polyvinyls, polypropylene,* and *polyethylene* are typical insulating materials used in flexible printed cables. Table 5.1 shows

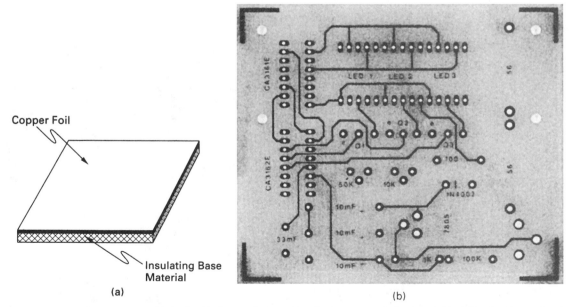

**Figure 5.1**  Rigid single-sided printed circuit boards: (a) copper-clad printed circuit blank; (b) processed single-sided printed circuit conductor pattern.

*Printed Circuit Board Materials*                                                                            **61**

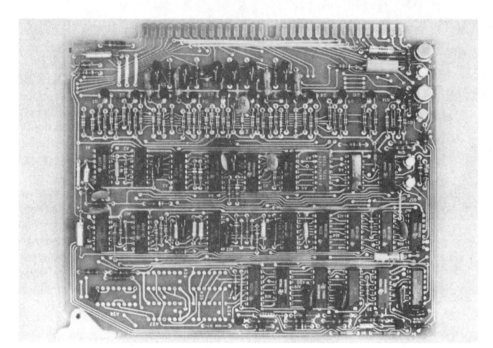

**Figure 5.2** Assembled double-sided rigid printed circuit board.

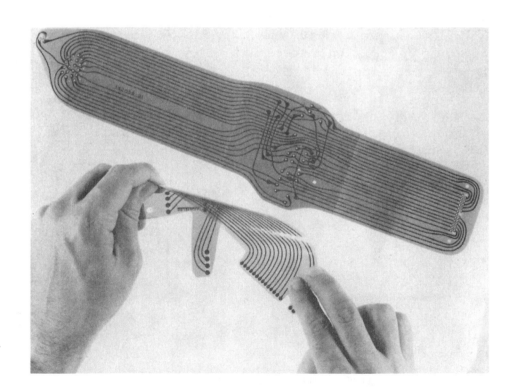

**Figure 5.3** Flexible printed circuit cables. *Courtesy of DuPont Company.*

**TABLE 5.1**

Insulation Characteristics for Flexible Printed Cables (Courtesy of Insulfab Plastics, Inc.)

| | TFE Fluorocarbon | TFE Glass Cloth | FEP Fluorocarbon | FEP Glass Cloth | Polymide | Polyester | Polychlorotri-fluoroethylene | Polyvinyl Fluoride | Polypro-pylene | Polyvinyl Chloride | Poly-ethylene |
|---|---|---|---|---|---|---|---|---|---|---|---|
| Specific gravity | 2.15 | 2.2 | 2.15 | 2.2 | 1.42 | 1.395 | 2.10 | 1.38 | 0.905 | 1.25 | .93 |
| Square inches of 1 mil film per pound | 12,800 | 13,000 | 12,900 | 13,000 | 19,450 | 21,500 | 12,000 | 20,000 | 31,000 | 22,000 | 30,100 |
| Service temp. deg. C (Minimum) | −70 | −70 | −225 | −70 | −250 | −60 | −70 | −70 | −55 | −40 | −20 |
| (Maximum) | 250 | 250 | 200 | 250 | +250 | 150 | 150 | 105 | 125 | 85 | 60 |
| Flammability | Nil | Nil | Nil | Nil | Nil | Yes | Nil | Yes | Yes | Slight | Yes |
| Appearance | Translucent | Tan | Clearbluish | Tan | Amber | Clear | Clear | Clear | Clear | Translucent | Clear |
| Thermal expansion × 10 inches/inch/deg. F | 70 | Low | 50 | Low | 11 | 15 | 45 | 28 | 61 | — | — |
| Bondability with adhesives | Good | Good | Good | Good | Good | Good | Good | Good | Poor | Good | Poor |
| Bondability to itself | Good | Poor | Good | Good | Poor | Poor | Good | Good | Good | Good | Good |
| Tensile strength PSI @77°F | 3,000 | 20,000 | 3,000 | 20,000 | 20,000 | 20,000 | 4,500 | 8,000 | 5,700 | 3,000 | 2,000 |
| Modulus of elasticity PSI | 80,000 | 3 | 70,000 | 3 | 430,000 | 550,000 | 200,000 | 280,000 | 170,000 | — | 50,000 |
| Volume resistivity ohms-cm | $2 \times 10$ | 10 | 10 | 10 | 10 | $1 \times 10$ | $1 \times 10$ | $3 \times 10$ | 10 | $1 \times 10$ | $1 \times 10$ |
| Dielectric constant 10-10 cycles | 2.2 | 2.5/5 | 2.1 | 2.5/5 | 3.5 | 2.8-3.7 | 2.5 | 7.0 | 2.0 | 3-4 | 2.2 |
| Dissipation factor 10-10 cycles | .0002 | .0007/.001 | .0002 | .0001/.001 | .002/.014 | .002-.016 | .015 | .009-.041 | .0002/.0003 | .14 | .0006 |
| Dielectric strength (5 mils thickness) volt/mil | 800 | 650/1600 | 3,000 | 650/1600 | 3,500 | 3,500 | 2,000 | 2,000 750v/mil | .125 in thk | 800 | 1,500 |
| Chemical resistance | Excellent | Excellent | Excellent | Excellent | Excellent | Excellent | Excellent | Good | Excellent | Good | Excellent |
| Water absorption, % | 0 | .10/.68 | 0 | .18/.30 | 3 | 0.5 | 0 | 15 | .01 | .10 | 0 |
| Sunlight resistance | Excellent | Excellent | Excellent | Excellent | Excellent | Fair | Excellent | Excellent | Low | Fair | Low |

a comparison of the various characteristics for common insulations. Since various methods and adhesives are used for the internal bond of these laminates, no one method of insulation removal will suffice for all types of these cables. The insulation is usually stripped by *friction wheel strippers, knife-edge strippers, chemical dipping,* or *melting* by a concentration of hot air. Because flexible printed cables have a large flat conducting area, they can handle larger currents than conventional wire cable. Figure 5.4 shows the approximate current capacities and derating values when a multiple number of conductors are laminated within a common cable.

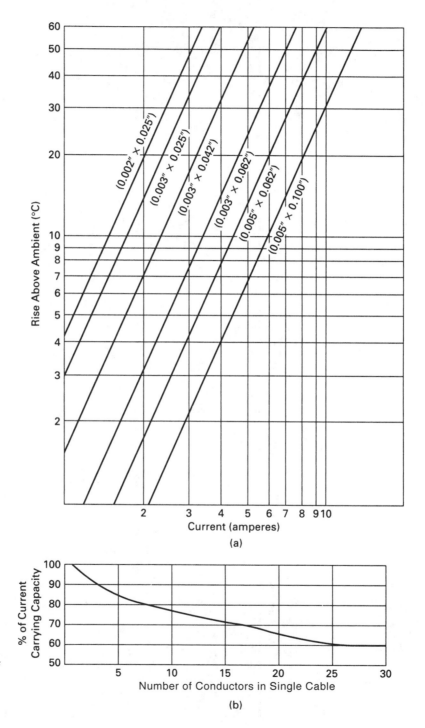

**Figure 5.4** Conductor Temperature Rise and Current Derating for Flat Cables: (a) Rise in Temperature Above Ambient as a Function of Current; (b) Approximate Derating Curve for Flat Cable.

Printed circuit boards of the rigid variety will be discussed exclusively in the remaining sections of this text since they are by far the most common. The insulating or resin material that provides the support for the conductor pattern and components for these boards is either a *thermosetting plastic* or a high-temperature-resistant *thermoplastic polymer.* Our discussion will be limited to the thermosetting plastics since they are used almost exclusively in rigid boards.

Base materials for printed circuit boards are laminates containing a *reinforcement* such as *paper, glass,* or *fabrics,* including *cotton cloth, asbestos, nylon cloth,* and *glass cloth.* These laminates are formed under pressure and heat. The thermosetting plastic resins common to the manufacture of printed circuit boards are *phenolic, epoxy, melamine, silicone,* and *Teflon.* Silicon-glass is mechanically inferior and more expensive than either phenolic, epoxy, or melamine resin boards. Teflon is used primarily for microwave applications and is approximately fifteen times more expensive than phenolic and at least five times as costly as the epoxy and melamine resin types. The phenolic types are less expensive than the epoxy resins and for many applications are both electrically and mechanically suitable. In general, the epoxy types possess electrical and mechanical characteristics superior to phenolic or melamine. In most cases, the reliability of a particular type of board will by far outweigh the cost factor.

To aid the technician in selecting the most suitable base material for a specific application, there follows a description of some of the electrical, mechanical, and chemical qualities of the more common base materials (see Table 5.2). The classifications are in accordance with NEMA (National Electrical Manufacturers Association) standards. These standards designate phenolic resin printed circuit boards as *X, XX,* or *XXX,* which represent the amount of resin present, the amount increasing with the number of *X*s. In general, the more resin content, the more improved are the electrical and mechanical characteristics of the board. A *P* suffix together with the *X* designation denotes that the phenolic is capable of being punched, but only at elevated temperatures, where a *PC* suffix indicates cold-punch qualities.

***Grade XX.*** This grade of paper-base phenolic possesses fair electrical characteristics and good mechanical properties. It machines well but is not recommended for punching. It offers little abrasive action to drills and, as a result, does not dull these bits easily. This grade is used by many manufacturers as a reference for determining the costs of other laminate types. For example, if the price index of nylon cloth phenolic is 3.5, this indicates that its cost is that much more than grade *XX,* which has a cost index of 1 (see Table 5.3).

***Grade XXP.*** This type is similar in electrical and mechanical properties to grade *XX* and it can be punched at increased temperatures of between 200 and 250°F (93 to 121°C).

***Grade XXX.*** This material possesses some improved electrical and mechanical properties over the *XX* type and is recommended for use at radio frequencies. It is available in a *self-extinguishing* form, meaning that it will not support combustion if ignited. This grade possesses fair *dimensional stability* (i.e., it maintains its size within close tolerances under environmental changes such as temperature and moisture).

***Grade XXXP.*** This grade is recommended for most general-purpose applications and is similar to grade *XXX,* but possesses an additional advantage in that it can be punched at elevated temperatures. Frequencies up to 10 MHz are permitted with this grade.

***Grade XXXPC.*** This is similar in electrical and mechanical properties to grade *XXXP* board and has higher insulation resistance and lower water absorption. These qualities make this type of board particularly useful for high-humidity applications. In addition, this material can be punched at temperatures of 70 to 120°F (21 to 49°C). For these reasons, *XXXPC* printed circuit boards are widely preferred over grades *XX* and *XXP.* However, when punching and moisture are not considerations, grade *XXXP* is generally selected.

## TABLE 5.2a
Insulating Material: Grades, Uses, and Properties (*Courtesy of Insulfab Plastics, Inc.*)

| Description Characteristics—Applications | NEMA Grade | Conforms to Military Spec. MIL-P | Color* | Standard Thicknesses (others available) | Mechanical Times 1000 | | | | |
|---|---|---|---|---|---|---|---|---|---|
| *These are AVERAGE and not GUARANTEED properties. Guaranteed values are NEMA or applicable MIL-P specifications.* | | | | | Compr. Str. (Lb./Sq.In.) ½" Face | Flexural Str. (Lb./Sq.In.) 1/16" Face—Lengthwise | Flexural Str. (Lb./Sq.In.) 1/16" Face—Crosswise | Tensile Str. (Lb./Sq.In.) 1/16" Lengthwise | Tensile Str. (Lb./Sq.In.) 1/16" Crosswise |
| Paper/Phenolic—General purpose. For panels, contractors, terminal blocks—high strength, low cost | X | | N.B.C. | .010–2.0 | 38.0 | 28.0 | 20.0 | 20.6 | 16.0 |
| Paper/Phenolic—For applications requiring good mechanical strength and dim. stability with elec. properties secondary | XP | | N.B.C. | .020–2.0 | 35.0 | 30.0 | 25.0 | 20.0 | 16.0 |
| Paper/Phenolic—Excellent cold punching, high strength | XPC | | N.B.C. | .015–.250 | 42.0 | 23.0 | 20.0 | 17.0 | 12.7 |
| Paper/Phenolic—An economy grade offering excellent shear and punch characteristics at room temp. | XPC | | N.B.C. | .031–.125 | 32.0 | 24.0 | 20.0 | 12.0 | 10.0 |
| Paper/Phenolic—Good mechanical, electrical, and machining properties | XX | 3115, PBG | N.B. | .010–2.0 | 34.0 | 22.0 | 19.0 | 16.0 | 13.0 |
| Paper/Phenolic—Excellent punching, high mechanical, and high electrical. Best all purpose | XXP | | N.B.C. | .015–.250 | 41.0 | 23.0 | 18.0 | 16.3 | 13.8 |
| Paper/Phenolic—Excellent electrical grade used where dim. stability or min. cold flow is essential | XXX | 3115, PBE | N.B. | 0.10–2.0 | 35.0 | 20.0 | 17.0 | 13.1 | 10.5 |
| Paper/Phenolic—Best quality paper base. High insulation resistance, low loss material | XXXP | 3115, PBEP | N.B. | .031–.250 | 28.0 | 19.0 | 17.5 | 12.0 | 9.5 |
| Paper/Phenolic—High quality, cold punching material. High insulation resistance | XXXPC | 3115, PBE-P | N.B. | .031–.50 | 30.0 | 19.0 | 15.0 | 14.0 | 11.0 |
| Cotton Cloth/Phenolic—General purpose. Excellent mach. and punching. For panels, wedges, gears, wear strips | C | 18324 15035, FBM | N.B. | .031–10.0 | 41.0 | 22.0 | 18.0 | 12.1 | 9.80 |
| Cotton Cloth/Phenolic—Best canvas base electrical grade. Excellent acid resistance. For panels, steam valve discs | CE | 15035, FBG | N.B. | .031–3.0 | 39.0 | 21.0 | 19.0 | 13.5 | 9.80 |
| Cotton Cloth/Phenolic. A finer weave than Grade C. Better for fine punching, close machining. Mechanical usage | L | 15035, FBI | N.B. | .020–4.0 | 39.0 | 21.5 | 18.0 | 11.9 | 7.9 |
| Cotton Cloth/Phenolic—Best insulation of cloth grades. Excellent resistance to moisture, mild acids, alkalies | LE | 15035, FBE | N.B. | .010–3.0 | 38.0 | 22.0 | 17.0 | 15.0 | 10.6 |
| Fine Weave Cotton Cloth/Phenolic, with graphite added. Non-insulator, used for thrust bearings, pump bearings | Linen Graphite | 5431 (AER.) | Green-Black | .008–6.0 | 37.0 | 20.0 | 16.0 | 15.0 | 11.0 |
| Asbestos Paper/Phenolic—Suitable for continuous use at 150° C. Good mechanical, poor electrical strength | A | 8059 (USAF) | N.B. | .015–1.0 | 45.0 | 28.0 | 21.0 | 18.0 | 13.0 |
| Asbestos Cloth/Phenolic-High heat resist., dimensional stab., low water absorp., low thermal expan., good impact resist. | AA | 8059 (USAF) | N. | .047–4.0 | 40.0 | 30.0 | 19.0 | 9.2 | 8.1 |
| Nylon Cloth/Phenolic—Excellent electrical and good mechanical properties under humid conditions | N1 | 15047-NPG | N. | .015–1.125 | 40.0 | 13.5 | 10.5 | 10.0 | 6.0 |
| Glass Cloth/Phenolic—Used in applications where stability of elec. and mech. properties is essential | G3 | | N. | .031–2.0 | 55.0 | 20.0 | 18.0 | 23.0 | 18.0 |
| Glass Cloth/Melamine—Excellent mechanical and electrical properties. Good arc resistance, flame retardance | G5 | 15037, GMG | Gray-Brown | .008–3.5 | 70.0 | 50.0 | 43.0 | 39.0 | 36.0 |
| Glass Cloth/Silicone—Suitable for high temp. (Class H) applications. Low loss electrical properties | G7 | 997, GSG | White | .010–2.0 | 62.0 | 38.0 | 32.0 | 33.0 | 31.0 |
| Glass Cloth/Melamine—An arc and flame resistance laminate with outstanding elec. prop. to meet MIL-P15037C, GME | G9 | 15037C, GME | Gray-Brown | .008–3.5 | 62.0 | 83.0 | 61.0 | 48.0 | 35.0 |
| Glass Cloth/Epoxy—High ins. resist., very low absorp. Highest bond strength of glass laminates. High stability in humidity | G10 | 18177, GEE | Green | .010–2.0 | 50.0 | 70.0 | 55.0 | 49.0 | 38.0 |
| Glass Cloth/Epoxy—High flexural strength retention at elevated temperatures. Extreme resistance to solvents | G11 | 18177, GEB | Green | .010–2.0 | 58.0 | 70.0 | 65.0 | 48.0 | 37.0 |
| Paper/Epoxy—Designed for optimum balance of mechanical and electrical properties, with excellent machinability | FR-3 | 22324, PEE | Ivory | .031–.50 | 32.0 | 26.5 | 22.0 | 12.0 | 9.0 |
| Glass Cloth/Epoxy—A flame resistant epoxy laminate with excellent machining and electrical properties | FR-4 | 18177, GEE | Green | .010–.125 | 55.0 | 82.0 | 75.0 | 48.0 | 37.0 |
| Glass Cloth/Epoxy—High strength retention at elevated temperatures. Self-extinguishing | FR-5 | 18177, GEB | Green | .010–1.5 | 50.0 | 58.0 | 52.0 | 49.0 | 36.0 |

Sheet Sizes: Range from 36″ to 48″ wide × 36″ to 120″ long, depending on Grade and Thickness. Consult nearest Insulfab office for additional information.

*Colors: Symbols N, B, or C refer to Natural, Black, or Chocolate. "N" ranges from light tan to dark brown. Other referenced colors are "Natural for grade." MIL-P specifications approve Natural colors only.

| Mod. of Elas. (Lb./Sq. In.) 1/16" Lengthwise (×10⁶) | Mod. of Elas. (Lb./Sq.In.) 1/16" Crosswise | Izod Impact (Ft.Lb./In. Notch) 1/16" Build-Up Edge—Lengthwise | Izod Impact (Ft.Lb./In. Notch) 1/16" Build-Up Edge—Crosswise | Bonding Strength 1/2≤ (lbs.) | Rockwell Hardness M Scale | Diel. Str. (V/Mil) Perp. to Lam. 1/16" S/S @ 23° C | Diel. Breakdown KV Par. to Lam. 1/16" S/S @ 23°C | Dissipation Factor @ 1 Mc. 1/16" As Received | Dissipation Factor @ 1 Mc. After 24 Hr. in Water 1/16" @ 23°C | Diel. Constant @ 1 Mc. 1/16" As Received | Diel. Constant @ 1 Mc. After 24 Hr. in Water 1/16" @ 23°C | Continuous Operation Temp. Limit Deg. C | Coef. Of Expansion In./In./C° Lengthwise ×10⁻⁵ 1/16" | Coef. Of Expansion In./In./C° Crosswise ×10⁻⁵ 1/16" | Deformation Under Load 1/16" (Percentage) @ 70°C | Thermal Conductivity BTU/Hr./Sq. Ft./In. Thick/F° | Water Absorp. % 24 Hr. @ 23°C 1/16" Thick | Specific Gravity | NEMA Grade |
|---|---|---|---|---|---|---|---|---|---|---|---|---|---|---|---|---|---|---|---|
| 1.70 | 1.40 | 1.00 | 0.80 | 1030 | 102 | 580 | 45 | 0.042 | 0.047 | 5.4 | 5.6 | 120 | 2.0 | 2.5 | 0.98 | 1.6 | 1.70 | 1.38 | X |
| 1.50 | 1.10 | 0.70 | 0.60 | 1100 | 110 | 550 | 62 | 0.054 | 0.080 | 6.5 | 7.5 | 120 | 2.0 | 2.5 | 0.60 | 1.6 | 3.00 | 1.44 | XP |
| 1.00 | 0.80 | 0.65 | 0.60 | — | 93 | 470 | 65 | 0.046 | 0.078 | 5.1 | 6.5 | 120 | 1.6 | 3.0 | 1.69 | 1.7 | 5.00 | 1.36 | XPC |
| 1.40 | 1.00 | 0.90 | 0.75 | — | 85 | 550 | 73 | 0.051 | 0.073 | 5.1 | 5.8 | 120 | 1.6 | 3.0 | 2.00 | 1.7 | 2.60 | 1.37 | XPC |
| 1.26 | 0.88 | 0.60 | 0.50 | 900 | 102 | 625 | 60 | 0.040 | 0.046 | 5.0 | 5.4 | 120 | 1.1 | 2.6 | 0.93 | 1.7 | 1.53 | 1.35 | XX |
| 1.10 | 0.80 | 0.80 | 0.70 | — | 105 | 555 | 65 | 0.036 | 0.044 | 5.0 | 5.3 | 120 | 1.5 | 1.7 | 1.59 | 1.7 | 1.25 | 1.36 | XXP |
| 1.36 | 0.90 | 0.60 | 0.50 | 960 | 110 | 745 | 68 | 0.034 | 0.036 | 5.0 | 5.4 | 120 | 1.7 | 3.5 | 0.56 | 1.8 | 0.78 | 1.35 | XXX |
| 0.62 | 0.47 | 0.5 | 0.5 | — | 98 | 740 | 90 | 0.028 | 0.030 | 3.8 | 3.9 | 120 | 1.06 | 1.21 | 1.64 | 1.7 | 0.40 | 1.27 | XXXP |
| 0.70 | 0.60 | 0.60 | 0.50 | 1050 | 104 | 740 | 70 | 0.028 | 0.032 | 4.1 | 4.3 | 120 | 1.1 | 1.2 | 1.67 | 1.7 | 0.40 | 1.28 | XXXPC |
| 1.05 | 0.79 | 2.80 | 2.40 | 2100 | 103 | 310 | 40 | 0.053 | 0.089 | 5.2 | 5.8 | 120 | 1.8 | 2.2 | 1.00 | 2.3 | 2.00 | 1.33 | C |
| 1.14 | 1.00 | 2.50 | 2.20 | 1800 | 104 | 400 | 60 | 0.045 | 0.050 | 5.0 | 5.6 | 120 | 1.8* | 2.2 | 1.30 | 2.3 | 1.52 | 1.35 | CE |
| 1.00 | 0.80 | 1.70 | 1.40 | 1820 | 105 | 300 | 50 | 0.055 | 0.070 | 5.8 | 6.5 | 120 | 2.2 | 3.5 | 1.20 | 2.2 | 1.60 | 1.33 | L |
| 0.91 | 0.75 | 2.00 | 1.60 | 1820 | 108 | 430 | 60 | 0.045 | 0.050 | 5.6 | 6.0 | 120 | 2.2 | 3.5 | 0.94 | 2.2 | 1.20 | 1.33 | LE |
| 0.84 | 0.70 | 1.87 | 1.15 | 1700 | 105 | — | — | — | — | — | — | 120 | 2.2 | 3.5 | 0.96 | 2.9 | 2.00 | Linen 1.40 | Graphite |
| 1.30 | 1.10 | 3.00 | 2.80 | 1250 | 100 | 120 | 5 | — | — | — | — | 150 | 1.0 | 1.3 | 1.06 | 2.3 | 1.00 | 1.56 | A |
| 1.42 | 1.06 | 5.00 | 3.50 | 2100 | 99 | 90 | 10 | — | — | — | — | 170 | 0.9 | 1.4 | 1.30 | 3.5 | 1.20 | 1.64 | AA |
| 0.40 | 0.35 | 4.00 | 2.50 | 1500 | 85 | 330 | 70 | 0.030 | 0.040 | 3.7 | 3.8 | 120 | 3.0 | 5.0 | 3.40 | 2.2 | 0.50 | 1.17 | N1 |
| 1.70 | 1.50 | 8.00 | 6.00 | 1010 | 103 | 500 | 60 | 0.018 | 0.090 | 4.9 | 8.0 | 150 | 1.0 | 1.5 | 0.30 | 1.8 | 2.10 | 1.73 | G3 |
| 2.10 | 1.90 | 13.00 | 10.00 | 1800 | 123 | 395 | 34 | 0.011 | 0.065 | 6.1 | 7.2 | 150 | 1.0 | 1.1 | 0.60 | 2.4 | 2.20 | 1.90 | G5 |
| 1.80 | 1.60 | 16.00 | 16.00 | 850 | 102 | 400 | 70 | 0.002 | 0.003 | 4.1 | 4.2 | 250 | 1.0 | 1.0 | 0.25 | 1.4 | 1.77 | 0.10 | G7 |
| 3.20 | 2.20 | 15.00 | 13.00 | 2100 | 115 | 400 | 65 | 0.012 | 0.015 | 7.0 | 7.2 | 150 | 1.5 | 1.8 | 0.50 | 2.4 | 0.68 | 1.92 | G9 |
| 1.90 | 1.80 | 10.00 | 8.00 | 2300 | 105 | 510 | 65 | 0.021 | 0.022 | 4.5 | 4.6 | 130 | 1.0 | 1.5 | 0.26 | 1.8 | 0.20 | 1.80 | G10 |
| 2.50 | 2.40 | 11.00 | 9.00 | 2100 | 114 | 600 | 60 | 0.018 | 0.022 | 4.8 | 5.0 | 150 | 1.0 | 1.4 | 0.10 | 1.8 | 0.20 | 1.77 | G11 |
| 1.00 | 0.90 | 0.70 | 0.65 | — | 95 | 550 | 65 | 0.031 | 0.032 | 4.3 | 4.4 | 120 | 1.3 | 2.5 | 1.50 | 1.6 | 0.40 | 1.42 | FR-3 |
| 2.80 | 2.50 | 11.00 | 9.00 | 2100 | 107 | 500 | 65 | 0.016 | 0.018 | 4.2 | 4.4 | 130 | 1.2 | 1.5 | 0.25 | 1.8 | 0.20 | 1.93 | FR-4 |
| 2.50 | 2.40 | 13.00 | 9.00 | 2000 | 109 | 490 | 60 | 0.013 | 0.019 | 4.5 | 4.6 | 150 | 1.0 | 1.5 | 0.10 | 1.8 | 0.13 | 1.95 | FR-5 |

**TABLE 5.2b**

Physical Properties of Common PC Board Grades (*Courtesy of Insulfab Plastics, Inc.*)

| Physical Properties | MIL-P-18177 MIL-P-13949 | G-10 | FR-4 | 1226F |
|---|---|---|---|---|
| Water Absorption % 029/23 . . . . . . . . . . . . . . . . . . . . . . . . . . . | 0.35 (max.) | 0.14 | 0.13 | 0.13 |
| Specific Gravity. . . . . . . . . . . . . . . . . . . . . . . . . . . . . . . . | — | — | — | — |
| Rockwell Hardness M Scale . . . . . . . . . . . . . . . . . . . . . . . . . . . . . . | — | — | — | — |
| Flexural Strength | | | | |
| Lengthwise, condition A-lbs./sq. in.. . . . . . . . . . . . . . . . . . . . . . . | 50,000 (min.) | 79,000 | 73,100 | 71,000 |
| Crosswise, condition A . . . . . . . . . . . . . . . . . . . | 40,000 (min.) | 60,000 | 65,100 | 63,500 |
| Tested 150°C after 1 hour 150°C | | | | |
| Lengthwise. . . . . . . . . . . . . . . . . . . . . . . . . . . . | | | | |
| Crosswise. . . . . . . . . . . . . . . . . . . . . . . . . . . . | N/A | N/A | N/A | N/A |
| Tensile Strength, psi | | | | |
| Lengthwise, condition A . . . . . . . . . . . . . . . . . . . . . | — | 54,370 | 54,470 | 59,170 |
| Crosswise, condition A . . . . . . . . . . . . . . . . . . . . . | — | 45,565 | 45,065 | 44,765 |
| Compressive Strength, flatwise, psi. . . . . . . . . . . . . . . . . . | — | 33,033 | 33,033 | 33,033 |
| Izod Impact, edgewise, ft. lbs./in., 1/8 in. thick | | | | |
| Lengthwise, condition A . . . . . . . . . . . . . . . . . . . . | 7.0 (min.) | 13.3 | 13.3 | 13.3 |
| Crosswise, condition A . . . . . . . . . . . . . . . . . . . . | 5.5 (min.) | 8.5 | 8.5 | 8.5 |
| Bond, strength, condition A lbs., 1/2 in. thick . . . . . . . . . . . . | 2000 | 2560 | 2560 | 2560 |
| condition D 48/50 . . . . . . . . . . . . . . . . . . . . . . . . | 1600 | 2370 | 2370 | 2370 |
| *Electrical Properties* | | | | |
| Dielectric Strength, parallel, KV (360 sec.) . . . . . . . . . . . . . . | 35 (min.) | 55 | 55 | 55 |
| Condition A . . . . . . . . . . . . . . . . . . . . . . . . . . . | 30 (min.) | 50 | 50 | 50 |
| Condition D48/50 . . . . . . . . . . . . . . . . . . . . . . | | | | |
| Dissipation Factor, 1 MHz Condition D24/23 . . . . . . . . . . . . . . . . . . . . . . . . | 0.030 (max.) | 0.020 | 0.021 | 0.021 |
| Dielectric Constant, 1 MHz Condition 24/23. . . . . . . . . . . . . . . . . . . . . . . . . | 5.4 (max.) | 4.6 | 4.8 | 4.8 |
| Insulation Resistance, megohms, condition C96/35/90. . . . . . . . . . | | $.76 \times 10^7$ | $2.38 \times 10^7$ | $2.2 \times 10^7$ |
| Arc Resistance, seconds, condition D48/50. . . . . . . . . . . . . | 60 (min.) | 75 | 82 | 78 |
| Volume Resistivity, megohms cms, condition C96/35/90. . . . . . . . | $10^6$ (min.) | $8.2 \times 10^7$ | $5.5 \times 10^8$ | $2.3 \times 10^8$ |
| Surface Resistivity, megohms, condition C96/35/90 . . . . . . . . . . | $10^4$ (min.) | $4.9 \times 10^5$ | $2.10 \times 10^6$ | $2.2 \times 10^6$ |
| Flame Resistance, seconds . . . . . . . . . . . . . . . . . . . . . . . | 15 (max.) | NA | 3 | 4 |
| UL94 Flame Classification UL file—E45456 A & B . . . . . . . . . . . | | | UL94-VO | |
| Maximum Operating Temperature, °C continuous . . . . . . . . . . . . | | 130 | 130 | 130 |
| Blister Resistance, 1/16" × 1" × 1" NEMA LI-1971. . . . . . . . . . | Excellent | Excellent | Excellent | Excellent |
| Resistance to Plating Solution: trichlorethylene vapors and etching solutions. . . . . . . . . . . . . . . . . . . . . . . . | Excellent | Excellent | Excellent | |

## TABLE 5.3
### Insulating Material Comparison Chart (Courtesy of Insulfab Plastics, Inc.)

| Properties | X | XP | XPC | XX | XXP | XXX | XXXP | XXXPC | C | CE | L | LE | A | AA | N1 | G3 | G5 | G7 | G9 | G10 | G11 | GP0-1 | GP02 | FR2 | FR3 | FR4 | FR5 |
|---|---|---|---|---|---|---|---|---|---|---|---|---|---|---|---|---|---|---|---|---|---|---|---|---|---|---|---|
| Insulation Resistance | 5 | 2 | 3 | 7 | 5 | 7 | 7 | 10 | 1 | 2 | 1 | 3 | 1 | * | 10 | 1 | 2 | 2 | 6 | 10 | 10 | 1 | 1 | 10 | 10 | 10 | 10 |
| Dielectric Str. Perp. to Laminations | 9 | 6 | 8 | 10 | 8 | 9 | 8 | 10 | 1 | 6 | 1 | 6 | 2 | * | 8 | 8 | 8 | 5 | 8 | 10 | 6 | 6 | 5 | 10 | 10 | 10 | 10 |
| Dielectric Str. Par. to Laminations | 1 | 6 | 2 | 6 | 8 | 8 | 8 | 10 | 1 | 4 | 1 | 6 | 1 | * | 10 | 2 | 4 | 6 | 6 | 6 | 6 | 5 | 5 | 9 | 9 | 10 | 10 |
| Dielectric Losses Radio Frequency | * | 1 | 1 | 2 | 2 | 3 | 3 | 5 | * | 1 | 1 | 1 | * | * | 6 | 2 | 5 | 10 | 4 | 6 | 6 | 6 | 5 | 5 | * | 6 | 6 |
| Dielectric Losses Power Frequency | 1 | 1 | 1 | 2 | 3 | 3 | 3 | 4 | 1 | 2 | 1 | 3 | * | * | 9 | 2 | 6 | 10 | 3 | 10 | 9 | 6 | 6 | 4 | 4 | 10 | 10 |
| ARC Resistance | 1 | 1 | 1 | 1 | 1 | 1 | 1 | 1 | * | 1 | 1 | 1 | * | * | 1 | 1 | 8 | 10 | 9 | 5 | 5 | 3 | 2 | 5 | 5 | 5 | 5 |
| Dielectric Properties Stability | 1 | 3 | 2 | 5 | 5 | 7 | 6 | 8 | 1 | 3 | 1 | 4 | * | * | 10 | 3 | 5 | 6 | 9 | 8 | 8 | 3 | 3 | 8 | 8 | 8 | 8 |
| Mechanical Strength | 5 | 2 | 2 | 3 | 3 | 2 | 4 | 2 | 4 | 3 | 4 | 3 | 2 | 4 | 2 | 6 | 8 | 4 | 9 | 10 | 10 | 8 | 8 | 2 | 3 | 10 | 10 |
| Impact Strength | 3 | 3 | 2 | 1 | 2 | 1 | 2 | 1 | 4 | 3 | 3 | 2 | 2 | 5 | 4 | 8 | 8 | 7 | 9 | 10 | 10 | 8 | 8 | 1 | 1 | 10 | 10 |
| Bond Strength | 1 | 3 | * | 1 | 1 | 2 | 2 | * | 9 | 9 | 7 | 7 | 1 | 9 | 9 | 1 | 9 | 1 | 7 | 10 | 8 | 6 | 6 | * | * | 10 | 10 |
| Water Absorption | 1 | 2 | 2 | 3 | 5 | 6 | 5 | 8 | 3 | 5 | 3 | 5 | 4 | 4 | 9 | 4 | 4 | 9 | 7 | 10 | 10 | 6 | 6 | 8 | 9 | 10 | 10 |
| Dimensional Stability Due to Moisture | 1 | 2 | 2 | 3 | 4 | 5 | 3 | 6 | 1 | 3 | 1 | 3 | 5 | 7 | 5 | 7 | 8 | 10 | 8 | 10 | 10 | 6 | 6 | 6 | 6 | 10 | 10 |
| Heat Resistance | 3 | 3 | 2 | 2 | 2 | 2 | 2 | 2 | 3 | 3 | 3 | 4 | 7 | 8 | 3 | 8 | 9 | 10 | 9 | 6 | 9 | 5 | 6 | 2 | 3 | 6 | 9 |
| Dimensional Stability Due to Temp. | 5 | 2 | 5 | 3 | 3 | 2 | 2 | 2 | 4 | 3 | 4 | 2 | 8 | 9 | 1 | 10 | 9 | 10 | 10 | 7 | 8 | 5 | 5 | 2 | 3 | 7 | 8 |
| Specific Gravity | 1.36 | 1.36 | 1.36 | 1.36 | 1.36 | 1.32 | 1.33 | 1.34 | 1.36 | 1.35 | 1.36 | 1.35 | 1.55 | 1.58 | 1.15 | 1.67 | 1.90 | 1.73 | 1.95 | 1.80 | 1.80 | 1.74 | 1.75 | 1.38 | 1.42 | 1.90 | 1.93 |
| Machinability | 4 | 4 | 6 | 5 | 6 | 5 | 4 | 5 | 7 | 7 | 8 | 7 | 4 | 1 | 8 | 2 | 2 | 1 | 2 | 4 | 3 | 3 | 3 | 5 | 7 | 4 | 4 |
| Punching | 4 | 10 | 9 | 5 | 9 | 3 | 7 | 8 | 8 | 7 | 10 | 8 | 3 | 2 | 8 | 3 | 3 | 3 | 2 | 3 | 2 | 3 | 3 | 8 | 9 | 5 | 5 |
| Flame Retardancy | 1 | 1 | 1 | 1 | 1 | 1 | 1 | 1 | 1 | 1 | 1 | 1 | 5 | 1 | 1 | 1 | 10 | 10 | 10 | 1 | 2 | 1 | 7 | 7 | 8 | 10 | 10 |
| Acid Resistance | * | * | * | 2 | 2 | 2 | 2 | 1 | 4 | 6 | 4 | 6 | 2 | * | 6 | 4 | 2 | 8 | 2 | 4 | 6 | 4 | 4 | 1 | 1 | 4 | 4 |
| Alkali Resistance | * | * | * | 2 | 2 | 1 | 2 | 1 | 1 | 2 | 1 | 2 | 1 | 3 | * | 3 | 4 | 4 | 4 | 5 | 7 | 3 | 3 | 1 | 1 | 5 | 5 |
| Res. to Oxidizing Agents | * | * | * | 2 | 2 | 1 | 2 | 1 | 1 | 1 | 1 | 1 | 1 | 1 | * | 2 | 4 | 8 | 4 | 4 | 7 | 3 | 3 | 1 | 1 | 4 | 4 |
| NEMA Grade | X | XP | XPC | XX | XXP | XXX | XXXP | XXXPC | C | CE | L | LE | A | AA | N1 | G3 | G5 | G7 | G9 | G10 | G11 | GPO-1 | GPO2 | FR2 | FR3 | FR4 | FR5 |
| Military Type | | | | PBG | PBE | PBE | PBEP | PBEB | FBM | FBG | FBI | FBE | | ASB | NPG | | GMG | GSG | GME | GEE | GEB | | | | PEE | GEE | GEB |
| Base | Paper | Paper | Paper | Paper | Paper | Paper | Paper | Paper | CRS Fab. | CRS Fab. | Fine Fab. | Fine Fab. | Paper | Paper | Nylon Cloth | Cont Glass Cloth | Cont Glass Cloth | Cont Glass Cloth | Cont Glass Cloth | Cont Glass Cloth | Cont Glass Cloth | Mat Glass | Mat Glass | Paper | Paper | Cont Glass Cloth | Cont Glass Cloth |
| Type Resin | PHE | PHE | PHE | PHE | PHE | PHE | PHE | PHE | PHE | PHE | PHE | PHE | PHE | PHE | PHE | PHE | MEL | SIL | MEL | EPOX | EPOX | POLY | POLY | PHE | EPOX | EPOX | EPOX |
| A.I.E.E. Insulation Class | A | A | A | A | A | A | A | A | A | A | A | A | B | B | A | B | B | H | B | B | F | B | B | A | A | B | F |
| Comparative Price | .8 | .9 | .9 | 1.0 | 1.0 | 1.2 | 1.3 | 1.4 | 1.3 | 1.3 | 1.7 | 1.7 | .9 | 2.6 | 3.5 | 2.5 | 2.2 | 5.8 | 2.4 | 3.1 | 3.1 | .7 | .8 | 1.1 | 2.9 | 3.1 | 3.6 |

Use only for general selection. Consult complete data sheets for final decision. Price comparisons are only a guide. Properties improve in quality from 1 to 10.

PHE: Phenolic—MEL: Melamine—SIL: Silicone—POLY: Polyester

*Indicates non-significant values.

**Grade G-5.** This is a melamine-glass base material with excellent electrical and mechanical properties. It is the least expensive glass-cloth base but it is one of the most abrasive grades and, as a result, is very seldom used for general-purpose printed circuit applications.

**Grade G-10.** Manufactured with a glass-cloth reinforcement and epoxy resin, this grade possesses superior electrical and mechanical properties when compared to phenolic boards. Because of its high flexibility, low dielectric losses and moisture absorption, and high bond strength between the base and the conducting foil, this laminate is widely used and is recommended for frequencies up to 40 MHz.

**Grade G-11.** Also employing a glass-cloth reinforcement with epoxy resin, this grade is similar in characteristics to grade G-10 but is more flexible at higher temperatures. In addition, it is more resistant to acids, alkali, and heat than G-10, but is more difficult to machine. The cost of both of these grades is the same.

**Grade FR-3.** This grade is designed for optimum balance of electrical and mechanical properties and is manufactured with paper reinforcement impregnated with an epoxy resin. It is considered superior to grade XXXPC because of its higher arc resistance, lower moisture absorption, and improved dimensional stability qualities. It is used almost exclusively as a self-extinguishing type in the higher frequency ranges, up to 10 MHz.

**Grade FR-4.** This grade is identical to grade G-10 with the addition of flame retardants. For this reason, it is currently the type most often specified for applications in minicomputers, CB radios, and military and aerospace instrumentation equipment.

The information presented for these various types and grades of laminates is included with additional information in Tables 5.2a, 5.2b, 5.3, and 5.4. These tables provide a complete comparison of a wider range of electrical, mechanical, and chemical properties of the described grades, in addition to many other base materials and their properties. All of these properties should be evaluated when selecting the most suitable base material for a particular application.

**TABLE 5.4**

Copper-Clad Properties (*Courtesy of Insulfab Plastics, Inc.*)

| Copper-Clad Properties | | | | | | |
|---|---|---|---|---|---|---|
| | | Bond Strength Lbs./Inch Width | | | | |
| | | 1 oz. | | 2 oz. | | Hot Solder Resist. |
| NEMA Grade | Mil-P Type | Ave. | Min. | Ave. | Min. | @ 500°F. Sec-Min |
| | 13949B | | | | | |
| XXXP | PP | 10 | 9 | 12 | 10 | 10 |
| XXXP | | 10 | 9 | 12 | 10 | 10 |
| FR-2 | | 10 | 9 | 12 | 10 | 10 |
| XXXP | | 10 | 9 | 12 | 10 | 15 |
| | 22324 | | | | | |
| FR-3 | PEE | 10 | 9 | 12 | 10 | 25 |
| | 13949B | | | | | |
| G-10 | GE | 11 | 10 | 13 | 12 | 40 |
| | 13949B | | | | | |
| FR-4 | GF | 11 | 10 | 13 | 12 | 20 |
| | 13949B | | | | | |
| G-11 | GB | 9 | 8 | 10 | 9 | 40 |
| FR-5 | | 9 | 8 | 10 | 9 | 30 |

Single- and double-sided copper-clad laminates are available in sheets that range from 26 to 36 inches wide by 36 to 48 inches long. Smaller, precut dimensions are also available at extra cost. The thicknesses of the laminates, including the copper foil, range from $\frac{1}{32}$ to $\frac{1}{4}$ inch in increments of $\frac{1}{32}$ inch.

A characteristic of all printed circuit boards is their tendency to *warp* and *twist*. These conditions are the result of the two dissimilar materials (base and copper foil) bonded together to form the laminate. After etching, however, the boards will significantly straighten out because of the large amount of copper loss. Warp and twist describe the deviation of the board from a straight line fixed by two points at the extremities of the board. To determine the degree of warp, this line is taken along the longest edge. Twist is evaluated along a diagonal line between opposite corners of the board. These values of warp and twist are expressed in percentage of board or diagonal length, with the thinner boards tending to warp more than the thicker ones. Deviations range from approximately 12% for a thickness of $\frac{1}{32}$-inch to 5% for a $\frac{1}{4}$-inch thickness. Double-sided boards reduce these values by at least a factor of 2. These deviations of the boards become important considerations when designing for plug-in type circuits, wherein only extremely small amounts of warp or twist can be tolerated.

## 5.3 CONDUCTING FOIL

Copper is the principal type of foil used in the manufacture of printed circuits. For special applications, nonstandard materials such as *aluminum, steel, silver,* and *tin* foils are used. Copper foil has the advantages of high conductivity, excellent soldering characteristics, low cost, and ready availability in a variety of widths. The copper used for printed circuits is at least 99.5% pure in order to maintain its high-conductivity properties.

The manufacturers of copper foil for printed circuit applications employ two basic production methods. These methods are termed *rolled foil* and *electrodeposited foil*. Rolled foil is formed from refined blocks of pure copper. The foil thus produced is largely free of pinholes and imperfections and has a smooth surface on both sides. Made in widths not exceeding 24 inches and thicknesses less than 0.001 inch, this rolled film has excellent tensile-strength properties. Because of its smooth surfaces, however, rolled foil requires special bonding treatments to enhance the *wetting* between the adhesive and the foil and to remove contaminants.

Electrodeposited foil, used almost exclusively in the manufacture of printed circuit boards, is produced by the *plating* of the film from solutions of *copper sulfate* or *copper cyanide* on a revolving stainless steel drum from which the foil is continuously stripped. The inner surface of the resulting film exhibits a smooth finish, whereas the outer surface is coarse, thereby promoting improved bonding with the increase of surface area. The thickness of the foil is controlled by the solution concentration and the electrical and mechanical parameters of the plating process. Thicknesses of less than 0.001 inch and widths in excess of 5 feet are obtainable by this method.

Foil thicknesses, regardless of the manufacturing process, are specified for printed circuit boards in *ounces* of foil per *square foot*. Foils of 1 ounce per square foot have an approximate thickness of 0.0014 inch, whereas 2- and 3-ounce foils have thicknesses of 0.0028 and 0.0042 inch, respectively. Other available thicknesses are shown in Table 5.5, which gives film thickness with the thickness tolerance.

The selection of a thickness of copper for a particular circuit application is determined principally by the amount of current that it must be able to handle. This current-carrying capability depends primarily on the thickness and width of the conductor path, in addition to the temperature. Owing to the larger radiating conductor surface area, the current capacities of printed circuit boards can easily approach 50 amperes. Table 5.6 shows the maximum current capacities for several conductor width and foil thicknesses to aid in the proper weight selection.

**TABLE 5.5**

Copper Foil Thickness and Tolerance

| Weight (oz/ft$^2$) | Normal Thickness (in.) | Tolerance (in.) |
|---|---|---|
| $\frac{1}{2}$ | 0.0007 | ±0.0002 |
| 1 | 0.0014 | +0.0004 |
|  |  | −0.0002 |
| 2 | 0.0028 | +0.0007 |
|  |  | −0.0003 |
| 3 | 0.0042 | ±0.0006 |
| 4 | 0.0056 | ±0.0006 |
| 5 | 0.0070 | ±0.0007 |

**TABLE 5.6**

Maximum Recommended Current-Carrying Capacity for Various
Conductor Widths and Thicknesses

| Conductor Width (in.) | Current (A) | | | |
|---|---|---|---|---|
|  | $\frac{1}{2}$-oz Foil | 1-oz Foil | 2-oz Foil | 3-oz Foil |
| 0.005 | 0.13 | 0.50 | 0.70 | 1.00 |
| 0.010 | 0.50 | 0.80 | 1.40 | 1.90 |
| 0.020 | 0.70 | 1.40 | 2.20 | 3.00 |
| 0.030 | 1.00 | 1.90 | 3.00 | 4.00 |
| 0.050 | 1.50 | 2.50 | 4.00 | 5.50 |
| 0.070 | 2.00 | 3.50 | 5.00 | 7.00 |
| 0.100 | 2.50 | 4.00 | 7.00 | 9.00 |
| 0.150 | 3.50 | 5.50 | 9.00 | 13.00 |
| 0.200 | 4.00 | 6.00 | 11.00 | 14.00 |

## 5.4 BONDING

The manufacturers of printed circuit boards choose the proper adhesives to bond the conducting foil to a particular base material in order to produce copper-clad laminates that will suit the needs of a variety of applications. The characteristics of *bond strength, hot solder resistance,* and *chemical resistance,* as listed in Tables 5.3 and 5.4, largely depend on the particular adhesive used by the manufacturer for bonding and therefore will influence the technicians' choice of boards. Of these characteristics, bond strength is most important. This affects the reliability of a printed circuit board. If the foil should lift from the base during manufacture or in service, repair is virtually impossible. The bond must withstand the stresses involved in processing, fabricating, and servicing the printed circuit. These stresses include chemical attack from etching solutions; physical forces introduced through shearing, bending, twisting, shock, and vibration; and thermal shock during the soldering process. When shearing large sheets of single-sided copper-clad laminate into smaller blanks, the foil side should be facing *upward* with the insulating base material against the shear table. By cutting through the foil side first, there is less tendency of tearing the foil away from the edge of the blank. In addition, care must be exercised that the shear blade be allowed to return *slowly* to its normal position as it passes the newly cut edge of the board. If the blade were allowed to spring back quickly, it might snap the edge of the board just enough to crack the base material and upset the bond along that edge.

The two most common types of adhesives used in printed circuit bonding are *vinyl-modified phenolics* and *modified epoxies.* Although both adhesives display excellent bonding qualities, the epoxy resins are superior in this respect. The manufacturing process of bonding involves the application of a uniform thickness of adhesive on the foil, after which it is forced-air dried and laminated to the base material in a press. Once the adhesive has cured, the excess foil is trimmed around the edges of the board. Finally, the foil is cleaned to remove all films and oxides resulting from the manufacturing process.

## 5.5 BOARD SELECTION

Selecting the most appropriate printed circuit board grade for a particular application depends primarily on the mechanical and electrical requirements of the circuit. The information contained in Tables 5.2, 5.3, and 5.4 will dictate which grade of board is most suitable for a specific application. An evaluation of the various board characteristics will be made with reference to selecting the most appropriate grade of base material for an amplifier's printed circuits. This example will serve as a guide to illustrate the important factors that need to be considered for proper board selection.

*Example.* A printed circuit board grade is to be selected to meet the requirements of an audio power amplifier with circuit specifications as follows:

*Type of construction:* prototype

*Maximum continuous current:* 1.6 amperes

*Maximum voltage:* 36 volts

*Frequency:* audio range (maximum 20 kHz)

*Board mounting:* corners secured by machine screws to metal chassis

*Hardware mounting:* swaged terminals and machine screws

*Machine operations:* drilling and shearing only

*Soldering:* hand

Grades of printed circuit boards containing silicone or Teflon resins are not considered because of their higher cost and high-frequency characteristics that are far beyond the requirements of the amplifier. Melamine resin boards are also not considered because of their extremely abrasive characteristics, which make them difficult to drill.

One important consideration for comparison and selection is *hot solder resistance.* This characteristic expresses thermal bond strength as a function of time and temperature. Since the boards will be hand-soldered, it is difficult to accurately control soldering time and temperature. Therefore, a board grade with a moderate-to-high value of hot solder resistance should be selected. Table 5.4 compares some common boards. The *FR* and the *G* grades are found to be superior to the phenolic grades listed.

*Bond strength* also needs to be considered in terms of the mechanical stresses that result from the fabrication of the board. Tables 5.2, 5.3, and 5.4 show that the epoxy-glass resins have a higher bond strength than all other grades, with *FR-4* and *G-10* having high comparable values of bond strength.

Warp and twist are of no great concern, since all of the boards in the amplifier circuit are to be secured to the chassis element with machine screws and nuts.

The *machinability* of the board is another important mechanical consideration. As shown in Table 5.3, the phenolics can be drilled easier than the epoxy-glass types. Finally, the base material's *mechanical strength* must be taken into account. Mechanical strength and machinability are inversely related. The epoxy glass is stronger than the phenolics. Since mounting holes will be formed in the corners of each board and terminals will be swaged to serve as circuit interconnections, fracturing of the base material is a necessary consideration. Although the machining of the glass-base epoxy materials is more difficult than that of the phenolics, this characteristic is not as significant as mechanical strength of the laminate in the selection process. Machinability refers to the ease or difficulty of drilling or punching and the abrasive action of the material, which affect the selection of tools and methods of machining. Because no punching or machining other than drilling will be performed on any of the boards used in the amplifier, the mechanical superiority of glass-base epoxy outweighs the small difference in machinability characteristics.

In considering the electrical characteristics, as shown in Tables 5.2, 5.3, and 5.4, *all* the board materials listed for the amplifier circuits have parameters that are in excess of those required. The parameters of *dielectric constant* and *dissipation factor* are the most important for high-frequency applications. Therefore, since the amplifier will operate in the low-frequency (audio) range, these parameters need not be considered here. For dc to low-frequency applications, the parameters that must be considered are *arc* and *tracking resistance, insulation resistance,* and *dielectric breakdown voltage.*

Arc resistance is the ability of a base material to prevent undesired conducting paths created by an arc between portions of the existing conductor pattern. Tracking resistance describes the formation of undesired conductor paths *through* the insulating material. These characteristics need not be considered at low-power and low-voltage applications, such as the 36-volts maximum of the amplifier.

Insulation resistance, expressed in megohms, is the measure of the total *leakage* resistance among adjacent conductors. Because of the extremely high ohmic values, as listed in Table 5.4, this parameter does not become a consideration for this application.

Dielectric breakdown voltage is a function of the spacing between conductors and, as such, need not be considered here for purposes of board material selection. This property is discussed under conductor spacing in Chapter 6.

Based on the mechanical and electrical requirements of the amplifier circuit and comparing the properties of the various laminates in the tables provided, grades *XXXP, FR-4, G-10,* and *G-11* appear to be the best choices to obtain a realistic compromise. In our selection process, we first eliminate grades *XXXP* and *G-11* because their bond strength and machinability values are lower. Also, grade *XXXP* is not as strong mechanically as *FR-4* and *G-10.*

For our amplifier circuit, either grade *FR-4* or *G-10* is suitable since both have comparable properties for this application. However, even though it is of no concern to our amplifier circuit, grade *FR-4* does have one additional advantage over grade *G-10* in that it has a much higher *flame retardancy* value, as seen in Table 5.3. Since the cost of these two grades is about the same, we will select grade *FR-4* for our amplifier circuit.

One final consideration that also influences the selection of the *FR* or *G* series over the phenolic types is the ease of tracing circuits on the completed printed circuit board. As shown in Table 5.2a, the phenolics are available in *opaque natural, brown,* or *black* colors. The brown color is used almost exclusively with printed circuit boards. The type *FR* or *G* grades are generally *translucent* with a *green tint.* This translucent characteristic is extremely helpful in inspecting circuits. It is difficult to compare component and conductor patterns with a schematic diagram when the base material is opaque. By holding the translucent board up to a strong light, the relationship between components and conductor pattern is easily seen.

With the base material selected, the final step is the choice of the most suitable base thickness and copper foil weight. As mentioned previously, warp and twist present no problem in this amplifier package. Since all boards will be firmly secured to the chassis element, no other mechanical parameters need to be considered. For these reasons, $\frac{1}{16}$-inch base material will be used. For the smaller boards, $\frac{1}{32}$-inch material may be used; but for the sake of consistency and to avoid board fracture, especially around swaged terminals and mounting holes, all boards in this package will be $\frac{1}{16}$-inch.

In selecting the foil thickness, the major consideration is the amount of current that the circuit is expected to handle. The maximum circuit current is 1.6 amperes. From Table 5.6, for a typical conductor width of 0.030 inch, the maximum current capacity for 1-ounce copper foil is approximately 1.9 amperes. This standard 1-ounce foil, which will easily meet the circuit current demands, is therefore selected. In summary, the solution to the problem of board selection for the amplifier package is a *$\frac{1}{16}$-inch, 1-ounce, FR-4, single-sided copper-laminated printed circuit board.*

The printed circuit board having been selected, the technician must next prepare the necessary artwork for transforming the schematic diagrams of the various circuits into conductor patterns that will be processed later. In the following chapter we develop the techniques for producing printed circuit conductor pattern artwork.

## EXERCISES

### A. Questions

5.1 List the advantages of printed circuit boards over conventional wiring methods.

5.2 What does the acronym NEMA represent?

5.3 Describe the meaning of NEMA suffixes *X, P,* and *C.*

5.4 What is the temperature rise of a 0.005- by 0.062-inch conductor if the current through it is 4 amperes?

5.5 Using Table 5.2a, select the laminate that provides the optimum balance of mechanical and electrical properties.

5.6 Referring to Table 5.3, match the grades of insulation independently in terms of the least and most desirable characteristics of cost, machinability, punching quality, water absorption, bond strength, and arc resistance.

5.7 Using Table 5.6, select the minimum conductor width for $\frac{1}{2}$-, 1-, 2-, and 3-ounce copper foil if the maximum dc current rating is to be 2.5 amperes.

5.8 What is the advantage of using flexible printed cables as compared to round cables?

5.9 What causes pc boards to warp and twist?

5.10 Referring to the tables provided in this chapter, determine the optimum pc insulating material and minimum conductor thickness and width to meet the following requirements: 1-ampere dc current, a frequency of 20 kHz, drilling and shearing operations only, wire-wrap terminals, and finger-type connectors for lead and conductor path connections.

### B. True or False

Circle *T* if the statement is true, or *F* if any part of the statement is false.

5.1 Grade *FR-4* pc boards have a greater flame retardancy than grades *G-10* or *G-11.*   T   F

5.2 As the copper foil weight in ounces per square foot doubles, the foil thickness will also double.   T   F

5.3 The more *X*s in NEMA standards for designating phenolic resin in pc boards, the higher the resin content and the better the punching quality.   T   F

5.4 Foil thicknesses are designated in ounces per square inch.   T   F

5.5 When shearing blanks of single-sided pc boards, the copper foil side should be facing upward.   T   F

5.6 The current-carrying capacity of a printed circuit conductor path is dependent on temperature, foil thickness, and path width.   T   F

5.7 For any given foil thickness, as path width increases, current capacity decreases.   T   F

5.8 Mechanical strength and machinability of pc boards are inversely related.   T   F

5.9 A *PC* suffix on the NEMA standards indicates that the pc board may be punched at elevated temperatures.   T   F

## C. Multiple Choice

Circle the correct answer for each statement.

5.1 Printed circuit board material manufactured from (*epoxy, phenolics*) offers little abrasive action during drilling.

5.2 Printed circuit board material used for microwave applications uses a base material made from (*Teflon, epoxy*).

5.3 The principal foil material used in the manufacture of pc boards is (*aluminum, copper*).

5.4 A copper foil having a thickness designation 2 oz/ft$^2$ has a thickness of (*0.0014, 0.0028*) inch.

5.5 The maximum recommended current-carrying capacity of a pc board conductor made from $\frac{1}{2}$-oz/ft$^2$ copper foil with a width of 0.050 inch is (*1.5, 2.5*) amperes.

5.6 The copper used to manufacture the foil on pc boards must be at least (*80%, 99.5%*) pure.

5.7 The most commonly used pc board is the (*flexible, rigid*) type.

## D. Matching

Match each item in Column A to the most appropriate item in Column B.

| COLUMN A | COLUMN B |
|----------|----------|
| 1. *XXXPC* | a. Copper purity |
| 2. Double-sided | b. Friction wheel stripper |
| 3. 99.5% | c. Epoxy resin |
| 4. *G-10* | d. 0.0014 inch |
| 5. Multilayer PCB | e. Two foil layers |
| 6. 1 oz/ft$^2$ | f. Cold-punching quality |
| 7. Flexible PCB | g. Flame retardant |
| 8. *FR-4* | h. Four foil layers |

# 6 Single-Sided Printed Circuit Board Design: The Preliminary Sketch

## LEARNING OBJECTIVES

*Upon completion of this chapter on the preliminary sketch of a single-sided printed circuit board design, the student should be able to*

- Assemble the basic materials required to prepare a preliminary sketch of a single-sided printed circuit board.
- Assemble the required information necessary to prepare a preliminary sketch.
- Establish a two-sided viewing format for the layout sketch.
- Be familiar with sketching common component and device outlines.
- Select the appropriate sizes of pad diameters.
- Determine on-center distances between pads.
- Properly sketch pad outlines on grid intercepts.
- Determine the optimum positioning of components.
- Become familiar with component layout for automatic insertion applications.
- Properly route conductor paths for electrical interconnections.

## 6.0 INTRODUCTION

After the engineer has completed the circuit schematic diagram, the printed circuit designer converts this into drawings and artworks that will result in a functional packaging arrangement that will be fabricated into a PCB. The *component layout sketch*, which includes conductor path routing, is completed first.

Component layout sketches are scaled two-dimensional representations showing the size, shape, and position of all components, devices, and hardware that will appear on the finished board, as well as the overall size and shape of the board. Also included on the layout sketch is the routing of all conductor paths that form the electrical connections specified in the schematic. In addition, *terminal pads* are shown at each lead access hole. These pads will be used to electrically connect (by soldering) all component and device leads to the circuit pattern. The component layout sketch is a composite view showing *both* sides of the PCB. It is therefore essential that these drawings be carefully constructed so that each side is clearly distinguishable from the other.

The component layout sketch is typically developed in two stages: the preliminary sketch and the finished drawing. Depending on the complexity of the circuit, the experienced designer may not need to prepare the sketch but may draw the finished drawing directly using *computer-aided drafting,* which is discussed in Chapter 7. In general, however, most electronic systems are of such complexity that they do require a preliminary working sketch to serve as a guide in preparing the final drawing. It is rare, and unnecessary, for the beginning designer to attempt a finished drawing without the aid of a sketch. The sketches not only help in establishing the overall layout plan but also result in a more technically correct drawing with fewer oversights and errors.

This chapter contains the basic information required to produce a preliminary sketch. The development of a finished component layout drawing is discussed in Chapter 7. The information provided is in three major categories: (1) how to draw the body outlines for components and devices, (2) how to use the schematic diagram to position the components and devices for a single-sided board design, and (3) how to route the electrical interconnections properly so that the circuit will function in accordance with the circuit schematic.

## 6.1 MATERIALS REQUIRED FOR DEVELOPING A COMPONENT LAYOUT "WORKING SKETCH"

The term *sketch* often implies a freehand drawing, made to no particular scale. We will not be using the term in this common form for the generation of the "working sketch" for a PCB component layout drawing. This type of sketch is drawn freehand and to a 2:1 scale. It needs to be emphasized that the working sketch is a guide to be followed closely when producing the finished component layout drawing. It is therefore essential that scale and thoroughness be maintained in all phases of generating this preliminary working sketch.

The basic implements required to develop the preliminary sketch are (1) pencils, (2) erasers, (3) decimal scales, (4) decimal and fractional circle templates, and (5) ruled grid paper.

The pencils used may be mechanical pencils with thin leads, 0.5 or 0.7 millimeter, or standard wooden drafting pencils with a lead hardness of at least 2H. For standard drafting pencils, appropriate sharpeners are necessary to maintain their points. For single-sided board design, two lead colors are required, the standard black and a contrasting color such as blue or red. The eraser used should be compatible to the leads and the type of paper used.

For making measurements, a rule with major graduations of 0.1 ($\frac{1}{10}$) inch is required. Fractional scales do not lend themselves to component layout drawings and should be avoided. The length of the rule will depend on the overall size of the design. Generally, a length of 12 or 18 inches is suitable.

Decimal and fractional plastic circle templates are used to draw terminal pads and round-bodied components. The templates selected should have hole sizes ranging from 0.1 to 1.00 inch and from $\frac{1}{32}$ to about 2 inches.

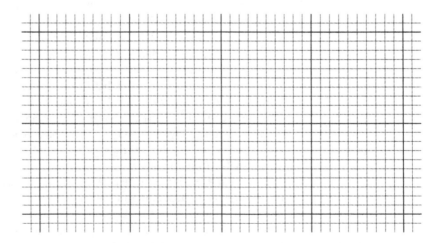

**Figure 6.1** 10-division-per-inch grid system.

The drawing media on which the sketches are to be produced should be a good-quality grid paper in a size large enough to easily accommodate the overall size of the scaled drawing. The color of the grid lines is preferably blue, although black is acceptable. The grid is a series of equally spaced vertical and horizontal lines and serves as an invaluable aid in the positioning of components and the placement of pads and conductor paths. The grid spacings should be 0.10 inch for both the vertical and horizontal lines. Grids with spacings of more than 0.1 inch should not be used in printed circuit layout. Figure 6.1 shows a piece of 10-division-per-inch graph paper with highlighted vertical and horizontal grids on every 1-inch increment.

## 6.2 BASIC INFORMATION FOR BEGINNING THE LAYOUT SKETCH

To begin producing the layout sketch, the PCB designer must have information available relative to the circuit and the parts. Primarily, a detailed circuit schematic with all parts identified either on the circuit drawing or on an accompanying parts list is initially required. An example of a detailed schematic is shown in Fig. 6.2, which is a standard common-emitter transistor amplifier. The parts list provided with the schematic identifies each of the components.

In addition to the circuit schematic, each of the parts or data sheets, which provide the physical size and shape of all components, devices, and hardware items, needs to be available. Also, electrical data sheets for all devices showing the pin arrangements are necessary.

Typically, the design engineer will provide the draftsperson with additional information regarding electrical and mechanical restrictions, such as the size and shape of the board, positioning of critical components, forbidden areas of the board for certain parts placement, required slots or cutouts, and placement of external interconnections.

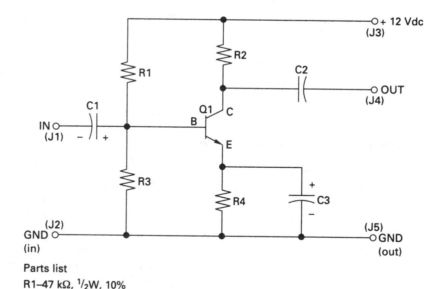

**Figure 6.2** Standard common-emitter transistor amplifier circuit schematic with parts list.

Parts list
R1–47 kΩ, $1/2$W, 10%
R2–2.7 kΩ, $1/2$W, 10%
R3–2.2 kΩ, $1/2$W, 10%
R4–180 Ω, $1/2$W, 10%
C1–1.0 µF, 15 WVDC Electrolytic Capacitor
C2–0.33 µF, Disk Capacitor
C3–100 µF, 3 WVDC Electrolytic Capacitor
Q1–2N3053, NPN Transistor,
   TO–5 Case Style

## 6.3 FORMAT OF THE LAYOUT SKETCH

Printed circuit boards are typically flat, rigid surfaces on which components are mounted and are electrically connected. For single-sided boards, the components, devices, and hardware are normally mounted on the *insulated* side, termed the *component* side, or *side 1*. The terminal pads and conductor paths are processed into the copper foil on the opposite side, referred to as the *circuit* or *foil* side, or *side 2*. Because the layout sketch represents a composite view of both sides, a direction of viewing the board must be adopted before layout can begin. It is common practice for PCB designers to view the layout from the component side of the board. This is because the components are considered to be the dominant elements in a circuit. In addition, the intended layout must not only result in a functional system but also should have the components well arranged so that the board is appealing to the eye. To achieve this, the components should be uniformly spaced using all of the available board area. By careful parts placement, the resulting board will have a balanced distribution of components with no crowding and/or open areas that are devoid of parts. It can be seen from this discussion that the logical viewpoint for laying out a PCB is the component side. Thus, initial attention will be focused on the placement of the components with the routing of the conductor paths to be considered after a balanced parts layout has been achieved.

With the direction of viewing established, it must be noted that, even though the components are positioned first, conductor routing must be considered at each step; that is, the conductor paths throughout the component arrangement must be routable. Thus the components and devices must be placed in such a position as to allow them to be conveniently connected electrically by a conductor pattern on the foil side of the board. To avoid confusion in distinguishing each of the board sides on the layout, a color-coded system needs to be established. For example, all detail appearing on the component side of the board, such as component and device bodies, hardware, and outside board edges, will be drawn using *black* lead. All detail appearing on the circuit side of the board, such as terminal pads and conductor paths, will be drawn using *blue* lead. Because this book does not have multicolor illustrations, all conductor paths in this chapter are shown as single, solid black lines connecting the circles that represent the terminal pads. It should, therefore, not be difficult to differentiate the linework on one side from that on the other.

Finally, the scale to which the drawing will be made must be considered. Industrial layouts are typically drawn to either a 2:1 or 4:1 enlarged scale. A 2:1 scale is used in this chapter to show techniques used to make enlarged scale drawings. Scale drawings are necessary to reduce small unavoidable layout errors.

## 6.4 DRAWING COMPONENT AND DEVICE BODY OUTLINES

With all of the necessary information available and the format established, we can now begin developing the working sketch. Before the components can be positioned onto the sketch, we must first learn how they are shown as two-dimensional body outlines that replace the graphic symbols in the circuit schematic. These outlines will be drawn to a 2:1 scale and will be shown so as to represent the components and devices as they will appear on the layout drawing, viewed from the component side.

The circuit schematic shown in Fig. 6.2 will be used to illustrate the drawing of the component body outlines. In the parts list of the amplifier, there are four $\frac{1}{2}$-watt resistors (R1, R2, R3, and R4), one tubular electrolytic capacitor (C1), one ceramic disk capacitor (C2), one vertically mounted electrolytic capacitor (C3), one npn transistor (Q1), and five external connections [IN, GND(in), +12 Vdc, OUT, and GND(out)].

The exact dimensions of each component must first be determined. The axial-lead-style $\frac{1}{2}$-watt resistors are found to have a diameter of 0.160 inch and a length of 0.416 inch with lead diameters of 0.032 inch. This style of component is mounted to the board surface by first bending the two leads at right angles at a specified distance from the ends of the body. The leads are then passed through drilled access holes with sufficient length to make connection to the terminal pads on the circuit side of the board. It is common practice when drawing the body outline of axial-lead components not to show the lead access holes or the lengths of leads between the body ends and the hole. Rather, since this is a composite view, only properly positioned terminal pads that will accept each lead are drawn beyond the ends of the body outline. The diameter of the pads and the distance between them must first be determined.

Terminal pad diameters for *unsupported* holes (not plated-through) on single-sided boards are dependent on the diameter of the lead to be inserted, the diameter of the drill hole, and the amount of copper foil required to remain about the pad after the hole is drilled. To provide the necessary clearance for our resistor leads, which have a 0.032-inch diameter, the drill hole should be from 0.010 to 0.020 inch larger in diameter. This would require a hole size of diameter between 0.042 and 0.052 inch. It is general practice for unsupported holes that the terminal pad diameter be at least 0.040 inch larger than the drill hole. For our example, a pad diameter of between 0.082 and 0.092 inch would be acceptable on a 1:1 scale. Commonly available pad diameters in this range are 0.075, 0.080, 0.093, and 0.100 inch in diameter. It should be noted that after a board has been processed, the bonding of the conductor paths and terminal pads is related to the amount of foil remaining after the hole has been drilled. For this reason, it is recommended that the size of the terminal pad be made as large as possible. Making them too large, however, reduces the available space for conductor path routing. The final size selection thus becomes a trade-off between conductor spacing and maximum diameter of terminal pads.

Additional factors that must be considered at this time are (1) reducing the number of different hole sizes, which results in fewer drill bit changes during fabrication, and (2) selecting only a small number of terminal pad diameters to satisfy all lead diameters in the design, thus reducing the required inventory of pad sizes. Every effort should be made to select the fewest drill and pad sizes that will satisfy the total range of hole and pad size requirements.

In selecting pad diameters for single-sided boards, the following guidelines are recommended. For leads with diameters of less than 0.020 inch, a 1:1 scale pad diameter of 0.075 inch may be used. Lead diameters of between 0.020 and 0.040 inch may effectively use a pad diameter of 0.100 inch.

The on-center distances between pads is determined by both the component body length and the lead diameter. Since an individual pad must be provided for each lead, one is shown at each end of the body outline. The preferred distance that the leads should extend outward from the component body before the 90-degree bend is made is 0.060 inch. To determine the overall bend distance X, a bend radius should be added. The minimum bend radius for each lead should be approximately 0.030 inch for lead diameters up to 0.030 inch, and 0.060 inch for lead diameters of 0.031 to 0.050 inch. Lead diameters greater than 0.050 inch should have a bend radius of two times the lead diameter. Refer now to Fig. 6.3. It can be seen that the formula for calculating the on-center distance (S) for terminal pads is

$$S = 2X + B$$

where X is the bend distance (extension distance + bend radius) and B is the body length. From the discussion above, and with reference to Fig. 6.3, the *minimum* value of terminal pad spacing for axial-lead-style components using maximum lead diameter is 2(0.060 + 0.060) plus the body length (B).

Referring again to the $\frac{1}{2}$-watt resistor shown in Fig. 6.3. and using the maximum values given for X and B in the accompanying table, we calculate the value of S as follows:

$$S = 2X + B$$
$$= 2(0.060 + 0.060) + 0.416$$
$$= 0.240 + 0.416 = 0.656 \text{ inch}$$

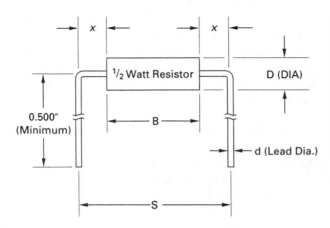

| Dimension | Inches | | Millimeters | |
|---|---|---|---|---|
| | MIN | MAX | MIN | MAX |
| B | 0.344 | 0.416 | 8.74 | 10.6 |
| D (DIA) | 0.115 | 0.160 | 2.92 | 4.06 |
| X * | 0.090 | 0.120 | 2.29 | 3.05 |
| d (Dia.) | 0.026 | 0.036 | 0.66 | 0.91 |
| S | 0.524 | 0.656 | 13.3 | 16.7 |

* Depends on application. Minimum values for dense layouts.

**Figure 6.3** Tabulating dimensional information on detail drawings.

This figure would normally be increased to 0.70 inch so that the on-center pad dimensions will fall on grid.

The terminal pads shown in Fig. 6.4a have been drawn on grid with an on-center dimension of 1.4 inches and diameters of 0.2 inch at 2:1 scale. Although the pads are shown in black, they should be drawn with blue pencil so as to more readily distinguish their true position as being on the foil side of the board.

Note in Fig. 6.4 that no center lines are used in drawing the terminal pads. Simply drawing the outline of the pad with a circle template is sufficient. The underlying grid should be used as a guide to locate and center the pad. It is important to center the pads on grid intercepts and construct the component body outlines off-grid as required by their dimensions. For precise artwork generation and board fabrication, it is more important for the pads to be positioned accurately on grid than it is for the component body outlines.

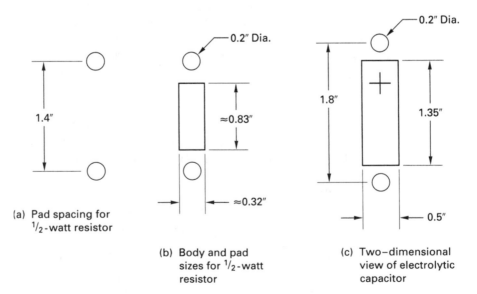

**Figure 6.4** Drawing the two-dimensional component view of axial-lead resistors and capacitors.

(a) Pad spacing for ¹/₂-watt resistor

(b) Body and pad sizes for ¹/₂-watt resistor

(c) Two-dimensional view of electrolytic capacitor

When mounted to the component side of a PCB, axial-lead components normally rest flush against the insulated board surface. As such, their two-dimensional body outlines, as viewed from the component side (side 1), would appear as rectangles. Using the adopted color-code system, all of the linework for side 1 would be drawn with black lead. Since our layout will be drawn to a 2:1 scale, each full-size dimension must be multiplied by a factor of 2 before drawing the body outlines. For a $\frac{1}{2}$-watt resistor, the maximum dimensions for the body length and diameter given in the table of Fig. 6.3 would be doubled, resulting in a rectangle drawn as close to 0.32 inch $\times$ 0.83 inch as the grid will permit. This is shown in Fig. 6.4b. Note that the grid lines provide a useful guide for measuring and drawing the body outlines by simply counting squares. When using a 0.1-inch grid to show the outline of this resistor, the body is to be laid out to a rectangle of just over 3 squares (0.32 inch) by a little more than 8 squares (0.83 inch). Note also that the body outline is neatly drawn and reasonably centered between the two terminal pads using a straightedge even though the layout is termed a *sketch*.

Capacitor C1 in Fig. 6.2 is the electrolytic type with axial leads and with a body diameter of 0.25 inch. Its body outline, shown in Fig. 6.4c, is drawn similar to that of a resistor, the only differences being the dimensions of the lead diameter, body diameter, and length. Because its maximum lead diameter is slightly less than that of a $\frac{1}{2}$-watt resistor (0.031 versus 0.032 inch), the diameter of its pads may be the same as those used with the resistor (i.e., 0.2 inch at 2X scale). Also, since the lead diameter is 0.031 inch, the bend radius will be 0.060 inch, the same as that for the resistor. The total body length of the electrolytic capacitor is 0.625 inch. Therefore, the on-center pad spacing is $S = 2(0.060 + 0.060) + 0.625 = 0.865$ inch. To be on grid, this becomes 0.90 inch between pad centers at a 1:1 scale. As shown in Fig. 6.4c, the body outline is centered between 1.8-inch spaced pads and is shown as a rectangle of approximately 0.5 $\times$ 1.35 inches at a 2:1 scale. Note the plus sign (+) inside the body outline near one end. Because electrolytic capacitors are polarized, they need to be correctly assembled onto the board. The plus sign serves as a key for proper installation.

Another component commonly used in electronic circuits is the radial-lead ceramic-type *disk* capacitor, which is shown in Fig. 6.5a properly mounted to a PCB. Capacitor C2 in Fig. 6.2 is a disk type. It is seen from its side view that the body is approximately circular and that it has two unpolarized leads that extend directly through the board without bends. Above the side view shown in Fig. 6.5a is the two-dimensional shape of the disk capacitor as viewed from the top or as it would appear on the component layout drawing and viewed from side 1. Accurately drawing this complex shape, which tapers almost to points at its outside edges with maximum body width at its center, is time-consuming and unnecessary for PCB layout. Rather, the less difficult shape of a rectangle has been adopted even though it does not exactly represent the body shape.

To construct the component view of disk capacitor C2, the on-center lead spacing is first measured and found to be 0.375 inch. At 2X scale, this dimension becomes 0.750 inch. This is then increased to 0.80 inch for the on-grid locations of pad centers. The lead diameters of the component are 0.025 inch. With this information, the pads, at 2X scale, will be drawn with 0.20-inch diameters on grid. This is shown in Fig. 6.5b. Remember that these pads would normally be drawn with blue pencil. The diameter $w$ and thickness $t$ of the capacitor are next measured and found to be 0.590 and 0.156 inch, respectively. At 2X scale, these dimensions become 1.18 and 0.312 inch. The two-dimensional view of the component is thus represented by a rectangle that is slightly over 3 squares by approximately 12 squares and centered between the terminal pads. Note in Fig. 6.5c that the terminal pads are on grid and are *inside* the body outline. This is not always the case. Other types of orientations of body outlines to terminal pads are shown for different sizes of disk capacitors in Fig. 6.5d.

Capacitor C3 in Fig. 6.2 is an electrolytic type with a tubular shape having two radial leads protruding from the *same* end of the body. This style is designed to be mounted

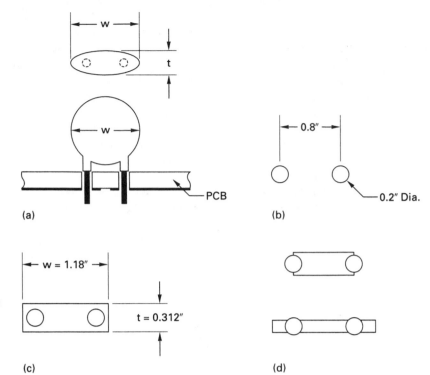

**Figure 6.5** Drawing the two-dimensional component view of a disk capacitor: (a) viewed from the top and from the side; (b) terminal pads drawn on grid; (c) pads shown inside the body outline; (d) orientation for different sizes of disk capacitors.

vertically onto the PCB with its leads extending straight into the access holes. This orientation is similar to that of the disk capacitor except that the base of the electrolytic type is seated flush with the surface of the board. This is shown in Fig. 6.6a. As with other PCB-mounted components, the pad positions and size will first be determined. The 1:1 on-center lead spacing and lead diameters are found to be 0.2 and 0.025 inch, respectively. Thus the pads will be drawn on grid at 2X scale with a 0.4-inch on-center dimension and diameters of 0.2 inch. This is shown in Fig. 6.6b. The case diameter is measured to be 0.335 inch, and at 2X scale it becomes 0.67 inch. Using a 0.7-inch hole in a circle tem-

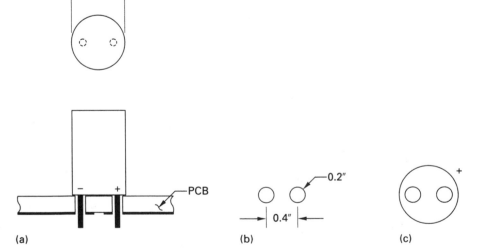

**Figure 6.6** Drawing the two-dimensional component view of an electrolytic capacitor with radial leads: (a) top and side views; (b) terminal pads drawn on grid; (c) body outline drawn around the terminal pads.

plate, the body outline is centered and drawn around the two terminal pads (see Fig. 6.6c). As seen from the component view, this capacitor appears as a circle since it is vertically mounted. Because it is polarized, a plus sign is placed beside the appropriate pad outside of the body outline to ensure correct orientation onto the board.

The five external connections to the points labeled IN, GND(in), +12 Vdc, OUT, and GND(out) in Fig. 6.2 will each be made with turret terminals for our example design problem. Pictorial, top, and side views of a turret assembly are shown in Fig. 6.7.

Turret terminals are a common method of providing points for soldering wires to make electrical connections between the PCB circuit and other external points. Each of the five turret terminals will require a pad in order to make electrical connections to a path on the circuit side of the board. The two-dimensional representation of a turret terminal is quite simple, as shown by Fig. 6.8a. The inner circle represents the diameter of the turret section, the next larger circle the shoulder, and the outer circle represents the terminal pad. Actually, all that is required is to show the terminal pad diameter for each of the terminals. This is shown in Fig. 6.8b. Appropriate labeling on the layout sketch will make it clear which of the pads will be used for making external connections requiring turret terminals. In addition, their larger diameters and locations near the board edges make them easy to identify.

To serve as an example, we will use a turret terminal with a shank diameter of 0.1 inch. If an access hole of 0.01 inch greater than the shank diameter is selected for ease of installation, a drill diameter of 0.11 inch would be required. Refer to Appendix II for a table of common drill bit sizes. It can be seen from the table that a No. 35 drill bit would be used for this hole. To satisfy the requirement of selecting a pad diameter that will be 0.040 inch

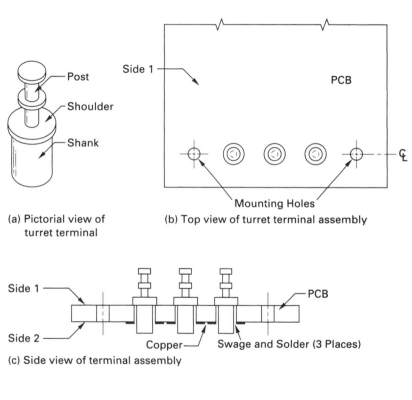

**Figure 6.7** Turret terminal detail.

(a) Pictorial view of turret terminal

(b) Top view of turret terminal assembly

(c) Side view of terminal assembly

**Figure 6.8** Drawing the two-dimensional representation of a turret terminal pad: (a) top view of turret terminal; (b) terminal pad only.

*Single-Sided Printed Circuit Board Design: The Preliminary Sketch*　　**85** larger than the drilled hole, a pad with a 1:1 scale diameter of 0.150 inch, having a 2X scale diameter of 0.300 inch, would be required. This is shown in Fig. 6.8b. This size pad would be shown for all external connections requiring the installation of turret terminals having a shank diameter of 0.1 inch.

The final component of Fig. 6.2 to be drawn is the transistor Q1. The body outline and pad arrangement for transistors are more complex in their drawing than the components discussed thus far. Although all of the critical dimensions can be measured with a scale, this is a type of component where dimensional information is best obtained from the manufacturer's specification sheets. Refer to Fig. 6.9a for the manufacturer's outline drawing of the TO-5 transistor case specified for Q1. Note that the maximum lead diameter is 0.019 inch. Therefore, a 1:1 scale pad diameter of 0.075 is required. When drawing the pads to a 2X scale, the diameter used will be 0.150 inch.

The transistor leads are numbered 1 (*emitter*), 2 (*base*), and 3 (*collector*). These leads are positioned in a triangular arrangement on a 0.2-inch-diameter dotted-lined circle with the base lead centered but elevated a distance of 0.1 inch above the in-line leads of the emitter and the collector. The emitter lead is closest to the case tab and the collector lead is farthest away from the tab with a separation of 0.2 inch. The three terminal pads that will accept the transistor leads for straight entry into the board are drawn with 0.15-inch diameters at a 2X scale and laid out at the dimensions shown in Fig. 6.9b. Note that the arrangement of the pads is inverted from that of the bottom view shown for this case style. This is necessary to represent the true orientation of the device as viewed from the

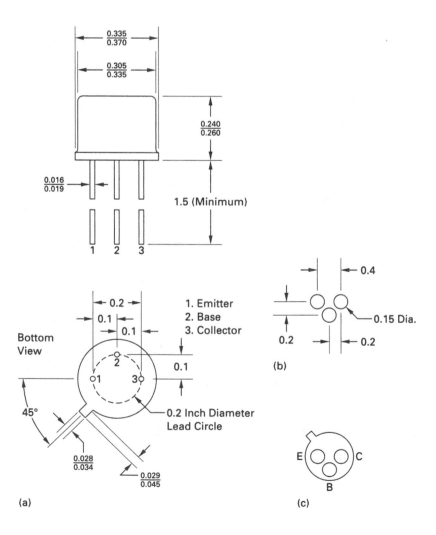

**Figure 6.9** Drawing the two-dimensional component view of a TO-5 style transistor using manufacturer's specifications.

(a)

(b)

(c)

top of the case as it is positioned on the component side of the board. Care must always be exercised in arranging pad positions for devices from the component view since the manufacturer's specifications commonly provide lead designations from the lead side or bottom view of the device.

Referring again to the specifications for the TO-5 case, the largest body diameter is specified as 0.37 inch at a 1:1 scale, or approximately 0.75 inch at a 2X scale. This circular body is drawn concentric to the dotted-line lead circle. For ease of drawing, the device tab is shown as a square approximately 0.1 by 0.1 inch and is positioned closest to the pad numbered 1 (emitter). This square is drawn outward from the body outline at a 45-degree angle to the center of the horizontal grid line, which passes through the centers of the emitter and collector pad circles.

The complete outline drawing of the TO-5 case with leads coded and with the identification tab is shown in Fig. 6.9c. Note from this view that the emitter, base, and collector are read in a counterclockwise direction from the tab.

It is not possible to demonstrate here the drawing of body outlines for all styles of components and devices and it is hardly necessary. As emphasized in this section, the basic objective is to draw the top view of the part as seen from the component side of the board. As much detail as necessary is shown to illustrate the important features of the part. Drawing excessive detail not only may be a waste of time but also may cause confusion.

In summary, terminal pads are first positioned and drawn on grid after their scaled dimension (in our case, 2X) and spacing have been determined. The component body outline is then drawn in its relative position to its pads. The following sequence should be followed: (1) calculate the pad diameter from the lead diameter and the hole size; (2) determine the 2X pad spacing for the component or device; (3) position and draw the 2X diameter terminal pad on grid for each lead; and (4) center and construct the body outline in its proper position. Finally, because the pads will be processed from the copper foil on the circuit side of the board, they should be drawn in a contrasting color, normally *blue*. The body outlines, which appear on the component side of the board, should be drawn with *black* lead.

## 6.5 POSITIONING COMPONENTS ON THE LAYOUT SKETCH

Now that you have become familiar with the drawing of common components and devices typically mounted on a PCB, we will begin the discussion of the most creative and interesting aspect of printed circuit design. This is the positioning of the parts on the board and the routing of the electrical connections (conductor paths) between terminal pads in accordance with the requirements of the circuit schematic. The typical concern faced by beginning designers is that after several parts have been tentatively positioned on the board, connecting paths between some of their pads will be unroutable. That is, there may be no possible routes available to run a conductor path between pads that must be connected without crossing another drawn path that is not associated with it. This problem, however, will be eliminated as proper layout skills are developed. This also emphasizes the need for drawing the preliminary sketch prior to preparing the final component layout. The sketch is used to resolve any problems of components being positioned so as to create unroutable paths.

It needs to be understood initially that printed circuit design is a creative task and, as such, does not lend itself to any single "correct solution." There are many acceptable layout practices, some with obvious reasons for specific component positioning and some where the reasons are somewhat obscure but are based on past practices that have proven successful. Your unique designs may be just as acceptable and correct as long as the basic guidelines initially established are not violated.

For our first design, we will construct a preliminary sketch of the amplifier circuit shown in Fig. 6.2 using the skills for drawing component body outlines developed in the preceding section. A statement of the design problem follows.

*Design Problem.* Using a 2:1 scale and a 0.1-inch grid system, design a component layout sketch of the amplifier circuit of Fig. 6.2. There are no restrictions on final board size or shape except that the finished layout maximizes the total board area. There are also no restrictions on the position or orientation of any external connections. All terminal pads are to be drawn on grid. The component view should show all necessary labeling using black lead to show the component side and blue lead to represent all pads and conductor paths on the foil side of the printed circuit board.

It is seen from the design problem that the designer has few restrictions, especially in component positioning and external connections. This is not typical, but it is felt that your first design should provide layout practice without rigid requirements.

We begin the design with the circuit schematic shown in Fig. 6.10, which is essentially the same as that shown in Fig. 6.2 with the addition of specific points identified with letters and numbers to serve as guides for locating portions of the circuit as they are discussed.

Next, a sketch, such as that shown in Fig. 6.11, is made of the overall board size with usable component area. The overall board outline is indicated by penciled corner brackets. A solid continuous line, called the *component border,* is drawn inside the overall board outline. This line shows the usable area on the board for placing components. The margin around the edges of the board is called the *forbidden area.* This space is reserved for drilling holes to accommodate hardware for mounting the board without interfering with components or conductor paths. For this design, an initial board size of 5 × 6 inches is selected. This size is at a 2:1 scale.

There is no easier way to begin the component layout sketch than to position components and hardware inside the component border as they appear on the schematic. This is termed the *schematic viewpoint* and will provide ease of conductor path routing. Rearrangement of initial component placement is usually necessary to obtain a well-balanced layout having a uniform distribution of parts in the usable area with no crowding or large wasted spaces. Precut component outlines are commercially available to aid in rapidly obtaining optimum layouts for final sketching. These precut outlines, called *dolls* or *puppets,* can be conveniently positioned and repositioned until the desired layout is achieved. It is possible, by using the computer-aided drafting techniques discussed in Chapter 7, to produce component outlines that can be cut into dolls.

Using the schematic viewpoint, the dolls are placed onto the 0.1-inch grid inside the component border. The result may appear as shown in Fig. 6.12. All component body out-

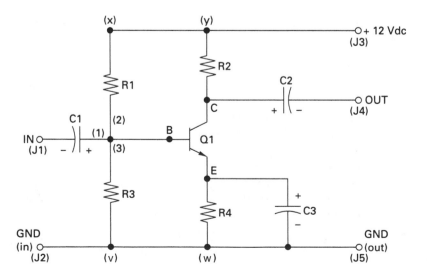

**Figure 6.10** Schematic labeled for use as a guide for parts placement and conductor path routing.

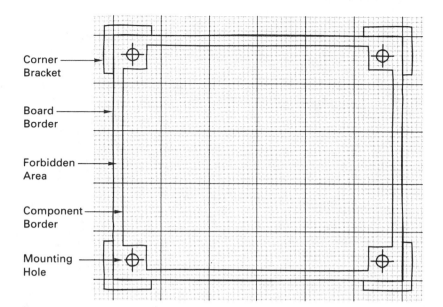

**Figure 6.11** Initial border layout sketch.

lines and terminal pads are to a 2:1 scale. All of the parts are labeled with the same designations as on the schematic. Note that the parts are positioned approximately as they appear in the schematic. For example, R1 is directly above R3 and to the left of R2, which is above Q1. Some components, like capacitor C1, require special orientation. This capacitor must be positioned with its negative side, which is not labeled, closest to the external terminal marked J1 and its positive terminal, labeled +, facing toward Q1. The input, output, ground, and power terminals are positioned toward the ends of the board, which follows their position on the schematic. This method of solving the problem of arranging components to make path routing for input and output connections with the least amount of difficulty is termed the *peripheral viewpoint*.

Another method of ensuring optimum component location is called the *central viewpoint* and is similar to the schematic viewpoint. This viewpoint, however, focuses attention

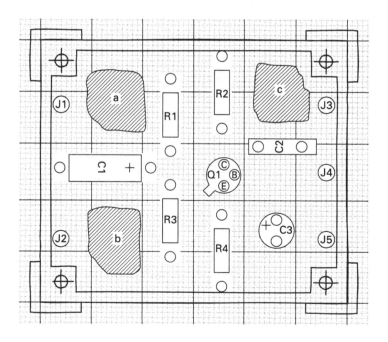

**Figure 6.12** Initial component layout using dolls and the schematic as a guide.

on the placement of passive components that support the operation of an active device such as Q1. The initial location of Q1 should be central to the resistors and capacitors that support its operation as shown in the schematic.

Observe that there are no crossover leads in the circuit schematic. Since our parts placement follows that of the schematic, it is possible to make all of the electrical connections as shown in the schematic. Even though the design is routable as shown, it does not meet the specified requirement of maximizing the total board area. The bold outline drawn around the positioned dolls in Fig. 6.12 represents the component border. Note the large unused areas of the board. These areas are shown as a, b, and c. This gives the appearance that the board is much larger than required for this design. To overcome this problem, we will slightly reposition some of the dolls until a more reasonable balance is achieved. Compare Fig. 6.12 with Fig. 6.13. To begin, C1 will be rotated 90 degrees with the GND (in) and J1 terminal pads positioned directly above and below this capacitor while still remaining centrally positioned. Also, note that for better component balance, moving Q1 to the right will allow R2 and R4 to be brought inward so that they will be adjacent to R1 and R3. This will reduce the overall width of the board. C3 can then be placed to the right of R4 bringing it directly under Q1. Finally, C2 is rotated 90 degrees and positioned to the right of R2 and above Q1. Wherever possible, centers of component bodies should be aligned as was done with R1, R2, R3, and R4. There are trade-offs, however, in reducing the total required space and improving the overall appearance of the board. For example, C1 is positioned midway between the top and bottom edges of the board. C2, Q1, and C3 are spaced vertically to avoid needless crowding, but this does not allow C2 to center on R2 nor C3 to center on R4. These types of trade-offs in rearranging components help to optimize the overall layout regarding uniform distribution. Note also that no body outlines are positioned on a diagonal to the horizontal and vertical grid lines. This is to prevent the generation of wasted board space. In addition, any component placed at a diagonal would look out of place compared to the other vertically and horizontally placed parts. The final component layout incorporating all of the techniques discussed is shown in Fig. 6.13.

Even though parts have been rotated and repositioned, two important points need to be emphasized. First, no part has been moved any great distance from its initial position in the schematic. The resulting path routing distances are thus as short, or in some cases

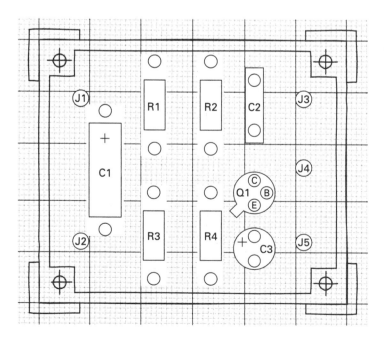

**Figure 6.13** Finalized component and hardware positioning.

shorter, than in our initial parts placement. Keeping interconnecting paths as short as possible is normally a prime requirement of printed circuit design. Second, each component move or rotation has resulted in reducing the amount of required board space, generating a more compact layout, which is also required in a good design.

Once the components have been repositioned to result in an improved layout, the external connections on both ends of the board can then be moved inward to reduce the required length of the board. After the final component and hardware positions are obtained, each doll is carefully traced with a pencil and removed from the layout. Care must be taken to sketch the pads associated with each doll in the correct location. Remember that all pad centers are to be placed on grid.

We now summarize how optimum component placement is accomplished. We first establish an initial point of reference by positioning the parts as they appear in the circuit schematic. The schematic is a concise representation of all electrical connections between parts. It contains minimum crossovers and, as such, part placement based on this scheme also results in minimum crossovers. The parts may then be rearranged to (1) reduce path routing problems, (2) result in a more compact design to reduce wasted board space, and (3) improve the appearance of the board by arriving at a uniform distribution of parts. In general, conductor path routing is simplified when the component layout parallels that of the circuit schematic.

## 6.6 LAYOUT CONSIDERATIONS FOR AUTOMATIC ASSEMBLY

When designing a component and conductor pattern layout of a pc board that is to be assembled by automatic insertion, component placement is critical and is governed by specific design criteria. To make automatic assembly cost effective, the designer must standardize the components selected for the circuit. By reducing the number of different size and shape axial- and radial-lead components, the variety of lead access hole spacings will be kept to a minimum. It is also important to arrange component positions on as few axes as possible, with rows and columns having uniformity of lead spacing and orientation. The optimum layout is that in which all components have the same lead span between terminal pads and are placed in columns and rows having all of their axes parallel. When more than one span is part of the layout, grouping of components of similar span where possible increases assembly efficiency by reducing worktable travel. See Appendix XVI.

In most systems, the insertion head does not rotate. Because of this, component axes placed perpendicular to others reduce assembly efficiency since the board must be rotated 90 degrees to install these parts. In addition, no component axis should be positioned at a diagonal to any board edge.

As much as possible, component layouts should orient keyed components uniformly. For example, all tabs on radial lead ICs or notches on dual-in-line package ICs that key pin 1 should face the same direction as well as placing the ICs in precise columns and rows. See Appendix XVI.

For automatic insertion designs, boards must be provided with two tooling holes from which all lead access holes are placed on a fine grid in increments of 0.025, 0.050, and 1.00 inch. These tooling holes are later used to position the board on the automatic insertion machines. As in other designs, pad centers must be precisely located on grid.

Lead access hole sizes for automatic insertion are, in general, made 0.015 inch larger than those made for manual component assembly. This means that the diameters of the terminal pads should be increased by 0.030 inch to keep the same amount of copper surrounding the lead access holes. This surrounding copper is called the *annular ring* and is generally given a minimum-width specification, such as 0.012 inch.

In Chapter 16, Fig. 16.20 shows an automatic insertion head together with finger-to-lead and finger-to-body clearance for parallel-positioned axial components. The typical length and width clearances required for each guide finger are 0.100 and 1.30 inch, respectively. When axial components are placed side by side, as in Fig. 16.20a, essentially no clearance is needed. However, to locate insertion holes on grid, the body center-line spacing of adjacent components is rounded off to the nearest 0.100 inch. When positioned end to end, a minimum of 0.100 inch between insertion holes is required. See Appendix XVII.

The pad centers for adjacent ICs should have a minimum spacing of 0.150 inch. When they are positioned end to end, the spacing can be extremely small—as small as two times the manufacturer's body tolerance. See Appendix XVI.

## 6.7 CONDUCTOR PATH ROUTING

The final process in the completion of the component layout sketch is to interconnect correctly the terminal pads associated with each of the components and devices in accordance with the circuit schematic. To continue in our layout of the amplifier circuit shown in Fig. 6.10, conductor paths between the pads shown in Fig. 6.13 will be drawn.

The width of the conductor paths is primarily related to (1) the maximum current they are expected to handle, (2) circuit density, and (3) fabrication limitations. Because it is the purpose of this chapter to introduce the basic techniques for routing conductor paths, we will focus only on this topic. (Refer to Chapter 5 for determining conductor path width and to Chapter 7 for the criteria for the spacing of these paths.)

Conductor paths are drawn on the sketch with a straightedge and are shown as single lines, typically 0.5 or 0.7 mm thick. Blue lead should be used to indicate that these paths are on the foil side of the board together with the pads. Since the conductor paths form the electrical connections, they must originate at a pad or at the intersection of other paths and terminate at a pad or intersect with another path. The conductor paths should be drawn on grid lines whenever possible since the pads are centered on grid. They should extend toward the pads as if to bisect them. For clarity, the conductor line drawn should cut across the pad circumference and extend approximately halfway to its center point. See pad (a) in Fig. 6.14. Changes in path direction are drawn in 45-degree angle steps, as shown in the paths between pads (a) and (b) and between pads (b) and (d) in Fig. 6.14. When routing a conductor path that must intercept another path for electrical connection, it is drawn only at right angles, as shown in the path labeled (e) in Fig. 6.14. Finally, no more than four paths should originate or terminate at one terminal pad. Space permitting,

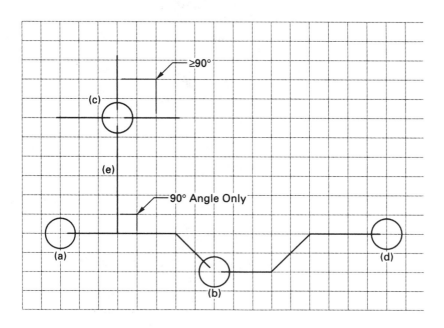

**Figure 6.14**
Techniques for routing conductor paths.

paths entering a pad should be separated by an angle equal to or greater than 90 degrees. See pad (c) in Fig. 6.14.

The task of conductor path routing begins with drawing as many of the short interconnect points as possible, leaving the longer paths until later. The circuit schematic of Fig. 6.10 shows an electrical connection between points $x$ and $y$. A single, solid blue line is drawn horizontally between the pads labeled $x$ and $y$ to form this connection. This is shown in Fig. 6.15. As each connection is made on the layout sketch, it should be checked off or lined out on the circuit schematic. This is done to ensure that all connections have been correctly made in accordance with the schematic. The procedure is repeated for the connecting points labeled $v$ and $w$.

A grouping of short connections is labeled as 1, 2, 3, and B on the schematic. This indicates that a single electrical connection is to be made from the + side of C1 (point 1) to the bottom pad of R1 (point 2) and the top pad of R3 (point 3) as well as to the base of Q1 (point B). Points 2 and 3 are first connected with a single vertical line. The connection between point 1 can then be made to either point 2 or 3 since they are now shown as electrically connected. However, it is recommended that this connection be made to the pad of point 2 since it is closer to the + side of C1 and thus will result in a shorter path (see Fig. 6.15). Note that this path is partially routed through the body outline of C1. This is normal practice since the conductor path (in blue) is on the foil side of the board and the component body is on the component side, which prevents mechanical or electrical interference problems. Even though conductor paths are generally made to follow grid lines, long diagonal paths are acceptable as long as they contribute toward high circuit density and leave no large unused board areas. To complete this connection, a line is drawn to the base pad of Q1, which is perpendicular to the path between points 2 and 3. It is important to note that this path was positioned approximately midway between the bottom pads of R1 and R2 and the top pads of R3 and R4. Following is an explanation of this design criterion. The conductor paths and terminal pads represent copper conductors that are separated by an insulating path or gap. As such, conductor path positioning is always planned so as to maintain as uniform an insulating gap as possible between adjacent paths and terminal pads as allowed by proper grid positioning. The insulating space between conductor paths and pads and

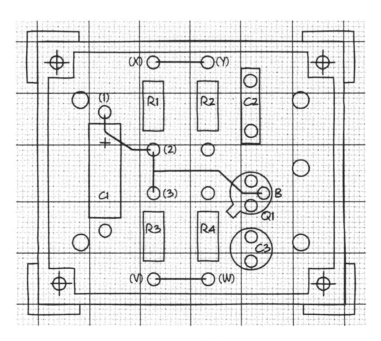

**Figure 6.15** Routing conductor paths using the schematic as a guide.

between adjacent paths is of major concern in PCB layout. Needless crowding between conductors is to be avoided. A conductor path may have unequal insulation spaces on either side as it is routed between other conductors, but this should not result in a severe difference in their proportions. Conductor paths must not be routed along the very edge of the board. A space must be allowed between the outside edge of the path and the board edge to prevent damage to the path when the board is processed into its finished size.

The finished layout sketch is shown in Fig. 6.16 properly labeled and coded. Following is a summary of the basic steps used in conductor path routing:

1. Route conductor paths along vertical and horizontal grid lines.
2. Change path direction by using short 45-degree line segments.
3. Draw conductor path lines so that they extend slightly inside the pad.
4. Avoid crowding.
5. Begin path routing by drawing the shortest and most direct routes.
6. Keep track of each path drawn by checking it off on the circuit schematic.
7. Draw path-to-path intersections as perpendicular lines.
8. Keep angles of paths entering common pads equal to or larger than 90 degrees.

Because of the complexities and difficulties inherent in the design of printed circuits, a flow chart is provided in Fig. 6.17 to aid the technician by showing the logical sequence of steps necessary to generate a component and conductor layout sketch.

The next chapter is devoted to demonstrating computer-aided drafting using a computer system and a computer-aided design (CAD) software package to convert the component and conductor pattern layout sketch into highly accurate component layout and conductor pattern scaled drawings.

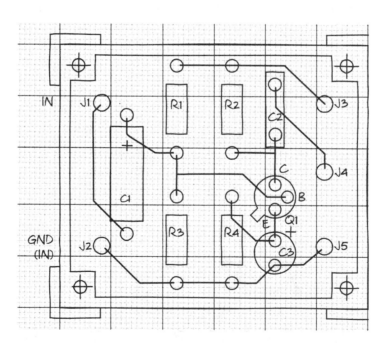

**Figure 6.16** Final component and conductor pattern layout sketch.

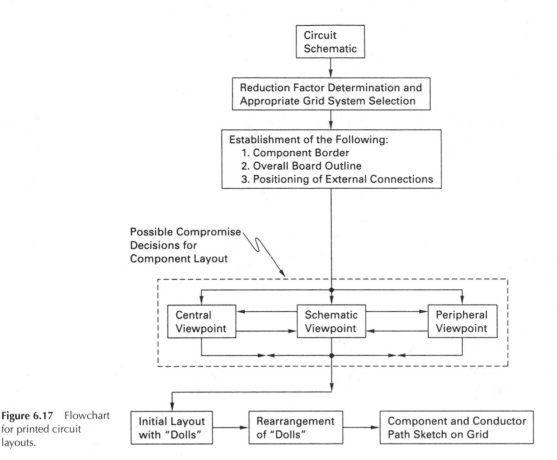

**Figure 6.17** Flowchart for printed circuit layouts.

The flowchart contains the following boxes connected by arrows:

Circuit Schematic

↓

Reduction Factor Determination and Appropriate Grid System Selection

↓

Establishment of the Following:
1. Component Border
2. Overall Board Outline
3. Positioning of External Connections

Possible Compromise Decisions for Component Layout

Central Viewpoint ← → Schematic Viewpoint ← → Peripheral Viewpoint

↓

Initial Layout with "Dolls" → Rearrangement of "Dolls" → Component and Conductor Path Sketch on Grid

## EXERCISES

### A. Questions

6.1   What features are shown in a layout sketch (figure 6.16)?

6.2   Why is the layout sketch described as a composite view?

6.3   What are the recommended grid system and scale to be used for layout sketches?

6.4   Why are body outlines referred to as two-dimensional views?

6.5   Briefly describe the focus of each of the three viewpoints used to obtain optimum component positioning for routing conductor paths.

6.6   Why are terminal pads drawn on grid?

6.7   To what does the term *forbidden area* refer?

6.8   Give two examples of employing specific design criteria to make automatic assembly cost effective.

6.9   How much larger are the lead access hole diameters made for component assembly by automatic insertion than by manual insertion?

6.10   How are path entries to terminal pads and changes in path directions made?

## B. True or False

Circle *T* if the statement is true, or *F* if any part of the statement is false.

| | | | |
|---|---|---|---|
| 6.1 | Layout sketches are drawn at a 2:1 scale on a 0.01 grid system. | T | F |
| 6.2 | Dolls are precut component outlines at a 2:1 scale that show the shape and size of the part with terminal pad requirements of location and size. | T | F |
| 6.3 | Resistor outlines include a keying symbol for proper orientation. | T | F |
| 6.4 | Conductor paths must not be drawn through component body outlines. | T | F |
| 6.5 | The component and conductor pattern layout allows both sides of the pc board to be viewed simultaneously. | T | F |
| 6.6 | A separate terminal pad must be provided for each component lead to be soldered to the conductor pattern. | T | F |
| 6.7 | Conductor paths are drawn only horizontally or vertically. | T | F |
| 6.8 | The width of the conductor path is determined by the amount of current it is to carry. | T | F |
| 6.9 | The angle between paths entering any terminal pad should not be less than 45 degrees. | T | F |
| 6.10 | Designing a pc board for automatic assembly uses the same criteria as those for manual assembly. | T | F |

## C. Multiple Choice

Circle the correct answer for each statement.

6.1    The overall board outline is delineated by (*corner brackets, forbidden area*) on the layout sketch.

6.2    The recommended format for a single-sided layout sketch is a composite view of both sides of the board where the conductor side is considered to be the (*bottom, top*) view.

6.3    When positioning dolls, a (*balanced, symmetrical*) layout is achieved if there is a uniform distribution of components with no large areas of wasted space.

6.4    The recommended pad diameter for lead diameters between 0.020 and 0.040 inch is (*0.080, 0.10*) inch.

6.5    The on-center distance between terminal pads for a $\frac{1}{2}$-watt resistor when $X = 0.090$ inch and $B = 0.344$ inch is (*0.60, 1.10*) inch.

6.6    The outer circle shown in the top view of an installed turret terminal represents the terminal (*shoulder, pad*).

6.7    The metal tab that is used to key a TO5-style transistor is closest to the (*collector, emitter*) lead.

6.8    The initial location of an active component is best accomplished by using a (*peripheral, central*) viewpoint.

6.9    As conductor paths are routed to terminal pads, direction changes should be made at (*45°, 90°*) angles.

6.10    The surrounding copper remaining after a terminal pad is drilled is called the (*annular ring, shoulder*).

## D. Matching Columns

Match each item in Column A to the most appropriate item in Column B.

| COLUMN A | COLUMN B |
|---|---|
| 1.  Grid system | a.  Axes |
| 2.  Precut component outlines | b.  Peripheral |
| 3.  Corner brackets | c.  Turret terminal |
| 4.  Keying | d.  Conductor path |
| 5.  Automatic insertion | e.  Doll |
| 6.  Shoulder | f.  +, −, metal tab |
| 7.  Routing | g.  Schematic |
| 8.  Balance | h.  0.1 inch and 1.0 inch |
| 9.  Parts list | i.  Overall board outline |
| 10.  Viewpoint | j.  Component placement |

## E. Problems

6.1 Using the appropriate templates and poster paper, trace and cut 2:1 scale dolls for all resistors, capacitors, and the TO5-style transistor described in this chapter. Make a new layout sketch of Fig. 6.2 with all input and output terminals on the right side of the pc board. The order from top to bottom is to be IN, OUT, +12 Vdc, and GND.

6.2 Construct a component layout and conductor pattern for the series feedback regulator circuit shown in Fig. 6.18. All input and output connections are to be located on one edge of the pc board and designed for finger-type connection. The maximum size of the pc board is to be 3 by $3\frac{1}{2}$ inches. The printed circuit is fabricated in Problem 10.2 with component assembly and soldering performed in Problems 16.2 and 17.2, respectively.

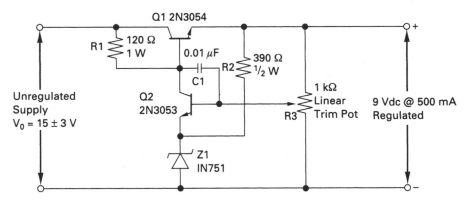

**Figure 6.18** Series feedback regulator.

6.3 Design 2:1 scale component and conductor pattern layouts for the tachometer circuit given in Fig. 6.19. Maximum pc board size is to be $1\frac{1}{2}$ by 2 inches with provision for a suitable pc connector on one of the 2-inch sides of the board. All components and external connections are to be identified and keyed where necessary. The silk screens are constructed in Problem 10.1 and employed in Problem 10.3. Printed circuit board component assembly and soldering are undertaken in Problems 16.3 and 17.3, respectively. The pc board and meter are mounted in Problem 25.3. The harness for interconnections is constructed in Problem 26.5. An optional double-sided circuit design may be considered for this problem.

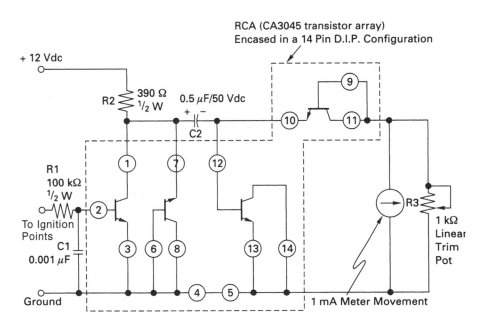

**Figure 6.19** Tachometer.

# 7 Drawing PCB Artworks with Computer-Aided Drafting Techniques

## LEARNING OBJECTIVES

*Upon completion of this chapter on drawing printed circuit board artworks using computer-aided drafting techniques, the student should be able to*

- Access the main menu and establish the name for a new PCB component layout drawing and conductor pattern artwork.
- Select the SETTINGS submenu from the ROOT menu and establish the units, grid, snap, and limits for a new drawing.
- Enter the DRAW menu and use the LINE function to define the component border and corner brackets for a new PCB design.
- Use the DRAW menu functions to draw 1:1 scale two-dimensional component outlines for resistors, capacitors, devices, and integrated circuits.
- Use the TEXT and DTEXT editor to create and position component labels such as $R_1$, $C_2$, and $IC_1$.
- Draw a 1:1 scale component and conductor pattern artwork using layers to display, individually or as composite layer combinations, corner brackets with component layout, donut pattern for solder mask, and component marking mask artwork.
- Save and print the 1:1 scale artwork layers for PCB processing.

## 7.0 INTRODUCTION

The demand continues to grow for engineering drawings and PCB layouts to be done using computer-aided drafting (CAD) techniques. This industrial trend became prevalent in the early 1980s and the technology is now invading education. Two of the more popular software packages used in industry, and currently being adopted by some colleges and universities for their computer laboratories, are AutoCAD and VersaCAD. These software packages are very powerful commercially available programs with an astounding array of options, but they are relatively expensive.

To be compatible with this trend, we have selected AutoCAD Release 12 as the software package to illustrate this technology. It is used in this chapter to create the component layout drawing and conductor pattern artwork for the amplifier circuit designed in Chapter 6 and hand sketched in Figure 6.16. Using AutoCAD, we describe the primary operations of drawing lines, rectangles, circles, donuts, conductor patterns, as well as scaling, labeling, making modifications, saving, and printing files.

*CAUTION*: When using a computer station for long periods of time, ergonomics should be applied. The following may help keep you more comfortable at your computer.

1. To avoid eye discomfort, viewing distance should be made appropriate for screen size. This is generally 12 to 24 inches for 15-inch screens—more for larger

screens. The monitor should also be adjusted so that the top of the screen is at eye level.

2. Adjust screen brightness and contrast so that characters and other displayed features are brighter than the background.

3. Use a document holder that is adjusted to the same height as your computer screen.

4. Place the keyboard directly in front of you, and where your wrist and forearm may be supported. Position the mouse pad within comfortable reach beside the keyboard.

5. Use an adjustable chair that is easily moved, that provides lower back support, and whose height may be adjusted so that hips and knees are level, with lower arms roughly horizontal with the keyboard.

6. Position the chair so that the wrists are straight—not flexed.

7. Be sure that the work station has adequate leg room and a foot rest if needed for hip/knee alignment.

8. Organize your bench and work space so that operations are within easy reach. Reaching behind your shoulder line or above shoulder height should be avoided.

The section on safety at the beginning of the book should be reviewed for more specific information.

## 7.1 INEXPENSIVE CAD SOFTWARE

The list of software packages for drafting is large and expanding. The selection process can be difficult. While a commercial software package like AutoCAD might be available at the university computer center, low-cost drafting packages are also needed for student home use. Writing in the Spring 1993 issue of the ASEE *Journal of Engineering Technology,* Professor Martin Pike of Purdue University Programs at Kokomo details the process he followed for selecting a low-cost drafting package in his article "Inexpensive CAD Software for Student Home Use." According to this article, studies have shown that many engineering students routinely purchase computers for their own unrestricted personal use and to do homework assignments. Most students select an IBM-compatible computer because of its flexibility and compatibility with university equipment. Furthermore, students often request information on low-cost, reasonable substitutes for AutoCAD and VersaCAD that can be used to complete drafting assignments at home.

The conclusions from this article indicate that a low-cost shareware program called Draft Choice or the very economical student edition of a commercial program such as Auto-Sketch are good choices for student home use. Cited by Professor Pike as reasons for these selections are (1) the ease of installation, (2) standard dimensioning techniques for linear systems, and (3) an interactive menu system similar to AutoCAD and VersaCAD. However, if transferring files is important, AutoSketch is the choice, since its files can be saved in a file format that makes them transportable to a university-installed AutoCAD program.

Selecting the most appropriate CAD software package is becoming more difficult as the number of available packages increases. However, the selection and use of one of the available CAD software packages has now become a virtual necessity.

## 7.2 AUTOCAD SETUP FOR PRINTED CIRCUIT BOARD DESIGNS

To use AutoCAD to create a PCB component layout drawing, we must first access the main menu to begin a new drawing. A name is given to the new drawing at the appropriate main menu prompt. Then we proceed with a routine setup procedure for the drawing. The setup involves the selection of units, grid, snap, and limits. Each of these important first steps is considered in detail in this section on AutoCAD setup.

The authors understand that many are still using earlier revisions of AutoCAD. For this reason we continue to include the material that describes the manipulating of specific commands and selections when using earlier AutoCAD revisions in Sections 7.2.1 through 7.7. Since the basic printed circuit board design and layout concepts as presented in this chapter are adaptable to the most recent CAD software packages, this material also remains unchanged. Students who are currently using Windows verions of the latest, more user-friendly software packages such as AutoCAD Release 14 or Release 2000, can accomplish their designs in a more productive fashion. Since it takes less time to produce the CAD art-works with Windows version software, these students can concentrate more on the material in this chapter that deals with the basic pcb layout concepts, which can result in gaining a quicker, more focused approach to understanding the concepts used to produce a finished pcb layout design.

**7.2.1 AutoCAD Main Menu**   If you are using earlier AutoCAD releases you may have to access the AutoCAD subdirectory on the hard drive (C:) of your computer by keying in the following underlined keystrokes:

$$C:\backslash\underline{CD\backslash ACAD<Enter>}$$

Then at the ACAD subdirectory prompt type:

$$C:\backslash ACAD>\underline{ACAD<Enter>}$$

You are presented first with the main menu for AutoCAD Release 10 and 11. This menu is shown in Fig. 7.1. Note that the main menu has a series of selections, most of which are self-explanatory. Selecting 0 allows you to exit AutoCAD. Items 1 to 4 are related to creating/editing/plotting new drawings, and items 5 to 9 are utility options. At the bottom of the main menu you are prompted to enter your selection.

Since we wish to create a new drawing, we select item 1:

$$Enter\ selection:\underline{1<Enter>}$$

The name of the new drawing is entered after the next prompt:

$$Enter\ NAME\ of\ Drawing:\underline{PCB-1<Enter>}$$

Note that a valid name consists of no more than eight letters and/or numbers. AutoCAD next loads the drawing editor and presents a graphics screen and the screen menu.

**7.2.2 AutoCAD Graphics Screen**   The AutoCAD graphics screen and screen menu are shown in Fig. 7.2. Using a mouse, you can move up or down the screen menu (right side of the graphics screen) to highlight an item. Then pressing the mouse's left (pick) button will

**Figure 7.1**   The main AutoCAD menu has 9 options. Selecting option 1 allows you to create, edit, and plot a new drawing. Options 5 to 9 are utilities options.

```
Main Menu

    0.   Exit AutoCAD
    1.   Begin a NEW drawing
    2.   Edit an EXISTING drawing
    3.   Plot a drawing
    4.   Printer Plot a drawing

    5.   Configure AutoCAD
    6.   File Utilities
    7.   Compile shape/font description file
    8.   Convert old drawing file
    9.   Recover damaged drawing

Enter selection:
```

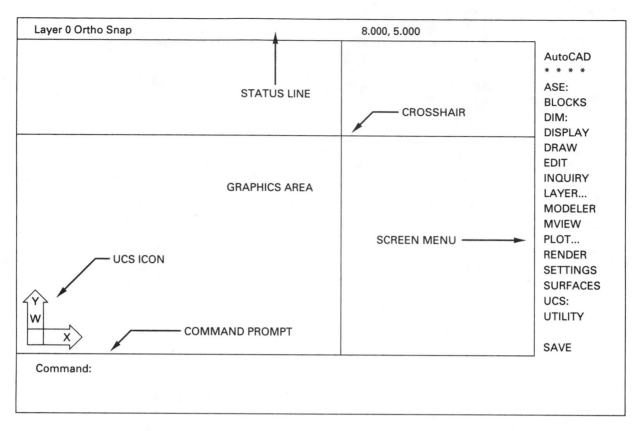

**Figure 7.2** AutoCAD's graphics screen includes a screen menu (right), status line (above), command prompt (below), user coordinate system (UCS, lower left), and graphics area.

select the action you want AutoCAD to perform. This first screen menu is called the ROOT menu. Selecting any item from the ROOT menu will call a submenu of new items. We use the mouse to select SETTINGS from the ROOT menu in the next section to establish our new drawing's (PCB-1) units, grid, snap, and limits.

The center of Fig. 7.2 is the graphics area and will display the PCB layout as the drawing is developed. At the top is a status line. This line identifies the layer currently being displayed, whether the ORTHO and SNAP functions (discussed in later sections) are active, and the exact X, Y coordinate location of the cursor.

At the bottom left edge of the graphics screen, a command line is shown. This command line is used to display and enter information.

### 7.2.3 Settings, Grid, Snap, and Limits    To begin a new drawing, type **NEW** at the command prompt followed by <Enter>:

Command: <u>new <Enter></u>

Type a file name, no more than 8 letters or numbers. When the dialog box appears, note that the cursor's crossed lines change to an arrow pointer, which is moved with the mouse to make selections in the dialog box. The file name selected in Sec. 7.2.1, PCB-1, is used and appears beside the blinking bar in the New Drawing Name dialog box displayed on the screen:

Command: <u>PCB-1 <Enter></u>

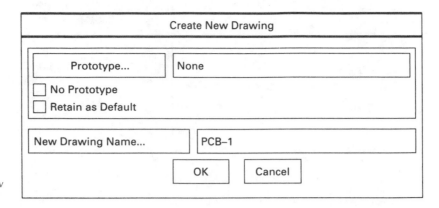

**Figure 7.3** Create New Drawing dialog box.

Move the arrow pointer and click the OK box with the mouse, or simply use <Enter> (see Fig. 7.3).

The PCB-1 drawing will be printed on a standard sheet of $8\frac{1}{2}$ by 11-inch white paper. This paper will be oriented with its long axis along the horizontal. To establish the setting for PCB-1, highlight the SETTINGS submenu in the ROOT menu with the mouse pointer and select it. In this text we identify a highlighted menu item with bold caps followed by <Enter> to indicate the mouse selection or pick button:

**SETTINGS**<u>&lt;Enter&gt;</u>

You are presented with the first page of the SETTINGS submenu. Since UNITS is on the next page of this submenu, highlight NEXT:

**NEXT**<u>&lt;Enter&gt;</u>

***Units.*** To format the units for PCB-1, highlight UNITS and pick it with your mouse:

**UNITS:**<u>&lt;Enter&gt;</u>

A series of options will be displayed, allowing us to tailor how AutoCAD interprets units. The first is the UNITS Report formats shown in Fig. 7.4a. Note that there are five options and that at the bottom command line, the default value is given in < > brackets. The <1>, indicating scientific units, was the default value. Since we want to express every dimension in decimals, we select 2 as follows:

Enter choice, 1 to 5 <1>: <u>2&lt;Enter&gt;</u>

Next you are asked to select the degree of displayed precision. We enter 3 at the prompt

Number of digits to right of decimal point (0 to 8) <4>:
<u>3&lt;Enter&gt;</u>

(see Fig. 7.4a).

AutoCAD is capable of displaying angular measurements in several formats, as shown in Fig. 7.4c. The default is option 1, decimal degrees. Choose this method of measurement by keying in <Enter> at the command prompt, since the displayed default value is also 1:

Enter choice, 1 to 5 <1>: <u>&lt;Enter&gt;</u>

Again we are asked to select the degree of displayed precision. We choose the default value 0, at the prompt, indicating that fractional degrees are not necessary (Fig. 7.4c).

Angles are measured with respect to a zero reference, with either a clockwise (CW) or counterclockwise (CCW) direction for positive angles. Normal convention dictates 0 degrees

```
Command: _UNITS Report formats:    (Examples)

        1.  Scientific              1.55E+01
        2.  Decimal                 15.50
        3.  Engineering             1'-3.50"
        4.  Architectural           1'-3½"
        5.  Fractional              15½

With the exception of Engineering and Architectural
formats, these formats can be used with any basic unit of
measurement. For example, Decimal mode is perfect for
metric units as well as decimal English units.

Enter choice, 1 to 5 <2> : <Enter>
```

(a)  Decimal dimensioning is used for PCB designs.

```
Number of digits to right of decimal point (0 to 8) <4>
:3 <Enter>
```

(b)  A displayed precision of 0.001 inches is suitable.

```
Systems of angle measure:            (Examples)

        1.  Decimal degrees           45.0000
        2.  Degrees/minutes/seconds   45d0'0"
        3.  Grads                      50.0000g
        4.  Radians                    0.7854r
        5.  Surveyor's units          N45d0'0" E

Enter choice, 1 to 5 <1> : <Enter>
```

(c)  All angles will be in units of degrees.

```
Number of fractional places for display of angles (0 to 8)
<0> : <Enter>
```

(d)  Fractional degrees are not necessary for most basic PCB
     designs.

```
Direction for angle 0:

    East        3 o'clock = 0
    North      12 o'clock = 90
    West        9 o'clock = 180
    South       6 o'clock = 270

Enter direction for angle 0 <0> : <Enter>
```

(e)  All angles are measured from 3 o'clock reference

```
Do you want angles measured clockwise? <N> : <Enter>
```

(f)  All angles are measured counterclockwise.

**Figure 7.4**  Units is the first necessary setup to begin a PCB drawing.

at 3 o'clock with positive angles measured CCW from that reference. We choose the normal convention in Figs. 7.4d and e, <0>: <Enter> and <N>:<Enter>, respectively.

*Grid and Snap.*  At the command line prompt press <F1>. Then return to the first page of the SETTINGS submenu by first highlighting and then selecting LAST:

**LAST**<Enter>

Next highlight GRID and select it:

**GRID**<u><Enter></u>

From our discussion of grid selection (see Sec. 6.1) a 0.05-inch grid is suitable for the PCB-1 drawing to be designed in this chapter. Therefore, at the GRID command line shown in Fig. 7.5a enter 0.05 followed by <u><Enter></u>. The grid is turned on or off using the F7 key.

SNAP is an AutoCAD function. It is the smallest dimension the cursor can resolve. If set to the grid setting above, the cursor can only be positioned on the grid, in both the X and Y directions. The snap is set to the grid spacing by typing 0.05 after the SNAP command line shown in Fig. 7.5b. To set SNAP equal to GRID, highlight the grd=snap function on the GRID menu:

**grd=snap**<u><Enter></u>

Then at the command line enter 0.05 followed by <u><Enter></u>. The SNAP function can be switched on or off using function key <F9>. The command line on the graphics screen indicates the on or off state of the **SNAP** function, <Snap on> or <Snap off>.

*Limits.* Sizing the plotting paper completes the AutoCAD setup of a new drawing. Drawing size is set in the LIMITS submenu. Return to the first page of the SETTINGS submenu by highlighting LAST:

**LAST**<u><Enter></u>

Next highlight LIMITS and select it:

**LIMITS**<u><Enter></u>

Figure 7.6 illustrates that coordinates 0.000, 0.000 identify the lower left corner of the paper. Press <u><Enter></u> and the upper-right-hand corner defaults to 12 by 9 inches, where the longer length is along the X axis. To resize to 4 by 4 inches, type 4, 4 after the <12.000, 9.000> prompt. Function key F6 controls the X, Y coordinate information on the status line. This is done by entering Zoom, followed by entering All to activate the coordinates for the limits selected.

---

**Figure 7.5** The Grid and Snap functions can be set equal using the grd=snap function.

> Command: '_GRID
> Grid spacing(X) or ON/OFF/Snap/Aspect <0.000> : <u>0.05</u>
> <u><Enter></u>
>
>   (a) Grid spacing is set at this command prompt.
>
> Command: _SNAP
> Snap spacing or ON/OFF/Aspect/Rotate/Style <0.100> : <u>0.05</u>
> <u><Enter></u>
>
>   (b) Snap spacing is set at this command prompt.

**Figure 7.6** Setting the limits in AutoCAD.

> Command: '_LIMITS
> Reset Model space limits:
> ON/OFF <Lower left corner> <0.000,0.000> : <u><Enter></u>
>
> Upper right corner <12.000, 9.000> : <u>4,4 <Enter></u>

## 7.3 DRAWING THE COMPONENT BORDER AND CORNER BRACKETS

As detailed in Chapter 6, the size of a printed circuit's component border is determined after considering several factors. For the amplifier component and conductor pattern layout shown in Fig. 6.16, we need a 2.1- by 2.5-inch PCB to support the components and conductor pattern. The overall board edge and component border are defined by corner brackets. We begin the component layout drawing (PCB-1) for the amplifier by drawing the 2.1- by 2.5-inch rectangular component border and defining it with corner brackets.

### 7.3.1 Defining the Component Border
Enter the DRAW submenu by first highlighting DRAW in the SETTINGS menu and then choose this selection with the mouse pick button:

**DRAW:**<u>\<Enter\></u>

Then select LINE:

**LINE:**<u>\<Enter\></u>

Using F6 and the COORDS displayed on the status line, we center our 2.5 by 2.1 component border in the 4- by 4-inch grid defined in the previous section. Using \<Ortho on\> (F8) and \<Snap on\> (F9), move the cursor until it is positioned at location 0.750, 1.000, as indicated on the status line. Set this point by using the pick button on the mouse. Next move to location 0.750, 3.100 and select this point with the mouse. You will note a straight line between these two points. Continue the component border construction by moving to location 3.250, 3.100 and then 3.250, 1.000, setting the points each time with the mouse pick button. Complete the box by moving back to the starting location 0.750, 1.000. Set this point with the pick button of the mouse and terminate the line construction with a space bar \<SP\>.

Figure 7.7a illustrates the component border just drawn with the coordinates of the corners shown. In the next section we position the corner brackets at the four corners of the component border.

### 7.3.2 Defining the Corner Brackets
To locate a 0.1-inch.-wide, half- by half-inch corner bracket at each corner of the PCB, we use the Zoom and Window features of AutoCAD accessible from the pull-down VIEW menu. Moving the cursor up to the status line will display 9 menus. These pull-down menus are (1) FILE, (2) ASSIST, (3) DRAW, (4) CONSTRUCT, (5) MODIFY, (6) VIEW, (7) SETTINGS, (8) RENDER, and (9) MODEL. Move the cursor to highlight the VIEW menu and select it with the mouse pick button:

**VIEW**<u>\<Enter\></u>

Figure 7.7b shows the VIEW menu options for Zoom; from these options select Zoom and Window with the mouse. The command line will prompt for the first corner. Locate the cursor at the lower-left corner of a viewing window that will expose one corner of the component border. Press the mouse pick button. Then move to the upper-right corner of the viewing window and select with the pick button (see Fig. 7.7c). The command lines for the Zoom and Window functions are given in Fig. 7.7d.

To draw the corner brackets, enter the DRAW menu and again select LINE. Construct a half- by half-inch bracket, 0.1 inch wide, using the steps outlined in the last section on drawing a component border. To exit Zoom and Window, again highlight the VIEW pull-down menu and select and pick Zoom and All to return to the full view.

Repeat the steps outlined in this section to construct corner brackets at the edges of the component border. Figure 7.7e shows the PCB component border with half-inch corner brackets.

---

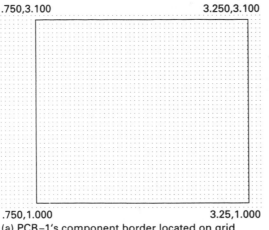

.750,3.100                     3.250,3.100

.750,1.000                     3.25,1.000

(a) PCB–1's component border located on grid

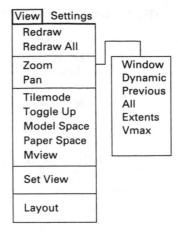

(b) Zoom, Window, and Zoom All are just three selections in the View pull-down menu

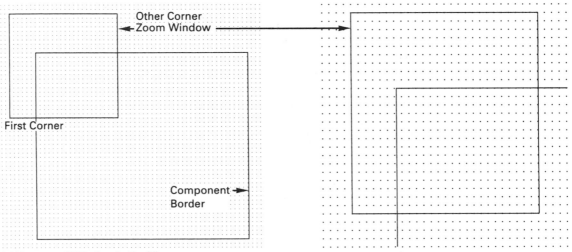

Other Corner
Zoom Window

First Corner

Component
Border

(c) Zoom and Window in the View menu

```
Command:  _zoom
All/Center/Dynamic/Extents/Left/Previous/Vmax/Window/⟨Scale(X/XP)⟩ : _window
First corner:  Other corner:
```

(d) Command line for the Zoom and Window functions

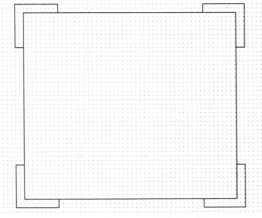

(e) Component border (2.5 × 2.1 in.) with corner brackets ($\frac{1}{2}$ × $\frac{1}{2}$ in. long × 0.1 in. wide)

**Figure 7.7**   The Line function in the DRAW menu is used to construct both the component border and corner brackets. Corner brackets are drawn using the DISPLAY menu's Zoom and Window functions.

## 7.4 DRAWING PCB COMPONENTS

The component and wiring sketch of the amplifier developed in the last chapter is now converted into a detailed 1:1 scaled component layout drawing. This two-dimensional drawing is viewed from the component side of the PCB and includes component bodies, lead access holes (circles with a diameter equal to the donut diameter of the conductor pattern artwork), and component and device labeling. Circuit wiring is also part of the component layout drawing. Using the component border developed in the last section and the sketch shown in Fig. 6.16, we begin by drawing the components in their approximate locations and to exact (1:1 scale) size.

**7.4.1 Axial-Lead Components** As an illustration, we draw a two-dimensional representation of a half-watt resistor (for this exercise, the snap will have to be changed to 0.025 inch). This component measures 0.15 inch in diameter, is 0.8 inch long, and the center of the lead access hole is 0.10 inch from the end of the body. Begin by entering the DRAW menu and calling the Line function. Then using the procedure outlined in Section 7.3.1 draw a rectangle using the grid as a guide and the dimensions given above (see Fig. 7.8a). Next highlight and select the Circle function in the DRAW menu:

CIRCLE: &lt;<u>Enter</u>&gt;

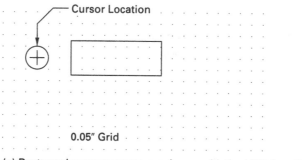

(a) Rectangular components are drawn with the LINE function

Command:__CIRCLE 3P/2P/TTR/ ⟨Center point⟩: Diameter/⟨Radius⟩:__D
Diameter: 0.1 ⟨Enter⟩

(b) The CIRCLE command line prompts for the diameter of the circle

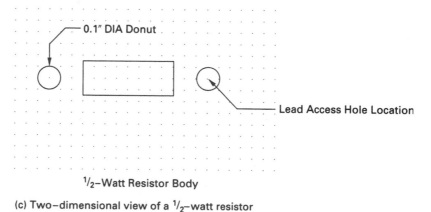

**Figure 7.8** Rectangular components are generated using the Line and Circle functions in the DRAW menu.

(c) Two–dimensional view of a ¹/₂–watt resistor

Select CEN, DIA: which indicates center of the circle and its diameter:

$$\mathbf{CEN,DIA:}<\underline{Enter}>$$

At the command line prompt, <center point>, locate the cursor in the center of the component body, 0.075 inch from either edge of the component body, and 0.10 inch from the end of the body. Press the mouse pick button. Then at the command line prompt for circle diameter, enter 0.1 followed by <Enter> (see Fig. 7.8b for the CIRCLE command line).

Repeat the drawing instructions for circles to add a second lead access hole circle. Figure 7.8c shows the final component configuration for a half-watt resistor.

**7.4.2 Radial-Lead Components**    Figure 7.9 shows a two-dimensional view of several radial-lead components. Note that they can all be adequately represented by drawing either a rectangle or circle for the body outline and circles to represent the terminal pad areas or donuts. Therefore, the techniques outlined in Sec. 7.4.1 can also be used to construct these components.

**7.4.3 Devices and Integrated Circuits**    Figure 7.10 shows the two-dimensional view of a few typical devices and integrated circuits (ICs). Note again that the bodies of these devices are represented by line segments and again circles are used to show the donuts. The student should gain some experience by duplicating the component and device outlines illustrated in this section.

**7.4.4 Final Component Layout and Labeling**    The AutoCAD drawing shown in Fig. 7.11a completes the component layout for Fig. 6.16. Note that all components and de-

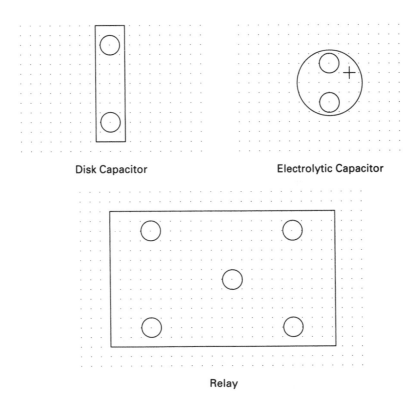

Disk Capacitor                    Electrolytic Capacitor

**Figure 7.9**  Several radial-lead devices.

Relay

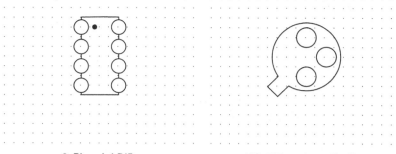

**Figure 7.10** Some standard devices and IC packages.

8–Pin mini DIP          TO–5 Transistor Outline

vices are positioned on grid. We next add descriptive labels to the components by using the text editor. At the command line prompt, type

TEXT &lt;Enter&gt;

You will be prompted to locate the starting point of the text:

Justify/Style/&lt;Start point&gt;:

Move the cursor to locate the lower-left edge of the letter/number and select this point with the mouse pick button. When prompted for the letter/number height, it is recommended that you pick a minimum of 0.1 inch. Therefore, at the command line prompt, type

**Height&lt;0.200&gt;**0.1&lt;Enter&gt;

Normally labeling is done at a zero rotation angle:

**Rotation angle &lt;0&gt;** &lt;Enter&gt;

The component layout drawing with all parts identified is shown in Fig. 7.11b. In the next section we add the conductor pattern or wiring traces.

**7.4.5 Adding the Wiring Traces**    Before adding the wiring traces on the component layout drawing, check the status line for the condition of the Ortho (orthogonal) function. This function can be switched on and off using the F8 function key. With the Ortho function on, lines can be drawn only at right angles to one another and parallel to the edges of the working drawing. Switching the Ortho function off, however, allows line segments to take any angle. Therefore, use F8 to switch Ortho off before adding the wiring traces to the component layout drawing.

Figure 7.12 shows the finished component layout drawing for the amplifier circuit with all wire traces added. The illustrated wiring trace segment from point a to b and then to point c is completed by first entering the DRAW menu and selecting the Line function. Then at the command line prompt

Command: LINE From point:

use the mouse to locate the donut center at point a and select it with the pick button. At the next command line prompt

To point:

move the cursor to point b, selecting it with the pick button. Finally, at the next prompt

To point:

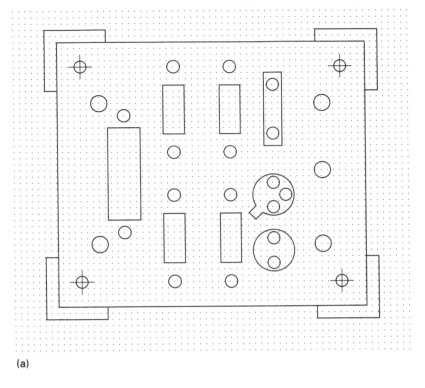

(a)

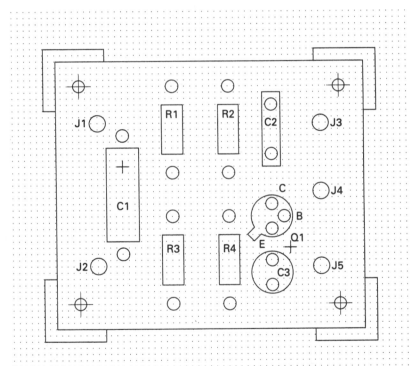

**Figure 7.11** Final component layout using AutoCAD: (a) components positioned; (b) components with labels.

(b)

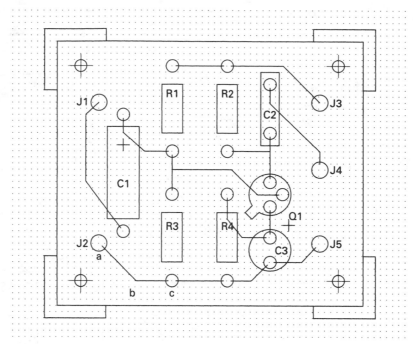

**Figure 7.12**
Component layout with conductor path routing complete.

again move the cursor, this time to point c, select it by pressing the pick button, and press the pick button one more time to terminate this trace segment. All wiring traces are added to the drawing in a similar fashion.

## 7.5 SAVING THE COMPONENT LAYOUT DRAWING

During the length of time it takes to make a complete set of PCB drawings using AutoCAD, the work should be periodically saved in the hard drive of your computer. This is to avoid losing all work completed due to either the computer hanging up or a power outage. The drawing is saved with the original filename that was specified when the new drawing was originated. It is possible to periodically save your work and remain in AutoCAD. At the command prompt, type SAVE followed by <Enter>:

Command: SAVE<Enter>

Depress the Ctrl and C keys simultaneously to return to the command prompt if necessary. When SAVE is used, a dialog box, as shown in Fig. 7.13, appears on the screen. The original filename, PCB, is typed at the command prompt and appears in the File: box. The arrow pointer is then moved to the OK box to return to the drawing.

## 7.6 CONDUCTOR PATTERN ARTWORK

Complete sets of printed circuit board drawings are made using AutoCAD layering. Layers can be thought of as a stack of transparent pieces of paper that are used to form a system of drawings. Each layer is drawn separately from any other layer. The layers can be viewed individually or in composite combinations. This gives the ability to maintain precise registration between all layers used. Layers can also be printed individually or in composite combinations.

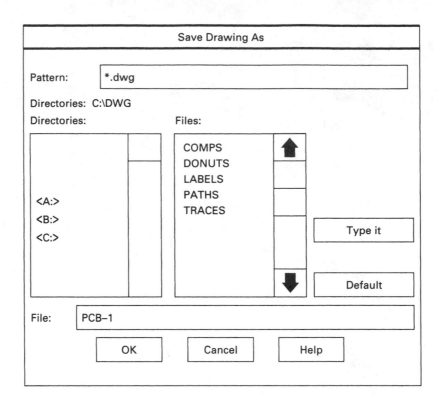

**Figure 7.13** Dialog box used to save drawings.

To access Layers, pick Settings at the top of the screen. The pull-down menu shown in Fig. 7.14 will appear on the screen. Now pick Layer Control . . . in the menu and the Layer Control dialog box shown in Fig. 7.15 will be displayed. Each layer to be drawn is given a name. The name is typed in the box with the blinking cursor, followed by <Enter>. Once entered, pick the New box with the arrow pointer. The layer name will then appear in the Layer Name listing section of the Layer Control dialog box. The layers to be viewed are turned on by picking the On box. Drawing entities will be produced only on the current layer even with other layers on in a composite view. The current layer must also be on for drawing entities to be seen on the screen as they are made. To make the layer to be drawn current, select the layer name with the pick arrow. The name will become highlighted. Be sure that this layer is also turned on. Now pick the Current box. This returns you to the drawing screen with the current layer name displayed at the left side of the status line. The component layout shown in Fig. 7.11a was drawn on layer 0. The labeling was produced on layer 2 with layer 0 on as a guide for positioning the literal detail. Therefore, Fig. 7.11b is the resulting display with layers 0 and 2 turned on. A composite print of these layers is often desired when a marking mask is required (see Sec. 7.8). Wiring traces are drawn on layer 4. The results in Fig. 7.12 are with layers 0, 2, and 4 turned on.

The conductor pattern artwork is the 1:1 scaled drawing that will be used to fabricate the pc board. It consists of 0.1-inch-wide solid (filled) corner brackets to define the finished size of the board, 0.1-inch-diameter donuts at every lead access hole, 0.125-inch-diameter donuts at all terminal pads, and 0.025-inch-wide conductor paths for the wiring traces. The donuts and the conductors will also be drawn on separate layers (layers 1 and 3, respectively). It should be noted that AutoCAD automatically arranges Layer names in alphabetical order.

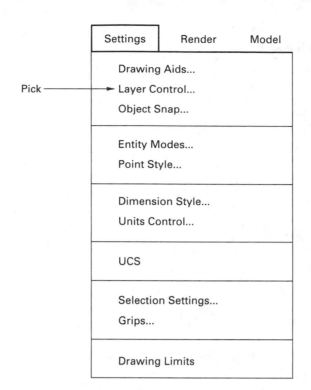

Pick ──────▶

Figure 7.14  Settings
pull-down menu.

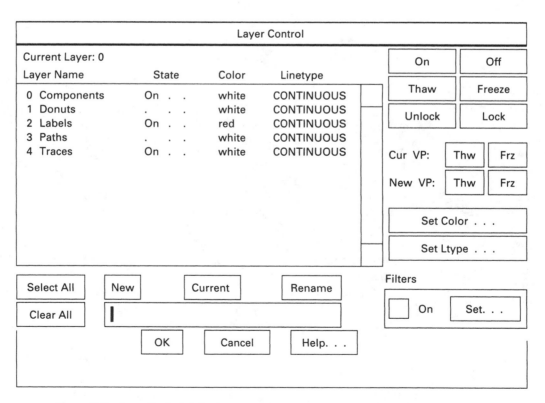

**Figure 7.15**  Layer Control dialog box.

***Donuts.*** We begin to generate the conductor pattern artwork by first making a NEW layer called Donuts (layer 1) current. Layers 0, 2, and 4 are left on. Next enter the DRAW menu and highlight DONUT:

`DONUT:`<u>`<Enter>`</u>

Then at the command prompt for Inside diameter, enter zero, since we will be using solid donuts:

`Inside diameter <0.5000>:` <u>`0`</u> `<Enter>`

At the Outside diameter prompt, enter 0.1 inch, the diameter selected for this project:

`Outside diameter <1.000>:` <u>`0.1 <Enter>`</u>

Then AutoCAD will prompt for the center of the donut. Using the mouse, locate the cursor at the center of one of the lead access holes and push the pick button to draw a solid, filled donut. Repeat this process until all of the lead access holes are filled. Then change the donut size to 0.125 inch for filling the terminal pads by again entering the DRAW menu and selecting Donut to be prompted for donut size desired. Figure 7.16a shows all of the donuts located. The donuts are often placed on a separate layer in case a solder mask is to be made (see Sec. 7.8).

***Conductors.*** To draw the conductors, a NEW layer called Paths (layer 3) is made current. Layers 1 and 4 are turned on as a guide for drawing the paths to the correct donuts. Conductor patterns are created using the Poly line or Pline function. This function is used to draw lines with any defined width. To draw a conductor path, select DRAW from the screen menu. Now pick NEXT near the bottom of the menu. The continuation of the DRAW menu is displayed. Finally, select PLINE: at the top of this menu:

`PLINE:`<u>`<Enter>`</u>

The software command line then prompts for the starting point. From Point: use the mouse to locate the center and select one of the donuts requiring a conductor path. You will be prompted with the following command line:

`Arc/Close/Halfwidth/Length/Undo/Width/<Endpoint of line>:`

Since we want to specify the width of the poly line (conductor path), type <W> followed by <Enter>. When prompted for the starting and ending width, type 0.025 inches:

`Starting width <0.0500>` <u>`0.025<Enter>`</u><br>
`Ending width <0.0250>` <u>`<Enter>`</u>

At this point you are again presented with the command line shown above, where <Endpoint of line> information is required. Use the mouse to locate the transition point of the line or the center of a donut. Selecting this location with the pick button will generate the line. Repeat this process until all conductor paths are drawn. Figure 7.16b shows the completed conductor pattern artwork for the amplifier circuit designed in the previous chapter.

It should be noted that any lettering that is to be produced in copper on the circuit side of the pc board must appear in backward lettering. This is due to the fact that the circuit side is viewed through the component side layer when the conductor pattern layout is generated. A reversal in orientation of this view is required when a photo is made of the conductor pattern artwork. The backward lettering ensures that the correct view is used when processing the board (see Fig. 7.16b).

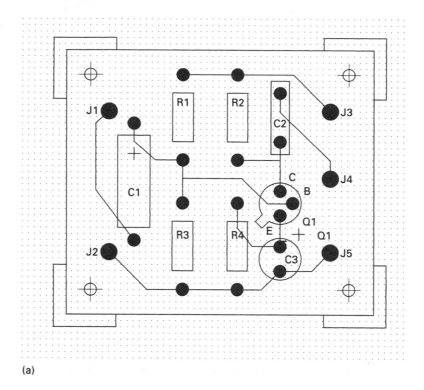

(a)

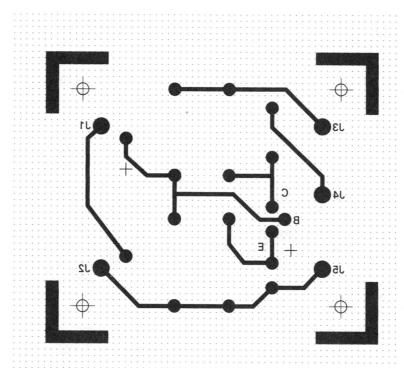

**Figure 7.16** Conductor pattern layout: (a) donut layer with layers 0, 2, and 4 turned on; (b) conductor path layer 3 with donut layer 1 on and all other layers turned off.

(b)

To form backward lettering, first select the DRAW pull-down menu from the top of the screen. Select Text in this menu and then select Set Style from the smaller dialog box that appears beside the DRAW menu. The Select Text Font dialog box now appears on the screen. Roman Duplex and the OK box are picked with the arrow pointer to return to the drawing screen. The following are the continuous sequence of AutoCAD prompts and the response to each one after the <defaults>:

```
Command: `_style Text style name (or ?) <ROMAND>:
romand
Existing style.
Font file <romand>: romand Height <0.0000>: .15
<Enter>
Width factor <1.0000>: <Enter>

Obliquing angle <0>: <Enter>
Backwards? <N> Y <Enter>
Upside-down? <N> <Enter>
Vertical? <N> <Enter>

ROMAND is now the current text style.
Regenerating drawing.
```

After entering Draw and DText from the menu, all characters will appear backward-reading.

***Corner Brackets.*** The sequence of steps outlined in the previous section to draw conductor paths is also used to draw the corner brackets. However, the width should be set to 0.1 inch for this detail (see Fig. 7.16b).

***Conductor Pattern Arrangements.*** When positioning conductor paths, several considerations need to be noted to avoid improper conductor pattern arrangements:

1. A separate terminal pad must be provided, with its center on grid, for each component lead to be soldered to the conductor pattern. In cases in which large conductor areas are used instead of individual conductor pads, each component lead must still be provided with its own access hole.

2. Whenever possible, conductor paths should be along the vertical and horizontal lines of the grid system. Irregular paths are acceptable only if they simplify the conductor pattern.

3. Attention should be focused on the conductor width and spacing. The cross-sectional area of the conductor path is determined in part by the amount of current it will be expected to handle. As discussed in Chapter 5, once the foil thickness has been selected, the minimum conductor width allowable is determined as outlined in Sec. 5.3 with the use of Table 5.6. Once this width has been established, the minimum spacing between conductors must next be determined. Table 7.1 lists recommended conductor spacings for boards that will not be encapsulated (sealed) when completed. These spacings are *minimum* for a wide range of ac or dc voltages and take into consideration the *dielectric breakdown* voltage characteristics. These characteristics refer to that voltage at which flashover (arc) among conductor paths occurs. For optimum circuit reliability, the components should be positioned in such a manner that the conductor width and spacing are not less than the minimums specified throughout the entire conductor pattern. In this phase of layout, the design must not be compromised. Once these minimum values have been determined, any reduction in either one for the purpose of preferred component repositioning for convenience or for the sake of appearance is unacceptable.

**TABLE 7.1**

Recommended Minimum Conductor Spacing

| Voltage Between Conductors Dc or Ac Peak (V) | Minimum Spacing |
|---|---|
| *Sea Level to 10,000 Feet* | |
| 0–150 | 0.025 inch |
| 51–300 | 0.050 inch |
| 301–500 | 0.100 inch |
| Greater than 500 | 0.0002 (inch per volt) |
| *Over 10,000 Feet* | |
| 0–50 | 0.025 inch |
| 51–100 | 0.060 inch |
| 101–170 | 0.125 inch |
| 171–250 | 0.250 inch |
| 251–500 | 0.500 inch |
| Greater than 500 | 0.001 (inch per volt) |

An example that illustrates the use of Tables 5.6 and 7.1 for the purpose of selecting an appropriate conductor width and spacing follows. Assume the following specifications:

*Maximum continuous current:* 1.6 amperes

*Maximum voltage:* 36 volts

Table 5.6 shows that a conductor width of 0.030 inch will safely handle 1.9 amperes using 1-ounce copper foil. This width allows an acceptable margin of safety. The spacing for the specified voltage requirements can be determined from Table 7.1, which shows that the minimum spacing for a 0- to 150-volt range is 0.025 inch. Therefore, no conductor path on the finished board can be closer to any other path than this minimum. The solution to this problem, therefore, is

*Conductor width:* 0.031 inch

*Conductor spacing:* 0.025 inch

Note that these specifications apply to the *finished* pc board at a 1:1 scale.

4. Finally, terminal area shapes and conductor path configurations must be evaluated so that only recommended artwork forms will be employed. Figure 7.17 shows various terminal shapes and path layouts with the preferred pattern configurations, along with those that should be avoided. The use of those configurations that are not recommended may introduce problems such as undesirable voltage distribution along ground paths, nonuniform solder flow onto the conductor pattern (especially at terminal areas), and weakening of the foil bond.

## 7.7 PRINTING THE CONDUCTOR PATTERN ARTWORK

To print the conductor pattern artwork layer, all other layers must be turned off. The PLOT command is entered at the command line:

Command: Plot <u>\<Enter\></u>

The Plot dialog box now appears on the screen as shown in Fig. 7.18a. Before attempting to plot the conductor pattern artwork, the following entries are established to obtain 1:1 scale plots on $8\frac{1}{2}$- by 11-inch paper.

In the Paper Size and Orientation block, Inches is picked using the arrow pointer. Next the Size box is picked and a Paper Size subdialog box is displayed on top of the Plot Configuration dialog box (see Fig. 7.17b). Paper size A is selected followed by

| Recommended | | Not Recommended | |
|---|---|---|---|
| Pattern Design | Comments | Pattern Design | Comments |
| | Uniform distribution of solder around lead access hole | | Non-symmetrical pad area around lead access hole will result in a non-uniform solder flow |
|  | Double filleted entries result in symmetrical solder fillets |  | Non-uniform pad area will result in a non-uniform solder fillet |
|  | Allows symmetrical solder fillets |  | Uniform solder flow is prevented around outside edges of pad |
|  | Pattern allows for symmetrical solder fillets |  | Non-uniformity in solder flow due to large center area of pattern |
|  | Proper path entry |  | Inferior entry results in sharp conductor path edges detracting from appearance and hampering good solder flow |
|  | 45° elbow maintains good foil bond |  | Weakened foil bond due to sharp exterior edge of elbow which could create problems during etching |
| 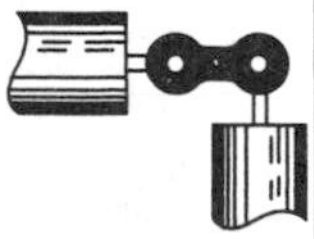 | For symmetrical soldering an individual access hole for each component lead is required |  | Excessive pad area is removed to accommodate multiple leads resulting in foil bond weakening |
|  | Shortest possible path length |  | Excessive pad length and sharp angle at entries |
| 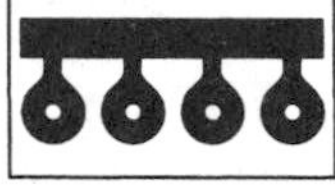 | Lower voltage drops due to heavy buss |  | High voltage drops along thin conductor paths |

**Figure 7.17**
Recommended conductor pattern configurations.

picking the OK box to return to the Plot Configuration dialog box. Now pick the Rotation and Origin box, and their subdialog box (Fig. 7.17c) appears. Pick 0 for rotation and 0.000 for both X and Y origins followed by clicking OK. To obtain a 1:1 scale print, type the number 1 in both the Plotted Inches and Drawing Units boxes. By picking Full and then Preview, the position of the artwork will be shown. If the position needs to be changed, the origin values selected are simply changed. The Plot is now accomplished by making an <Enter>. Plotting commands automatically appear at the command prompts (see Fig. 7.19). However, another <Enter> has to be made after the instruction to Position Paper in Plotter.

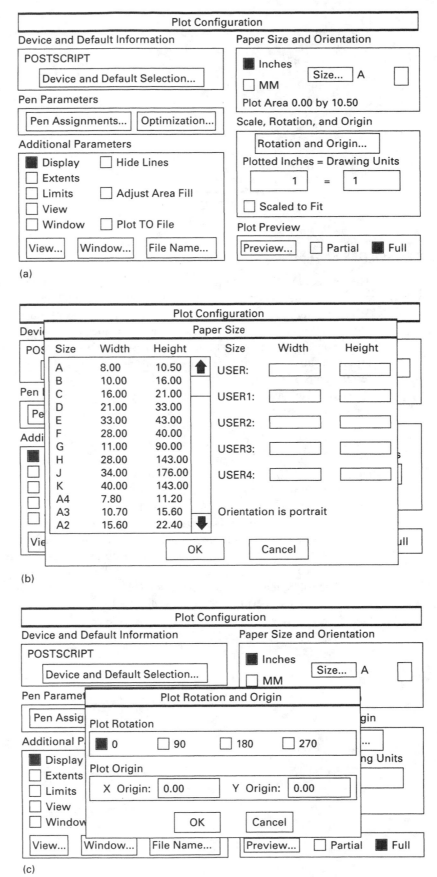

**Figure 7.18** Dialog boxes: (a) Plot Configuration dialog box; (b) Paper Size subdialog box; (c) Rotation and Origin subdialog box.

```
Command: '_plot Effective plotting area: 8.50 wide by 8.70
high
Position paper in plotter.
Press RETURN to continue or S to Stop for hardware setup
Regeneration done 100%

108 raster data scan lines to be sent.
Sending scan line: 32
Scan lines complete.

Plot complete.
```

**Figure 7.19** Plotting commands that appear during printing of the conductor pattern artwork.

## 7.8 SILK-SCREEN SOLDER AND MARKING MASKS

The procedures and techniques thus far discussed for designing component layouts and conductor pattern drawings represent the minimum requirements for producing pc boards. To provide for more efficient soldering, a *solder mask* may be employed. Solder masks consist of corner brackets and duplicate-size donuts located on the solder mask layer as they appear on the conductor pattern layer. These donuts represent the terminal pads on the processed board to which component leads are soldered. The solder mask artwork is referred to as a *positive pad master.*

To form the solder mask over the conductor pattern on a processed pc board, a *silk screen* is necessary. This silk screen is prepared from a 1:1 scale photographic negative of just the donuts and their positions as they appear on the conductor pattern layout. Therefore, using AutoCAD, a donuts layer positive artwork is generated in addition to the component and conductor layers. The silk screen is produced by using the techniques considered in detail in Chapter 10. All of the areas seen as white on the positive solder mask artwork will be clear of film gelatin when the screen is processed. The silk screen solder mask becomes a negative artwork screening tool and must be of the same size as the donut layer. When the conductor side of the board is positioned under the silk screen, the application of an appropriate solder resistant substance through the screen will coat all surface areas of the board, including the conductor paths, with the solder stop-off film but will leave the terminal pads to be soldered exposed. A typical solder mask is shown in Fig. 7.20.

The solder mask will protect the conductor side of the board during and after the solder process since it will not be removed. This mask is applied *after* the board has been

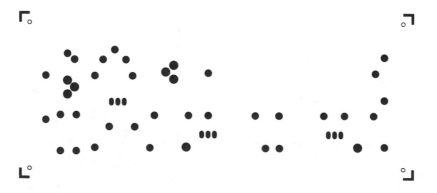

**Figure 7.20** Positive artwork solder mask.

etched as described in Chapters 10 and 13. Solder will alloy only with the exposed bare copper terminal pads and provide sound electrical connections for component leads. A solder mask is used almost exclusively for automatic soldering methods such as wave soldering (described in Chapter 17). Since only the terminal pads are exposed, less solder is required, thereby reducing weight and cost. Without the mask, the entire conductor pattern would be alloyed with the solder.

A *marking mask* is employed to aid in component assembly. This mask shows component outlines, without terminal pads, their position, and identification numbers. When processed into a silk screen, the mask will allow this detail to be printed onto the component side of the pc board. The component outlines are drawn using AutoCAD techniques previously discussed in this chapter. Component identifications such as $R_1$, $R_2$, $C_1$, $C_2$, and $Q_1$, should be in accordance with the schematic diagram. Component outlines on the marking mask layer must also be the same size as those appearing on the component layer.

The marking mask is processed into a silk screen with the screen becoming a *negative* marking mask tool. The application of a suitable epoxy paint will produce the desired detail on the component side of the board. A completed marking mask for a pc board is shown in Fig. 7.21.

At this point, a differentiation should be noted between the lettering detail shown on the marking mask and that shown on the conductor pattern layer. Backward-letter orientation, as previously mentioned, must be used on the conductor pattern layer. Forward-oriented lettering must be used on the marking mask layer to obtain the correct letter orientation when the component side of the board is screened through the marking mask. The application of letter symbols on both sides of the board aids in identifying components and rapidly locating test points when these needs arise.

The 1:1 scale artworks generated by the techniques discussed in this chapter are camera ready when printed and are reproduced into photographic negatives and positives by processes discussed in Chapter 9.

To exit AutoCAD, simply type QUIT at the command prompt.

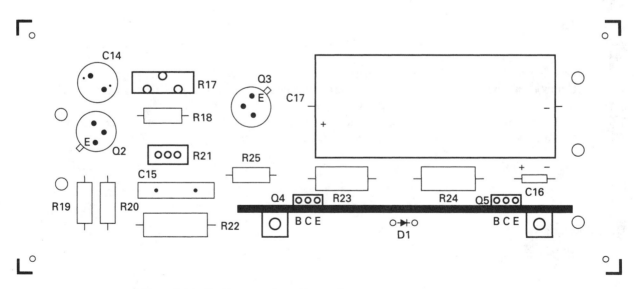

**Figure 7.21**    Positive artwork marking mask.

## EXERCISES

### A. Questions

7.1    Explain how to access the main AutoCAD menu.
7.2    List the four settings that are made before starting a new drawing.
7.3    What keys are used to turn on or off Grid and Snap?
7.4    Describe the information that is displayed on the status line.
7.5    Outline the procedure for assigning a filename to a NEW drawing.
7.6    What are typical Grid and Snap settings for drawing a 1:1 scale layout on AutoCAD?
7.7    How are individual drawing layers established?
7.8    Discuss the procedures for entering forward and backward lettering on a layer.
7.9    Describe the detail that normally appears on the component conductor, marking mask, and solder mask layers.
7.10   Discuss when work should be saved and how this is accomplished.

### B. True or False

Circle *T* if the statement is true, or *F* if any part of the statement is false.

| | | | |
|---|---|---|---|
| 7.1 | To begin a NEW drawing in AutoCAD, the Grid, Snap, Units, and Limits are set first. | T | F |
| 7.2 | No more than eight characters can be used for a file name. | T | F |
| 7.3 | Decimal dimensioning is used for pc board layouts. | T | F |
| 7.4 | Grid and Snap settings should be set to the same value. | T | F |
| 7.5 | Corner brackets are drawn inside the screen X and Y Limits setting. | T | F |
| 7.6 | Component outlines are drawn by selecting Pline from the DRAW menu. | T | F |
| 7.7 | Work is saved both periodically as the drawing is developed and when all layers are completed. | T | F |
| 7.8 | The inside diameter selected for donuts used to generate a conductor pattern and a solder mask is 0.010 inch. | T | F |
| 7.9 | A terminal pad must be provided for each component and device lead. | T | F |
| 7.10 | Marking masks reduce the amount of solder required on a pc board. | T | F |

### C. Multiple Choice

Circle the correct answer for each statement.

7.1    The status line is found at the (*top, bottom*) of the graphics screen.
7.2    When selecting the UNITS report, (*Decimal, Engineering*) format is chosen.
7.3    To turn the Grid on or off, the (*F6, F7*) key is depressed.
7.4    Drawing corner brackets is aided by using (*Zoom All, Zoom and Window*) from the VIEW pull-down menu.
7.5    Components such as (*disk capacitors, resistors*) are constructed with their leads having axial orientation to the component body.
7.6    It is important to locate (*component bodies, terminal pads*) on grid when drawing the component layout.
7.7    DText is used when labeling on the conductor layer to obtain (*backward, forward*) letter orientation.
7.8    Drawing entities will be produced only on the (*Current, New*) layer.
7.9    The amount of (*current, voltage*) determines the minimum spacing between conductor paths.
7.10   The solder mask layout is a (*negative, positive*) artwork.

### D. Matching

Match each item in Column A to the most appropriate item in Column B.

|   | COLUMN A | | COLUMN B |
|---|----------|---|----------|
| 1. | Settings | a. | Backwards |
| 2. | Zoom     | b. | Solder mask |

| | | | | |
|---|---|---|---|---|
| 3. | DRAW menu | | c. | Limits |
| 4. | Current | | d. | Save As |
| 5. | DText | | e. | Layer |
| 6. | Voltage | | f. | Window |
| 7. | Plot | | g. | Resistor |
| 8. | Donuts | | h. | Spacing |
| 9. | Prototype | | i. | Pline |
| 10. | Axial | | j. | Preview |

## E. Problems

7.1     Measure the body size, lead size, and lead spacing for several different components and devices. Convert your measurements to make 2:1 scale drafting aids using AutoCAD. Print a drafting aid for each part drawn.

7.2     Draw 1:1 scale component and conductor pattern layouts using AutoCAD of the hand sketch made for the tachometer in Problem 6.3. Make individual layers and print finished drawings.

7.3     Using AutoCAD, draw and print a 1:1 scale marking mask, including corner brackets, for the tachometer component layout generated in Problem 7.2.

7.4     Draw a 1:1 scale donut layer for the tachometer using the conductor layer generated in Problem 7.2 and print the solder mask with corner brackets for the design.

7.5     Referring to the 14-pin dual-in-line test jig configuration shown in Fig. 7.22, draw a 1:1 scale conductor pattern layout of the device terminal pad layout and paths 1 through 14 with the donuts shown.

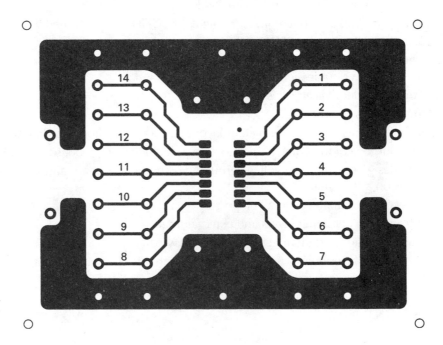

**Figure 7.22** Fourteen-pin dual-in-line test fixture.

# 8 Double-Sided Plated-Through-Hole and Multilayer Printed Circuit Artwork and Design Concepts

## LEARNING OBJECTIVES

*Upon completion of this chapter on artwork design and layout of double-sided plated-through-hole and multilayer printed circuit boards, the student should be able to*

- Understand the basic differences between single-sided boards and double-sided plated-through-hole pc boards.
- Make interconnections between both sides of a double-sided pc board by using plated-through holes.
- Better understand how to use component bodies to bridge paths for greater flexibility.
- Create a circuit schematic using the EAGLE schematic module.
- Perform a computer-aided electrical rules check (ERC).
- Translate a schematic into a pc board layout using the EAGLE board layout editor.
- Create conductor paths using the EAGLE autorouter module.
- Understand the color code system adopted by industry to distinguish conductor patterns and other detail on both sides of a double-sided pc board layout.
- Perform a computer-aided design rules check (DRC).
- Output finished layouts or schematic.
- Become familiar with basic conductor pattern layout considerations for plated pc boards.
- Understand the purpose of including a thief area on a double-sided pc board layout for plated boards.
- Understand the design concept of a voltage/ground type multilayer board and show layer-to-layer interconnections on a drawing.
- Visualize all layers simultaneously in a composite view.
- Draw connections between inner layers and external fingers.
- Properly provide antipads and thermal pads on inner layers.
- Design four-layer multilayer boards using the guidelines in this chapter and Chapters 6 and 7.

## 8.0 INTRODUCTION

As electronic systems have become more complex, the requirements of higher-density packaging have partially been met with the widespread use of double-sided printed circuit boards and the rapidly growing importance of multilayer and surface-mount pc boards (see Chapters 14 and 18). Double-sided board fabrication requires the preparation of at least two layouts, one for each side of the pc board. This, of course, adds to the time and costs of producing a double-sided board. Precise alignment of all terminal pads between both sides of the board is required for each lead access hole. This alignment problem is not a consideration in the preparation of single-sided boards since there is only one conductor pattern on one side of the board. In addition, double-sided boards require many more steps in fabrication, which increases their cost as compared to single-sided boards. Even though they are more costly and time-consuming to fabricate, double-sided pc boards allow more components to be packaged in less space than single-sided boards and thus meet the demands for high-density electronic packaging.

Appendix XVIII presents a packaging feasibility study for the most appropriate type of pc board (i.e., single-sided, double-sided, or multilayer) to design for analog circuits of varying complexity. Also see Appendix XIX, which presents a packaging feasibility study for the most appropriate pc board to design for digital circuits.

*CAUTION:* When using a computer station for long periods of time, ergonomics should be applied. The following is a list of things to do that may help you keep more comfortable at your computer.

1. To avoid eye discomfort, viewing distance should be made appropriate for screen size. This is generally 12 to 24 inches for 15-inch screens but should be more for larger screens. The monitor should also be adjusted so that the top of the screen is at eye level.

2. Adjust screen brightness and contrast so that characters and other displayed features are brighter than the background.

3. Use a document holder that is adjusted to be at the same height as your computer screen.

4. Place the keyboard directly in front of you and position the mouse pad within comfortable reach beside the keyboard where your wrist and forearm may be supported.

5. Use an adjustable chair that is easily moved, that provides lower back support, and that adjusts for height so that hips and knees are level, with lower arms roughly horizontal with the keyboard.

6. Position the chair so that the wrists are straight and not flexed.

7. Be sure that the work station has adequate leg room and a foot rest if needed for hip/knee alignment.

8. Organize your bench and workspace so that operation can be performed within easy reach without having to extend beyond the point of comfort, and reaching behind your shoulder line or above shoulder height is unnecessary.

The section on safety at the beginning of the book should be reviewed for more specific information.

## 8.1 BASIC DOUBLE-SIDED PRINTED CIRCUIT BOARDS

A double-sided pc board differs from a single-sided board in that both sides are processed with conductor patterns and terminal pads. The arrangement of conductor paths and pairs of terminal pads to form a double-sided pc board is shown in Fig. 8.1. As with single-sided boards, all components are mounted only on one side. This side is called the *component side,* or *top*. Component leads pass through previously drilled and plated lead-access holes that have been processed through each pair of terminal pads on either side of the board. The component lead is *clinched,* or bent over, onto the terminal pad that is on the *circuit side,* or *bottom,* of the board. Note in Fig. 8.1 that the lead access hole drilled for the component lead also passes through a terminal pad on the component side of the board. Unlike single-sided designs, conductor paths may extend away from terminal pads on both sides of the board. Electrical interconnections between conductors on both sides of the board are made with *plated-through holes* (PTH), which become an integral part of the overall conductor pattern in the fabrication process.

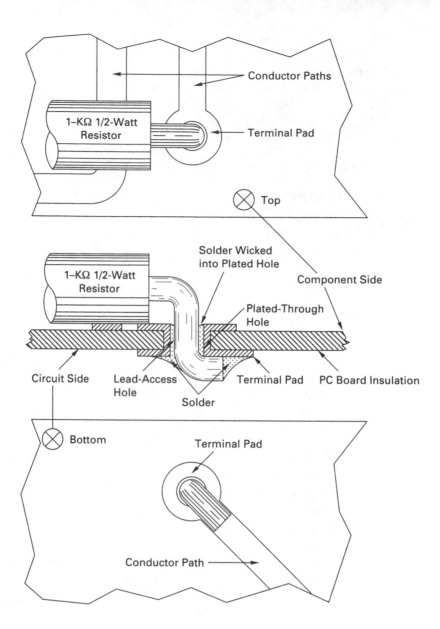

**Figure 8.1** Double-sided PTH pc board conductor path and terminal pad arrangement.

When it is required only to make an interconnection between both sides of a double-sided pc board, a *via hole* is used. Also referred to as feed-throughs, via holes are not intended to accept component leads. For this reason, the terminal pads and drill hole sizes are small. Typically, a via hole pad size has a finished diameter of 0.050 inch with a drill hole size of 0.030 inch. The top and side view of a via hole would appear identical to that shown in Fig. 8.1 without the component lead, and in Fig. 14.3 for a multilayer board.

The electrical interconnections between both sides of a pc board are formed by first drilling all lead-access holes in the bulk stock. Through a series of processes in chemical baths, all these holes are then plated with copper to form a continuous electrical connection between the copper surfaces of both sides of the board. Each copper surface is then imaged with the intended conductor pattern design. Further chemical processes result in additional copper and then solder to be plated on all the drilled hole walls. These processes, as well as those required to complete a double-sided plated-through-hole pc board, are discussed in detail in Chapters 11, 12, and 13.

The circuit side of the board can be considered a conventional single-sided layout. The conductors and terminal pads on the component side serve to extend the amount of conductor path routing. With both sides of the board used to route electrical interconnections, less space is required for conductor paths on the bottom side, thus allowing for greater component density.

One consideration involved in the fabrication of a double-sided pc board is the component bodies that must rest on the conductor paths, which are on the top of the board. Most component and device cases, such as resistors, capacitors, plastic-cased transistors, and integrated circuits, are electrically insulated from their leads. The bodies of these components may safely contact a conductor path without electrical interference. However, some metal-cased integrated circuits and transistors have one lead attached to the case. These packages must not be allowed to come into contact with any of the conductor paths on the top side of the pc board. Electrical contact is avoided by mounting these components slightly above the board surface to create an insulating air gap.

## 8.2 DOUBLE-SIDED PC BOARD LAYOUT

The generation of a 1:1 scale component and conductor pattern layout for a double-sided board is accomplished by an entirely different manner than those procedures described in Chapters 6 and 7. The major differences are the elimination of using layout dolls, the elimination of making layout sketches by hand, and the substitution of a software drawing package by CadSoft Computer, Inc., called EAGLE 2.6, in place of AutoCAD.

EAGLE 2.6 is capable of producing 1:1 scale layouts for extremely complex single-sided, double-sided, and multilayer pc boards. However, a basic relaxation oscillator circuit, shown in Fig. 8.2, will be used as the vehicle to allow a greater focus on the manipulation of this software drawing package while gaining an in-depth understanding of layout concepts.

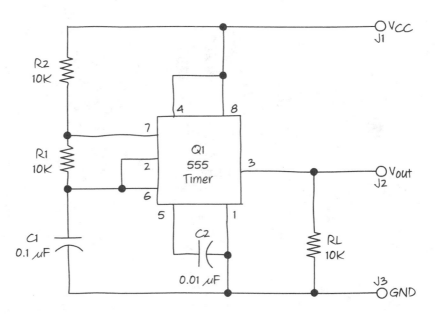

**Figure 8.2** Hand sketch of a basic relaxation oscillator circuit.

## 8.3 EAGLE 2.6 SOFTWARE FOR PC BOARD LAYOUT

EAGLE 2.6 is a powerful software drawing package optimized for printed circuit board layout and design. The complete package supplied by CadSoft Computer, Inc., contains a *schematic module,* board *layout editor,* and an *autorouter module.* Also available as part of the software is a large selection of components and devices, stored in a series of libraries as either *package macros* for pc board layouts or *schematic symbols* for schematic drawings. EAGLE 2.6 also supports up to 255 drawing layers. Layers below 100 have been predefined by the software manufacturer. For example, layer 1 represents the conductor paths called *tracts* on the *top* (component) side, and layer 16 the *tracts* on the *bottom* (circuit) side of a pc layout. Some layers are assigned functions such as, *silk screen top side* (layer 21), *silk screen bottom side* (layer 22), *pads* (layer 17), and *vias* (layer 18). Layers above 100 can be defined by the user with the LAYER command.

**8.3.1 The EAGLE 2.6 Main Menu**  Figure 8.3 illustrates EAGLE's *main menu,* containing a group of some of the available commands. Any action in EAGLE is started by calling one of these commands. Commands can be entered via the keyboard or selected with a mouse by first highlighting the command and then pressing the select (left) button. In certain EAGLE commands, pop-up menus are activated from which parameters for the initial command can be selected.

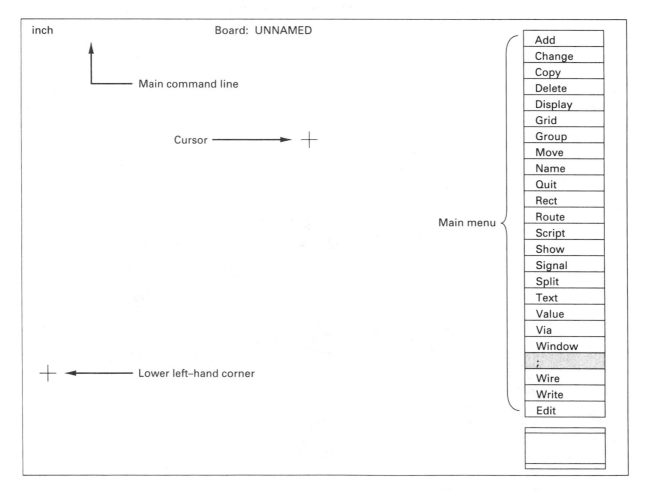

**Figure 8.3**  EAGLE start-up display with main menu.

Twenty-four of the most common commands are displayed on the right side of the screen when the main menu is in view, and the semicolon (;), highlighted in red, is the main terminator within the EAGLE software program. A main command line, indicating both dimensional units (*inch*) and drawing environment (*Board: UNNAMED*), appears at the top of the screen, and a cross (+) identifies the lower-left corner of the drawing.

**8.3.2 Creating a Circuit Schematic Using the Eagle Schematic Module** The relaxation oscillator shown in Figure 8.2 will be used to illustrate how a circuit schematic is generated in EAGLE and then how pc board component and conductor pattern layouts are created from the schematic. Create a new schematic by first highlighting the EDIT command within the main menu and selecting it with the mouse. From the pop-up menu, select first SCHEMATIC and then NEW. At the prompt, enter 555 Timer as the name for this new file, followed by a carriage return <CR>:

**enter name:** <u>555 Timer <CR></u>

The command line of the main menu will be updated to

**Sheet: 555 TIME.1/1**

indicating the first page of a schematic with 555 TIME as the file name.

To place schematic symbols for a schematic on the drawing, you must first choose the appropriate library in which the symbol resides. There are some 37 individual libraries in EAGLE 2.6. Each library is grouped according to function (ECL, MEMORY, 68000, etc.) or manufacturer (Maxim, Zilog, etc.). For the relaxation oscillator of Fig. 8.2 we find the 555 timer symbol in the DEMO library, and the three resistors and two capacitor symbols in the DISCRETE library listed as RESUS-7,5 and CAP-7,5 respectively. The +VCC and ground (GNDA) symbols are found in the SUPPLY library.

To select and place a resistor symbol, such as R1, first type USE to display the pop-up menu of libraries. Select DISCRETE to load this library and then highlight the ADD command from the main menu and choose RESUS-7,5 with the mouse. As you move the mouse onto the screen you will drag the symbol for R1 along. If desired, rotate the symbol 90 degrees counterclockwise by pressing the right mouse button once, 180 degrees by pressing twice, etc. When the resistor is positioned correctly, press the select (left) mouse button. Repeat this procedure of library selection (USE), symbol selection (ADD), and symbol positioning (MOVE, ROTATE) until all elements are located on the drawing.

To label the value of each element, highlight and select the VALUE command. Then use the left mouse button to select the symbol, and at the pop-up prompt type the component value. For example, the value of R1 can be set to 10K by

**R1: RESUS-7,5** <u>10K <CR></u>

This procedure can be repeated to rename elements using the NAME command.

Electrically connecting the symbolic elements with the wiring (nets or airwire) requires the use of the NET and JUNCTION commands. Using the mouse, first highlight and then select the NET command from the main menu. Click on the lead end of an element and drag the mouse until it is positioned where desired and then double-click to terminate. Repeat until all elements are wired electrically as desired. To form the nets and component leads into connections, highlight the JUNCTION command. At the intersection of each net and component lead (pins), click the mouse select button to establish an electrical connection. Figure 8.4 illustrates the final schematic layout of the relaxation oscillator generated with the EAGLE 2.6 schematic module.

The final step in developing a circuit schematic is to perform an electrical rules check (ERC) to look for open pins and connections, and so on. By typing ERC, EAGLE will check

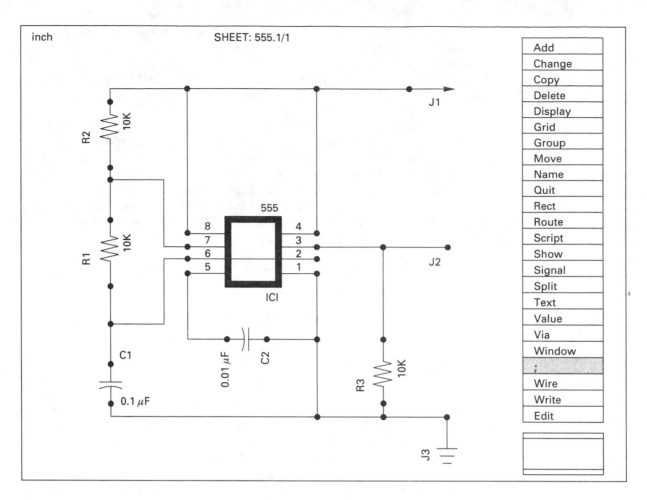

**Figure 8.4** Finalized schematic of relaxation oscillator.

and display errors and warnings. The results of an ERC test are displayed below the main command line. A typical report might read as follows:

```
inch                      Sheet: 555 TIME.1/1
ERC: finished. 0 errors, 1 warning (C:\EAGLE\555 TIME.ERC)
```

### 8.3.3 Translating a Schematic into a PC Board Layout

The circuit schematic developed in the last section can be converted into a pc board layout using EAGLE's BOARD command and the layout rules discussed in Section 8.2. Begin the process by first typing the BOARD command. All the components, along with their airwire connections (RATS-NEST), will appear beside a blank pc board outline (see Fig. 8.5a).

Then highlight the MOVE command, and with the mouse select each component individually and move it into the pc board outline. When each component is approximately positioned, click on the select button. Repeat this process until all parts are within the pc board outline.

To obtain a more detailed image of all the parts and their relative location, select WINDOW from the main menu. Use the mouse to locate (left button) the lower-left and upper-right corners of the whole image containing all the parts. Then selecting or typing the semicolon (;) will window the image to best highlight all the parts (see Fig. 8.5a). Use the

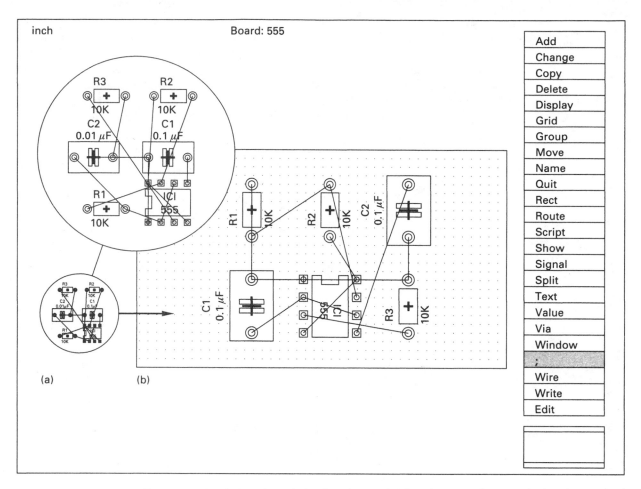

**Figure 8.5** Translating schematic into board: (a) EAGLE board command start-up display; (b) optimized parts placement with RATS-NEST.

MOVE and ROTATE commands and employ the central, schematic, and peripheral viewpoints discussed in Chapter 6 to optimize the parts placement within the pc border. Once parts placement is optimized, the net positions are optimized by using the RATS-NEST command (see Fig. 8.5b).

### 8.3.4 Creating the Conductor Paths with an Autorouter
The autorouter module supplied with the EAGLE 2.6 software employs a ripup and retry algorithm that tends to create PCB layouts that are 100% routed without additional commands or any other manual intervention. This router can be configured to minimize the number of vias used on a double-sided board and also to smooth the traces.

The autorouter is accessed by typing AUTO; then from the AUTOROUTER SETUP menu, select Start. Figure 8.6 shows the complete autorouted layout for the relaxation oscillator. Although individual layers can be turned on or off, the screen display of the autorouted pc board appears in Fig. 8.6a as a composite view with all of the layers turned on. A multi-color coding system has been adopted by industry to avoid confusion and misinterpretation of track patterns on different layers. The interpretation when observing the composite view is as follows: The layout is a top view with the first layer being that of the components. The tracks under the parts on the component side are shown with their pad connections and are presented in the color red, forming the next layer of the composite view. Still viewing the board from the top, the track pattern and pad connections on the bottom of the pc board would be "seen" as though one were looking through the board. Red and blue tracks may "cross" each other since each color represents conductor paths on different layers separated by the insulating board material. Because of the absence of colored illustrations in this book, a legend is given in Fig. 8.6b to aid in the interpretation of the autorouted layout shown in Fig. 8.6a. Blue tracks are shown as solid black paths and red tracks are shown as parallel lines.

Additional features such as corner brackets and mounting holes would be added. Nomenclature on the top side of the board is shown with forward-reading orientation. Some pc boards may require labeling on the bottom side of the board. If any nomenclature is added to the bottom side of the board, it must be added in backward-reading orientation on that layer to obtain forward orientation when viewing the actual bottom side of the pc board.

The final layout should be checked for spacing errors, short circuits, track widths, pad diameters, and so on, using the design rule check (DRC) command. The results of

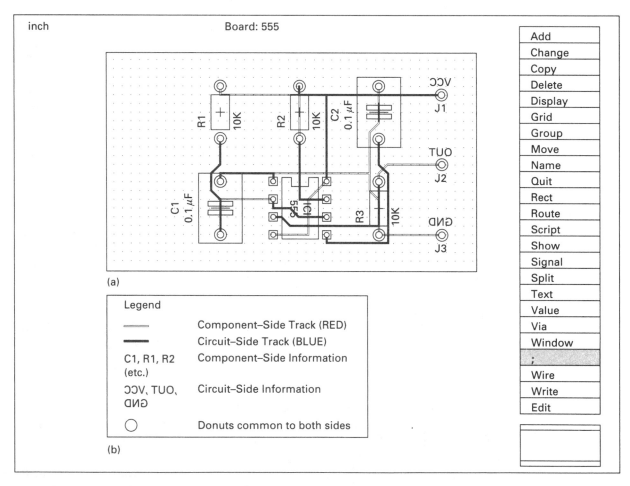

**Figure 8.6**  Composite view of double-sided pc board layout of relaxation oscillator: (a) top (red layer) and bottom (blue layer) track patterns; (b) legend for interpreting layer information.

a DRC test are displayed below the main command line. A typical report might read as follows:

```
inch                                    Sheet: 555 TIME
DRC: Top: finished. no errors!
DRC: Bottom: finished. no errors!
```

**8.3.5 Creating Outputs with XPLOTS**   A program called XPLOT is used to output finished layouts or schematics. XPLOT allows the plotting of any drawing layer individually or in combination with other layers. XPLOT is invoked only from the DOS prompt. Therefore use the WRITE command to save the layout and QUIT to the DOS command prompt. Then type

```
XPLOT options filename layer
```

It should be noted that the information presented here on EAGLE's 2.6 software is the barest minimum information necessary to illustrate its use on a simple pc board layout. This software is quite powerful and can best be learned by reference to the CadSoft Company, Inc., user manuals. Information pertaining to requirements regarding conductor layers and patterns for multilayer pc boards and boards employing surface mount devices (SMD) is presented in Chapters 14 and 18, respectively.

## 8.4  LAYOUT CONSIDERATIONS FOR PLATING

Chapter 13 describes the process of pattern-plating both copper and 60/40 tin–lead alloy onto the conductor path areas of both sides of the pc board. Many of the problems associated with an improper balance of plated metals can be attributed to the component and conductor pattern layouts. If a double-sided pc board is to be electroplated to a uniform thickness and proper alloy, it must meet the following specifications:

1. The total conductor pattern areas to be plated on each side of the board must be as close to equal as possible.
2. The conductor pattern on a given side of the board must be uniformly and evenly distributed throughout.

Strict adherence to the specifications above is essential if the correct 60/40 tin–lead alloy is to be deposited during the plating process. An imbalance of conductor pattern areas between the two sides results in the side with the less area to be tin-rich–plated and the side with more area to be lead-rich–plated. If the ratio of alloys plated on either side is significantly different from the optimum 60/40 ratio, proper soldering of the board will not be realized.

If the conductor pattern is not uniformly and evenly distributed across either side of the board, the area with the heavy concentration of copper conductors will be low in tin-content plating, while the area with the light concentration of copper conductors will be plated with an alloy having high tin content. This will again result in soldering difficulties.

One final provision that must be included in the conductor pattern layout phase for boards that will be electroplated is an area of solid copper just outside the corner brackets of the pc board. This area is called the *thief,* or *robber,* area and its purpose is to direct some of the high current density at the edge of the board toward itself during the plating process. This will result in a more consistent 60/40 alloy on both sides of the pc board. The thief area is included in the conductor pattern layouts and appears as a wide border just outside the corner brackets. At 1:1 scale, the thief area on the CAD artworks should be from $\frac{1}{2}$ to 1 inch wide. Figure 8.7 shows the conductor patterns for both sides of the pc board with the inclusion of the thief area.

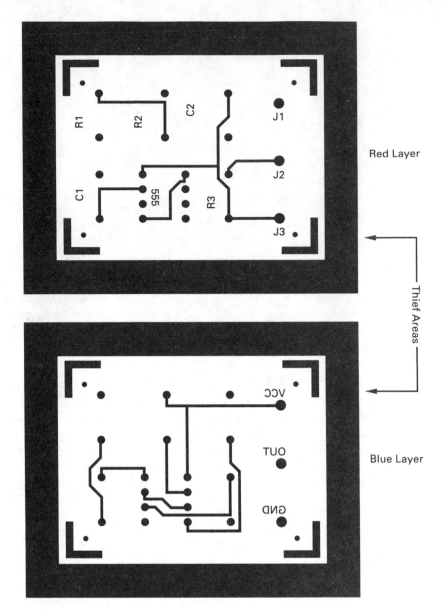

**Red Layer**

**Thief Areas**

**Blue Layer**

**Figure 8.7** Printed track patterns with the addition of thief areas.

The following sections, 8.5 through 8.10, present the concept and design of multilayer boards and will complete layout and design information given in this text for single-sided, double-sided, and multilayer printed circuit boards.

## 8.5 DESIGNING A MULTILAYER BOARD

In Chapters 5, 6, 7, and the beginning of this chapter, we discussed the proper selection of printed circuit materials, laminate characteristics and selection, copper foil properties, single-sided and double-sided PCB designs, and the problems and restrictions involved in their layout. Prior to beginning the layout of a board, some of the criteria that needed to be determined included the selection of conductor widths and spacings, center-to-center lead spans, and pad diameters. In addition, design feasibility studies given in Appendix XVIII and Appendix XIX show how to determine package density. This aided in the decision of what level of design (single-sided, double-sided, or multilayer) was required. In the remaining sections of this chapter, we treat dense packages that require a multilayer design. Although it is true that multilayer designs will generally allow a denser packaged circuit, we will see that this is not the only consideration in the selection process. In general, however, density levels greater than 25 to 30 holes/in$^2$ for analog circuits and 1.5 to 1.8 EDIPs/in$^2$ for digital circuits should be considered to be laid out on a multilayer board (MLB).

Another advantage of using a multilayer design over a double-sided board is easier routing of conductor paths on the outer layers. Taking a four-layer board as an example, one inner layer is used to distribute power to the circuit and the other inner layer serves as ground or common reference connections. In a double-sided layout, space would have to be allowed for power and ground interconnections. With these designed onto the inner layers of a multilayer board, the outer layers need include just signal paths, which allow easier routing of denser packages.

Digital circuits designed onto a multilayer board with a large copper ground plane results in a reduction of some forms of noise that is commonly generated in double-sided boards. Thus fewer decoupling capacitors are required to effectively protect the ICs.

Multilayer boards also have the advantage of the option of using one of the inner layers to dissipate large amounts of heat that may be generated from on-board components. Heat dissipation is becoming an increasingly difficult problem as higher-density designs are laid out on PCBs. As the EDIPs/in$^2$ value increases, so does the generated amount of heat per square inch increase. If the increase is high enough, elaborate and expensive methods of heat removal need to be incorporated into the design to optimize circuit performance.

One must also consider the disadvantages of a multilayer board compared to a double-sided design. Most important, multilayer boards are more costly to manufacture than a typical double-sided board having plated-through holes. This cost factor is two to three times higher. In addition, a multilayer board requires a great deal more time to lay out and to manufacture. As a general rule, unless specific advantages unique to a multilayer design can justify its use, consideration should be given to a dense double-sided board that will not affect cost and production time as severely.

When a multilayer design is required, specific rules and methods of illustration are necessary for the layout designer to convey the correct information to the manufacturer. In the following sections we present MLB design and layout techniques. A four-layer power/ground board will be used to demonstrate the layout of a multilayer board. The information presented will allow more complex designs to be laid out. It will show how electrical connections are made to the inner layers, as well as how to insulate between component leads and inner layers where isolation is required as they pass through the board. The use of *thermal relief pads* for electrical connections and *antipads* for isolation is discussed. Finally, a simple IC timing circuit will be used to demonstrate a complete four-layer power/ground design. This circuit was selected only to show methods of layout and in no way suggests that a multilayer board is required for this design. The four-layer power/ground configuration is perhaps the least difficult to design and the least expensive to manufacture when compared to more complex multilayer boards. These advantages make this the most popular of all the multilayer boards.

## 8.6 LAYER-TO-LAYER INTERCONNECTIONS IN AN MLB

The *multilayer board* (MLB) is made up of a combination of equivalent single- and/or double-sided PCBs bonded together to form one integral board having two outside layers of conductive foil in addition to one or more internal layers. All of these conductive surfaces are used during the packaging design to form the required electrical interconnections. Multilayer boards are classified by (1) the number of conductive layers, and (2) the type of internal layer circuitry.

Multilayer boards are most commonly described by a number (3, 4, 5, etc.) that corresponds to the number of conductive layers. An exploded view of a typical four-layer MLB is shown in Fig. 8.8. Note that there are two outer conductive layers, similar to standard double-sided PCBs, and two inner layers, thus the number 4. If the inner layers of foil are processed with conductor patterns (i.e., conductive paths terminating in pads), the board is designated a *signal* MLB. Where the inner layers are used basically as unbroken sheets of copper foil, the board is termed a *voltage/ground plane* MLB. To familiarize the technician with the most common multilayer design and fabrication, we will discuss the voltage/ground plane MLB.

A four-layer voltage/ground MLB is shown in Fig. 8.9. The top side, labeled layer 1 (or component side), appears identical to a single-sided board processed with terminal pads and conductor paths. The bottom side, layer 4 (or circuit side), also appears identical to a single-sided PCB with terminal pads and conductor paths processed into the copper foil. The two inner conductive layers (labeled layers 2 and 3) are made from a laminate with copper foil on both sides, similar to a double-sided board. The inner layer patterns are processed before the completed MLB is formed. After the MLB has been

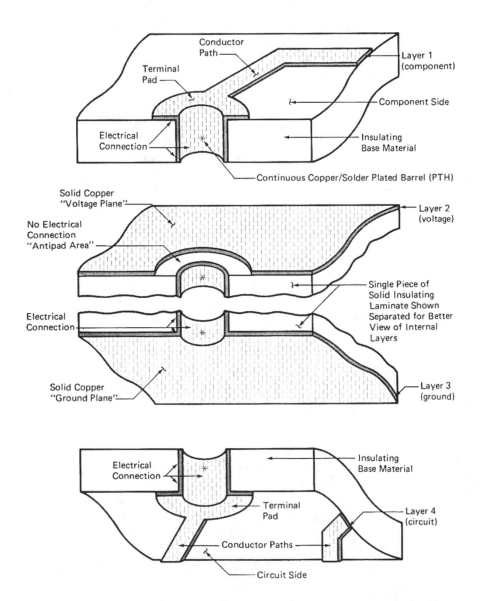

**Figure 8.8** Exploded view of a voltage/ground type MLB.

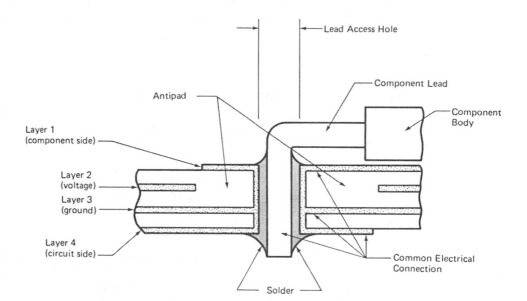

**Figure 8.9** Cross-sectional view of the voltage/ground type multilayer board.

completed, the plated-through holes are processed through all layers in a similar manner as that used in double-sided boards. In Fig. 8.8, the laminations and the barrel of the plated-through hole are shown split for illustration of the inner copper layers 2 and 3. Layer 3 (*ground*) is a solid sheet of copper extending from the plated hole. A component lead soldered into this hole will make electrical contact with layer 3 in addition to both outer pads on layers 1 and 4. Note, however, that layer 2 (voltage) has been processed with an insulation gap or clearance around the lead access hole so that its copper foil does not extend to the plated barrel. Thus a component lead inserted through this hole and soldered to its terminal pads on layers 1 and 4 will *not* make electrical contact with layer 2. This clearance provided around lead access holes in copper planes where no electrical contact between component leads and inner conductive layers is to be made is termed an *antipad* area.

All multilayer boards will have inner conductive layers with some antipads for isolation from plated-through holes as well as connections to others. Refer again to Fig. 8.9. Note that the barrel of the plated-through hole connects the conductive copper on layers 1, 3, and 4, while the antipad isolates the foil of layer 2 from the PTH. When the component lead is soldered, the solder will fill the barrel and electrically connect the lead to layers 1, 3, and 4 but not to layer 2 because of the antipad area.

As an application of the use of a voltage/ground MLB, refer to the digital circuit shown in Fig. 8.10. This circuit is constructed with four 14-pin DIP integrated circuits, each package of which designates pin 14 as the power connection and pin 7 as the ground connection. If this circuit was to be packaged using a voltage/ground MLB, the board would be fabricated as shown in Fig. 8.11. The voltage plane (layer 2) would have antipads about all lead access holes except for those pins numbered as 14. No antipads would appear at all holes for the number 14 pins so that the power plane would extend to the plated-through hole for electrical connection to these leads. Similarly, antipads would be required about all lead access holes of the ground plane (layer 3) except for those connected to the number 7 pins of each DIP package, which are to be grounded. The resulting MLB provides power and ground connections on layers 2 and 3, respectively. All signal conductor paths can then be designed onto the two outer surfaces of the board.

The multilayer board is the most complex and costly to design and fabricate, but it provides distinct advantages over the double-sided PCB. When voltage and ground connections

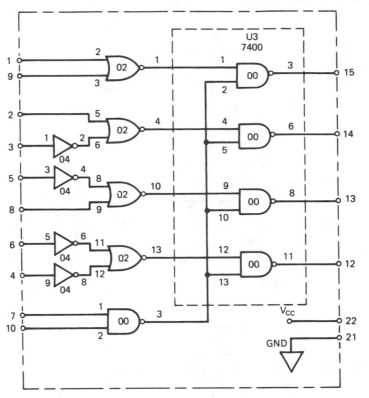

**Figure 8.10** A typical logic circuit consisting of 7400 series TTL gates.

VCC for all ICs at PIN 14
GND for all ICs at PIN 7

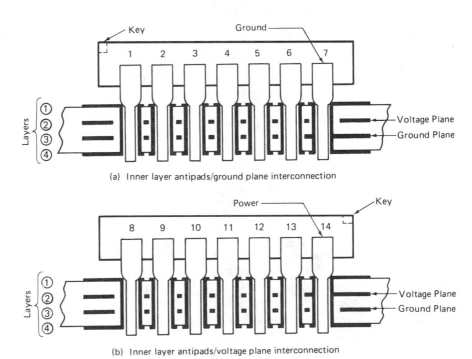

(a) Inner layer antipads/ground plane interconnection

(b) Inner layer antipads/voltage plane interconnection

**Figure 8.11** Cross sections of a voltage/ ground style multilayer board for a DIP integrated circuit.

are made to inner layers, more flexibility and space are available to the designer in routing signal paths on the outer layers. Further, inner layers may also be used for signal paths in the design of high-density systems. Other advantages of the MLB are that they can provide electrical shielding with the use of an internal ground layer as well as internal heat sinking where required. Because the multilayer board plays a prominent role in today's high-density electronic packaging, it is essential that the PCB designer and the technician gain familiarity with this type of design.

Figure 8.8 showed an exploded view of a voltage/ground type MLB. Cross section and cutaway views are now shown in Figure 8.12 to better illustrate a four-layer board cutting across a plated-through hole. Once again, the top layer is the *component side* and is labeled *layer 1*. On this layer, a typical component lead hole and pad connected to a conductor path are shown. The bottom layer is the *circuit side,* labeled *layer 4*. A pad is also shown on this layer. Note that the pads shown in layers 1 and 4 are oriented in a similar manner to that of a double-sided board; that is, the two outer layers are electrically connected by a copper-plated barrel surrounding the hole wall that is formed during the manufacturing process. In the view shown in Fig. 8.12a, it is not evident whether there are any

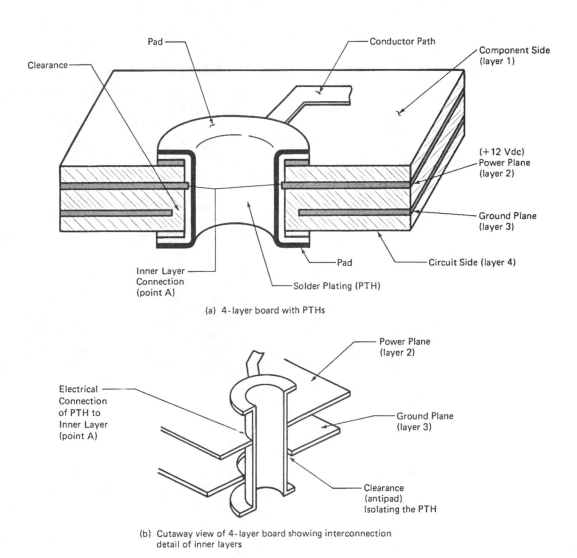

(a) 4-layer board with PTHs

(b) Cutaway view of 4-layer board showing interconnection detail of inner layers

**Figure 8.12**   Cross-sectional views of layer-to-layer interconnections in a multilayer PCB.

conductor paths connected to the pad of layer 4. This fact is not pertinent to our discussion at this point.

Note that references made to Fig. 8.12a so far do not differ greatly from those of a double-sided board with a plated-through hole. We will now consider the inner layers of the cross section shown in Fig. 8.12a. The *+12-Vdc power plane,* labeled *layer 2,* is shown electrically connected to the outer layer pads by means of the plated-through hole. This connection is shown as point A, which is the edge of the copper plane of layer 2. It was exposed during the drilling operation prior to the plating of the hole. Note that with +12-Vdc power connected to layer 2, it will also be electrically connected to the pads on layers 1 and 4 and to any conductor paths that may be connected to these pads.

The *ground plane* shown in Fig. 8.12a is labeled *layer 3.* Note that no electrical connection exists between this layer and the barrel of the plated-through hole. The clearance shown to result in this electrical isolation is made with an *antipad,* which is discussed in the next sections, 8.7 and 8.8.

The cross-sectional view shown in Fig. 8.12b is the same as that shown in Fig. 8.12a, but with *all of the insulating board material removed,* exposing only the individual layers of copper. The view more clearly illustrates the association of the pads, antipads, plated-through hole, and the power and ground planes of this four-layer configuration.

## 8.7 LAYER REPRESENTATION ON THE COMPONENT LAYOUT DRAWING

From our discussion in the preceding section on layer-to-layer interconnections on a processed multilayer board, it is seen that there are four separate and distinct layers of circuitry. Conventionally, the outer layers are labeled layer 1 (component side) and layer 4 (circuit side). The inner layers are labeled layer 2 for the power plane and layer 3 for the ground plane. Thus the "hot" terminal of the power supply (either + or −) is electrically connected to layer 2. This will result in every lead access hole that contacts layer 2 (see Figs. 8.12a and b) being electrically connected to dc power. Similarly, layer 3 must be electrically connected to the GND terminal of the power supply. Accordingly, every lead access hole contacting layer 3 is electrically connected to ground. (This is not shown but would simply be the reversal of layer 2 and 3 connections to the plated-through hole as shown in Fig. 8.12b.)

While laying out a four-layer board, the designer must be able to visualize four layers of conductive planes simultaneously as a composite drawing. This is not as difficult as it may appear initially. Each layer will be constructed separately but will ultimately require layer-to-layer registration. There is one of three options of making electrical connection combinations for every plated-lead access hole. These options are shown in Fig. 8.13a and are designated as *case A, case B,* and *case C.* For each layout possibility, each of the four layers will be shown independently. The symbolic representation for the layout of copper pads for each of the following three cases is shown in Fig. 8.13a.

### Case A

A component lead passing through an access hole is to be connected to +dc (power). In the layout, a pad would be drawn on both outer layers 1 and 4. This is a requirement of all layouts. (Conductor paths may or may not be required on both the component side and the circuit side. Recall that paths on the component and circuit sides are shown in this book as parallel and solid lines, respectively. In an actual drawing, they would be drawn as solid color-coded lines to indicate the appropriate side of the board.) To electrically connect to

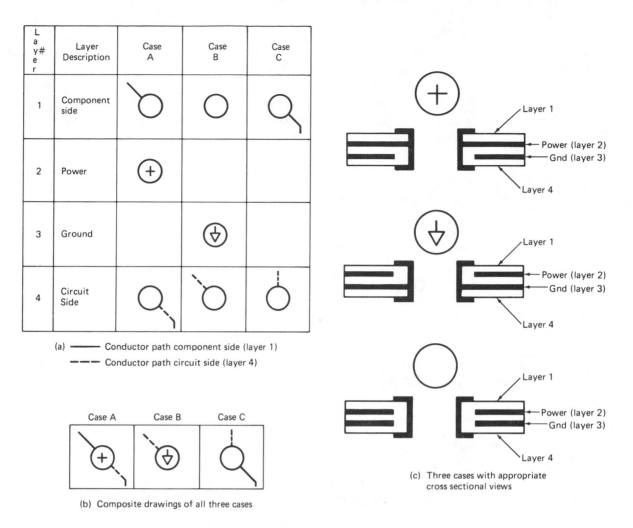

| Layer # | Layer Description | Case A | Case B | Case C |
|---------|------------------|--------|--------|--------|
| 1 | Component side | | | |
| 2 | Power | | | |
| 3 | Ground | | | |
| 4 | Circuit Side | | | |

(a) ——— Conductor path component side (layer 1)

——— Conductor path circuit side (layer 4)

| Case A | Case B | Case C |
|--------|--------|--------|

(b) Composite drawings of all three cases

(c) Three cases with appropriate cross sectional views

**Figure 8.13** Coding pads for the three possible layer-to-layer interconnections.

power, layer 2 is provided with a round pad symbol that represents a thermal relief pad and is drawn with a + symbol inside the circle. This indicates that this pad makes electrical connection to layer 2, which is the power plane of the layout. Because no electrical power connection is to be made to the ground plane, no pad is shown on layer 3. The absence of a pad indicates that clearance (antipad) is required between the ground plane and the plated-through hole.

### Case B

A component lead passing through the access hole is to be connected to ground. See Fig. 8.13a. This would be shown as a pad drawn on both outer layers (1 and 4) and appropriate conductor paths added where required. The electrical connection to ground is shown with a round thermal relief pad drawn on layer 3, and the *ground*, or *common reference*, symbol added within the circle. Since no electrical connection is to be made to layer 2, no pad or symbol is shown on this layer. Again, the absence of a pad indicates the need for an antipad to isolate the plated-through hole from layer 2.

**Case C**

A component lead passing through an access hole is not to be electrically connected to either the power or ground planes but is to connect directly from layer 1 to layer 4. The symbol for this connection is shown in Fig. 8.13a. This type of connection is typical in a double-sided board design. Pads are shown on both outer layers (1 and 4) but no pads or symbols are drawn on either inner layer. This indicates the requirement of antipads to isolate the plated-through hole from both the power and the ground planes.

Note in all of the three cases presented, no pad or symbol is required when no connection is to be made to an inner copper plane. With the absence of a pad, it is understood that an antipad is required for isolation of the plated-through hole.

The symbols listed in Fig. 8.13a are shown as if they were drawn on four separate sheets of vellum (i.e., one sheet for each layer). They are shown in composite form in Fig. 8.13b, which is how they would appear on a single sheet of vellum to represent all four layers.

Using the convention of symbols shown in Fig. 8.13a, the composite drawings of Fig. 8.13b are interpreted as follows:

*Case A:* Path connections—layers 1 and 4. Thermal relief pad—layer 2. Antipad— layer 3.

*Case B:* Path connection—layer 4. Thermal relief pad—layer 3. Antipad—layer 2.

*Case C:* Path connections—layers 1 and 4. Antipads—layers 2 and 3.

For each hole to be drilled into the multilayer board, the designer simply relates it to one of the three cases above to determine how it will be drawn on the conductor pattern layout. Each hole symbol will thus show which plated-through hole will connect to the appropriate inner layer to satisfy the power and ground connections as dictated by the circuit schematic. Remember that for representing plated-through holes that are to be connected only to layers 1 and 4, the round pad drawn will have no symbol associated with it, which indicates no inner layer connections. The cross-sectional views representing the three cases just described are shown in Fig. 8.13c.

The symbols used to represent layer representation on the component layout drawing may be more conveniently read if provided with a coded system. For example, it may be much more convenient to represent the three cases of connections shown in Fig. 8.13b with different geometric shapes (e.g., round, square, and rectangular). Another form of coding is using a different color for each of the three cases. Regardless of the system of coding used, a legend should accompany the component layout drawing so that there will be no confusion in interpretation.

In all cases in the layout of multilayer boards, the power plane and the ground plane must be electrically isolated. One of the first tests that a board is subjected to after it has been manufactured is to ensure that this isolation has not been violated. This test is referred to as *continuity test,* or *ring out.* An ohmmeter is used to ascertain if there is infinite resistance (open circuit) between the power plane and the ground plane, which indicates that no electrical connection exists. This test is performed by first attaching one lead of the ohmmeter to a plated-through hole (or finger edge connector) that connects the power inner layer to a point on one of the outer layers. The other meter lead is attached to another plated-through hole, or finger, which connects the ground layer to a point on one of the outer layers. Any finite value of resistance indicated on the ohmmeter renders the board defective.

Figure 8.14a shows how electrical connections between inner layers and external fingers are drawn on the component layout. This figure shows the internal power plane (layer 2) electrically connected to the fingers labeled 1 and 2 on the component side of the board. Two via holes are shown inside the power plane edge limit and drawn with the appropriate

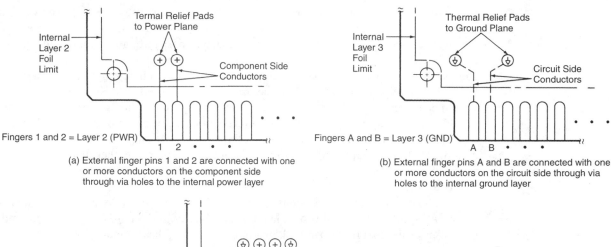

**Figure 8.14** Finger connections to internal power and ground layers.

(a) External finger pins 1 and 2 are connected with one or more conductors on the component side through via holes to the internal power layer

(b) External finger pins A and B are connected with one or more conductors on the circuit side through via holes to the internal ground layer

(c) Four-layer composite view of finger to inner power and ground planes connections

symbol. Two conductor paths (shown as solid lines) are drawn from the pads to the tops of fingers 1 and 2. In low-current applications, it may be necessary to connect only one finger to the power plane using only one via hole. Where current requirements are high, it is common practice to use several fingers and via holes.

Connections from the ground plane (layer 3) to fingers A and B are shown in Fig. 8.14b. Again, two via holes with the appropriate symbols are drawn above the fingers and inside the ground plane edge limit. Two conductor paths (shown as dashed lines) are then drawn on the circuit side to the tops of fingers A and B.

Note that Fig. 8.14a, showing layers 1 and 2, and Fig. 8.14b, which shows layers 3 and 4, were drawn separately for the sake of clarity of explanation in connecting fingers to inner layers. On a component layout drawing, this would be drawn as a composite view showing all four layers on one sheet of paper. This composite view is shown in Fig. 8.14c, which is simply the result of superimposing Fig. 8.14a on top of Fig. 8.14b.

The number of fingers connected to inner layers for any design is determined by the amount of current that is to be supplied to the board. Of course, the number of fingers connected to the power plane should be the same as that connected to the ground plane.

## 8.8 CLEARANCE WITH ANTIPADS AND INTERCONNECTIONS WITH THERMAL RELIEF PADS

It has been shown that a clearance, or insulation gap, is required on all inner layers around plated-through holes that are not to be connected to power or ground planes. For each plated hole, at least one, or at most two, clearance pads (or *antipads*) will be required on the inner-layer artworks. One form of these antipads is shown in Fig. 8.15a. The copper inner layer of the foil plane is electrically isolated from the inner copper pad (through which the lead drill hole will pass) by a circular insulating gap, shown as a shaded area. This gap can vary in width from 0.010 to 0.025 inch. It can be seen that where an antipad is used, the plated-through hole will pass through the inner pad but will be electrically isolated from the foil plane. A table of typical values of inner pad diameters with accompanying maximum finished (after plating) hole diameters is shown in Fig. 8.15b. Antipads are selected on the basis of the finished hole size. Their diameter should be a minimum of 0.040 inch larger than the hole size. This is to provide a degree of additional safety in the manufacturing phase in cases of misregistration of the inner layers. Even with a 0.040 inch clearance, if the misregistration is severe, the result could be unwanted electrical connections throughout the board. Remember that an antipad is required at each lead access or via hole on each inner layer where *no* electrical connection is to be made.

To make electrical connections to an inner layer represented by pads having a *plus* or *ground* symbol, a special pad called a *thermal relief pad,* or *spider pad,* is required. One form of thermal relief pad is shown in Fig. 8.15c. (These are also available in preformed shapes.) This pad is composed of an inner pad diameter, which is electrically connected to the copper foil plane of the inner layer by *connecting ribs.* These ribs have the same width as the surrounding insulation gap and vary from 0.010 to 0.025 inch. The design of this style of pad also provides high-resistance paths to heat flow from the inner pad to the foil plane. During the soldering process, a lead passing through a plated hole and the inner pad heats quickly and alloys properly with the solder inside the hole. This is the result of not allowing the heat being generated at the lead to be rapidly sinked from

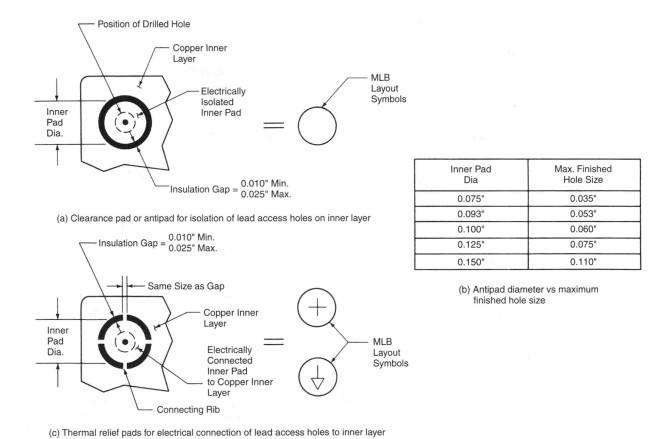

| Inner Pad Dia | Max. Finished Hole Size |
|---|---|
| 0.075" | 0.035" |
| 0.093" | 0.053" |
| 0.100" | 0.060" |
| 0.125" | 0.075" |
| 0.150" | 0.110" |

(a) Clearance pad or antipad for isolation of lead access holes on inner layer

(b) Antipad diameter vs maximum finished hole size

(c) Thermal relief pads for electrical connection of lead access holes to inner layer

**Figure 8.15**  Antipad and thermal relief pad configurations.

the plated hole and pad to the large copper plane because of the high thermal resistance provided by the connecting ribs. Thus maximum heat is maintained at the lead/hole interface and results in sound soldered connections. The inner pad diameter of the thermal relief pad should be at least 0.040 inch larger than the finished plated hole size. Remember that a thermal relief pad is placed at each lead access hole on each layer where an electrical connection is to be made.

It needs to be emphasized that the antipad and the thermal relief pad configurations shown on the left side of Figs. 8.15a and c are illustrated only to provide the PCB designer with a better understanding of their function. These configurations are produced with the use of a computer-aided drafting package such as AutoCAD. On the component layout drawing, they are shown as circles, with or without symbols, as illustrated on the right side of Figs. 8.15a and c.

## 8.9 SETUP CONSIDERATIONS FOR MULTILAYER BOARDS

Many of the points discussed in the layout of densely packaged double-sided PCBs also apply to that of multilayer boards. For example, the determination of the lead span dimensions between pad centers for axial lead components are based on the same criteria. In addition, those procedures for the selection of outer layer pad diameters, component spacing, and conductor width and spacing also apply to the layout of multilayer boards. For this reason, they will not be repeated here. In developing the multilayer component layout drawing in the following section, these criteria will be applied.

In addition to those layout considerations, which are common to both double-sided and multilayer designs, there are several others that are unique to multilayer boards: the grid and scale, defining the usable board area on each layer, layer-to-layer registration, and legends and codings used on drawings.

Positional accuracy to result in precise layer-to-layer registration in multilayer boards is more critical than in the layout of double-sided designs. In extremely high multilayer packaging densities greater than 25 holes/in$^2$ or 1.5 EDIPs/in$^2$, registration is much more challenging to the designer. Grid and Snap settings of 0.050 and 0.025 will be necessary when using CAD.

Corner brackets drawn on layers 1 and 4 again define the edges of the finished board. For the inner layers, a typical limitation is that the edges of the copper foil planes are no closer than 0.10 in. ($1\times$ scale) to the finished board edges. This is to ensure total sealing of the inner layers and also to prevent layer delamination. In addition, inner foil clearance normally must be provided around all mounting and hardware assembly holes. To define the limits of the foil planes and the required mounting hole clearance for the inner layers, an outline of the inner copper border is drawn on the layout. This is shown as the *inner layer foil limit* in Fig. 8.16.

For precise layer-to-layer registration of the four individual boards, a minimum of three registration and alignment targets are required. They are positioned on the layout outside the finished board outline. These also are shown in Fig. 8.16.

Of prime concern in any layout is the necessity of conveying all required information to eliminate the possibility of misinterpretation. For multilayer designs, there are three types of information required on the component layout drawing. Similar to double-sided layouts, pad and conductor path legends are shown. The third legend, which is unique to multilayer designs, is one that codes all the layers by number, name, connections (symbols, colors, or pad shapes), and the final foil thickness of each. The appropriate legends are shown together with the basic outline of a rectangular multilayer board in Fig. 8.16. Note that the board outline includes corner brackets, borders, mounting holes, targets, reduction scale, usable layout area, and the inner layer foil limit shown as dashed lines. This outline will be used in the next section to illustrate the techniques used to draw a multilayer component layout design.

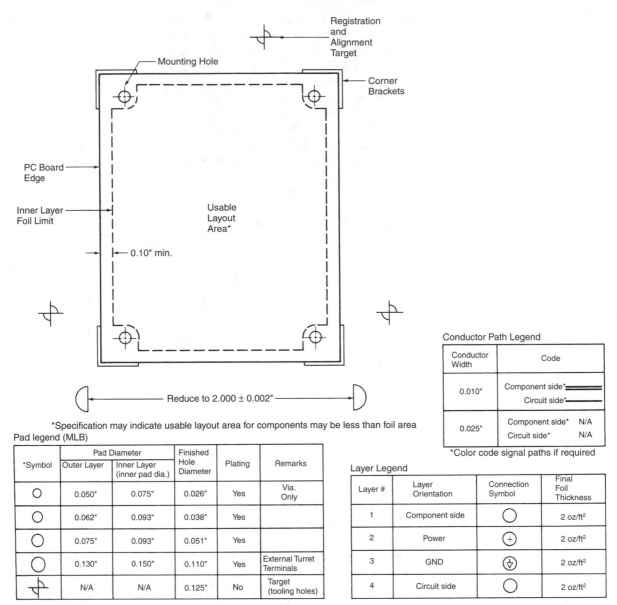

**Figure 8.16** Mechanical specifications for multilayer pc board design.

Registration and Alignment Target

Mounting Hole

Corner Brackets

PC Board Edge

Inner Layer Foil Limit

Usable Layout Area*

0.10" min.

← Reduce to 2.000 ± 0.002" →

*Specification may indicate usable layout area for components may be less than foil area

Pad legend (MLB)

| *Symbol | Pad Diameter | | Finished Hole Diameter | Plating | Remarks |
|---|---|---|---|---|---|
| | Outer Layer | Inner Layer (inner pad dia.) | | | |
| ○ | 0.050" | 0.075" | 0.026" | Yes | Via. Only |
| ○ | 0.062" | 0.093" | 0.038" | Yes | |
| ○ | 0.075" | 0.093" | 0.051" | Yes | |
| ○ | 0.130" | 0.150" | 0.110" | Yes | External Turret Terminals |
| ⊕ | N/A | N/A | 0.125" | No | Target (tooling holes) |

*If required color code for pad size

Conductor Path Legend

| Conductor Width | Code |
|---|---|
| 0.010" | Component side*▬▬▬▬<br>Circuit side*━━━ |
| 0.025" | Component side* N/A<br>Circuit side* N/A |

*Color code signal paths if required

Layer Legend

| Layer # | Layer Orientation | Connection Symbol | Final Foil Thickness |
|---|---|---|---|
| 1 | Component side | ○ | 2 oz/ft² |
| 2 | Power | ⊕ | 2 oz/ft² |
| 3 | GND | ⊕ | 2 oz/ft² |
| 4 | Circuit side | ○ | 2 oz/ft² |

## 8.10 FOUR-LAYER MULTILAYER DESIGN

To demonstrate the basic layout techniques for designing a multilayer board, we will use the simple schematic of a relaxation oscillator shown in Fig. 8.17. This circuit employs a 555 IC timer chip packaged in a TO-99 style 8-pin case. Power is applied between the +12-Vdc terminal and the ground terminals of the IC. Current flows through R1 and then the diode CR1, which is properly biased for conduction. This current charges capacitor C1. When the voltage across C1 reaches the predefined trigger level of the 555 timer, the capacitor quickly discharges through resistor R2 and into pin 7 of the IC. This charge/discharge cycle of capacitor C1 applied to the IC produces a square-wave output across the load resistor R3 that is connected to pin 3. The output waveform is measured across the *out* and *ground* terminals.

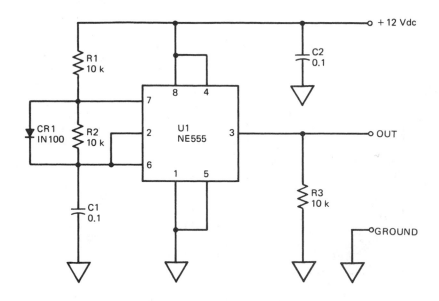

**Figure 8.17** Circuit schematic of a simple NE555 IC timer/oscillator.

All resistors are 1/2 watt ± 10% unless otherwise specified.
All capacitors are in microfarads unless otherwise specified.

Using the layout practices established in previous chapters, we will design a multi-layer board for the circuit of Fig. 8.17 using the board outline and accompanying legends shown in Fig. 8.16. As usual, the layout begins with the parts placement, using the schematic as a guide for initial component positioning. Attention is given to achieving a uniform distribution and balance of parts throughout the usable board area. This component layout is shown in Fig. 8.18a. Note that all parts are provided with a pad for each lead and that all parts are coded with both their appropriate designators and values.

At this point, we proceed with the routing of conductor paths as though the layout were for a double-sided design. Signal paths having 0.010-inch widths as per the conductor path legend of Fig. 8.16 will be routed along the circuit and component sides (layers 1 and 4). Note in the conductor path legend that a conductor width of 0.025 inch is not applicable (N/A) in this design. Sometimes in double-sided designs, this conductor width is specified for power and ground paths. Because we will be making power and ground connections only to inner layers 2 and 3, respectively, no conductor paths will be required for these connections. Thus the only routing of conductors that is required for a multilayer board is that for signal paths on layers 1 and 4. These are shown in Fig. 8.18b. Note that no outer-layer conductor paths are routed to the external terminal pad labeled +12 Vdc or to any component lead requiring a connection to +12 Vdc. Similarly, no outer-layer paths are routed to the external ground terminal pad or to any component lead requiring a connection to ground. In short, no conductor path is used to make any power and ground connections.

The final phase of our design will be to code the appropriate component and external terminal pads for inner layer power and ground connections using the layer legend of Fig. 8.16. Referring to the circuit schematic of Fig. 8.17, we begin by making all the power connections. Pins 8 and 4 of U1 must be connected to power, as are the top ends of resistor R1, capacitor C2, and the external turret terminal labeled +12 Vdc. The pads associated with these points will be connected to the inner power layer (layer 2). Using the symbol +, all of these pads are coded to show that they are to be connected to layer 2 as well as to both outside layers.

Recall that all pads shown on the outer layers (1 and 4), which have no plus or ground symbols, are to have no electrical connection to either power or ground. This means that pairs of antipads will be used on layers 2 and 3 with these uncoded pads.

*Double-Sided Plated-Through-Hole and Multilayer Printed Circuit Artwork and Design Concepts* **147**

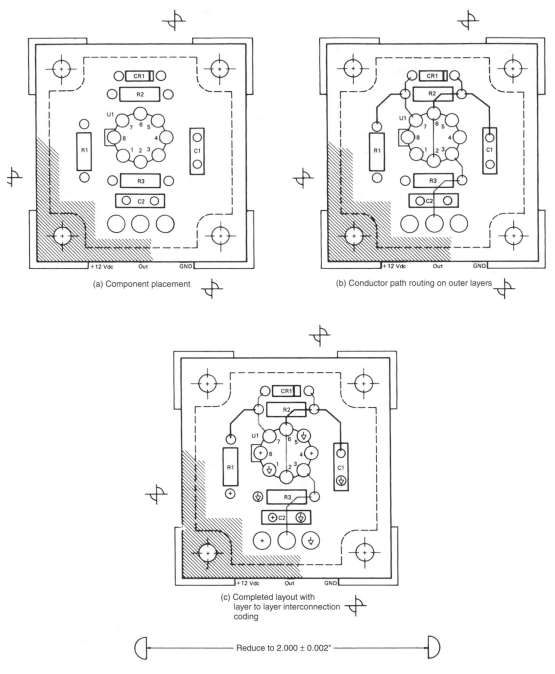

(a) Component placement

(b) Conductor path routing on outer layers

(c) Completed layout with layer to layer interconnection coding

Reduce to 2.000 ± 0.002"

**Figure 8.18** Sequence of multilayer board design.

In like fashion, ground connections are made to layer 3. The schematic shows that pins 1 and 5 of U1 as well as the external ground pad and the bottom ends of capacitors C1 and C2 and the resistor R3 are to be connected to ground. These points are coded on the drawing with a ground symbol inside of each pad. The completed component layout drawing for the multilayer board is shown in Fig. 8.18c.

This chapter completes the treatment of printed circuit component and conductor pattern design.

Next, we will describe the sequential processing steps that are performed to fabricate single-sided, double-sided, plated-through-hole, and multilayer boards. Because these processing steps are involved and require a great deal of detailed explanation, they will be spread over the next six chapters. Chapter 9 begins the process by converting the printed track layers into silver halide film negatives and positives and diazo film positive phototools. Chapter 10 describes the processing steps for fabricating single-sided pc boards and producing silk screen marking and solder masks. The method of applying these masks to a pc board is also covered. Chapter 11 presents the drilling of the bulk stock and its preparation for copper deposition followed by a flash plating process. Chapter 12 discusses the conductor pattern imaging process using dry film photo resist and the diazo phototools. Chapter 13 completes the fabrication of the double-sided plated-through-hole pc board by considering copper pattern plating, solder plating, photo resist stripping, etching, and infrared solder reflow. Finally, Chapter 14 presents multilayer printed circuit board fabrication.

## EXERCISES

### A. Questions

8.1 Describe specifically how interconnections are made between the circuit side and the component side of a double-sided plated-through-hole pc board.

8.2 What is the color-coding scheme for double-sided conductor pattern layouts?

8.3 When placing literal and numerical information on double-sided layouts, what is the convention in regard to forward- and backward-reading directions?

8.4 What command is used to have an automatic electrical check made of the schematic diagram drawn on the screen?

8.5 To what does the term *net* refer?

8.6 After the schematic is translated into an optimized component layout, what command optimizes the net positions?

8.7 When autorouting is completed, what command obtains an automatic test that all mechanical and electrical requirements are being met?

8.8 To achieve uniform plating, what two layout specifications must be adhered to in circuit board design?

8.9 What is the purpose of the thief area?

8.10 What might be a possible cause of tin-rich plating?

8.11 How does a voltage/ground MLB differ from a double-sided pc board?

8.12 How are antipads and thermal pads used?

### B. True or False

Circle *T* if the statement is true, or *F* if any part of the statement is false.

8.1 The bottom side of a double-sided pc board is also referred to as the circuit side. **T** **F**

8.2 A PTH forms a continuous metal path between copper on the top and bottom sides of a double-sided pc board. **T** **F**

8.3 Bodies of components, such as resistors and integrated circuits, must not contact conductor paths on a pc board. **T** **F**

8.4 On a composite view of both sides of a double-sided pc board, conductor paths must not be allowed to cross each other. **T** **F**

8.5 When using CAD, component and conductor pattern layouts must be drawn to a 2:1 scale. **T** **F**

8.6 In the layer color-coding system for double-sided pc boards, red represents the tracks on the component side. **T** **F**

| | | |
|---|---|---|
| 8.7 | Nets are optimized using the electrical rules check after autorouting is completed. | **T** **F** |
| 8.8 | The purpose of the thief area is to direct some of the high current density at the edge of the board toward itself during the plating process. | **T** **F** |
| 8.9 | Antipads have uninterrupted 0.010 to 0.025-inch insulation gaps. | **T** **F** |
| 8.10 | MLB design should be considered for density levels greater than 25 to 30 holes/in$^2$ for analog circuits and 1.5 to 1.8 EDIPs/in$^2$ for digital circuits. | **T** **F** |

## C. Multiple Choice

Circle the correct answer for each statement.

8.1 All the literal and numerical information should be (*forward-*, *backward-*) reading on the component side and (*forward-*, *backward-*) reading on the circuit side of a double-sided pc board.

8.2 The circuit side of a double-sided pc board is the (*top*, *bottom*) side.

8.3 Airwire positions are optimized by using the command (*ERC*, *RATSNEST*).

8.4 Tracks having the color (*blue*, *red*) represent conductor paths on the circuit side of the board.

8.5 An area of a pc board having a heavy concentration of conductor paths will receive a (*high-*, *low-*) tin content when solder plated.

8.6 One of the voltage/ground MLB inner layers is used for (*power*, *signal*) connections.

8.7 Pin 7 on a 7400 series TTL gate is connected to the (*power*, *ground*) layer.

8.8 MLB thermal pads employ (*gaps*, *ribs*) to prevent rapid heat sinking away from the PTH and pad to the large copper planes during the soldering process.

8.9 The inner layer foil limit lies (*inside*, *outside*) the mounting hole locations.

8.10 There are (*three*, *four*) combinations of inner layer connections represented by the convention of inner layer connection symbols.

## D. Matching

Match each item in Column A to the most appropriate item in Column B.

| COLUMN A | | COLUMN B | |
|---|---|---|---|
| 1. | Thief area | a. | Output |
| 2. | Airwire | b. | PTH |
| 3. | Circuit side | c. | Track |
| 4. | Via | d. | Junction |
| 5. | Autoroute | e. | Short circuit |
| 6. | XPLOT | f. | Plating |
| 7. | DRC | g. | Blue |
| 8. | Insulation | h. | Uncoded pad |
| 9. | Layer 4 | i. | Ground |
| 10. | Inner layer | j. | Antipad |

## E. Problems

8.1 Using CAD, reproduce the component layout in Fig. 8.18a on a 2.7 by 2.7-inch pc board. This layout will be used to test your understanding of the different methods of making inner layer connections using antipads and thermal pads in an MLB design. To aid in drawing this figure, use the following dimensional information.

1. Draw all resistors having body sizes of 0.5 inch long and 0.2-inch diameters with 0.1-inch diameter pads spaced 0.7 inch apart.

2. The capacitors are to have body sizes of 0.2 inch by 0.6 inch with 0.1-inch diameter pads spaced 0.4 inch apart.

3. The diode should have a length of 0.35 inch and a diameter of 0.15 inch with 0.1-inch diameter pads spaced 0.6 inch apart.

4. The pad pattern for device U1 is to have 0.15-inch pads on a 0.7 lead circle and a 0.15 by 0.2-inch key at pin number 8.

5. There are to be four 0.2-inch mounting holes located 0.3 inch inside of each edge at the corners of the board.

6. The foil limit is to be 0.2 inch inside all edges of the board with a 0.3-inch radius from hole centers to skirt each mounting hole position.

7. The hatched forbidden area is to be made around the entire periphery of the board and extend 0.1 inch inside the foil limit.

8. Refer to Fig. 8.15 to make antipads and thermal pads. Draw both connected and isolated inner pads having a 0.1-inch diameter with their center pads being 0.015 inch in diameter. Conducting ribs and insulation gaps are to be made 0.025 inch wide.

9. Draw conductor paths on both sides of the board 0.010 inch wide.

8.2 Determine the number of conventional pairs of pads, not counting mounting holes, needed on layers 1 and 4, and the number of antipads and thermal relief pads required on layers 2 and 3. Refer to Figs. 8.13, 8.17, and 8.18.

8.3 Add just the outer layer conductor paths to the drawing generated in Problem 8.1 using Figs. 8.18b and c that are needed, other than power and ground, to satisfy the electrical connections of the timer/oscillator circuit shown in Fig. 8.17. Replace the double-lined paths on the circuit side of the board, as shown in Fig. 8.18b, with 0.010 inch filled polylines. Color code conductor paths on layer 1 as red and on layer 4 as blue. When routing conductor paths on layers 1 and 4, be sure that no path is routed to an external terminal pad labeled +12 Vdc or to any component lead requiring +12 Vdc. Similarly, no outer layer path is to be routed to any external terminal pad or component lead requiring an electrical connection to ground.

8.4 Draw the power and ground layers, properly employing antipads and thermal relief pads that will satisfy the required power and ground innerlayer connections of the timer/oscillator circuit to complete the four-layer artworks. Be sure that all four layers are in perfect registration when viewed simultaneously on the screen.

8.5 After all four layers have been completed in Problems 8.1 through 8.3, first display only layers 1, 2, and 4. Then perform an electrical check using the circuit schematic in Fig. 8.17, and symbol codes in Figs. 8.13b and c and Fig. 8.18c to see that all antipads and thermal pads have been correctly positioned to satisfy power connection requirements. Next view layers 1, 3, and 4 to see that all ground connections have been properly provided, using the same figures as a guide, and that there are no short circuits between layers 2 and 3.

# Unit

# THREE

## Printed Circuit Board Fabrication and Inspection

# 9 PCB Photographic Processes

## 9.0 INTRODUCTION

Until recently, 2:1 and 4:1 scaled *taped artwork masters* were produced by manual means on clear acetate sheets with opaque tape. Today most new PCB designs are generated using CAD software packages such as those illustrated in Chapters 7 and 8 of Unit 2. The *artwork master* that results from these new software tools is almost always a 1:1 scaled conductor pattern. Most CAD software is capable of producing negative phototools directly. If a Gerber-type photoplotter is available, all software can produce a high-quality 1:1 scaled positive print (black conductor pattern on white paper) with a laser printer.

The best method to photographically process old 2:1 or 4:1 scaled taped artwork masters or a new 1:1 scaled CAD-generated artwork master into *working tools* to be used to fabricate the finished printed circuit board is with a copy camera. The resulting positive or negative photographic films are commonly known as *phototools*.

This chapter begins with a discussion of the setup, use, and adjustments of a typical copy camera for photographically reducing old taped artworks or processing new CAD artworks into phototools. This is followed by a presentation of films, chemicals, and bath preparations as well as film processing. Also included is the technique of 1:1 scaled contact printing to obtain image reversals. The chapter concludes with information on diazo phototools for double-sided printed circuit board fabrication.

The accuracy and quality of the phototools described in this chapter will determine the quality of the resulting printed circuit board. For this reason, this chapter is sufficiently detailed so as to emphasize careful attention to each of the processes described. Properly prepared phototools can avoid pc board fabrication problems.

*CAUTION:* The copy camera described in this chapter is equipped with four 800-watt quartz iodide lamps. While using this equipment, **do not** look at the illuminated lamps. Shield your eyes from the lamps when exposures are being made.

The diazo developing unit discussed in this chapter uses ammonia that is heated between 200° and 220°F (93 to 104°C). Be sure this unit is used in a well-ventilated station that is working properly to be protected from inhaling ammonia fumes.

A chemical hazard exists in the materials presented in this chapter. Approved face shields (goggles) in compliance with ANSI Z87.1-1989 and chemical-resistant clothing and footwear and neoprene or nitrile gloves must be worn while performing any of these operations. Also, the section on safety outlined at the beginning of the book should be reviewed for more specific information.

## 9.1 PHOTOGRAPHIC IMAGING OF THE ARTWORK MASTERS

Commercial photocopy cameras are capable of accepting artwork masters in excess of 48 by 60 inches. These masters may be reduced into positives or negatives by a factor of 5:1, 4:1, 2:1, or 1:1. This reduction is dependent on the type of lens available on the camera. One-to-one reversals of CAD-generated artwork masters can also be made with the copy camera.

Artwork reductions made on a copy camera can be produced to a tolerance of ±0.002 inch, although ±0.005 inch is the tolerance typically specified. The results of artwork exposure with a copy camera have high-quality line resolution, high contrast between light and dark areas, and negligible imperfections such as pinholes (voids) in the opaque areas of the processed film.

A typical commercial copy camera used for artwork reduction in the printed circuit board industry is shown in Fig. 9.1. This is a nuArc Copy Camera Model SST1418, which is capable of handling a film image size of up to 14 by 18 inches and artworks as large as 21 by 25 inches if front-lighted and 15 by 19 inches for back-lighted work. When used with the appropriate film, this camera will provide excellent-quality exposures that are ideally suited for small- to medium-size pc boards.

The nuArc Model SST1418 copy camera will be described here to illustrate photographic procedures and typical camera operation. Other cameras used for printed circuit photography will involve similar operating procedures but may have distinctive differences so as to require the operator to consult the manufacturer's instructions for any operating variations. The nuArc camera has eight basic components: (1) transparent glass copyboard with cover glass and translucent rear panel for backlighting; (2) four 800-watt quartz iodide lamps; (3) lens with diaphragm control, electric shutter, and cover; (4) bellows; (5) vacuum-frame film holder with vacuum pump and foot switch; (6) focusing plate; (7) two scale-adjustment control wheels; and (8) control console with focus, scale illumination, and master switches and lamp timer.

To operate this camera for reducing old taped artwork masters, the procedure is as follows. First, the lens cover is removed. The copyboard is then rotated to the horizontal position for artwork loading. This is done by pulling out the frame release handle and gently lowering the frame. The cover glass is then unlatched and raised until the internal lock in the cover support arm engages to hold the cover open. The artwork to be processed is placed onto the transparent base of the copyboard and centered (see Fig. 9.2). To aid in obtaining consistent uniformly developed exposures, a transparent-to-opaque *photographic step exposure table* is placed over a clear area of the artwork but outside the corner brackets so that it does not become a part of the resultant phototool. This is also shown in Fig. 9.2. (A detailed discussion of the use of a step table is given in Sec. 9.2, on processing film.)

With the artwork, together with the step table, positioned on the copyboard, the cover glass is lowered and locked into place. Raising the copyboard to its locked vertical position will establish the correct orientation for exposure. For maximum contrast between light and dark areas of the artwork, backlighting is the preferred method. This is shown in Fig. 9.3. Note that the lamps are directed at approximately 45-degree angles to the rear translucent light-diffusing panel of the copyboard. With this arrangement, the artwork is uniformly and brilliantly illuminated without *hot spots* or reflections during film exposure. For new CAD-generated artwork masters, the lamps are directed at approximately 45-degree angles to the front of the copyboard. This is necessary since the artwork master is usually printed on white paper.

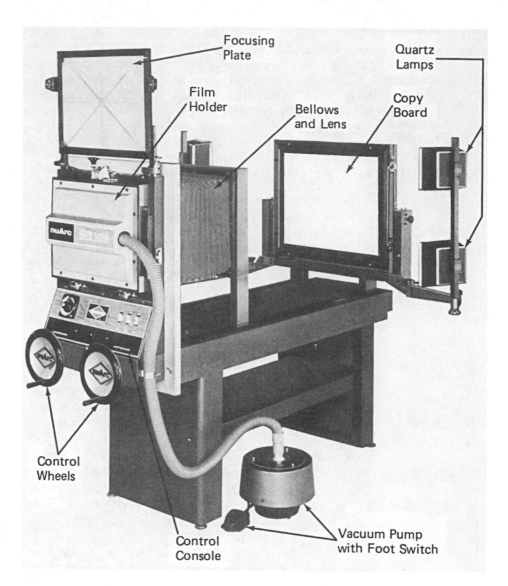

Figure 9.1 Horizontal copy camera. *Courtesy of nuArc Company, Incorporated.*

All adjustments for photoreduction are done at the rear of the camera. Power to the control console is first applied by depressing the tops of the *master power* rocker switch and the *scale tape light* switch. The scale reduction settings are then made. Also located at the rear of the camera are two control wheels for adjusting both the reduction scale and the focus of the artwork image. One control wheel is labeled *lensboard* and the other is labeled *copyboard*. Scale adjustment using these control wheels is made in conjunction with scale markings on two stainless steel calibration tapes that are viewed through 2× magnifying lenses located at the center of the control console above the wheels. To adjust for a 2:1 scale reduction, the *copyboard* control wheel is rotated until the accompanying calibration tape reads *50*. Clockwise rotation of this wheel moves the copyboard *away* from the lens while a counterclockwise rotation moves it *toward* the lens. The *lensboard* control wheel is then adjusted so that its calibration tape also reads *50*. Clockwise rotation of this wheel moves the lens *toward* the copyboard and counterclockwise rotation moves it *away* from the copyboard. With both calibration tapes set at *50,* the camera is set for a 2:1 artwork reduction. The controls used for setting the camera to the desired reduction scale are shown in Fig. 9.4. Note that new CAD-generated artwork masters, printed at a 1:1 scale, require both tapes to be set at *100*.

**Figure 9.2** Loading taped artwork master and step table onto camera's copyboard.

**Figure 9.3** Correct position of quartz lamps for backlighting of copyboard. *Courtesy of nuArc Company, Incorporated.*

To check the accuracy of the tape setting as well as the focus, the vacuum frame film holder is first opened by releasing the latch. The frame is then lowered to the horizontal position. The ground-glass focusing plate, which is stored above the film holder, is next unlatched and swung downward until it locks into the viewing position (see Fig. 9.5). Under safe light conditions, a visual check of the image may now be made. By depressing the top of the *timer/focus* rocker switch, the camera lights will turn on and the artwork image will be seen on the focusing plate, which has horizontal as well as vertical scale markings for making general measurements. However, best results can be achieved in checking the accu-

**Figure 9.4** Detail of the control console and adjustment control wheels. *Courtesy of nuArc Company, Incorporated.*

racy of the setting on the reduction scales by using a *precision transparent polyester scale,* such as the Bishop Graphics Accuscale, placed flat on the focusing glass in line with the reduction marks. Any fine adjustment required is then made only with the lensboard wheel by rotating it slightly clockwise or counterclockwise until exact alignment of the projected artwork scale reduction marks and the precision scale markings result on the focusing glass.

After the accuracy of the artwork reduction image has been verified, the focusing glass may be raised and locked into its vertical storage position above the film holder. The *timer/focus* switch should then be positioned to *timer* and the *tape* switch moved to *off* by depressing the bottoms of these rocker switches. In this mode, the exposure lamps will be off as well as the scale illumination lamps.

Proper film exposure is a function of the *amount* of light that strikes the film and the *length* of exposure time. The amount of light is set by the aperture opening within the lens system, the size of which is designated by *f/stops.* The maximum amount of light is transmitted through the lens using the widest aperture opening (lowest f/stop number). The nuArc Model SST1418 camera has an f/stop range of f/10 to f/22. It has been determined empirically that an aperture setting of f/16 will result in high-quality artwork phototools. This opening is set with the pointer using the upper scale shown in Fig. 9.6.

Exposure time is controlled by an electrical timer that is synchronized with the shutter mechanism of the lens. Refer again to Fig. 9.4. The timer pointer is set to the desired number of seconds of exposure and is activated by depressing the center button (see Fig. 9.7). Immediately, the lights are illuminated and the lens shutter is opened. The film is exposed for the preset time, at the end of which the timer activates the switching that turns off the lamps and closes the shutter. After each exposure, the timer will recycle to the preset position in preparation for another time exposure. With the optimum aperture opening of f/16, a timer setting of 10 seconds results in high-quality negatives ideally suited to pc artwork processing. A flow chart of the photographic reduction procedures using the nuArc SST1418 copy camera is shown in Fig. 9.8.

**Figure 9.5** Focusing plate positioned to inspect artwork image.

**Figure 9.6** Lens with diaphram percentage control. *Courtesy of nuArc Company, Incorporated.*

**Figure 9.7** Time of film exposure controlled with an electrical timer. *Courtesy of nuArc Company, Incorporated.*

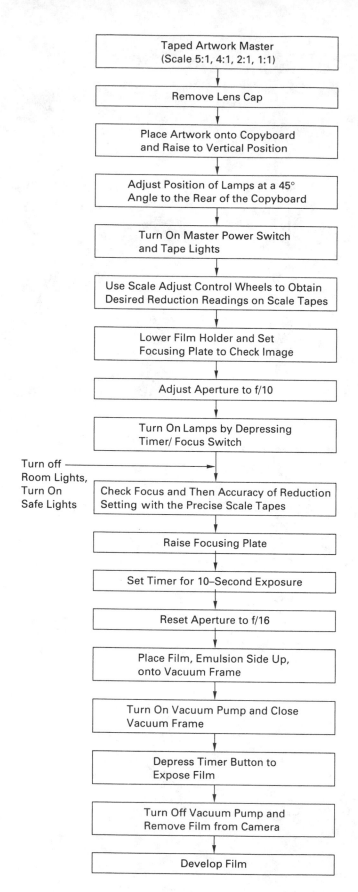

Taped Artwork Master
(Scale 5:1, 4:1, 2:1, 1:1)

Remove Lens Cap

Place Artwork onto Copyboard
and Raise to Vertical Position

Adjust Position of Lamps at a 45°
Angle to the Rear of the Copyboard

Turn On Master Power Switch
and Tape Lights

Use Scale Adjust Control Wheels to Obtain
Desired Reduction Readings on Scale Tapes

Lower Film Holder and Set
Focusing Plate to Check Image

Adjust Aperture to f/10

Turn On Lamps by Depressing
Timer/ Focus Switch

Turn off Room Lights, Turn On Safe Lights

Check Focus and Then Accuracy of Reduction
Setting with the Precise Scale Tapes

Raise Focusing Plate

Set Timer for 10–Second Exposure

Reset Aperture to f/16

Place Film, Emulsion Side Up,
onto Vacuum Frame

Turn On Vacuum Pump and Close
Vacuum Frame

Depress Timer Button to
Expose Film

Turn Off Vacuum Pump and
Remove Film from Camera

Develop Film

**Figure 9.8** Flow diagram for photographic reduction of artwork master.

## 9.2 EXPOSING AND PROCESSING PHOTOGRAPHIC FILM

The nuArc SST1418 copy camera or other similar unit used for photographic processing of artwork masters requires the use of suitable film. This film must possess high line contrast, reasonably fast processing time, and excellent dimensional stability. Several manufacturers have produced sheet film with these qualities, among them Kodak Kodalith Ortho Type 3 and Dupont Cronalith film. For purposes of illustration, the Kodak Ortho 3 film and its associated chemicals are discussed. If other film is used, the specific manufacturer's literature on film processing must be consulted.

The Kodak negative film is processed in the following sequence: *exposure, develop, stop, fix, wash,* and *dry.* Each of these steps will be described in detail.

To load the camera, a sheet of Kodak negative film is first removed from its light-tight box under *safe-light* conditions (Wratten 1-A or equivalent). Under normal lighting, this film would appear pale *red* but appears as light gray under the safe lamp. The *emulsion* side of the film (as opposed to the 0.004-inch-thick triacetate support backing side) is discernible by its lighter color. (If extremely high-dimensional stability and durability are required, 0.007-inch film is available at slightly higher cost.) The film is centered onto the vacuum frame film holder so that its emulsion side is facing upward toward the operator. This is shown in Fig. 9.9. The vacuum pump is then turned on by a foot switch. With the vacuum holding the film firmly in place, the film holder is raised to its vertical position and latched. The artwork to be reduced is positioned on the copyboard so that it is centered over the translucent backing. An exposure step-table having 21 transparent-to-opaque sequential blocks is taped onto the artwork outside the corner brackets. With the setup in place, the glass cover is secured and the copyboard is raised to its vertical position. Before making the exposure, the operator should check the f/stop setting, timer setting, and reduction scale tapes as well as the position of the lamps.

The film is exposed by depressing the timer *start* button shown in Fig. 9.7. After exposure, the film is removed from the vacuum frame and is ready to be developed.

Kodak film is developed by the use of two concentrated chemicals: *Kodalith Liquid Developer Solution A* and *Kodalith Liquid Developer Solution B.* Since these chemicals are

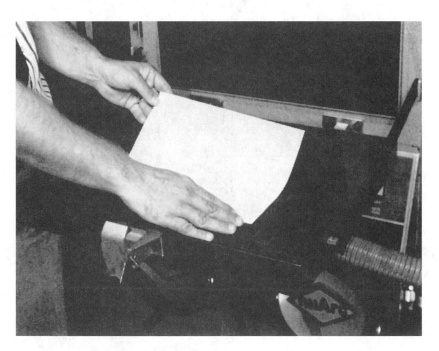

**Figure 9.9** Vacuum frame holds film in position for exposure. *Courtesy of nuArc Company, Incorporated.*

in liquid form, they require no special mixing equipment. For best results, they should be mixed with water that is at a temperature of between 65 and 70°F (18 to 21°C). (This water temperature is recommended for mixing all film baths.) The following five steps outline the preparation of the developer.

1. Determine the volume of developer required.
2. Divide this total in half to determine the amount of diluted developer A and diluted developer B required.
3. Divide this figure by 4 to find the amounts of both concentrates A and B needed.
4. Add three equal parts of water to concentrate A as determined in step 3; repeat for concentrate B.
5. Mix diluted A to diluted B to yield the total volume as determined in step 1.

The concentrates A and B are never added together. They must be diluted before mixing.

An example for the preparation of the developer will aid in understanding the various steps. In this example, it is required that *1 quart of Kodalith Developer be prepared.* Following are the five steps just presented:

1. Total volume required—1 quart (32 ounces).
2. Subvolume—32/2 = 16 ounces of diluted A and 16 ounces of diluted B.
3. 16A/4 = 4 ounces of A concentrate; 16B/4 = 4 ounces of B concentrate.
4. 4 × 3 = 12 ounces of water. Four ounces of A plus 12 ounces of water = 16 ounces of diluted A solution; 4 ounces of B plus 12 ounces of water = 16 ounces of diluted B solution.
5. Sixteen ounces of A diluted solution added to 16 ounces of B diluted solution yields 32 ounces (1 quart) of working developer.

The prepared solution is placed in a plastic developing tray. Under safe-light conditions, the exposed film with the emulsion side *up* is placed in the tray and constantly agitated until the film is developed completely. At a developing solution temperature of 68°F (20°C), the development time is between 2 and $2\frac{1}{2}$ minutes, the time decreasing as the temperature increases. At 74°F (23°C), the developing time is $1\frac{1}{4}$ to 2 minutes and at 80°F (27°C) reduces to 1 to $1\frac{1}{2}$ minutes.

A chart showing approximate developing time in minutes for specific bath temperatures is shown in Fig. 9.10. Proper developing time is critical. Depending on how long

| Bath Temperature | Developing Time |
|---|---|
| 68°F | 2 min. 15 sec. |
| 69°F | 2 min. 10 sec. |
| 70°F | 2 min. 7 sec. |
| 71°F | 2 min. 3 sec. |
| 72°F | 1 min. 58 sec. |
| 73°F | 1 min. 52 sec. |
| 74°F | 1 min. 49 sec. |
| 75°F | 1 min. 45 sec. |
| 76°F | 1 min. 40 sec. |
| 77°F | 1 min. 37 sec. |
| 78°F | 1 min. 32 sec. |
| 79°F | 1 min. 26 sec. |
| 80°F | 1 min. 22 sec. |
| 81°F | 1 min. 18 sec. |
| 82°F | 1 min. 15 sec. |

**Figure 9.10** Developing time as a function of bath temperature.

the film is allowed to remain in the developing bath, it can be under- or overdeveloped. An explanation of the developing process will aid in understanding the reason for paying close attention to the required time. When the film is initially exposed in the camera, certain portions remain underexposed because the black conductor pattern on the copyboard absorbs light, whereas the surrounding white area reflects light. As a result, the underexposed film portions form a *latent* image of the conductor pattern. When the film is developed, it yields a white conductor pattern image on a black background. During the developing process, the total emulsion surface is softened and feels oily to the touch. For this reason, care needs to be exercised to avoid scratching this surface. If the film is allowed to remain in the developer for an excessive amount of time, the entire emulsion surface will become totally black. The film negative should be agitated in the developer solution long enough for a clear-cut image to appear with the desired contrast. Optimum contrast can be gauged by continuous visual inspection of the device outline configuration. Attention should be focused on the smallest lead holes to be developed. When they appear as distinct sharp black dots, the developing process should be terminated *immediately* by quickly placing the film into a *stop bath*. If lead holes are not present, continuous comparison of the depth of color in the black areas of the developing film between *both* sides is made. As soon as both sides appear to have equal degrees of blackness, the film is ready to be placed into the stop bath.

Determining proper developing time by the methods just described is appropriate for experienced technicians. For the beginner, however, a more exacting method may be used. This method takes into account the light source and exposure time, lens setting and reduction scale, developing solution bath temperature, and time in the developing bath. It compensates automatically for any variations in these details to result in a correctly developed film. A Stouffer chart, shown in Appendix VII, is placed outside the corner brackets of the artwork on the copyboard of the camera. As the developing process progresses, the blocks in the step table will begin to blacken, starting with the transparent block, step 1. By monitoring the step chart, the film is removed from the developing solution and placed into the stop bath at the point where the *sixth* step of the table has completely blackened with the seventh, and perhaps the eighth and ninth steps, partially blackened. This method of film developing assures consistent results regardless of any variations in the exposure setting. The appearance of a properly developed step table is shown in Fig. 9.11.

Whichever technique of film developing is employed, care should be exercised to avoid the possibility of under- or overdeveloping. Immediately after the film is properly developed, it is quickly placed into a *stop bath*. This stop bath is made up of 1 part of *glacial acetic acid* to 32 parts of water. To serve as an example of this mixture, to prepare 330 milliliters of stop bath, 10 milliliters of glacial acetic acid is mixed with 320 milliliters of water.

The developer is completely neutralized in the stop bath when the oily characteristics of the film emulsion no longer exist. This neutralization process will take approximately 1 minute. There is no visible change in the image. The negative is gently agitated in this stop bath for approximately 1 minute and then is placed in a tray containing *Kodak Rapid Fixer (Solution A)* with a *hardener (Kodak Solution B bath for fixer)* at 68°F (20°C) for approximately 1 to 2 minutes. The black areas of the film are insoluble in the fixer, but the white emulsion of the conducting pattern image will dissolve in the solution. The negative must be agitated in the fixer solution for a sufficient period of time for all the emulsion in the conducting pattern area to be dissolved, exposing the underlying transparent backing. A black tray should be used because it permits the complete dissolving of the white emulsion to be more easily detected. The visual impression is a totally black film. The hardener is included in the fixer solution to stabilize the remaining emulsion.

To prepare 1 gallon of fixer bath, 1 quart of solution A is added to 2 quarts of water at 65 to 70°F (18 to 21°C). Three-and-one-half fluid ounces of solution B (hardener) is stirred into the solution. Finally, sufficient water is added to make the total solution

**Figure 9.11** Properly developed 21-step table for Kodalith Ortho Type 3 film (settings: f/16, 10 seconds, 2:1 reduction).

equal to 1 gallon. A negative film being removed from the fixer solution is shown in Fig. 9.12.

After the fixing process, the film negative is washed in a spray of water at approximately 70°F (21°C) for several minutes to remove all traces of the fixing solution (Fig. 9.13). Once the film has been washed, it can be allowed to dry by simply suspending it with a clamp attached to an overhead line. This method is not recommended, however, since the surfaces of the negative tend to waterspot. To prevent this spotting, a squeegee or photographic sponge may first be used to remove the excess water and then the negative should be allowed to dry. The film is now completely processed into a negative of the printed circuit pattern. The negative should be handled carefully because it will scratch easily. Because of the involved nature of the photographic process, a flow diagram is provided in Fig. 9.14 to guide the technician through the individual steps.

**Figure 9.12**
Developing sequence for reduced negative.

**Figure 9.13**  Final film rinse.

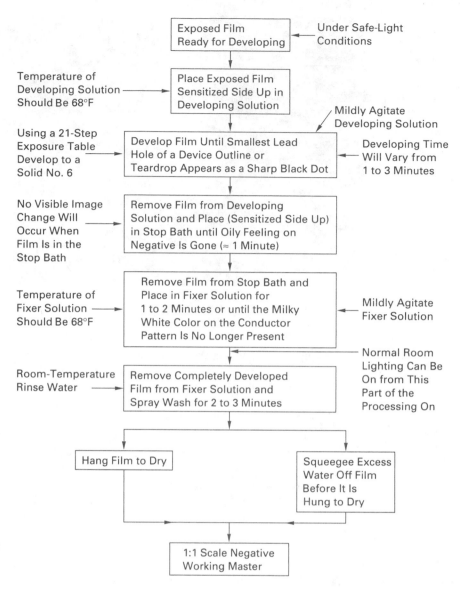

**Figure 9.14** Flow chart for developing negative film.

The flow chart contains the following boxes and annotations:

Exposed Film Ready for Developing ← Under Safe-Light Conditions

Temperature of Developing Solution Should Be 68°F → Place Exposed Film Sensitized Side Up in Developing Solution

Using a 21-Step Exposure Table Develop to a Solid No. 6 → Develop Film Until Smallest Lead Hole of a Device Outline or Teardrop Appears as a Sharp Black Dot ← Mildly Agitate Developing Solution / Developing Time Will Vary from 1 to 3 Minutes

No Visible Image Change Will Occur When Film Is in the Stop Bath → Remove Film from Developing Solution and Place (Sensitized Side Up) in Stop Bath until Oily Feeling on Negative Is Gone (≈ 1 Minute)

Temperature of Fixer Solution Should Be 68°F → Remove Film from Stop Bath and Place in Fixer Solution for 1 to 2 Minutes or until the Milky White Color on the Conductor Pattern Is No Longer Present ← Mildly Agitate Fixer Solution

Room-Temperature Rinse Water → Remove Completely Developed Film from Fixer Solution and Spray Wash for 2 to 3 Minutes ← Normal Room Lighting Can Be On from This Part of the Processing On

Hang Film to Dry

Squeegee Excess Water Off Film Before It Is Hung to Dry

1:1 Scale Negative Working Master

## 9.3 CONTACT PRINTING

When the artwork masters have been photographically reduced into negative transparencies, the required 1:1 scale *positives* of the marking mask negatives can now be processed. If the film used in the initial reduction process yields a negative transparency, one more operation will generate the positive transparencies. To produce a positive from a reduced artwork negative is in fact to obtain a positive transparency of the original artwork master that is reduced to the proper 1:1 size scale for processing a pc board marking mask. (The process of converting a positive transparency of a marking mask into a silk screen is discussed in detail in Sec. 10.5.)

The processing of a 1:1 scaled positive from the same-size negative is best accomplished by *contact printing* using the same Kodak Kodalith Ortho Type 3 film used for the reduction of the original artworks. An arrangement showing a marking mask being contact-printed is shown in Fig. 9.15. This system consists of a vacuum frame, pump, lamp with adjustable intensity, and exposure timer controls.

**Figure 9.15** Negative and film being exposed with contact printer.

For contact printing, a sheet of Ortho 3 film is removed from its light-tight box under safe-light conditions. It is placed, emulsion side up, on the black nonreflective mat in the vacuum frame. The reduced negative is then placed on top of the film and the glass cover of the vacuum frame is lowered and locked into place. The vacuum pump is switched on to maintain pressure of the negative against the film. Good contact is achieved when 25 inches Hg or greater is registered on the vacuum gauge. The contact printer lamp is positioned approximately 3 feet directly above the film to be exposed. For exposure, the intensity control of the light source is switched to *intensity level 2* and the timer is set at 10 seconds. The *start* button is then depressed to activate the light and the timer.

After exposure, the vacuum pump is switched off and the film is removed from the frame. Still under safe-light conditions, the film is processed through the same developer, stop bath, fixer, and water rinse procedures as described in the preceding section. The flow diagram of Fig. 9.14 used to process negative film is also applicable for contact print processing.

## 9.4 DIAZO PHOTOTOOL FOR DOUBLE-SIDED PRINTED CIRCUIT BOARDS

For processing a double-sided pc board with plated-through holes, two 1:1 scale positives are required. These *silver halide* positives are generated from contact-printing the negatives, as described in the previous section. Silver halide positives, however, are not suitable for proper registration, as will be explained here. Correct registration of the 1:1 phototools on both sides of a previously drilled panel is accomplished by aligning the solid terminal pads of the phototools over the holes. The aim is to obtain uniform annular rings about all holes. Because the conductor pattern on silver halide positives is 100% opaque, the drilled holes cannot be viewed through the pads on the positive to obtain proper registration. For this reason, phototools must be made from the 1:1 silver halide positives, which have a lower visual density (light-transparent color) of the opaque conductor pattern, yet be opaque to ultraviolet radiation. These positives will allow the drilled holes to be viewed through the pads for proper registration.

One type of film available for producing this special phototool is a light-sensitive *diazo emulsion* film on a polyester base. Diazo film, whose emulsion is sensitive to ammonia gas and is commonly used in the pc board industry, makes direct image transfer (without reversal) when contact printed with standard silver halide positives. The contact printing is accomplished under gold safe-light conditions using the same ultraviolet light-exposure units used for photo-resist exposure (see Chapter 12). The film is then dry-developed in hot ammonia gas.

To illustrate the process of producing two diazo phototools (one for the component side and one for the circuit side), we will use *Dynachem Type G-2 Trans Opaque* film. This diazo film is supplied on a dimensionally stable 0.007-inch polyester base. It should be stored at 50°F (10°C) or below when not in use in order to extend its shelf life. It is always processed in a gold safe-light environment such as that found in the photo-resist exposure area of a pc board fabrication facility.

To make a 1:1 contact-printed diazo positive from a silver halide positive, a sheet of G-2 film is first placed in the vacuum frame of an exposure unit (similar to the one shown in Fig. 9.16) on a black nonreflective surface. The film is positioned with the emulsion side facing *upward*. In this position, the notch on the sheet of film will be in the upper-right corner. The silver halide positive is then placed on top of the diazo film with its emulsion side *down*. The frame cover is closed and the vacuum pump is turned on, which will bring both films in tight contact with each other before the frame is slid under the exposure lamps. Exposure time is determined empirically. It has been found that with the 400-watt ultraviolet light source shown in Fig. 9.16, proper exposure is completed in under 2 minutes. Those areas on the diazo film that are exposed to the light through the transparent areas of the silver halide positive will remain clear when developed in ammonia gas. Those areas under the opaque conductor pattern of the silver halide positive are unexposed to the light. In the diazo developing process, these areas became an amber color, which is transparent to visible light but opaque to ultraviolet light.

After the diazo film has been exposed, it is removed from the vacuum frame and developed in a gold-lighted area using a *Dynachem Diazo G-2 Film Processor,* which is shown in Fig. 9.17. This unit is a simple conveyor having hard-rubber rollers that transport the film slowly through an environment of ammonia gas heated to 200 to 220°F (93 to 104°C). By simply introducing the film into the unit, total developing time is approximately 1 minute.

When the diazo film has been developed, the amber conductor pattern image will become visible (see Fig. 9.18). Since it is now possible to see through the solid terminal pads, proper registration of the film over a predrilled board can be effectively accomplished.

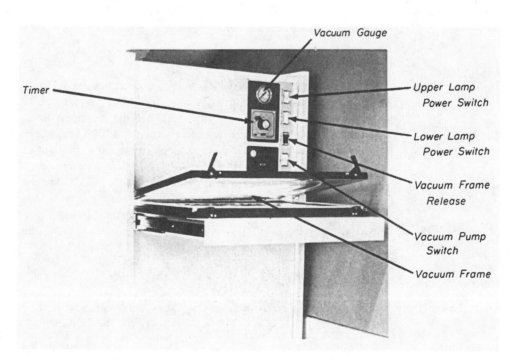

**Figure 9.16** Exposure unit for double-sided pc boards.

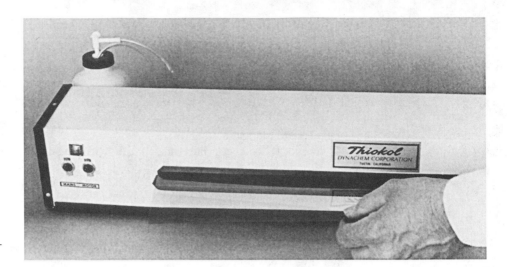

**Figure 9.17** Exposed film being developed with the Dynachem diazo developing unit.

**Figure 9.18** Diazo film phototools with double-sided masking tape.

Since the processed films will be used as phototools to image pc boards, they must have areas that will effectively block ultraviolet light and other areas that will transmit UV energy. This is true for both silver film and diazo film. With diazo film, the amber areas must block UV light while the clear portions transmit UV energy. Accurate measurements of the ability of the film areas to block or transmit UV light can be obtained easily and quickly with the use of a Macbeth Transmission Densitometer (see Fig. 9.19). To use the instrument, the two areas of film are alternately placed over the sampling area and readings are shown on a digital display. The opaque area that is to block UV light must show a reading of 4.00 or greater. The clear portion must produce a reading of 0.6 to 0.7 for diazo film and 0.02 to 0.06 for silver films to ensure effective transmission of UV energy.

The procedure for using the densitometer is as follows:

**Step 1.** The filter wheel is rotated to the UV position. This places the ultraviolet filter at the sampling aperture.

**Step 2.** The rocker switch located on the rear panel is turned on. Three figure Cs will appear on the digital display and the light in the density button will appear.

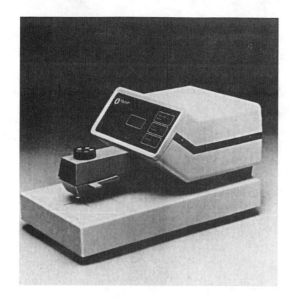

**Figure 9.19** The densitometer measures the UV filtering and transmission characteristics of phototools. *Courtesy of Macbeth, a division of Kollmorgen Corporation.*

**Step 3.** The sampling aperture arm and the zero button are pressed simultaneously, which causes 0.00 to appear on the display.

**Step 4.** The calibration film strip is placed under the sampling aperture and depressed. This should result in a reading of 3.00. If a different reading is displayed, the CAL TRANS potentiometer on the rear panel is adjusted for a reading of 3.00.

**Step 5.** The sampling aperture arm is released and the calibration film is removed.

**Step 6.** The diazo film's amber area (black for silver halide) is placed under the sampling aperture and depressed. A reading of 4.00 or greater should be displayed to indicate a properly exposed and developed film.

**Step 7.** The sampling aperture arm is released and the film is shifted to a clear area and the aperture is again depressed. For effective UV transmission, readings of 0.02 to 0.06 should be displayed for silver film and 0.6 to 0.7 for diazo film.

The densitometer is an important instrument in assuring the effectiveness of the photo tools to filter and transmit UV energy.

Final preparation of the phototools requires that the two sheets of diazo film first be cut to size. Two pieces of double-sided Scotch or masking tape are then placed on each phototool on the side that will contact the photosensitized pc board during exposure (Chapter 12). This is the side where the literal information appears backward-reading. The tape ensures that registration will be maintained between the phototool and the pc board throughout the exposure process.

With the completion of the diazo phototools, we will continue with the discussion of the sequential processes in the fabrication of both single- and double-sided pc boards. Because these processes are involved and require a great deal of detailed explanation, they will be spread over the next four chapters. Chapter 10 is devoted to the fabrication of single-sided pc boards using a simple print-and-etch technique and the negative phototools discussed in this chapter. Chapter 11 presents the drilling of the bulk stock and its preparation for copper deposition, followed by a flash-plating process. Chapter 12 discusses the conductor pattern imaging process using dry film photo resist and the diazo phototools described in this chapter. Chapter 13 completes the fabrication of a double-sided plated-through-hole pc board by considering copper pattern plating, solder plating, photo-resist stripping, etching, and infrared solder reflow.

## EXERCISES

### A. Questions

9.1 What type of camera is recommended for photoreducing artworks for printed circuits?

9.2 How should the step table be positioned on the copyboard with the artwork?

9.3 Which of the following f/stops will allow the transmission of the least amount of light: f/5.6, f/11, f/16, or f/22?

9.4 (a) What is the recommended temperature of the developing solution?
(b) If the developing solution temperature is 78°F, what is the estimated time for developing exposed film?

9.5 Explain the term *latent image*.

9.6 Briefly describe the three methods for determining proper developing time.

9.7 What is the purpose of the stop bath in the developing process?

9.8 What areas of the developed film dissolve in the fixer solution?

9.9 How does the developed film, after contact printing, compare to the artwork through which it was exposed?

9.10 Define the term *phototool*.

9.11 How are hot spots or reflections eliminated during film exposure?

9.12 What are the six processing steps for Kodak negative film?

9.13 What determines when the film should be removed from the fixer tray?

9.14 Why is diazo film chosen over silver halide film for phototools used to image double-sided pc boards?

### B. True or False

Circle *T* if the statement is true, or *F* if any part of the statement is false.

9.1 When producing a phototool, a 1:1 scaled image reversal occurs. **T F**

9.2 To obtain properly developed exposures, the step exposure table is placed over a clear area of the artwork inside the corner brackets. **T F**

9.3 Maximum contrast between light and dark areas of the artwork results when back-lighting is used on the camera. **T F**

9.4 Under safe-light conditions, the emulsion side of Kodalith film appears lighter than the triacetate backing side. **T F**

9.5 Using an f/stop setting of 16, Kodalith film is exposed for 2 to $2\frac{1}{2}$ minutes. **T F**

9.6 The emulsion around the conductor pattern dissolves when the film is placed into the fixer bath. **T F**

9.7 A stop bath is used before the fixer bath to prevent overdeveloping. **T F**

9.8 When contact printing, the film to be exposed is placed in the vacuum frame with the emulsion side facing up. **T F**

9.9 Reversal of the image does not result when contact printing with Kodalith film as it does when photoreducing on a camera. **T F**

9.10 Ultraviolet light will not pass through the areas of silver halide film but will pass through the amber-colored areas of diazo film. **T F**

### C. Multiple Choice

Circle the correct answer for each statement.

9.1 Using a copy camera for artwork master reduction, a phototool can be produced to a tolerance of (*±0.01, ±0.002*) inch.

9.2 Maximum contrast between light and dark areas of the artwork results with (*front-, back-*) lighting on the copy camera.

9.3 The recommended exposure setting for a copy camera is (*f/16 at 10 seconds, f/8 at 30 seconds*).

9.4    The emulsion side of film is placed facing (*up, down*) in the developing tray.

9.5    Developing time (*is, is not*) a function of developing solution temperature.

9.6    Neutralizing time in the stop bath is approximately (*1 second, 1 minute*).

9.7    The exposed film becomes a transparency in the (*fixer, stop*) bath.

9.8    Contact printing is used to obtain 1:1 scale (*negatives, positives*) of the marking mask.

9.9    (*Diazo, Silver halide*) film uses an emulsion that is developed in a dry ammonia gas.

9.10   Diazo film is exposed using (*ultraviolet, infrared*) sources.

## D.   Matching

Match each item in Column A to the most appropriate item in Column B.

| COLUMN A | | COLUMN B |
|---|---|---|
| 1.  Fixer solution | a. | Aperture opening |
| 2.  Contact printing | b. | 2 to 2 1/2 minutes |
| 3.  Developing process | c. | Glacial acetic acid |
| 4.  f/16 | d. | 1:1 scale reversal |
| 5.  Diazo | e. | Transparent film |
| 6.  Emulsion side of film | f. | Lighter color |
| 7.  Support backing of film | g. | Darker color |
| 8.  Stop bath | h. | Ammonia gas |

# 10

# Single-Sided PCB Processing: Print-and-Etch Technique

## LEARNING OBJECTIVES

*Upon completion of this chapter on the print-and-etch technique for processing single-sided circuit boards, the student should be able to*

- Apply photo resist to copper foil.
- Expose the photo resist.
- Develop the pc board and apply dye.
- Prepare etchant solution and etch a pc board.
- Use a wide-shank carbide-tip drill bit to drill the required holes in an etched pc board.
- Strip and completely clean an etched board.
- Process contact films for silk-screen printing.
- Be familiar with the various available types of silk screens and printing inks and paints and select the appropriate type for a given application.
- Prepare and use a silk screen for printing on pc boards.

## 10.0 INTRODUCTION

Of the many methods available for the processing of single-sided pc boards by the simple print-and-etch technique, the use of liquid photo resist (wet film) is perhaps the most economically feasible for prototype work. This technique requires inexpensive materials and equipment, thus making it ideally suited for this purpose. The topics discussed in this chapter include detailed discussions on circuit board cleaning, photosensitive resist application, etching, drilling, and the silk-screen process for preparing marking masks.

In the wet-film technique of circuit board processing, certain methods are more suitable for prototype work, whereas others lend themselves more readily to mass production applications. The selection of a specific process will depend on such factors as costs, quantity of boards to be fabricated, and labeling and solder mask requirements. The use of the various chemicals and materials for processing are included to aid in the selection of the most appropriate method. Also included is information on simple construction of some of the equipment necessary for the various phases of board fabrication.

*CAUTION:* A chemical hazard exists in the materials presented in this chapter. The processes discussed should be performed in adequately ventilated facilities and with applicable equipment having proper fume exhaust systems. Approved face shields (goggles) in compliance with ANSI Z87.1-1989 and chemical-resistant clothing and footwear and neoprene or nitrile gloves must be worn while performing any of these operations. Also, the section on safety outlined at the beginning of the book should be reviewed for more specific information.

For developing the conductor pattern onto the foil side of the copperclad pc board, a negative of the original conductor pattern artwork master is required. This negative becomes one of the basic elements used to process either one board for prototype work or hundreds of boards for mass production requirements.

The surface of the foil must first be treated so that selected areas of the copper can be protected from the etching solution, which will remove all the copper except the conductor pattern. The copper may be treated by several methods to accomplish selective foil pattern protection. The method to be considered here is *photosensitizing.* In this process, the foil is treated with a *photosensitive resist,* exposed through a 1:1 negative of the conducting pattern, and developed. The result will be a 1:1 positive conductor pattern produced on the surface of the foil, which is etchant resistant, whereas the undesired areas of the foil are unprotected and will be removed by the etching process.

Several chemicals are readily available to implement this photosensitized technique. The chemicals and procedures discussed in this section are those of the Eastman Kodak Company. Chemicals of other manufacturers may require processing and exposure times that are different from those discussed.

The process begins by cutting the necessary size copper-clad board from sheet stock. (See Chapter 5 for information on shearing pc boards.) The work must then be thoroughly cleaned before the photo resist is applied. This cleaning is essential in order to remove all contaminants, such as grease or copper-oxide film, which readily form on exposed copper. Unless these contaminants are removed, they will adversely affect the adhesion of the resist with the copper surface.

The copper is first chemically cleaned in a 5 to 10% by volume bath of hydrochloric acid and water. This bath will remove copper oxide and grease film for the improvement of photo resist adhesion. To prepare this bath, 75 milliliters of hydrochloric acid is slowly added to 925 milliliters of cold tap water in a container of Pyrex or acid-resistant plastic such as polyethylene or polypropylene. (*CAUTION:* Never add water to acid. This causes violent eruptions to occur.)

After the board is removed from the acid bath, the copper foil is then scrubbed with a 3M Scotch-Brite 96 pad while water-spray rinsing. This technique is shown in Fig. 10.1. Firm pressure should be applied with small overlapping circular motions. If the recommended cleaning pad is not readily available, any gentle abrasive household cleanser that contains no bleach is acceptable. Coarse abrasives should not be used because they would deeply scratch the soft copper foil. Proper cleaning removes contaminants without causing surface imperfections in the foil. Cleaning is sufficient when water flushed over the surface of the copper "sheets" and no water "beads" remain. Remaining grease contaminants will appear as dry spots from which the water tends to pull away. To complete the cleaning process, the foil is thoroughly rinsed with water and swabbed with a lint-free cloth to remove any remaining cleaner deposits. Once the copper is completely cleaned, all surface moisture must be removed. This is accomplished by first patting dry with a paper towel. Since moisture severely affects resist adhesion, a final drying operation is important. Total drying will be achieved if the board is placed in a forced-air oven for approximately 10 minutes at a temperature of 120°F (49°C). After the board is removed from the oven and allowed to cool to room temperature, the application of the photo resist can begin. The board must be handled carefully because fingerprints will contaminate the copper. The cleaning and drying process appears time-consuming and tedious, but it is only through a thoroughly clean and dry surface that a quality printed circuit will result.

The application of liquid photo resist is done by *spraying, dipping, roller coating, low-speed whirling,* or *flow coating.* The first four methods are common to industrial applications and result in a consistently uniform layer of resist. These methods require expensive specialized equipment and, for this reason, are not intended for prototype pc board fabrication. Flow coating requires no special equipment, and even though it does not produce a consistently uniform resist thickness, the results are acceptable.

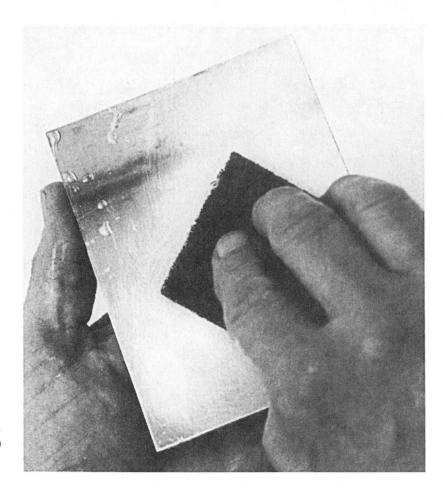

**Figure 10.1** Printed circuit board material contaminant removal is best accomplished with a Scotch-Brite pad and cold water flush.

Presensitized photo resist pc boards ready for imaging are commercially available in a variety of sizes. These are maintained in light-tight plastic bags.

Since the photo resist is sensitive to ultraviolet light, safe-light conditions are required in its use. Lighting should be provided by red safe light (Wratten 1-A or equivalent) or fluorescent tubes covered with gold or amber plastic sleeves with the ends sealed with black vinyl electrical tape to prevent light leaks.

Kodak KPR liquid resist lends itself well to flow-coating applications. It is clear in color and is applied directly to the pc board without thinning. For ease of application, a convenient amount of liquid resist may be transferred into a small bottle or other dispenser. Figure 10.2 shows the application of Kodak resist with the use of a dropper. The resist is initially applied onto the center of the pc board. The board is then rotated and tilted in a circular motion so as to allow the resist to flow outward toward all four edges, thereby covering the entire surface of the copper. With this process completed, one corner of the board is tilted and placed into the mouth of the resist container to drain off all excess resist. The board is then placed in a horizontal position to dry at room temperature in a dust-free area. In approximately 10 minutes, the board is tack-free.

To ensure successful imaging with proper resist adhesion to the copper surface, a baking cycle is required. The baking removes any residual resist solvent that may remain after being air-dried. Trapped solvent will prevent the conductor pattern image from being fully protected during the etching process.

Baking can be accomplished by either one of two methods. The first method is to place the board into an oven for 20 minutes at a temperature of 176°F (80°C) or a maximum of 10 minutes at 250°F (120°C). The second method uses a 275-watt ruby-faced in-

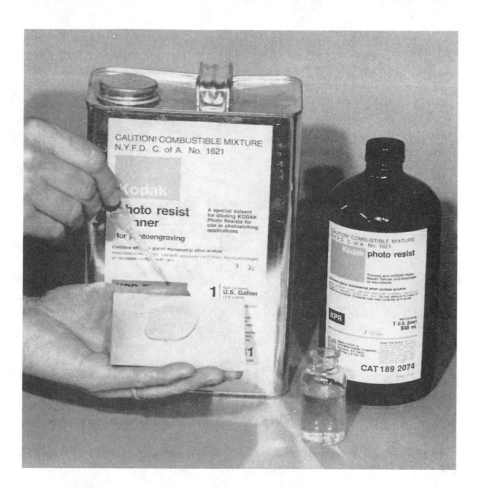

**Figure 10.2** Flow-coating a pc board.

frared heating lamp positioned 12 to 14 inches above the board. This process will require 20 to 30 minutes to eliminate resist solvent. In any type of baking cycle, extreme care must be exercised to avoid heat or light fog affecting the film. This would degrade the film and cause later processing problems. The resist surface temperature should not exceed 250°F (120°C). If using a lamp, reflectors should be opaque with no ultraviolet leaks.

After the photo resist hardens in the baking cycle, it becomes sensitive to light. For this reason, care must be exercised to protect it from exposure to any intense light when removing the board from the oven or from under the lamp.

The board should be allowed to cool to room temperature prior to initiating the imaging process. Large-diameter 16-mm film cans provide a convenient means of storing sensitized boards for short periods of time.

Kodak KPR photo resist can be exposed with a variety of sources rich in ultraviolet light, such as carbon-arc lamps, pulsed-xenon lamps, and unfiltered ultraviolet fluorescent lamps. A commercial exposure unit, shown in Fig. 10.3, uses high-pressure mercury vapor lamps and is ideally suited for photo resist imaging.

Exposure time will depend on such factors as light intensity, type of light source, distance of the source from the resist surface, and thickness of the resist layer. Optimum exposure can best be determined by trial and error.

It is essential that the light source be provided with a shield that directs the light onto the work and at the same time protects the eyes from exposure to the light rays. Light in the ultraviolet range is dangerous to the eyes and prolonged exposure must be avoided. If a light shield is not available, protective glasses should be worn. The pc board is placed onto the copyboard with the sensitized side *up* under safe-light conditions. The reduced negative of

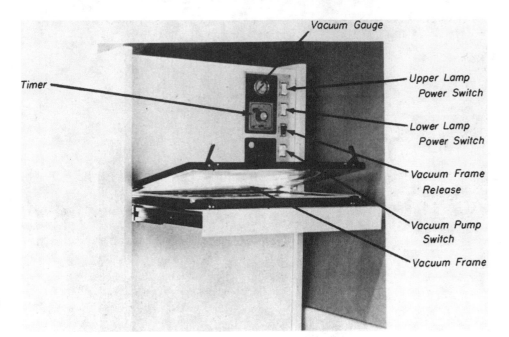

**Figure 10.3** Exposure unit for photo resist imaging.

the conductor pattern is positioned directly onto the sensitized surface. It is held in direct contact with and flat against the pc board by the weight of the glass. The border delineation marks on the negative must be aligned with the corners of the board. In addition, the negative is positioned so that all literal and numerical designations are readable. Care must be exercised to ensure that the sensitized surface will be exposed through the correct side of the negative. If an orientation error is made and the board is later etched, it will be impossible to correct the conductor pattern.

Those photo resist areas exposed to the light through the clear conductor pattern are polymerized and will become insoluble to the developer and thus remain on the copper. Those areas under the opaque portions of the negative are not exposed and remain soluble to the developer. This photo resist will wash away, leaving a photo resist mask on the surface of the copper that duplicates the desired positive conductor pattern.

The exposed board is developed with the use of a Pyrex or stainless steel tray that contains at least 1 inch of Kodak KPR developer. The board is placed into the tray with the imaged copper side facing up to prevent damage to the photosensitized emulsion. The amount of time that the board remains in the developer solution is critical. Normally, approximately 2 minutes at room temperature is sufficient to completely remove all of the resist areas that have not been polymerized by the light sources. The tray should be gently rocked for the full developing time. During the entire developing process, the remaining resist remains clear but a faint image of the conductor pattern becomes visible when the board is viewed at an angle under normal lighting conditions.

After the board is developed, it is removed from the tray, being handled only by the edges since the resist is soft and can be easily damaged. Both board surfaces are then flushed with cold tap water at high volume and low pressure for at least 30 seconds on each side.

The developed resist image is next inspected to determine if there are any imperfections, such as pinholes, bridges, or open paths. This inspection process makes use of Kodak KPR dye. The developed and rinsed board is dipped into a bath of this blue-colored dye for approximately 30 seconds. It is then removed and immediately flushed with cold tap water. This is shown in Fig. 10.4. The dye will adhere only to the photo resist remaining after the developing process.

**Figure 10.4** Conductor pattern inspection is aided with the use of dye.

The board is next dried in an oven for 10 minutes at 250°F (120°C) or under a lamp at 6 to 8 inches for 10 minutes. In no case should the resist surface temperature exceed 250°F. After the drying cycle, the board is allowed to cool to room temperature.

The board can now be inspected and any errors or pattern imperfections that may be present can be corrected by using a sharp, pointed knife. If there are any breaks in the photo resist covering of the desired conductor pattern at this point, they can be repaired by the application of *etchant ink resist* (non–light-sensitive). This resist is carefully applied with a fine artist's brush to close the break. It must overlap the photo resist on either side of the break while maintaining width. This technique is shown in Fig. 10.5. There is a number of asphaltum-based materials available from graphic arts suppliers that may be

**Figure 10.5** Application of etchant resist to repair defective pattern.

used for this purpose. Some may require dilution to a workable consistency. The solvent used should be in the turpentine family. Touch-up pens are also available to close breaks in conductor paths.

After the conductor pattern touch-up has been completed, another baking cycle is recommended to remove the solvent from the resist. A 20-minute cycle at approximately 200°F (93°C) is sufficient to remove all solvents.

The photosensitizing procedure just discussed is an effective method of preparing a pc board for etching in prototype work. A flow diagram to aid the technician in following the many steps in this process is shown in Fig. 10.6.

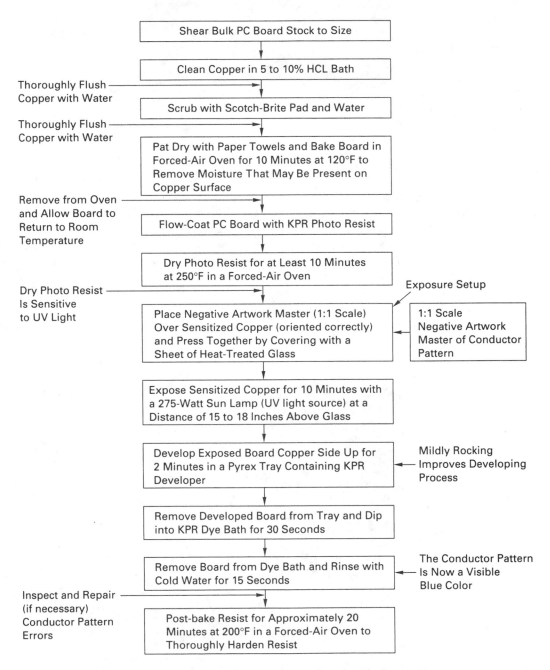

**Figure 10.6**  Flow diagram for sensitizing printed circuit boards with photo resist.

## 10.2 CIRCUIT BOARD ETCHING

*Etching* is the process of attacking and removing the unprotected copper from the pc board to yield the desired conductor pattern. Several commercial etchants and techniques are available for processing pc boards. These materials and methods attack the unprotected copper yet do not affect the adhesive, supporting laminate, and photo resist. The most common chemicals used by industry as an etchant are *ferric chloride, ammonium persulfate, chromic acid,* and *cupric chloride.* Of these, ferric chloride is commonly used since it is less expensive and potentially the least dangerous.

Methods of etching include *tray rocking, tank etching,* and *spray etching.* Tray rocking is the simplest system, consisting of a tray of etchant attached to a powered rocking table. Tank etching involves dipping the work into a vat containing the etchant, which is maintained at an optimum temperature. Spray etching, although a much faster process, calls for etchant to be pumped under pressure onto the surface of the pc board. This method requires an elaborate and expensive system. These three methods will be described using ferric chloride as an etchant.

Ferric chloride ($FeCl_3$) is an etchant that will react with the unprotected areas of the copper surface and produce copper ions, which are soluble in solution. This chemical is available premixed to the correct concentration. Ferric chloride crystals are also available if liquid storage is a problem but require premixing and concentration testing.

During the etching process, the concentration weakens because the soluble cupric and ferric ions precipitate out of solution in the form of a sludge that tends to settle on the bottom of the etching vat. When the etching time becomes excessively long, the used etchant should be discarded and replaced with a fresh solution.

Ideal etching conditions require that the etchant be heated to a temperature of between 100 and 130°F (38 to 54°C). Temperatures above this range should be avoided because of the fumes that are generated. Although etchant temperature is not critical, the activity of the etching process is increased when the solution is heated above room temperature. With the temperature maintained reasonably constant, the length of time required for etching will be consistent. However, because of the weakening of the solution, etching time will progressively increase.

In addition to heating, efficient etching requires that the etchant be continuously agitated to allow "fresh" solution to flow over the copper. A simple means of achieving this agitation is to use a Pyrex tray filled with etchant and placed on top of a *rocker table.* Rocking motion is imparted to the table by means of a small *motor, cam, connecting arm,* and off-center *pivot* arrangement. The pc board is placed copper side *up* in the tray. Only one board should be etched at one time since, because of the rocking motion, they tend to come in contact with each other. This could cause some of the etchant resist to be scraped off of a surface. As the table is rocked, a wave is generated in the etching solution that travels back and forth across the tray and board surface, keeping the major portion of any sludge formed off the copper surface. There are two disadvantages in using this arrangement. First, no method is provided for heating the etchant unless an infrared heat lamp is used, in which case etchant temperature is difficult to control. Second, some sludge is continuously carried over the surface of the board, which reduces some of the effectiveness of the etchant. However, outweighing these disadvantages is the fact that this system is relatively simple and inexpensive and suitable for prototype work. A 5- by 5-inch square of 1-ounce single- or double-sided pc board can be completely etched in approximately 30 minutes.

A more elaborate system for agitating the solution is shown in Fig. 10.7. These systems generally consist of a *vat* (Pyrex or other suitable nonmetallic material), *heating tape, variable transformer, air feed,* and *thermometer.* The temperature of the etchant contained in the vat is controlled by varying the voltage level delivered to the heating tape by the transformer. This temperature is monitored by a thermometer placed in the solution in proximity to the boards being etched. The etchant is circulated (agitated) by means of air

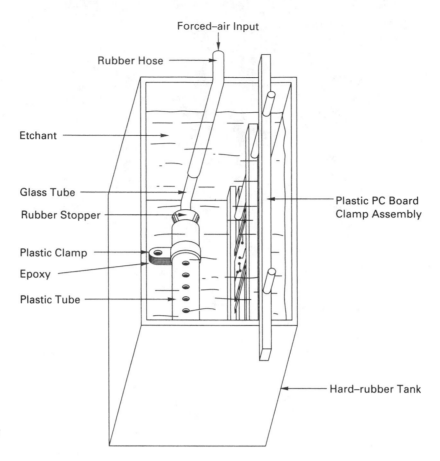

Forced–air Input

Rubber Hose

Etchant

Glass Tube

Rubber Stopper

Plastic Clamp

Epoxy

Plastic Tube

Plastic PC Board
Clamp Assembly

Hard–rubber Tank

**Figure 10.7** Forced-air etching system.

pumped into the vat under low pressure and escaping through a perforated plastic tube placed along one side of the vat. The sludge (copper and ferric ions) settles to the bottom of the vat, allowing relatively sludge-free etchant to continuously circulate across the surface of the work. The boards to be etched are suspended in the solution from noncorrosive rods and clamps with their copper side facing the source of the air bubbles. This method, although more expensive, can etch a 5- by 5-inch 1-ounce board in approximately 15 minutes.

Spray etching involves jet spraying the surface of the board with pressurized nozzles. Systems such as that shown in Fig. 10.8 also are available with a heating element to decrease etching time. With such a system, etching times of between 1 and 2 minutes can be realized without difficulty. The etching cycle is controlled by a mechanical timer. Several precautions should be pointed out. The thermostat setting should not exceed the manufacturer's recommended maximum limits to avoid damage to the equipment. In addition, spray etchers of this type may be hazardous if the cover is not properly seated prior to operation to completely contain the etchant within the system.

When the etching is completed, the board is rinsed under water for approximately 30 seconds and patted dry. This washing process will not remove the photo resist, which is allowed to remain on the board to protect the copper during drilling and punching operations. Aprons should be worn to protect clothing, and tongs used when transferring etched boards from the etchant to the initial rinse. After this washing and drying process, the boards are now prepared for drilling and punching.

**Figure 10.8** Industrial etching system. *Courtesy of Chemcut Corporation.*

## 10.3 PRINTED CIRCUIT BOARD DRILLING AND PUNCHING

For drilling printed circuit boards, a specially constructed drill bit is available. This bit, shown in Fig. 10.9, consists of a short length of twist bit with an enlarged shank of approximately $\frac{1}{8}$ inch. Printed circuit bits used for drilling through epoxy-resin boards are made of tungsten carbide. Although carbide is a brittle material, it does not dull as quickly

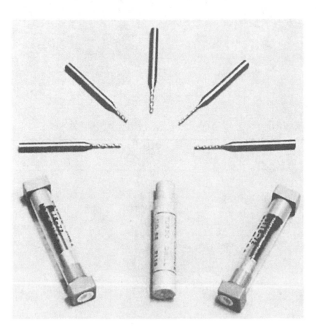

**Figure 10.9** Special carbide-tip printed circuit twist drills.

as HS drills. Carbide withstands the heat better at speeds of 15,000 to 80,000 rpm, which are typical in CNC (computer numerical control) drilling. Automated drilling is used on double-sided printed circuit boards for the plated-through-hole (PTH) process discussed in Chapter 11.

Drilling is always performed from the *copper* side of the processed board for the following reasons: (1) to use the small etched center holes in terminal pads, which aid in bit alignment (it is extremely difficult to locate hole positions for drilling from the reverse side of the board, especially if the insulating base material is opaque), and (2) to minimize the possibility of pulling the copper foil away from the base (Fig. 10.10). The maximum lead diameter for some common components and devices together with the recommended drill sizes for lead access holes are shown in Table 10.1.

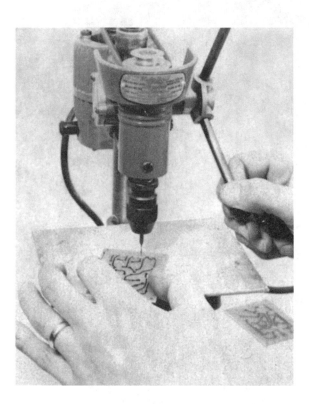

**Figure 10.10** Printed circuit board drilling with wide-shank carbide-tip drill bit.

**TABLE 10.1**

Lead Hole Drilling Reference for Printed Circuits

| Component or Device | Lead Diameter (in.) | Drill Size | Decimal Equivalent of Drill (in.) |
|---|---|---|---|
| 1/8-watt resistor | 0.016 | #75 | 0.0210 |
| 1/4-watt resistor | 0.019 | #72 | 0.0250 |
| 1/2-watt resistor | 0.027 | #66 | 0.0330 |
| 1-watt resistor | 0.041 | #64 | 0.0469 |
| 2-watt resistor | 0.045 | #55 | 0.0520 |
| Disk capacitor | 0.030 | #64 | 0.0360 |
| TO-5 case style | 0.019 | #72 | 0.0250 |
| TO-18 case style | 0.019 | #72 | 0.0250 |
| DO-14 case style | 0.022 | #70 | 0.0280 |
| 77-02 plastic power transistor | 0.026 | #67 | 0.0320 |
| TO-99 (8-pin IC) | 0.019 | #72 | 0.0250 |
| TO-116 (14-pin DIP) | 0.023 | #69 | 0.0292 |

Punching should also be performed from the copper side for the same reasons just mentioned. If necessary, the grade of board used may require elevated temperatures for punching (see Sec. 5.2). Where necessary, an oven should be used for heating just prior to the punching operation. The punch and die should be sharp for best results. Before additional processing can be undertaken, the board must be completely cleaned of all photo resist, dye, and etchant impurities.

## 10.4 CONDUCTOR PATTERN SURFACE CLEANING

To prepare the etched copper surface for additional processes, the photo resist and the dye are removed by dipping the board in a bath of Kepro PRSK photo resist stripper for 1 minute. The board is then removed from the bath and scrubbed with a 3M Scotch-Brite 96 pad while water-spray rinsing. This scrubbing is essential to remove any photo resist residue. If not removed, this residue can cause problems in soldering leads to terminal pads. The board is given a final water rinse for about 30 seconds.

Etching in a ferric chloride solution produces contaminants on the copper foil, which will adversely affect soldering or plating. These salts of ferric chloride require an additional cleaning process. The following procedure for their removal is recommended. The contaminating salts remaining on the copper are made water soluble by dipping and mildly agitating in a bath of Lonco Copperbrite No. 48 HT. A final water rinse for 1 minute will remove all residues from the copper surface. After this final rinse, the board is wiped dry with a lint-free cloth.

## 10.5 SILK-SCREEN PRINTING

A relatively inexpensive method of reproducing identical pc boards is by the silk-screen printing technique. This process utilizes ink resist that is applied through the stencil of the desired pattern that has been affixed to a fine-mesh silk screen. (Nylon or wire mesh can also be used.) Although the silk-screen method is extensively used for processing pc conductor patterns, the illustrative descriptions provided will involve preparing marking and solder masks. The end result will be stencils that will yield acceptable tolerances of line definition. Artwork for silk-screen processing can be generated by CAD techniques as discussed in Chapter 7.

The necessary materials to produce a silk-screen stencil are a silk or nylon screen mounted on a printing frame with hinged clamps, stencil film, film developer blackout, rubber squeegee, paint or solder resist, and the artwork master processed into a 1:1 scale positive or negative transparency. These materials and the procedures for producing and using a silk screen for the marking mask shown in Fig. 10.11 will be discussed as an illustrative example.

The contact films used in the development of silk-screen stencils are normally extremely slow-speed types with respect to exposure time and, as a result, do not require darkroom facilities. The film selected for this example is *Ulano Super Prep*. Even though the film speed is slow, it is recommended that this film be used in subdued light avoiding direct sunlight or high-output fluorescent lamps. The 1:1 scale positive transparency, as shown in Fig. 10.11, is used to implement this film.

A piece of plain white paper, larger than the artwork, is first placed flat to serve as a background. The Super Prep film, allowing at least a 1-inch border on all sides, is placed on top of the white sheet with the sensitized (emulsion) side of the film positioned *against* the white paper. (The sensitized side of the film is determined by observing the *dull* finish as opposed to the glossy side of the polyester backing.) The exposure unit shown in Fig. 10.3 will be used for this example. The 1:1 scale positive marking mask transparency is then placed on *top* of the film and held firmly in place with the glass printing frame. Exposure times will vary from 5 to 15 minutes, depending on such factors as film type and light source. Optimum exposure times can be experimentally determined

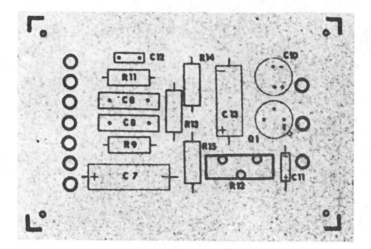

**Figure 10.11** Positive 1:1 scale marking mask.

and, once established, should not vary appreciably. Those areas of the film emulsion exposed to the light source will become insoluble in warm water after the developer bath. Those areas protected from the light by the opaque portions of the positive will dissolve in warm water after the developer bath. After exposure, the film is placed in a tray of developer with the emulsion side *up* for between 1 and 2 minutes. One end of the tray should be slowly raised and lowered to allow the solution to agitate gently over the surface of the film.

The developer for Ulano Super Prep film is available in powder form and packaged in premeasured packets labeled *A* and *B*. To prepare the developing solution, one packet of each of the powders is dissolved together in 16 ounces of water at a temperature of approximately 65 to 70°F (18 to 21°C). This developer is light sensitive and should be prepared in subdued light and stored in an amber-colored bottle. The solution has an effective lifetime of approximately 24 hours, after which time it begins to deteriorate. One indication of deterioration is the appearance of wrinkles on the polyester support of the film during the developing process. When this occurs, the developer should be discarded and replaced with a fresh mixture.

After the film has been in the developer for the necessary amount of time, it is removed and immediately placed in a tray of water at a temperature of approximately 100°F (38°C). The emulsion side of the film is again facing upward to prevent any smearing of the softened emulsion from the bottom of the tray. Gentle rocking of the water bath will remove the soluble portions of the emulsion (those areas protected during exposure). Approximately 1 minute is all the time required for the water bath to process the film. The film being processed in the water is shown in Fig. 10.12.

The film is next removed from the warm-water bath and rinsed with cold water for about 30 seconds to remove any remaining emulsion from those areas of the negative film that must be completely transparent. (*CAUTION:* Extreme care must be exercised when handling the film after it is taken from the warm-water bath since the remaining emulsion is soft and tacky and is easily smeared. The cold-water rinse is best applied by placing the film on a sheet of glass with the polyester support backing against the glass. The emulsion side can then be gently rinsed under a cold-water tap.)

The silk-screen negative stencil of the original artwork master can now be obtained by firmly pressing the soft emulsion of the film against the surface of the silk screen. After the entire emulsion side has been pressed against the silk screen, both sides of the screen are gently blotted with soft absorbent paper towels to remove any excess water (Fig. 10.13). A block-out material, such as *Ulano No. 60 Water Soluble Fill-in,* is then

**Figure 10.12** Warm water used to process film after developing.

spread with a small brush over the unused areas of the screen bounding the stencil. The block-out confines all excess paint or resist to the upper surface of the screen during printing (Fig. 10.14).

The screen is then allowed to dry thoroughly. If a small fan is available, drying time will require only approximately 30 minutes. After the screen has dried, the transparent backing of the film is peeled from the screen, which is then carefully inspected by tilting it back and forth while holding it up to an intense light. No residual film should remain on those areas of the screen that are not blocked and appear as translucent. This residual film will appear to glisten and, if not removed, will impair smooth transfer of

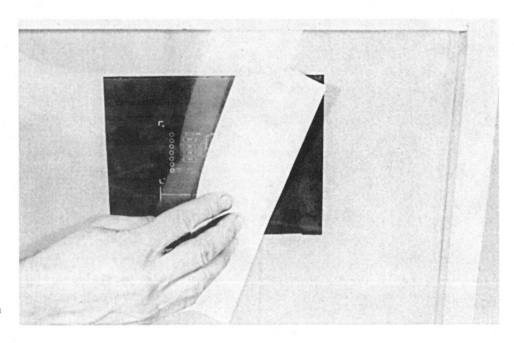

**Figure 10.13** Removing excess moisture from both sides of a silk screen.

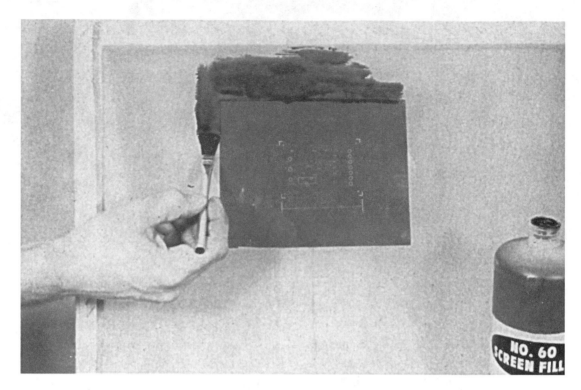

**Figure 10.14**   Block-out applied to unused areas of screen.

the paint or resist through the screen stencil during its application. This film is removed with standard mineral-oil spirits and a stiff brush. The emulsions forming the stencil or the block-out are resistant to these solvents and will not be affected by this cleaning process. The use of water, however, should be avoided because these emulsions are water soluble.

The stencil just developed is now ready for use. Before proceeding, however, some consideration needs to be given to the type of screen to select for a particular application. In this section, the concern centers around screens for marking and solder masks. Therefore, a moderate degree of control must be maintained on positioning paint or resist flow through the screen in order to effect acceptable line definition.

Screens are classified in two categories, *mesh number* and *durability.* Mesh numbers for silk screens, generally running from No. 1 to No. 25, refer to the number of filaments per inch. A No. 1 mesh screen has approximately 30 filaments per inch and is considered very coarse. Mesh No. 25 represents a fine screen of approximately 196 filaments per inch. The more filaments per inch, the greater the control of paint or resist flow through the screen and consequently the sharper the resulting pattern. For extremely critical line definition, *monofilament* nylon mesh screens with as many as 283 filaments per inch are also available. For applying epoxy paint for marking masks and solder resist for solder masks, a screen with a mesh number of 20 will produce adequate results. This mesh is also suitable for conductor pattern stencils.

The durability of the screen is a measure of its ability to withstand continued reuse. Durability classifications are *X, XX,* and *XXX,* with the latter used only if many reproductions are to be made. For prototype work, screens with a durability classification of *X* are adequate. However, if the screen is to be continually stripped clean and reprocessed for other stencils, *XX, XXX,* or wire mesh screens should be considered.

To facilitate the silk-screen labeling process, it is first necessary to provide a base and a set of clamp hinges to accommodate the frame and also to provide positive positioning of

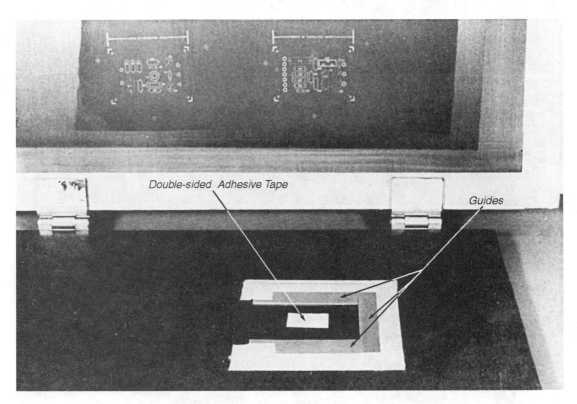

**Figure 10.15** Register guides to aid in board alignment.

the surface to be labeled. These aids are shown in Fig. 10.15. The frame is clamped to the printing base by means of the clamp hinges in such a manner that the stencil is closest to the base as the frame is swung down into the printing position. With the frame properly clamped, the pc board is placed on the printing base under the frame and its corners are aligned with the border delineation marks on the stencil. Once the desired alignment is achieved, three small *register guides* cut from either cardboard or metal, no thicker than the board, are positioned against any three board edges. This procedure is also shown in Fig. 10.15. The register guides are secured in place with masking tape and help hold the work in position during printing. They also allow for rapid positioning of identical workpieces when many boards are to be processed.

Once the work and stencil are properly aligned, a liberal amount of epoxy-type paint is poured across the left-hand margin of the marking mask stencil from top to bottom. A hard-rubber squeegee, preferably one that will span the entire stencil, is brought down onto the screen behind the pool of paint. With a sweeping motion, the squeegee is drawn across the stencil, spreading the paint as it moves. This technique is shown in Fig. 10.16a. Familiarity with this technique will enable the technician to completely label the entire work surface with a single sweeping motion.

After the stenciling is completed, the screen is raised from the work (Fig. 10.16b). Small boards may tend to cling to the stencil when it is raised. Usually, they will slide slightly before they fall, thus causing the labeling to smear. This problem can be avoided by using double-sided adhesive tape placed on the underside of the work when it is first positioned on the base.

When labeling has been completed, the paint must be allowed to dry thoroughly before handling. Depending on the type of paint used, drying time will vary from 2 hours to a complete curing time of 4 days. Drying time can be reduced by baking according to

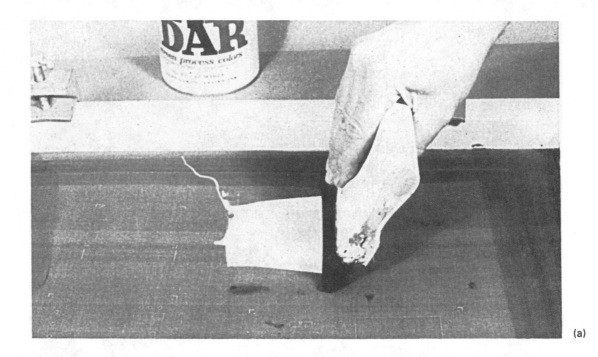

(a)

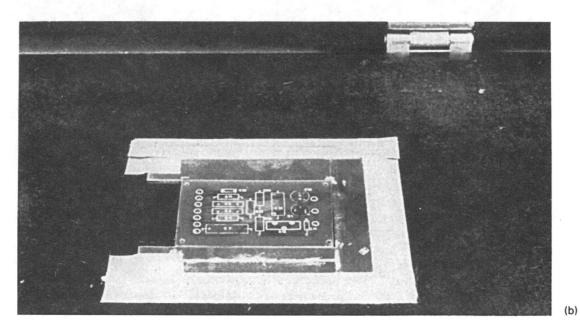

(b)

**Figure 10.16** Use of silk screen for board marking: (a) a single stroke of the squeegee labels the board; (b) results of the screening process.

paint manufacturers' specifications. Certain lacquers and vinyls used in the silk-screen process for labeling will air dry in 5 to 15 minutes. No difficulty will be encountered in obtaining any desired labeling characteristics in terms of color, durability, or finish because there are hundreds of different types of paints, enamels, and lacquers available from graphical suppliers.

When the labeling process has been completed, the excess paint can be removed from the margins of the stencil with a small spatula and stored. The remaining paint is then washed off with an appropriate solvent. If the existing stencil is no longer to be used, the screen can be reclaimed for further stencil use by washing it in warm water. This will dissolve the water-soluble emulsion and blockout. Difficult areas can be gently scrubbed with a soft-bristle brush.

The discussions above have been limited to silk-screen printing for producing marking masks. This process is also used for processing solder masks. For this purpose, the procedures for developing the stencil are identical except that a 1:1 *negative* is used initially to generate the silk-screen film. Also, instead of using an epoxy paint, a solder resist, such as *Lambert No. 184-10-V green solder resist,* is applied through the screen.

Silk-screen stencils and finished boards with solder masks applied are shown in Fig. 10.17. Because of the complexity of the silk-screen process, a flow diagram is provided in Fig. 10.18 to help the technician follow the various steps.

After the pc boards have been processed, the components and hardware must be mounted and leads soldered securely to the conductor pattern. The procedures and equipment necessary for this phase of construction are discussed in the succeeding chapters.

**Figure 10.17**   Solder masks with boards screened.

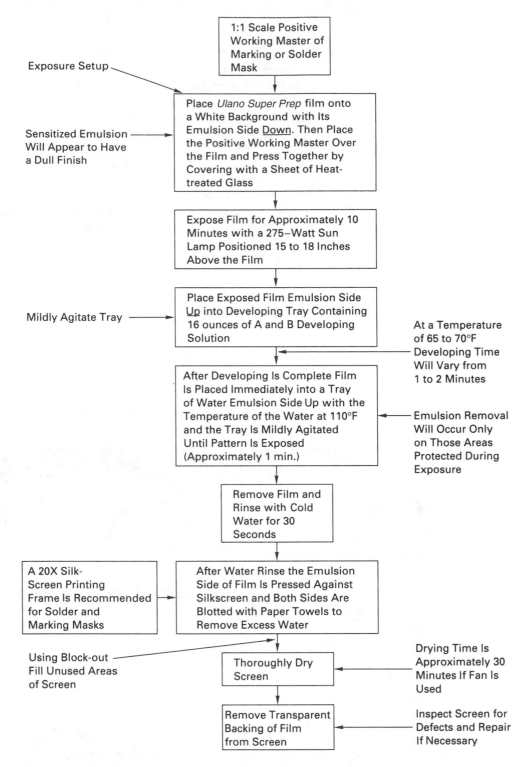

**Figure 10.18**  Flow diagram for developing silk-screen stencils.

## EXERCISES

### A. Questions

10.1      What is the purpose of photo resist?

10.2      How should pc board copper surfaces be cleaned prior to the application of photo resist?

10.3      List the various methods of applying photo resist.

10.4      Define the term *etching*.

10.5      What would be the result if a positive phototool was used to contact print the photosensitized board instead of a negative phototool?

10.6      What is the correct orientation for the negative when contact printing the photosensitized board?

10.7      Define the term *polymerize*.

10.8      What precaution must be observed when developing exposed photo resist?

10.9      After the application of liquid photo resist, why are the pc boards dried in an oven?

10.10     Determine the required drill sizes for $\frac{1}{4}$-watt resistors, disk capacitors, and DO-14 case leads. Which component requires the largest hole size?

### B. True or False

Circle *T* if the statement is true, or *F* if any part of the statement is false.

| | | | |
|---|---|---|---|
| 10.1 | Etching efficiency increases with agitation and decreased temperature. | **T** | **F** |
| 10.2 | Before drilling a pc board, each terminal pad is center punched. | **T** | **F** |
| 10.3 | The flow coating of photo resist requires no special equipment. | **T** | **F** |
| 10.4 | Ferric chloride is commonly used as an etchant. | **T** | **F** |
| 10.5 | Drilling of a pc board is performed from the opposite side of the copper conductor pattern. | **T** | **F** |
| 10.6 | The contact films used in the development of silk-screen stencils do not require that they be used under safe-light conditions. | **T** | **F** |
| 10.7 | Some grades of pc board require elevated temperatures prior to punching. | **T** | **F** |
| 10.8 | Etchant resist is removed by scrubbing the board with a Scotch-Brite pad. | **T** | **F** |
| 10.9 | Developer solution for Ulano Super Prep film is light-sensitive. | **T** | **F** |
| 10.10 | A silk screen with a mesh number of 2 is ideally suited for conductor pattern stencils. | **T** | **F** |

### C. Multiple Choice

Circle the correct answer for each statement.

10.1      Light-sensitive photo resist is exposed with a(n) (*black, infrared*) light.

10.2      When preparing a bath of hydrochloric acid cleaner the (*water, acid*) is slowly added to the (*water, acid*).

10.3      The drilling of a processed pc board is performed from the (*insulation, copper*) side.

10.4      Kodak KPR liquid resist is (*clear, amber*) in color.

10.5      To expose photo resist, the negative is positioned so that literal and numerical designations are (*forward, reverse*) reading.

10.6      An exposed board to be developed is placed into the developing tray with the imaged copper side facing (*up, down*).

10.7      After the touch-up process has been completed, the board is baked to remove (*gases, solvents*) from the resist.

## D. Matching

Match each item in Column A to the most appropriate item in Column B.

| COLUMN A | COLUMN B |
|---|---|
| 1. Copperbrite | a. Clean copper foil |
| 2. Sun lamp | b. Rubber squeegee |
| 3. Water "sheeting" | c. 2 minutes |
| 4. Etchant | d. Contamination remover |
| 5. Photo resist developing | e. 3600 angstroms |
| 6. Silk screening | f. 30 filaments/inch |
| 7. Mesh number | g. Ferric chloride |

## E. Problems

10.1 Prepare silk screens for the marking and solder mask layouts designed for the tachometer circuit of Problem 6.3. (The screens are used in Problem 10.3.) Obtain the meter movement specified for the circuit. Using a commercially available tachometer as a calibration reference, redesign the meter scale and construct a silk screen for the new scale.

10.2 Construct a pc board from the artwork master for the series feedback regulator circuit designed in Problem 6.2 from $\frac{1}{16}$-inch FR-4 board with 1-ounce copper foil. Components are assembled and soldered in Problems 16.2 and 17.2, respectively.

10.3 Fabricate a pc board from the artwork master for the tachometer circuit designed in Problem 6.3 from $\frac{1}{16}$-inch *XXXP* board with 1-ounce copper foil. Using the silk screens fabricated in Problem 10.1, construct the meter scale, mark the pc board, and mask the conductor pattern for soldering. Components are assembled to the pc board in Problem 16.3.

# 11

# Double-Sided PCB Processing: Electroless Copper Deposition and Copper Flash Plating

## LEARNING OBJECTIVES

*Upon completion of this chapter on double-sided printed circuit board copper deposition and flash plating, the student should be able to*

- Stack and shear pc boards to their correct size.
- Understand the major components and performance characteristics of production CNC drill/router machines.
- Understand the equipment and processes involved in an electroless copper deposition line.
- Leach all tanks and accessories in an electroless copper deposition line.
- Properly mix, maintain, and control all the solutions of the electroless copper deposition line.
- Successfully operate an electroless copper deposition line.
- Understand the equipment and processes involved in copper flash plating.
- Properly mix, maintain, and control the solution of the flash plating bath.
- Flash-plate copper onto pc boards.

## 11.0 INTRODUCTION

In Chapter 7, a number of design requirements for the generation of artwork suitable for double-sided pc boards was presented. Recall that major emphasis was placed on electrically interconnecting circuitry on the component side of the board to that on the circuit side. For prototype applications, where only a few boards of any specific type are processed, this side-to-side interconnection may be accomplished by first wave-soldering the processed board on the circuit side and then hand-soldering the leads on the component side to its circuitry. Although this method is time-consuming, it is adequate for obtaining the desired results.

Volume production manufacturing requires that side-to-side interconnections be made an integral part of the pc board in the fabrication process. This interconnection technique, adopted by high-volume pc board manufacturers, is the *plated-through-hole* (PTH) process, which lends itself to mass-production applications and results in a much higher degree of finished board control. Because it is an extensive multistepped process, the discussion of double-sided plated-through-hole pc board fabrication will extend into the following two chapters. The following chapter descriptions provide an overview of this complex process.

This chapter begins with a detailed discussion of shearing pc board stock to size and then drilling all of the required holes. Through a series of chemical baths, a very thin layer of copper is then deposited electrolessly over the entire board surface covering all exposed areas, including the edges and the walls of the drilled holes. This layer of copper results in the initial electrical interconnection between the copper foil on both sides of the board through the plating of the walls of each drilled hole. To increase the thickness of the copper deposited on the walls of the holes, the board is further treated in a *copper flash plating* process, which is a standard electrolytic method employing a copper plating bath and electric current.

In Chapter 12 the fabrication process continues with the application of dry-film photo resist to image the copper foil on both sides of the board with the desired conductor patterns. (Double-sided conductor pattern artwork is presented in Chapter 7.) The board is completed in Chapter 13 by first additionally electroplating the exposed conductor pattern and hole walls with copper followed by a solder electroplating process. This solder plate serves as the etching resist and ensures a highly solderable pc board. After all of the plating is completed, the dry-film resist is stripped from the board followed by etching, reflowing, and finally routing to size.

As can be seen from this introduction, the fabrication of a double-sided plated-through-hole pc board requires many sequential processes, some of which demand exacting procedures. For this reason, each of the processes are described in detail.

*CAUTION*: A chemical hazard exists in the materials presented in this chapter. The processes discussed should be performed in adequately ventilated facilities and with applicable equipment having proper fume exhaust systems. Approved face shields (goggles) in compliance with ANSI Z87.1-1989 and chemical-resistant clothing and footwear and neoprene or nitrile gloves must be worn while performing any of these operations. Also, the section on safety outlined at the beginning of the book should be reviewed for more specific information.

## 11.1 ELECTROLESS COPPER DEPOSITION: GENERAL INFORMATION

The discussion of double-sided plated-through-hole fabrication begins with the selection of the appropriate pc board stock. For our purposes, we have selected grade FR-4 double-sided board stock having a thickness of 0.059 inch and a copper thickness of 1 ounce per square foot ($oz/ft^2$) on each side. This thickness is most often specified, although thinner copper thicknesses are becoming more popular for fine-line applications (i.e., less than 10-mil path widths and spaces). The stock is first sheared into the correct panel size to accommodate the required conductor pattern artwork (see Chapter 7).

Two precisely located tooling holes are then punched into each panel as well as into a layer of backup material and entry material. All these layers are then stacked and pinned together for drilling. A pinned lay-up of panels for drilling is shown in Fig. 11.1. The bottom backup panel is a piece of $\frac{1}{8}$-inch XP hardboard material. Three double-sided copper clad pc boards are shown stacked on top of the bottom panel. The top drill-entry material is a thin sheet of aluminum. The purposes of the entry and backup panels are to (1) prevent burring of the drilled holes, (2) clean and cool the drill bit to prevent tearing hole walls, and (3) protect the underlying surface of the drilling machine worktable. The layers are held firmly together with steel stacking pins especially made for this purpose. The pins are pressed into the previously drilled tooling holes.

A typical multispindle CNC (computer numerical control) drilling machine, used to automatically drill separate stacks of panels simultaneously, is shown in Fig. 11.2. These machines are also used to route individual pc boards to size and shape. The major components of a CNC drill/router machine are (1) computer controller; (2) granite slab base; (3) moving worktable, servomotor drive, and feedback system; and (4) directly programmable 15,000- to 80,000-rpm spindles. The massive base is required to resist the immense inertial forces of the rapidly moving worktable as well as to support the entire machine structure.

Typical drilling performance characteristics of CNC drilling machines are (1) $x$ and $y$ table positional speeds of up to 500 inches per minute, (2) positioning accuracies of $\pm 0.0002$ inch with repeatability of $\pm 0.0001$ inch, (3) drill hole accuracy of $\pm 0.001$ inch, (4) production speeds of 400 "hits" (drilled holes) per minute, and (5) automatic drill size changes in

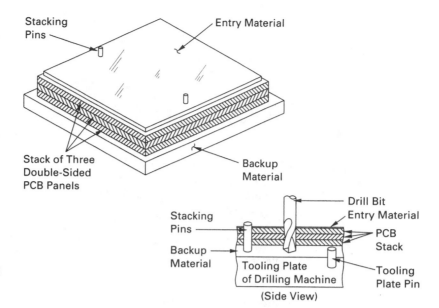

**Figure 11.1** Stacking and pinning setup used on CNC machines for multipanel drilling.

**Figure 11.2** Automatic driller/router machine. *Courtesy of Excellon Automation, Division of Excellon Industries, an Esterline Company.*

10 seconds. The CNC machine is programmed to drill multiboard panels in a step-and-repeat mode and to take optimum paths between holes to further reduce drilling time.

After the boards have been drilled, they are unstacked and mechanically deburred on both sides using 400-grit sandpaper. These boards are now ready for the *electroless copper deposition* process. This process will deposit a layer of copper approximately 20 to 30 millionths of an inch ($\mu$inches) thick on all board surfaces, that is, the copper foil on both sides, the board edges, and most important, around the inner glass/epoxy surface of each drilled hole. To accomplish this, the drilled boards are first secured in specially designed racks and then processed through 16 chemical baths and water rinses. In this plating line, the chemical makeup of baths, the solution temperatures, the immersion time rate of board agitation, the air agitation of solution, and the quality of water rinses are rigidly controlled. After the

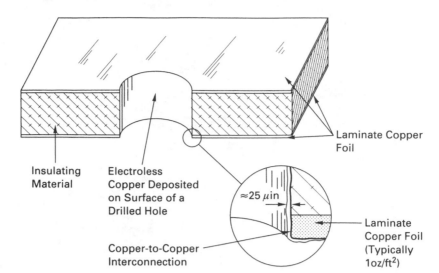

**Figure 11.3** Cross-sectional view of a double-sided plated-through-hole board after copper deposition.

Laminate Copper Foil

Insulating Material

Electroless Copper Deposited on Surface of a Drilled Hole

≈25 $\mu$in

Copper-to-Copper Interconnection

Laminate Copper Foil (Typically 1oz/ft$^2$)

boards have been processed through the 16-bath copper deposition line, they will exhibit a slightly brownish copper appearance throughout, that is, on the copper foil surfaces, the edges, and walls of the drilled holes. A cross-sectional view of a board after it has been processed with the copper deposition baths is shown in Fig. 11.3.

## 11.2 ELECTROLESS COPPER DEPOSITION LINE: MECHANICAL CONSIDERATIONS

A typical electroless copper deposition line is shown in Fig. 11.4a. It consists of 16 stations, or bays, all of which are the same size. In the line shown, each bay is designed to accommodate a 10-gallon (37.85-liter) bath. Thus their physical dimensions are approximately 6 inches wide, 24 inches deep, and 24 inches high. The bays are constructed of a welded rigid plastic material such as stress-relieved polypropylene with PVC plumbing. Each bay is fitted with a PVC ballcock drain valve leading into a central drain manifold to facilitate solution removal. A high-torque low-rpm motor is mounted to the left side of the deposition line. The motor is coupled through a cam and a 1-inch-square fiberglass drive bar to a pair of 1-inch-square rails. These rails are positioned on Teflon guides and run the length of the line along the top outside edges. The racked boards are lowered into each of the baths with the arms of the rack resting on the rails. A typical rack is shown in Fig. 11.4b. The motor-driven rails provide a transverse motion of the racked boards, which forces the solution through the holes as the boards travel back and forth in the baths. The rate of travel should be relatively slow, not to exceed 12 complete cycles per minute with a 2- to 4-inch stroke cycle.

The deposition line can be divided into two major bay groupings: (1) water rinse stations and (2) chemical treatment stations. There are eight of each of these groupings in the line. Refer to the deposition line shown in Fig. 11.4a. The first bay on the left is the only one required to be operated above room temperature. It is fitted with a 1000-watt stainless steel heater that is sufficient to raise the temperature of 10 gallons of solution to the required 150°F (66°C) in a reasonable amount of time. A temperature control with a probe is used to maintain this temperature and prevent overheating of the solution. Bay 14 in Fig. 11.4a is the electroless copper bath and is fitted with a length of PVC tubing connected to an aquarium pump. The tubing extends to the bottom of the bay. The bubbles generated from the pump aid in stabilizing the electroless copper solution.

Each of the rinse stations is equipped with an overflow drain near the top rim of the bay. This is shown in Fig. 11.4a. Rinse water is supplied from a mixing valve that regulates the temperature to a central PVC manifold water system that feeds each rinse station. From this system, a $\frac{1}{2}$-inch PVC water pipe is routed through the rear wall of

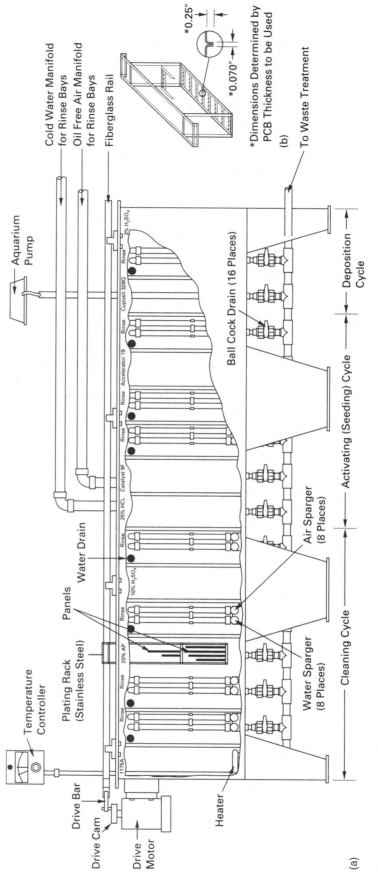

**Figure 11.4** Standard 16-bay electroless copper deposition line: (a) detailed front-sectional view; (b) 316 stainless steel rack for copper deposition.

each rinse bay, down the back of the tank, and through the bottom center of the tank. Several small, equally spaced holes are drilled into the bottom section of the pipe. In this way, clean water enters continuously from the bottom of the tank and moves upward, finally leaving the bay through the drain at the top. This results in contaminants, which are generated from rinsing the racked boards, being forced upward and out into the drain. For optimum results, the rinse-water temperature should be maintained between 60 and 70°F (16 and 21°C). Sufficient volume of water to have a bath turnover of from 5 to 10 times the capacity of the tank per hour should be supplied. This will maintain the level of rinse-water contamination, even for a constantly working line, within acceptable limits. For the relatively small deposition line shown in Fig. 11.4a, which has eight 10-gallon rinse bays, a turnover of 10 times per hour for each bay would require a water supply of 800 gallons per hour.

To reduce the large volume of water required and to result in more effective board rinsing than that provided by rack movement alone, a PVC air sparger is installed in each of the rinse stations. These 1/2-inch-diameter pipes are L-shaped and enter the top of the bay through a hole at the rear of the tanks. They run downward and along the bottom of the tanks. Each sparger has equally spaced holes drilled along the bottom of its total horizontal length. This pipe is returned to a central manifold, which in turn is connected to an oil- and dust-free blower system equipped with airflow control valves. Air entering each rinse bay at the rate of 1 1/2 to 2 cubic feet per minute for each square foot of tank surface will result in greatly increasing the solution movement. This provides more effective rinsing with minimum water input required.

The rack used with the copper deposition line is made from 316-type stainless steel and is designed so that it will rinse freely and entrap no chemicals. The rack shown in Fig. 11.4b has a two-tier arrangement of slotted spacers. The slots are approximately 1/2 inch apart, 1/4 inch deep, and 20 thousandths of an inch wider than the thickness of the pc boards to be processed. Boards mounted onto the rack are therefore slightly off vertical, which results in good solution-to-hole movement and more effective rinsing. The rack is fitted with front and back extensions so that it may be positioned onto the work rod agitator.

In this section we have described the mechanical aspects of a typical low-production, hand-operated deposition line common to many printed circuit fabrication shops. To aid in efficiency, several improvements can be made to this basic line. For example, the addition of a hoist to raise and lower the rack into the baths would increase productivity. Also, the installation of conductivity meters in the rinse stations to monitor the level of contamination would reduce production time by sounding an alarm when levels of contamination exceed preset limits.

In the next section we describe the procedures for starting up a new deposition line and also the chemical baths involved in the copper deposition process.

## 11.3 ELECTROLESS COPPER DEPOSITION: CHEMICAL PROCESSES

After the construction of a new copper deposition line, all bays, racks, and hardware must be thoroughly *leached* (chemically cleaned) to remove all contaminants prior to filling any of the tanks with chemical baths or rinse water. All heaters, plumbing, and spargers should be installed so that they may be leached at the same time. The leaching process is required only for startup purposes, that is, before the chemicals are added to the tanks for the first time. A detailed leaching process is given in Appendix IX. The student should refer to this appendix before proceeding to the next section. It should be emphasized that even though a new tank may look clean, it is probably the single largest source of bath contaminations. Those contaminations, such as mold release film, are typically the cause of many copper deposition failures. For this reason it is absolutely essential that all parts that will come into contact with any of the chemical baths be properly leached.

The basic 16-bay deposition line as described in Sec. 11.2. and shown in Fig. 11.4a can be divided into three major categories, described by the functions of each bath. Viewing the line from left to right in Fig. 11.4a, these categories are (1) bays 1 through 7 for *cleaning*, (2) bays 8 through 13 for *activating*, and (3) bays 14 through 16 for *copper deposition* and *holding* (see Fig.11.5). During the cleaning cycle, the copper surfaces of the pc board as well as the hole walls and board edges are cleaned and conditioned to accept activation and subsequent copper deposition. The *activation* process "seeds" (deposits) particles of precious metal (palladium) onto the surfaces of the nonconductive hole walls in order to provide a site for subsequent copper deposition. In the *copper deposition* process, these sites are deposited with 20 to 30 millionths of an inch (20 to 30 μinches) of copper resulting in a completely metalized board capable of being *flash* electroplated with copper (Sec. 11.5) to a thickness of approximately 0.000125 inch (125 μinches).

To describe the complete deposition process in detail, a standard *low-speed electroless copper process* developed by the Shipley Company, Inc. is shown in flowchart form in Fig. 11.5 and will be discussed in detail. This is called the Cuposit 328Q process and will be described using the 16-bay deposition line discussed in Sec. 11.2 and shown in Fig. 11.4a. Recall that each of the bays has a capacity of 10 gallons.

As shown in Fig. 11.5, the boards are first racked in accordance with the discussion of Sec. 11.2. Remember that during this entire process, the rails supporting the plating rack are driven back and forth to ensure the proper movement of solution through the holes in the board.

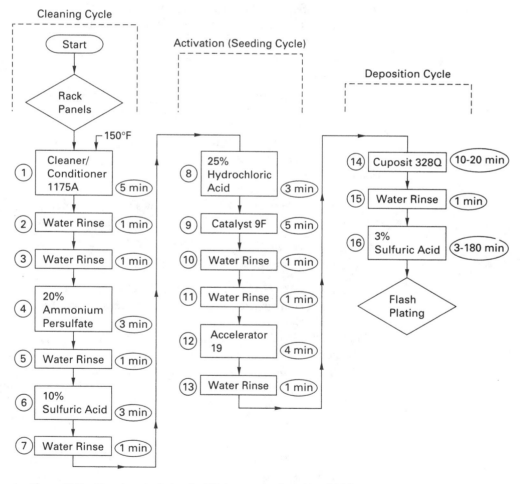

**Figure 11.5** Flowchart depicting the Shipley copper deposition 328Q process.

*Double-Sided PCB Processing: Electroless Copper Deposition and Copper Flash Plating*   **199**

Each of the baths shown in Fig. 11.5 will now be described.

***Bay 1: Cleaner/Conditioner 1175A.*** This is a strongly alkaline bath designed to (1) clean oils, stains, and oxides from the copper surfaces, and (2) condition the glass/cloth hole wall surface to accept the activation (seeding) layer. This 2.5% by volume bath is prepared by adding 0.25 gallon (946 milliliters) of 1175A concentrate to 9.75 gallons (36.9 liters) of tap water to result in 10 gallons of working solution. With the bath maintained at an operating temperature of 150°F to a maximum of 180°F (66 to 82°C), the racked boards are loaded into the solution for 3 to 5 minutes.

***Bay 2: Water Rinse.*** The rack is removed from bay 1, allowed to drain, and then rinsed in this room-temperature ($\approx 70°F$) water bath for 1 minute. Both air and water spargers are used in this and subsequent rinse baths to prevent chemical drag-out from one bath to the next.

***Bay 3: Water Rinse.*** The rack is removed from bay 2, allowed to drain, and then again rinsed for 1 minute in room-temperature water.

***Bay 4: 20% Ammonium Persulfate.*** To ensure a good copper-to-copper bond (electroless to foil), a light etch is required on the copper foil to coarsen the surface and also to remove any contamination. This *micro-etch* bath is prepared by adding 15 pounds of ammonium persulfate crystals to 10 gallons of tap water. (This is a ratio of 1 1/2 pounds of AP per gallon of water.) The solution is stirred with a leached paddle until all the crystals are dissolved. Finally, 10 ounces (300 milliliters) of reagent or chemically pure (CP) grade sulfuric acid ($H_2SO_4$) is added to complete the working solution.

The rack is removed from bay 3, allowed to drain, and then placed into the ammonium persulfate micro-etch bath, which is maintained at room temperature, for 3 minutes. After this process, the boards should exhibit a uniform mat salmon color over all copper surfaces.

***Bay 5: Water Rinse.*** The rack is removed from bay 4, allowed to drain, and rinsed in room-temperature water for 1 minute.

***Bay 6: 10% Sulfuric Acid.*** This bath is necessary to ensure that all complex ammonium persulfate crystals embedded into the copper surface from the micro-etch bath are removed. To make 10 gallons of 10% solution of sulfuric acid, 1 gallon (3.785 liters) of reagent-grade sulfuric acid is slowly added to 9 gallons (34 liters) of water.

The rack is removed from bay 5, allowed to drain, and then loaded into the 10% sulfuric acid bath, maintained at room temperature, for 2 to 3 minutes.

***Bay 7: Water Rinse.*** The rack is removed from bay 6, allowed to drain, and then rinsed in room-temperature water for 1 minute.

***Bay 8: 25% Hydrochloric Acid.*** This bath will further prevent the next bath (bay 9) from being contaminated with any of the previously used chemicals. To make 10 gallons of 25% hydrochloric acid solution, 2.5 gallons (9.46 liters) of reagent-grade hydrochloric acid is slowly added to 7.5 gallons (28.39 liters) of distilled water.

The rack is removed from bay 7, allowed to drain, and loaded into the 25% hydrochloric acid solution for 1 to 5 minutes. This bath is also operated at room temperature. No rinse is required after this process since this solution is compatible with that of bay 9.

***Bay 9: Catalyst 9F.*** This is a proprietary colloidal solution of palladium-tin in hydrochloric acid. In this solution, the palladium is surrounded and protected by stannic tin molecules. The purpose of this bath is to "seed" (implant) palladium onto the nonconductive glass/epoxy wall surfaces of the drilled holes. This is necessary so that these surfaces will accept the subsequent electroless copper deposition process. To prepare a

10-gallon bath of Catalyst 9F, 4 gallons (15.14 liters) of either reagent-grade or CP-grade hydrochloric acid is slowly added to 4 gallons of distilled water and stirred to mix. To this solution, 2 gallons (7.57 liters) of Catalyst 9F concentrate is added and again stirred.

The rack is removed from bay 8, allowed to drain, and then loaded into the Catalyst 9F bath for 3 to 10 minutes. This bath is operated at room temperature.

**Bay 10: Water Rinse.** The rack is removed from bay 9, allowed to drain, and then rinsed for 1 minute in room-temperature water.

**Bay 11: Water Rinse.** The rack is removed from bay 10, allowed to drain, and again rinsed for 1 minute in room-temperature water.

**Bay 12: Accelerator 19.** Accelerator 19 is a proprietary solution designed to dissolve the tin molecules that were deposited in bay 9. This exposes the palladium sites to achieve uniform copper deposition. The Accelerator 19 bath also increases the life of the subsequent copper bath by minimizing drag-in of Catalyst 9F. To prepare a 10-gallon bath of Accelerator 19, 1.666 gallons (6.306 liters) of concentrate is added to 8.333 gallons (31.54 liters) of distilled water.

The rack is removed from bay 11, allowed to drain, and then loaded into the Accelerator 19 bath, operated at room temperature, for 4 to 8 minutes.

**Bay 13: Water Rinse.** The rack is removed from bay 12, allowed to drain, and then rinsed in room-temperature water for 1 minute.

**Bay 14: Cuposit 328Q Electroless Copper Bath.** Cuposit 328Q is also a proprietary solution designed to deposit approximately 25 $\mu$inches of copper onto properly *activated* surfaces. To prepare 10 gallons of Cuposit 328Q solution, add 1.25 gallons (4.73 liters) of Shipley Copper Mix 328A to 7.5 gallons (28.39 liters) of distilled water and stir thoroughly. To this mixture, 1.25 gallons of Copper Mix 328Q is added and again stirred well. Finally, 0.25 gallon (946 milliliters) of Copper Mix 328C is added and stirred. This bath is more stable if gentle air bubbling is allowed to agitate the solution to reduce the formation of cuprous oxide.

The rack is removed from bay 13, allowed to drain, and then loaded into the Cuposit 328Q bath for a period of time ranging from 8 to 20 minutes, depending on the bath temperature and solution concentration. These time and concentration relationships are shown in Fig. 11.6. The typical range of bath operating parameters is shown on the three-dimensional surface response curve. The plating time varies from a low of 8 minutes if the bath is at its optimum copper concentration of 100% and at its highest recommended temperature of 78°F (point *a*) to as much as 20 minutes if the bath is operated at its low limit of 65°F with a copper concentration of 80% (point *h*). The shaded area of the response curve bounded by points *a, d, h,* and *e* shows the effects that both temperature and copper concentration have on plating time.

Plating time can additionally be determined from either of the simplified two-dimensional curves also shown in Fig. 11.6. The time-versus-percent copper concentration graph relates plating time as a function of % copper for each of four constant temperatures. The time-versus-bath temperature graph relates plating time as a function of temperature for two constant copper concentrations (80 and 100%).

**Bay 15: Water Rinse.** The rack is removed from bay 14, allowed to drain, and then rinsed in room-temperature water for 1 minute.

**Bay 16: 3% Sulfuric Acid.** This bath is intended as an alkaline neutralizer and also serves as a storage tank to prevent the electroless deposited copper on the processed boards from rapidly oxidizing, which would occur if they were exposed to air. To prepare 10 gallons

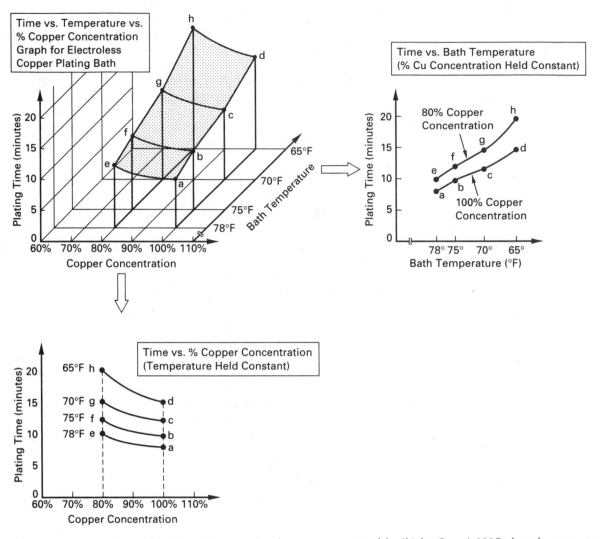

**Figure 11.6** Three-dimensional surface response curve of the Shipley Cuposit 328Q electroless copper deposition bath. This graph demonstrates the effect on plating time as both bath temperature and percent copper concentration are varied. All data presented are to obtain approximately 25 μinches of electroless copper.

of 3% sulfuric acid, 0.3 gallon (1.46 liters) of reagent- or CP-grade sulfuric acid is slowly added to 9.7 gallons (36.71 liters) of water.

The rack is removed from bay 15, allowed to drain, and then loaded into the 3% sulfuric acid bath for a period from 3 minutes up to 3 hours. This bath is operated at room temperature.

After the boards have been processed through the 16-bay electroless copper deposition line, they are completely coated with approximately 25 μinches of copper. The next process is to immediately panel plate or flash plate the boards to build up the copper thickness on the hole walls an additional 0.001 inch (100 μinches). This will ensure electrical integrity between the copper foil surfaces on opposite sides of the board through the plated holes.

Before introducing the copper flash plating process, we complete the discussion of electroless copper deposition by considering visual inspections for bath monitoring and bath replenishments necessary to maintain quality control.

# 11.4 ELECTROLESS COPPER DEPOSITION: PROCESS MONITORING AND CONTROL

To maintain the deposition line in good chemical balance, several monitoring operations must be undertaken on a routine basis. Most often, monitoring is performed by visual interpretation of bath color or by the amount of dwell time that the boards are processed through any given bay. For the Shipley low-speed electroless copper deposition line described in the preceding section, each of the baths is considered in terms of their monitoring and process control.

*Water Rinse.* Rinse water should always be maintained at temperatures above 60°F with a rate of flow from 5 to 10 bath volumes per hour. If air agitation is provided, the lower rate of flow may be used with good results. To maximize the effectiveness of the rinse water, a *solution conductivity* or *resistivity* meter should be used. A specially designed test probe leading to the meter is placed directly into the rinse water. (The readings of either one of these types of meters is simply the reciprocal of the other.) These meters measure the contamination in the rinse water by monitoring the increase of conductivity. As the level of contamination rises, the purity of the water is degraded resulting in higher values of conductivity. Conductivity is measured in units of microsiemens per centimeter ($\mu$S/cm). Ideally, the conductivity meter may be set to alert the operator when contaminants have degraded the rinse water to a conductivity reading of approximately 3500 $\mu$S/cm. For this reading, the rinse water should be discarded and replaced with fresh water.

*Cleaner/Conditioner 1175A.* Tap water is used to replenish the solution due to evaporation loss caused by the high operating temperature of this bath. The strength of the bath can be checked visually. If *light* oxides are not removed from the copper surface after a dwell time of 1 minute in the bath, the solution should be discarded and replaced with a fresh batch. Another method of monitoring the effectiveness of this bath is to keep account of the number of square feet of pc boards that have been processed. The bath should be replaced after 200 square feet of surface area per gallon of solution is processed. For our 10-gallon tank, this would translate into 2000 square feet of board surface.

*20% Ammonium Persulfate.* This solution should be replaced when its color becomes medium blue. In addition, the solution strength can be monitored by visual inspection of the board surfaces. If a uniform pink or salmon appearance is not displayed over the entire copper surfaces on both sides of the board after the proper dwell time in the bath, the solution should be replaced.

*10% Sulfuric Acid.* This solution should be replaced when its color becomes light blue.

*25% Hydrochloric Acid.* This solution should be replaced when its color becomes pale green.

*Catalyst 9F.* This solution will exhibit a dark brown-to-black color at room temperature when freshly prepared. A crude check on this bath is to observe if the color turns to a pale amber. This is an indication that the solution has not been properly maintained and should be replaced. Another check on this solution is to observe the boards after they have been loaded for the proper dwell time. A uniform tan color should appear on all insulation surface edges for FR-4 and G-10 grade boards. This bath should be maintained at a strength of at least 10% by periodic replenishments made according to the schedule shown in Fig. 11.7. The procedure for determining the concentration of Catalyst 9F is as follows:

1. Maintain the bath level by additions of concentrated hydrochloric acid (reagent or CP grade).
2. Remove a 20-milliliter sample of the bath. In a separate beaker, mix 45 milliliters of distilled water and 15 milliliters of reagent-grade hydrochloric acid. This constitutes 60 milliliters of a 25% by volume solution of hydrochloric acid.

| Catalyst 9F Replenishment Schedule for a 10-Gallon Bath | |
| --- | --- |
| Bath Strength Determined by Comparison Against Color Standards | Additions* of Catalyst 9F Concentrate |
| 100% | None |
| 90% | 750 ml |
| 80% | 1500 ml |
| 70% | 2250 ml |
| 60% | 3000 ml |
| 50% | 3750 ml |

Maintain Concentration Above 70%

*Note: Remove an equal volume of bath before making additions. Mix bath completely after additions are made.

**Figure 11.7**
Recommended replenishment schedule for the Catalyst 9F bath.

3. Mix the 20-milliliter sample and the 60 milliliters of 25% hydrochloric acid solution into a bottle that is the same size and shape of the color standard bottles that are available from the manufacturer and, in our case, from Shipley Co., Inc. The mixture is covered and the solution shaken well. The color of the sample is then compared with the color standards that indicate various bath strengths. This comparison is made by first spacing the color standard bottles approximately 2 inches apart in good lighting (near a window works well). The prepared sample bottle is then placed between adjacent standard bottles and moved until its color most closely matches one of the standards. The percent concentration of the sample is equal to the value written on the cap of the matching standard bottle.

4. Replenish the bath by first removing a bath volume equal to the amount of Catalyst 9F to be added according to the replenishment schedule shown in Fig. 11.7.

*Accelerator 19.* The level of this bath should be maintained with distilled water only. A crude check on the strength of the solution is to observe when it turns cloudy or blue. At that time it should be immediately replaced. The amount of board processing can also be used as a measure of solution control. After 300 square feet of pc board stock per gallon of solution has been processed, the bath should be replaced.

*Cuposit 328Q.* Shipley's low-speed electroless copper bath is deep blue in color. The copper concentration is controlled by using a set of blue color standards, which are obtained from the manufacturer. The color-matching procedure is similar to that described for the Catalyst 9F solution. The procedure for determining the copper concentration is as follows:

1. Measure out 89 milliliters of Cuposit copper mix color indicator, which is obtained from the manufacturer.

2. Add 11 milliliters of electroless copper bath to the above. The solution is thoroughly mixed and placed in the same size and shape bottle as the color standards. The prepared sample is then compared against the blue color standards to determine the copper concentration of the bath.

3. Replenish the bath by first removing a bath volume equal to the amount of 328A and 328Q concentrates according to the replenish schedule shown in Fig. 11.8.

It should be emphasized that the bath should be replenished to the 100% level for each morning startup. During the day, sufficient work should be processed through the bath so that it is depleted to 70% copper concentration or below for overnight storage.

| Cuposit 328Q Replenishment Schedule for a 10-Gallon Bath | | |
|---|---|---|
| Copper Concentration of Bath Determined by Color Standards | 328A Additions* | 328Q Additions* |
| 100% | None | None |
| 90% | 475 ml | 475 ml |
| 80% | 950 ml | 950 ml |
| 70% | 1425 ml | 1425 ml |
| 60% | 1900 ml | 1900 ml |
| 50% | 2375 ml | 2375 ml |

Maintain Bath Above 70% for Daily Operation

*Notes: 1. Remove an equal volume of bath before making additions.
2. Replenish bath by adding the correct amount of 328A concentrate first, mixing the bath completely and then adding the correct amount of 328Q concentrate. Again stir bath completely to mix solution.

**Figure 11.8**
Recommended replenishment schedule for the Cuposit 328Q electroless copper bath.

The three primary elements in the Cuposit 328Q solution are *copper sulfate, sodium hydroxide,* and *formaldehyde.* The copper sulfate provides the copper metal for deposition, the sodium hydroxide controls the pH (alkalinity) of the solution, and the formaldehyde acts as a reducing agent during the plating process. Even though there are other elements in the solution, these three are mainly expended during the plating process and must be replenished. As was previously discussed, the copper concentration is monitored and controlled through the use of color standards. Both the sodium hydroxide and the formaldehyde can be monitored and controlled through chemical analysis with the use of a pH meter and a titration procedure. A less exacting but effective monitoring and control method is by visual inspection of the processed boards according to the following procedure. Over a weekend of bath inactivity, the copper concentrate of the solution is first checked and replenished to operating strength using the color standards as previously described. A pc board is processed in the bath and it is inspected to see if the copper appearance is *pink* in color. If not, sodium hydroxide and formaldehyde must be added. The Shipley Company supplies sodium hydroxide as a concentrate, which is labeled Cuposit Z, and formaldehyde concentrate, labeled Cuposit Y. These concentrates are added to the solution in 200-milliliter steps. First, 200 milliliters of Cuposit Z are added followed by 200 milliliters of Cuposit Y. Again, a board is processed and inspected. If it is pink in color, no further additives are required. If it does not appear as pink, additional Cuposit Z and Cuposit Y are added in 200 milliliter steps and another board is processed and inspected. This procedure continues until a processed board appears as pink.

*3% Sulfuric Acid.* This bath must be in a pH range of 1 to 2. The solution should be replaced when the pH level falls outside this range.

Monitoring and controlling of the copper deposition line may first appear tedious and difficult. However, with a short period of experience, you will find that it will become quite simple and will require only a modest amount of time each day.

## 11.5 FLASH PLATING: GENERAL INFORMATION

The preceding section described how drilled pc boards were completely metalized with a thin ($\approx 25$ $\mu$inches) deposit of electroless copper over all surfaces and the walls of each hole. These processed boards are allowed to remain in the 3% sulfuric acid solution in preparation of a copper *flash* plating, which will build up the copper thickness to approximately

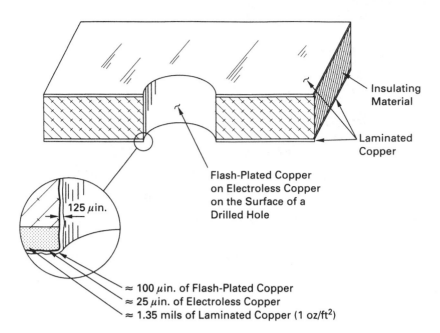

**Figure 11.9** Cross-sectional view of a double-sided, plated-through-hole board after flash plating.

Insulating Material

Laminated Copper

Flash-Plated Copper on Electroless Copper on the Surface of a Drilled Hole

125 μin.

≈ 100 μin. of Flash-Plated Copper
≈ 25 μin. of Electroless Copper
≈ 1.35 mils of Laminated Copper (1 oz/ft$^2$)

0.000125 inch (125 μinches). This plating is required to prevent the electroless copper deposited on the hole walls from oxidizing, which would result in voids, or nonconducting areas, after subsequent board processing.

Flash plating is accomplished in a plating tank containing a copper sulfate plating solution. The pc board is suspended in this solution and electrically connected to the negative side of a power supply. The positive side of the power supply is connected to a large copper *anode,* which is also suspended in the plating solution. An electric current is passed through the plating tank causing copper ions from the plating solution to be attracted to the board and be deposited onto all copper surfaces. These copper ions are supplied to the plating solution by the corrosion of the copper anode. A thickness of 125 μinches of copper can be electroplated by this method in a little more than 4 minutes. The result of flash plating is shown in Fig. 11.9. Note that the buildup of copper is continuous and uniform over the entire copper surfaces of the board in addition to the hole walls.

## 11.6 COPPER PLATING TANK AND PLATING ACCESSORIES

This section describes the plating tank, plating rack, and the power supply for a typical prototype plating process. A rack for plating printed circuits holds either one or more boards in a fixed position in the plating solution. It must be capable of handling the large currents required in the plating process and not be easily affected by the plating solution. A basic plating tank setup is shown in Fig. 11.10a. The rack contacts the cathode work rod of the plating tank and is held securely with a wing nut or thumb screw. Tight contact between the thumb screw and cathode reduces resistance and holds the rack rigidly to the rod. The lower section of the rack connects to the copper foil of the pc board and is held firmly by another thumb screw or setscrew.

Plating racks are often fabricated from square copper rod stock, because this material is capable of handling 1000 amperes per square inch of cross-sectional area. A commercially fabricated plating rack capable of handling at least four boards at one time is shown in Fig. 11.10b. Because most of the rack assembly will be in contact with the plating solution, it is protected with a pestasol coating to prevent plating buildup. The contact points for the boards are fabricated of 300 series stainless steel. A smaller custom single board rack is shown in Fig. 11.10c. It is constructed of copper stock and stainless steel screws. Subsequent discussions on plating will make use of this rack.

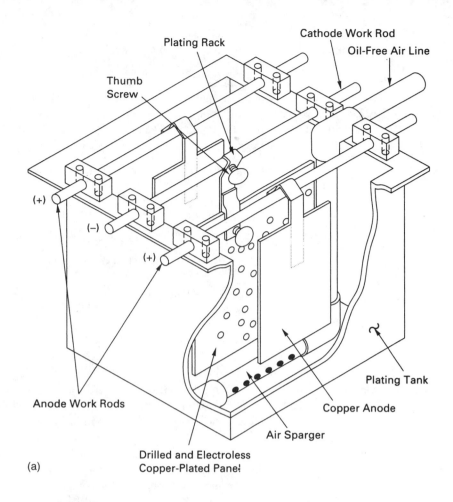

Plating Rack

Cathode Work Rod

Oil-Free Air Line

Thumb
Screw

(+)

(−)

(+)

Anode Work Rods

Plating Tank

Copper Anode

Air Sparger

Drilled and Electroless
Copper-Plated Panel

(a)

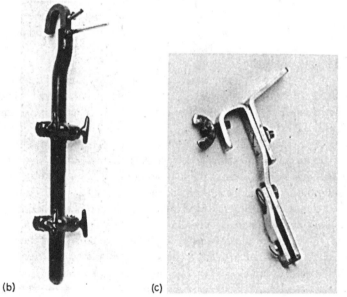

**Figure 11.10** Plating tank and accessories for electrolytic copper plating: (a) prototype copper plating tank; (b) commercial plating rack; (c) custom single-board rack.

**(b)**

**(c)**

Because the copper will be electroplated from a sulfate plating solution, the recommended material for the tank is stress-relieved polypropylene or polyethylene plastic.

A tank size of 12 inches long by 12 inches wide by 12 inches deep will be adequate for most prototype work. This tank will have a solution capacity of approximately 6 gallons (22.7 liters) when the plating solution is 9.75 inches from the bottom. The exact volume can be calculated by dividing the number of cubic inches of solution by 231 cubic inches per gallon. The following example will demonstrate this calculation.

*Example:* Determine the exact volume of plating solution for a 12-inch by 12-inch by 12-inch tank.

$$\text{Volume of solution (cubic inches)} = \text{length} \times \text{width} \times \text{depth of solution}$$

$$= 12 \text{ inches} \times 12 \text{ inches} \times 9.75 \text{ inches}$$

$$= 1404 \text{ cubic inches}$$

$$\text{Solution volume (gallons)} = \frac{\text{volume of solution (cubic inches)}}{231 \text{ cubic inches/gallon}}$$

$$= \frac{1404 \text{ cubic inches}}{231 \text{ cubic inches/gal}} = 6.08 \text{ gallons}$$

A 12-inch-deep tank with approximately 10 inches of plating solution (2 inches of clearance to the rim of the tank) will allow the insertion of the anodes and the racked pc board to be plated without causing the solution to spill over the top of the tank.

The tank is fitted with two anode work rods and one cathode work rod. These rods are made of 3/8-inch-diameter solid copper rod stock. This diameter will easily support the weight of the copper anodes and the plating rack with pc boards as well as handle the large currents used in the plating process. The work rods are held securely in position with pairs of Lucite or plastic blocks fastened onto opposite ends of the tank rim. The anode work rods are positioned over the tank parallel to one another and 1 inch inward from the sides of the tank. The cathode work rod is positioned parallel to the anode rods and centered between them. The on-center spacing between either anode work rod and the cathode work rod is 5 inches (see Fig. 11.10a). These distances must remain constant between the surfaces of the board and the anodes for proper plating.

For uniform plating, effective solution movement through the holes in the board is required. The amount of time required to electroplate a given thickness of copper can be reduced with more vigorous solution movement. To accomplish this, an air agitation system is built into the plating tank (see Fig. 11.10a). The source of air should be an oil-free, low-pressure blower delivering approximately 1.5 to 2 cubic feet per minute of air for each square foot of plating tank surface area. The air is conveyed to the plating solution through a 1/2-inch PVC sparger system. Note in Fig. 11.10a that the sparger pipe is positioned directly below the pc board to be plated. A series of 1/16-inch-diameter holes is drilled into the bottom of the sparger pipe at 45-degree angles to its center line. The sparger is positioned approximately 1 inch above the bottom of the tank. Air entering the pipe exits through these small holes and is deflected off the bottom of the tank. The upward motion of these expanding air bubbles causes vigorous solution movement across the outside surfaces of the board as well as through all the drilled holes.

The negative terminal of a plating power supply is electrically connected to the cathode work rod. Both anode work rods are electrically connected together (anode strap) and then connected to the positive terminal of the power supply (see Fig. 11.11). All electrical connections should be made with AWG No. 12 stranded wire protected with a chemical-resistant plastic insulation.

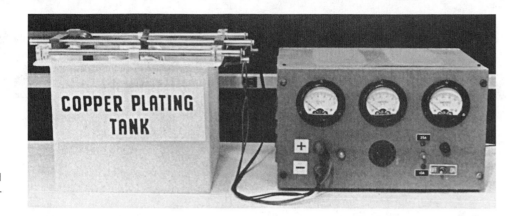

**Figure 11.11** Typical prototype copper plating setup.

The plating power supply should have the following characteristics:

1. Continuous adjustment of output voltage from 0 to 6 volts dc.
2. Current capacity in excess of plating current requirements (a maximum of 15 amperes is acceptable for the plating tank illustrated).
3. Percent ripple at full load current should be equal to or less than 5%.
4. A meter to monitor the dc output voltage at the + and − terminals.
5. A meter to monitor the load current.

The circuit schematic for a typical plating power supply is shown in Fig. 11.12.

The types of material used for the anode, together with its shape and surface area, will affect the quality of the resulting plating onto the pc board surface. Only *phosphorized* copper anodes should be used, since they will corrode properly in an acid copper-plating solution. They should contain 0.04 to 0.06% phosphorus by weight. Copper anodes designated as OHFC or pure copper anodes should not be used.

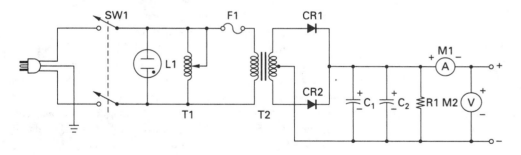

**Figure 11.12** Schematic diagram of plating power supply.

Parts List
SW1-DPST switch
L1-neon lamp
T1-2-A 0- to 120-V ac 0.25-kVA autotransformer
F1-1.2-A 250-V, SLO-BLO fuse
T2-Stancore P-6433 5.0 V CT at 15-A transformer
CR1-CR2-IN3900, 20 A at 100 PIV
C1, C2-24,000 μF at 35-WVDC electrolytic capacitor
R1-1 kΩ, 1/2 W, 10%
M1-0- to 25-A meter
M2-0- to 10-V meter

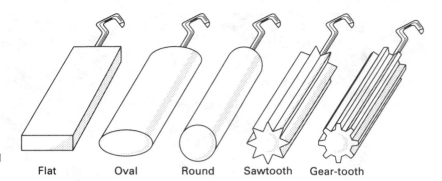

**Figure 11.13** Standard anode shapes.

Flat     Oval     Round     Sawtooth     Gear-tooth

Anodes are available in various shapes and in a wide range of sizes (Fig. 11.13). The standard shapes are *flat, oval, round, sawtooth,* and *gear tooth.* A flat anode is preferable in a small plating tank since it results in more uniform current density. The oval, sawtooth, and other special shapes are designed to increase the surface area of the anode. The criterion for selecting an anode is its surface area relative to the area of the pc board to be plated. For the plating of copper, an anode-to-cathode surface area ratio of approximately 1:1 is recommended. This means that the surface area of the anode facing the pc board in the plating tank should have a surface area approximately equal to the one side of the pc board that it is facing. Uniform plating through the holes of double-sided pc boards will result if the overall exposed copper path area is equal on both sides of the pc board. Double-sided plating is accomplished with the use of two anodes, one positioned on each side of the pc board, both having identical surface areas. This will simplify and optimize the plating process.

The anodes must be provided with a hook whose length will allow the anode to extend from the anode work rod down into the plating solution yet allow the top of the anode to protrude slightly above (1/4 inch) the surface of the solution. The length of the anode should be such that it does not extend below the bottom of the pc board more than 1 inch.

To illustrate the calculation of the correct size of an anode for a specific application, we will assume the requirement of processing a 5- by 5-inch card. The area of copper that will be exposed to the plating solution is determined by multiplying the length $L$ of the board by its width $W$. This will yield the overall area in square inches of exposed copper surface for one side. For our example, this results in 5 inches by 5 inches, or 25 square inches. This figure is used to determine the anode area required for plating as well as the plating power supply current density (Sec. 11.8) to obtain optimum plating.

There are two restrictions involved in the selection of the anode size required for plating: (1) Its surface area must approximately equal that of the cathode, and (2) it cannot extend more than 1 inch below the pc board. Selecting an anode whose length is 6 inches (1 inch longer than the board) will satisfy the length criterion. From the range of standard available widths, a 5-inch-wide anode will result in an area of 30 square inches, which is slightly larger than that of the cathode (25 square inches). Since each side of the pc board has equal areas, two 5- by 6-inch anodes will meet all of the requirements for flash plating our example board even though the ratio of anode area to work area is slightly larger than the optimum 1:1 proportion.

Prior to the placement of the anodes into the plating tank, proper procedures must be followed to avoid contaminating the plating solution. All plating solutions are extremely susceptible to organic contamination. For this reason, the anodes must be properly cleaned and correctly *bagged.* Anode degreasing is accomplished by scrubbing the surface with a brass bristle brush and cold tap water. The anode is then rinsed thoroughly and dipped into a 10% by volume solution of sulfuric acid for approximately 10 minutes. After this acid dip, the anode is ready to be bagged.

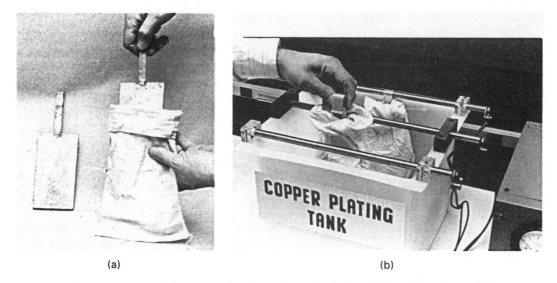

(a)                                        (b)

**Figure 11.14**  Anode bag preparation: (a) anode and leached anode bag; (b) anode installation into copper plating tank.

The anode bags used should be made of Dacron or polypropylene (never cotton) and should be 2 inches wider and 2 inches longer than the anode. The bag must be *leached* (purified) of all sizing or organic contaminations that it was exposed to in its manufacturing process. Leaching is accomplished in three steps. First, the bag is boiled in water for at least 1 hour. It is then removed from the boiling water and rinsed in cold water. Finally, the bag is soaked for 24 hours in 1/2% by volume of Copper Gleam PC solution (see Sec. 11.7). One gallon of this solution will be sufficient for two bags of the size described. To mix 1 gallon of this solution, 190 milliliters of Copper Gleam PC is added to 3594 milliliters of water. The treated bags are then tied around the cleaned anodes. This is shown in Fig. 11.14a. The anodes are now ready to be installed into the plating tank (Fig. 11.14b), which, together with all the hardware and plumbing, must be thoroughly leached prior to adding the plating solution and the anodes. The leaching process is the same as that described in Sec. 11.3 for the electroless copper deposition line. Again, leaching is required only when the plating solution is added for the *first* time. Recall that detailed information for leaching is given in Appendix IX and should be reviewed before proceeding to the next section.

## 11.7  COPPER SULFATE PLATING SOLUTION

After the plating tank and the accessories have been leached, the plating solution is prepared. For our purposes we have selected a copper sulfate solution, since it requires only one principal additive and is thus easy to control. A key ingredient in this plating bath is manufactured by the Lea Ronal Company and is sold with the trade name of Copper Gleam PC. The use of this solution results in a bright copper deposit that is ideally suited for plating pc boards.

The copper sulfate plating solution is made up of the following:

1. Copper sulfate pentahydrate liquid
2. Sulfuric acid (reagent grade)
3. Hydrochloric acid (reagent grade)
4. Copper Gleam PC additive/brightener
5. Distilled water

Highly pure copper sulfate provides the source of copper in the plating solution. The most convenient use of this chemical is in liquid form. Sulfuric acid is added to the solution to improve its conductivity and also aids in anode corrosion. Reagent-grade sulfuric acid is used to prevent contamination from entering the plating cell. A small amount of hydrochloric acid is added since chlorides aid in brightening the plating. The Copper Gleam PC provides a bright ductile copper deposit over the entire board surface. The formulation of 10 gallons of Copper Gleam PC copper sulfate plating solution is given in Table 11.1, which shows the required amount of each ingredient and the order of mixing.

To prepare a plating bath for the 6-gallon tank described previously, Table 11.2 lists the ingredients and order of mixing. The amounts of each ingredient were derived by simply taking 6/10 of those amounts listed in Table 11.1. The 6-gallon bath is prepared by first placing 3.72 gallons (14.08 liters) of distilled water into the leached tank. Next, 1.68 gallons (6.36 liters) of liquid copper sulfate pentahydrate, having a copper concentration of 2.25 pounds of copper per gallon, is added to the water. The solution is stirred slowly until it is thoroughly mixed. The 0.6 gallon (2.27 liters) of sulfuric acid is then added, stirred, and the solution is allowed to cool to room temperature. Finally, 2.7 milliliters of hydrochloric acid and 0.03 gallon (114 milliliters) of Copper Gleam PC are added and the solution is thoroughly mixed. The height of the solution should be marked on the outside of the tank to note any decrease in level due to evaporation. Any significant decrease of solution should be replaced so as to return it to the original level in the tank. The bath should be maintained at room temperature and loosely covered with sheet plastic when not in use to protect the solution from dust and other foreign matter.

**TABLE 11.1**

Formulation of 10 Gallons of Copper Plating Solution with Copper Gleam PC Brightener

| Order of Mixing Ingredients | Compound | Amount | |
|---|---|---|---|
| 1 | Water (distilled or deionized) | 6.2 gal | 23.5 liters |
| 2 | Copper sulfate pentahydrate (2.25 lb copper/gal) $CuSO_4\ z5H_2O$ | 2.8 gal | 10.6 liters |
| 3 | Sulfuric acid (reagent grade) | 1.0 gal | 3.8 liters |
| 4 | Hydrochloric acid (reagent grade) | — | 4.5 mL |
| 5 | Copper Gleam PC | 0.05 gal | 190 mL |
| | Total | ≈ 10.05 gal | ≈ 38.1 liters |

**TABLE 11.2**

Formulation of 6 Gallons of Copper Plating Solution with Copper Gleam PC Brightener

| Order of Mixing Ingredients | Compound | Amount | |
|---|---|---|---|
| 1 | Water (distilled or deionized) | 3.72 gal | 14.08 liters |
| 2 | Copper sulfate pentahydrate (2.25 lb copper/gal) $CuSO_4\ z5H_2O$ | 1.68 gal | 6.36 liters |
| 3 | Sulfuric acid (reagent grade) | 0.6 gal | 2.27 liters |
| 4 | Hydrochloric acid (reagent grade) | — | 2.7 mL |
| 5 | Copper Gleam PC | 0.03 gal | 114 mL |
| | Total | ≈ 6.03 gal | ≈ 22.83 liters |

All plating solutions require periodic chemical analysis. The techniques of chemical analysis for the type of bath just described are beyond the scope of this book. However, since the amount of solution described in Table 11.2 is relatively small and not significantly expensive, the entire bath should be replaced if plating problems are encountered. A fresh batch of solution can be used successfully for a reasonable length of time if extreme care is exercised to minimize the introduction of organic contaminations into the bath. Detailed chemical analysis procedures for the bath described may be obtained from the Lea Ronal Company.

Before pc boards can be plated, the anodes must be conditioned (electrolyzed) after they have been cleaned, bagged, and hung into the plating bath. The conditioning process causes a film to be formed on the anodes and involves simply hanging sample copper-clad panels from the cathode rod into the plating solution. The power supply is adjusted to a value that will result in a current density of 10 amperes per square foot of surface area on the sample panels used. A suitable anode film is formed in approximately 4 hours.

## 11.8 COPPER SULFATE FLASH PLATING

With the pc boards that have been processed through the electroless deposition line stored in the 3% sulfuric acid holding tank and the plating tank fully prepared, the flash plating process can begin. The air supply is turned on and the rack of boards is removed from the holding tank. They are immediately placed into a rack, such as the ones shown in Fig. 11.10, and are ready for the electroplating process. The power supply is turned on and set just slightly above its zero reading. The racked boards are placed into the plating solution and swished back and forth for 15 seconds before electrical connection is made to the cathode rod. This 15-second dip will remove any minor oxides that may have formed, thereby reactivating the surfaces for electroplating. The rack holding the pc board is then lowered and hooked onto the cathode rod and secured with the thumb screw (refer to Fig. 11.10a).

The current density recommended by the manufacturer for flash plating is 30 amperes per square foot ($\approx$ 208 milliamperes per square inch) of pc board surface. *However,* it is important to note that this high-current density initially subjected to the 20 $\mu$inches of electroless copper deposited on the hole walls *would cause its destruction.* It is therefore essential that one begin the plating process by applying one-fifth the recommended current density, that is, 6 amperes per square foot ($\approx$ 42 milliamperes per square inch), for 2 minutes and then an additional 4 minutes at the recommended rate of 30 amperes per square foot.

The power supply is to deliver 30 amperes of current for every square foot of copper that is exposed to the plating solution. In Sec. 11.6 it was determined that our example panel has a surface area of 25 square inches. Since this is a double-sided board, the total surface area exposed to the plating solution is twice this amount, or 50 square inches. We will now calculate the amount of current that the power supply is to deliver for the first 2 minutes (i.e., at one-fifth the recommended rate of 208 milliamperes per square inch) as follows:

initial current setting

$$= \frac{1}{5} \text{ recommended current density} \times \text{total surface area}$$

$$= \left(\frac{1}{5}\right) \frac{208 \text{ milliamperes}}{\text{square inch}} \times 50 \text{ square inches}$$

$$= 2080 \text{ milliamperes} (\approx 2 \text{ amperes})$$

When the power supply is set at 2 amperes, you will note that the output voltage is low (< 1 volt).

After 2 minutes at the 2-ampere rate, the power supply should be readjusted for the recommended current density for an additional 4 minutes. The final current setting is calculated as follows:

$$\text{final current setting}$$

$$= \text{recommended current density} \times \text{total surface area}$$

$$= \frac{208 \text{ milliamperes}}{\text{square inch}} \times 50 \text{ square inches}$$

$$= 10{,}400 \text{ milliamperes}$$

$$= 10.4 \text{ amperes } (\approx 10 \text{ amperes}).$$

Recall that the power supply, specified in Sec. 11.6, has a current capacity of 15 amperes, which is well in excess of the requirements of our plating application.

Thickness of the copper plated onto the surface of the pc board is dependent on (1) the current density and (2) the plating time. For the copper sulfate bath operated at 30 amperes per square foot, copper is deposited at the rate of 100 $\mu$inches every 3 1/2 to 4 minutes. To result in a thickness of 125 $\mu$inches, the required plating time would be slightly more than 4 minutes.

After the flash plating process has been completed, the power supply is turned off and the rack is loosened from the cathode work rod and immediately removed from the tank. The pc board is then thoroughly rinsed in cold tap water for at least 1 minute and then dried. Inspection of the plating should reveal a smooth and bright deposit of copper on all surfaces, including the walls of all the drilled holes.

The next phase in the processing of the pc board is to photosensitize both of its sides so that the conductor patterns may be imaged. Chapter 12 presents the procedures for photosensitizing and imaging in preparation for the final processing steps discussed in Chapter 13.

## EXERCISES

### A. Questions

11.1 The formula

$$\text{gallons} = \frac{L \times W \times H}{231}$$

is used to calculate the capacity, in gallons, for rectangular-shaped tanks where $L$, $W$, and $H$ are in inches. Calculate the capacity of tanks having the following dimensions in inches:

| Tank | L | W | H |
|------|-----|-----|-----|
| 1 | 24 | 6 | 24 |
| 2 | 12 | 12 | 12 |
| 3 | 96 | 24 | 48 |

11.2 Calculate the amount of rinse water required in gallons per hour for a 12- by 18- by 36-inch rinse tank with a turnover rate of six times per hour.

11.3 Explain the function and determine the makeup of a 20-gallon ammonium persulfate bath.

11.4 Discuss the function and solve for the makeup of a 60-gallon bath of Catalyst 9F.

11.5    Describe the function and calculate the makeup of an Accelerator 19 bath for a 6- by 30- by 60-inch tank.

11.6    Give the function and determine the makeup of a 328Q bath for an 8- by 16- by 36-inch tank.

11.7    Using the graph of time versus percent copper concentration shown in Fig. 11.6, determine the plating time required at 75°F. Next, using the graph of time versus bath temperature, find the plating time required with a bath having an 80% copper concentration.

11.8    Determine the resistance in ohms for a conductivity reading of 2500 $\mu$S/cm using the formula

$$R = \frac{1}{\text{conductivity } (\mu\text{S/cm})}$$

11.9    Calculate the anode size for copper plating a 6- by 9-inch double-sided pc board.

11.10   What plating bag size should be used for a 4- by 20-inch copper anode?

11.11   Discuss the function and solve for the makeup of a 35-gallon copper sulfate plating bath using Copper Gleam PC as the brightener.

11.12   Determine the initial power supply setting for current and then the recommended current density for a 7- by 12-inch double-sided pc board for copper flash plating.

11.13   What is a *plated-through hole?*

11.14   What is the purpose of electroless copper deposition?

11.15   How are anode bags leached, and for what purpose?

## B.   True or False

Circle *T* if the statement is true, or *F* is any part of the statement is false.

11.1    When drilling stacks of pc boards on CNC machines, a thin entry sheet of aluminum is placed on top of the stack.                                                                T   F

11.2    The PTH process requires that pc boards be drilled after the electroless copper deposition and copper flash plating processes are completed.                                      T   F

11.3    The electroless copper deposition process will deposit a layer of copper approximately 20 to 30 thousandths of an inch thick on board surfaces.                                  T   F

11.4    Catalyst 9F is used to seed particles of palladium onto the surfaces of hole walls to provide a site for copper deposition.                                                     T   F

11.5    Air spargers are used in rinse tanks to improve the efficiency of rinsing pc boards.         T   F

11.6    Flash plating is used to prevent deposited electroless copper from oxidizing.                T   F

11.7    For copper plating, the surface area of the anode must be approxmately equal to that of the cathode.                                                                           T   F

11.8    The current density for flash plating is 30 amperes per square inch of pc board surface.     T   F

11.9    Anode bags should be made of cotton.                                                         T   F

11.10   All heaters, plumbing, and spargers used with a copper deposition line should be leached at the same time prior to the startup use of the line.                               T   F

## C.   Multiple Choice

Circle the correct answer for each statement.

11.1    CNC multipanel drilling uses (*aluminum, XP hardboard*) as entry material and (*aluminum, XP hardboard*) as bottom backup material.

11.2    The electroless copper deposition process is performed (*before, after*) the drilling operation.

11.3    The air spargers installed in each of the rinsing bays of the electroless copper deposition line (*increase, decrease*) the volume of water used.

11.4    The electroless copper deposition line uses eight water rinse cycles, which require the water to be maintained at a temperature of (*70°F, 150°F*).

11.5    Palladium is seeded onto the insulated material of the pc board in the (*Accelerator 19, Catalyst 9F*) bath.

11.6    Approximately 25 μinches of copper is deposited on an activated surface in the (*Catalyst 9F, Cuposit 328Q*) bath of the electroless deposition line.

11.7    The holding tank for board storage after the copper deposition process contains a (*3%, 30%*) concentration of sulfuric acid.

11.8    The purpose of copper sulfate in the electrolytic flash plating tank is to (*improve conductivity, provide a source of ions*).

## D.   Matching

Match each item in Column A to the most appropriate item in Column B.

| COLUMN A | COLUMN B |
|---|---|
| 1.  3% sulfuric acid | a.   $30\ A/ft^2$ |
| 2.  CNC drill/router | b.   Seeding |
| 3.  Catalyst 9F | c.   Micro-etch |
| 4.  Copper anodes | d.   Positive terminal |
| 5.  Electroless copper deposition | e.   Storage bath |
| 6.  Current density | f.   125 μinches of copper |
| 7.  Ammonium persulfate | g.   400 hits/minute |
| 8.  Flash plating | h.   25 μinches of copper |

# 12 Double-Sided PCB Processing: Imaging

## LEARNING OBJECTIVES

*Upon completion of this chapter on the imaging of double-sided pc boards, the student should be able to*

- Know the processes involved in double-sided imaging.
- Chemically and mechanically clean pc boards.
- Be familiar with dry-film photo resists.
- Use the laminator to apply dry-film resist to pc boards.
- Properly register the diazo phototools onto both sides of a pc board.
- Expose dry-film photo resist.
- Use a radiometer to monitor exposure time.
- Prepare developing solution and spray-develop pc boards.
- Inspect and touch up any defects in developed pc boards.

## 12.0 INTRODUCTION

In Chapter 11, the first of many phases in the processing of double-sided plated-through-hole pc boards was discussed. The panels to be processed were stacked and pinned together and all the holes were drilled with an automatic multispindle CNC machine. After drilling, the panels were unpinned, mechanically deburred, and then processed through cleaning and drying. They were then vertically racked in specially designed stainless steel racks and processed through a 16-bath electroless copper deposition line, which plated approximately 25 $\mu$inches of pure copper on all board surfaces including the edges and the walls of the drilled holes. To ensure a continuous copper surface without voids, especially on the hole walls, the boards were finally passed through a flash-plating process, which electroplated an additional 100 $\mu$inches of copper onto all surfaces.

With the boards plated with approximately 125 $\mu$inches of copper, the next step in their processing is *dry-film photo resist imaging*. The boards are first mechanically cleaned and then chemically treated to activate the flash-plated copper. Both sides of the boards are then laminated with a light-sensitive dry-film photo resist. The light-sensitive surfaces are exposed through positive-working phototools of the conductor patterns (see Chapter 7). These images are next developed, which exposes just the copper patterns of conductors and terminal pads of the artworks. These patterns, as well as hole walls, will subsequently be electroplated with copper and then 60/40 tin–lead solder (Chapter 13).

Recall that in the print-and-etch technique for processing single-sided boards (Chapter 10), the photo resist employed was a light-sensitive liquid. Although this photo resist is suitable for simple prototype work, it does not lend itself to plated-through-hole applications and thus has been superseded by the newer dry-film types. There are two reasons for this. First, the solvent-type developers required to process liquid resists are becoming increasingly more difficult to dispose of properly because of environmental considerations. Second, it is difficult to accurately control the thickness of liquid resists, resulting in lower yields of flaw-free boards. Both of these problems are overcome with the use of dry-film photo resists. Uniform thickness results in more predictable exposure time and little, if any, pinhole touch-up is required. Dry-film resists that can be developed and stripped in aqueous-based chemicals are also available. These resists can be applied simultaneously to both sides of a double-sided board, lending themselves ideally to this application. In addition, dry-film resists can more easily withstand almost any electroplating bath solution and, as a result, allow the application of copper from a copper sulfate bath as well as 60/40 tin–lead solder from a fluoborate bath. In this chapter we describe the imaging process using only dry-film photo resists.

*CAUTION:* A chemical hazard exists in the materials presented in this chapter. The processes discussed should be performed in adequately ventilated facilities and with applicable equipment having proper fume exhaust systems. Approved face shields (goggles) in compliance with ANSI Z87.1-1989 and chemical-resistant clothing and footwear and neoprene or nitrile gloves must be worn while performing any of these operations. Also, the section on safety outlined at the beginning of the book should be reviewed for more specific information.

## 12.1 PROCESSES INVOLVED IN DOUBLE-SIDED IMAGING

The basic steps required to image a double-sided pc board are shown in the flowchart of Fig. 12.1. The flowchart is accompanied by illustrations that show the results of the sequential steps described in the imaging process. Subsequent sections of this chapter detail each of these steps.

The imaging process begins with boards that have been drilled, deposited with electroless copper, and flash-plated (Chapter 11). Both sides of these panels are then chemically and mechanically prepared to accept a uniform lamination coating of dry-film photo resist. With the resist laminated on both sides, it is exposed with two diazo phototools (Chapter 7), one for the component side and one for the circuit side. Phototool alignment will be accomplished using the underlying hole pattern on the board. Recall that diazo phototools show the conductor pattern images as translucent (amber), which aids in this alignment.

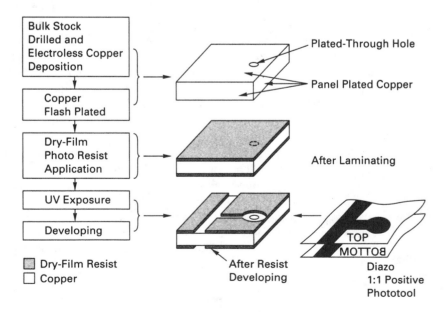

**Figure 12.1** Basic flowchart with pictorial sequence of the imaging process.

Also, the amber pattern is opaque to ultraviolet light, which is used to expose the resist. The translucent quality of the diazo film allows each pattern to be registered in perfect alignment with the drilled holes on either side of the board due to the fact that they are clearly visible through the amber pattern. This eliminates the need for any special registration pinning arrangement between the two phototools.

After exposure through the two diazo phototools, the board is developed to produce an image such as that shown in Fig. 12.1. Note that after developing, the resist is removed from only the conductor pattern, terminal pads, and holes. These will all be copper- and solder-electroplated. All the copper that is to be removed remains covered with the photo resist.

## 12.2 COPPER CLEANING FOR DRY-FILM APPLICATION

Optimum dry-film resist adhesion is accomplished only if the copper surface is properly cleaned. The cleaning cycle outlined in Fig. 12.2 will produce copper surfaces that are completely degreased, slightly acidic, and thoroughly dry. Careful attention to this cleaning cycle cannot be emphasized enough. Most problems that develop in subsequent processing steps can often be directly attributed to improper cleaning.

First, an acid dip is required to leave the copper surface in a slightly acid state to improve resist adhesion. The acid dip uses CP (chemically pure) or reagent-grade sulfuric acid. One part of acid is *slowly* added to 9 parts of cold tap water. This mixture will yield a 10% by volume acid dip solution. (***CAUTION:*** *Never add water to acid*. This causes violent eruptions to occur.) The container used for the dip tank can be of Pyrex or an acid-resistant plastic such as polyethylene or polypropylene. Plastic gloves should be worn whenever handling acid solutions. The pc board is dipped into the acid solution for approximately 2 minutes while gently rocking the container (Fig. 12.3a). The pc board is then removed from the acid dip and completely rinsed on both sides for at least 30 seconds.

Cleaning of the copper surfaces is best accomplished with a mechanical scrubber, such as that shown in Fig. 12.3b. This scrubber consists chiefly of a series of motor-driven hard-rubber rollers positioned horizontally along the length of the unit. Panels to be cleaned are placed on these rollers and are transported into the unit at a speed of approximately 4 to 5 feet per minute. Inside the unit and above the rollers is a cylindrical scrubbing brush (320 grit) that rotates in a direction counter to that of board travel. The vertical position of the brush is adjusted by a small handwheel located at the top-right side of the machine. After the boards enter the scrubber, water flows over their surface as they pass under the brush. The copper surface (one side only) is thus mechanically cleaned of grease, oils, and foreign particles. The small flow of water on both the board and the brush acts as a coolant and also flushes away contaminants. As the board moves from under the brush, it is sprayed with water to further aid in removing foreign matter. It finally passes between a series of *air knives* (parallel air tubes having a narrow slit along their lengths), which direct streams of

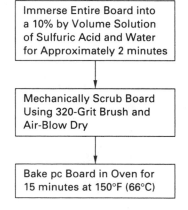

**Figure 12.2** Chemical and mechanical cleaning procedure for bulk pc board stock prior to resist application.

Immerse Entire Board into a 10% by Volume Solution of Sulfuric Acid and Water for Approximately 2 minutes

Mechanically Scrub Board Using 320-Grit Brush and Air-Blow Dry

Bake pc Board in Oven for 15 minutes at 150°F (66°C)

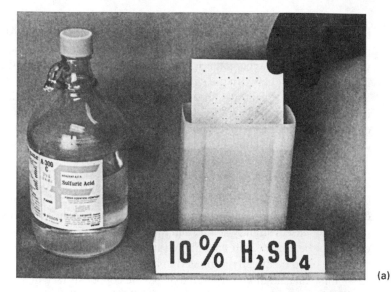

(a)

(b)

**Figure 12.3** Cleaning cycle for copper in preparation for dry-film lamination: (a) chemical cleaning by an acid dip; (b) Somaca scrubber used to mechanically clean copper.

high-pressure air over both surfaces of the board simultaneously to promote drying. Because the scrubber described cleans only one side of a board at a time, a double-sided board must be turned over and run through a second time. Larger and more elaborate scrubbers are available that will clean, rinse, and dry both sides of pc boards simultaneously.

The cleaned board is then placed vertically into a forced-air oven for approximately 15 minutes at a temperature of 150°F (66°C). This prebake cycle just prior to resist application accomplishes two functions: (1) The elevated temperature will remove any moisture from the board, which would adversely affect resist adhesion, and (2) by raising the temperature of the copper surfaces to 150°F, improved resist adhesion results. Timing and temperature in this heating phase are somewhat critical. Prolonged exposure to heat will cause surface oxides to be redeposited onto the copper surfaces. Although dry-film resist can be applied successfully over a *light* oxide film, this should be avoided. In addition, increasing the oven temperature to shorten the prebake cycle should also be avoided, since higher temperatures do not improve resist adhesion and can degrade the copper bond. After this prebake cycle, the board is ready for dry-film resist application.

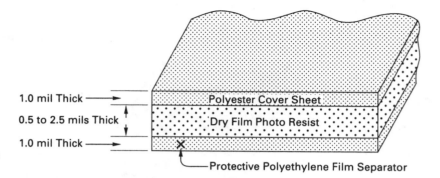

**Figure 12.4**
Composition of dry-film resist.

1.0 mil Thick ⟶ Polyester Cover Sheet

0.5 to 2.5 mils Thick ↕ Dry Film Photo Resist

1.0 mil Thick ⟶ ✕

Protective Polyethylene Film Separator

## 12.3 DRY-FILM PHOTO RESIST

The use of dry-film photo resists in the manufacture of printed circuit boards is a process that has been used extensively since the early 1970s. A typical composition of the dry-film resist is shown in Fig. 12.4. The resist is a light-sensitive photopolymer, dry in nature, and available in thicknesses of from 0.5 to 2.5 mils. Widths range from 3 to 25 inches that are available on continuous rolls of 100, 250, and 400 feet. The resist is sandwiched between two layers of plastic. One side is protected with a polyester cover sheet and the other side is covered with a polyethylene separator sheet. These plastic sheets are intended to protect the dry film from handling damage and also to provide a gastight seal against the oxygen in the air, which would quickly degrade the film.

As with any other light-sensitive resist, correct handling is crucial. Because these dry films are sensitive to light in the ultraviolet range, they should be stored and used only in an area illuminated with gold fluorescent tubes or lighting that does not emit ultraviolet light. Dry films are also sensitive to extreme heat and should therefore be stored at temperatures below 70°F (21°C).

Dry-film resists are categorized as either solvent- or aqueous-developing types. With the demands being placed on the industry for more and better environmental controls, the aqueous types are being used more extensively. Although dry-film resists are available from several manufacturers, all are similar in makeup and also in the fabrication processes required. For our purposes, we have elected to use Laminar AX, which is an aqueous-developing dry film, blue in color, manufactured by the Dynachem Corporation of California. This resist will withstand both the copper sulfate and the tin–lead fluoroborate plating solutions that will be used in the pattern-plating phase of processing described in Chapter 13. The optimum size of resist selected for our purposes is a thickness of 2 mils and a width of 8 inches, which will cover the bulk stock described in this chapter.

## 12.4 DRY-FILM LAMINATOR

Dry-film photo resists are applied to the clean and slightly acidic copper surfaces with a *laminator*. A typical laminator capable of processing boards up to 12 inches wide is shown in Fig. 12.5. This laminator applies the dry film simultaneously to both sides of the pc board. Dry-film resists are applied under pressure at elevated temperatures with a specific feed-through rate. A typical laminator consists of (1) a pair of hard-rubber pressure rollers, (2) a pair of take-up rollers, (3) a pair of heat shoes, (4) a pair of temperature sensors for each shoe, (5) a variable-pressure adjustment control, and (6) a variable-motor-speed control to adjust the feed rate of the dry film.

Dry-film threading of the laminator for double-sided pc boards is shown in Fig. 12.6. Two rolls of dry film, one for each side of the pc board, are used. The protective

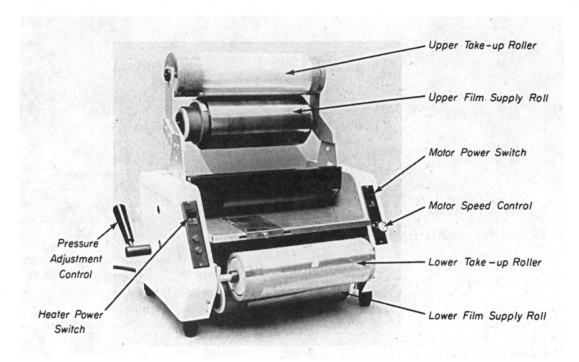

**Figure 12.5** Prototype laminator for dry-film photo resist application.

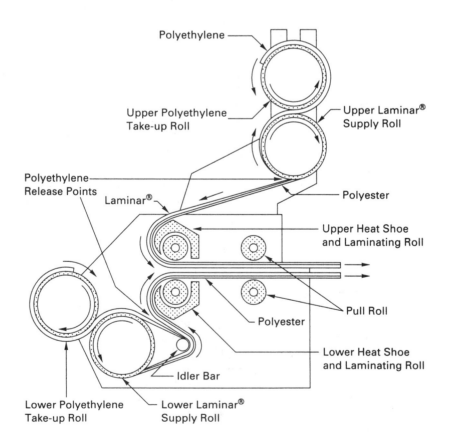

**Figure 12.6** Typical threading diagram for laminator.

polyethylene separator sheets are removed from both rolls of dry film and fed into both the upper and lower take-up rollers. The upper and lower rolls of dry film are fed between the hard-rubber rollers with exposed dry film in contact with each other. The opposite sides of both rolls of film, together with the polyester cover sheets, are in contact with both heat shoes. As the pc boards are fed between the rolls of dry film, the take-up rollers remove, on a continuous basis, the polyethylene separator sheets just prior to the film contacting the pc board. The film is heated as it passes over the heat shoes and is laminated onto the copper surfaces by the pressure of the hard-rubber rollers as the pc board moves through the laminator.

Successful application of Dynachem Laminar AX to prepared copper surfaces requires the following laminator specifications: (1) upper and lower heat shoe temperatures set to $235 \pm 10°F$, (2) laminating feed rate of approximately 5 linear feet per minute, and (3) moderate pressure setting of the hard-rubber rollers. With the laminator's parameters correctly set, the application of the dry-film photo resist onto the pc board can now be performed.

## 12.5 DRY-FILM PHOTO RESIST APPLICATION

In Sec. 12.2, the pc board was placed into an acid dip, machine scrubbed and dried, and prebaked in an oven. These heated boards should be laminated immediately upon removal from the oven. Cold boards would act as heat sinks and remove heat from the dry film, resulting in poor resist adhesion. For this reason, the oven should be placed close to the laminator so that no delay will result between removing the board from the oven and placing it into the laminator.

With the laminator controls set to the settings specified in Sec. 12.4 and allowing the laminator time to reach the preset temperatures, approximately 1 foot of dry film is run through the laminator. This is a precautionary measure that should be done any time a heated laminator has remained idle for any length of time. The reason for doing this is that the first foot of film on the rolls has been in close contact with the heat generated by the heat shoes of the laminator and, as a result, has been degraded by the long exposure to elevated temperatures. Film damaged from excessive heat results in *heat fog*. A pc board processed with heat-fogged film will not process successfully in the subsequent steps.

After the one foot of film has passed through the laminator and with the film feed still on, the board is removed from the oven (held by the edges only) and fed immediately into the laminator by pushing it gently up to the moving rollers. The board contacting the rollers will be pulled into the laminator, where heated dry-film resist will be applied simultaneously to both sides of the copper surface. This process is shown in Fig. 12.7a.

As the board with the applied resist leaves the laminator, it is removed by cutting off the film either with the aid of the slide cutter mounted at the rear of the laminator or with a single-edge razor blade (Fig. 12.7b). Efficient laminating can be realized if two people operate the unit: One person feeds the boards into the laminator while the other cuts and removes the processed boards.

A laminated double-sided pc board is shown in Fig. 12.8. The board is sandwiched between two layers of dry film, each protected by a polyester cover sheet. After the board has been laminated, it is extremely important not to disturb or remove the polyester cover sheets (even at the corners). These sheets serve as "gas seals," preventing oxygen in the atmosphere from contacting the underlying resist film and causing it to polymerize (i.e., change it chemically into heavier molecules). The excess film is carefully trimmed as close as possible to the board edges, using a sharp razor blade. This technique is shown in Fig. 12.9.

The temperature of the laminator must be carefully maintained at $235 \pm 10°F$. At elevated temperatures, resist blistering occurs, and at temperatures lower than the specified range, a breakdown in resist adhesion to the copper surface can result. If the laminator is in continuous use, careful monitoring of the two thermometers is essential to minimize problems.

(a)

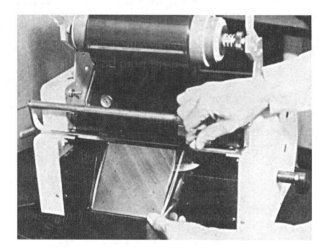

(b)

**Figure 12.7** Dry-film laminating technique: (a) pc board fed into laminator; (b) laminated pc board removed using rear film cutter.

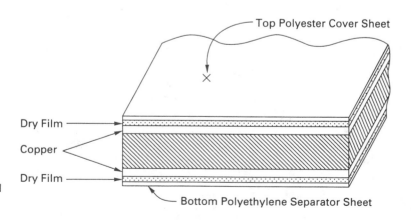

Top Polyester Cover Sheet

Dry Film

Copper

Dry Film

Bottom Polyethylene Separator Sheet

**Figure 12.8** Laminated double-sided pc board.

**Figure 12.9** Trimming excess dry film from edges of laminated pc board.

After laminating, a normalizing period of 30 minutes before exposure is recommended by the manufacturer. This 30-minute period allows the board and dry film to return to room temperature, resulting in an improvement of the adhesion properties of the film. Because the film is extremely sensitive to ultraviolet light, the normalizing interval must be in a gold-light area.

## 12.6 EXPOSURE OF DRY-FILM PHOTO RESIST

The dry-film resist is now ready to be exposed to an ultraviolet light source. The amount and quality of ultraviolet light contacting the resist are critical to obtain results that are consistently reliable. Too short an exposure time will result in an image that is not completely developed. Too long an exposure time will result in resist brittleness and improper resist adhesion to the copper surface. Resist adhesion problems usually become apparent in the cleaning cycle prior to plating and during the plating operation itself.

Any point-light source rich in ultraviolet light is suitable for exposure of Laminar AX resist. Excellent results are obtainable with an exposure unit equipped with high-pressure mercury vapor lamps. The exposure unit shown in Fig. 12.10a is a small prototype unit suitable for double-sided exposures. This unit consists of two 400-watt high-pressure mercury vapor lamps, one above and one below the vacuum frame. Also included in this unit are the controls to activate the vacuum pump, a vacuum gauge, and a resettable timer to control the exposure time. The technique of exposure is somewhat a function of the equipment available. The following explanation relates directly to the exposure unit shown in Fig. 12.10a but is similar to most fixed-light source units. The main power is turned on and both upper and lower lamps are energized. The unit is then allowed to warm for at least 30 minutes before exposing any boards. A warm-up time is required of all exposure units equipped with high-pressure mercury vapor lamps to allow the lamps to operate at full light intensity.

Proper exposure time for a particular resist and exposure unit must be determined empirically. For any exposure unit using a Stouffer 21-step exposure chart, the procedure for this determination is detailed in Appendix VIII. With the correct exposure time specified for the Laminar AX resist film, a *radiometer,* shown in Fig. 12.10b, is used to measure the UV light energy that strikes the film during exposure. The Dynachem Model 500 UV Integrating Radiometer shown measures the light energy in units of millijoules per square centimeter. The direct liquid crystal digital readout allows for accurate and rapid monitoring.

The Dynachem radiometer is approximately 5 inches in diameter and only $\frac{3}{8}$ inch in thickness. This geometry permits its use in any area of the exposure frame, even in a

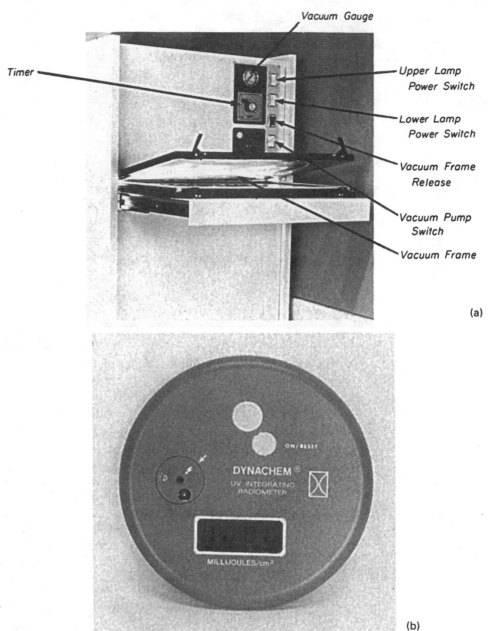

Vacuum Gauge

Timer

Upper Lamp
Power Switch

Lower Lamp
Power Switch

Vacuum Frame
Release

Vacuum Pump
Switch

Vacuum Frame

(a)

DYNACHEM ®
UV INTEGRATING
RADIOMETER

ON/RESET

MILLIJOULES/cm²

(b)

**Figure 12.10** Exposing photo resist film: (a) exposure unit for double-sided pc boards; (b) radiometer.

vacuum frame. The procedure for using the radiometer to test the upper lamp of an exposure unit is as follows:

*Step 1.* The film manufacturer's recommended exposure time is set with the Stouffer chart positioned under a clear area of the phototool. For a silver film phototool and Laminar AX resist, this exposure time is about 30 seconds under a 400-watt mercury vapor lamp.

*Step 2.* The radiometer is placed face up in the vacuum frame and the ON/RESET button is pressed. This will result in the display showing four zeros.

*Step 3.* The sensor area of the radiometer (a $\frac{1}{2}$-inch-diameter white circle) is covered with a clear portion of the phototool.

***Step 4.*** The vacuum frame is closed and the exposure unit energized for the time obtained in step 1.

***Step 5.*** At the completion of the exposure cycle, the radiometer display will read the standard exposure in millijoules per square centimeter. For the Laminar AX film, this reading should be 55 mJ/cm$^2$ for the exposure time determined by the use of the Stouffer chart.

***Step 6.*** The procedure is repeated to test the lower lamp of the exposure unit. The radiometer is positioned face down on top of a clear area of a phototool placed on the bottom glass surface of the vacuum frame.

The radiometer will automatically turn off in about 45 seconds if not exposed to UV light. The Dynachem unit is battery operated and is equipped with a charger. It is rated for 8 hours of continuous use per full charge.

The radiometer is an essential tool for obtaining optimum imaging. It accurately monitors correct exposure time by indicating problems resulting from a decrease in UV energy caused by lamp output, voltage drop, dirty reflectors, or changes in the clear areas of the phototool. An undetected decrease in UV energy can result in imaged boards having resist patterns that will not withstand the plating and etching processes.

After the laminated pc board has normalized to room temperature, the circuit-side diazo phototool is aligned over the board using the underlying holes for registration. Two pieces of double-sided tape are placed on the phototool in the thief area (where they will not show) to contact the pc board. When this is completed, the board is turned over and the component-side phototool is registered and taped into position. The setup for exposure is shown in Fig. 12.11a. The phototools and pc board are placed onto the glass portion of the vacuum frame (Fig. 12.11b). Note the small wire lying in the vacuum frame. One end of the wire is positioned

(a)

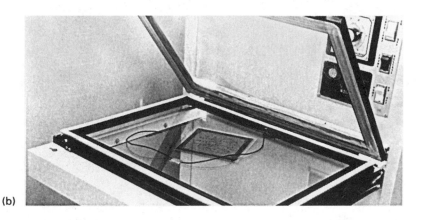

**Figure 12.11**
Preparation for dry-film exposure: (a) alignment of phototool with laminated board; (b) proper position of phototool on vacuum frame.

(b)

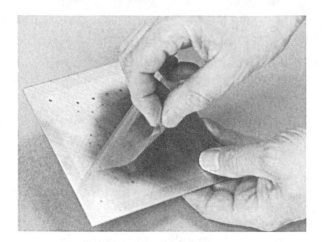

**Figure 12.12** Removal of protective polyester cover sheet after exposure.

across one corner (thief area) of the board and the other end is placed into the vacuum port. This ensures rapid air removal from the frame so that the total surface areas of all elements will be pressed tightly together. For exposure, the plastic cover is lowered and locked into place over the setup. Engaging the vacuum pump will create a vacuum between the glass and the plastic sections of the frame forcing both the top and bottom positives into immediate contact with the polyester cover sheets of the dry film. A good seal is indicated by a reading of 25 psi for the unit illustrated in Fig. 12.10a. A good seal having been achieved, the frame is pushed into the exposure unit between the upper and lower lamps and exposed for the preset interval on the timer control. For the exposure unit shown, an exposure time of approximately 30 seconds, as determined with the step table, results in correct exposure for Laminar AX photo resist to obtain a resist image that will withstand the subsequent plating processes.

As with any negative-acting resist, exposure of the film to ultraviolet light causes the following to occur. All those areas of the resist protected from the light (those areas appearing as amber conductor patterns on the diazo positive) will remain unchanged chemically and will be completely removed in the developing phase. Those areas that are exposed to the light (all transparent areas) are polymerized and will not be removed in the developing phase.

After the film is properly exposed, the vacuum frame will be automatically ejected from the unit. The vacuum pump is then turned off and both the exposed pc board with the phototool are removed from the unit. At this point, a faint image of the conductor pattern is discernible through the transparent gas seal on the resist. This is the result of a visible indicator dye that was added to the film by the manufacturer. Again, the exposed pc board needs to be allowed to normalize to room temperature for approximately 30 minutes. The protective gas seal should remain in place and undisturbed during the normalizing period.

One side of a correctly exposed double-sided pc board with its protective polyester cover sheet partially removed (after the 30-minute normalizing period) is shown in Fig. 12.12. This board is now ready for developing.

## 12.7 SPRAY DEVELOPING

Spray developing, similar to other photographic processes, should be done in a gold-lighted area to prevent ultraviolet light from destroying the imaged boards. Spray developing is accomplished by spraying, under high pressure, a mildly alkaline developer solution onto the exposed pc board for a specific length of time and at a specific temperature. This temperature should be within the range of 80 to 90°F (27 to 32°C). Higher developing temperatures will decrease developing time, but may degrade the quality of the resist pattern remaining on the pc board. Too low a temperature will increase the required developing time. The prototype spray developer shown in Fig. 12.13 will completely develop a pc board in approximately 45 seconds at a temperature of 85°F (29°C) with new developer solution. Development time is determined by the "freshness" of the developing solution and its temperature. Optimum developing can be determined visually—all of the resist that was not exposed to the ultraviolet light is completely removed, leaving clear copper. A properly developed pc board is shown in Fig. 12.14. In mass-production conditions, an additional 15

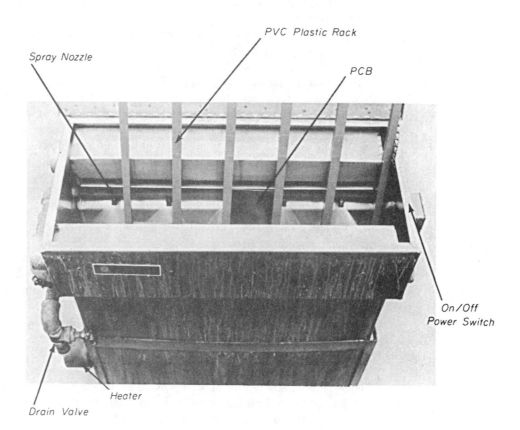

**Figure 12.13** Spray developing a pc board.

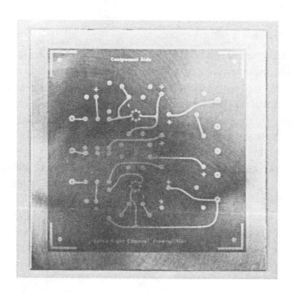

**Figure 12.14** A pc board after developing.

seconds is added to the developing time to compensate for the fact that the more boards that are developed, the more resist goes into the developing solution resulting in longer time.

A developer that can be used for Laminar AX resist is a 1.5% by weight solution of trisodium phosphate (TSP) and tap water. For this solution, 0.125 pound (57 grams) of TSP per gallon of water is required. This solution is cost-effective but results in a relatively slow developer. Improved developing efficiency can be achieved with Dynachem's concentrated developer solutions for Laminar AX resist. These developers are under proprietary trade labels of Concentrate Developer KB-1A and KB-1B. An example of preparing developing solution will make use of the spray developer shown in Fig. 12.13. This developer has a sump capacity of 10 gallons minimum and 15 gallons maximum.

To prepare the developing solution, 12 gallons of cold tap water are placed into the developer sump. A measure of 1875 milliliters of Concentrate Developer KB-1A and 1875 milliliters of Concentrate Developer KB-1B is added to the water. To suppress the formation of excessive foam during the developing operation, the addition of Dynachem Antifoam No. 40 at the rate of one milliliter per gallon of developer will complete the solution. To the approximately 13 gallons of solution in the sump, 13 milliliters of Antifoam No. 40 are added. This solution will result in rapid and effective pc board developing. It should be mentioned at this point that with this developing solution, if developing time exceeds $1\frac{1}{2}$ minutes at a spray temperature of 85°F (29°C), the solution has been saturated with developed Laminar AX. It should be discarded and a fresh batch of developer prepared. In addition, the pH level of the solution should be monitored with a pH meter to avoid obtaining less durable resist images. The developer should be replaced when the pH level drops to a reading of 10.

In preparation for developing, the immersion heater temperature control is adjusted to 85°F (29°C) and at least 1 hour should be allowed to pass for the solution to reach this operating temperature. After the solution has reached its operating temperature, the board to be developed, with both protective polyester cover sheets removed, is placed in a suitable rack and suspended into the spray. A slow up-and-down motion improves the uniformity of developing, as does rotating the entire rack 180 degrees in the spray at least once for each batch of boards to be developed.

After developing, the board is immediately removed from the rack and given a rinse under a cold-water tap for at least 30 seconds on each side. Any developer solution remaining on the pc board after rinsing is neutralized with a 30-second dip in a diluted sulfuric acid solution. A 5% by volume solution is recommended. This solution is prepared by slowly adding 5 parts of CP sulfuric acid to 95 parts of tap water. After this neutralizing dip, the board is rinsed for 30 seconds and then allowed to air-dry.

## 12.8 IMAGE TOUCH-UP

After the board has been properly developed and air dried, it is critical that the images on both sides be carefully inspected for any defects. This inspection is best accomplished with the aid of a 5× magnifier fitted with a 22-watt fluorescent lamp. The entire image and the surrounding film-resist areas should be examined. Any specks of undeveloped photo resist in the pattern image should first be removed. These will appear as shiny and almost clear areas on the copper pattern. These particles can be removed by carefully scraping with the tip of a scalpel.

The boards should also be examined to observe if there are any voids or breaks in the surrounding film resist. If not corrected, these voids will leave unwanted copper exposed, which will result in conductor paths or terminal pads being shorted in the electroplating process. These areas are easily repaired with the use of a special fiber-tip ink-resist pen. Two types of dry-film touch-up pens are available. One is coded with a *red* top and is made to be used with photo resist films that can be stripped in an alkaline solution. This is the type that would be used with the Laminar AX film. The other touch-up pen is coded with a *blue* top and is used with photo resists that must be stripped with a solvent. It is important that the touch-up ink selected be removed with the same stripper used to remove the dry-film resist on which it is used. This will avoid additional processing steps. By simply filling in any breaks with the resist ink, possible shorts are avoided. Before the board is further processed, all touch-up ink must be allowed to dry completely.

After the board has been inspected and any necessary repairs performed, it is ready for the remaining phases of processing to complete its fabrication. These processes, discussed in Chapter 13, are pattern plating, solder plating, dry-film stripping, etching, and solder fusion followed by cutting the board to final size.

## EXERCISES

### A.  Questions

12.1    What is the purpose of baking the mechanically cleaned copper panels at 150°F for 15 minutes prior to dry-film application?

12.2    List three precautions in the handling of dry film.

12.3    What are the three laminator settings and their values for dry-film laminating?

12.4    Explain the term *heat fog* and how it is avoided.

12.5    Explain the term *gas seal*.

12.6    Why does the manufacturer of dry film specify a 30-minute normalizing period after dry-film lamination?

12.7    Explain the procedure for determining proper exposure for dry-film resist using a Stouffer chart.

12.8    Calculate the required amount of trisodium phosphate to make a 20-gallon developing bath for Laminar AX film.

12.9    How much Antifoam 40 should be added to a 20-gallon developing bath?

12.10    What is the purpose of the neutralizing bath after boards have been developed and what is its makeup?

12.11    After the pc board has been exposed through the diazo phototool, from which areas is the resist removed by the developing process?

12.12    What is the approximate vacuum gauge reading that will result in a good seal in the exposure unit?

12.13    In how much time can a pc board be developed in a spray developer having a developer solution temperature of 85°F?

### B.  True or False

Circle *T* if the statement is true, or *F* if any part of the statement is false.

12.1    The copper surfaces of pc boards are chemically cleaned in a 20% by volume solution of sulfuric acid and water for about 2 minutes.    **T    F**

12.2    When mixing acid solutions, the water is slowly added to the acid.    **T    F**

12.3    Prior to dry-film laminating, the cleaned pc boards are baked for approximately 15 minutes at 150°F.    **T    F**

12.4    The laminating feed rate is approximately 15 linear feet per minute.    **T    F**

12.5    Heat fog is caused by leaving the pc boards in the oven for too long.    **T    F**

12.6    Immediately after a pc board is laminated, the polyester cover sheets are removed and the board is allowed to normalize for approximately 30 minutes.    **T    F**

12.7    Developing solution for Laminar AX should be used at a temperature of approximately 85°F and replaced when the pH level drops to 10.    **T    F**

## C. Multiple Choice

Circle the correct answer for each statement.

12.1    To prepare a 10% solution of sulfuric acid, one part of acid is slowly added to (*9, 10*) parts of cold water.

12.2    Chemical cleaning of a pc board requires (*alkaline, acid*) followed by a mechanical scrubbing.

12.3    The recommended laminating temperature of dry film photo resist is (*235°F, 361°F*).

12.4    After the photo resist laminating process is completed, a normalizing time of (*30 minutes, 30 seconds*) is required prior to further pc board processing.

12.5    The developing solution for Laminar AX requires (*sulfuric acid, trisodium phosphate*).

12.6    The development of Laminar AX is accomplished by (*spraying, dipping*) with a solution temperature of (*60°F, 85°F*).


## D. Matching

Match each item in Column A to the most appropriate item in Column B.

|  | COLUMN A |  | COLUMN B |
|---|---|---|---|
| 1. | Diazo | a. | 21-step exposure table |
| 2. | Dry-film photo resist | b. | Developing process |
| 3. | Trisodium phosphate | c. | Phototool |
| 4. | Stouffer chart | d. | TSP |
| 5. | Antifoam No. 40 | e. | 30 minutes |
| 6. | Normalizing time | f. | Polyester cover sheet |

# 13

# Double-Sided PCBs: Final Fabrication Processes

## LEARNING OBJECTIVES

*Upon completion of this chapter on the final fabrication processes for double-sided pc boards, the student should be able to*

- Determine the copper image area of a conductor pattern.
- Set up a cleaning line to prepare pc boards for copper pattern plating.
- Plate copper onto the conductor pattern of pc boards.
- Prepare a tank and accessories for tin–lead pattern plating.
- Prepare the tin–lead plating solution.
- Solder-plate the conductor pattern of pc boards.
- Prepare resist the stripping solution and remove the resist from pc boards.
- Prepare a modified ammonium persulfate solution to use with a spray etcher to etch pc boards.
- Use an oven to reflow solder on pc boards.
- Understand an alternative method of fabricating single- and double-sided prototype PCBs using a mechanical protoboard system that employs Quick Circuit™ drilling, engraving, milling, and routing techniques.
- Understand the application concept of using Quick Circuit™, Quick Press™, and Quick Plate™ multilayer board technology to produce prototype MLBs.

## 13.0 INTRODUCTION

In Chapter 12, 1:1 diazo phototools were used to image the laminated pc board, which had previously been completely covered with a 125-$\mu$inch layer of copper (Chapter 11), to form the conductor path and terminal pad patterns. The result of this imaging is shown as the first sequential step in the processing flowchart of Fig. 13.1.

The next phase of fabrication is the *pattern-plating* process. This involves plating of the imaged pattern with additional copper followed by a layer of tin–lead alloy (solder). The pattern-plating process first adds on an additional thickness of approximately 0.001 inch (1 mil) of copper onto all the exposed conductor pattern as well as the inside surfaces (barrels) of all the drilled holes. This will ensure continuous metal paths having good electrical integrity of all through-hole connections from one side to the other. The solder (60/40 tin–lead alloy) is then electroplated over the conductor pattern and the hole barrels to a thickness of approximately 0.0003 inch (0.3 mil). The solder serves a dual purpose: first as an etchant resist and subsequently as a fused protective coating against oxidation for the entire pattern. This solder coating improves board shelf life and promotes improved soldering to the pattern. Refer again to Fig. 13.1 to see the results of pattern plating.

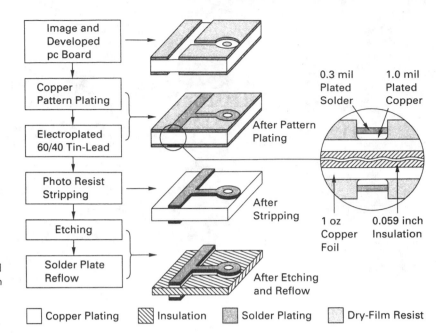

**Figure 13.1** Basic flowchart with pictorial sequence of the pattern plating process, including etching and reflow.

After the image has been pattern plated, the dry-film photo resist, which is used as a plating *stop-off* (plating resist to all areas of unimaged copper) during the plating process, has served its purpose. It may now be chemically stripped (removed) from the board, thereby exposing the underlying layer of copper that was flash plated onto the board's foil. (See the next sequential step in Fig. 13.1.) This is the copper that was not pattern plated and thus exposed to the etchant, which will remove it completely from the board. The copper image, which is protected by the pattern-plated solder (now serving as an etchant stop-off), will not be attacked by the etchant and will thus remain on the board. Note in Fig. 13.1 that all conductor paths, terminal pad areas, and all connections made between both sides of the board through plated holes remain after the etching process.

The final process to complete the conductor pattern is called *solder reflow*. As a result of the stripping and etching process, the solder plating on the conductor pattern image becomes contaminated and must first be chemically treated and then heated to its melting point. This heating process causes the solder to alloy (fuse) with the underlying copper, resulting in a much denser and less porous metal that will be less susceptible to oxidation and will enhance the component soldering phase of construction. The processed panels are finally routed (contoured or profiled) to shape to complete their fabrication.

The topics presented in this chapter begin with describing several methods of determining the pattern area to be plated for purposes of setting the correct plating current density. Since a detailed discussion on plating copper was presented in Chapter 11, it is referred to in this chapter only in consideration of copper pattern plating. This is followed by describing the cleaning cycles and a detailed discussion of the solder plating process. Included is a description of etchants that are used with solder as a stop-off. Solder reflow is considered using prototype bench equipment. Finally, board routing is described to profile the panel into a finished double-sided pc board.

*CAUTION:* A chemical hazard exists in the materials presented in this chapter. The processes discussed should be performed in adequately ventilated facilities and with applicable equipment having proper fume exhaust systems. Approved face shields (goggles) in compliance with ANSI Z87.1-1989 and chemical-resistant clothing and footwear and neoprene or nitrile gloves must be worn while performing any of these operations. Also, the section on safety outlined at the beginning of the book should be reviewed for more specific information.

## 13.1 DETERMINING THE IMAGE AREA TO BE PATTERN PLATED

One of the first steps in preparing to pattern plate is to precisely determine the area of exposed copper to be plated. This is necessary so that the power supply used in the process will be correctly set to result in high-quality plating. Three methods can be used to determine the pattern area: (1) the *calculation* method, (2) the *weight loss* method, and (3) using automatic area calculator equipment. Each of these methods has its advantages and disadvantages and each is discussed here.

To illustrate the methods of determining the image area, we will use the 5- by 5-inch card that was used in Sec. 11.6 as an example in the copper plating process. Recall that the conductor patterns laid out on each side of the board were made as close to equal as possible. Thus, by determining the pattern area on one side, it is assumed that it is the same on the other side.

We first demonstrate the calculation method. The thief-area calculation presents no special problem, since it is normally a square or rectangular border outside the conductor pattern area. The thief area is calculated as follows (Fig. 13.2). The area of the bulk stock is first determined in square inches. The thief area is then obtained by subtracting the available conductor pattern area from the bulk stock area:

thief area (square inches per side) = $(A \times B) - (C \times D)$

(where dimensions $A$, $B$, $C$, and $D$ are in inches). For the example card, the dimensions are

$$A = 5.0 \text{ inches}$$
$$B = 5.0 \text{ inches}$$
$$C = 4.0 \text{ inches}$$
$$D = 4.0 \text{ inches}$$

Therefore,

thief area (square inches per side) = $(5 \times 5 \text{ inches}) - (4 \times 4 \text{ inches})$
= (25 square inches) − (16 square inches)
= 9 square inches

The determination of the area of the conductor pattern is not quite as simple. To calculate the copper surface area of the conductor pattern that will face each anode, the area of each of the shaped configurations (i.e., donuts, IC patterns, paths, etc.) used in the lay-

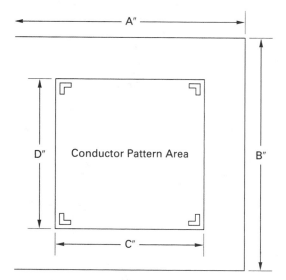

**Figure 13.2**
Calculation of thief area. Thief area = $(A \times B) - (C \times D)$.

out must first be determined. Table 13.1 is provided to aid in this calculation. Many standard patterns used in pc layouts are listed together with their areas in square inches. In addition, several common conductor path widths are given, together with their square-inch area per linear inch. The information given in Table 13.1 is for a 1:1 scale.

The following procedure is used with Table 13.1 to determine the square-inch area of a conductor pattern for *one* side of a double-sided layout: (1) Count the number of donuts of each size used in the layout and multiply their numbers by the appropriate square-inch areas listed for the various terminal pad sizes, which will determine the total area of just the donuts; (2) count and add the total area of each special layout configuration, such as IC patterns; and (3) determine the total length in inches of conductor paths and multiply this length by its area per linear inch. The addition of items 1, 2, and 3 will produce the total conductor area for one side of the pc board. The example shown in Table 13.2 of the foregoing procedure will illustrate its use.

To calculate the total area to be plated, simply add the previously calculated thief area to the copper surface area of the conductor pattern. For one side of the card:

$$\text{total area to be plated} = \text{thief area} + \text{conductor pattern area}$$
$$= (9 \text{ square inches}) + (\approx 1.00 \text{ square inch})$$
$$= 10 \text{ square inches}$$

## TABLE 13.1
Areas of Some Standard PC Patterns

| Pattern | Approximate Copper Area (in.$^2$) |
|---|---|
| *Terminal pad diameter (in.)* | |
| 0.150 | 0.018 |
| 0.125 | 0.012 |
| 0.100 | 0.008 |
| 0.09375 | 0.007 |
| 8-pin mini-DIP IC | 0.035 |
| 14-pin DIP IC | 0.06 |
| 16-pin DIP IC | 0.07 |
| 8-pin round IC | 0.038 |
| 10-pin round IC | 0.039 |

| Conductor Width (in.) | Area (in.$^2$) per Linear Inch | Area (in.$^2$) per 10 Linear Inches |
|---|---|---|
| 0.020 | 0.020 | 0.2 |
| 0.025 | 0.025 | 0.25 |
| 0.030 | 0.030 | 0.3 |
| 0.040 | 0.040 | 0.4 |
| 0.050 | 0.050 | 0.5 |
| 0.060 | 0.060 | 0.6 |

## TABLE 13.2
Pattern Area Calculation Example at 1:1 Scale (One Side)

| Pattern | Number Inches in Length | | Area of Each (in.$^2$) | | Total Area (in.$^2$) |
|---|---|---|---|---|---|
| $\frac{1}{8}$-inch donuts | 8 | × | 0.012 | = | 0.096 |
| $\frac{3}{32}$-inch donuts | 52 | × | 0.007 | = | 0.364 |
| 8-pin ICs | 2 | × | 0.038 | = | 0.076 |
| $\frac{1}{32}$-inch path ($\approx 0.030$) | 16.8 | × | 0.030 | = | 0.504 |
| | | | | Total | 1.04 |

The results of our calculation indicate that there are approximately 10 square inches of conductor pattern area to be plated on each side of the pc board. This value will be used to set the correct power supply setting to obtain the proper plating density (Secs. 13.3 and 13.6).

The apparent advantage to the *calculation* method is that it requires no special equipment. But even though it yields fairly accurate results, it is extremely time-consuming, especially when dense designs are involved.

The *weight loss* method of determining pattern area calculation initially requires the pc board stock to be processed, having 1 ounce per square foot (oz/ft$^2$) of copper foil on each side, accurately cut to its initial size. In the case of our pc board, this is a 5- by 5-inch panel. The panel is next heated at 150°F (65°C) for at least 1 hour to expel any moisture that may be present. It is then precisely weighed to within 0.01 gram on an analytical scale. This initial weight is recorded. It has been found that the panel weight for the board is 53.724 grams. We will call this weight $W_1$. Using the two positives and the imaging techniques discussed in Chapter 12, the panel is processed through dry-film lamination, exposure, and developing. Since positives are used, the *exposed* copper image is that of the total conductor pattern and thief areas. Etching the board at this point will remove just those exposed copper areas on both sides of the board. The panel is then rinsed, dried, again heated to 150°F for 1 hour to remove moisture, and then weighed. This new weight measurement is the panel *without* the copper forming the conductor pattern and thief areas. For our example panel, this weight is 49.78 grams. This second weight measurement will be called $W_2$. The weight of copper removed from both sides of the board is given by

$$W \text{ (total pattern)} = W_1 - W_2$$
$$= 53.72 - 49.78 = 3.94 \text{ grams}$$

To determine the area in square inches represented by the weight of copper removed from the board, we simply multiply the total weight loss figure by 5.079. This number represents the area in square inches per gram of 1-oz/ft$^2$ copper. Thus the total pattern area on the example panel is calculated as follows:

$$\text{total area (square inches)} = \frac{5.079 \text{ square inches}}{\text{gram}} \times 3.94 \text{ grams}$$
$$= 20.0 \text{ square inches}$$

Remember that this is the pattern and thief areas on *both* sides of the board. It can be seen that, as determined by the calculation method, the pattern and thief area of each side is approximately 10 square inches.

The weight loss method is less tedious than the calculation method, but it does require an accurate scale, proper preparation of the sample, and precise measurements to obtain correct results.

The third method of determining the pattern area uses commercial equipment made for this purpose, such as the Kahn area calculator, shown in Fig. 13.3. This instrument is capable of measuring the area of a pattern on either a positive or a negative phototool to an accuracy of ± 2%. Further, the phototool may be either silver halide or diazo film. In using this area calculator, all that is required is that it first be calibrated with the appropriate artwork that comes with the instrument. The Kahn area calculator consists of a lightproof box with a fixed-intensity light source in its drawer section. Above this source is a set of light-sensitive resistors (photoconductors) whose resistance is proportional to the intensity of light to which they are exposed. Calibration of the instrument is in units of square inches.

To measure the pattern area using the Kahn area calculator, the phototool is first placed in the center of the light table section of the drawer. The entire surrounding area is then blanked out with scrap pieces of opaque film so that all that is visible is the conductor pattern and thief area of the phototool. The drawer is closed and the digital readout, when

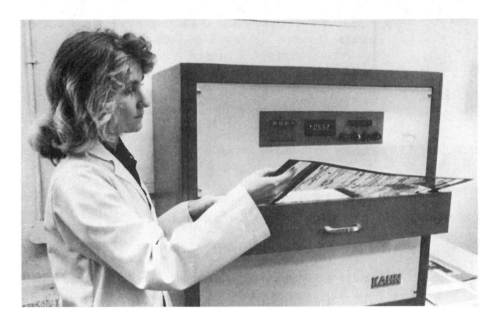

**Figure 13.3** Operator using area calculator. *Courtesy of Kahn Instruments, Inc.*

stabilized, displays the numerical value of the conductor pattern area directly in square inches. When using 1:1 scale positive or negative phototools, no corrections for scale enlargements or reductions are necessary. The use of an area calculator is the most rapid method of determining pattern area—and the most expensive.

## 13.2 PATTERN PLATING: GENERAL INFORMATION

With the conductor pattern imaged onto the panel, it is ready to be pattern plated. Since only the conductor pattern and thief areas are exposed to the plating solutions, they will be selectively plated with an additional amount of copper followed by a layer of solder. The process of pattern plating is similar to that of flash plating as discussed in Chapter 11; that is, the panel is first racked and chemically cleaned. In addition, the pattern-plated copper is deposited at the same current density of 30 amperes per square foot ($\approx$ 208 mA/in$^2$) per side. However, since only the conductor images and thief areas are exposed to the copper pattern-plating solution, the power supply setting will be lower than that used in flash plating the entire panel surface area. Plating time will depend on many variables, but, in general, it takes approximately 50 to 60 minutes to pattern plate an average of 1 mil of additional copper that will be deposited on all conductors, terminal pads, and barrels of all holes. This additional layer of copper in the hole barrels is required to ensure good electrical conductivity between the component-side and circuit-side terminal pads.

After the pattern plating is completed, the panels are removed from the plating tank and passed through another cleaning cycle to prepare them for solder plating. The solder plating is accomplished in a 60/40 tin–lead alloy bath. The same pattern areas are exposed to the solder plating solution as were exposed to the copper plating solution. Since solder is deposited at an average current density of 20 amperes per square foot ($\approx$ 140 mA/in$^2$) the power supply setting will be even lower than that used for copper pattern plating. To produce a good etch resist as well as to provide for an extended shelf life after solder reflow, a minimum average thickness of 0.0003 inch (0.3 mil) of solder is required. At the recommended current density of 20 amperes per square foot, this thickness of solder can be deposited in approximately 10 minutes.

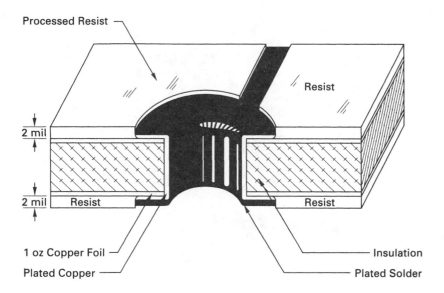

**Figure 13.4** Cross-sectional view of a pattern-plated hole.

Labels in figure: Processed Resist; Resist; 2 mil; 2 mil; Resist; Resist; 1 oz Copper Foil; Plated Copper; Insulation; Plated Solder

The cross-sectional view of a plated-through hole after it has been pattern plated with copper and solder is shown in Fig. 13.4. Note that all plated metal is deposited within the boundaries of the resist image. In the next section, detailed procedures for pattern plating copper and solder are discussed.

## 13.3 PATTERN PLATING OF COPPER

The pattern plating of copper onto an imaged panel begins with an effective chemical cleaning cycle. This cleaner should properly prepare the exposed copper surfaces to accept the additional layer of copper but must not degrade the dry-film resist on the panel. A cleaning cycle recommended where Dynachem Laminar AX dry-film resist is used to image the board is shown in Table 13.3. Boards to be plated are first placed in racks such as those shown in Fig. 11.10. The plating rack should make good electrical contact between the thief area of the board and the plating rack thumb screw.

The cleaning line for a board such as the 5- by 5-inch panel can be set up in laboratory beakers or small plastic tanks (polyethylene or polypropylene). All chemicals and solutions should be handled with extreme care in a well-ventilated area. In addition, a sink and water must be available for rinsing. Such a laboratory cleaning line is shown in Fig. 13.5. The board is processed through the following baths in the order presented.

### TABLE 13.3

Recommended Cleaning Cycle Prior to Copper Plating
for Dynachem Laminar AX Photo Resist

---

1. *Dip board into 20% by volume solution of Dynachem LAC-41 and water for 5 minutes at 160 ± 10° F (71°C). (Slowly agitate.)*
2. *Rinse board in warm tap water for 1 minute.*
3. *Rinse board in cool tap water for 1 minute.*
4. *Place board in ammonium persulfate, $(NH_4)_2S_2O_8$, micro-etch solution for 30 seconds at 70°F (21°C). (Slowly agitate.) Micro-etch solution—1 ½ pounds (680 grams) of ammonium persulfate crystals per gallon of tap water.*
5. *Rinse board in cold tap water for 1 minute.*
6. *Dip board in 3% by volume solution of sulfuric acid, $H_2SO_4$, and water for 2 minutes.*
7. *Electroplate.*

---

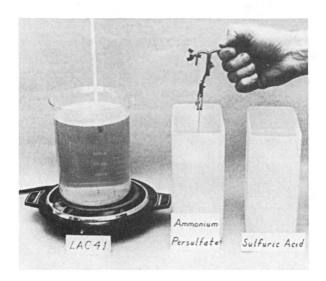

**Figure 13.5** Cleaning line (water rinses not shown).

**Step 1.** A proprietary solution of Dynachem LAC-41 will be used to remove oxides and greases from the exposed copper area of the conductor pattern and the thief area. This solution is selected because it will not adversely affect the resist or its bond to the copper surface. To prepare this cleaner, 1600 milliliters of water are placed into a 2000-milliliter beaker. To this beaker, 400 milliliters of concentrated LAC-41 solution are added and stirred. This yields a 20% by volume working bath of cleaning solution. The beaker is then placed onto a hot plate and the solution brought up to and stabilized at a temperature of 160°F (71°C). A laboratory thermometer is used to monitor this temperature. The racked board is then placed into the heated solution for 5 minutes, during which time the board is slowly rocked back and forth to allow fresh solution to pass across the exposed copper areas.

**Step 2.** The board is removed from the LAC-41 solution and rinsed completely under warm tap water for at least 1 minute.

**Step 3.** While still at the sink, cold tap water is turned on and the board is rinsed for an additional minute. This assures that all the cleaning solution is removed from the copper surfaces.

**Step 4.** An ammonium persulfate (AP) etching bath is the next phase of the cleaning cycle. This bath is prepared by adding 12 ounces of AP crystals to 2000 milliliters of warm water. To achieve uniform etching, it is absolutely essential that the board be slowly rocked back and forth continuously for approximately 30 seconds. A visual inspection of the conductor paths should show a smooth matte finish that is salmon in color. The proper operation of this micro-etch solution is critical to maintain good-quality boards. If the panel is left in this bath too long, the attack of the etchant solution on the hole walls will result in voids being generated. In the worst case, the entire 125-$\mu$inch thickness of copper on the barrels could be etched away. On the other hand, if the boards are not left in the bath long enough, the etch will not completely remove the photo resist particles that may be locked into the copper. This would result in poor adhesion of the plated metal.

**Step 5.** The board is removed from this micro-etch bath and rinsed under cold tap water for at least 1 minute to remove the majority of the ammonium persulfate residue from the board surface.

**Step 6.** To remove all traces of ammonium persulfate residue and to neutralize the copper surfaces, the board is dipped into a diluted solution of sulfuric acid and rocked slowly for a minimum of 2 minutes to a maximum of 20 minutes (holding tank). This solution is prepared by adding 60 milliliters of CP sulfuric acid to 1940 milliliters of water. This mixture will yield a 3% by volume solution of sulfuric acid.

**Step 7.** After the final 3% sulfuric acid dip, the rack with the board is transferred directly into the plating tank for the copper pattern-plating process.

It is recommended that all cleaning solutions used be prepared with either CP (chemically pure) or reagent-grade acids. In small quantities, all solutions used in the cleaning line should be changed daily to minimize the possibility of cleaning problems. The use of large quantities makes daily discarding and preparation time consuming and economically unfeasible. To determine the effectiveness of the cleaning solutions and procedures, a modified water-break test should be used periodically. A test pc board (no pattern necessary) is passed sequentially through all the cleaning steps as outlined above. After the board is removed from the final sulfuric acid bath, the acid should "sheet" across the entire copper surface for a minimum of 30 seconds with no detectable "water breaks," which is an indication of the presence of contaminants not successfully removed. A cleaning cycle that passes this test is effective in preparing the copper surfaces for quality plating.

The process for pattern plating copper employs the same plating tank and copper plating solution as described in Secs. 11.6 and 11.7, respectively. These sections should be reviewed before proceeding further.

After the board has been treated through the chemical cleaning and micro-etching baths, the rack is removed from the 3% solution of sulfuric acid. The board is then placed directly into the plating tank and the rack is secured to the cathode work rod with the air sparger activated and the power supply turned on to a low setting. With the board in the tank, the power supply may then be adjusted to the correct current setting. As discussed previously, the manufacturer's recommendation for optimum current density is 30 amperes per square foot ($\approx 208$ mA/in$^2$) of copper area that is exposed to the plating solution. Recall that it was determined that the conductor patterns and thief areas on *both* sides of our 5- by 5-inch panel had a total copper surface area of approximately 20 square inches. The plating power supply setting is thus found according to the following relationship:

$$\text{current setting} = \text{current density} \times \text{total surface area}$$
$$= \frac{208 \text{ milliamperes}}{\text{square inch}} \times 20 \text{ square inches}$$
$$\approx 4200 \text{ milliamperes} \approx 4.2 \text{ amperes}$$

It will be noted that with the power supply set at 4.2 amperes the output voltage is low (less than 1 volt) for this plating solution.

The required thickness of pattern-plated copper is an average of 1 mil. Plating thickness is a function of the current density and the plating time. The plating rate for varying current densities is shown in Table 13.4. Note that this table is based on 100%

**TABLE 13.4**

Plating Rate for Copper Sulfate Solution Based on 100% Cathode Efficiency

| Thickness | | Plating Time (minutes) at Several Current Densities (A/ft$^2$) | | | |
|---|---|---|---|---|---|
| in. | mils | 20 A/ft$^2$ | 25 A/ft$^2$ | 30 A/ft$^2$ | 40 A/ft$^2$ |
| 0.0005 | 0.5 | 27 | 21 | 18 | 13 |
| 0.0006 | 0.6 | 32 | 25 | 21 | 16 |
| 0.0007 | 0.7 | 37 | 30 | 25 | 18 |
| 0.0008 | 0.8 | 42 | 34 | 28 | 21 |
| 0.0009 | 0.9 | 48 | 38 | 32 | 24 |
| 0.001 | 1.0 | 53 | 42 | 35 | 26 |
| 0.00125 | 1.25 | 67 | 54 | 44 | 33 |
| 0.0015 | 1.50 | 80 | 64 | 53 | 40 |

cathode current efficiency. Because this is not practical, it has been found that the plating time listed in Table 13.4 should be increased by approximately 25% to result in the specified plating thickness. For example, to obtain our specified copper thickness of 1 mil with a current density of 30 amperes per square foot, the ideal time of 35 minutes should be multiplied by 1.25, which increases the time to approximately 45 minutes.

After the panel has been plated, the power supply is reduced to a low setting ($\approx \frac{1}{2}$ ampere) and the rack is removed from the cathode work rod. The board is immediately placed into a water rinse tank for at least 1 minute and then inspected for plating quality. There should be no *pits, frostings,* or *hazy regions* appearing on the imaged areas.

With copper pattern-plating completed and the board rinsed, the panels are prepared for solder plating by dipping them into a 10% by volume solution of fluoboric acid at room temperature. This bath is prepared by adding 200 milliliters of concentrated fluoboric acid to 1800 milliliters of water and stirring. The panels should be slowly agitated for 1 minute in this bath. They are now ready for the solder pattern-plating process.

## 13.4 TIN–LEAD PLATING TANK AND ACCESSORIES

In the preceding section, the conductor patterns and thief areas of the pc board were pattern plated with an additional 1 mil thickness of copper in preparation for solder plating. Prior to the discussion of this plating process, however, the construction of a prototype plating tank will first be presented. You will note that this tank is similar to that used for flash plating and described in Sec. 11.6.

Because the tin–lead alloy will be electroplated from a fluoborate plating solution, the recommended material for the tank is stress-relieved polypropylene or polyethylene plastic. Glass-silicated materials and titanium are not suitable for fluoborate solutions and their use should be avoided.

A tank size of 12 inches long by 12 inches wide by 12 inches deep will be adequate for most prototype work. This tank will have a solution capacity of approximately 6 gallons when the plating solution is 9.75 inches from the bottom. As before, the exact volume can be calculated by dividing the number of cubic inches of solution by 231 cubic inches per gallon. A 12-inch-deep tank with approximately 10 inches of plating solution (2 inches of clearance to the rim of the tank) will allow the insertion of the anodes and the racked pc board to be plated without causing the solution to spill over the top of the tank.

The tank is fitted with two anode work rods and one cathode work rod. The anode work rods are made of $\frac{3}{8}$-inch-diameter solid copper rod stock. This diameter will easily support the weight of the tin–lead anodes as well as handle the currents used. These anode work rods are held securely into position with pairs of Lucite or plastic blocks fastened onto opposite ends of the tank rim. The rods are positioned over the tank parallel to one another and 1 inch from the sides of the tank. The cathode work rod made from the same $\frac{3}{8}$-inch-diameter copper stock is positioned parallel to the anode work rods and centered between them. The spacing between either anode work rod and the cathode work rod is 5 inches. The cathode work rod rests on a pair of plastic rollers attached to opposite ends of the tank rim and centered between the Lucite anode rod support blocks. This allows movement of the cathode rod parallel to the stationary anode rods. The cathode work rod must be capable of back-and-forth travel to provide agitation of the pc board in the plating solution. This type of rod motion will produce "knife agitation" of the racked pc board through the solution and is recommended for tin–lead alloy plating provided that the rate of motion is not violent. For tin–lead alloy plating, the recommended distance of travel of the pc board in the solution is from 3 to 6 feet in 1 minute. To realize this optimum range of travel distance, a high-torque motor operating at 12 rpm is mounted to the tank rim. A drive wheel of approximately 2 to 3 inches in diameter is assembled to the motor's shaft and the cathode work rod is linked to the drive wheel at a distance of 1 inch from the center of the motor's shaft. The linkage of the cathode work rod to the motor *must* be made with a nonconductive material, preferably plastic. As the motor rotates the drive wheel through one revolution, the cathode work rod will be pushed forward,

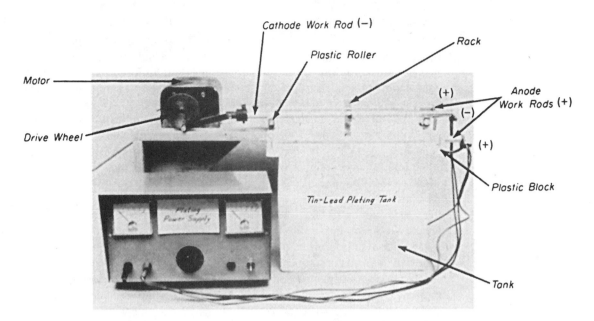

**Figure 13.6** Typical plating tank and accessories for tin–lead solder plating.

pulled back, and then pushed forward again for a total distance of 4 inches. With a 12-rpm motor, the total distance traveled will be 48 inches, or 4 feet, in 1 minute. This distance is within the recommended range for "knife agitation" used for tin–lead alloy plating. A typical plating tank fabricated as described in this section is shown in Fig. 13.6.

The negative terminal of a plating power supply is electrically connected to the cathode work rod. Both anode work rods are electrically connected (anode strap) and then connected to the positive terminal of the power supply. All electrical connections should be made with AWG No. 12 stranded wire protected with a chemical-resistant plastic insulation.

The plating power supply should have the same characteristics as that used for flash plating as described in Sec. 11.6. It can, however, have a lower current rating since pattern plating requires much less current than does flash plating the entire panel area.

The type of material used for the anode, together with its shape and surface area, will affect the quality of the alloy that is plated onto the pc board. Because the tin–lead alloy to be plated requires a metal ratio of 60% tin and 40% lead, anodes with this same ratio should be utilized. The anodes must also be free of oxides and have low metal impurities. Anode material that meets all these specifications is high-purity Vaculoy. Anodes made with this material will corrode evenly in the plating solution, produce less sludge, and improve bath stability.

Recall from Sec. 11.6 and Fig. 11.13 that the standard anode shapes are *flat, oval, round, sawtooth,* and *gear tooth.* For prototype applications, the flat configuration is preferred. The criterion for selecting an anode is its surface area relative to the area of the pc board to be plated. For the plating of tin–lead alloy, an anode-to-cathode surface area ratio of approximately 2:1 is recommended. This means that the surface area of the anode facing the pc board in the plating tank should have a surface area approximately twice as large as the one side of the board it is facing. You will recall from Chapter 7 that uniform plating on double-sided pc boards will result if the overall exposed conductor path area is made to be equal on both sides of the board. Double-sided plating is accomplished with the use of two anodes, one positioned on each side of the pc board, both having identical surface areas. This will simplify and optimize the plating process and a uniform coating of 60/40 tin–lead alloy will be plated on both sides of the pc board.

The anodes must be provided with a hook whose length will allow the anode to extend from the anode work rod down into the plating solution yet allow the top of the anode to protrude slightly above ($\frac{1}{4}$ inch) the surface of the solution. The length of the anode should be such that it extends from 1 to 2 inches below the bottom edge of the pc board, which is suspended from the cathode work rod. This will result in a higher current density on the thief area of the pc board and tends to improve the uniformity of alloy composition in the conductor-path area. Thus the two restrictions that are placed on the selection of the anode size are that (1) it has a surface area from one to two times larger than that of the pattern facing it and (2) it is from 1 to 2 inches longer than the pc board bulk stock. We will again use our example panel to select the correct anode size for solder pattern plating. Recall that the exposed conductor pattern and thief area on each side of the 5- by 5-inch card is 10 square inches.

First, an anode is selected whose length is 6 inches (1 inch longer than the 5-inch bulk stock). If the anode width is chosen to be 4 inches, the anode surface area will be 6 inches $\times$ 4 inches $=$ 24 square inches, which is within the range of from one to two times larger in area than the area of the pattern to be plated. Since each side of the layout has approximately equal areas, two 4- by 6-inch anodes will meet all of the requirements for plating the pc board.

Prior to the placement of the anodes into the plating tank, proper procedures must be followed to avoid contaminating the plating solution. Tin–lead fluoborate plating solutions are extremely susceptible to organic contaminations. For this reason, the anodes must be properly cleaned and correctly *bagged.* Anode degreasing is accomplished by scrubbing the surface with a brass bristle brush and cold tap water. The anode is then rinsed thoroughly and dipped into a 15% by volume solution of fluoboric acid for approximately 10 minutes. After the acid dip, the anode is ready to be bagged.

The anode bags used should be made of either polypropylene or dynel (*but never cotton*) and should be 2 inches wider and 2 inches longer than the anode. The bag must be *leached* (purified) of all organic contaminations that it was exposed to in its manufacturing process. Leaching is accomplished by boiling the bag in water for at least 1 hour at 150°F (66°C). The bag is then removed, rinsed in cold water, and tied around the cleaned anode. The anodes are now ready to be installed into the plating tank. Recall that the anode cleaning and bag leaching procedure is similar to that used for the anodes and bags in the copper plating process discussed in Sec. 11.6.

In addition to the anode preparation, the plating tank itself must be thoroughly leached of all contaminations prior to the addition of the plating solution. Tank leaching should be done with all the hardware and electrodes removed from the tank. Leaching of the tank need be done only before the plating solution is added to the tank *for the first time.* Detailed information for leaching the tank is given in Appendix IX. The student should refer to this appendix before proceeding to the next section.

## 13.5 TIN–LEAD PLATING SOLUTION

After the plating tank has been properly leached, the preparation of the plating solution may begin. For our purposes, a high-throwing power formulation will be used to electrodeposit 60/40 tin–lead alloy onto the copper conductor pattern surface. The solution described is a high-acid, low-metal (tin and lead) mixture formulated under the trade name HI-THRO, manufactured by the Allied Chemical Corporation in Morristown, NJ, under U.S. Patent 3,554,878. The plating bath is made up of the following chemicals and compounds:

1. Stannous fluoborate concentrate, 51.0% by weight
2. Lead fluoborate concentrate, 51.0% by weight
3. Fluoboric acid, 49% by weight
4. Boric acid
5. Stabilized liquid peptone solution (available from Allied Chemical Corporation)
6. Water

Stannous and lead fluoborate are used to provide the source of metal in the plating bath. Because variations in the metal content of the plating bath will adversely affect the content of the alloy deposited, both the stannous and lead fluoborate additives must be carefully controlled when preparing the bath so that it will contain 60% by weight of tin as a metal and 40% by weight of lead as a metal.

The fluoboric acid aids in increasing the conductivity of the bath and improves the grain quality of the deposit by making it finer and smoother. Peptone also improves the fine-grained deposit and inhibits the formation of "trees," which are small growths of deposit extending away from the conductor pattern. Boric acid aids in maintaining stability by minimizing the decomposition of the fluoborates in the solution.

The chemicals and compounds necessary to prepare a 10-gallon bath of HI-THRO tin–lead fluoborate plating solution are listed in Table 13.5. The table includes the required amounts and the order of mixing the ingredients. For purposes of preparing a plating bath for the 6-gallon tank previously discussed, Table 13.6 lists the amount of each ingredient and the order of mixture. The required amount of each of the ingredients was derived by taking $\frac{6}{10}$ (0.6) of the amounts listed in Table 13.5.

The 6-gallon plating solution is prepared by first placing one-half of the required amount of water (0.96 gallon or 3633 milliliters) into the plating tank. To 1500 milliliters of hot water, 1.11 pounds (503.5 grams) of boric acid are added and stirred until it is completely dissolved. This solution is then added to the plating tank and stirred. The following are then stirred into the plating tank in the order given: 0.126 gallon (480 milliliters) of lead fluoborate solution, 3.6 gallons (13.62 liters) of fluoboric acid, 0.27 gallon (1020 milliliters) of stannous fluoborate solution, and 0.102 gallon (390 milliliters) of stabilized peptone. The remaining water (2137 milliliters) is added to the tank to complete the bath solution. The solution is thoroughly stirred and allowed to settle overnight before use.

The height of the solution in the tank should be marked on the *outside* to note any decrease due to evaporation. Any significant decrease of solution should be replaced so as to return it to the original level in the tank. The bath should be maintained at room temperature and loosely covered with sheet plastic when not in use to protect the solution from dust and other foreign matter.

## TABLE 13.5
Formulation of 10 Gallons of Tin–Lead Plating Solution

| Order of Mixing Ingredients | Compound | Amount | |
|---|---|---|---|
| 5 | Stannous fluoborate concentrate, 51% | 0.45 gal | 1.70 liters |
| 3 | Lead fluoborate concentrate, 51% | 0.21 gal | 0.80 liters |
| 4 | Fluoboric acid, 49% | 6.00 gal | 22.70 liters |
| 2 | Boric acid | 1.85 lb | 839 grams |
| 6 | Stabilized peptone solution | 0.17 gal | 0.64 liters |
| 1,2,7 | Water (deionized) | 3.20 gal | 12.10 liters |
| | Total solution | 10.03 gal | 37.94 liters |

## TABLE 13.6
Formulation of 6 Gallons of Tin-Lead Plating Solution

| Order of Mixing Ingredients | Compound | Amount | |
|---|---|---|---|
| 5 | Stannous fluoborate concentrate, 51% | 0.270 gal | 1.02 liters |
| 3 | Lead fluoborate concentrate, 51% | 0.126 gal | 0.48 liters |
| 4 | Fluoboric acid, 49% | 3.600 gal | 13.62 liters |
| 2 | Boric acid | 1.110 lb | 503.50 grams |
| 6 | Stabilized peptone solution | 0.102 gal | 0.39 liters |
| 1,2,7 | Water (deionized) | 1.920 gal | 7.27 liters |
| | Total | 6.018 gal | 22.78 liters |

All plating solutions require periodic chemical analysis. In alloy plating, the ratio of metals deposited must be maintained within narrow limits. The techniques of chemical analysis for the type of bath just described are beyond the scope of this book. However, since the amount of solution described in Table 13.6 is relatively small and not significantly expensive, the entire bath should be replaced if plating problems are encountered. A fresh batch of solution can be used successfully for a reasonable length of time if extreme care is exercised to minimize the introduction of organic contaminations into the solution. Detailed chemical analysis techniques may be obtained from Allied Chemical Corporation for HI-THRO plating solution.

## 13.6 60/40 TIN–LEAD PATTERN PLATING

After the board has been copper pattern plated, rinsed, and dipped into the 10% fluoboric acid solution, it is ready for the solder plating process. The power supply is first turned on and set to a low reading (0.5 ampere) to activate the plating solution. The racked board is then removed from the fluoboric acid bath and placed into the solder plating tank. The rack is secured onto the cathode work rod with the thumbscrew and the power supply is adjusted to the correct current density. The current setting determines the proper alloy plating and needs to be properly set. The manufacturer recommends that the optimum current density be 20 amperes per square foot ($\approx 140$ mA/in$^2$) for this plating solution. In Sec. 13.1 it was calculated that the conductor patterns and thief areas on both sides of the example board each have a surface area of 10 square inches. Therefore, the plating power supply must be set as follows:

$$
\begin{aligned}
\text{current} \quad &= \text{current density} \times \text{total surface area} \\
&= \frac{140 \text{ milliamperes}}{\text{square inch}} \times 20 \text{ square inches} \\
&= 2800 \text{ milliamperes} = 2.8 \text{ amperes}
\end{aligned}
$$

It will be noted that when the power supply is set at 2.8 amperes, the output voltage is low (in the order of 1 volt or less) for this plating solution.

The thickness of the plating onto the pc board is a function of the current density and the plating time. For the HI-THRO alloy bath, 60/40 tin–lead is plated at a rate of 0.0001 inch (0.1 mil) every 2.3 minutes. To achieve a plating thickness of 0.3 mil requires 6.9 minutes, and 0.5 mil requires 11.5 minutes of plating. A minimum of 0.3 mil of plating thickness is required for the tin–lead alloy to provide an effective etching resist. A plating time of 7 minutes has been established as optimum to provide the pc board with the minimum required plating thickness.

After the pc board has been plated, the power supply is turned down to a few milliamperes to keep the tank "live," the motor driving the cathode rod is turned off, and the

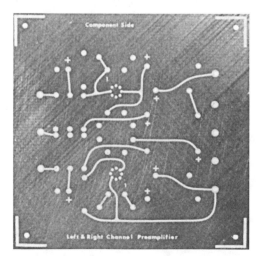

**Figure 13.7** Properly processed PTH board through copper and solder electroplating.

rack is removed from the cathode work rod. Finally, the power supply is turned off. The board is then thoroughly rinsed in cold tap water for at least 1 minute and then dried. Inspection of the plating should reveal a smooth, fine-grained deposit with a light dull-gray surface having no apparent dark streaks. A properly plated pc board is shown in Fig. 13.7.

## 13.7 RESIST STRIPPING

The Laminar AX resist previously applied to the pc board may now be removed by chemical stripping since it has served its purpose. The Dynachem Corporation provides an alkaline solution that will remove the resist by simple immersion. For Laminar AX resist, 400K stripper is available as a concentrate. The stripping solution is prepared by mixing 1 part of 400K stripper to 9 parts of water in a beaker. The stripping solution is then heated on a hot plate to 130°F (54°C). The temperature should be checked with a laboratory thermometer. Immersion of the pc board into the stripper for approximately 2 minutes will dissolve the resist in the hot water. The board is then removed from the stripper and rinsed first in hot tap water and then in cold tap water for at least 1 minute each to completely remove any remaining film residue. It is recommended that the board surface be gently scrubbed with a nylon bristle brush while under the hot-water rinse. Since the stripper solution is highly alkaline, chemical-resistant gloves should be worn.

## 13.8 ETCHING

After stripping the resist from the processed pc board, it is ready to be etched. The copper conductor pattern is protected with a 0.3-mil-thick layer of 60/40 tin–lead alloy while the exposed surrounding areas are unprotected and can be removed by etching. Etching is accomplished in a similar manner as described in Chapter 10. However, ferric chloride, which was used in Chapter 10 as the etching solution, cannot be used because it will attack the tin–lead alloy as well as the exposed copper. The tin–lead alloy plating acts as the etch "resist" and an etchant solution must be used that will be effective in attacking the exposed copper without adversely affecting the plating. Four common etching solutions can be used for this purpose: alkaline ammonia, chromic-sulfuric acid, and peroxide/sulfuric or modified ammonium persulfate.

*CAUTION:* Both the alkaline ammonia and the chromic-sulfuric acid solutions produce noxious and poisonous fumes, especially when heated. Peroxide/sulfuric solution requires constant monitoring and control. When using these solutions as etchants, extreme caution and good ventilation are required. For these reasons, alkaline ammonia, chromic-sulfuric acid, and peroxide/sulfuric are limited to industrial applications.

The modified ammonium persulfate solution is much safer to use and is the best choice for prototype applications. This etchant solution is made by dissolving 2 pounds (0.91 kilogram) of ammonium persulfate crystals per gallon (3.785 liters) of deionized water at a temperature of 120°F (49°C). Phosphoric acid is then added in the amount of approximately 50 milliliters per gallon of etching solution. The formulation of 10 gallons of modified ammonium persulfate etchant is shown in Table 13.7.

**TABLE 13.7**
Formulation of 10 Gallons of Ammonium Persulfate Etchant

| Compound | Amount |
|---|---|
| Deionized water | 10 gal |
| Ammonium persulfate crystals | 20 lb (18.2 kg) |
| Phosphoric acid | 500 mL |

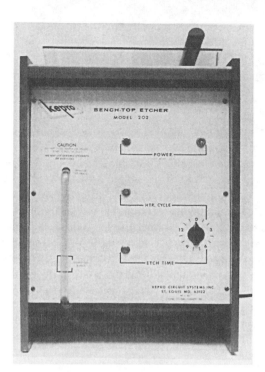

**Figure 13.8** Spray etcher for both single- and double-sided pc boards.

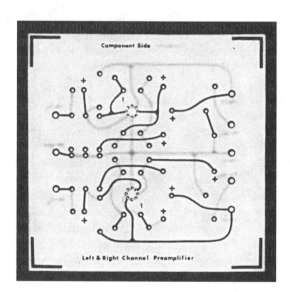

**Figure 13.9** A pc board after etching.

Etching double-sided pc boards is best accomplished with the use of a spray etcher, such as that shown in Fig. 13.8. This etcher allows both sides of the pc board to be etched at the same time. The spray etcher is operated at a temperature of 110 to 120°F (43 to 49°C). The time required to accomplish the etching is a function of the temperature and the amount of copper in the solution. A 1-ounce copper board should satisfactorily be etched in 3 to 5 minutes. As more boards are etched, additional copper goes into the solution and a point will be reached where etching time will exceed 6 minutes. At this point, the etchant should be replaced with fresh solution. Where small volumes are involved, it is advisable to prepare a fresh batch daily.

After the board has been etched, it is rinsed in cold tap water and thoroughly dried with paper towels. The etched pc board is shown in Fig. 13.9.

Once the pc board has been etched, it is ready for final processing: drilling, shearing, component assembly, and component soldering. However, an interim step is recommended at this stage of processing. This process is called *fusion,* sometimes referred to as *reflow,* and involves heating the 60/40 tin–lead alloy electroplated onto the copper surface to a temperature sufficient to melt the plating and fuse it to the underlying copper. The following advantages are realized through this process: (1) improved solderability, (2) extended shelf life, (3) elimination of solder overhang, which may have resulted after the etching process, (4) a check to observe if de-wetting of the copper surface is a problem, and (5) a rough check on the metal ratio of plating. Items 1, 2, and 3 are self-explanatory. Items 4 and 5 will be discussed in detail. During the pattern-plating process, the 60/40 tin–lead metals have been plated onto the copper surface. The plating has not been alloyed with the copper but is adhered through a metallic bond. For alloying to occur, the solder must be elevated to its melting temperature.

If the solder has been plated to a clean copper surface, the fused solder will appear as bright and shiny over the entire conductor path areas. After fusion, an indication of an improper bond is a nonuniform surface showing nodules of solder. This de-wetting condition is a probable indication of an unclean copper surface and the board should be rejected to avoid further problems during the component soldering process, whether it be by hand or wave methods. The fusion process, then, is a quality-control check on the cleaning process. If de-wetting occurs after fusion, the cleaning process needs to be improved and/or corrected.

Fusion can also be used as a means of roughly checking the tin–lead ratio being plated. Reference to the tin–lead fusion diagram in Fig. 17.1 shows that at temperatures between 361 and 400°F, the range of tin–lead solder alloys that will melt are 55/45 through 60/40 to 80/20. If the plating has not melted and formed an alloy with copper after the fusion process, the metal content of the plating bath is out of balance and should be replaced.

Fusion is accomplished in an infrared drawer-type oven, similar to the one shown in Fig. 13.10. This style of oven consists of a wire belt drawer with infrared heating elements positioned above and below, a preheat timer, a high-intensity timer, a cycle button, and an exhaust blower system. The recommended time settings for $\frac{1}{16}$-inch pc board stock are 15 seconds for the preheat timer and 9 seconds for the high-intensity timer.

Before the pc board is placed into the oven drawer, it must be fluxed with a solder-fusing liquid. This liquid is applied to all areas of the pc board by hand brushing, spraying,

**Figure 13.10** Bench-type solder fusion oven. *Courtesy of Glo-Quartz Ovens, Inc.*

or roller coating. The fusing liquid should be a mildly acidic flux that is completely water soluble. One such flux is Solder Flow No. 2111A available from Glo-Quartz Ovens, Inc. The purpose of the flux is to prevent air from contacting the molten solder, to minimize the formation of oxides, to hold any impurities in suspension so that they will not interfere with the alloying process, and to reduce the possibility of de-wetting.

After the pc board has been properly fluxed, it is inserted into the drawer of the oven. The oven door is closed and, with the two timers set, the cycle button is activated. During the preheat cycle ($\approx$ 15 seconds) the board is elevated to a temperature of approximately 200°F (93°C) to reduce the possibility of thermal shock. At the end of the preheat cycle, the high-intensity cycle ($\approx$ 9 seconds) automatically increases the board's temperature to approximately 400°F (204°C), melting the plate alloy and fusing it to the copper surface. The board is allowed to cool for a few minutes before removing it from the oven. The result of fusion is a conductor pattern having a bright and shiny appearance.

Because Solder Flow No. 2111A is a mildly acidic solution, the residue remaining on the copper surfaces must be completely removed. This is best accomplished by briskly brushing both sides of the pc board in warm water using a nylon bristle brush. A final rinse under cold tap water for 30 seconds on each side followed by paper-towel drying completes this phase of the pc board processing.

## 13.10 MECHANICAL METHODS USED TO FABRICATE PROTOTYPE PCBS

There is an alternative method available to PCB users for obtaining prototype circuit boards besides the multistep chemical processing techniques discussed in Chapters 9 through 13. In many instances, the PCB user may have to send circuit board designs to a circuit board manufacturer that uses wet-processing techniques when only one or two prototype boards are needed. The turnaround time is often lengthy and very costly when only a few boards are needed. There is a software-controlled prototyping system available, designed by T-Tech, Inc., that solves many PCB user problems. The system is used to make single- and double-sided analog, digital, RF/microwave prototype boards and its software is compatible with standard CAD output files. The system, Quick Circuit™, is an X–Y milling/drilling table that can perform drilling, engraving, milling, and routing operations and was developed specifically for research and design laboratories requiring prototype quantities. One of the Quick Circuit™ systems, model 5000HS with a software-controlled, high-speed spindle, is shown in Fig. 13.11. Models 5000HS and 7000 will accept Gerber™, DXF, HPGL, and Excellon™ files. See Appendix XXI for software options associated with using the Quick Circuit™ system.

The standard Quick Circuit™ model 5000 has a spindle speed range of 8,000 to 23,000 rpm, while models 5000HS and 7000 machines have software-controlled, high-

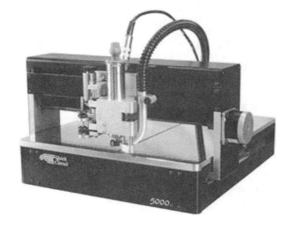

**Figure 13.11** The Quick Circuit™ System 5000 HS. *Courtesy of T-Tech, Inc. Atlanta, GA, www.t-tech.com*

speed spindles that can be set anywhere from 5,000 to 60,000 rpm. Vendors of high quality microwave materials often recommend spindle speeds of up to 60,000 rpm when using smaller end mills. The high-speed spindle allows for faster machining, longer tool life, and the use of thicker board materials. See Appendix XXII for Quick Circuit™ models 5000, 5000HS, and 7000 mechanical and electrical specifications, computer requirements, and additional capabilities. Models 5000 and 5000HS have work areas of 10 inches by 11 inches (254 mm × 279 mm); model 7000 has a work area of 13 inches by 19 inches (330 mm × 480 mm).

Quick Circuit™ systems produce accurate, economical protoboards in a fraction of the time of other methods of simulations. Quick Circuit™ systems can produce mechanically engraved prototype circuit boards the same day you complete your design, and will function identically to printed circuit boards produced using conventional plating and acid-etch processes.

Another machine with a larger work area is also available from T-Tech, Inc. The ProtoDrill™ shown in Fig. 13.12 is a single-spindle drilling and routing machine with three-axis control and a work area of 24 inches by 24 inches by 2 inches (610 × 610 × 2 mm). This system is capable of drilling up to three standard circuit panels at one time with speeds in any axis of up to 300 inches per minute (7.5 meters/minute) and an accuracy of ± 0.001 inch/1 mil (0.025 mm). The ProtoDrill™ can be driven from any Windows 95/98/NT capable PC. The ProtoDrill accepts standard 0.125-inch diameter, 1.5-inch length tooling. Its spindle speeds are software controlled and vary from 5,000 rpm to 50,000 rpm. Holes as small as 0.008 inch (0.2 mm) can be drilled. See Appendix XXIII for mechanical and electrical specifications and computer requirements.

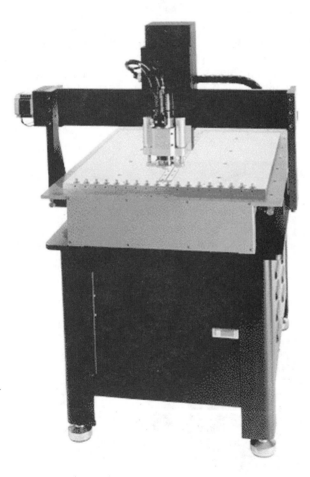

**Figure 13.12** The single-spindle drilling and routing ProtoDrill machine. *Courtesy of T-Tech, Inc., Atlanta, GA, www.t-tech.com*

There are six major steps involved in producing a protoboard from circuit board CAD designs using a Quick Circuit™ system. The following describes each of these steps.

**Step 1.** The CAD design is completed by generating a Gerber™ format file for each circuit layer. In addition, an Excellon™ format file must also be generated for all drilling information. These two files contain the same data that would be sent to a traditional board house to have costly prototypes manufactured. With your existing CAD package and this data, the Quick Circuit™ system is then capable of producing a circuit board prototype. Fig. 13.13 illustrates software being used to obtain data necessary to complete a total CAD design package.

**Step 2.** Next, software conversion is performed. In addition to being a full-featured Gerber™, Excellon™, and HPGL™ editor, the Isolation software is used to create drilling and milling data for the Quick Circuit™ CNC table. From your CAD files, the Isolation software computes the isolation path outlining each electrical net on your circuit board (see Fig. 13.14). The isolation paths become the mill paths that the Quick Circuit™ CNC table uses to mechanically engrave the desired conductor pattern into the copper foil surface of the protoboard.

**Step 3.** Next, board material is selected and simple machine setup undertaken. Standard FR4/G10 is commonly chosen. Also, for tarnish resistance and ease of soldering, solder-plated board material may be selected. Aluminum clad backup and drill entry materials are also used. A variety of board, backup, and entry materials are available. To begin the setup of the three elements on the machine, aluminum clad backup material is placed beneath the circuit board material being used for the prototype board. The backup material is used to prevent hole-wall damage, drill breakage, and drilling into the top of the CNC table top and to extend tool life by dissipating heat away from the drill tip. Backup material with an adhesive side to assist in securing flat, thin laminate materials is also available. Entry material, which consists of two thin sheets of aluminum with a cellulose interior layer, is placed onto the drill-entry side of the circuit board material. The use of the entry material helps to reduce entry burrs, minimizes drill wander, aids in making small drill holes associated with microvias, provides better front to back registration, and extends drill

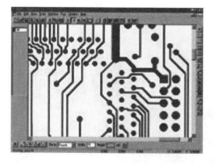

**Figure 13.13** Working with your existing CAD package. *Courtesy of T-Tech, Inc., Atlanta, GA, www.t-tech.com*

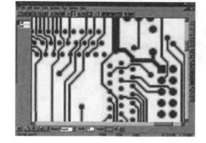

**Figure 13.14** Using isolation software for outlining conductor paths. *Courtesy of T-Tech, Inc., Atlanta, GA, www.t-tech.com*

life by also absorbing heat from the drill bit. Once the three elements are properly positioned, their tooling holes are drilled, pinned together, and affixed to the CNC table. See Fig. 13.15. The table's precision pin, bushing, and groove features ensure perfect registration when the material is turned over for milling the bottom of the circuit board.

*Step 4.* Once the board material with entry and backup panels have been pinned to the CNC table, precise drilling and milling operations are performed. The Quick CAM™ software that drives the Quick Circuit™ CNC table is easy to use. First all holes are drilled, then the isolation milling is performed. Fig. 13.16 shows the milling operation being performed. Quick Circuit's high resolution and repeatability allows trace widths as small as 0.004 inch (0.1 mm). Quick Circuit's precision allows you to prototype high-density digital, microwave, and surface mount circuit boards. See Appendix XXIV for tooling information about carbide drill bit and endmill selection and styles used for these two operations that are available from T-Tech, Inc. Also refer to Appendix XXV for the recommended feed rates (inches per minute) for milling tools.

*Step 5.* Once the protoboard has been drilled and its isolation paths milled, contour routing to profile the board to finished size is done. Quick Circuit™ contour routes circuit boards of virtually any shape and size. Figure 13.17 shows the contour routing operation being performed. Interior routing is just as easy. Large "drill" holes can also be routed instead of drilled. This is a unique feature of the QuickCAM™ software that saves time and money by no longer needing excessive inventories of large drill bits. An additional Quick Circuit™ routing capability is used for de-paneling individual boards from multiboard panels. See Appendix XXIV for tooling information about carbide routers and Appendix XXV for recommended feed rates and spindle speeds for routers.

*Step 6.* Once the board has been contour routed to final size and shape, the last major step in producing your Quick Circuit™ protoboard is the assembly and soldering of components onto the board. As was previously mentioned, the populated Quick Circuit™ protoboard will function identically to a printed circuit board produced through a conventional plating and acid-etch process.

**Figure 13.15** Setup is easy. *Courtesy of T-Tech, Inc., Atlanta, GA, www.t-tech.com*

**Figure 13.16** Milling an isolation path. *Courtesy of T-Tech, Inc., Atlanta, GA, www.t-tech.com*

**Figure 13.17** Contour routing the drilled protoboard to finished size. *Courtesy of T-Tech, Atlanta, GA, www.t-tech.com*

## 13.12 PROTOBOARD RF/MICROWAVE AND MULTILAYER TECHNOLOGY

Since the Quick Circuit™ models 5000HS and 7000 are high-speed circuit board plotters (5,000 to 60,000 rpm), they are capable of manufacturing precision RF/microwave circuit board prototypes on a variety of substrates, including PTFEs and ceramic loaded substrates. These machines can drill holes as small as 0.2 mm (8 mils) and are capable of milling traces as narrow as 0.1 mm (0.004 inch or 4 mils) spaced as closely as 0.1 mm (4 mils) apart. Straight-flute end mills are typically required for RF/microwave applications in order to ensure rectangular cuts into the dielectric. These systems can also create RF/microwave circuits and tune them with exceptional accuracy (0.006 mm/0.00025 inch) and repeatability (0.006 mm/0.00025 inch). Fig. 13.18 shows a variety of RF/microwave prototype boards.

T-Tech multilayer technology, using the equipment shown in Fig. 13.19, provides a quick and economical way for designers, engineers, and technologists to produce multilayer (MLB) protoboards. T-Tech's Quick Circuit™ machine is complemented by a full line of multilayer equipment. The combination of a high-speed spindle Quick Circuit™, a Quick Plate™ electroplating system (power supply, electroplating tank with accessories and a rinse tank), and a Quick Press™ to laminate MLBs having four layers and up, as

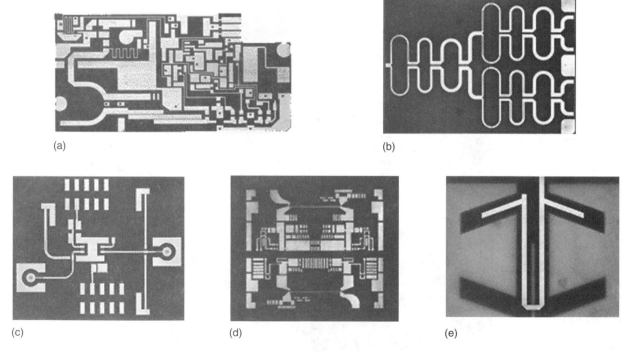

(a)

(b)

(c)

(d)

(e)

**Figure 13.18** Examples of protoboards used in RF/microwave applications. *Courtesy of T-Tech, Inc., Atlanta, GA, www.t-tech.com*

shown in Figure 13.19a, gives flexibility to design complex circuits that only MLBs can accomplish. Figure 13.19b shows a light-duty press that is recommended for making MLB protoboards up to four layers. It should be noted that unlike single- or double-sided protoboards, MLBs require electroplating and pressing processes.

An overview of the procedure is as follows: T-Tech's Quick Circuit™ is used to first mill the inner-layer antipad and thermal relief patterns of the designed MLB. See Section 14.1. Then the additional outer layers are pressed onto the inner layers using the Quick Press™ and associated laminating materials. See Section 14.2. After the laminating process, holes are drilled using Quick Circuit™. The next process, using a specially formulated emulsion and the Quick Plate™ system, is to condition the hole walls to accept electroplated copper to form plated-through-holes to establish the desired interlayer electrical connections. Finally, the Quick Circuit™ is used once again to mill the circuit conductor pattern design onto the outer layers. The finished product is a multilayer board with electroplated through-holes that is ready to be tested for the desired pattern of continuity. Fig. 13.20 shows a cross section of an electroplated through-hole using T-Tech's multilayer technology. In the following chapter, conventional multilayer printed circuit board processes that are also used to produce MLBs are presented.

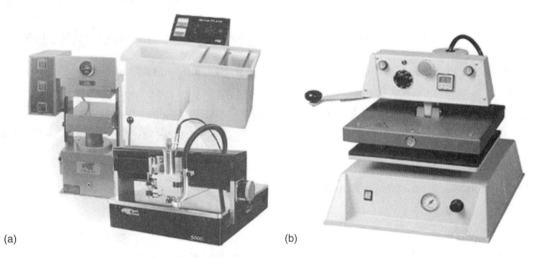

(a)      (b)

**Figure 13.19** Equipment used to produce MLB protoboards: (a) Quick Circuit™, heavy-duty Quick Press™ and Quick Plate™; (b) light-duty Quick Press™ used for protoboards up to four layers. *Courtesy of T-Tech, Inc., Atlanta, GA, www.t-tech.com*

**Figure 13.20** Cross-section of a plated-through-hole using T-Tech's multilayer technology. *Courtesy of T-Tech, Inc., Atlanta, GA, www.t-tech.com*

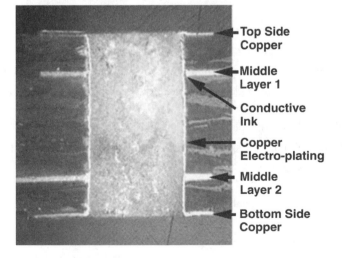

Top Side Copper

Middle Layer 1

Conductive Ink

Copper Electro-plating

Middle Layer 2

Bottom Side Copper

# EXERCISES

## A. Questions

13.1 What are the three methods used to determine the image area of a pc board?

13.2 Calculate the thief area of a 6- by 9-inch board that has a 4- by 7-inch image area.

13.3 Solve for the image area, in square inches, for one side of a double-sided board design having 15 resistors with 0.125-inch pads, 20 sixteen-pin DIP ICs, 5 eight-pin round ICs, and 85 inches of 0.040-inch conductors. Use Table 13.1 as a guide.

13.4 Using the values obtained in Questions 13.2 and 13.3, find the total area to be plated for (a) one side and (b) both sides of the pc board.

13.5 Determine the total area of 1 ounce per square foot copper to be pattern plated (double-sided board) if $W_1$ is 185.63 grams and $W_2$ is 179.21 grams.

13.6 Calculate the proper current settings to pattern plate copper onto a double-sided board having a total area of (a) 121 square inches and (b) 3.6 square feet.

13.7 With reference to Table 13.4, which is based on an ideal of 100% cathode efficiency, calculate the *practical* time to plate 1.5 mils of copper with a current density of 30 amperes per square foot.

13.8 Using the values obtained in Questions 13.4 and 13.5, calculate the correct power supply setting for solder plating.

13.9 How much does each of the plating processes (copper pattern plating and solder plating) add to the thickness of the metal surfaces?

13.10 What reduction in hole diameter size results from pattern and solder plating?

13.11 What are the recommended current densities for electroplating copper? For electroplating solder?

13.12 Briefly describe the difference between flash plating and pattern plating.

13.13 After the pattern plating and solder plating have been completed, how is the resist film removed from the pc board?

13.14 Define the term *solder fusion*. By what other name is this process known?

13.15 What are the advantages to the use of the solder fusion process?

## B. True or False

Circle *T* if the statement is true, or *F* if any part of the statement is false.

13.1 In the pattern plating process, the dry-film photo resist acts as a plating stop-off and the pattern-plated solder acts as an etchant stop-off.     T    F

13.2 The solder reflow process causes the solder to alloy with the underlying copper conductor pattern.     T    F

13.3 The thief area of a 6- by 9-inch pc board having a 1-inch-wide border is 29 square inches.     T    F

13.4 The optimum current density for plating solder is 30 amperes per square foot.     T    F

13.5 To provide an effective etching resist, a minimum thickness of plated solder is 0.003 inch.     T    F

13.6 For the plating of tin–lead alloy, an anode-to-cathode surface area ratio of approximately 3:1 is recommended.     T    F

13.7 The anode and cathode work rods in a plating tank are made from solid copper rod stock.     T    F

13.8 The most accurate method of determining the image area to be pattern plated is the *calculation* method.     T    F

13.9 The optimum thickness of pattern-plated copper is 3 mils.     T    F

13.10 Ferric chloride cannot be used to etch solder-plated pc boards.     T    F

## C. Multiple Choice

Circle the correct answer for each statement.

13.1    The three methods used to determine pattern area are calculation, using automatic equipment, and (*weight gain, weight loss*).

13.2    It takes approximately (*60 seconds, 60 minutes*) to pattern plate 1 mil of copper.

13.3    To produce an effective etch resist, a minimum of (*0.3 mils, 30 μinches*) of solder is required.

13.4    For pattern plating, the pc board is securely attached to the (*anode, cathode*) rod of the plating tank.

13.5    The time required to pattern plate 60/40 tin–lead to the minimum required thickness is approximately (*7 seconds, 7 minutes*).

13.6    Dynachem 400K solution should be (*heated, cooled*) prior to resist stripping.

13.7    Etching a pattern plated pc board should be accomplished with a solution of (*ammonium persulfate, ferric chloride*).

13.8    A de-wetting condition exhibited by pc boards after the reflow process is the result of a (*good, poor*) copper surface cleaning cycle.

## D. Matching

Match each item in Column A to the most appropriate item in Column B.

| COLUMN A | | COLUMN B | |
|---|---|---|---|
| 1. | Ammonium persulfate | a. | Stripper |
| 2. | Peroxide/sulfuric | b. | $30 \text{ A/ft}^2$ |
| 3. | Copper pattern plating | c. | Fusion |
| 4. | Solder pattern plating | d. | Cleaner |
| 5. | 400K | e. | Etchant |
| 6. | LAC-41 | f. | Micro-etch |
| 7. | Reflow | g. | $20 \text{ A/ft}^2$ |

# 14

# Multilayer PCB Fabrication

## LEARNING OBJECTIVES

*Upon completion of this chapter on the fabrication of multilayer printed circuit boards, the student should be able to*

- Understand the basic structure of multilayer boards.
- Know the raw materials required to fabricate multilayer boards.
- Use a laminating fixture to pin and register multilayer boards.
- Fabricate the inner layers of multilayer boards.
- Apply a black oxide treatment to the inner-layer core.
- Properly prepare a layup book in preparation of pressing a multilayer board.
- Use a cold press cycle to laminate raw multilayer boards.
- Drill multilayer boards.
- Apply an epoxy-smear treatment to drilled boards.
- Complete the fabrication of multilayer boards.

## 14.0 INTRODUCTION

Since the early 1960s, there has been an ongoing demand on the packaging designer to incorporate more electronics into a smaller space in order to reduce both system size and overall weight. With the density of the double-sided pc board being pressed beyond its limits, the industry's solution is the *multilayer board* (MLB).

The multilayer board consists of two outer layers of circuitry that sandwich two or more inner layers. The outer layers contain conductor paths and terminal pads and are identical in appearance to standard double-sided boards. Each of the inner layers serves a specific circuit function, as is explained in subsequent sections. Each of the board layers is separated from the others by sheets of insulating material. The layers are ultimately laminated together under heat and pressure to form a single functional board that is capable of providing more dense circuitry than are double-sided boards.

One method of classifying MLBs is in the manner in which the layers are connected electrically. The most common method of layer-to-layer interconnection is to drill clearance holes (usually lead access holes but via holes are also used) and employ a plated-through-hole technique similar to that used on double-sided boards.

Multilayer boards are also classified by the manner in which the inner layers are used. If conventional circuitry (i.e., terminal pads and conductor paths) is to be designed on one or more inner layers, it is referred to as a *signal* MLB. On the other hand, if the inner layers are predominantly solid conductive copper planes, they are classified as *power/ground* or *voltage/ground* MLBs. For the latter classification, one inner layer is used exclusively for making all circuit *ground* interconnections, while the other is used for making all the *dc voltage* interconnections.

In addition to reducing both space and weight, multilayer designs are also used in high-frequency circuits. The use of the large copper planes as ground greatly reduces *crosstalk* and other forms of interference common to high-frequency circuits. At microwave frequencies, MLBs can be designed to maintain a uniform impedance necessary for proper circuit performance.

The solid copper planes of MLBs may also serve as effective heat sinks in circuits in which excessive heat is generated. For this purpose, the solid planes are used as the outer layers and the inner layers are used for signal paths. In this type of design, the two outer layers may also be used for power and ground while also serving to remove heat from the circuit.

Although the MLB overcomes many design problems, it is the most expensive type of pc board to manufacture. This is primarily due to the fixed setup costs, more processing steps, and critical controls and inspections, which often result in reduced yields. For these reasons, the use of a multilayer design should be made only after other alternatives and trade-offs have been considered.

The following topics are treated in this chapter: (1) artworks for MLBs, (2) multilayer material, (3) inner-layer core manufacture, (4) layup processes, (5) MLB press, (6) drilling and board preparation, and (7) standard plated-through-hole processing.

*CAUTION:* A chemical hazard exists in the materials presented in this chapter. The processes discussed should be performed in adequately ventilated facilities and with applicable equipment having proper fume exhaust systems. Approved face shields (goggles) in compliance with ANSI Z87.1-1989 and chemical-resistant clothing and footwear and neoprene or nitrile gloves must be worn while performing any of these operations. Also, the section on safety outlined at the beginning of the book should be reviewed for more specific information.

## 14.1 ARTWORK FOR MULTILAYER BOARDS

High-quality artwork is absolutely essential for the manufacture of MLBs. The resulting board can be no better than the artwork supplied to the manufacturer. This artwork includes at least one 1:1 scale film on a 0.007-inch polyester backing for each layer of the MLB. For a typical four-layer voltage/ground MLB, positive and negative films are required. These are shown in Fig. 14.1. Note that the conductor paths (signal traces) appear on only the two outer layers: the component side (layer 1) and the circuit side (layer 4). Film *positives* are provided for these layers. Layer 2 is positioned below the component side and serves as the *voltage* or *power* plane. Layer 3 is the *ground* plane and is between the power plane and the circuit side (layer 4). Film *negatives* are providing for the power and ground layers.

In viewing each film shown in Fig. 14.1, the dark-shaded areas of layers 1 and 4 are to be considered as *copper* on the finished board and the unshaded areas as *insulation* (i.e., devoid of copper). For the inner layers (2 and 3), the dark-shaded areas are to be considered as *insulation* on the finished board and the unshaded areas as *copper.*

Layer-to-layer registration is accomplished by aligning the three staggered targets as each film is placed upon the other in the layup. As can be seen in Fig. 14.1, both the power and ground inner layers are largely solid planes of copper.

If no electrical connection is to be made between an inner layer and an outer layer, an *antipad* (clearance area of insulation) is provided around each clearance hole. This antipad is a round or square insulation gap, devoid of copper, that is at least 0.050 inch larger than the diameter of the hole drilled in that area (see Fig. 14.2a). This insulation gap ensures that there will be no electrical connection between that layer and the plated barrel of the hole.

If an electrical connection is to be made to one of the inner layers, a *thermal relief pad* is used (see Fig. 14.2b). The connection is made by drilling a hole through the solid inner pad. This hole will be plated and will connect to the two *splines* (copper paths) of the pad. The geometry of the thermal relief pad prevents the occurrence of inferior solder connections in the

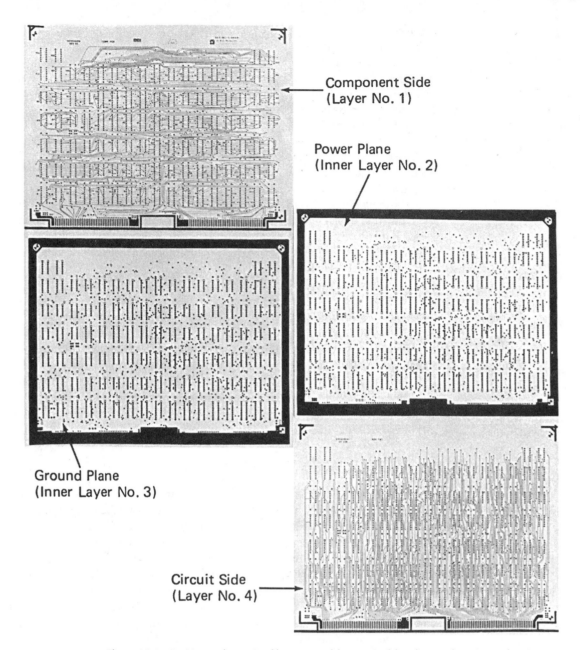

**Figure 14.1**  Positive and negative films are used for a typical four-layer voltage/ground MLB.

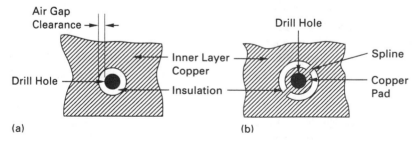

**Figure 14.2** Special pad geometries are used on inner layers to provide isolation or interconnection with hole barrel: (a) antipad; (b) thermal relief pad.

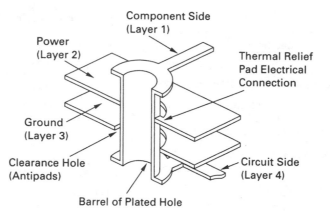

Power
(Layer 2)

Component Side
(Layer 1)

Thermal Relief
Pad Electrical
Connection

Ground
(Layer 3)

**Figure 14.3** Cutaway view showing a properly registered PTH in an MLB.

Clearance Hole
(Antipads)

Circuit Side
(Layer 4)

Barrel of Plated Hole

wave-soldering process. Direct connection of a plated-through hole to a solid copper plane could cause the copper to act as a heat sink, drawing heat away, and result in a poor connection.

A cutaway view of a properly registered MLB is shown in Fig. 14.3. It is essential that the following be included in all MLB artworks: (1) all layers clearly marked, (2) corner brackets, (3) reduction scale and tolerances, and (4) tooling or mounting holes clearly identified.

## 14.2 RAW MATERIALS USED IN MULTILAYER BOARD FABRICATION

In the design of a standard four-layer voltage/ground MLB, a finished thickness of 0.059 ± 0.005 inch is a typical specification. Due to electrical circuit considerations, different inner-layer thicknesses of insulation as well as their specific location relative to the outer layers may be a design requirement. For these reasons, the laminate suppliers provide stock with dielectric (insulation) thicknesses of 0.002 inch (2 mils) to 0.031 inch (31 mils) in increments of 0.001 inch (1 mil).

A common insulating material used is a fiberglass-cloth epoxy-based material that conforms electrically and mechanically to either grade G-10 or FR-4. (Grades of insulating base material are discussed in Chapter 8.) This material is completely cured and is called *C-stage material.* Polyimide materials are also available. C-stage laminate is available with copper foil bonded to one or to both sides. The copper thickness varies from ultrathin 1/8 ounce per square foot (oz/ft$^2$) (0.0002 inch or 5 microns) to 10 oz/ft$^2$ (0.014 inch or 350 microns). Laminates clad with copper on both sides are called *cores* and are often used to form the inner layers of an MLB. When used in this way, a minimum foil thickness of 2 oz/ft$^2$ (0.0028 inch or 70 microns) is recommended. C-stage laminates with copper foil on only one side are called *cap sheets* and are typically used to form the outer layers of an MLB. The foil thickness on cap sheets used on MLBs ranges from 1/2 oz/ft$^2$ (0.0008 inch or 20 microns) for fine-line work to 1 oz/ft$^2$ (0.0014 inch or 35 microns) for normal conductor widths and spacings.

A typical MLB specification for a four-layer voltage/ground design is the following: core thickness of 0.031 inch with 2 oz/ft$^2$ copper foil on both sides and two cap sheets having a laminate thickness of 0.008 inch (8 mils) with 1 oz/ft$^2$ copper foil on one side. An exploded view of this type of MLB is shown in Fig. 14.4.

To bond the cap sheets to the processed circuitry on the inner-layer core, several layers of adhesive material are sandwiched between the foil layers to form the finished assembly as shown in Fig. 14.4. This adhesive is called *prepreg* or *B-stage* material. It is composed of thin sheets of fiberglass cloth impregnated with an epoxy resin that is semicured. Complete curing of prepreg material is accomplished under elevated pressure and temperature in a multilayer press.

Because prepreg material is semicured and hydroscopic, it will absorb moisture quickly. For this reason it must be carefully stored. Some manufacturers store prepreg material under refrigeration at temperatures ranging from 10 to 50°F (−12 to 10°C) in a relative

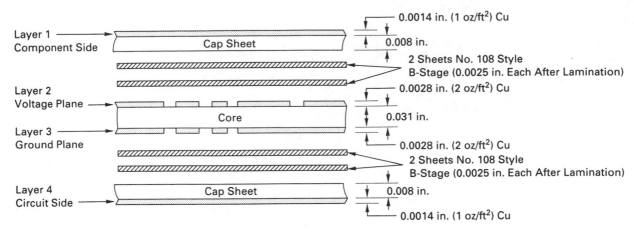

**Figure 14.4** Exploded side view of a four-layer MLB.

humidity of 50% or less. Others prefer storage in a dry gas atmosphere at or slightly above ambient temperature. Both methods of storage are effective. Because prepreg material can be degraded by extended storage, date codes should be incorporated with the inventory.

Before prepreg material is to be used, it is recommended that it be given a mild pre-bake cycle or placed in a vacuum chamber. These processes will remove any moisture or volatiles (solvents) that may have been absorbed while in storage. The handling of prepreg material requires that white gloves be worn at all times.

Six commonly used styles of prepreg bonding sheets together and their important characteristics are listed in Table 14.1. Note that the cloth thickness ranges in nominal value from 0.001 to 0.007 inch. With the resin added, the "as received" thickness is approximately 0.002 inch larger than the cloth alone, ranging from 0.003 to 0.008 inch for the styles listed. The *cured thickness* column in Table 14.1 lists the amount of thickness reduction expected after bonding is completed in the multilayer press. This column is used to determine the finished thickness of the MLB. We will use the example MLB shown in Fig. 14.4, together with Table 14.1, to determine its finished thickness. Using two sheets of Style 108 prepreg to bond the component side cap sheet to the core and another two sheets to bond the circuit side cap sheet to the core, the resulting cured thickness for each bond will be 0.005 inch ($2 \times 0.0025$), or a total of 0.010 inch added to the overall thickness of the boards. The overall thickness of the finished MLB will thus be the sum of (1) core thickness (0.031 inch), (2) two 2-oz/ft$^2$ inner-layer copper thicknesses at 0.0028 inch per layer, (3) two cap sheets at 0.008 inch each (insulation thickness), and (4) the bonding thickness of 0.010 inch. It can be seen that the finished MLB for our example will be approximately 0.062 (1/16) inch.

Referring again to Table 14.1, note that four additional columns of information are provided to aid in the selection of the style of prepreg material to be used. An explanation of each of these columns follows.

**Resin Content.** This is a measure of the amount of epoxy resin present in a given sample of prepreg and is specified in *percent by weight* of the total amount of prepreg. The resin content is an important selection criterion. If this content is too low, a resin-starved bond line will result. With too high a resin content, the resulting bond line will be too thick on the processed inner layers.

**Resin Flow.** This is a measure of the amount of epoxy resin present in a given sample that will be flowing during the press cycle. Given as *percentage of average resin content,* the flow decreases as the amount available decreases. The concern here is that too little flow can cause too thick a bond line with voids, while too high a flow can result in a resin-starved bond line.

**TABLE 14.1**

Epoxy System Multi-Flow Prepreg Bonding Sheets (G-10/FR-4)*

| Glass Cloth | | Bonding Sheet | | | | | |
|---|---|---|---|---|---|---|---|
| | Nominal | Thickness | | Resin | Resin | | Volatiles |
| Style | Thickness | As Rec'd | Cured | Content (%) | Flow (%) | Gel (sec) | (% max.) |
| 104 | $0.001 \pm 0.0002$ | $0.003 \pm 0.0015$ | $0.0015 \pm 0.0005$ | 70–85 | 40–55 | 90–130 | 0.75 |
| | | | | | 40–55 | 220–280 | |
| 106 | $0.0015 \pm 0.0003$ | $0.0035 \pm 0.0015$ | $0.002 \pm 0.0005$ | 65–80 | 40–55 | 90–130 | 0.75 |
| | | | | | 40–55 | 220–280 | |
| 108 | $0.002 \pm 0.0005$ | $0.0045 \pm 0.0015$ | $0.0025 \pm 0.0005$ | 60–80 | 35–50 | 90–130 | 0.5 |
| | | | | | 40–55 | 220–280 | |
| 113 | $0.003 \pm 0.0005$ | $0.0055 \pm 0.0015$ | $0.0035 \pm 0.0005$ | 50–70 | 30–45 | 90–130 | 0.5 |
| | | | | | 35–50 | 220–280 | |
| 116 | $0.004 \pm 0.0005$ | $0.006 \pm 0.002$ | $0.0043 \pm 0.0005$ | 45–65 | 20–35 | 90–130 | 0.5 |
| | | | | | 25–40 | 220–280 | |
| 7628 | $0.007 \pm 0.0005$ | $0.008 \pm 0.002$ | $0.006 \pm 0.001$ | 38–45 | 15–25 | 90–130 | 0.5 |
| | | | | | 15–30 | 220–280 | |

*Typical glass styles and resin parameters shown. A wide specifications range is necessary to encompass military specifications and customer process variations. Significantly tighter tolerances are maintained on off-the-shelf bonding sheets. All are tested to MIL-G-55636. Multi-Flow prepreg bonding sheets are furnished 38 inches wide in continuous lengths or sheeted to order.

**TABLE 14.2**

Suggested Fill Chart for Epoxy and Polymide Multi-Flow Prepreg Bonding Sheets*

| Copper Weight Thickness | | Glass Style Recommended Plies | | | | |
|---|---|---|---|---|---|---|
| Copper Weight $(oz/ft^2)$ | Copper Thickness (in.) | 104 | 106 | 108 | 113 | 116 |
| 1 | 0.0014 | 2 | 2 | 2 | 2 | 2 |
| 2 | 0.0028 | 2 | 2 | 2 | 2 | 2 |
| 3 | 0.0042 | 3 | 3 | 3 | 2 | 2 |
| 4 | 0.0056 | 4 | 4 | 4 | 3 | 3 |
| 5 | 0.0070 | 5 | 5 | 4 | 4 | 4 |
| 6 | 0.0084 | 6 | 5 | 5 | 5 | — |
| 7 | 0.0098 | 7 | 6 | 6 | 5 | — |
| 8 | 0.0112 | 8 | 7 | 6 | 5 | — |
| 9 | 0.0126 | 9 | 8 | 7 | — | — |
| 10 | 0.0140 | 10 | 9 | 8 | — | — |

*Fill chart based on hypothetical circuitry with 30% copper density.

***Gel Time.*** This is a measure of how long the resin remains molten before gelling or solidifying. Within the range 220 to 280°F (104 to 138°C), the gel time is typically constant from 90 to 130 seconds.

***Volatiles.*** In the manufacture of prepreg, solvents are used to control the viscosity of the resin. These volatiles are only partially driven off since the resin is semicured. Those solvents remaining are expressed as a percent by weight of the total amount of prepreg. Volatiles can cause vapor bubbles or voids between layers. For this reason, a prebake cycle of B-stage material is important to drive off any remaining solvents.

Laminate manufacturers provide useful information to aid the designer and manufacturer of MLBs in selecting the appropriate material. For example, refer to Table 14.2, which lists the recommended number of bonding sheets of B-stage material for specific glass styles and copper foil thicknesses of the core sheet. Using this table and referring to Fig. 14.4, it can be seen that two sheets of Style 108 prepreg between layers is recommended to properly bond and fill the spaces in etched 2 $oz/ft^2$ copper foil.

## 14.3 LAMINATING FIXTURES AND PINNING

One of the most critical phases in the manufacture of MLBs is the precise layer-to-layer registration of all terminal pads. Because these pads are used for inner-layer interconnections, they must be in exact alignment during inner-layer imaging and in the layup in order to successfully drill and plate the MLB. Recall that Fig. 14.4 shows the side view layup of a typical MLB. Precise layer-to-layer registration of the phototools used to manufacture this board is accomplished with a pinning system that is keyed to a set of laminating fixture plates. These plates are used to hold all layers and materials fixed in close registration tolerance during the laminating press cycle.

A typical set of 12- by 12-inch laminating fixtures used for manufacturing MLBs is shown in Fig. 14.5. The bottom plate (Fig. 14.5a) is made from 0.250-inch hot-rolled steel stock having a coefficient of thermal expansion similar to that of the pc board stock. This plate is ground to a precision flatness with both surfaces parallel to one another. The plate is fitted with four hardened steel tooling pins (Fig. 14.5b) having a diameter of 0.250 inch and made to "snug" fit into four hardened steel bushings. For the MLB layup shown in Fig. 14.4, the length of the pin should be 0.500 inch. It is essential that the length of the pin be less than the total thickness of the top and bottom plates plus that of the finished thickness of the MLB. If the pin is longer than this combination, it will cause damage to the press. Since the positions of the four toolings pins are accurately laid out, precise alignment of all artwork phototools with these pins will result in exact layer-to-layer registration.

The top laminating fixture (Fig. 14.5c) fits over the bottom plate with the MLB layup sandwiched between them for pressing. The top plate is also made from the same 0.250-inch hot-rolled steel stock with both flatness and parallelism specifications identical to those of the bottom plate. This plate is provided with four slots to accept the tooling pins for proper registration and to allow for material expansion. The width of each slot is made slightly larger than the 0.250-inch-diameter tooling pin it is to accept.

The 12- by 12-inch laminating fixture just described is used for registration purposes in the following manner. The MLB raw stock (core, cap sheets, and B-stage) is first cut to 11 by 11 inches. All the stock is then slotted with the same configuration as on the top plate of the fixture and in precise alignment with the pins of the bottom plate. The slots in the MLB material are made with an air-actuated punch, which automatically aligns and forms the four slots simultaneously.

Note in Fig. 14.5c that opposite slots in the top plate are in the same direction and that adjacent slots are perpendicular to one another. The reason for this slotting configuration is to allow for heat expansion in the press. When heated in the press, both the pinned pc board stock and the bottom plate expand along the x and y planes. Slots A and B in the top plate and in the board material allow space for expansion in the x direction. The edges of pins C and D hold the exact center of the board stock between pins A and B in a fixed position, which allows the expansion error to be split in half on either side of this center. In a similar fashion, slots C and D allow expansion in the y direction. Again, the expansion error due to heat is split in half on either side of the center, which is held in a fixed position by the edges of pins A and B.

With the four slots punched into the pc board stock, the phototools for each layer of the MLB are punched with the same four slots in the same orientation. These slots are also aligned with the tooling pins of the laminating fixture. In this way all phototools are precisely registered to one another for layer-to-layer alignment and all material is registered to the phototools through these series of four slots. In the following section, the core layers of an MLB are fabricated using the alignment techniques just discussed.

## 14.4 INNER-LAYER FABRICATION

After the slots have been punched into the core and cap sheets for positioning onto the laminating fixture, the next phase in MLB fabrication is to subject all the C-stage raw stock to an initial oven-baked cycle to reduce internal stresses, thereby improving the dimensional

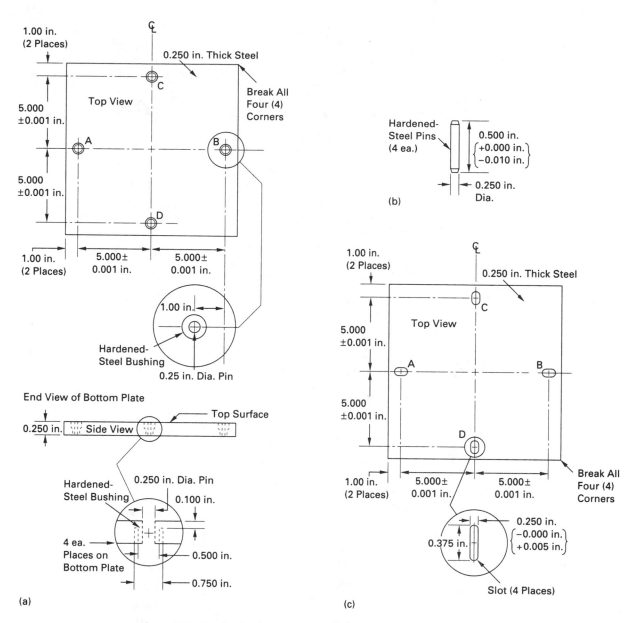

**Figure 14.5** Laminating fixture geometry for pinning and registering an MLB layup: (a) bottom caul plate; (b) tooling pin; (c) top caul plate.

stability of the material. A 6- to 8-hour bake at 300°F is recommended for this cycle. During this baking cycle, the material, which is very thin, will tend to soften. For this reason, the sheets should be racked rather than laying one on top of the other. Racks with vertical slots will support the edges of the sheets and prevent sagging.

When the C-stage material has returned to room temperature, it can be processed into the inner-layer voltage and ground patterns. We will use the layup of the MLB shown in Fig. 14.4 as an example. Both the voltage and ground patterns of the inner layers are processed simultaneously. The flowchart of Fig. 14.6 shows the sequential processes involved. With the exception of the last step (black oxide treatment), you will note that the flowchart is similar to that of the print-and-etch processes discussed in earlier chapters. These processes should be reviewed before continuing in this section.

Both copper surfaces of the core are first chemically cleaned in a 10% $H_2SO_4$ acid dip. This is followed by a water rinse. The core is then mechanically scrubbed and air-knife dried.

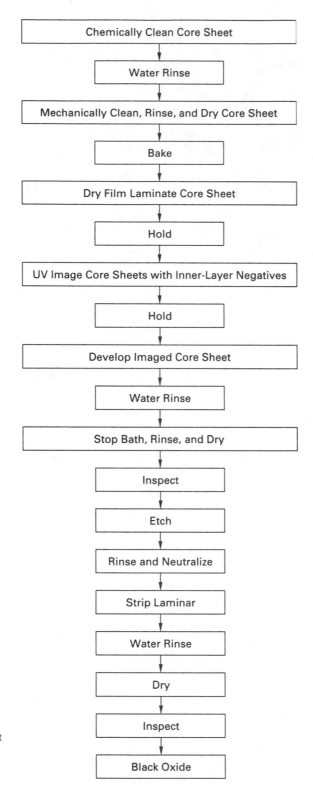

**Figure 14.6** Flowchart for inner-layer fabrication by the print-and-etch process.

In preparation for dry-film photo resist laminating, the core is baked at 150°F for 15 minutes. Proper racking of the core during this bake cycle is important to prevent sagging. The core is removed from the oven and immediately laminated on both sides with 1-mil Dynachem Laminar AX photo resist (see Chapter 12). The excess resist is trimmed along all edges. All tooling holes are carefully cleaned of resist with a sharp scalpel. After laminating, a hold time of 30 minutes is allowed to improve photo resist adhesion to the copper surfaces.

To establish the proper layer-to-layer registration between the inner layers, the voltage and ground-layer phototools are secured to the core with two *Carlson pins* (see Fig. 14.7). The pins are 0.250 inch in diameter and approximately 0.055 inch long. The phototools are positioned one above and one below the slotted core stock and pinned into registration.

Once registered, the core is processed through the print-and-etch sequences. After exposure, the core is again allowed a holding time of 30 minutes, after which it is developed. Laminar AX resist is developed by first removing the gas seals from both sides of the exposed core sheet. Both sides are then spray-developed for approximately 45 seconds at a developing solution temperature of 85°F (29°C). The developing process is immediately followed by immersing the core sheet into a stop bath of 10% sulfuric acid to neutralize the developing action. The image should be developed to a solid 4 on the Stouffer table.

After the core has been developed and neutralized, it is ready to be etched. Any of the common etchants, such as ferric chloride, ammonium persulfate, peroxide/sulfuric, or ammoniated etchants may be used. Once the core has been etched, it is immediately rinsed in tap water. Depending on which type of etchant solution is used, an appropriate neutralizing step may be necessary. The circuit features now appear on the voltage and ground layers. Since the photo resist is no longer required, it is stripped from both sides of the core. Time in the stripper bath must be held to an absolute minimum to reduce the amount of methylene chloride that is absorbed by the epoxy substrate. A final water rinse follows the stripping process and the core is allowed to dry.

To improve laminated bonding to the large copper areas of the inner layers, the core is treated with a *black oxide* solution. (Recall that this treatment was not discussed in the detailed description of the print-and-etch technique presented in previous chapters.) This commercially available solution chemically provides a heavy cupric oxide coating, which increases the copper surface area by roughening the copper. A flowchart showing the process for a typical black oxide treatment is shown in Fig. 14.8. The core is first mechanically cleaned using a conveyorized cleaning machine such as that shown in Fig. 12.3b. The core is then racked and passed through a series of eight baths as described here.

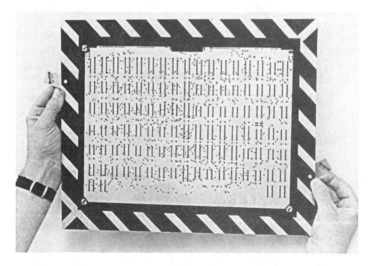

**Figure 14.7** Inner layer registration is established through the use of Carlson pins.

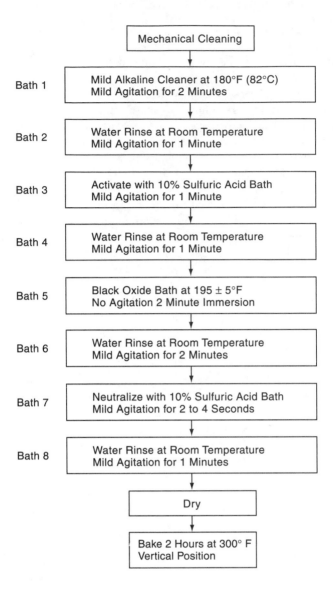

```
                    ┌─────────────────────────┐
                    │   Mechanical Cleaning   │
                    └─────────────────────────┘
                                 ↓
         ┌───────────────────────────────────────────┐
 Bath 1  │  Mild Alkaline Cleaner at 180°F (82°C)     │
         │  Mild Agitation for 2 Minutes              │
         └───────────────────────────────────────────┘
                                 ↓
         ┌───────────────────────────────────────────┐
 Bath 2  │  Water Rinse at Room Temperature           │
         │  Mild Agitation for 1 Minute               │
         └───────────────────────────────────────────┘
                                 ↓
         ┌───────────────────────────────────────────┐
 Bath 3  │  Activate with 10% Sulfuric Acid Bath      │
         │  Mild Agitation for 1 Minute               │
         └───────────────────────────────────────────┘
                                 ↓
         ┌───────────────────────────────────────────┐
 Bath 4  │  Water Rinse at Room Temperature           │
         │  Mild Agitation for 1 Minute               │
         └───────────────────────────────────────────┘
                                 ↓
         ┌───────────────────────────────────────────┐
 Bath 5  │  Black Oxide Bath at 195 ± 5°F             │
         │  No Agitation 2 Minute Immersion           │
         └───────────────────────────────────────────┘
                                 ↓
         ┌───────────────────────────────────────────┐
 Bath 6  │  Water Rinse at Room Temperature           │
         │  Mild Agitation for 2 Minutes              │
         └───────────────────────────────────────────┘
                                 ↓
         ┌───────────────────────────────────────────┐
 Bath 7  │  Neutralize with 10% Sulfuric Acid Bath    │
         │  Mild Agitation for 2 to 4 Seconds         │
         └───────────────────────────────────────────┘
                                 ↓
         ┌───────────────────────────────────────────┐
 Bath 8  │  Water Rinse at Room Temperature           │
         │  Mild Agitation for 1 Minutes              │
         └───────────────────────────────────────────┘
                                 ↓
                    ┌─────────────────────────┐
                    │           Dry           │
                    └─────────────────────────┘
                                 ↓
                    ┌─────────────────────────┐
                    │  Bake 2 Hours at 300° F │
                    │  Vertical Position      │
                    └─────────────────────────┘
```

**Figure 14.8** Flowchart for black oxide treatment.

***Bath 1.*** A 2-minute dip in a mild alkaline cleaner at an operating temperature of 180°F (82°C) to remove fingerprints, greases, and oils.

***Bath 2.*** A 1-minute water rinse at room temperature.

***Bath 3.*** A 1-minute dip in a 10% solution of sulfuric acid, which serves as an activator.

***Bath 4.*** A 1-minute water rinse at room temperature.

***Bath 5.*** Black oxide solution that is heated to 190 ± 5°F. Rapid immersion into this bath is essential so that the clean copper does not begin to oxidize from the heat generated by the solution. The core remains in this bath for approximately 2 minutes and should not be agitated.

***Bath 6.*** A 1-minute water rinse at room temperature.

***Bath 7.*** A 10% sulfuric acid bath to neutralize hydroxyl ions present after the black oxide treatment. If not removed, these hydroxyl ions will adversely affect the B-stage material by causing voids to occur in the MLB. The core should remain in this bath for 2 to 4 seconds.

**Bath 8.** A 1-minute water rinse at room temperature.

After the core has been processed through the eight baths and dried, the copper surfaces will appear as a uniform gray-black color on both sides.

The final preparation of the core prior to the laminating process is another oven-bake cycle for 2 hours at 300°F. When processing several cores, they must not be stacked one on top of the other, but rather separated and positioned vertically in the oven. In addition, care should be taken to prevent scratching the black oxide surfaces. White cotton gloves should be worn to prevent oils and greases from contaminating these surfaces. The final bake cycle is required to remove water and solvents that were absorbed in the previous processing steps. This removal is essential to eliminate voids due to moisture absorption by the core.

## 14.5 MULTILAYER LAYUP PROCESS

After the inner layers of an MLB have been processed through the black oxide treatment and final bake cycle, they are ready to be pinned together with the two outside layers and other elements to complete the layup of the multilayer *book,* which will then be pressed into the raw MLB. The press cycle is discussed in the following section.

The layup on an MLB should be performed in a dust-free, positive-pressure room that is environmentally controlled at 70°F and 50% RH. All technicians working in the layup room should wear white cotton gloves.

The assembly of the MLB book begins by cleaning and inspecting the bottom laminating plate (caul plate). The work surface should be freed of all foreign matter by cleaning with lint-free cloth saturated in trichlorethylene. The tooling pins are next installed into the bottom fixture plate. The side view of a layup sequence to press two identical MLBs between one set of 12- by 12-inch caul plates is shown in Fig. 14.9. (Typically, four to five MLBs are pressed in a book at one time.) Referring to Fig. 14.9, the construction of the MLB book is performed in the following sequence with each step numbered in the figure.

1. The cleaned bottom caul plate, with the tooling pins installed flush to the bottom surface of the plate, is placed on a worktable.

2. A sheet of mold-release material is cut so that its dimensions are approximately 1 inch larger than the caul plate on all four sides. For a 12- by 12-inch plate, the mold release sheet would be cut to a dimension of 14 by 14 inches. This prevents sticking between the pc boards, the caul plates, and separator plates. Various materials, such as Tedlar, PTFE sheeting, kraft paper, or chipboard, are used as mold-release material. These materials will not contaminate the MLB with agents of their own. With a deep-throated hand-operated paper punch, four 1/2-inch-diameter holes are punched into the release sheet, aligned with the four tooling pins. The mold-release sheet is then positioned over the bottom caul plate.

3. With the tooling pins used for alignment, the 11- by 11-inch bottom cap sheet (layer 4) is placed over the mold-release sheet with its copper surface facing downward. (Recall from Sec. 14.4 that this cap sheet has been stress-relieved for 6 hours at 300°F.)

4. Two sheets of Style 108 B-stage material (as recommended in Table 14.2) are cut to the same size as the raw stock (11 by 11 inches). One-half-inch holes are also punched into this material, again to align with the tooling holes. The prepreg sheets should be allowed to stabilize for at least 24 hours at room temperature at no more than 50% RH prior to being installed in the layup. A *J-type thermocouple* is inserted into a bond line, between the two prepreg

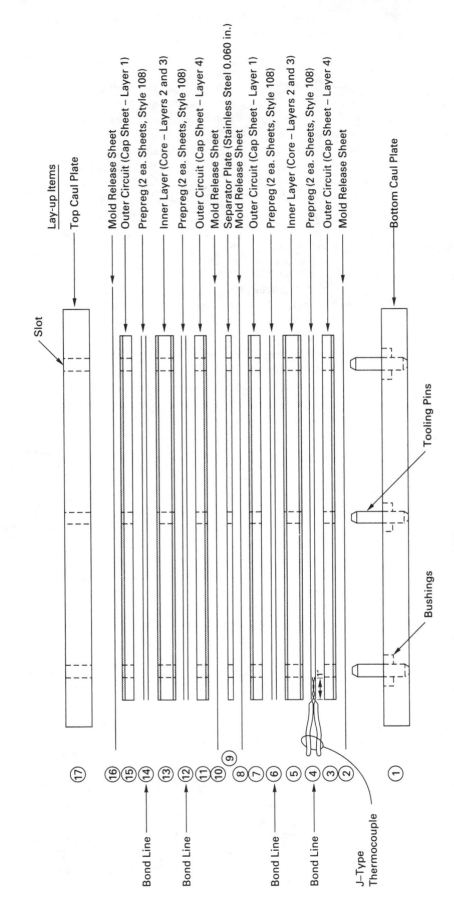

**Figure 14.9**  Exploded side view of a book having a layup to press two MLBs.

270

sheets approximately 1 inch from the outer edge. This will be used to monitor the thermal profile of the press cycle as discussed in Sec. 14.6.

5. A previously processed and black-oxide-treated core (inner layers 2 and 3) is pinned to the assembly using the slotted holes and tooling pins for alignment. The assembly instructions should be consulted for the correct orientation of the core. In our assembly, layer 3 (ground) will face downward.

6. Two sheets of Style 108 prepreg material are placed above the inner layer in a similar fashion to that described in step 4.

7. A cap sheet (layer 1) is positioned above the prepreg material and aligned using the tooling holes. The solid copper surface of layer 1 is made to face upward.

8. Another 14- by 14-inch sheet of mold-release material is prepared similar to that described in step 2 and placed above the top cap sheet. This completes the layup for one of two MLBs to be placed between the same pair of laminating fixture plates.

9. To separate the two MLBs in the laminating fixture, a 12- by 12-inch stainless steel separator plate is positioned above the mold-release material described in step 8. This separator plate has a typical thickness of 0.060 inch and it is provided with four oversized holes aligned with the tooling pins.

10–16. The procedures described in steps 2 through 8 are repeated to assemble the top MLB with the exception of omitting the J-type thermocouple.

17. After the top MLB has been assembled above the stainless steel separator plate, the top caul plate of the fixture is assembled over the top mold-release sheet using the tooling pins and slots for alignment.

The layup book for two MLBs just described is now ready to be pressed into raw MLBs. This assembly can be made to accommodate up to five MLBs in a single book by the addition of a separator plate and mold-release material for each additional MLB and increasing the length of the tooling pins as required.

## 14.6 MULTILAYER COLD PRESS CYCLE

The application of both pressure and temperature is required to laminate the book shown in Fig. 14.9 to result in two identical MLBs. This lamination process is accomplished with a multilayer press and by subjecting the MLBs to a *cold press cycle.* (This term is derived from the fact that the pressing cycle is initiated at room temperature.) A basic unit for laminating MLBs consists of a *single-opening press* with two *platens* that are electrically or steam heated to the recommended temperature. These platens are also equipped with cooldown coils. The primary requirements of the platens are that they be perfectly flat and possess thermal uniformity throughout their surfaces. Pressure is applied pneumatically or hydraulically. Controls that set, control, and monitor temperature, pressure, and time are also a requirement of the laminating press. A typical small-production MLB press is shown in Fig. 14.10.

Because the controls are not identical on different presses, a general operating procedure is described here. Pressing a book such as that discussed in the preceding section begins by cutting four sheets of 90-pound-weight kraft paper or chipboard to a size that is at least 2 inches larger all around than the laminating fixture plates. For our example 12- by 12-inch caul plates, the sheets should be approximately 16 by 16 inches. These layers are used to keep the platens clean, but also serve to slightly adjust the thermal profile of the book as the press cycle is initiated. The number of sheets applied determines the rate at which heat is imparted from the platens to the book to fit the recommended thermal profile. For our example press cycle, we will use two sheets of kraft paper positioned on the bottom platen. The book is placed onto the paper and centered on the platen. The two remaining sheets of kraft

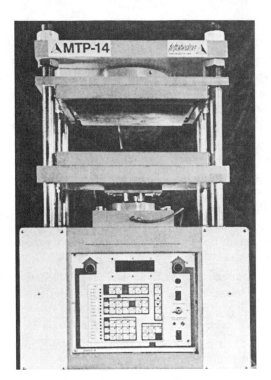

**Figure 14.10** Small production MLB press. *Courtesy of Tetrahedron Associates, Inc.*

paper are placed over the top caul plate. This arrangement of the book and the kraft paper positioned in the press is shown in Fig. 14.11.

Before the press cycle is initiated, the thermocouple wires extending outward from the book should be connected to the appropriate terminals of a chart recorder. Typically, the constant wire connects to the − terminal and the iron wire to the + terminal of the recorder. The thermocouple leads are shown in Fig. 14.11.

With power applied to the recorder and to the laminating press, both platens are set to 345 ± 5°F. The timing controls should be set as follows:

*Initial heat-up:* 30 minutes

*Cure:* 1 hour

*Cool-down:* typically 15 to 30 minutes

This timing represents an overall press cycle of approximately 2 hours. A typical thermal profile that occurs inside the book obtained from the thermocouple and the chart recorder is shown in Fig. 14.12.

With all of the controls set, the press platen opening (space between the top and bottom platens) is closed and the pressure adjustment control is immediately set to apply 350 psi pressure to the caul plate surface. Our 12- by 12-inch plates have a surface area of 144 square inches. Therefore, the required pressure $P$ is

$$P = 144 \text{ square inches} \times 350 \ \frac{\text{pounds}}{\text{square inch}} = 50{,}400 \text{ pounds}$$

This pressure of approximately 25 tons should be set and maintained throughout the press cycle.

Refer again to Fig. 14.12 to observe the temperature profile over the entire press cycle. As the temperature of the platens rises from approximately 70°F toward 350°F, it passes through 200 to 210°F, where the Style 108 B-stage material begins to flow. To maintain good resin flow, the stack temperature should be controlled at a rate of 8 to 10°F rise per minute from 200 to 300°F. During this resin-flow cycle, air is forced out of the inner layer,

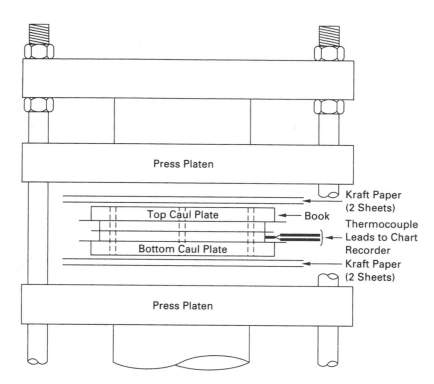

**Figure 14.11** Placement of book between plates in preparation for press cycle.

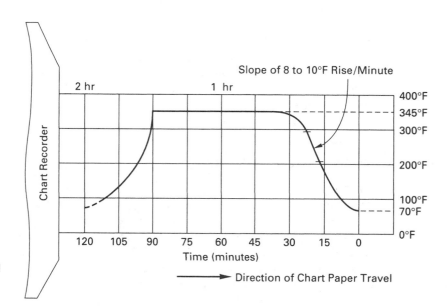

**Figure 14.12** Thermal profile of a multilayer cold press cycle.

filling in all of the tiny voids between circuit features with molten resin. As the stack temperature rises above 280 to 290°F, the resin stops flowing and begins to solidify. The cure continues with increasing temperature. After approximately 60 minutes, resin stiffening occurs, and molecular cross-linking results in a homogeneous mass.

After the cure cycle is complete (1 hour), the cool-down cycle begins. The book is *frozen* by bringing the temperature of the platens down from approximately 345°F to below 100°F in 15 to 30 minutes. The recommended rate of temperature change during the cool-down cycle is again 8 to 10°F per minute.

With the cool-down cycle complete, the book is removed from the press. The pins are then extracted from the caul plate and the MLBs are removed. The mold-release and kraft paper sheets are discarded and any resin seepage around the edges of the stack is trimmed off. To reduce any internal stresses that have occurred during the pressing operation, the MLBs are weighted with steel plates and baked for 2 hours at a temperature of 250°F.

## 14.7 DRILLING MULTILAYER PRINTED CIRCUIT BOARDS

The raw MLBs that were laminated in the preceding section now appear as double-sided pc board raw stock containing tooling holes. These boards are now ready to be drilled on an NC or CNC drilling machine. Recall from Chapter 11 that the panel is pinned with backup material and a thin sheet of aluminum as top drill-entry material.

The drill operator is provided with the proper drill program for the MLB together with instructions on the stack-up (i.e., number of MLBs to be stacked and drilled simultaneously). Typically, MLBs are drilled *one-up* (a single board at a time). However, depending on the thickness and the amount of hole-position tolerance that must be maintained, several boards may be stacked and drilled together.

One of the most serious problems that faces the MLB manufacturer is called *epoxy smear,* which is the result of the drilling operation. As the drill bit passes through the copper and epoxy-glass material, it generates heat (above 350°F) due to friction as it cuts through the copper. This heat melts some of the epoxy inner layer, and the molten resin is dragged across the entire barrel of the hole as the bit is withdrawn. As it cools, the epoxy forms an insulating layer that will prevent electrical interconnection between the inner and outer layers by way of the plated-through hole. Although it is possible to reduce the amount of epoxy smear, it cannot be completely eliminated due to the nature of the drilling process. Epoxy smear generation is dependent on the drill bit geometry, the speed and feed rate adjustment, the number of hits before bit replacement, the flatness and type of entry material, the type of backup material, and the amount of board cure before drilling. The use of resharpened drill bits should be avoided. It can be seen that many factors contribute to the generation of epoxy smear and it can be held to a minimum by carefully controlling these factors. However, a post-drilling smear removal process is necessary to improve the reliability of inner-layer interconnection in the plated-through-hole process. This removal process is discussed in the following section.

## 14.8 EPOXY SMEAR REMOVAL AND ETCHBACK

After the board has been drilled, it is subjected to an epoxy smear removal process before continuing with its fabrication. This process should have the characteristics to produce three specific results. The *first* is the complete removal of the epoxy residue resulting from the drilling operation. The *second* is to produce a positive *etchback* of the cured epoxy between all layers of copper. This etchback is typically specified as a minimum of 0.0002 inch to a maximum of 0.003 inch. The etchback process produces an inner layer *foot* to obtain a reliable three-point electrical contact that will result during the plating process (see Fig. 14.13). The *third* is to etch the fiberglass material of the inner layers and condition (roughen) the epoxy surface to improve electroless copper-plating adhesion.

Although there is no universally accepted process for accomplishing the three characteristics outlined, we present a de-smear/etchback process that is a generally accepted standard and currently used in the industry. This process is shown in flowchart form in Fig. 14.14. Each of the steps is discussed with reference to this flowchart.

***Bath 1.*** The boards are racked horizontally and immersed in concentrated sulfuric acid at 80°F for 15 to 30 seconds. Extreme care must be exercised when working with this concentrated acid bath. This bath will remove the epoxy smear and etchback the inner-layer epoxy material. The degree of removal is a function of time, temperature, acid concentration, and the amount of agitation of the board. Increasing the time above 30 seconds does not improve etchback, apparently due to the fact that fresh acid cannot contact with the epoxy sur-

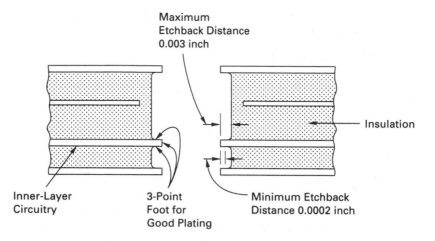

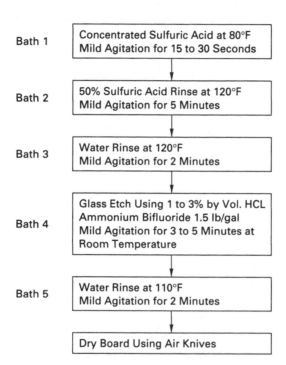

**Figure 14.13** Etchback produces a superior surface geometry for the formation of a PTH.

**Figure 14.14** Flowchart for smear removal and etchback process.

**Bath 1**

| Concentrated Sulfuric Acid at 80°F |
| Mild Agitation for 15 to 30 Seconds |

**Bath 2**

| 50% Sulfuric Acid Rinse at 120°F |
| Mild Agitation for 5 Minutes |

**Bath 3**

| Water Rinse at 120°F |
| Mild Agitation for 2 Minutes |

**Bath 4**

| Glass Etch Using 1 to 3% by Vol. HCL |
| Ammonium Bifluoride 1.5 lb/gal |
| Mild Agitation for 3 to 5 Minutes at |
| Room Temperature |

**Bath 5**

| Water Rinse at 110°F |
| Mild Agitation for 2 Minutes |

| Dry Board Using Air Knives |

face. Increasing the temperature does, however, reduce the viscosity of the acid and improves its solvent action. When the concentration of acid is diluted to below approximately 92%, the solvent action is severely impaired and the bath should be replaced. Mild up-and-down board agitation improves solution flow in the holes and improves etchback.

**Bath 2.** The boards are removed from the sulfuric acid bath, allowed to drain, and then immersed in a 50% sulfuric acid rinse at 120°F for a minimum of 5 minutes. This bath will prevent epoxy residue from being redeposited in the holes. The board is mildly agitated in this bath.

**Bath 3.** This is a water rinse heated to 120°F. The board is immersed in this bath and mildly agitated for 2 minutes. This will flush the holes free of any remaining residue.

**Bath 4.** This glass-etching bath is made up of 1.5 pounds per gallon of ammonium biflouride and 1 to 3% by volume of hydrochloric acid and is operated at room temperature. The board is mildly agitated in this bath for 3 to 5 minutes. This bath will etch away the protruding glass fibers.

**Bath 5.** The board is placed in a final warm-water (110°F) rinse and mildly agitated for 2 minutes.

Finally, the board is dried and is ready for the succeeding fabrication processes.

## 14.9 FINAL MULTILAYER BOARD PROCESSING

With the completion of the smear removal and etchback processes, the MLB is next processed in an *identical* manner to that of a double-sided pc board. These processes will not be repeated here, but chapters can be referenced where detailed information of each is provided. The MLB is first passed through an electroless copper deposition line to initially form the plated-through holes (Chapter 11). This is followed by copper flash plating (also Chapter 11). The two outer layers (layers 1 and 4) are then laminated with dry-film photo resist (Chapter 12). Copper and solder pattern plating is next applied to the total pattern (Chapter 13), which completes the formation of the plated-through holes. The board is etched (Chapter 13) to result in the final conductor patterns on layers 1 and 4. Upon completion of the etching process, the board is profiled to size and shape (Chapter 13), which results in a finished multilayer pc board.

## EXERCISES

### A. Questions

14.1   Explain the basic difference between a signal MLB and a voltage/ground MLB.

14.2   What are the advantages of using solid copper planes in MLB construction?

14.3   Define the purpose of (a) an antipad and (b) a thermal relief pad.

14.4   (a) If 3 oz/ft$^2$ inner-layer copper is used in an MLB design, how many layers of Style 104 prepreg material should be used to obtain a good bond with a cap sheet? Refer to Table 14.2. (b) For the same MLB design, how many layers of Styles 106, 108, 113, and 116 should be used?

14.5   Using Table 14.1, determine the cured thickness (i.e., after pressing) for three sheets of Style 113 prepreg material.

14.6   A four-layer power/ground MLB is to be fabricated with a 0.020-inch core having 4 oz/ft$^2$ copper on both sides. Two outer-layer cap sheets, each having an insulation thickness of 0.020 inch, are to be bonded to the core using Style 113 prepreg material. Determine (a) the number of layers of B-stage material to be used to bond each cap sheet and (b) the overall thickness of the cured MLB.

14.7   Explain the purpose of the slotted design in the top laminating fixture.

14.8   Explain the purpose and use of Carlson pins for inner-layer fabrication.

14.9   What is the purpose of the black oxide treatment?

14.10  List the elements in their correct order to assemble a single four-layer MLB book.

14.11  In what portion of the press cycle (see Figure 14.12) and between what temperatures does resin flow occur?

14.12  How does etchback improve the condition of the barrels in an MLB to obtain good-quality electrical interconnections?

14.13  What is the difference between B-stage and C-stage materials?

14.14  What are *cap sheets,* and where are they used?

14.15  How is the thermal profile of the press cycle monitored?

14.16  What is *epoxy smear,* and what is its cause?

## B. True or False

Circle *T* if the statement is true, or *F* if any part of the statement is false.

| | | | |
|---|---|---|---|
| 14.1 | MLBs are classified as power/ground if the outer layers are predominantly solid conductive copper planes. | **T** | **F** |
| 14.2 | For a four-layer voltage/ground MLB, positive phototools are required for the outer layers and negative phototools for processing the inner layers. | **T** | **F** |
| 14.3 | Thermal relief pads and antipads are found only on internal layers of an MLB. | **T** | **F** |
| 14.4 | B-stage material is used to form the outer layers of an MLB. | **T** | **F** |
| 14.5 | If there is too little resin flow during the press cycle, a resin-starved bond line will result. | **T** | **F** |
| 14.6 | Caul plates are made from steel stock whose coefficient of thermal expansion is the same as the pc board stock. | **T** | **F** |
| 14.7 | Black oxide solution is used to improve laminate bonding to the copper areas of the inner layers of an MLB. | **T** | **F** |
| 14.8 | Only one MLB at a time can be pressed between a set of caul plates. | **T** | **F** |
| 14.9 | The required pressure applied to an MLB in the press cycle is approximately 25 tons. | **T** | **F** |
| 14.10 | Epoxy smear is caused by the heat generated from the drilling operation. | **T** | **F** |

## C. Multiple Choice

Circle the correct answer for each statement.

14.1 Completely cured pc board material is called (*B-stage, C-stage*), while partially cured material is called (*B-stage, C-stage*).

14.2 Cap sheets are typically pc board material with copper foil on (*one, two*) sides.

14.3 Improved laminate bonding between the large copper areas of the inner layers results by treating the core material with (*etchback, black oxide*) solution.

14.4 The cold press cycle in the fabrication of MLBs requires a 1-hour cure cycle at a temperature of (*345°F, 145°F*).

14.5 Typically, MLBs are drilled (*one, two, three*) at a time.

14.6 Laminating fixtures for MLB fabrication are made from (*stainless, hot-rolled*) steel, which approximates the coefficient of thermal expansion of pc board stock.

## D. Matching

Match each item in Column A to the most appropriate item in Column B.

| COLUMN A | COLUMN B |
|---|---|
| 1. Caul plate | a. Etchback |
| 2. Prepreg | b. Inner-layer interconnection |
| 3. Mold release | c. Double-sided pc board |
| 4. Antipad | d. B-stage |
| 5. Epoxy smear | e. Two hours |
| 6. Thermal relief pad | f. Laminating fixture |
| 7. Cold press cycle | g. Clearance area |
| 8. Core material | h. Kraft paper |

# 15 Quality Assurance Inspection Procedures for PCBs

## LEARNING OBJECTIVES

*Upon completion of this chapter on techniques of quality assurance procedures for printed circuit boards, the student should be able to*

- Understand the difference between destructive and nondestructive testing.
- Use an inspection magnifier lamp to inspect a pc board visually.
- Repair breaks in conductor paths.
- Repair bridges between pattern features.
- Use a comparator to measure pattern features.
- Visually recognize hole-plating defects.
- Use precision instruments to measure pc boards.
- Use an ohmmeter to perform basic electrical testing.
- Be acquainted with automated test equipment.
- Microsection, inspect, and evaluate specimens from a pc board using a high-powered microscope.
- Use a microscope to take precision measurements.

## 15.0 INTRODUCTION

The manufacture of a printed circuit board, whether it be a single-sided, double-sided, or multi-layer design, requires many sequential processing steps, expensive equipment, and a considerable amount of materials and chemicals. Because of the numerous processes, some of them critical, a finished pc board is prone to a host of faults. Problems such as defective materials, incorrect processing sequence, errors in chemical timing or temperature, and poor workmanship are not uncommon and never completely unavoidable. A fabricated pc board represents a significant cost to the manufacturer. Many detected defects are either correctable or acceptable and may not render a board useless. In addition, the components that are to be installed onto the board represent further major costs. For this reason, it is not economically sound to load a board that has not been inspected and tested to determine its acceptability. In this chapter we establish inspection procedures that will determine the quality of a processed board before the component assembly and final soldering processes are initiated.

It is not the intent of this chapter to define the absolute criteria for accepting or rejecting a processed pc board. These criteria are determined by the manufacturer of the product based on the system's functionability and the standards of acceptability as defined by the pc board industry. The objective of this chapter is to identify the common features of a processed board to be inspected and to show the most appropriate method of determining the relative quality of these features. In many cases, a common problem is in interpreting the results obtained in the inspection procedure.

In general, inspection procedures for pc boards can be separated into two major categories: (1) *nondestructive* testing and (2) *destructive* testing. As the terms imply, the nondestructive test will not damage the board, which is functional after inspection. This is the most cost-effective method of board testing. However, to accurately determine the quality of manufacture, a destructive test is necessary.

Nondestructive tests presented in this chapter are (1) visual, (2) mechanical, and (3) electrical. *Visual inspections* are those that can be made with the naked eye or with the aid of a magnification device. *Mechanical inspections* primarily involve the checking of measurements of length, width, thickness, and hole diameters. *Electrical inspections* can involve such tests as checking for short or open circuits on conductor patterns or within the inner layers of multilayer boards. Some of these tests use sophisticated computer systems that completely check the electrical pattern for any defects. To properly test a pc board, all three methods, visual, mechanical, and electrical, should be employed. The treatment of destructive testing will include the preparation and interpretation of microsectioned samples of plated-through holes.

*CAUTION:* When removing specimens with a punch press from printed circuit boards, keep fingers clear from the punch and die to avoid injury. Care must also be taken to keep hands and fingers away from the sharp blade when inserting or withdrawing samples from the clamp on the diamond saw machine. A chemical hazard exists in materials presented in this chapter regarding specimen etching. The processes discussed should be performed in adequately ventilated facilities and with applicable equipment having proper fume exhaust systems. Approved face shields (goggles) in compliance with ANSI Z87.1-1989 and chemical-resistant clothing and footwear and neoprene or nitrile gloves must be worn while performing any of these operations. Also, the section on safety outlined at the beginning of the book should be reviewed for more specific information.

## 15.1 VISUAL INSPECTION

Even though the finished pc board can be inspected with the naked eye, an inspection magnifier lamp is an ideal aid. This instrument more readily allows the detection of flaws and results in less eye strain. The magnifier lamp has a circular 22-watt fluorescent tube that encircles an interchangeable magnifying lens. Typical lens powers are 3, 4, 5, 8, 10, and 12 diopters (1.75 to 4.00×), which provide shadow- and distortion-free viewing. The lower-powered lenses are used for viewing a reasonably large area, while the higher-powered lenses are used for closer scrutiny of a smaller area or feature. See Table 15.1 for a comparison of lens power and magnification.

An inspection magnifier lamp used to inspect a pc board is shown in Fig. 15.1. The types of surface defects readily visible under a magnifying lamp are *broken* circuits and *bridged* circuits. Broken circuits are viewed as obvious interruptions in conductor paths,

**TABLE 15.1**

Comparison of Lens Power and Magnification

| Diopters | Magnification | Object Distance from Lens (in.) |
|---|---|---|
| 3 | 1.75 × | 13 |
| 4 | 2.00 × | 10 |
| 5 | 2.25 × | 8 |
| 8 | 3.00 × | 5 |
| 10 | 3.50 × | 4 |
| 12 | 4.00 × | 3 |

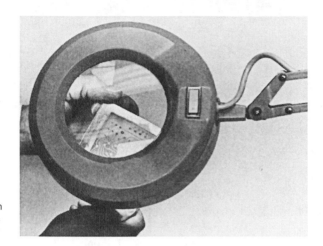

**Figure 15.1** Inspection of pc board under magnification.

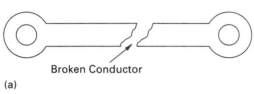

**Broken Conductor**

**(a)**

**Figure 15.2** Typical conductor defect and repair: (a) open circuit; (b) one method to repair open circuit.

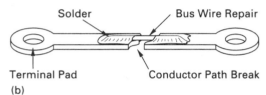

Solder     Bus Wire Repair

Terminal Pad     Conductor Path Break

**(b)**

such as that shown in Fig. 15.2a. These may be the result of improper handling of the artwork master or the phototools, opaque particles allowed to remain on the phototools, incomplete developing of the resist film, or scratched pattern-plated solder before the etching process. This type of defect must be repaired. If the number of such defects is not extensive and if it is permissible by company policy, the break can be repaired with bus wire. A piece of bus wire cut long enough to be laid along and span the broken path serves as a conductive bridge. Flux is applied to the paths on each side of the break, and with the wire held in place it is soldered to the broken ends of the conductor paths. The result of such a repair is shown in Fig. 15.2b. The flux should be completely removed after the wire has been soldered. (*Note:* This type of repair should be undertaken after the pc board is completely assembled and soldered.)

Bridged circuits appear as unwanted connections between any features in the conductor pattern (see Fig. 15.3). These *shorts* are again primarily due to improper handling of the artwork masters, the phototools, or the imaged board. A scratch in the photo resist can also cause these bridged circuits to form. If permissible, these defects can be easily repaired by removing the unwanted metal (solder over copper plating over laminate copper) down to the insulation surface with a scalpel.

When visual inspection requires the precise measurement of the conductor pattern features, an important device used for this purpose is the 10× magnifier called a comparator. This has a precisely scaled reticule such as the one shown in Fig. 15.4a. As shown in Fig. 15.4b, the scale has 100-mil (0.1-inch) major divisions with 25-mil (0.025-inch) subdivisions and 5-mil (0.005-inch) minor divisions. The comparator is designed with a transparent 360-degree base to admit a maximum amount of light. It is used by simply centering its base directly over the feature to be measured. By careful aligning the reticule on the

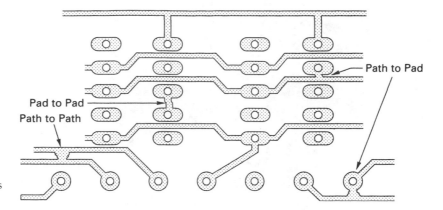

**Figure 15.3** Examples of circuit bridges.

feature, its size can be precisely determined to within 5 mils and a good reading to within 1 or 2 mils can be estimated.

The use of a comparator is shown in Figs. 15.4c, d, and e. Figure 15.4c shows the conductor path width being measured. Note that the width distance falls precisely on a minor subdivision, resulting in an exact measurement of 15 mils (0.015 inch). Figure 15.4d shows a comparator being used to measure the annular ring of a pad. Here the result is less precise since the dimensions of the feature span between minor divisions. However, a reading of approximately 7 to 8 mils (0.007 to 0.008 inch) can be closely estimated. Note also in Fig. 15.4d that there is some misregistration of the pattern since the terminal pad is not centered on the hole as seen by the nonuniformity of the annular ring. Another example of the use of a comparator is the measurement of the air gap between conductor pattern features. This is shown in Fig. 15.4e, where a conductor path-to-terminal pad spacing is read as less than 5 mils. A minimum air gap of this dimension may be questionable as to its acceptability.

A severe imperfection on double-sided pc boards that can be detected by visual inspection is defects in hole wall plating. These *voids* are the absence of both solder and copper in random areas of the barrel (see Fig. 15.5a). These are often the result of excessive cleaning prior to pattern plating, incorrect processing of the deposition line, or a variety of other processing problems. Voids in a plated-through hole may well render the board useless. Usually, one or two random and isolated voids can be considered acceptable, but if they appear in the same hole, they most likely exist in all holes and in all boards processed in that batch.

Another form of void, also shown in Fig. 15.5a, is a *rim void* or *barrel crack,* which occurs at the interface of the surface of the hole and the top or bottom ends of the barrel. These cracks are totally unacceptable since they break the electrical continuity between both sides of the board. Voids can often be detected with the naked eye by viewing the holes at a 45-degree angle to the board surface. A comparator may also be used to enhance the view. It is also placed at a 45-degree angle to the board surface (see Fig. 15.5b).

A special magnifier, used exclusively for the inspection of plated-through holes, is shown in Fig. 15.5c. This 8× magnifier has a center prism that splits the image of the inside of the hole into nine views that simultaneously show the entire surface area of the barrel without moving the instrument. It is simply placed flush to the surface of the board and centered on the hole to be inspected. Thus a 360-degree undistorted view of the barrel is achieved without rotating the board or the instrument.

Inspection of plated-through holes also detects foreign matter, such as dust and dirt, which may have resulted from drilling, or from nodulation (rough surface) caused by poor plating processes. Anything on the hole walls other than a smooth and continuous layer of metal is an indication of processing problems starting with the drilling operation.

Highly sophisticated optical equipment is also available to improve the inspection of a pc board. One such instrument is shown in Fig. 15.6a. It is designed to project a magnified

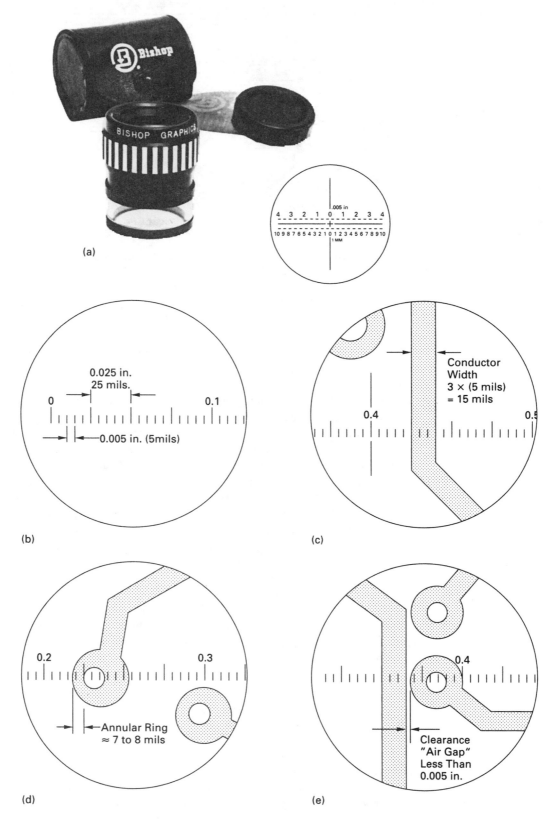

**Figure 15.4** Measurements that can be made using a comparator: (a) 10× comparator, *courtesy of Bishop Graphics, Inc., Westlake Village, CA;* (b) scaled reticule; (c) conductor width measurement; (d) annular ring measurement; (e) air gap measurement.

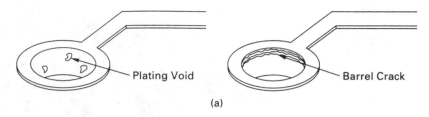

Plating Void — Barrel Crack

(a)

(b)

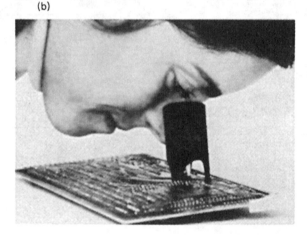

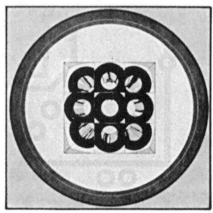

**Figure 15.5** Plating defects and methods of visual inspection: (a) voids; (b) inspection using comparator at a 45-degree angle; (c) optical hole inspector. *Courtesy of ALCHEMITRON.*

view of the pc board onto a large screen. It includes a microscope to permit even closer examination of any selected feature. This inspection system is equipped with an indexing *x–y* table onto which the board to be viewed is placed. The entire area of the board can be rapidly and systematically scanned at as much as 25× magnification. Sufficient space is available to allow some minor repairs to be made within the viewing area. One of the advantages of this type of inspection system is that it can aid in viewing conductor pattern defects that may not easily be detected by eye or with the aid of low-magnification devices. Some of these defects, together with those previously mentioned, are shown in Fig. 15.6b. In addition to inspecting finished pc boards, these optical inspection systems may also be used to check and correct artworks and phototools. This greatly aids in preventing a number of board defects that might otherwise occur.

(a)

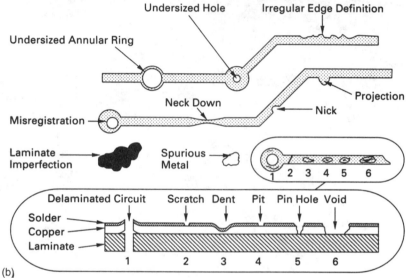

**Figure 15.6** Defects are more readily found by magnified optical inspection: (a) optical inspection, *courtesy of Circuit Equipment Corporation, Baltimore, MD;* (b) conductor pattern defects.

## 15.2 MECHANICAL MEASUREMENTS

Mechanical measurements on pc boards may be grouped into two general categories: (1) the measurement of board length, width, and thickness, and (2) hole size measurement of both plated and unplated holes. The length and width of the finished board should be measured with precision instruments such as the dial caliper. This caliper is typically accurate to 0.001 inch, which is suitable for making board measurements. The results of the dial caliper readings should be compared with those of the dimension drawing of the board to determine if they are within the specified tolerance. For single- and double-sided boards, the length and width dimensions are critical, especially if the board is to slide into grooved guides, shown in Fig. 25.18, so that its fingers fit into an edge-line connector. On the other hand, if the board is designed to simply be installed into a chassis, such as that shown in Fig. 25.19, its dimensions are far less critical.

In the case of a multilayer board, the *thickness* dimensions are extremely important if the fingers are to fit snugly into a connector. To avoid mechanical or electrical problems, the board must not be thicker or thinner than the specified tolerance. Thickness measurements can also be accurately made with the dial caliper or with a micrometer. These measurements should be taken in the finger area to determine correct fit into the connector. To obtain the most accurate reading, only the *insulation* thickness should be measured, not a portion of the board that contains any copper pattern. If thickness measurements of all four corners of a multilayer board show a wide variation, the problem is usually the result of a fault in the press platens.

For the accurate measurement of unplated holes having straight sides (i.e., uniform top-to-bottom diameters) a *dial-indicating tapered-pin hole gauge,* such as the one shown in Fig. 15.7a, may be used. To measure the hole's diameter, the tapered pin is first inserted into the hole. The base of the gauge is then brought down flush to the insulation surface of the board (see Fig. 15.7b). The dial indicator is calibrated to read the hole diameter directly as a result of reading the diameter of the tapered pin at its top entry point. The gauge shown in Fig. 15.7a will provide a hole-diameter reading to an accuracy of 0.001 inch. Four different gauges of this type will give a full range of measurements from 0.010 to 0.330 inch, which represent drill bit sizes from No. 80 (0.35 mm) to 3/8 inch (9.5 mm).

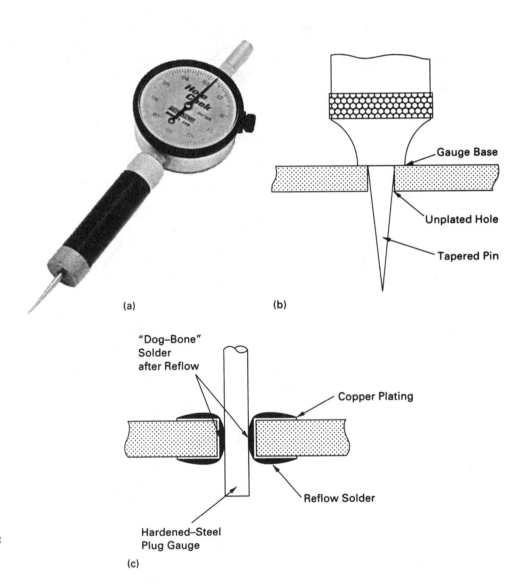

**(a)**  **(b)**

**(c)**

**Figure 15.7**
(a) Tapered-pin hole gauge, *courtesy of ALCHEMITRON;* (b) measurement technique for unplated holes; (c) measurement technique for plated holes.

A cross-sectioned view of a drilled hole that has been plated and reflowed is shown in Fig. 15.7c. Note the *hourglass* or *dog-bone* effect of the plated metals. A major concern in plated holes is that the specified component leads fit into the holes without any degree of difficulty. Accurate measurement will determine this in addition to ensuring that proper plating thickness has been achieved. To measure the minimum diameter of these nonuniform barrels, a set of untapered plug gauges, having accurate and uniform diameters, must be used. Typical sets of plug gauges for this application, such as those available from the Meyer Gauge Company,* consist of a series of 2-inch microfinished hardened-steel pins. One set ranges in size from 0.011 to 0.060 inch in diameter, increasing in increments of 0.001 inch; another set ranges from 0.061 to 0.250 inch, increasing in the same increments. The size of pin that when inserted into the hole will meet extremely slight resistance or drag indicates the hole's diameter. These pins must be used with care. If even moderate pressure is required to insert a pin into the hole, damage to the plating could result.

## 15.3 ELECTRICAL MEASUREMENTS

One of the primary concerns in a finished pc board is that no unwanted shorts or opens exist. For relatively uncomplicated designs, visual inspection, as previously discussed, is suitable for detecting these faults. A great deal of difficulty is encountered, however, in the inspection of high-density double-sided and multilayer boards. For these levels of designs, it is beyond effective visual inspection, even if aided with optical equipment. After initial visual and mechanical inspection, computer-based test equipment designed to automatically test the entire circuit for proper continuity is employed. Other electrical instruments are also used to measure the average thickness of the copper deposited on the barrels of plated-through holes.

In multilayer boards, severe misalignment of inner layers may cause, for example, a power layer to be shorted to a ground layer (see Fig. 15.8a). Note that proper alignment of

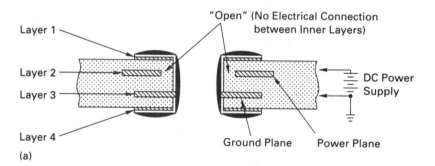

(a)

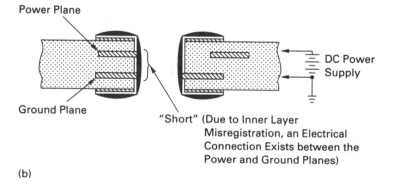

**Figure 15.8** Testing for an internal short circuit in an MLB: (a) properly aligned inner layers; (b) misregistration of power plane.

(b)

*Meyer Gauge Co., 230 Burnham St., South Windsor, CT 06074.

inner layers 2 (power) and 3 (ground) results in the power layer connected to the barrel and an open circuit existing between it and the ground layer. Thus, a power source connected across these two layers will not be shorted. The problem resulting from the misalignment of layer 3 (ground) is shown in Fig. 15.8b. Here, ground makes electrical connection to the barrel, resulting in its being shorted to layer 2 (power). Because the power and ground layers are connected through the barrel, a power source connected to these points will be placed across a direct short.

For the reasons just stated, one of the most basic electrical tests made on multilayer boards is to check for *inner-layer shorting*. This type of test is not intended to measure the degree of misalignment but determines the acceptance or rejection of a board. The only instrument required is an ohmmeter. After zeroing the meter, one test probe is brought into contact with an outer-layer terminal of the board, which is connected through a plated hole to layer 2 (power). The second meter probe is made to contact an outer-layer terminal pad that connects to layer 3 (ground) through another plated hole. A reading of *infinite ohms* (i.e., an open circuit) should be read on the ohmmeter, which indicates an acceptable board. This reading shows no electrical connection existing between layers 2 and 3. Power can safely be applied to this board. If, however, the ohmmeter reads *zero* ohms (i.e., a short circuit between power and ground layers), the board must be rejected because of inner-layer misregistration. Any power applied to these inner layers will result in the supply being shorted, causing fuses to blow, circuit breakers to trip, and so on.

Recent demand for higher-density boards and the number of layers required in one board has resulted in it becoming impossible to test, with 100% accuracy, the electrical integrity of the total pattern. The use of the most sophisticated visual inspection techniques coupled with mechanical tests and electrical checking with meters is found to be unsuitable for high-density, ultra-fine-line boards. Not only is this type of testing extremely time-consuming, but only 50 to 70% of faults in a multilayer board are detectable, due to human limitations. To overcome these shortcomings and to develop test equipment more compatible with the state of the art, manufacturers have developed a new concept in the electrical testing of bare boards (i.e., boards that have no parts mounted). This automated test equipment (ATE) uses a computer that can be programmed to quickly and repeatedly test 100% of the conductor pattern on any level of the board. It can locate faults (shorts or opens) and identify the type of fault. *Bare-board testing* equipment is classified as either *dedicated* or *universal* systems. They are intended for the testing of large volume work and not for prototype applications.

A bare-board pc board test system is shown in Fig. 15.9a. It consists of a microprocessor-based computer connected to a multiple-contact test fixture, or head, through a wire cable system. Additional equipment required is an external vacuum pump to activate the test head, and a printer to provide a readout of the test results. The test head is manufactured to test a specific circuit board pattern (dedicated) or designed to test any pattern (universal). A universal test fixture is far more expensive than the dedicated type. For both test fixture designs, the head consists of a series of spring-loaded test probes or *nails*. They are press-fitted into an insulated base plate in a pattern that aligns each nail to make contact with each terminal pad of the conductor pattern. Several styles of test probes are shown in Fig. 15.9b. The probe with the *serrated* top is the one used in the tester shown in Fig. 15.9a. This style makes a firm electrical contact on its assigned terminal pad. The other end of the probe is connected to a lead in a cable that is connected to the computer. The cable provides the electrical connections between the individual terminal pads and the microcomputer.

A second insulated panel rides on guide pins secured to the base plate. This panel is provided with a clearance hole for each contact point (i.e., pad) and is fitted with tooling pins for registering the board to the *bed of nails* (see Fig. 15.10). After the pc board is positioned onto the tooling pins, a sheet of surgical rubber is placed over the head to create a vacuum seal on the board. When the vacuum system is activated by the computer, the board is forced downward onto the bed of nails to make electrical contact with each terminal pad.

A dedicated test head has test pins located below each terminal pad for a specific conductor pattern only. On the other hand, a universal test head has a test probe positioned on

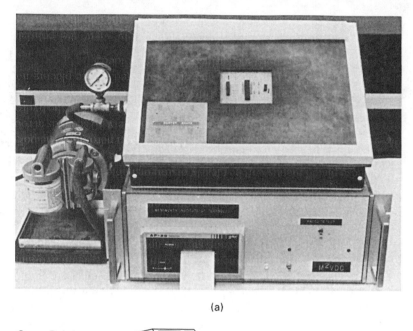

(a)

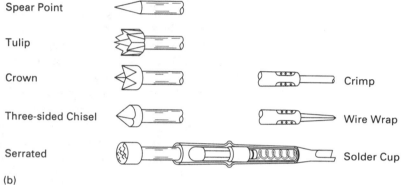

Spear Point

Tulip

Crown             Crimp

Three-sided Chisel       Wire Wrap

Serrated          Solder Cup

(b)

**Figure 15.9** Electrical testing of bare boards: (a) bare-board test system; (b) styles of contact pins.

a grid spaced at 0.1-inch intercepts. Large universal test systems may have a matrix of as many as 40,000 probes (20- by 20-inch grid). With this type of system, any board design may be tested as long as its terminal pad centers are located on a 0.1-inch grid and the tooling holes on the board are made to align with the tooling pins on the fixture. To test a specific conductor pattern on a universal test system, a plexiglass plate with a hole image that matches the terminal pad locations is positioned between the probe matrix and the board. This allows only those probes that align with pads to make contact with the board, while the other probes are blanked out.

The bare-board tester, shown in Fig. 15.9a, has two modes of operation. These are the *learn* mode and the *test* mode. When power is initially applied to the microcomputer, the system is in the *learn* mode. A pc board having a predetermined correct conductor pattern is placed on the test head. The *press to test* button is depressed, which activates the vacuum system and causes the board to be pressed against the probes. The microcomputer then samples each test probe and stores in its memory the number assigned to each probe, its terminal pad position, and the correct conductor path interconnection sequence. The microcomputer has thus sampled each test probe and *learned* the electrical continuity map of the correct master board. This board is then removed from the test fixture and the system is ready to electrically test boards that have been processed with the same conductor pattern as the master board. A board to be tested is placed on the fixture and the *press to test* button is again activated. The contacting probes feed information to the computer, which compares the conductor pattern of this board with that of the master board stored in memory.

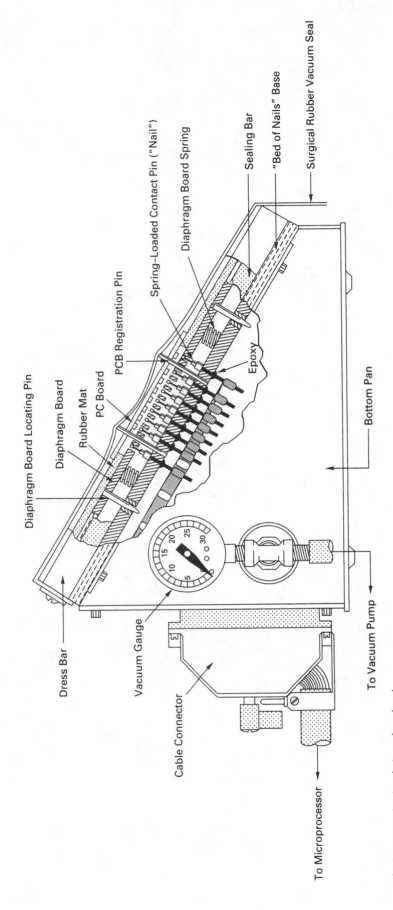

**Figure 15.10** Sectional view of test head.

289

INSTRUCTIONS FOR
LOADING DATA

THE MICROCOMPUTER
SYSTEM IS READY TO
LOAD DATA FROM THE
MASTER PC KEYBOARD.

PLACE THE MASTER PC
BOARD ON THE BED OF
NAILS AND DEPRESS
THE "PRESS TO TEST"
SWITCH.

**Results of learn mode**

WENTWORTH INSTITUTE
OF TECHNOLOGY

ELECTRONIC SHOP
BARE BOARD TESTER

THE PC BOARD NOW
BEING TESTED HAS
NO WIRING ERRORS.
JOB WELL DONE.

**Results of correct PC board**

WENTWORTH INSTITUTE
OF TECHNOLOGY

ELECTRONIC SHOP
BARE BOARD TESTER

6 WIRING ERRORS

| DRIVE<br>PIN | ERROR<br>POINT | TYPE OF<br>ERROR |
|---|---|---|
| 12-- | 13--- | SHORT |
| 12-- | 59--- | SHORT |
| 13-- | 39--- | SHORT |
| 39-- | 59--- | SHORT |
| 100-- | 110--- | OPEN |
| 106-- | 110--- | OPEN |

**Results of PC board
with six errors**

(a)

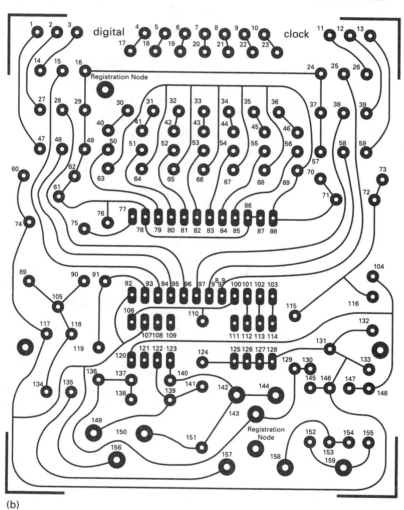

**Figure 15.11** Bare-board conductor pattern code and results of automatic testing: (a) bed of nails printouts; (b) terminal pad identification.

(b)

The results of this test are printed on a paper tape that indicates if the unit under test is acceptable (no shorts or opens) or if there are any faults. Any shorts or opens are specified together with their locations by identifying the assigned numbers of the terminal pads involved (see Fig. 15.11a). To most efficiently make use of the tape readout in locating any faults on the board, a layout of the conductor pattern, together with the assigned terminal

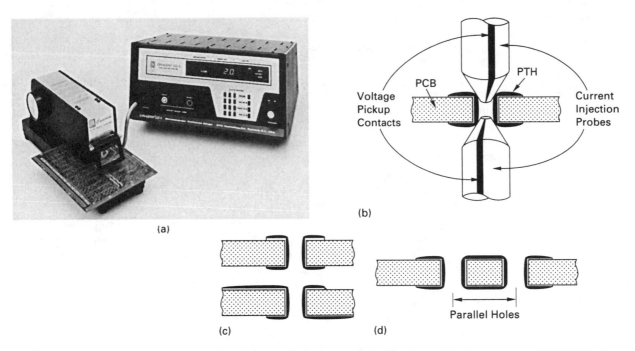

**Figure 15.12** Nondestructive testing of plated-through holes: (a) Caviderm, *courtesy of UPA Technology, Inc.;* (b) self-centering conical test probes; (c) isolated PTHs do not result in parallel paths; (d) parallel holes closer than 1/4-inch spacing will result in incorrect readings.

pad identification numbers, is essential. This is shown in Fig. 15.11b. It can be seen that a bare-board testing system is quick and repeatable with a consistent degree of accuracy.

Automatic test equipment is also available that will quickly and nondestructively measure the average thickness of the copper plated onto the barrel of a plated-through hole. One such system is shown in Fig. 15.12a. This computerized tester is called the *Caviderm CD7* and is manufactured by UPA Technology, Inc. The Caviderm makes a microresistant ($\mu$R) measurement that can be used to nondestructively test both the quality as well as the plating thickness of a plated-through hole. The equipment consists of two self-centering conical contacts that touch the top and bottom rim of the hole (see Fig. 15.12b). The *current injection cones* apply a precisely known current (usually 20 mA) through the copper plating that makes up the hole wall. The resulting voltage drop developed across the hole is detected by the *voltage pickup contacts*. Electronically, the voltage developed is divided by the current injected to result in micro-ohms of resistance ($R = V/I$). The computer processes this information and translates it automatically into barrel thickness, shown on a digital display in mils. This method of testing actually measures the *average* resistance or *average* thickness of the barrel. For this reason, any voids, cracks, or thin spots will strongly influence the reading. The net result will be lower values of wall thickness displayed than those specified or that would result if these defects were not present. In testing plated-through holes in either double-sided or multilayer pc boards, the measurement is essentially the copper thickness and is not really affected by the solder layer above the copper. The measurement can, however, be misleading if not performed correctly. For measurements to be accurate, an *isolated* hole must be used. Shown in Fig. 15.12c, an isolated hole is either one that has no circuit etches (top or bottom) leaving the hole, or one having interconnecting terminations that do not result in a parallel circuit. An inaccurate reading will also result if the hole to be measured forms a parallel circuit with a hole closer than 1/4 inch (see Fig. 15.12d).

The Caviderm CD7 tester is self-calibrating. All that is required to initiate the measurement is to key into the computer the board thickness and the hole diameter.

All of the pc board inspection and test techniques discussed to this point are nondestructive; that is, the board has not been physically damaged in any way. These tests are essential for ensuring a degree of quality control. However, nondestructive testing falls short in providing the product user or the manufacturer a complete assurance of the quality of the finished pc board. This assurance can result only through a destructive cross-sectioning test procedure called *microsectioning.* In this type of test, a sample board from a production lot of double-sided or multilayer boards is destroyed in order to accurately view the interior walls of a plated-through hole to determine its quality. Microsection testing can also be performed using a specially prepared sample, called a *coupon,* which does not require that a finished board be sacrificed. This nondestructive microsectioning test will be discussed in conjunction with specimen selection procedures.

Visual inspection of the cross-sectional view of the barrel of a typical plated-through hole requires the use of a microscope having a minimum of 50× magnification. This highly magnified view allows the electrical integrity of the interconnection points between the top and bottom ends of the barrel to be more accurately evaluated. Plated-through holes are the most critical processes in the fabrication of a pc board. Continuous metal must be formed through the hole and over the original copper outer surface.

Microsectioning aids in evaluating a variety of plated-through-hole characteristics, including barrel copper and solder thickness, voids, barrel cracks (rim voids), nodulation (rough plating), annular ring, conductor width, multilayer board registration, epoxy smear, and etchback. (These defects and their appearance are discussed later in this section.) The visual evaluation of these characteristics provides the end product user with a more definitive representation of the quality of the finished board. Used as a means of quality control, microsectioning tests are an early warning indicator to the manufacturer in determining potential processing problems. If allowed to go undetected, these problems can dramatically reduce the yield of acceptable boards.

The following steps outline the microsectioning process:

1. *Specimen selection and removal.* A sample plated-through hole is selected and then removed from the board with a punch.

2. *Cross-sectional cutting.* The cross-sectional side view of the hole is prepared for inspection by sawing through the center of the hole using a diamond saw.

3. *Specimen mounting.* The cross-sectional view is positioned in a mold that is next filled with an epoxy mounting material and allowed to harden. The epoxy-mounted sample is then removed from the mold.

4. *Sanding and polishing.* The surface to be inspected is sanded and then finely polished to a smooth finish.

5. *Specimen etching.* The area to be inspected is lightly etched to remove any smearing of the two metals (copper and tin–lead) so that they may be clearly distinguishable.

6. *Microscope inspection.* The specimen is now ready to be viewed on an inverted microscope.

Each of these steps will be discussed in detail followed by an evaluation of several illustrations showing typical plated-through-hole faults.

***Specimen Selection and Removal.*** The selection of a sample specimen may be made from the image area of a pc board or from a specially prepared coupon within the thief area (see Fig. 15.13). Those selected from the image area will result in the board being destroyed. Samples taken from the thief area will not damage the board, allowing it to be returned to the usable lot. It is thus apparent that there is a substantial cost advantage in selecting samples from coupons in the thief area. The disadvantage, however, is that the thief area is the section of the board having the highest current density. This will result in a hole

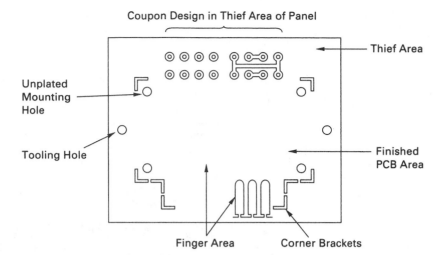

**Figure 15.13** Test coupon provided in thief area eliminates board destruction.

having a larger plating thickness than one appearing in the image area. Either sample will show essentially the same degree of plating quality and registration.

Which of the sample areas to select becomes a trade-off of cost versus inspecting a more representative sample (i.e., taken from the image area). This trade-off usually results in inspecting for plating quality and registration by using thief area coupons, and occasionally destructively sampling a hole from the pattern area to monitor copper and solder thickness.

*Specimen Removal.* When removing the plated-through-hole sample from the pc board, extreme care must be exercised to avoid damaging the plating. Any deformities or cracks resulting from its removal will render the sample useless in evaluating the quality of the plating. Specimen holes may be removed by sawing, shearing, or punching. On 0.059-inch boards, punching is the method most often used. A specially prepared strain relief punch and standard die, as shown in Fig. 15.14a, are required. The diameter of the punch should be at least 1/2 inch with a 3/8-inch-diameter relieved center having a depth of 0.1 inch. This arrangement can be used in a single-station punch press, or a standard lever-operated turret punch such as that shown in Fig. 21.11. To remove the sample, the board is positioned in the machine between the punch and die as shown in Fig. 15.14b. With the specimen hole centered under the relieved section of the punch, it is removed from the board with no damage resulting to the barrel plating. This method of specimen removal is quick, easy, and effective.

*Cross-sectional Cutting.* With the specimen removed, it is next cut through the hole slightly below its center line using a precision cutoff saw fitted with a diamond-coated abrasive blade. A typical saw used for this purpose is shown in Fig. 15.15a. The sample is firmly clamped into the cutoff arm in a precise position over the blade so as to bisect the hole. A micrometer adjustment is used for this positioning. The cutoff arm is weighted with approximately 75- to 100-gram weights in order to apply an adequate downward pressure of the sample against the saw blade during the cutting operation. The speed of the blade rotation is set to about 75 rpm. (This corresponds to a setting of 3.5 on the unit shown in Fig. 15.15a.) The sample is then gently lowered until it rests against the rotating blade. The bottom edge of the blade is continuously immersed in a mineral oil cooling bath in order to avoid the generation of heat during the cutting process, which could adversely affect the appearance and characteristics of the hole plating. With this type of cutting arrangement, bisecting of the sample takes between 2 and 3 minutes and results in an extremely smooth-cut edge. After the cut is complete, the saw will automatically shut off. The sample is then withdrawn from the clamp and washed in alcohol to remove all traces of the coolant. The cut sample is shown in Fig. 15.15b.

*Specimen Mounting.* To prepare the specimen for sanding and polishing, it is encapsulated in an appropriate mounting compound. This begins by positioning the sample

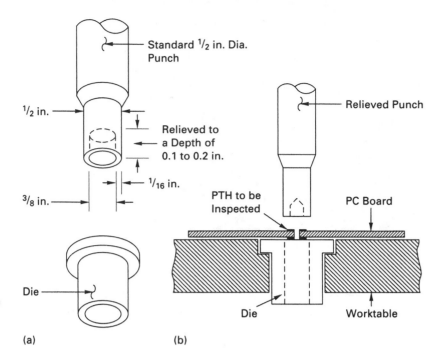

**Figure 15.14** PTH specimen removal by punching: (a) specifications for relieved punch for specimen removal; (b) removing sample PTH for inspection.

Standard 1/2 in. Dia. Punch

1/2 in.

Relieved to a Depth of 0.1 to 0.2 in.

1/16 in.

3/8 in.

Die

**(a)**

Relieved Punch

PTH to be Inspected

PC Board

Die

Worktable

**(b)**

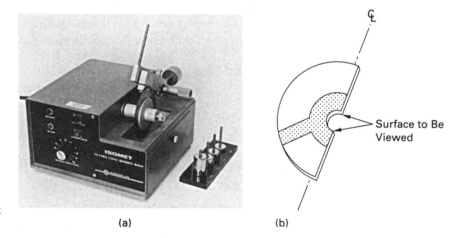

**Figure 15.15** Cutting specimen to expose cross section of PTH with a diamond saw: (a) diamond saw; (b) cut specimen.

Surface to Be Viewed

**(a)**

**(b)**

against a piece of glass with the hole to be viewed (i.e., the cut edge) facing the glass. It is secured in this position with a retaining spring, as shown in Fig. 15.16a. A plastic mounting ring, which has been sprayed with a silicon-base mold-release agent, is then centered around the sample. To prevent damage to the outer-layer plating during the sanding and polishing process, the mounting ring is filled with an encapsulating compound. This also provides support to the cut edges of the plated hole. The encapsulating material must be a cold-mounting compound that will harden quickly. Materials that require high pressure and elevated temperatures are unsuitable for pc board inspection applications since they can damage the plating. A product called Quick MOUNT* is an effective cold-application compound consisting of a powder and liquid resin. To mix this material, 1 part by volume of the

*Trademark of Fulton Metallurgical Products Corp., Saxonburg, PA.

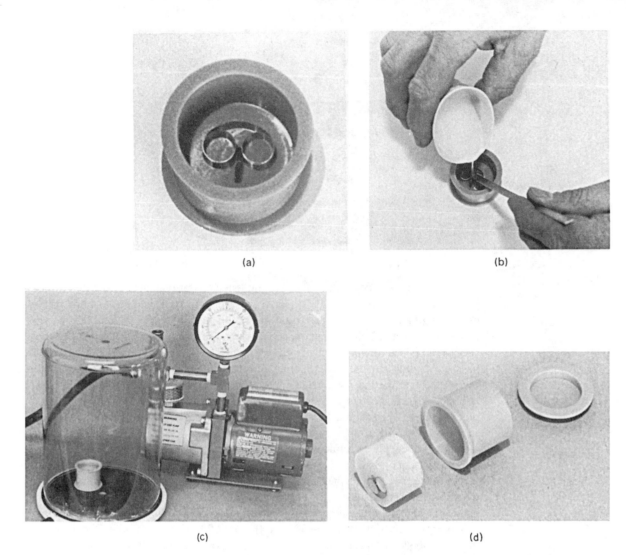

(a)

(b)

(c)

(d)

**Figure 15.16** Specimen mounting in preparation for sanding and polishing: (a) plastic mounting ring with specimen; (b) mounting compound must cover specimen and retaining ring; (c) vacuum pump may be used to remove bubbles generated in the mounting compound; (d) completely cured sample.

liquid resin is poured into a paper cup and 2 parts of the powder are added. This is stirred with a spatula for at least 1 minute to result in complete mixing. The compound is slowly poured into the mounting ring until it is almost full. To prevent the sample from moving while the compound is being added, it may be held in place with a thin rod (see Fig. 15.16b). The filled mounting ring is then allowed to stand undisturbed for a minimum of 30 minutes at room temperature. The mixing of cold-mounting compounds may cause air bubbles to be generated throughout the material. These air bubbles should be removed since they can become lodged around the cut edges of the hole and provide no support of the plating, causing possible misinterpretation of the examination results. The air bubbles may be effectively removed from the compound by placing the glass plate and sample into a vacuum bell jar immediately after the mounting ring has been filled. The vacuum pump is then activated for a maximum of 10 to 15 seconds. The sample is allowed to remain in the bell jar for 30 minutes to allow the compound to cure (see Fig. 15.16c).

After the curing cycle, the sample is removed from the bell jar and from the glass. Finger pressure is applied to the plastic mounting ring to remove the sample (see Fig. 15.16d). The mounted sample is now ready to be sanded and polished.

**Figure 15.17**
Semiautomatic sanding unit for initial sample preparation.

***Sanding and Polishing.*** This is a multistep operation that begins with a coarse sanding and is completed to a highly polished surface. Although sanding is often done by hand on *lapping* blocks (four stationary sanding blocks ranging from 240 to 600 grit), semiautomatic sanding and polishing machines, such as the one shown in Fig. 15.17, are widely used. When used for sanding, four different disks (240, 320, 400, and 600 grit) having adhesive backings are applied to separate wheels. Sanding begins with the 240-grit wheel being placed into the machine. The mounted sample is then placed into a retaining ring that is attached to a motor-driven arm. Steel weights totaling approximately 200 grams are placed on top of the sample to provide a good degree of downward pressure. This is necessary to prevent the sample from rocking, which would result in a nonflat surface. A small stream of water flows continuously from a nozzle to prevent overheating of the sample during the sanding operation. The machine is set at a rotating speed of 300 rpm for the 240-grit wheel. The arm to which the sample is attached moves in a back-and-forth direction as the wheel spins. This results in a figure-8 sanding pattern that duplicates that of sanding by hand. The sample is sanded on this coarse wheel for approximately 2 minutes, after which it is successively sanded on the 320-, 400-, and 600-grit wheels in an identical manner to that described. For these less coarse grits, however, the rotating speed of the wheel is increased to 600 rpm and the sample is again sanded for approximately 2 minutes on each wheel. It is essential that the sample be thoroughly rinsed under cold tap water between each change of grit wheel to remove any remaining coarser grit particles from its surface.

After the sample has been sanded on the 600-grit wheel, it is rinsed with water and inspected for flatness. In addition, the sanded surface should display only 600-grit sanding scratches. With the sanding process completed, the sample is ready to be polished.

Polishing of the sample is done in two phases: (1) *diamond* polishing (coarse) and (2) *alumina* micropolishing (fine). Coarse polishing with a diamond abrasive compound will remove all scratches produced by the 600-grit sanding process and will result in extremely fine surface scratches that are barely detectable by the naked eye. Fine polishing with an alumina compound removes these minute scratches and results in an ultrasmooth surface, free of any visual blemish.

The semiautomatic machine shown in Fig. 15.17 is also used for polishing. Coarse polishing requires a napless polishing cloth, such as a Texmet* disk, to be initially applied to the wheel. This cloth is secured to the wheel by its self-sticking adhesive backing. A diamond paste, such as Metadi* diamond compound, which has a maximum particle diameter of 1 micron (0.001 mm), is spread sparingly in a star arrangement over the polishing cloth (see Fig. 15.18a). The most appropriate lubricant and extender for diamond polish is a lapping oil, such

---

*Trademarks of Buehler Ltd., Evanston, IL.

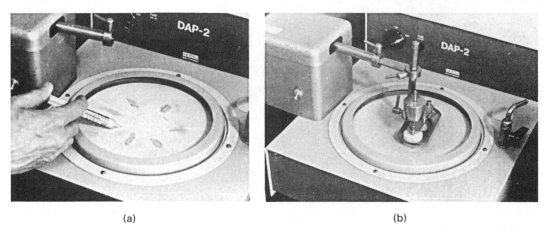

**Figure 15.18** Semiautomatic polishing: (a) application of diamond paste to Texmet polishing cloth; (b) diamond polishing at 250 rpm.

as Metadi fluid. A small amount of this oil is added to the diamond paste and polishing cloth. The polishing machine is next turned on and set to a rotating speed of 250 rpm. The sample is placed in the retaining ring and weighted as described previously (see Fig. 15.18b). Note that water is not used in this polishing process. Satisfactory diamond polishing is completed in 2 to 4 minutes with this arrangement. With the completion of this coarse polishing process, the sample is removed from the machine and rinsed thoroughly with tap water. Contact with fingers on the polished surface should be avoided while handling the sample.

Fine polishing requires a napped polishing cloth, such as AB Microcloth,* to be applied to a second polishing wheel. This cloth also has a self-sticking adhesive backing. The diamond polishing cloth and wheel are removed from the machine and replaced with the second wheel with the napped cloth. Fine polishing is accomplished with a 0.3-micron (0.0003-mm) alumina abrasive, such as AB Alpha Polishing Alumina,* in a water slurry. A small pool, approximately the size of a half-dollar coin, of this abrasive is deposited on the polishing cloth. This alumina abrasive requires no lubricants or extenders since they are contained in the slurry. Fine polishing requires a wheel rotation speed of 600 rpm and the sample again weighted with the same weights. The final polishing process requires only approximately 1 minute to result in an ultrasmooth polished sample. After completion, it is removed from the machine and rinsed in tap water, again avoiding finger contact with the polished surface. The viewing surface appears smooth and shiny and is ready for the final step in preparing the sample for inspection.

***Specimen Etching.*** The polishing processes cause a slight smearing of the copper and the solder platings of the microsectioned sample. To remove the smears so that each of these plated metals will be clearly distinguishable for inspection, a light etching is required. The liquid etch for this purpose is prepared in a petri dish by mixing 3 parts of 26% ammonium hydroxide with 1 part of 3% hydrogen peroxide. The polished surface of the sample is immersed in this mixture for approximately 30 seconds at room temperature. The ammonium hydroxide will attack the smeared copper, while the hydrogen peroxide will attack the solder to result in a microfinished sample ready for microscope inspection. The sample should be thoroughly rinsed and blow-air dried before placing it on a microscope. Finger contact with the polished surface must be avoided.

***Microscope Inspection.*** To effectively view the polished specimen, a standard inverted-type metallurgical microscope, such as the one shown in Fig. 15.19, is recommended.

---

*Trademarks of Buehler Ltd., Evanston, IL.

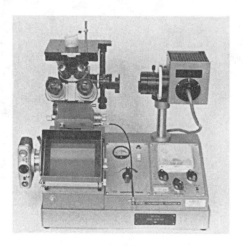

**Figure 15.19**
Metallurgical inverted microscope.

This microscope consists of (1) stage/clamp and stage adjustments, (2) eyepieces, (3) five objective lenses on a rotating nosepiece, (4) light source and internal adjustable power supply, (5) 35-mm and Polaroid camera mount, (6) internal exposure meter, and (7) large-format viewing screen. An inverted type of microscope is recommended so that the polished surface of the sample can be placed against the stage and viewed through the objective lens located below the stage. The cross-sectional area of the hole is centrally positioned in the center of the stage and held firmly in place with a spring clip as shown in Fig. 15.19. For inspection of all parts of the sample, the stage may be moved left to right and front to back by vertical and horizontal *stage control knobs*. Concentric coarse- and fine-focusing knobs adjust the height of the stage.

The microscope can be used with *minocular* or *binocular* eyepieces with a *filar*. Filars are available with magnifications of 5×, 10×, and 15× and are used together with the magnification of the objective lens to determine the overall viewing magnification. Objective lenses have typical magnifications of 2.5×, 5×, 10×, 20×, 40×, and 100×. The revolving nosepiece holds five objective lenses for convenience in quickly changing magnifications. It can thus be seen that with the microscope described, magnifications can range from as low as 12.5× (2.5 × 5) to a high of 1500× (15 × 100).

The source of illumination is a tungsten filament lamp whose intensity is controlled by an internal power source. Filters are also available to enhance various aspects of the viewed sample. The microscope shown in Fig. 15.19 has provisions for mounting 35-mm as well as Polaroid cameras and has built-in exposure and color temperature capabilities. In addition, it also has a large-format viewer that enables the magnified image to be inspected on a translucent glass plate.

***Specimen Evaluation.*** Examination of the specimen typically begins with a low magnification, such as 50×, in order to view the overall hole appearance and quality of fabrication. If an imperfection is detected or if a plating thickness measurement is to be taken, the sample is then viewed under a magnification of upward of 200×.

A sample may possess a variety of characteristics or conditions that are all subject to interpretation. Because these interpretations are subjective in most cases, they should be made only by experienced personnel to determine the processing causes of observed defects. In evaluating microsections, we will discuss only those major items that are of concern to both the pc board manufacturer and the end user.

A microsection of a properly processed double-sided board is shown in Fig. 15.20a; that of a correctly processed multilayer board is shown in Fig. 15.20b. Observe the continuously smooth plated surfaces of both copper and solder with no roughness in the holes caused by drilling or the presence of foreign matter. The overall appearance should be of fine-grained copper having good registration.

Copper and solder thickness measurements inside a plated hole should be made at a minimum of three separate locations within the hole. In Fig. 15.21 copper thickness mea-

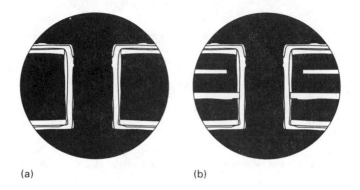

**Figure 15.20** Cross sections of plated-through holes: (a) double-sided; (b) four-layer MLB.

(a)                                    (b)

**Figure 15.21** Three measurement points to determine average plating thickness in a PTH.

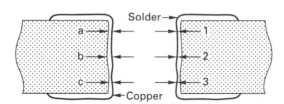

surements are taken at points *a, b,* and *c.* The results for evaluation are the average of these three readings. Readings of solder plating thickness (measured before reflow) are taken at points 1, 2, and 3 and again averaged.

Voids detected within the barrel are one of the most serious plating problems and, depending on their severity and number, may be subject to criteria for board lot rejection. A void is observed under a microscope as a break or absence in the plating that extends to the laminate base material. Such a void is shown in Fig. 15.22a. When a void is approximately the same lateral area of the hole, such as that shown in Fig. 15.22b, it is called a *circumferential,* or *rim,* void. This defect manifests itself as a break completely around the circumference of the hole resulting in an open circuit between layers. Rim voids are *always* cause for board lot rejection.

Another plating problem that is detectable in microsectioning evaluation is *nodulation,* which consists of severe plating irregularities or lumps appearing within the plated hole. This defect can result in additional problems, such as plating voids, reduced plating thickness in selected areas of the barrel, and reduced hole diameter. In severe cases, nodulation is cause for lot rejection. Severe nodulation of a plated-through hole is shown in Fig. 15.23.

Other measurements that can be easily made on a microsectioned sample are the minimum conductor width and the annular ring (see Fig. 15.24). The minimum conductor width measurement is shown as dimension *a.* It is a determination of the amount of undercutting that resulted from the fabrication processes. The measurement of annular ring is shown in Fig. 15.24 as dimension *b.* These measurements will detect both registration and processing problems that may exist.

For multilayer boards, inner-layer registration is critical and also can be determined on a microsectioned sample. A multilayer board with a severely misregistered inner layer is shown in Fig. 15.25. The amount of misregistration is found by first locating the center line (₵) of the hole. Dimensions *A* and *B* are then measured from this center line. Dividing the difference between dimensions *A* and *B* by 2 will yield the amount of misregistration of the inner-layer pad.

Another defect of plated-through holes in the multilayer boards is known as *epoxy smear.* This is the result of excessive heat generated in the drilling process, which causes the epoxy in the laminate to melt during drill penetration. When the drill is removed, it causes the epoxy to smear over the interface area of copper and laminate. This results in an insulated area

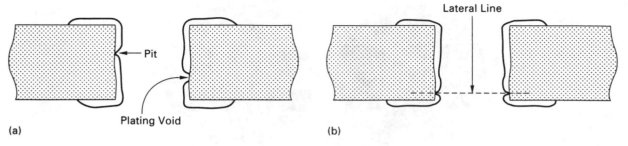

(a)

— Pit

Plating Void

Lateral Line

(b)

**Figure 15.22** Plating defects found in microsections: (a) plating pit and void; (b) rim void or circumferential void.

**Figure 15.23** Severe plating irregularity, termed nodulation.

$$\text{Misregistration} = \frac{A - B}{2} = \frac{34\text{ mils} - 26\text{ mils}}{2} = \frac{8\text{ mils}}{2}$$
$$= \underline{4\text{ mils}}$$

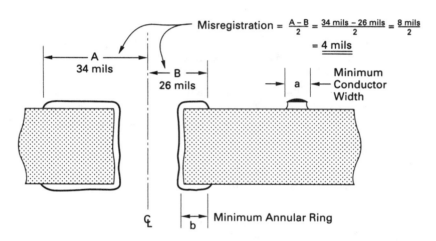

A
34 mils

B
26 mils

a

Minimum Conductor Width

Minimum Annular Ring

b

**Figure 15.24** Measurements made on a microsectioned PTH.

$$\text{Misregistration of Inner Layers (in mils)} = \frac{A - B}{2} = \frac{46\text{ mils} - 29\text{ mils}}{2} = \frac{17\text{ mils}}{2}$$
$$= 8.5\text{ mils}$$

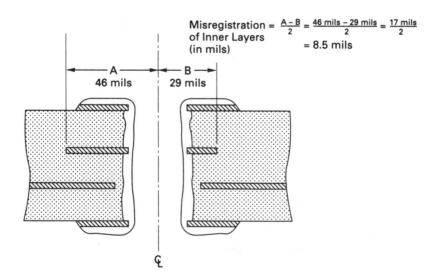

A
46 mils

B
29 mils

**Figure 15.25** MLB inner-layer misregistration measurement.

**300**   *Chapter Fifteen*

formed between the inner-layer copper and the plated barrel. Epoxy smear is detectable on the microsectioned sample and can be removed with the use of chemicals as discussed in Chapter 14. A multilayer board with epoxy smear on an inner layer is shown in Fig. 15.26.

*Etchback* is a chemical process used in the manufacture of multilayer boards. Its purpose is to remove a small amount of insulating material between inner layers before the holes are plated. This insulation removal provides a footing in order to obtain a reliable interconnection. Etchback is specified with a minimum and a maximum number, which range from 0.0002 to 0.003 inch, with typical ranges of 0.001 to 0.003 inch. If the amount of insulating material removed is insufficient, proper footing will not be achieved. On the other hand, an excessive amount removed will result in rough hole wall plating (see Fig. 15.27).

The six chapters in this unit detailed the industrial standard methods and procedures necessary to fabricate single-sided, double-sided, and multilayer printed circuit boards. In Unit 4, we turn our attention to PCB hardware and component assembly with information given on both hand and wave soldering techniques. Unit 4 also includes a discussion on surface-mount technology.

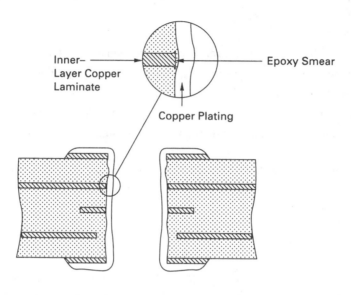

**Figure 15.26** Epoxy smear detected in microsection shows open-circuit defect.

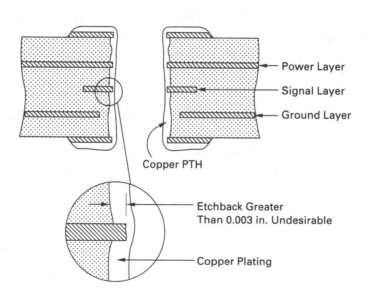

**Figure 15.27** Degree of etchback can be measured by microsectioning and is used for process control.

## EXERCISES

### A. Questions

15.1     What are the three types of tests required to completely inspect a pc board?

15.2     What is the magnification of a 5-diopter lens?

15.3     What type of equipment should be used to measure the width of a gold-plated finger?

15.4     Define the defect termed *rim void* and explain why it is unacceptable.

15.5     What caution should be exercised when using pin gauges to measure the diameters of plated-through holes?

15.6     Explain the difference between dedicated and universal bare-board testers.

15.7     What is the best type of equipment to use in measuring the plating thickness of a PTH?

15.8     List in order the six basic steps required to prepare and inspect a sample PTH microsection for evaluation.

15.9     Discuss the advantages and disadvantages of taking a sample PTH from the thief area compared to that taken from the imaged area.

15.10     What is the recommended method of removing a sample PTH from a board without damage in preparation for microsectioning?

15.11     Describe the procedure for specimen mounting.

15.12     List four major defects that are detectable through microsection evaluation.

15.13     What are the two types of surface defects that are readily visible under a magnifying lamp?

15.14     List three examples of circuit bridges.

15.15     What three common circuit pattern features can be measured with a comparator?

15.16     Describe a simple procedure for determining if an internal short circuit exists between power and ground planes of an MLB.

15.17     With reference to Fig. 15.25, calculate the inner-layer misregistration if dimension *A* is 50 mils and dimension *B* is 40 mils. Is this acceptable if the maximum allowable misregistration is 6 mils?

### B. True or False

Circle *T* if the statement is true, or *F* if any part of the statement is false.

| | | |
|---|---|---|
| 15.1 | As the lens power in diopters increases, the object distance from the lens decreases. | **T**    **F** |
| 15.2 | A 10× comparator is used primarily to detect flaws in etched conductor patterns. | **T**    **F** |
| 15.3 | Plating voids are acceptable if many of them appear in one processed hole. | **T**    **F** |
| 15.4 | A tapered-pin hole gauge is used to accurately measure a plated hole. | **T**    **F** |
| 15.5 | A universal test head has probes positioned on grids spaced at 0.1-inch intercepts. | **T**    **F** |
| 15.6 | The Caviderm nondestructively tests the plating thickness of a plated-through hole. | **T**    **F** |
| 15.7 | Test coupons used for microsectioning are located inside the corner brackets of a pc board. | **T**    **F** |
| 15.8 | Microsectional samples receive a final polishing phase using a 0.3-micron alumina abrasive. | **T**    **F** |
| 15.9 | A magnification of 600× is obtained when using a 15× filar with a 40× objective lens. | **T**    **F** |
| 15.10 | An etchback greater than 0.003 inch is required to result in proper footing of a PTH. | **T**    **F** |

## C. Multiple Choice

Circle the correct answer for each statement.

15.1     (*Destructive, Nondestructive*) testing is necessary to accurately determine the manufacturing quality of a pc board.

15.2     The most accurate annular ring measurements are made with a (*comparator, dial indicating caliper*).

15.3     Hole diameters on pattern-plated pc boards are more accurately measured with a (*tapered pin, plug gauge*).

15.4     Microsectioning is a (*destructive, nondestructive*) test procedure.

15.5     Bare-board testing of pc boards is intended for (*prototype, volume*) production.

15.6     A nondestructive test for measuring copper plating thickness is with the use of a (*Caviderm, bare-board*) tester.

15.7     The detection of epoxy smear requires the use of (*a Caviderm, microsectioning*).

## D. Matching

Match each item in Column A to the most appropriate item in Column B.

| COLUMN A | | COLUMN B |
|----------|-----|----------|
| 1. Microsectioning | a. | Detected in microsection |
| 2. Comparator | b. | Bed of nails |
| 3. Caviderm | c. | 10× magnifier |
| 4. Plug gauge | d. | Plating thickness measurement |
| 5. Etchback | e. | Destructive testing |
| 6. Bare-board tester | f. | Hole diameter |

Unit

# FOUR

## Printed Circuit Board Component Assembly and Soldering

# 16

# PCB Hardware and Component Assembly

## LEARNING OBJECTIVES

*Upon completion of this chapter on printed circuit hardware and component assembly, the student should be able to*

- Be familiar with the various available solder and solderless terminals used on pc boards.
- Install solder terminals onto pc boards.
- Be familiar with pc board connectors.
- Assemble components and devices onto pc boards.
- Heat-sink device leads prior to soldering.
- Install device sockets onto pc boards.
- Install heat-sinking devices onto pc boards.

## 16.0 INTRODUCTION

After a circuit design has been processed into a pc board, two major operations are necessary in order to produce a functional circuit: (1) assembling the individual components and hardware that comprise the finished board, and (2) soldering component leads to the etched copper foil conductors. This chapter is devoted exclusively to assembling components and hardware to pc boards.

In recent years, an emphasis has been placed on the development of reliable miniature devices and hardware that are adaptable to pc applications. Consequently, during the planning stages of a pc packaging problem, the technician is faced with the task of selecting the most suitable components and hardware for his or her specific purpose from an overwhelming variety of types and manufacturers. Coupled with this is the necessity of having available the technical information for mounting these parts onto the pc board. This chapter discusses the applications and assembly techniques of many of the common components and hardware associated with printed circuits.

Specific information for solder terminals and methods for their insertion into the board are considered. Also included are eyelets in addition to information on connectors and associated hardware for board-to-board or board-to-chassis interconnections. Basic techniques and tools necessary for mounting components such as resistors and capacitors are discussed. Included are methods of mounting devices such as diodes, transistors, ICs (integrated circuits), and FETs (field-effect transistors) with and without sockets. In addition, techniques for heat sinking these devices are discussed.

*CAUTION:* Prior to performing any cutting described in this chapter, approved safety glasses with sideshields in compliance with ANSI Z87.1-1989 must be worn. Care should be exercised to prevent ends of severed leads from becoming projectiles when using diagonal cutters. Also, the section on safety outlined at the beginning of the book should be reviewed for more specific information.

## 16.1 STANDARD SOLDER TERMINALS

One of the major problems encountered when designing and assembling a pc board is the means by which external connections are made. For this application, the selection of suitable hardware is almost unlimited and is determined by many design factors, some of which were discussed in Chapter 2.

One of the most economical and easily installed terminal types is a metal stud, mechanically and electrically secured to the board, on which clips for testing or wires for final assembly may be attached. These metal studs are termed *solder terminals*. These terminals are mounted through predrilled holes in the pc board's insulating material and copper conductor. They are then *swaged* to form a solid mechanical connection between the insulated side of the board and terminal pad. Finally, they are soldered to the copper conductor to provide a sound electrical connection.

A typical double-turret, single-ended terminal with its associated nomenclature is shown in Fig. 16.1a. These terminals are designated by *post style* and *size* (available in hollow or solid posts), *shank diameter,* and *shank length.* The shank length must fit the pc board thickness for which it is intended. These thicknesses generally range from 1/32 to 1/4 inch.

Solder terminals of the type shown are usually formed from tinned brass and are recommended for use with pretinned pc boards. This combination results in the proper flow of solder between the circuit pattern and the terminal, when heated with a soldering iron, to form a sound electrical connection with a minimum amount of solder applied to the joint.

To secure the solder terminals to the pc boards, they are swaged by one of two methods: *hand* or *pressure* staking. Both methods require that a hole be drilled into the pc board that is approximately 0.002 to 0.005 inch larger than the shank diameter of the terminal.

For hand staking, an *anvil,* such as that shown in Fig. 16.1b, is held firmly with a *toolholder* secured in a vise. The terminal to be swaged is placed, post *down,* into the anvil with the terminal *crown* seated. The hole in the pc board is positioned onto the terminal shank. When properly positioned over the terminal, the shank rim should protrude slightly out of the hole through the terminal pad. A simple toolholder can be fabricated from an appropriate-sized hex nut by first drilling out the threads to accommodate the tool diameter. A second hole is then drilled and tapped through the center of one of the flats for securing the tool with a set screw.

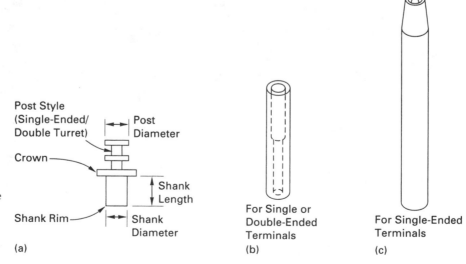

**Figure 16.1** Solder-type terminals: (a) solder terminal nomenclature; (b) anvil for swaging; (c) setting tool.

Post Style (Single-Ended/Double Turret)

Post Diameter

Crown

Shank Length

Shank Rim

Shank Diameter

(a)

For Single or Double-Ended Terminals

(b)

For Single-Ended Terminals

(c)

To swage the terminal, a *setting tool,* such as the one shown in Fig. 16.1c, is used. For single-ended terminals, a setting tool with a *solid face* that will "cup" the *shank rim* down firmly against the terminal pad is required. The swaging of double-ended terminals requires that both the anvil and setting tool be hollow. The setting tool hole diameter should be slightly larger than that of the terminal post for clearance. With the correct setting tool positioned directly over the terminal shank, it is firmly tapped with an 8-ounce hammer. This will cup the shank rim over against the copper conductor, forming a sound mechanical joint.

Pressure staking is the preferred method of securing solder terminals to pc boards because it allows for the least chance of fracturing the board. Hand staking imparts mechanical shocks to the board material and, unless extreme care is exercised, the base material could fracture. Pressure staking imparts a steady pressure to the terminal, eliminating the mechanical shocks and thus reducing the possibility of board fracture.

Pressure staking utilizes an *arbor press* such as that shown in Fig. 16.2a. The terminal is inserted into the anvil, exposing the shank onto which the hole of the pc board is positioned. With the appropriate setting tool attached, rotating the operating handle forward will bring the tool in contact with the terminal shank rim. Additional rotation will apply pressure to the terminal to complete the swaging. This operation is shown in Fig. 16.2a. Because excessive pressure could damage the anvil, setting tool, terminal, and the pc board, most arbor presses are equipped with a threaded anvil. This anvil is threaded into the press table to a height that will provide adequate clearance for the thickness of the pc board. Adjusting for the correct travel of the setting tool is made by trial and error. Optimum anvil setting is achieved when full forward rotation of the operating handle properly swages the terminal without excessive pressure being applied to the insulating material or terminal shank.

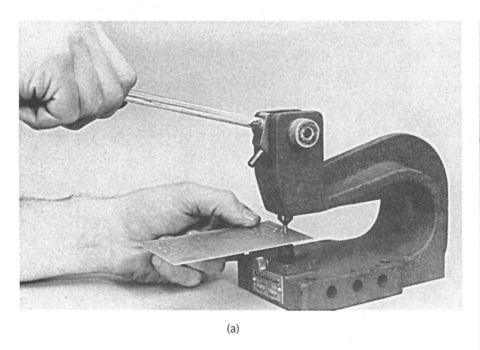

(a)

(b)

**Figure 16.2** Pressure staking solder terminals: (a) engaging setting tool with terminal using an arbor press; (b) properly swaged terminals.

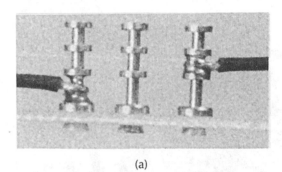

(a)

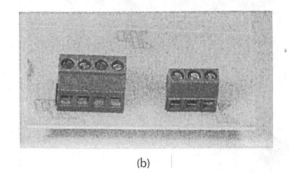

(b)

**Figure 16.3** Printed circuit board terminals: (a) solder, (b) solderless.

When a terminal shank is formed by a V- or funnel-type swage instead of being fully cupped, a *solder ring* may be slipped first over the shank where it protrudes through the board prior to the swaging operation. For this type of swaged shank, solder rings are recommended to ensure a uniform, reliable solder joint, especially if a board is to be hand-soldered.

Figure 16.2b shows the result of correct staking on a terminal. A properly swaged terminal shank will have the appearance of a perfectly smooth cupped eyelet with no fractures around the shaped rim. In addition, no "play" should be detected when pressing down or pulling up on the terminal post.

Terminals with leads soldered are shown in Fig. 16.3a. Figure 16.3b shows an alternative method of pc board wiring using solderless connectors. Techniques for properly orienting leads and soldering them to terminals are discussed in Chapter 26.

## 16.2 PRINTED CIRCUIT BOARD CONNECTORS

The preceding section introduced a simple means of providing terminal points on pc boards for external connections to chassis- and panel-mounted components. This section introduces basic and commonly available hardware and methods for *interconnecting* boards. Selecting the most appropriate connector for a specific application will depend on such factors as (1) cost, (2) mounting limitations, (3) required number of contacts, (4) board thickness, (5) vibration, (6) current and voltage requirements, and (7) required insertion and removal operations. These considerations must be examined before an intelligent choice of connector can be made.

Connectors are available for *board-to-board, board-to-wire, board-to-flexible cable,* and *board-to-chassis* interconnections. These can be separated into two general categories: board edge *plugs* and *receptacles.* The edge-receptacle or *finger*-receptacle-type connector consists of a molded insulator containing a number of female-type contacts that accept the edge of the pc board and mate with foil fingers to make the electrical connections. A typical edge connector is shown in Fig. 16.4a. An edge pin-style connector is shown in Fig. 16.4b, which contains either in-line or staggered contact pins swaged into the pc board and soldered to the terminal pads at the edge of the board. Assembly time is increased with this connector, but it is recommended when the unit is subject to high vibration. A two-piece

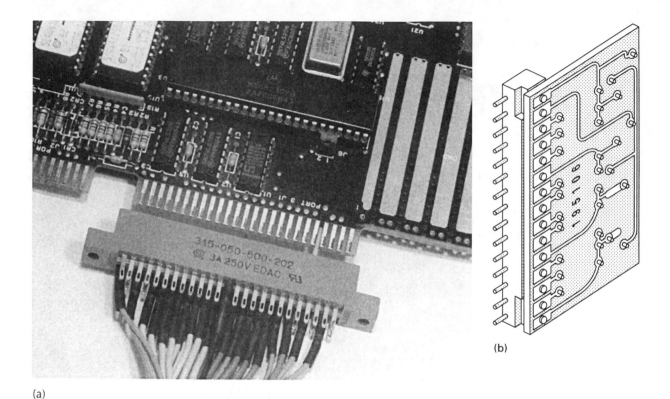

(a)

(b)

**Figure 16.4**  Printed circuit board connectors: (a) edge (finger); (b) pin style.

connector consists of a male plug and a female contact arrangement. One section is mounted to the pc board and the other section is attached to a chassis or another pc board. Electrical contact is made when these two connectors are mated.

Connectors are also specified in terms of female contact design. For the edge-receptacle type, which must contact the foil fingers extending to the edge of the board, two contact arrangements are necessary. The first is the type with contacts that mate with a row of fingers on a single-sided board, whereas the second mates with rows of fingers on *both* sides of a board.

The connectors just discussed are typically constructed of either *brass, phosphorous–bronze,* or *beryllium–copper.* When cost is a major factor, brass is the least expensive. It does, however, degenerate spring tension, owing to aging and continual insertion and removal of the pc board. Phosphorous–bronze contacts are superior in spring-retention characteristics, but they are slightly more expensive and have a higher contact resistance than brass. Beryllium–copper contacts, although more expensive, overcome all the disadvantages of the brass or phosphorous–bronze type. Since the materials used in the construction of the contacts for these connectors are relatively soft, they tend to wear under continual insertion and removal of the pc board. For this reason, they are plated with a gold alloy to a thickness of 20 to 100 millionths of an inch.

Typical contact terminations through the top or rear of the connector are shown in Fig. 16.5. The *eyelet* and *dip solder* types afford the most sound electrical connections and are recommended for use where vibrations are expected. *Wire-wrap* styles allow for more rapid assembly, with a resulting cost savings over the solder types.

Connectors have contact spacings that vary from 0.100 to 0.200 inch, the most common being an on-center spacing of 0.156 inch. The number of contacts per connector may vary from 6 to as many as 100 and will, naturally, depend on the number of contacts required on the pc board. Most manufacturers design their connectors to function equally well with either 1/16- or 1/32-inch-thick pc boards. Insertion-type connectors for thicker boards are also available.

**Figure 16.5** Typical connector terminations.

Eyelet          Dip Solder          Wire-Wrap

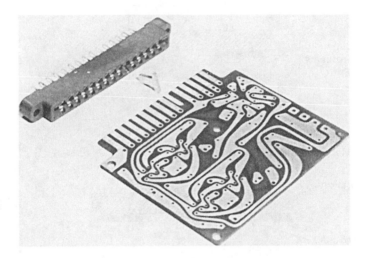

**Figure 16.6** Polarizing key used to correctly orient pc board in connector.

To eliminate the possibility of incorrectly inserting the pc board into the connector, especially in the case of double-sided boards, some method of connector polarization should be employed. To insert the board into the connector in the correct position, a plastic or metal *polarizing key* is inserted either between pairs of contacts or beside them in slots provided for this purpose (Fig. 16.6). For the board to be inserted into the connector when this type of keying is used, a notch in the leading edge of the board at the fingers must align with the key. The type of key that is positioned on the insulator beside contact pairs is sometimes preferable because it does not reduce the number of available contacts by one set. The leading end of the board that is inserted into the connector is chamfered at a 45-degree angle on both edges. This configuration allows the board to be inserted easily into the connector and reduces the possibility of the foil fingers lifting on boards that are frequently removed and replaced.

## 16.3 COMPONENT ASSEMBLY

The smaller components such as resistors and capacitors are first assembled. Under no circumstances should swaging be done on the board *after* any components have been assembled because of fragile components that might be damaged.

The components to be first considered are *axial lead* types (tubular shape having leads at each end). On single-sided boards, these components are normally mounted parallel to the pc board surface with the body of the component lying flush against the insulated side. The leads are bent at right angles to the component body and passed through the predrilled clearance holes to the foil for electrical connection. Once through the board, the leads are bent in the direction of the terminal pad entry. This operation keeps the components from slipping out of position as others are installed and ensures sound electrical connections after soldering. The proper methods of component assembly are shown in Fig. 16.7. Several considerations need to be taken into account before the components are assembled. First, only components that weigh 1/2 ounce or less and/or dissipate less than 1 watt should be considered for flush mounting against the board. Heavier components should be provided with some mechanical clamping arrangement, which is discussed later in this section. Components dissipating more than 1 watt should be mounted above the surface of the board (also discussed later in this section), since the heat generated from these components tends to weaken the foil bond.

*PCB Hardware and Component Assembly*　　　　　　　　**311**

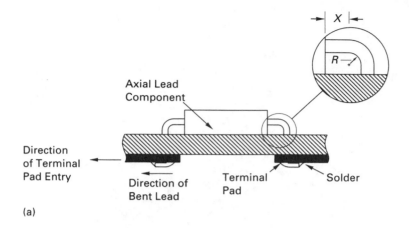

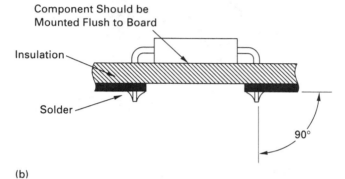

**Figure 16.7** Two methods of assembling axial lead components to printed circuit boards: (a) fully clinched leads; (b) leads with dead heads.

Second, to minimize stresses on the components and their leads, the following minimum specifications should be observed: (1) the minimum bend radius for all component leads should not be less than twice the lead diameter, and (2) this bend should begin no closer than 1/16 inch from the body or end bead of the component. These dimensions are shown in Fig. 16.7a, labeled *R* and *X*, respectively.

Finally, the leads must be properly bent *after* passing through the insulating material. One of two types of bending procedures may be adopted: the *fully clinched* lead arrangement, as shown in Fig. 16.7a, and a *dead head,* shown in Fig. 16.7b. The fully clinched lead is used when maximum rigidity of the component is desired. This technique also requires less solder to complete the electrical connection. Dead heads, as the name implies, are adopted if components may have to be removed later on. This technique makes it less difficult to remove a component after it has been soldered than one with fully clinched leads. Although the specifications provided do not conform exactly to any standards, they are realistic values that have proven to be effective in pc boards not exposed to severe environmental or physical stresses.

The leads of tubular components must be accurately bent *before* being inserted into the pc board. For standard or uniform spacing between terminal pad centers, the *universal bending block,* shown in Fig. 16.8, may conveniently be used. This type of bending block is usually graduated in 0.05- to 0.1-inch increments, with increasing on-center dimensions from 0.5 to 1.5 inches. The component body is positioned in the *center channel* with the leads resting in the appropriate pair of lead *slots* that most closely agree with the terminal pad on-center dimensions on the pc board. The leads are then bent over the edges of the block to properly form both the correct radius and on-center dimension.

If a bending block is not available, *tapered round-nose pliers,* such as those shown in Fig. 16.9, should be used to perform the lead-bending operation. Only pliers with smooth jaw surfaces should be used, to avoid scraping and nicking the component leads. A nicked lead could result in a complete fracture under operational stress or installation.

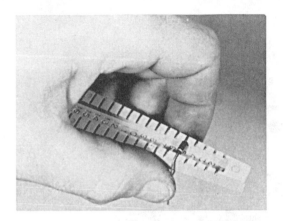

**Figure 16.8** Universal bending block.

To form the bend with this tool, the lead is gripped at the proper position between the tapered jaws. Holding the pliers stationary, the lead is bent at right angles to the component body (Fig. 16.9). Unlike bending tools, round-nose pliers do not allow for repeated accurate formation of on-center dimensions. With practice and care, however, the correct lead position in the tapered jaws can be determined quickly to achieve the desired bend radius and on-center dimensions.

After the component leads are bent, they are inserted into the correct holes in the pc board from the insulated side. The component body is held firmly against the surface of the board and the leads are cut using diagonal cutters, as shown in Fig. 16.10. (More information on these cutters is provided in Chapter 26.) For dead-head connections, the leads should be cut approximately 1/16 to 1/8 inch from the surface. If the leads are to be fully clinched, the jaws of a pair of long-nose pliers are used to force the lead firmly against the foil. Leads should never be clinched before cutting because the jaws of the diagonal cutter could gouge the copper conducting path.

Pneumatic cut-and-bend tools are commercially available for pc board assembly. These tools will, in one operation, bend the lead to the desired angle and cut it to the correct length.

The assembling of disk capacitors presents mounting problems due to their physical design. When possible, they should be mounted with a 1/8-inch space between the bottom edge and the insulation, and the leads passing straight through the board before bending. Since disk capacitors are circular, care must be taken to prevent the component from rotating during the lead clinching operation, which could upset the vertical and center positioning. Figure 16.11 shows the proper method of mounting a disk capacitor and the incorrect mounting method. It is advisable to fully clinch the leads of disk capacitors to secure these components firmly to the pc board prior to soldering.

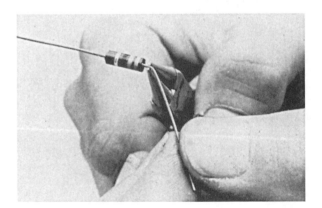

**Figure 16.9** Tapered round-nose pliers used to bend component leads.

**Figure 16.10** Excess lead length removed with diagonal cutters.

When space is critical on a pc board, axial lead components may be mounted vertically. This is accomplished by inserting one lead of the component, without bending, directly into the pc board. The other lead is bent 180 degrees around the body of the component and then passed through the board. This technique is shown in Fig. 16.12a. As can be seen from this figure, the mechanical security of even a small component mounted in this fashion is extremely poor. Under shock and vibration, stresses encountered between the component and the pc board could result in mechanical failure. To minimize this possibility, plastic supports, such as those shown in Fig. 16.12b, are used. They significantly improve mechanical security.

Physically large components (greater than 1/2 ounce) may present undue strain on their leads if not secured. A common type of support used to secure large components to pc boards is a clip-style holder provided with an expansion lug that secures the clip to the board by a simple press fit. This type of component clamp is desirable since it does not require any disassembly to remove the component.

The mounting of small trim-pots (screw-adjusted potentiometers) and thumb-pots (thumb-adjusted potentiometers) shown in Fig. 16.13 requires special consideration since their leads are in the form of rigid pins or tabs that may not be crimped or bent. These leads

**Figure 16.11** Disk capacitor assembly to pc boards.

Right          Wrong

**314**     *Chapter Sixteen*

**Figure 16.12** Vertical-mounting techniques of axial lead components: (a) unsupported mounting method; (b) plastic supports for vertical mounting of axial lead components. *Courtesy of Robison Electronics, Inc.*

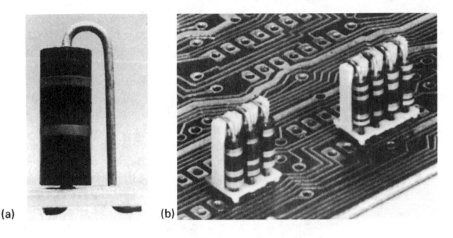

(a)  (b)

**Figure 16.13** Trim and thumb potentiometers.

must therefore be fed straight through when mounted. For mechanical security prior to soldering, the access holes for the pins are drilled for a press fit.

## 16.4 DEVICE ASSEMBLY

Assembling devices to pc boards presents special problems in terms of lead bending, aligning, inserting, crimping, and protecting the device before and during soldering. All of these points need attention if the finished pc board is to be reliable. Devices that are considered in this section for direct mounting on the pc board are *transistors, FETs, TO-5-style integrated circuits* (ICs), and *dual-in-line plastic pack ICs* (DIPs). Unique assembly problems for each of these devices will be considered separately. A large number of solid-state device sizes and case configurations have been standardized by the *Joint Electronic Devices Engineering Council* (JEDEC). Several diode and transistor case outlines are identified by numerical designations prefixed by *DO* or *TO* to indicate a specific outline. Although many power transistors and rectifiers do not follow any established standard for style and size, such as the JEDEC package identification system, the manufacturer's case numbers and specifications will readily provide the necessary device outline and spacing dimensions for assembly.

Before assembly of the devices is described, an important consideration regarding their protection will be discussed. Solid-state materials are extremely temperature sensitive, so precautions must be taken during assembly that will protect them from excessive heat while soldering. Heat is transferred from the soldering iron tip through the lead to the device and affects the lead bond. This thermal problem often dictates the choice of specific mounting arrangements for the device. Manufacturers often provide assembly information together with maximum allowable time at specific temperatures that the device can tolerate without ruining the lead bonds. For a 2N3053 transistor, for example, the manufacturer specifies that the maximum allowable thermal conditions while soldering are 10 seconds at 491°F (255°C). If these conditions are expected to be exceeded, some method of heat transfer is necessary. The most common device used is a *lead heat sink*. This aid is mechanically connected to the lead between the device case and the soldering iron tip. The heat sink absorbs and draws off excessive heat from the device during soldering. A common method of heat sinking is shown in Fig. 16.14. This is a commercially available heat sink that is made of aluminum and is spring-loaded. Aluminum, copper, or brass heat sinks are more efficient than steel pliers because these materials are better conductors of heat. As can be seen from Fig. 16.14, some space must be provided between the device case and the insulation side of the pc board, where sinking is necessary, to attach the heat sink. This obviously is a disadvantage in terms of mechanical security, because it prevents the device from being flush-mounted to the pc board. Heat sinking of leads is not normally necessary when wave soldering techniques (discussed in Chapter 17) are employed, but should be considered for hand soldering. Therefore, when assembling devices the technician must consider the necessity of lead heat sinking and provide for it where necessary.

A popular case configuration used for low-power applications below 1 watt is the *TO-5* style shown in Fig. 16.15a. As shown in Fig. 16.15b, this case configuration is relatively simple to assemble to pc boards. The wide *sealing plane* provides excellent mechanical security when mounted flush against the insulated side of the board. This type of mounting requires that lead holes in the board have the exact spacing and orientation as the transistor leads. Straight feed-through connections to the pc board are made, allowing for either dead head or fully clinched leads onto the foil side of the board. One possible disadvantage could result from this type of mounting. Flush mounting will cause contaminants to be trapped between the device body and the board while cleaning and removing flux. To avoid this, an insulated plastic spacer having riser buttons is placed between the device case and the board. The spacer allows solvents to be used to remove contaminants from under the device case.

The leads of the TO-5-style case initially are inserted most easily into both the spacer and the pc board by positioning the center lead over its respective lead hole. The case is then tilted at an angle to *raise* the other two leads above the surface of the board or spacer. This procedure is shown in Fig. 16.15b. Once this first lead is properly positioned, the case is rocked to align and insert another lead. The last lead is inserted in a similar manner. Once the leads have been inserted through the spacer and the pc board, the assembly can be firmly seated. The leads are bent over toward the conductor pad, cut to the desired length, and clinched similar to component leads, as discussed in Sec. 16.3. The transistor is now ready to be soldered.

Another popular case style is the *TO-92* package shown in Fig. 16.16. These in-line leads are best assembled by first orienting an end lead over its lead hole on the pc board. The middle lead is inserted next and finally the remaining lead is fed into its hole. The hole centers on the board are usually drilled farther apart than the lead orientation on the transistor to provide sufficient spacing between terminal pads because the leads are so closely

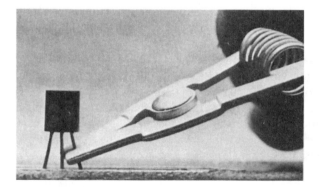

**Figure 16.14** Spring-loaded lead heat sink.

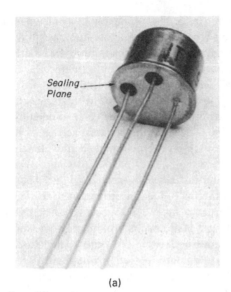
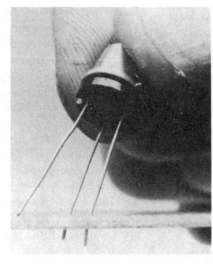

**Figure 16.15** TO-5-style case: (a) case configuration; (b) case is tilted for insertion of second lead.

(a)

(b)

arranged on this type of device. The transistor is mounted so that the height of the sealing plane is approximately 1/4 inch from the board surface. Lead heat sinking can then be easily employed. The completed assembly of a TO-92 transistor is shown in Fig. 16.16.

Some insulated-gate field-effect transistors (IGFETs) or metal-oxide semiconductor field-effect transistors (MOSFETs) require utmost care during assembly. These devices are extremely sensitive to the most minute static voltage at their gate terminal. For this reason, the manufacturers of these devices supply them with their leads packed in conductive foam, which shorts them together. This arrangement prevents any static charge from building up at the gate terminal, which could easily destroy the device. The leads should never be touched with the fingers and should be soldered only with a grounded soldering iron.

The *TO-99* style of case and lead configuration, shown in Fig. 16.17a, presents uniquely difficult assembly problems. Integrated circuits of this style have at least eight leads attached to a case with a diameter of 0.200 inch. The mounting problems arise when attempts are made to align and insert all eight leads simultaneously, since the manufacturer normally cuts all device leads to an equal length. Assembling these devices can therefore be simplified by *staircasing*—that is, by cutting each lead, starting at the *key lead,* succeedingly shorter than the next in a descending direction toward the sealing plane. The key

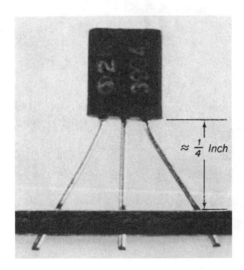

$\approx \frac{1}{4}$ *Inch*

**Figure 16.16** TO-92 transistor package assembled to a pc board.

lead is then used as the reference for lead alignment and placement into the pc board. The *shortest* lead must be cut to a length that will allow it to pass through the pc board and be easily soldered to the terminal pad. With the leads cut in the staircase fashion, the longest lead is inserted first into the proper hole without aligning the others. Each succeeding shorter lead is then inserted, *one at a time,* until all the leads are inserted and the component is properly mounted. The assembly of an IC with staircased leads is shown in Fig. 16.17b. An IC assembled with fully clinched leads and soldered is virtually impossible to remove without damaging the device. For this reason, dead heads are recommended if there is any chance that the device will have to be removed at a later time, which is sometimes the case in prototype work.

Assembling 14-pin *dual-in-line packaged (DIP)* ICs (TO-116 package) presents equally difficult problems in terms of lead insertion. Since the leads on these devices are too short to staircase, an alternative approach is necessary. The DIP is first tilted back with its leads positioned above the pc board and approximately over the lead holes. This step is shown in Fig. 16.18. Finger pressure is applied evenly to the outside edges of the first two opposite leads closest to the board until they align with their respective holes. These leads are partially inserted into the board. The device is then tilted down closer to the board surface until the next two adjacent pins are against the board. Again, finger pressure is applied to align and insert these leads. This technique is continued until all the leads have been inserted into the board. For large 40-pin DIP packages, a commercially available insertion tool, shown in Fig. 16.19, is helpful. Because the leads provided with the DIP are short, they are normally soldered while extending straight through the board. The DIP remains positioned during assembly as a result of the slight pressure exerted outward by the leads against the edges of their access holes. However, the package position should be checked just prior to soldering the board to ensure proper mounting.

(a)

(b)

**Figure 16.17** Method of assembly for TO-99 case style: (a) TO-99-style IC package; (b) staircased leads simplify their insertion into a pc board.

**Figure 16.18** Assembly of dual-in-line (DIP) package.

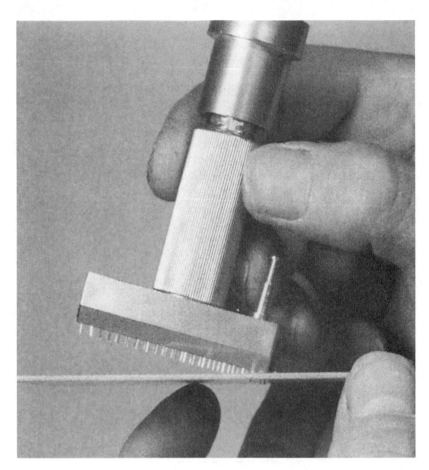

**Figure 16.19**
Assembly of a 40-pin dual-in-line (DIP) package using a commercially available insertion tool.

For large-volume production, automatic insertion equipment is used to rapidly assemble axial, radial, and dual-in-line components and devices to pc boards. As discussed in Sec. 6.6, automatic insertion is feasible only if the component layout is acceptable in terms of arrangement and spacing. Sufficient space between components must be allowed in the design for the insertion machine's head driver and guide fingers. With computer control, the pc board is positioned so that lead access holes are sequentially positioned under the insertion head. As each component is belt-fed into the head, its leads are formed just before the driver presses the body flush against the board surface. Automatic insertion of an axial lead component is shown in Fig. 16.20.

## 16.5 DEVICE SOCKETS

An alternative method of assembling devices to pc boards is with the use of commercially available *device sockets*. These sockets are obtainable in all the standard case configurations. Device sockets have two decided advantages over direct device soldering to the foil. First, the use of sockets allows devices to be removed and replaced easily. Second, lead heat sinking is unnecessary. The only disadvantage of sockets is the additional cost, a factor that must be emphasized throughout the fabrication of a prototype. Where cost is of prime importance, sockets should be used only for devices that may need occasional replacement because of aging or other anticipated design changes.

Some commercially available sockets for transistors and ICs are shown in Fig. 16.21. These sockets are provided with round pins for direct soldering into pc boards or with long square pins that serve the dual function of pc board mounting and wire-wrap lead assembly.

Another method of assembling devices that has all the advantages of sockets plus the additional benefit of versatility in positioning involves the use of *socket terminals* such as

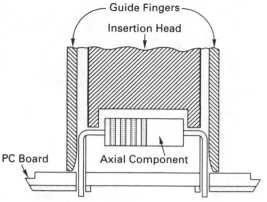

(a) Cross section of driver head and lead forming
guide fingers

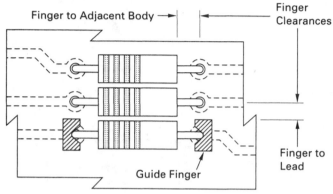

**Figure 16.20**
Automatic insertion of
axial lead component.

(b) Appropriate guide finger clearance

those shown in Fig. 16.22a. These individual terminals consist of a *wire-wrap stake* for interconnections at the base, a *four-leaf contact socket* arrangement for device lead insertion, and a *barb* on the side of the socket for positive positioning and securing the terminal when it is inserted into the pc board. To mount this type of socket, the appropriate diameter mounting holes are first drilled into the pc board in accordance with the socket lead orientation. The socket terminals are then inserted into the board and swaged into place with an arbor press. If desired, the pins can later be soldered to the copper foil of the pc board. High-density *DIP test panels* using the socket terminals are shown in Fig. 16.22b.

The wiring and testing of new IC packages, such as the 52-pin PLCC shown in Fig. 16.23, require special sockets. These sockets accept the pins of the new high-density chips and provide wire-wrap capability for rapid prototyping.

## 16.6  DEVICE HEAT SINKING FOR PRINTED CIRCUIT BOARD APPLICATIONS

Heat sinking devices on pc boards is usually less difficult than heat sinking chassis-mounted devices. When devices and heat sinks are assembled on the pc board, insulators are not necessary because the board itself is an insulator. This is not the case with chassis-mounted heat sinks (as was discussed in Chapter 2).

Several commercially available heat sinks used in low- and medium-power applications and the case styles to which they are adaptable are shown in Fig. 16.24. These heat sinks are fabricated from aluminum or beryllium–copper alloys. They are secured to the device case by either a press fit resulting from the spring tension of the sink material or by mechanical clamping techniques. *Silicon grease* (thermal joint compound) used where the

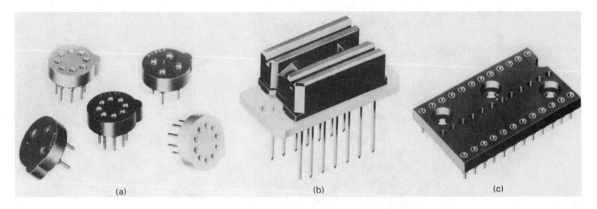

Figure 16.21 Transistor and integrated circuit sockets: 3-, 4-, 8-, 10-pin sockets for TO-5 case style; (b) 14-pin DIP socket with wire-wrap terminals; (c) 24-pin socket for large-scale integration (LSI). *Courtesy of Augat, Inc.*

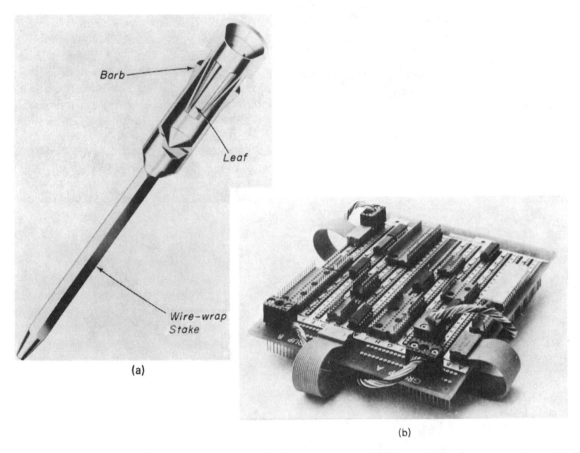

Figure 16.22 Socket terminal nomenclature and DIP panels: (a) wire-wrap socket terminal; (b) high-density test panels using wire-wrap socket terminals. *Courtesy of Augat, Inc.*

device case contacts the heat sink can reduce the thermal resistance between the two by as much as 100%. These sinks are finished in either a black anodized or natural unaltered metal surface. The black finish is more efficient in radiating heat away from the device case.

Techniques for heat sinking DIP-type ICs are shown in Fig. 16.25. The sink shown in Fig. 16.25a requires no hardware for mounting, since it is held by a friction fit between the base of the device and the top surface of the socket. For those DIP-type ICs that are to

**Figure 16.23** A modern 52-pin plastic lead chip carrier (PLCC).

(a)

(b)

**Figure 16.24** Printed circuit heat sinking for low- and medium-power devices: (a) sink for TO-5 and TO-18 case styles; (b) dual TO-5 sink. *Courtesy of International Electronic Research Corporation.*

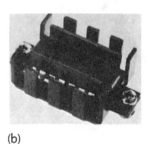

(b)

(a)

**Figure 16.25** Methods of heat sinking DIP integrated-circuit packages: (a) sinking for socket assembly, *courtesy of Astodyne, Inc., Wilmington, MA, subsidiary of Roanwell Corp.;* (b) assembly technique for dual surface heat sinking, *courtesy of Thermalloy.*

be mounted directly to the pc board without the use of a socket, the heat sink with its method of assembly as shown in Fig. 16.25b may be used. This arrangement allows heat to be removed from both the top and bottom surfaces of the device. The mounting used to secure the heat sink to the device is also used to secure the assembly to the pc board.

The selection of the proper heat sink is dictated by design factors discussed in Chapter 2. However, it is the responsibility of the technician to select the most appropriate type and style to meet the packaging requirements.

# EXERCISES

## A. Questions

16.1    What is the purpose of a polarizing key on connectors?

16.2    What is the difference between a full clinch and a dead head?

16.3    Why is it important to use pliers with smooth jaws when bending component leads?

16.4    What care should be taken when assembling disk capacitors to pc boards?

16.5    What is the advantage to staircasing device leads?

16.6    What are the advantages of using device sockets?

16.7    Why are heat sinks generally made with a black finish?

16.8    Why is heat sinking components on pc boards less difficult than heat sinking those that are chassis mounted?

16.9    Why are lead heat sinks used?

## B. True or False

Circle *T* if the statement is true, or *F* if any part of the statement is false.

16.1    Hand or pressure staking is employed when installing turret terminals.                **T    F**

16.2    Both anvil and setting tools must be hollow when swaging double-ended terminals.        **T    F**

16.3    Axial lead components dissipating up to 2 watts may be flush mounted on the pc board.    **T    F**

16.4    Dead heads are used if it is expected that the component may be removed at a later time.  **T    F**

16.5    Socket terminals can be used in place of device sockets.                                 **T    F**

16.6    A heat sink may be applied to the device case while soldering the leads.                 **T    F**

16.7    Some connectors are plated with a gold alloy to reduce wear.                             **T    F**

## C. Multiple Choice

Circle the correct answer for each statement.

16.1    Solder terminals are secured to the pc board by (*welding, staking*).

16.2    For maximum rigidity, components should be installed with (*full-clinch bends, dead heads*).

16.3    A major advantage of using device sockets is (*service, system reliability*).

16.4    A component lead bend should begin no closer than (*1/2, 1/16*) inch from the body end.

16.5    When a component is assembled into its mounting holes, the leads are bent (*toward, away from*) the conducting path at a (*90-, 30-*) degree angle.

16.6    When installing disk capacitors, their leads should be provided with (*full-clinch bends, dead heads*).

## D. Matching

Match each item in Column A to the most appropriate item in Column B.

| COLUMN A | COLUMN B |
|----------|----------|
| 1.  PCB connectors | a.  Arbor press |
| 2.  Solder terminal | b.  Silicon grease |
| 3.  Pressure staking | c.  TO-92 |
| 4.  Lead bend | d.  Bifurcated contacts |
| 5.  Transistor case style | e.  Dead head |
| 6.  Heat sink | f.  Metal stud |

## E. Problems

16.1 Cut two 1 1/2-inch lengths of 1/8-inch-diameter brass or copper rod. File a taper on one end and a flat side over the entire length of each piece. Remove the teeth from only the ends of the jaws of a small alligator clip and construct a lead heat sink similar to that shown in Fig. 16.26.

16.2 Mount the components on the pc board designed for the series feedback regulator in Problem 6.2 and constructed in Problem 10.2. (The assembled board is soldered in Problem 17.2.)

16.3 Install the components on the pc board designed for the tachometer in Problem 6.3 and fabricated and labeled in Problem 10.3. (The assembled board is soldered in Problem 17.3. The tachometer pc board and meter are chassis- or panel-mounted in Problem 26.5.)

Figure 16.26

# 17 PCB Soldering

*Upon completion of this chapter on printed circuit board soldering, the student should be able to*

- Understand the health hazards involved in the use of solder and flux.
- Understand the composition and temperature characteristics of soft solder.
- Know the major classifications of fluxes and their applications.
- Know the available soldering iron tip configurations and their temperature requirements for a variety of soldering applications.
- Use a soldering iron to solder pc board connections.
- Use a personal computer to obtain a thermal profile of a soldering iron.
- Be familiar with common solder joint deficiencies, their causes, and methods to correct them.
- Make an informed choice when selecting electronic grade specialty chemicals relating to their impact on the environment.
- Use de-soldering aids and tools.
- Understand the wave-soldering process and the basic components of a wave-soldering system.
- Know how to obtain a thermal profile of a wave soldering process.

## 17.0 INTRODUCTION

With the components properly assembled, the final process necessary to complete the construction of a pc board is to produce sound electrical connections between the leads and the foil pattern. These connections can be produced by several methods, the most common of which are *welding* and *soft soldering*. Welding requires the use of complex and expensive equipment and, as such, is rarely used in fabricating a prototype pc board. Soldering, on the other hand, can be performed quickly with less expensive equipment and results in excellent mechanical and electrical connections and protects the joint from oxidation. For these reasons, soldering will be discussed exclusively in this chapter.

*Soldering* is a metal solvent or chemical alloying action of the solder with the surfaces of the metal parts between which an electrical connection is formed. This completely metallic contact is produced by the application of soft solder with the heat of a soldering iron to the joint between the component lead and the terminal pad. The resulting connection is electrically sound, with the new alloy formed having different electrical and mechanical characteristics than either the solder or the metals joined. The soldering of pc boards requires the development of proper techniques if quality results are to be obtained. The two methods of soldering discussed in this chapter are *hand* and *wave* soldering. These methods are used extensively in pc board applications.

Specific information is also provided on the following topics: soft solder and its characteristics; the types of *flux* available and their application; the soldering iron, including the *tip*, *power rating*, and *tinning* processes; and hand-soldering techniques, with emphasis on the proper soldering of leads, swaged terminals, and conductor paths (tinning). Included is the technique of obtaining the thermal profile of a soldering iron with the use of a personal computer. A discussion of the characteristics of correct solder connections is presented to aid in visual inspection. *De-soldering* and the solvents used to remove the flux residues formed on soldered connections are also included. Finally, information on production-line methods of soldering, specifically wave soldering, is presented together with a method of obtaining a thermal profile of the wave soldering process.

*CAUTION:* Place a hot soldering iron into its holder when it is not in use. Keep hands and fingers away from the heating element and soldering iron tip. Be especially watchful that the heating element or tip does not come in contact with the power cord. Heat resistant gloves must be used when removing hot boards from wave-soldering machines. The processes discussed should be performed only in adequately ventilated facilities and with applicable equipment having proper fume exhaust systems. Be sure the fume exhaust system is turned on when using a wave-soldering machine. See Appendix XX regarding flux fumes, health issues related to flux fumes, and fume extraction.

A chemical hazard exists in the materials presented in this chapter. Approved face shields (goggles) in compliance with ANSI Z87.1-1989, chemical-resistant clothing and footwear, and neoprene or nitrile gloves must be worn while performing some of these operations. Also, the section on safety outlined at the beginning of the book should be reviewed for more specific information.

## 17.1 SOFT SOLDER AND ITS CHARACTERISTICS

Soft solder, used extensively in electronic equipment construction, is an alloy consisting principally of *tin* and *lead.* Soft solder is differentiated from hard solder by its tin content and lower melting point. The amount of tin contained in soft solder ranges from 50 to 70%. The tin–lead ratio determines the strength, hardness, and melting point of the solder.

Solder liquefies at temperatures between 361 and 621°F (183 to 327°C), the exact temperature depending on the tin–lead ratio. A metal such as copper, which has a melting point of 1981°F (1083°C), can be successfully alloyed with solder at temperatures well below this value because of the *solvent action* of solder when it is liquefied. At the melting point of solder, a thin film of metal is dissolved from the copper surface, forming an alloy and establishing an electrically continuous joint. The formation of this alloy between the metal and the solder has its unique physical properties, such as torsional, shear, and tensile strengths, which are different from those of either the solder or the metal. These properties will vary widely and will depend on the depth of alloying into the metal surface.

Pure tin melts at 450°F (232°C) (point *B,* (Fig. 17.1) and the melting point of pure lead is 621°F (327°C) (point *A*). When these two elements are combined, the melting point of the solder formed can be below that of either pure metal. A composition of 63% tin and 37% lead melts at 361°F (183°C), which represents the lowest melting point and most rapid transition from the solid to the liquid state of any other tin–lead ratio. This point, *C,* is also shown in Fig. 17.1. The 63/37 solder is termed the *eutectic* composition. All solder, with the exception of the eutectic composition, which melts sharply from solid to liquid, passes through a stage of softening between its solid and liquid stages known as the *plastic range.* A fusion diagram showing the temperature ranges for the various states of solder for all tin–lead combinations is shown in Fig. 17.1.

To determine the proper tin–lead ratio for a specific application, the function of the connection must be examined. The necessary properties that will dictate the composition of solder are mechanical resistance to fractures due to stress, ability to form a continuous metallic connection at low temperatures, and cost. These factors are discussed here.

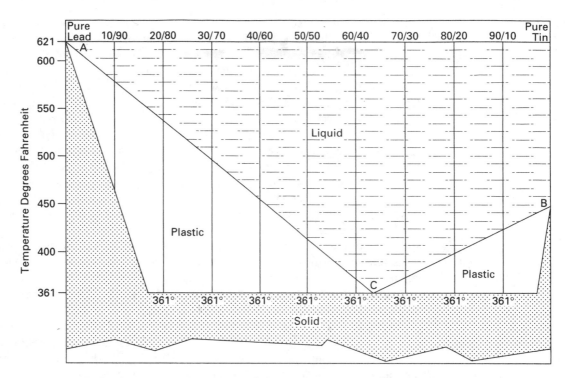

**Figure 17.1** Tin–lead fusion diagram. *Courtesy of Kester Solder, Division of Litton Systems, Inc.*

Since tin is more expensive than lead, solder with a higher tin content is more costly. It has been shown empirically that the highest joint resistance to stress exists with a 63/37 tin–lead ratio (eutectic solder). This concentration, therefore, affords the best alloying qualities in addition to the lowest melting point.

For most hand-wiring and printed circuit applications, solder with a tin–lead ratio of 60/40 is commonly used because of its excellent wetting action. *Wetting* is the term used to describe the ability of the solder to readily spread and alloy uniformly over the entire metal surfaces to be joined. Eutectic solder is sometimes selected to take advantage of the lowest and sharpest melting-point characteristics wherein maximum precautions are necessary to avoid component heat damage and upsetting the joint as it cools. If these two considerations are not critical, 60/40 solder is an excellent compromise.

Solder for electronic applications is available in bars, sheets, wire spools, and special forms, such as pellets, rings, and washers. For hand-wiring purposes, solder wire ranges from 0.030 to 0.090 inch in diameter. The larger sizes are used for general-purpose work and the smaller sizes for delicate soldering applications such as pc boards.

Solder wire is also available with a core containing flux (see Sec. 17.2) in specific amounts to promote sound solder connections. For this reason, flux-core solder wire is used almost exclusively for electronic applications.

## 17.2 FLUX

The interaction of metal parts with the atmosphere forms a thin layer of oxide on their surfaces. This oxidation increases as the metal is heated and will severely interfere with the solvent action of solder, thus preventing alloying and the formation of an electrically continuous joint. Consequently, the oxide must be removed. *Fluxes* are used for this purpose. They are chemical agents that aid in soldering by removing thin films of oxide present on the metal surfaces to be soldered. When applied to the joint, the flux attacks the oxides and suspends them in solution, where they float to the surface during the soldering process. When the joint is heated, the presence of flux also prevents further oxidation in addition to lowering the surface tension of the metals, thereby increasing the wetting action. It is important to remember that flux is not a cleaning agent for removing grease or other contaminants. Its sole function is to remove the oxide film. For optimum soldering results, the parts must be thoroughly cleaned before the flux is applied.

The flux in no way becomes a part of the soldered connection but aids in the process. Upon completion of the soldered joint, a flux residue appears on its surface, which contains the captured oxides. This residue should be removed with an appropriate solvent.

The ability to rapidly remove oxide films from metal surfaces constitutes *activity* of the flux. It would appear that a highly active flux is ideally suited for electronic construction since it would afford rapid alloying and thereby reduce the possibility of heat damage to the components. This, however, is not the case. Highly active fluxes may be corrosive at room temperature and, if allowed to remain as a residue, will deteriorate the conductor surfaces or reduce the resistance of the insulation between soldered connections. Corrosive damage to components may also occur, since some of the active residues will gradually spread as they absorb moisture from the atmosphere. Even with suitable solvents, there is no assurance of complete flux residue removal, especially around closely spaced terminals, connectors, or conductors on pc boards where complete solvent flushing is not possible.

There are three major classifications of flux: (1) *chloride* (inorganic salts), (2) *organic* (acids and bases), and (3) *rosin* fluxes. The chloride types are the most active (highly corrosive) fluxes. They absorb moisture from the atmosphere and strongly react with acid even at room temperature. Organic fluxes are slightly less active than the chlorides and are used mainly for confined areas in which fast soldering time is important and corrosion problems are not critical. Many of the organic fluxes are converted to an inert residue after thermal decomposition. They do not absorb moisture and are difficult to remove. For the reasons given above, chloride- and organic-type fluxes are not recommended for use in electronic construction. The rosin-type fluxes are used almost exclusively because of their noncorrosive characteristics at room temperature. They are corrosive at temperatures near the melting point of solder. Consequently, they attack the oxide film during the heating cycle but are inactive when room temperature recurs. Rosin fluxes are available with activating agents that greatly improve their activity. These activated rosin fluxes are much more corrosive than pure rosin when heated and present the appearance of an instantaneous melting, wetting, and flowing action of the solder. They are essentially as noncorrosive at room temperature as the pure rosin types and are often preferable if a higher degree of flux activity is dictated, such as in wave soldering.

Liquid flux may be applied by *wiping, dipping, spraying,* or *sponging.* Wiping or dipping methods are not extensively used. Uniform flux coatings are difficult to realize when wiping with a brush, and thorough wetting of all surfaces may not result when dipping because of air pockets or cavities created during this process. Another disadvantage of these two methods is that application of an excessive amount of flux requires extensive removal of residue after the soldering has been completed.

Spraying flux involves applying a fine mist to the joints. The use of this method ensures that leads and terminals are more uniformly and completely coated with flux. When spray methods of flux application are to be used, the manufacturer's literature should be consulted regarding the recommended spray gun, nozzle, pressure, and solvent to use.

The application of flux with a sponge is perhaps the most effective and least messy method. To apply the flux, the board is firmly pressed against a sponge that has been saturated with liquid flux.

When hand soldering, the proper amount of flux can best be applied with the use of *flux-core wire solder.* This form of solder contains a core of solid rosin flux in a single or multiple core. There is no significant advantage in using multiple-core solder since it is essentially the volume ratio (amount of flux to solder) that determines optimum soldering conditions. Core sizes are available that provide a ratio of rosin flux per unit volume of solder of 0.6 to 4.4%. These ratios can be obtained for any size of wire solder. Indications are that 60/40 rosin flux core solder with a diameter of 0.040 inch and 3.6% flux is ideally suited for hand soldering pc boards and other electronic precision work.

## 17.3 THE SOLDERING IRON

The soldering iron, shown in Fig. 17.2, consists of four basic parts: (1) *tip,* (2) *heating element,* (3) *handle,* and (4) *power cord.* Some soldering irons are equipped with either a cork finger grip or a heat deflector. These are designed to improve thermal insulation in the handle, which tends to become hot after prolonged use. In addition, the cork finger grip allows more comfortable use of the iron.

When power is applied to the soldering iron, the tip is heated by direct thermal contact with the heating element into which it is set. The solder is liquefied when it comes in contact with a tip that has reached its operating temperature.

A soldering iron is selected for a specific application by considering the following factors: (1) size and style of the tip, (2) tip material, (3) required tip temperature, and (4) tip-temperature recovery time. Where precise control of tip temperature is required, *temperature-controlled* soldering irons having tips with built-in temperature sensors are available. This type of iron is shown in Fig. 17.2. Temperature-controlled units allow tip temperature settings from 500 to 800°F (260 to 427°C). They maintain the set idling temperature to within seconds after each soldering operation. This type of temperature control ensures consistent soldering temperature conditions.

Selecting a tip style is somewhat a matter of personal preference. However, the shape of the particular tip used must provide the largest contact area to the specific connection for maximum heat transfer while minimizing the possibility of heat damage to surrounding leads or components. Several widely used tip styles are shown in Fig. 17.3. Each of the available tip configurations is designed for a specific soldering application. *Chisel*-style tips are commonly used for hand wiring and general repair work. The large flats of these tips allow large areas to be heated rapidly. The *turned chisel* tips lend themselves very well to soldering in confined areas, such as hand soldering components to double-sided pc boards. *Conical* tips are preferable for soldering high-density wiring, eyelets, and small heat-sensitive parts.

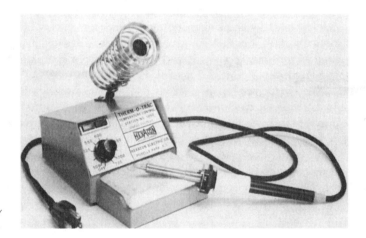

**Figure 17.2**
Temperature-controlled soldering iron. *Courtesy of Hexacon Electric Co.*

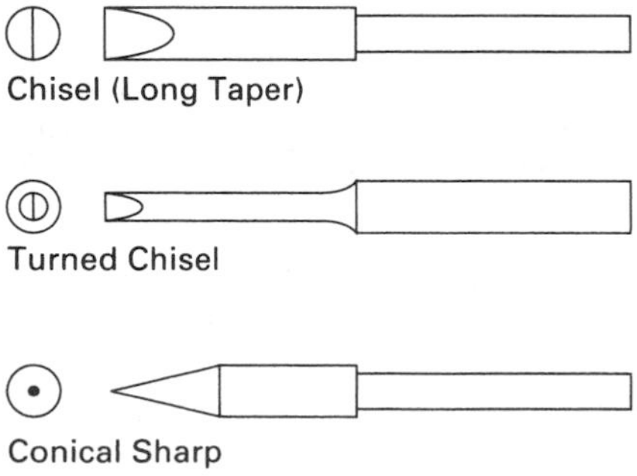

Chisel (Long Taper)

Turned Chisel

**Figure 17.3**  Widely used solder tip configurations.

Conical Sharp

*PCB Soldering*

Soldering iron tips are manufactured chiefly from copper and are available both in plated and unplated finishes. Typical platings for tips are *iron, gold,* and *silver.* These platings protect the copper tip from corrosion and pitting, which become pronounced over extensive periods of use. The plated tips should never be filed or cleaned with harsh abrasives that would remove the plating and expose the copper. The tips should instead be periodically cleaned by first dipping them cold into liquid rosin flux and then bringing them up to operating temperature to loosen any surface oxidation or contaminants (burned solder and flux residues) that may be present. Solder is then applied to the hot tip. Contaminated areas are easily detected by observing those areas that resist wetting by the solder. Sufficient solder has been applied when it begins to puddle. The tip is then wiped on a moist fine-pore cellulose sponge to remove the contaminants.

Tips should periodically be removed from the soldering iron because oxide will build up on the shank or threaded portion that fits into the heating element. If this oxide formation is not occasionally removed, the tip will seize in the element, making it difficult to remove and possibly damaging the element.

The required tip temperature is based on its application. For general-purpose soldering, such as to terminal strips and solder lugs, a tip temperature of 600 to 900°F (316 to 482°C) is sufficient. Printed circuit soldering requires fast heat transfer to the foil, yet excessive heat that could upset the foil bond must be avoided. For soldering to conductor patterns having widths of 1/32-inch or larger and for terminal pads having 3/32-inch diameters or larger, irons with tip temperatures of 800 to 850°F (427 to 454°C) are recommended. When soldering to more delicate pc boards or extremely fine wire (28 gauge or smaller), tip temperatures of 550 to 700°F (288 to 371°C) are suggested. However, when fine-component wire is to be soldered, especially in the case of heat-sensitive devices, a tip temperature of at least 900°F (482°C) is recommended (if lead heat sinks are not used) so that rapid heat transfer occurs with a minimum amount of tip contact time.

Irons are usually rated in *watts,* and are typically available from 20 to 60 watts for electronic applications. It is difficult to compare exact tip temperature with wattage rating since the length and the diameter of the tip largely influence tip temperature. For example, a 25-watt iron with a 3/16-inch tip diameter and 3/4-inch tip length will produce a temperature of approximately 680°F (360°C). Another 25-watt iron with a tip having the same 3/16-inch diameter but a 1 1/2-inch length will reach a temperature of approximately 605°F (318°C). As a rule, tip temperature decreases linearly, approximately 10°F for every 0.10-inch increase in tip length. Tip temperature is inversely related to tip diameter. However, this relationship is not linear, as is the case of length versus temperature. Therefore, empirical data, such as those provided in Table 17.1, must be consulted. This table shows comparative values of tip temperatures to specific wattage ratings for three common tip diameters. The values given in this table are for iron-clad chisel-style tips having a length of 3/4 inch.

A more general relationship of wattage to tip temperature is given in Table 17.2. This table provides a comparison of the common wattage ratings to a range of tip temperatures

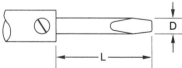

**TABLE 17.1**

Comparison of Wattage Rating and Tip Temperatures for Various Sizes of Solder Tips

| Diameter D (in.) | Length, L (in) | 20 W | 25 W | 30 W | 35 W | 40 W | 50 W | 60 W |
|---|---|---|---|---|---|---|---|---|
| $\frac{3}{64}$ | $\frac{3}{4}$ | 675°F | 725°F | 775°F | 825°F | 875°F | 925°F | 975°F |
| $\frac{1}{8}$ | $\frac{3}{4}$ | 640°F | 690°F | 750°F | 800°F | 860°F | 910°F | 960°F |
| $\frac{3}{16}$ | $\frac{3}{4}$ | 630°F | 680°F | 730°F | 790°F | 850°F | 900°F | 950°F |

that can be expected from any style or size tip. In general, approximately a 50°F change in tip temperature can be expected between the successive wattage ratings listed.

There is a useful technique for determining the approximate tip temperature of iron-clad tips by visual inspection. When the iron is heated to its *idling* (maximum operating) temperature, the flats of the tip are alternately wiped several times across the surface of a moist sponge. The color of the tip is observed 2 to 3 seconds after wiping. There is a relationship between the color obtained and the approximate tip temperature. This relationship is given in Table 17.3. Tip temperature can be periodically monitored using this technique.

*Recovery time* is the rate at which a tip will return to its idling temperature after transferring heat (tip cooling) to the work during soldering. In mass production applications wherein a rapid succession of soldered connections are to be made, fast recovery time is absolutely essential. It is not of major concern in prototype work, however, since the recovery rate is normally much greater than the rate at which soldered connections are made.

The following example illustrates how to evaluate the many available irons and tips to select the most appropriate one for a specific application. An iron will be selected for soldering a double-sided pc board having typical conductor widths of 1/32 inch with terminal pads of approximately 1/16 inch in diameter. Ideal tip temperature for pc board soldering is 800 to 850°F (427 to 454°C). As can be seen in Tables 17.1 and 17.2, a wattage rating of 35 watts will provide this necessary temperature with a 3/64- by 3/4-inch tip. The style of tip should permit ready access between components. A *conical chisel* design is selected since this configuration provides the desired contact area for component lead soldering to terminal pads. Finally, the iron-clad finish is chosen for durability. The tip selected for this example is quite versatile because of its thin chisel shape, which works well with both double- and single-sided boards.

In general, the soldering iron that produces the best performance over an extended period of time is one that has high heat conductivity and low heat loss when in contact with connections to, bring them up to soldering temperature. It should also have a cool, lightweight handle; a no-burn power cord with strain relief, and a simple method for removing the tip and disassembling the iron to replace the heating element.

**TABLE 17.2**

Soldering Iron Wattage Rating versus Range of Tip Temperatures

| Wattage Rating (W) | Range of Tip Temperatures (°F) |
|---|---|
| 20* | 550–750 |
| 25* | 600–800 |
| 30* | 650–850 |
| 35* | 700–900 |
| 40 | 750–950 |
| 50 | 800–1000 |
| 60 | 850–1050 |

*Commonly selected wattage ratings for electronic applications.

**TABLE 17.3**

Reference for Iron-Clad Tip Temperature by Visual Inspection

| Tip Temperature (°F) | Tip Color (2 to 3 seconds after sponge wipe) |
|---|---|
| 700 | Silver |
| 800 | Gold |
| 850 | Gold with streaks of blue at tip |
| 900 | Blue to purple |
| 1000 | Ash black |

## 17.4 THERMAL PROFILE OF A SOLDERING IRON

Modern personal computer technology has simplified the way an accurate thermal profile of a standard soldering iron is obtained. Two methods, typical of today's technology, are illustrated next.

### Using a Single-Channel Signal-Conditioning Module from Analog Devices, Inc.

Analog Devices, Inc., has introduced a new family of signal-conditioning modules that can connect a type J thermocouple temperature sensor directly into the serial port of a personal computer. This hardware, used in conjunction with a software package such as LABTECH ACQUIRE,* forms a complete data-acquisition system that can be used to accurately measure the thermal characteristics of a soldering iron.

The basic parts of this temperature measuring system are shown in Fig. 17.4a. The soldering iron tip is first prepared by drilling a hole equal in diameter to the width of the thermocouple. Heat is then applied and the drilled hole is tinned to provide good thermal contact with the thermocouple. The white and red leads of the type J thermocouple are connected directly to pins 3 and 2, respectively, of a 6B11 signal conditioning module. This module converts the analog input (thermocouple voltage) into a serial digital output that is accepted by the RS-232-C port of an IBM personal computer. Power for all signal conditioning is supplied from a +5-volt power supply module. The serial computer interface is made with a separate cable between the signal conditioning module and the computer.

The LABTECH ACQUIRE software package is menu driven and runs on computers compatible with the IBM PC, XT, and AT and provides the interface between the user and the instrumentation system. The software package can be installed onto a hard disc system by first creating a directory in DOS (Disc Operating System) in which ACQUIRE will run. This is done by typing from the C prompt

```
C:\DOS>MD\ACQUIRE<CR>
```

where C:> or C:\DOS> stands for the C prompt, MD for *Make Directory,* and <CR> for carriage return. Next, change the directory to ACQUIRE by typing

```
C:\DOS>CD ACQUIRE <CR>
```

where CD indicates *Change Directory.* A subdirectory within ACQUIRE is then made to save the setups. This is done by typing

```
C:\ACQUIRE>MD SETUP <CR>
```

To install the software, the ACQUIRE floppy disk is placed into drive A and the following is typed:

```
C:\ACQUIRE>COPY A*.*<CR>
```

When all files on the floppy disk are copied from drive A to drive C, the disk is removed and stored. To complete the software installation, ACQUIRE must know about the hardware configuration. From the C:\ACQUIRE prompt, type

```
C:\ACQUIRE>AQ<CR>
```

---

*LABTECH ACQUIRE is a registered trademark of Laboratory Technologies, Inc.

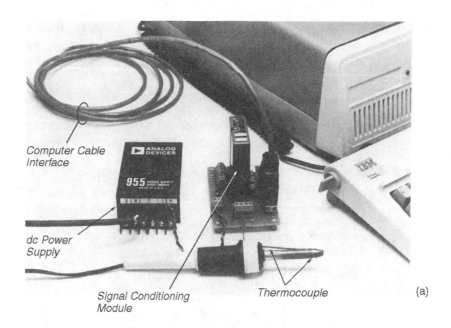

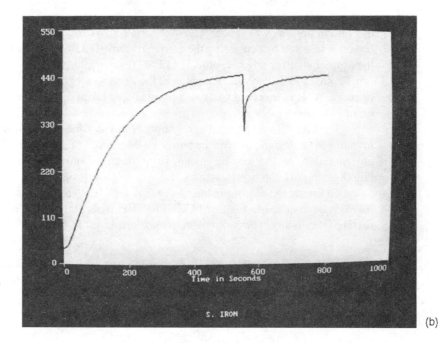

**Figure 17.4** (a) A single-channel, standalone, data acquisition system used to accept soldering iron tip temperature and input this information into a personal computer; (b) the thermal characteristics of a standard soldering iron.

which initiates ACQUIRE and the main menu appears as

```
CONFIGURE GO ANALYZE SAVE RECALL DELETE OPTIONS QUIT

set up data acquisition
```

The LABTECH ACQUIRE software is now loaded and ready for the hardware configuration. The thermal profile of a soldering iron will be used as an illustration. For this task, the only main menu options that will be used are CONFIGURE, GO, and QUIT. The ACQUIRE package has more versatility with additional menu options, such as ANALYZE, SAVE, RECALL, and DELETE. These are not discussed here.

To select from the main menu, press the left or right arrow key (⇄) until OPTIONS is highlighted in reverse video. By depressing the carriage return <CR> key, the OPTIONS MENU will be displayed as follows:

```
Current Value CGA (4 Colors)

                                               Options Menu


Display Type                              EGA (8 Colors)
Start on Key Press                              No
Hold Display on Screen                          Yes
Hardware Interface Device                   Demo Board
Analyze Drive, Path Name, File Name         C:\Lotus\123
Header Records in File?                         Yes
```

*Display Type* will be highlighted in reverse video. ACQUIRE inspects the system and displays the type of graphics adapter card present within the system.

The up or down (↑↓) arrow keys are used to move the options menu to highlight *Start on Key Press*. The *No* is changed to *Yes* by first depressing the F1 key. A pop-up menu with the *Yes No* options will appear. The up or down arrow key is used to highlight *Yes* and this option is selected by depressing the carriage return <CD> key. A *Yes* in this field will make the system wait for any key, other than ESC (escape), to be depressed before data acquisition begins. In this way, the beginning of the system receiving data can be controlled by the operator. A *No* in this field will result in data being taken the moment that the GO.EXE program is loaded.

The arrow keys are used to highlight *Hardware Interface Device*. This entry is used to change the designation for the hardware connected to the system. *Demo Board* and *6B* are included in the pop-up menu since there is only one 6B11 module attached. Highlighting 6B and depressing <CR> will select this option.

To return to the main menu, the escape (ESC) key is depressed. The left or right arrow (⇄) key is used to highlight CONFIGURE. This option is selected by depressing the carriage return key. The *Setup* menu will be displayed as follows:

```
Current Setting 1

                                               Setup Menu


Number of Analog Channels [1.4]                 1
Number of Digital Channels                      0
Time Stamp Date?                               Yes
Sampling Rate (Hz)                           10.000
Run Duration                                 30.000
Starting Method                            Immediate
Trigger Channel                                 1
Trigger Threshold                             0.000
Trigger Polarity                               High

File Name                                   TUTOR.PRN
Number of Windows                               1
Width of Windows in Seconds                  30.000
Window Color                                  Black
```

| Channel Number | 1 |
|---|---|

```
Channel Name                          Ch.1
Display in Window Number [1..1]        1.
Scale Factor                        1.000
Offset                              0.000
Minimum Display Value               0.000
Maximum Display Value              10.000
Trace Color                           RED.
```

The up and down arrow keys are used to move this setup menu down until *Sampling Rate (Hz)* is highlighted. The number on the menu is changed to 1 (one sample per second) by typing 1 <CR>. Next, *Run Duration* is highlighted and the setup number is changed to 800 <CR>. The system is now ready to take data once every second for 800 seconds, or about 13 1/3 minutes.

Dropping to *File Name,* IRON <CR> is entered. For *Window Width in Seconds,* the horizontal axis of the graph is matched to the run duration by typing 800 <CR>. At *Window Color,* the F1 key is depressed and a pop-up window with eight colors will appear on the screen. The up and down arrow keys are used to highlight white, which is selected by depressing <CR>.

Finally, *Maximum Display Value* is highlighted and the setup value is changed by typing 550 <CR>. The vertical axis is now limited to 550°C.

The configuration of the display is complete. To return to the main menu, press ESC. To initiate GO (i.e., load the GO.EXC program), <CR> is depressed when GO is highlighted in reverse video. Any key can then be depressed at the same time that the soldering iron is plugged in. As shown in Fig. 17.4a, data on tip temperature (°C) as a function of time (seconds) begin to be collected.

The thermal characteristic curve of the soldering iron is shown in Fig. 17.4b. This curve shows that the soldering iron reached a maximum tip temperature of 440°F in ~500 seconds. At ~500 seconds, the iron was wiped on a dampened sponge to simulate an actual soldering operation. Tip temperature dropped quickly to about 300°F, and recovery time to restore it to 440°F was about 200 seconds.

To exit ACQUIRE, press the ESC key to recall the main menu. QUIT is highlighted and the carriage return <CR> is keyed. A system similar to the one just described greatly simplifies the task of data collection and display.

### Using a General Purpose Data-Acquisition System from Hewlett-Packard, Inc.

The thermal profile of a soldering iron can also be created quickly by using any number of general purpose data-acquisition or computer-controlled measurement systems. One such system, manufactured by Hewlett-Packard, is illustrated in this section. This type of data-acquisition system consists of standard bench equipment, such as digital multimeters and oscilloscopes, function generators and dc power supplies, and a software package for programming and equipment control. The bench equipment is connected to, and communicates with, a personal computer along an interface bus called HPIB. The equipment is controlled with a software package, termed by the manufacturer HP VEE (Hewlett-Packard's Visual Engineering Environment). HP VEE is an object-oriented, graphical programming language, designed for creating automated measurement and test systems and solving engineering problems such as the one illustrated here.

A statement of the problem to be solved using this type of computer-based measurement system might be presented as follows: A typical 33 watt soldering iron, fitted with a chisel point tip, requires approximately 8 to 10 minutes to reach its maximum or operating temperature of approximately 400 to 450°C. Therefore, we need to use a thermocouple to collect data on tip temperature for 10 minutes (600 seconds), measuring the thermocouple's voltage in 5-second intervals, to correct and convert the measured thermocouple voltages to temperature, and to graphically display the soldering iron's temperature profile in degrees Celsius as a function of time in seconds.

As discussed earlier, a soldering iron is prepared by first drilling a small hole into the tip, installing a type J thermocouple, and then tinning the hole with solder to promote good thermal contact between the iron's tip and the measuring thermocouple. Next the *white* and *red* leads of the thermocouple are connected to the *positive* and *negative* leads, respectively, of a bench-type digital multimeter like the model hp34401a digital multimeter from Hewlett-Packard used for this test set-up.

The acquisition (measurement), manipulation, and display of thermocouple data versus time can all be controlled with software from the HP VEE main window. An object-oriented graphical program will next be written to measure the thermocouple's voltage every 5 seconds. This program will compensate for a reference-junction voltage error at room temperature, which is fixed at 25°C. A reference-junction voltage error will be generated by dissimilar metal connections at the digital multimeter (dmm) terminals. Finally, this program will mathematically convert the resulting corrected voltage data into degrees Celsius and display a graph of Temperature vs. Time on the computer monitor.

An object-oriented program to solve engineering problems like this one begins at the main HP VEE window. Figure 17.5 illustrates its four major areas: the *Title, Menu,* and *Tool Bars,* and the *Work Area.* From the *Menu Bar,* software items called *Objects* are selected from one of the seven menu options, positioned within the *Work Area,* and wired (programmed) to solve the engineering problem. Once a program is created it can be executed from the *Tool Bar* by clicking the *Run* button.

Our data acquisition problem will require the following four major software element groups:

1. A unit to take thermocouple measurements from 0 to 600 seconds in 5-second (delayed) intervals.
2. A unit (the hp34401a digital multimeter instrument display panel) to control the dmm and to measure the thermocouple's voltage.
3. A mathematical unit to add a reference junction correction voltage to the measured thermocouple voltage and convert this voltage data into temperature.
4. A unit to display the soldering iron's thermal profile of tip temperature vs. time.

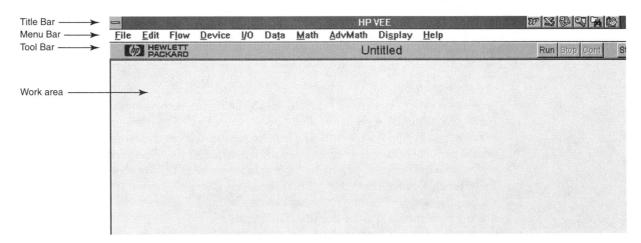

**Figure 17.5** The four major areas of the HP VEE main window are the *Title, Menu,* and *Tool Bars,* and the *Work Area.* Software *objects* are available from seven option menus: *Flow, Device, I/O, Data, Math, AdvMath,* and *Display.* (Reproduced with permission of Hewlett-Packard Company)

Fig. 17.6 shows the objects and programming (wiring) needed to solve this engineering problem; the following illustrates the steps to follow to create and execute this graphical program.

The first step in any graphical program is to call up all the needed software objects and locate them within the work area. For this experiment, we will initially need the *For Range* object. To select this object, move the cursor to *Flow* on the menu bar and click the left mouse button, then click *Repeat* in the *Flow* menu, and finally, click the *For Range* object. An outline of the object appears in the work area. Position this object in the upper-lefthand corner of the work area as shown in Fig. 17.6 and click the left mouse button one last time to fix or locate the object. Refer to the shorthand notation below to acquire this element.

$$Flow \longrightarrow Repeat \longrightarrow For\ Range$$

To change the range from the default setting, click the computer's left mouse button on the *Thru* window and enter 600 from the keyboard. Then either press Enter <CR> or click the left mouse button one more time to enter the new value of 600. In a similar fashion set the *Step* interval to 5 seconds. Next call up a *Delay* object, also from the *Flow* menu, and position it in the work area as shown in Fig. 17.6, using the shorthand notation:

$$Flow \longrightarrow Delay$$

To change the delay value from the default setting of 0 (seconds), click the number 0 and enter 5 from the keyboard. Then either press Enter (<CR>) or click the left mouse button one more time to enter the new delay value of 5. Add an hp34401a dmm software control panel to the work area, using:

$$I/O \longrightarrow Instrument \longrightarrow dmm\ (hp34401a\ @\ 722) \longrightarrow Get\ Instr$$

Figure 17.7 displays this instrument's display (control) panel in its "Open" or "Maximum" view. We want this dmm displayed in its maximum view for now because we must add a *Data Output* terminal to this object. With this added terminal we can transfer the measured data from the dmm to the input of the math function to follow.

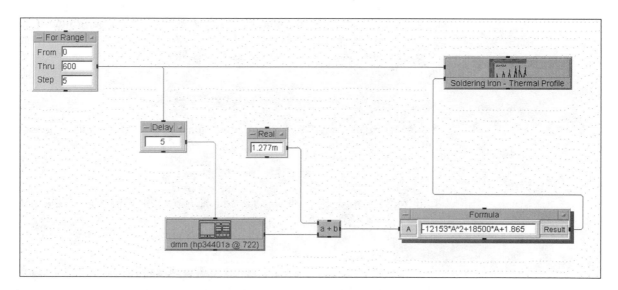

**Figure 17.6** The software objects and wiring needed to display soldering iron tip temperature vs. time. Note that the *For Range, Delay, Real,* and *Formula* objects are shown in "open" view and the *dmm* (instrument panel), *Add,* and *X vs Y Plot* objects are shown as *icons.* (Reproduced with permission of Hewlett-Packard Company)

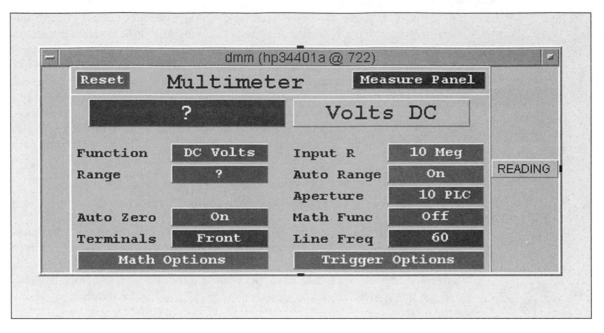

**Figure 17.7** An "Open" or "Maximum" view of the hp34401a instrument control panel with *Reading* terminal added to the right (*Data Output*) side of the object. The dmm is configured as a voltmeter to measure dc voltages. (Reproduced with permission of Hewlett-Packard Company)

To add an output terminal to the dmm, click the [-] button in the upper-left corner of the instrument's display panel only once. (See Fig. 17.7.) From the drop-down menu, select *Add Terminal* and then select *Data Output...Ctrl +A*. A large selection menu is presented. Scroll down until you can highlight *Reading (Real)*. When you click *OK,* an output terminal named *Reading* will appear on the right-hand side of the dmm instrument panel, using:

$$\text{Add Terminal} \longrightarrow \text{Data Output...Ctrl+A} \longrightarrow \text{Reading (Real)}$$

Finally, "Minimize" the hp34401a dmm into an *icon* by clicking the upper-righthand corner "Minimize" button. Reposition the minimized dmm icon labeled dmm (hp3440a @722) in the work area as shown in Fig. 17.6. To convert the measured thermocouple voltage to temperature, call up the following three HP VEE objects—*Real, a+b,* and *Formula*—and set each as indicated.

$$\text{Data} \longrightarrow \text{Constant} \longrightarrow \text{Real}$$

Set the constant to 0.001277, the reference junction correction voltage needed for a Type J thermocouple at 25°C. See Appendix XV.

$$\text{Math} \longrightarrow + \ -*/ \longrightarrow a + b$$

Call from the *Math* menu an object to add the correction voltage to the thermocouple reading.

$$\text{Math} \longrightarrow \text{Formula}$$

Then from the *Math* menu, click *Formula* and from the keyboard set a second-order polynomial to $-12153*A^2 + 18500*A + 1.8657$. This equation will linearize the output of the thermocouple and display the results in degrees Celsius. The development of this equation is given in Appendix XIV.

Finally to display the thermal profile of the soldering iron, call from the *Display* menu an *X vs. Y Plot* object, using:

```
Display  →  X vs Y Plot
```

Double-click this icon to create an "Open" view of this display. (See Fig. 17.8.) To scale the vertical axis for temperature, click the upper *Data* window and enter 500 and then set the lower Data window to 0. Repeat for the horizontal axis, setting the time limits between 0 and 600. Finally click the appropriate window and label the vertical and horizontal axes with suitable titles, such as those shown in Fig. 17.8.

Interconnecting (programming) the objects to solve this problem requires the following information. HP VEE object pins and terminals are universally configured as follows:

*Data Input*—on the left                     *Data Output*—on the right

*Sequence In*—at the top                    *Sequence Out*—at the bottom

As the names imply, *data* lines carry data between objects. When connected, *sequence* pins will dictate an execution order typically flowing from the top of the work area to the bottom. We will wire both data and sequence pins in this exercise.

Begin to program this experiment by connecting the data output pin of the *For Range* object to the *Sequence In* pin of the *Delay* object. This is accomplished by clicking the left mouse button near (*but not touching*) the *For Range, Data Output* pin, move the mouse pointer to the *Sequence In* pin of the *Delay* object, until this pin is highlighted, then click again. (Refer to Fig. 17.6.) In a similar fashion, connect the *Data Output* pin of the *Delay*

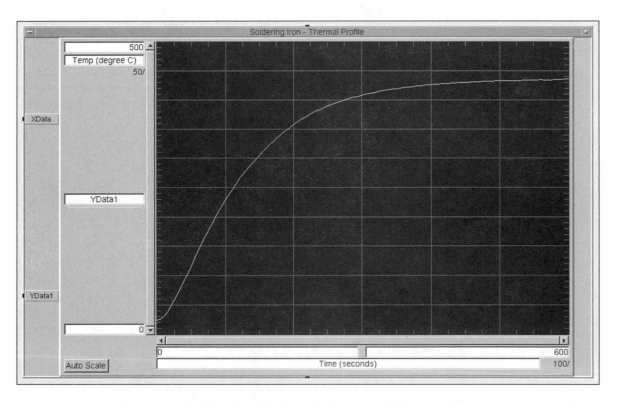

**Figure 17.8** The thermal profile of a standard 33-watt soldering iron is shown as an exponential rising plot of temperature as a function of time, eventually reaching an operating temperature of approximately 420°C. (Reproduced with permission of Hewlett-Packard Company)

*PCB Soldering*                                                                                      **339**

object to the *Sequence In* pin of the hp34401a digital multimeter instrument panel. When this program runs, the dmm will be delayed for 5 seconds for each count of the *For Range* object. Repeat this procedure until the objects are wired as shown in Fig. 17.6. When executed, this program performs as follows: Every time the *For Range* object increments, it "pings" or signals the *Delay* object. After a 5-second wait the *Delay* object "pings" the digital multimeter to make a thermocouple voltage measurement and output its reading to the math function that follows. Compensation for a room temperature reference junction of typically 25°C is accomplished by adding a correction voltage of 1.277mV (the output of a type J thermocouple at 25°C) to the multimeter reading. Finally the compensated thermocouple voltage is converted into units of degrees Celsius by performing this nonlinear second-order polynomial equation created in the *Formula* object:

$$\text{Temperature in °C} = -12153 * A^2 + 18500A + 1.8657$$

Finally, time data is applied to the *Xdata* input of the *X vs. Y Display* object, and temperature data to the *Ydata1* input. This program takes 10 minutes to run, makes 120 voltage measurements—each delayed by 5 seconds—and displays the results.

To execute this program, click the *Run* button in the Tool Bar at the same time power is applied to the soldering iron. Double clicking the *X vs. Y Plot* icon will present it in "open" view and allow the operator to see the thermal profile (heating cycle) of the soldering iron as it is generated. Figure 17.8 illustrates a complete thermal profile.

## 17.5 HAND SOLDERING PRINTED CIRCUIT BOARDS

Of prime importance in soldering is that the surfaces to be joined are clean and devoid of oxides. This is especially true for pc board soldering, since a poorly cleaned terminal area would prolong heating time excessively when soldering, which in turn could upset the foil bond. Not only should the terminals be cleaned, but it is imperative that liquid flux be applied to all joints prior to soldering even if terminal areas have been tinned. The flux may be applied to the pc board by any of the methods discussed in Sec. 17.2. However, wiping is preferable in prototype work when a solder mask is to be employed. This method allows the flux to be selectively applied only to those terminal areas that are to be soldered. The pc board is first secured in a *circuit board holder* and tilted to a horizontal position. Liquid flux is then applied with a fine artist's brush or toothpick onto the areas to be soldered. This arrangement is shown in Fig. 17.9. When all the surfaces to be soldered have been coated with flux, soldering may begin.

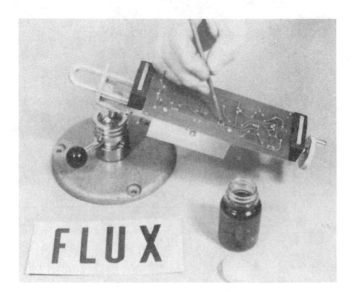

**Figure 17.9**
Application of liquid flux to selected areas.

To solder leads to pc boards, it is important that heat and solder be applied quickly and accurately. A small amount of solder is initially applied to the flat of the soldering iron tip (solder bridge) to promote rapid heat transfer to the joint (Fig. 17.10a). The entire flat of the tip surface is then placed in contact with the terminal area and lead simultaneously. The more surface area of the tip contacting the terminal pad and lead, the more efficient the transfer of heat (Fig. 17.10b). The joint is allowed to heat to the melting point of solder, usually taking from 1 to 2 seconds for most connections. Solder is then applied to the terminal pad and lead simultaneously and at a point on the opposite side of the lead from where the tip is contacting. The correct orientation of tip and solder is shown in Fig. 17.10c. As soon as the solder begins to flow, no more should be applied. Since solder follows heat flow, it will wet the entire connection and flow over the lead, forming a smooth contour of solder around the lead and terminal pad. The soldering iron is then removed and the joint allowed to cool. The joint must never be moved until the solder has completely solidified because any movement of the lead as it is cooling will prevent positive alloying from taking place. *It is important to apply only a minimum amount of solder sufficient to ensure proper alloying.* A properly formed solder joint is smooth and shiny in appearance. All surfaces must be completely wetted and the contour of the lead on the terminal pad clearly visible. A correctly soldered lead to a printed circuit terminal pad is shown in Fig. 17.10d. A precaution should be mentioned here. When soldering the individual terminal pads, it is important to keep track of those leads that must be heat sinked. Lead heat sinking is discussed in Sec. 16.4.

Excess solder and contaminated flux residue on a tip must be removed before another solder joint is formed. Cleaning is accomplished quickly and easily by wiping the tip on a moist sponge. Soldering should never be attempted with a tip covered with excess solder or contaminants if quality results are to be expected.

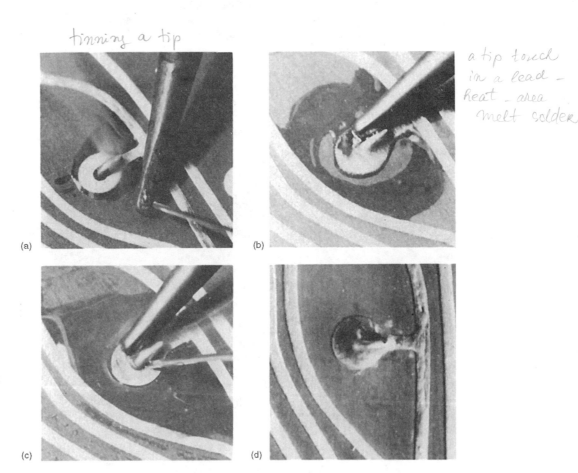

**Figure 17.10** Printed circuit board soldering: (a) forming a solder bridge; (b) heating solder pad and lead; (c) solder application; (d) correctly soldered terminal.

If soldering is required to comply with DOD-STD-2000-3 standards, reference should be made to Figs. 17.10, 17.11, and 17.12. The leads are formed perpendicular to the component side of the board and then passed through the board. The length of the clinched lead termination shown in Fig. 17.10a is cut to be equal to one pad diameter and the end of the lead is finally pressed firmly against the surface of the pad in perfect alignment with the conductor path as shown in Fig. 17.10a. As previously stated, by the proper positioning and application of the soldering iron tip and solder to the connection, as shown in Figure 17.10c, the solder should completely wet the surface of the clinched lead and the entire surface of the pad as shown in Fig. 17.10d. The wetting of the solder between the lead and the pad will form concave fillets where the contour of the lead formation is still readily visible. The resulting concave filleting of a properly soldered clinched lead termination, in accordance with DOD-STD-2000-3, is shown in Fig. 17.11 in three different views. These views show clinched leads soldered to single-sided pc boards and pc boards that are either double-sided without plated-through holes or with plated-through holes.

In some specifications, the lead termination at the pad is required to be an unclinched connection as shown in the cross-sectional views of Fig. 17.12. Once again, the lead is formed perpendicular to the surface of the board (component side). It is passed through the board and is allowed to extend only a small amount (1/32 inch) above this board surface prior to soldering. By using the proper positioning and application of the soldering iron tip

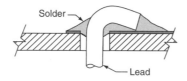

Single sided board

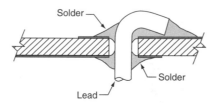

Double sided board (without plated-through hole)

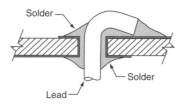

**Figure 17.11** Clinched lead termination, solder connection.

Double sided board (with plated-through hole)

and solder on opposite sides of the unclinched lead, similar to that shown in Fig. 17.10c for a clinched lead, the proper degree of wetting and filleting will result. The resulting filleting of a properly soldered unclinched lead termination, in accordance with DOD-STD-2000-3, is shown in Fig. 17.12 in three different views. These three views show unclinched leads soldered to double-sided pc boards with and without plated-through holes, interfacial (paths to pads on both sides of the board) and interlayer hole, and noninterfacial plated-through hole (no path to one of the pads). In the case of a single-sided board, the filleting would only be formed on the side of the board through which the lead protrudes.

Swaged terminals are soldered to pc boards in a similar manner to lead soldering, with one important exception. Since the terminal pad area for swaged terminals is larger than those used for lead connections, it is necessary to apply heat for a longer period of time. The flat area of the soldering tip is placed in contact with the swaged portion of the terminal and solder is applied to the terminal pad. When heated sufficiently, the solder will completely wet the terminal pad and flow uniformly around the swaged terminal in the area where it contacts the copper foil.

Since many pc boards do not employ solder masks, and if conductor pattern plating is not economically feasible, the technician may wish to tin the entire conductor area of the board to retard oxidation. This can be done during the soldering operation. Flux is first applied to the entire board surface. A small amount of solder is applied to the flat of the tip

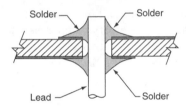

Double sided board (without plated-through hole)

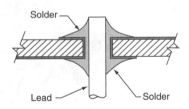

Plated-through hole interfacial and interlayer

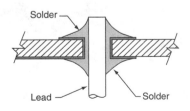

**Figure 17.12** Unclinched lead termination, solder connection.

Non interfacial plated-through hole

and the tip is slowly drawn along the conductor path. A thin layer of solder will be deposited on the surface of the copper. Several passes may be necessary to completely tin a conductor path. This will prevent excessive solder from forming in addition to minimizing foil bond damage. For this technique, a soldering iron with a tip temperature of between 500 and 600°F (260 and 316°C) should be used. The application of solder via the soldering tip is restricted to tinning conductor patterns. This technique is never employed for making soldered connections.

Properly soldered connections are uniform in appearance and their quality may be judged by visual inspection. Each solder joint may be inspected immediately upon soldering or after completion of the entire board. To aid in this inspection, a 5× to 10× magnifying glass may be used. Following is a list of deficiencies and their causes common to improperly soldered connections. These are shown in Fig. 17.13.

1. *Solder peaking.* Characterized by a sharp point of solder protruding from a connection. Peaking is caused by the rapid removal of heat before the entire joint has had an opportunity to completely reach its soldering temperature. The solder follows the hot tip as it is removed from the connection, resulting in this peaking condition. Reheating the joint will correct this deficiency.

2. *Incomplete wetting.* Occurs when portions of the soldered connections have not been alloyed with solder and are completely visible. This may be the result of both insufficient heat and solder. It may also be the result of contaminants on the soldering tip or terminal pad. Reheating the joint and applying additional solder will correct this fault if the condition is not caused by contaminants. In that case, desoldering and cleaning are necessary before a new joint is attempted. The terminal pad can be cleaned with a sharpened pencil-style typewriter eraser. (Desoldering techniques are discussed in Sec. 17.7.)

*[handwritten margin notes: Remove tip early or rapid remove of heat; insufficient heat and solder or contaminated tip and terminal pad]*

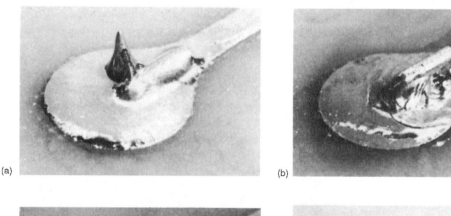
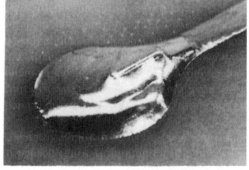

(a)      (b)      (c)      (d)

**Figure 17.13** Typical inferior solder connections: (a) solder peaking; (b) incomplete wetting; (c) excessive solder; (d) cold solder joint.

3. *Excessive solder.* Evident when the lead contour is not plainly visible. This is extremely undesirable since it prevents complete inspection of the alloying action of the solder, thus obscuring potential troubles. This condition can be rectified by removing some of the solder with the use of de-soldering aids.

4. *Cold solder joint.* An inferior connection easily detected by its dull-gray, grainy appearance, or as a cluster of solder that has not properly wetted all the surfaces. This is the result of applying insufficient heat, which prevents solder from alloying with the metal parts. Cold solder joints can generally be corrected by reheating the connections. If, however, the puddling of solder is caused by a connection that has not been thoroughly cleaned, it must be de-soldered and cleaned before it can be properly soldered.

*applying insufficient heat or not enough heat*

Flux residue remaining on the connection after soldering is undesirable. This residue presents a poor board appearance, and contaminants suspended in this residue could cause troublesome electrical leakage paths. If more active fluxes must be used, residue removal is doubly important to prevent corrosion. Flux removal can be accomplished by *hand brushing, dipping,* or with *ultrasonic equipment.*

Brushing with a stiff-bristle brush dipped in a flux solvent is often acceptable, but may not remove flux completely from inaccessible cavities characteristic of pc boards in which swaged eyelets and terminals are used. Dipping and mildly agitating the board in a solvent bath is preferable since it will improve solvent action in some areas that are not possible to reach by hand brushing (Fig. 17.14). A more efficient means of flux residue removal, especially when activated fluxes are used, is to use an ultrasonic tank containing the flux solvent. Several minutes in this system ensures the most thorough cleaning.

For most electronic soldering applications, rosin flux residue is the type most generally encountered. Isopropyl alcohol is an excellent solvent for this residue. As in the case of most solvents, certain hazards present themselves. Alcohol is flammable and should be used in a well-ventilated area and kept away from the skin. Finally, the remaining solvents are quickly evaporated by the use of an oil- and moisture-free air hose.

There is a number of aerosol cleaners available that are sometimes used instead of isopropyl alcohol. However, it is necessary to know, when choosing an aerosol cleaning solvent, what impact the product has on the environment, its regulatory compliance, and its

**Figure 17.14** Mild agitation for flux removal is accomplished by use of a rocker table.

characteristics. When ordering, catalog descriptions may include the content of substances that may deplete stratospheric ozone and/or may contribute to the formation of surface level ozone, along with performance data. Many acronyms have been adopted that identify the specific substances that are contained in the aerosol. These acronyms are CFC, CI. Solv, HCFC, HFC, HFE, and VOC. In addition, the numerical values of the content of these substances by percent of weight and the Ozone Depletion Potential (ODP) are given. ODP is determined in accordance with the Montreal Protocol and U.S. Clean Air Act of 1990 and is the relative value determined for the potential of a substance to deplete naturally occurring ozone in the stratosphere compared to trichlorofluoromethane gas, which has the highest depletion potential. As a reference, this gas has a depletion potential of 1.00. The definitions of the chemical compounds represented by each of the substance acronyms listed above, according to the *Contact East Advisor* article in *Contact East's* 1997 general catalog, are given as follows to allow one to make an informed choice of benefit when selecting an aerosol cleaner or other aerosol product:

CFCs (Chlorofluorocarbon) are compounds that contain carbon, chlorine, and fluorine, which are very stable as they collect in the troposphere. CFCs are not broken down easily in sunlight. Therefore, many will eventually reach the stratosphere where the chlorine can adversely react with the ozone, causing depletion.

CI. Solvs (Chlorinated Solvent) are compounds that contain hydrogen, oxygen, and nitrogen with the presence of at least one atom each of carbon and chlorine. Carbon tetrachloride and 111-trichloroethylene are two examples of chlorinated solvents that have substances that are reported to be both dangerous to breathe and to cause depletion of the ozone layer.

HCFCs (Hydrochlorofluorocarbon) are compounds composed of carbon, chlorine, fluorine, and hydrogen. HCFCs are less stable than CFCs in sunlight and most are broken down before reaching the stratosphere. In comparison to CFCs, HCFCs have ozone depletion values of only 2% to 10%.

HFCs (Hydrofluorocarbon) are compounds that contain carbon, fluorine, and hydrogen. These compounds do not contain chlorine and are broken down naturally in the lower atmosphere, and thereby are unable to contribute to the depletion of the ozone layer.

HFEs (Hydrofluoroether) are compounds that were developed by 3M. HFEs are low in toxicity and are nonflammable. Since the compounds contain only ethers, fluorine and hydrogen, they have short life in the atmosphere and are completely ozone-safe.

VOCs (Volatile Organic Compound), such as denatured and isopropyl alcohol, may contain chlorine, oxygen, and nitrogen atoms and at least one or more carbon atoms. They are found in the form of gases or volatile liquids that are less stable than CFCs and are broken down in the atmosphere with no danger of depleting ozone in the stratosphere. Certain compounds contribute to the formation of surface-level ozone, more commonly known as smog. VOCs fall in this category and the amount of their content is established by the weight percentage of the product's formulation that contributes to creating smog.

Most aerosols are now available having ODPs of 0.00 due to formulations used that have eliminated the harmful compounds that can cause ozone depletion. Even the more aggressive cleaners on the market have ODPs of only 0.03, 0.05, 0.07, and 0.10.

## 17.7 DE-SOLDERING PRINTED CIRCUIT BOARDS

When a lead is soldered to a terminal pad, it is difficult to remove without damaging the component or the terminal area. Several commercially available de-soldering aids, however, simplify this task. These aids are the *solder wick, de-soldering bulb, solder sucker, de-soldering tips,* and *extraction tools.*

A solder wick is made of finely woven strands of copper wire. This flattened wick is placed over the terminal area and lead to be de-soldered. This arrangement is shown in Fig. 17.15a. The soldering iron tip is placed in contact with the solder wick and pressed down against the connection. As the heat from the iron is transferred to the wick and connection, the solder will melt and flow in the direction of the heat transfer. The solder is thus trapped

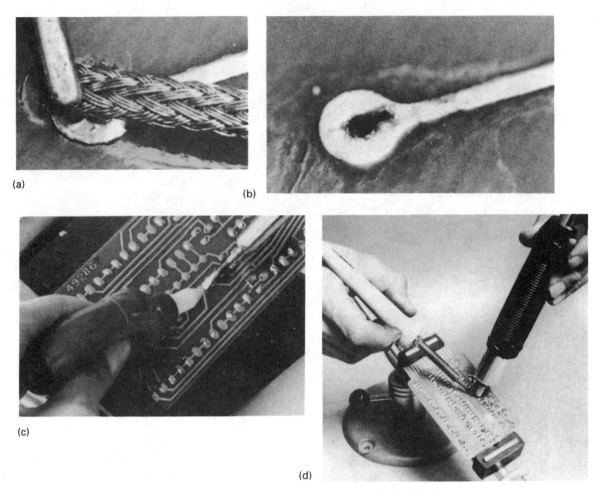

(a)

(b)

(c)

(d)

**Figure 17.15** De-soldering techniques: (a) position of solder wick and soldering tip; (b) de-soldered terminal pad using solder wick; (c) de-soldering bulb, *courtesy of Ungar, Division of Eldon Industries, Inc.;* (d) solder sucker.

by the solder wick as it flows up through the weave. The result is shown in Fig. 17.15b. The solder can be completely removed from the joint with this method. The used portion of the wick is discarded. Solder wick is available in rolls for many de-soldering applications. With the solder removed from the connection, the lead can be bent away from the terminal pad and the component easily withdrawn.

The de-soldering bulb and solder sucker can also be used for removing excess solder or for de-soldering component leads. Both tools have hollow Teflon tips with high heat resistance and will not scratch or mar delicate pc board conductors. The de-soldering bulb, shown in Fig. 17.15c, is employed by depressing the bulb and then placing the hollow tip alongside the soldering iron tip on the joint. As the solder begins to melt, the bulb pressure is released. The liquid solder is drawn up into the bulb by the suction. Both tools are then removed from the connection for inspection. If solder remains, the process must be repeated. The solder sucker shown in Fig. 17.15d is employed in basically the same manner except that the suction is produced by a spring-loaded piston. The piston handle is first pushed downward. The handle is then rotated to engage the release pin. As the solder begins to melt, the pin is disengaged and the solder is drawn up through the hollow tip as the piston snaps upward inside the tubular handle. Both tools are easily disassembled to remove the accumulated solder. These de-soldering tools do not generally remove sufficient solder

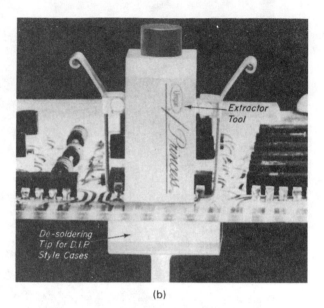

(a)                                                    (b)

**Figure 17.16**   Device de-soldering equipment: (a) de-soldering tips for TO-5 and DIP-style cases; (b) de-soldering DIP-style devices. *Courtesy of Ungar, Division of Eldon Industries, Inc.*

on the first attempt to allow components to be removed, thus rendering the solder wick the preferable method.

For removing ICs, the previously discussed de-soldering aids would have to be used on each lead individually, making it a time-consuming process. De-soldering *tips* specifically designed to simultaneously heat all the leads for removing devices quickly and easily are available for this purpose. De-soldering iron tips for TO-5-style cases and dual-in-line configuration ICs are shown in Fig. 17.16a.

The de-soldering tip is placed in contact with all the device leads simultaneously. As the leads are de-soldered from the foil side of the board, the device is removed from the component side with an extraction tool such as that shown in Fig. 17.16b. With this tool, the device is lifted away as the solder is melted. The remaining solder present on the terminal pad may obstruct the access holes and can be removed with any of the de-soldering aids discussed. The recommended type of extraction tool is one whose metal clamps grip the leads, thus providing some degree of heat sinking. It must be mentioned that to avoid damage lead heat sinking is just as important when de-soldering as it is during soldering.

## 17.8  WAVE SOLDERING

A much more elaborate system of production line pc board soldering is the process of *wave soldering*. This method is widely used in the industry because of its large soldering capacity; its ability to accurately regulate time and temperature exposure of the pc board; the scrubbing action of the solder wave, which aids in complete wetting and soldering of each joint area; and the consistently high-quality results obtained.

A typical wave-soldering system consists of three main sections: the *fluxer, preheater,* and *solder wave.* An integral part of the system is a conveyor, which is used to accurately control the movement of the pc board over the fluxer, the preheater, and the solder wave. Precleaning and postcleaning cycles of the pc board are not a part of the system and are performed separately.

A typical bench-type wave-soldering machine is shown in Fig. 17.17. The conveyor speed is adjustable from 1 to 15 feet per minute at angles of between 5 and 9 degrees with the surface of the bench. The board is introduced into the system by attaching it to the fingers of the conveyor. The circuit side is first run across a foam fluxer. At this station, a head of fine bubbles rises from the fluxer. These bubbles are typically generated by passing air through a submerged porous ceramic cylinder that is impervious to the flux. A chimney and brushes aid in uniform flux distribution. The bursting flux bubbles rise by capillary action and penetrate into plated-through holes. Rosin-based flux, such as Kenco No. 465 Resin Flux, having 25% or higher rosin content, is generally preferred in the wave-soldering process. This flux not only promotes uniform solder wetting and formation of a strong intermetallic bond with the copper, it also reduces the surface tension of the liquid solder. Surface tension reduction is important in preventing solder bridges and icicles (solder peaks) from forming.

After the board is fluxed, the conveyor continues its travel over a preheater station. The purpose of this station is threefold. First, preheating evaporates excess flux solvent on the board to prevent splattering and entrapped gases, which could cause pinholes in connections when the board is soldered. Second, it conditions the board and assembly against

**Figure 17.17**  Bench-type wave-soldering machine. *Courtesy of the John Treiber Company.*

thermal shock. Third, preheating the board helps to overcome the heat-sink effects on large components caused by the solder wave. Preheaters must be capable of providing a range of heat, as measured on the component side of the board, from 100 to 275°F (38 to 135°C). Overheating the flux must be avoided because it degrades its effectiveness by reducing its mobility.

With the board preheated, it is next conveyed over the solder wave, which quickly produces large numbers of high-quality and well-contoured solder connections simultaneously. These connections are both electrically and mechanically sound. The wave is formed by pumping molten eutectic solder (63/37) vertically upward through a solder nozzle having a large plenum chamber that rests in a solder pot. A high-capacity centrifugal pump propels the molten solder through the nozzle to form a standing wave. The most commonly used wave shape is bidirectional, that is, solder flows in two directions.

Bridges and icicles will usually result if the solder system has too narrow a solder wave and the boards are run on a horizontal conveyor (see Fig. 17.18a). A web of solder tends to extend beyond the wave in the direction of travel, which creates these problems. If the solder wave is narrow, inclining the conveyor to an angle of 5 to 9 degrees will increase solder *peel back* and therefore reduce the solder web and significantly reduce solder defects (see Fig. 17.18b). For best results, however, a solder system having a deep, wide wave (3 to 4 inches along the path of travel at 1 or more inches above the nozzle edges) and an inclined conveyor will virtually eliminate the formation of a solder web, which causes bridging and icicles (see Fig. 17.18c). Inclining the conveyor causes the board to be introduced onto the solder wave at a high solder-velocity point. This takes advantage of the natural

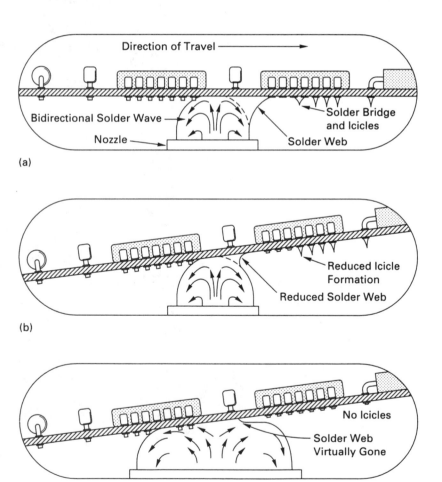

**Figure 17.18** Use of wide solder wave with oil and an inclined conveyor eliminates solder bridges and icicles: (a) narrow solder wave with horizontal conveyor; (b) narrow solder wave with inclined conveyor; (c) wide solder wave with inclined conveyor.

scrubbing action of the wave at this point in its velocity profile. A wider wave used together with an inclined conveyor also allows the soldered conductor pattern to exit the wave at a practically zero velocity point, which not only increases solder *peel back* but reduces heat transfer to the components being soldered.

To obtain the smoothest solder wave and minimum solder deposits having a bright, shiny appearance, the formation of dross (oxides of tin and lead) must be kept to an absolute minimum. One method of reducing the dross formation is to use an oil (soldering fluid) as a dross blanket in the solder pot. As solder is pumped, the oil forms a thin film on the surface of the wave. Another effective method is to intermix oil, such as Hollis No. 225 Soldering Fluid, by precision metering with the wave at the impeller of the pump. This causes the oil to disperse uniformly within the solder wave. In both methods, the oil reacts with the dross to form harmless tin and lead soaps without degrading the solder. In addition, the surface tension of the solder is reduced, which allows excess solder to drain more easily from the board. The soldered boards also become coated with the oil, which intermixes with the rosin flux and protects the soldered connections from oxidation. This mixture of oil and rosin is easier to remove than rosin alone in the cleaning process, assuring bright soldered joints.

Although wave soldering is an automated process, a degree of skill and knowledge of soldering is necessary to properly operate the system. This knowledge includes an understanding of the proper conveyor speed, type of flux, preheater temperature, width of the solder wave, solder temperature, the use of an oil blanket or oil intermix, and the established quality standards. When the pc boards have been assembled, soldered, and cleaned, they may be tested for operational performance and modifications can be made if necessary.

## 17.9 THERMAL PROFILE OF THE WAVE SOLDERING PROCESS

One essential function of a process engineer may be to evaluate and modify a wave soldering system in order to quantify both the magnitude and duration of temperatures that contact a pc board as it passes through the various operations. By tightly controlling the thermal profile of this process, high-quality results, with minimum rejects, will be obtained. The data acquisition systems described in Sec. 17.4 are ideally suited to evaluate the thermal characteristics of a pc board as it passes through the wave soldering system.

Typically, during the wave soldering process, a pc board will be subjected to temperature variations from about 25°C (room temperature) to as much as 250°C. For this reason, a type J thermocouple is selected and constructed by twisting and welding together iron and a constantan (40% nickel and 60% copper alloy) wire. This thermocouple is capable of accurate, repeatable measurements of temperatures from −184 to +760°C.

The thermocouple is first inserted into a small via hole of the board. (Refer to Chapter 8 for a description of a via hole.) The thermocouple must enter from the component side and should extend into the hole until it just begins to exit the circuit side. This arrangement will allow the temperature changes to be monitored and will simulate those changes to which a component lead would be subjected. The thermocouple extension leads pass above the component side of the board and to the input of the 6B11 signal conditioning module discussed in Sec. 17.4.

A thermal profile of a conventional wave soldering process obtained from the data acquisition system is shown in Fig. 17.19. Test panels are used to properly establish process parameters as follows: During the preheat or flux activation phase, a thermal rise of approximately 2°C/second is obtained by adjusting the conveyor speed within the typical range of 6 to 8 feet per minute. Optimum results are achieved when the circuit side surface temperature is about 183°C, the melting point of eutectic solder, as the board exits the preheater. As the board moves over the solder wave, additional heat is transferred to the board, which raises the circuit side surface temperature still higher. The board dwells in the molten solder wave for 3 to 5 seconds, causing the surface temperature to reach a maximum temperature of about 250°C. Control of the time that the board remains in the

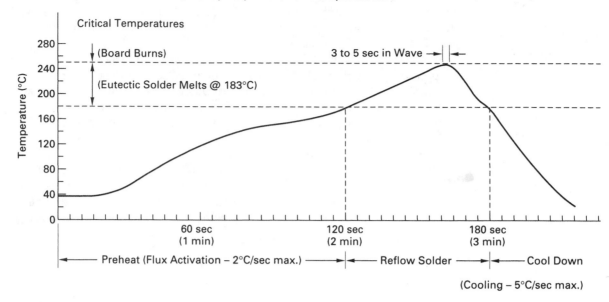

[Conveyer Speed – 6 to 8 Feet per Minute]

**Figure 17.19**   Thermal profile of a conventional wave soldering process.

wave is a function of the profile of the wave and the depth to which the board extends into the solder. Refer to Fig. 17.19. Test panels can be run and rerun to adjust the controls for optimum results. Finally, a maximum cool-down cycle of −5°C/second is recommended to minimize thermal shock and maximize joint strength.

The data acquisition system just described can be configured quickly with a personal computer and menu-driven software packages such as ACQUIRE, which are discussed in Sec. 17.4.

## EXERCISES

### A.   Questions

17.1   What are the two metals used to make solder, and what is the percentage of each in solder used for general-purpose electronic work?

17.2   Explain the term *eutectic solder.* What is its melting point temperature?

17.3   What type of flux is used for electronic work?

17.4   What is the purpose of flux?

17.5   What is the recommended method of cleaning a plated soldering iron tip?

17.6   What is the approximate temperature of a 3/4-inch-long tip having a diameter of 3/64 inch when used with a 25-watt heating element?

17.7   Explain the following terms: *cold solder, solder peak,* and *wetting.*

17.8   What is used to remove rosin flux residue from electrical connections?

17.9   What benefits can be derived from the generation of a thermal profile for a wave soldering process?

17.10   What is solder wick used for?

17.11   What is the major cause of solder bridge and icicle formations when wave soldering pc boards? How is this eliminated?

*determined in accordance with the Montreal Protocol and US Clean Air Act of 1990*

17.12   What does ODP stand for and how is it determined? *Ozone Depletion Potential*.

17.13   Give the acronyms for two specific substances in aerosols that have a high potential to cause depletion of ozone. *CFCs (chlorofluorocarbon) and Cl. Solvs (chlorinated solvent).*

17.14   How should clinched leads be formed onto pads? *The end of the lead is firmly pressed against the surface of the pad*

17.15   How should the soldering iron tip and solder be positioned relative to the clinched lead and pad to properly solder this type of connection? *The iron tip and solder a placed opposite side of the lead.*

## B.   True or False

Circle *T* if the statement is true, or *F* if any part of the statement is false.

17.1   Solder is an alloy of tin and lead. **(T)**   F

17.2   For most hand-soldering applications, 60/40 eutectic solder is used. **(T)**   F

17.3   Solder melts at a lower temperature than either pure tin or pure lead. **(T)**   F

17.4   Rosin flux with activating agents is noncorrosive at room temperature. **(T)**   F

17.5   Conical soldering iron tips are used when large areas must be rapidly heated. T   **(F)**

17.6   When hand soldering, the solder should be placed between the iron's tip and the lead to ensure proper wetting. T   **(F)**

17.7   A cold solder joint can be corrected by simply reheating the connection without the addition of more solder. **(T)**   F

17.8   Oil is used to reduce the formation of dross in a wave-soldering system. **(T)**   F

17.9   HFCs cause ozone depletion in the stratosphere. T   **(F)**

17.10   A lead may be clinched in any direction as it is pressed against the pad. T   **(F)**

## C.   Multiple Choice

Circle the correct answer for each statement.

17.1   Solder used in the electronics industry for component assembly consists chiefly of tin and (*copper,* *lead*).

17.2   Chloride-based fluxes are not used in the electronics industry because they are (*expensive,* *corrosive*).

17.3   Rosin-based fluxes are most active at (*room,* *high*) temperature.

17.4   For a constant soldering iron wattage rating, the tip temperature (*increases,* *decreases*) as the tip length increases.

17.5   A good solder joint results if a (*liberal,* *minimum*) amount of solder is applied to the connection.

17.6   After a solder joint has been completed, flux residue is removed with the use of a solvent such as (*alcohol,* *hydrochloric acid*).

17.7   The most common method used by industry for soldering large quantities of pc boards is (*hand,* *wave*) soldering.

17.8   HCFCs are (*less,* *more*) stable than CFCs in sunlight.

## D.   Matching

*Acid core flux for 60/40.*

Match each item in Column A to the most appropriate item in Column B.

| COLUMN A | COLUMN B |
|---|---|
| 1.   Dross  *e* | a.   Solder defect  ✓ |
| 2.   Eutectic  *f* | b.   Return to idle temperature  ✓ |
| 3.   Recovery time  *b* | c.   Flux solvent ✓ |
| 4.   Flux  *d* | d.   Rosin  ✓ |
| 5.   Rosin joint  *a* | e.   Oxidizing scum  ✓ |
| 6.  Solder wick - *g* | f.  63/37 ratio  ✓ |
| 7. Alcohol *c* | g.  desoldering tool ✓ |
| 8. PTH  *i* | h.  Concave ✓ |
| 9. Smog | i.  Interlayer ✓ |
| 10. Fillet *h* | j.  VOCs |

6. Solder wick      f.   63/37 ratio

7. Alcohol      g.   De-soldering tool

8. PTH      h.   Concave

9. Smog      i.   Interlayer

10. Fillet      j.   VOCs

## E. Problems

17.1    Lay out, shear, and bend an aluminum tray for holding a soldering iron tip cleaning sponge. The tray is to have inside dimensions of 3/4 inch by 3 1/2 inches by 3 1/2 inches. Tabs are not required on the inside corners. These corners are sealed in Problem 25.2 to retain excess moisture. Break all sharp edges with the appropriate file.

17.2    Solder the pc board for the series feedback regulator designed and constructed in Problems 6.2, 10.2, and 16.2.

17.3    Solder the pc board for the tachometer circuit designed and constructed in Problems 6.3, 10.3, and 16.3. Circuit board and meter installation are accomplished in Problem 25.3. Wiring is done in Problem 26.5.

# 18 Surface Mount Technology

## LEARNING OBJECTIVES

*Upon completion of this chapter on surface mount technology, the student should be able to*

- Apply the appropriate ergonomics when using a computer or an inspection microscope.
- Determine land sizes and footprints for many surface-mounted components and devices.
- Properly identify the side of a pc board on which surface-mounted and conventional through-hole components can be placed when both technologies are used on a common board.
- Select, test, and properly apply solder paste to surface-mounted components.
- Select, test, and properly apply adhesives to surface-mounted components.
- Be familiar with the wide variety of soldering methods used in surface mount technology.
- Hand solder surface-mounted components.
- De-solder surface-mounted components.
- Extend the magnification range of an inspection microscope.
- Become acquainted with the features of a variety of bench-top equipment and accessories used in prototype and small volume production of the Surface Mounted Technology (SMT) applications of stencil/screen printing, pick/place/dispense, reflow, soldering, and de-soldering.
- Understand the differences between SMT convection and conduction rework stations.

## 18.0 INTRODUCTION

Surface mounting stems from the technologies of hybrids where surface-mounted devices are soldered inside a ceramic or plastic package having through-hole leads. For these, it appears that the conventional methods of design and manufacturing have, to a large extent, reached the limit regarding factors of cost, reliability, size, and weight. With the emergence of surface-mounted technology (SMT), the generation of lighter, smaller, and more dense pc boards at lower costs and higher reliability has been realized.

Surface-mounted technology involves the bonding of components and devices directly to the surface of the board and connected directly to the conductor pattern, thus eliminating the requirement of drilling holes in the board.

The sizes of surface-mounted components (SMC) and devices (SMD) are very much smaller than their equivalents made for conventional through-hole installation. Because SMT is relatively new, all equivalencies are not available. In addition, there is a great deal of nonuniformity of standardization among manufacturers of SMT components and device packages. As a result, both conventional and SMT are required to be used on most pc board designs. With the improvements in the technology, more and more board area will be devoted exclusively to surface mounting.

*CAUTION:* When using soldering irons, place hot irons in holder when not in use. Keep hands and fingers away from the heating element and soldering iron tip. Be especially watchful that the heating element or tip does not come in contact with the power cord.

Heat resistant gloves must be used when removing hot boards from wave-soldering machines. The processes discussed should be performed in adequately ventilated facilities and with applicable equipment having proper fume exhaust systems. Be sure the fume exhaust system is turned on when using any of the types of soldering systems discussed in this chapter. See Appendix XX regarding flux fumes, health issues related to flux fumes, and fume extraction.

A chemical hazard exists in the materials presented in this chapter. Approved face shields (goggles) in compliance with ANSI Z87.1-1989 and chemical-resistant clothing and footwear and neoprene or nitrile gloves must be worn prior to performing any of these operations. Also, the section on safety outlined at the beginning of the book should be reviewed for more specific information.

Just as when using a computer station for long periods of time, ergonomics should be applied when using inspection magnifiers/lamps and microscope systems. This type of equipment should be used with ergonomically designed workstations that have height-adjustable tables with foot rests and height-adjustable seats with lumbar support. The following is a list of things to do that may keep you more comfortable at your inspection station when using either type of equipment.

1. When employing a magnifying lens/lamp, use both eyes as these systems are designed to be used as comfortably as a pair of eyeglasses.

2. Do not lean back from the lens to attempt to get greater magnification. Instead, position the lens at a distance of about 8" to 10" from your eyes so that you obtain maximum magnification without distortion.

3. Adjust seat and table height so that you will have a good working posture. By sitting in a correct ergonomic position, muscle fatigue can be avoided.

4. Use a lens that eliminates glare, such as Luxo Corporation's magnifier lenses that feature WAVE + PLUS™, which is an anti-reflective coating. This will eliminate the need to hunch over the magnifier to block the glare from overhead lighting, which is a cause of muscle fatigue and eyestrain due to reflections off the lens.

## 18.1 CONDUCTOR PATTERN LAYOUT

Surface-mounted conductor pattern layout does not lend itself to the relative ease of layout experienced with conventional through-hole design. This is largely due to the lack of standardization of SMT package designs and the wide variation in component tolerances in the industry. In addition, land pattern design is process-dependent, which means that the same design can experience different manufacturing defects depending on the process variables. (*Land pattern* is the term used in SMT to define the terminal pads to which the device leads are soldered.) This is brought about by the fact that each manufacturer has to establish specific land pattern designs that will function with their materials and processes. For this reason the land patterns presented in this section and the techniques for their design serve only as guidelines.

Common land patterns, also referred to as *footprints,* include *ceramic rectangular resistors and capacitors* (small outline RC and CC); *tantalum capacitors* (small outline molded plastic and welded-stub); *tubular passive components* (metal electrode leadless face or MELFs); *transistor chip patterns* (small outline transistors or SOTs); *small outline integrated circuits* (SOICs); *small outline J lead devices* (SOJ memory and SRAM packages); and *chip carriers* (plastic leaded chip carrier or PLCC, and leadless ceramic chip carrier or LCCC).

The land pattern design parameters for ceramic rectangular resistors and capacitors are shown in Fig. 18.1. The dimensions of component length ($L$), width ($W$), height ($H$),

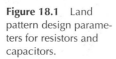

**Figure 18.1** Land pattern design parameters for resistors and capacitors.

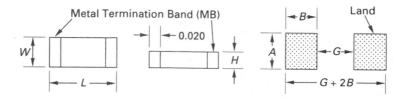

and the termination length of the metal bands (*MB*) are measured first. To determine the width of the land (*A*) and its length (*B*), the following relationships are used:

$$\text{For capacitors: } A_c = W + 0.01 \text{ inch}$$
$$B_c = H + MB$$
$$\text{For resistors: } A_r = W + 0.01 \text{ inch}$$
$$B_r = H + MB + 0.01 \text{ inch}$$

Resistors are much thinner than capacitors and require a longer land to prevent misalignment due to their lower mass. The space between the lands (*G*) is the same for both components and is determined as follows:

$$G = L - 2MB - 0.02 \text{ inch}$$

The overall length of the footprint would thus be *G* + 2*B*. Two examples of land patterns for ceramic resistors and capacitors are shown in Table 18.1.

### TABLE 18.1

Resistor and Capacitor Land Pattern Dimensions

| Component | A | B | G | G + 2B |
|-----------|-------|-------|-------|--------|
| R1206* | 0.065 | 0.060 | 0.070 | 0.190 |
| C1206† | 0.065 | 0.060 | 0.070 | 0.230 |

*$L = 0.130, W = 0.064, H = 0.030, MB = 0.020$

†$L = 0.130, W = 0.064, H = 0.030, MB = 0.020$

Dimensions are in decimals of an inch.

The procedure to determine the land patterns required for molded plastic and welded-stub tantalum capacitors is similar to that for ceramic resistors and capacitors, as shown in Fig. 18.2. Once the dimensional information is obtained for each component type, the following are used to determine the footprint dimensions and spacings:

$$\text{For molded plastic: } A_{mp} = W + 0.010 \text{ inch}$$
$$B_{mp} = H + MT$$
$$G = L - 2MT - 0.020 \text{ inch}$$
$$\text{For welded stub: } A_{ws} = W + 0.010 \text{ inch}$$
$$B_{ws} = H + MT$$
$$G = L - 2MT - 0.020 \text{ inch}$$

Two examples of tantalum capacitor footprints are shown in Table 18.2.

Surface-mounted tubular passive components (MELFs) are resistors, ceramic and tantalum capacitors, and diodes. These packages have metal end caps that are soldered to the lands. Because these packages are round and there is a tendency for them to roll out of position, a slot must be provided in the lands if they are to be soldered using the vapor reflow method. If the component is to be wave soldered, an adhesive is used, which eliminates the need for the slots.

The measurements used for the design of MELF patterns are shown in Fig. 18.3. The following are used to determine the footprints for this package style:

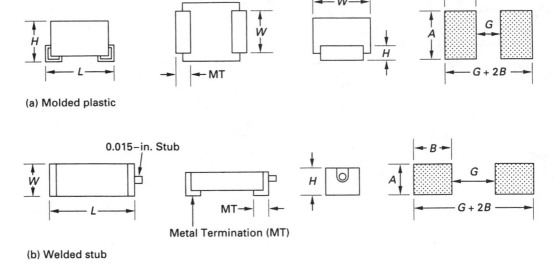

(a) Molded plastic

0.015–in. Stub

Metal Termination (MT)

(b) Welded stub

**Figure 18.2** Land pattern design parameters for tantalum capacitors.

## TABLE 18.2
### Tantalum Capacitor Footprint Dimensions

| Component | A | B | G | G + 2B |
|---|---|---|---|---|
| Molded plastic package B* | 0.090 | 0.060 | 0.050 | 0.170 |
| Welded stub package B† | 0.060 | 0.075 | 0.085 | 0.235 |

*L = 0.130, W = 0.080, H = 0.030, MT = 0.030, H = 0.075, W = 0.110.

†L = 0.160, W = 0.059, H = 0.045, MT = 0.030.

Dimensions are in decimals of an inch.

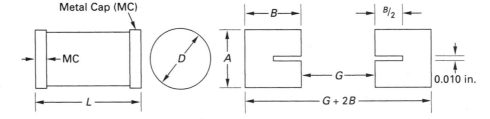

**Figure 18.3** Land pattern design parameters for MELFs.

$$A_{\text{MELF}} = \text{diameter } (D) + 0.010 \text{ inch}$$
$$B_{\text{MELF}} = \text{diameter } (D) + MC$$
$$G = L - 2MC - 0.020 \text{ inch}$$

If a notch is required, notch depth = $B/2$ and notch width = 0.010 inch. An example of MELF land pattern dimensions is given in Table 18.3.

Some of the most commonly used small outline transistor (SOT) chip patterns are the SOT23, SOT89, and the SOT143 packages. These footprints are shown in Fig. 18.4. When designing the land patterns for SOTs, the distance between land centers is generally made equal to the on-center dimensions of the device leads. At the same time, the land should be made 0.015 inch longer than the lead on each of its ends. Many footprints that will accommodate a variety of designs are available from graphic suppliers, such as Bishop Graphics.

**TABLE 18.3**

Sample of MELF Land Pattern Dimensions

| Component | A | B | G | G + 2B |
|-----------|-----|-----|-----|--------|
| MLL41* | 0.109 | 0.114 | 0.140 | 0.368 |

*L = 0.200, diameter = 0.099, MC = 0.015.

Dimensions are in decimals of an inch.

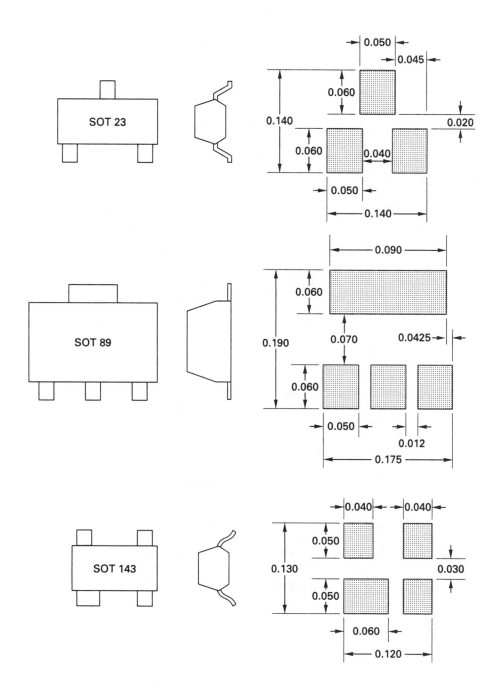

**Figure 18.4** Small outline transistor patterns.

The criteria for selecting commercially available footprints for small outline integrated circuits (SOICs) or R-packs are much less complicated than for those designs discussed previously. The land width and length should be 0.030 and 0.075 inch, respectively. The on-center spacing of the lands is typically 0.050 inch. This will result in about a 0.015-inch space between adjacent lands. To determine the space ($S$) between rows of lands, the relationship $S$ = body width − 0.01 inch is used.

The footprint of an SOIC, together with typical dimensions of various other packages with different lead counts, is shown in Table 18.4. The footprint design for plastic leaded chip carriers (PLCCs) and leadless ceramic chip carriers (LCCCs) is shown in Fig. 18.5. Note that the PLCC chip has leads that turn *under* the package. This formation is referred to as *J leads* as opposed to those typical of SOT packages, which are called *gull wings*. The footprint is determined by the following:

$$\text{on-center lead spacing} = 0.050 \text{ inch}$$
$$\text{land width} = 0.025 \text{ inch}$$
$$\text{land length} = 0.075 \text{ inch}$$
$$\text{dimension } A \text{ and } B = X + 0.030 \text{ inch (for PLCC)}$$
$$\text{dimension } A \text{ and } B = X + 0.070 \text{ inch (for LCCC)}$$

Note that for square chip carriers, dimension $A$ = dimension $B$. This is not true of rectangular packages. These dimensions for a variety of square and rectangular chip carrier footprints are given in Table 18.5.

Land Width = 0.030 in.
Land Length = 0.075 in.
Land On-Center Spacing = 0.050 in.
Spacing Between Lands = 0.015 in.

**TABLE 18.4**

SOIC and R-Pack Footprint Dimensions

| Package | L | W | S |
|---------|-------|-------|-------|
| S08 | 0.175 | 0.290 | 0.140 |
| S014 | 0.325 | 0.290 | 0.140 |
| S016 | 0.375 | 0.290 | 0.140 |
| SOL24 | 0.575 | 0.450 | 0.300 |
| R-pack 14 | 0.325 | 0.350 | 0.200 |
| R-pack 16 | 0.375 | 0.350 | 0.200 |

Dimensions are in decimals of an inch.

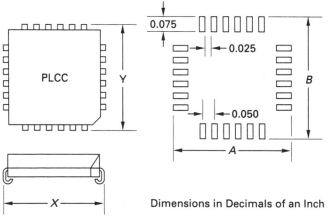

**Figure 18.5** PLCC and LCCC footprint dimensions.

Dimensions in Decimals of an Inch

Dimension A = X + 0.030 and B = Y + 0.030 for PLCC
Dimension A = X + 0.070 and B = Y + 0.070 for LCCC

Surface-mounted components are soldered to the pc board by either vapor phase, solder wave, or infrared processes that are discussed in Sec. 18.5. Variations in land pattern dimensions may be required for the same component for a different soldering process. One common problem encountered is the varied degree of solderability at termination ends. This can cause misalignment of components with footprints. Also, components such as resistors and capacitors, instead of remaining flat, may end up in a standing position, called *tombstoning*. If these components are positioned too close to each other, a large component may shadow a smaller one, causing insufficient solder or solder bridges to result. For these reasons, it is essential that all land pattern designs first be evaluated through the processing of test boards.

Some recommended spacing considerations for SO, SOJ, R-packs, and PLCC packages are shown in Fig. 18.6. These spacings also take into account automatic assembly and testing requirements.

**TABLE 18.5**

Overall Dimensions of PLCC and LCCC Land Patterns

| Package Style | Pin Count | Dimension A | Dimension B |
|---|---|---|---|
| Square | 16 | 0.370 | 0.370 |
| Square | 24 | 0.470 | 0.470 |
| Square | 44 | 0.720 | 0.720 |
| Square | 68 | 1.020 | 1.020 |
| Square | 124 | 1.720 | 1.720 |
| Rectangular | 22 | 0.350 | 0.550 |
| Rectangular | 32 | 0.520 | 0.620 |

Dimensions are in inches.

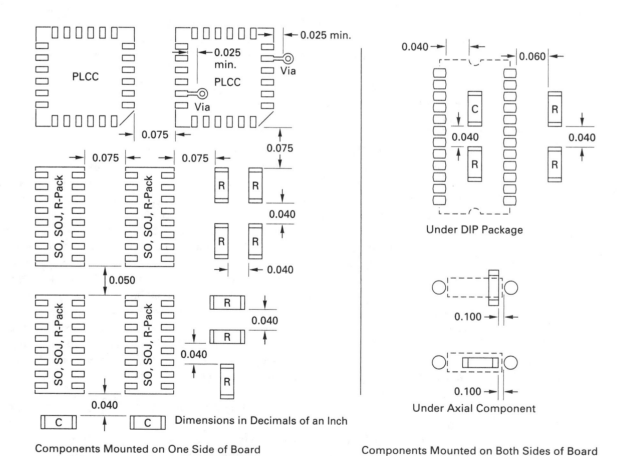

**Components Mounted on One Side of Board**

**Components Mounted on Both Sides of Board**

**Figure 18.6**  Minimum land-to-land or land-to-pad clearances.

*Surface Mount Technology*

The following societies may be consulted for additional design information: The Institute of Interconnecting and Packaging Electronic Circuits (IPC), Electronics Industries Association (EIA), and the Society of Manufacturing Engineers (SME).

## 18.2 SURFACE MOUNT ASSEMBLY

The surface mount packages discussed in the preceding section are shown in Fig. 18.7. Figures 18.7a through d show leadless resistor and capacitor packages with metal end caps or terminations, and Figs. 18.7e, f, g, and h illustrate SOT and SOIC packages with gull and modified gull wing leads. A leadless chip carrier with recessed edge conductors is shown in Fig. 18.7i, and Fig. 18.7j is a plastic chip carrier package having J leads.

As mentioned previously, a wide variety of components is not available for surface-mounted applications, resulting in pc board designs incorporating this technology together with conventional through-hole insertion. An exception to this is in the design of memory boards, where all required packages are available for surface mount.

As a result of the requirement for a combination of both technologies, SMT assemblies have been divided into three categories that describe their application. These are shown in Fig. 18.8. Type I assembly contains only surface-mounted components on a single-sided, double-sided, or multilayer pc board. Where components are placed on the bottom side, the tack of the solder paste applied to the lands prior to assembly is sufficient to hold them in place during board handling and baking prior to reflow soldering. If the reflow soldering is done by vapor-phase or infrared processes, the solder paste will continue to hold them in position. If wave soldering is used, all components must be held in place with an adhesive on any side that will be passed over the wave. Because many devices are heat sensitive, one side of the board may be vapor-phase reflowed while the less temperature-sensitive components on the other side of the board are wave soldered. This procedure is more common with type II and III assemblies that combine surface mount with through-hole designs. Type II assemblies place SMCs on one or both sides of the board with through-hole insertion components on the top only. Type III assembly places through-hole insertion components on the top side with SMCs on the bottom side.

Due to their size and delicate mechanical features, SMCs are mounted manually with the aid of a vacuum pipette or, for volume production, by automatic component placement. Prior to placement, solder paste and adhesive are applied as required. These processes are discussed in detail in Secs. 18.3 and 18.4, respectively. The two primary functions of automatic equipment for SMT assembly are picking up the components and placing them onto the correct lands. For this reason, this type of equipment is generally termed "pick and place." Assembly equipment having single or multiple pick and place vacuum heads are available. Components are placed sequentially with X, Y, and Z movements of the vacuum head. Automatic machines are usually fed with components that are bulk packaged on reels or in magazines called DIP sticks, which are plastic tubes that channel parts into the machine.

Two types of bench-top pick and place equipment used in prototype and small volume production are OK International's Single Arm Pick/Place/Dispensing System, shown in Fig. 18.9, and Automated Production Systems' (APS) Gold Place™ Manual Pick & Place and Dispensing System, shown in Fig. 18.10. OK International's system, shown in Fig. 18.9, has a single, low friction X, Y, and Z arm movement with infinite Theta (rotation) and is electrostatic discharge (ESD) safe. The spring-loaded adjustable board holder will accept PCBs up to 11" × 15 1/2" and has a movable arm rest. There is a built-in diaphragm pump having automatic vacuum switching with an LED indicator. The system operates on 120 VAC, 50/60 Hz, and requires shop air (up to 90 PSI). It is equipped with an air regulator and gauge. This dispenser features manual or timed mode, has adjustable fluid hold back, and an adjustable dispenser angle. The carousel is removable and allows for pre-kitting of components.

Automated Production Systems' equipment shown in Fig. 18.10 operates on 120 VAC, 50/60 Hz, and requires shop air (20 PSI minimum). It has both gross and fine X, Y, Z, and Theta movement and handles PCBs up to 16" × 24". This machine has a hand rest

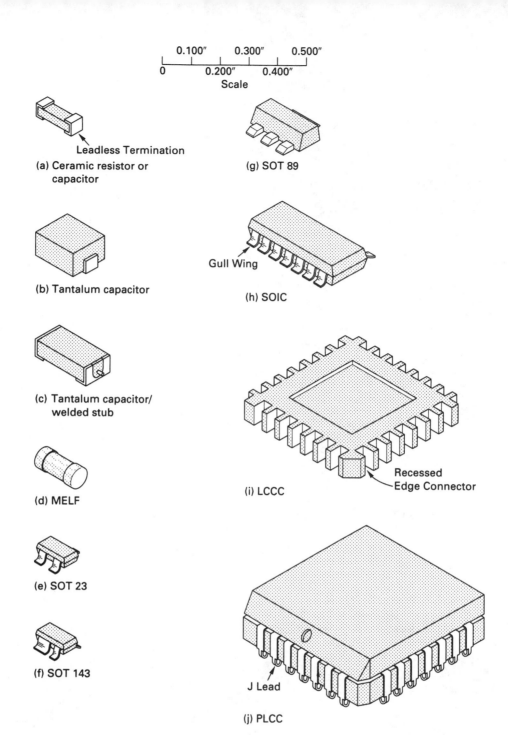

**Scale**

0.100"  0.300"  0.500"
0  0.200"  0.400"

**(a) Ceramic resistor or capacitor** — Leadless Termination

**(b) Tantalum capacitor**

**(c) Tantalum capacitor/ welded stub**

**(d) MELF**

**(e) SOT 23**

**(f) SOT 143**

**(g) SOT 89**

**(h) SOIC** — Gull Wing

**(i) LCCC** — Recessed Edge Connector

**(j) PLCC** — J Lead

**Figure 18.7** Surface mount component and device packages.

and arm assembly to alleviate operator fatigue by carrying the operator's hand throughout the entire assembly process. The vacuum tip is automatically toggled on and off during component pick and place, while the digitally timed dispenser accurately deposits adhesives, potting compound, or solder paste. This system operates in manual, semi-automatic, and automatic modes. The system also has ESD-safe bins that slide in close to placement. Optional stand, tape feeders, and stick feeders are also available.

When SMCs are manually installed, they should be handled by their bodies, not by their terminations or leads. In addition, care should be exercised to avoid smearing solder paste or adhesive onto the packages that would interfere with solder joint formation and result in short circuits.

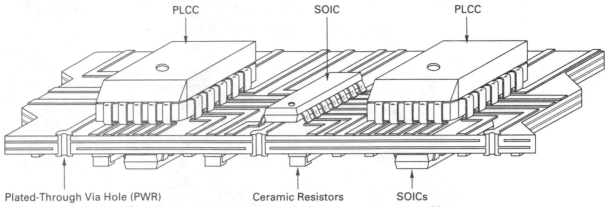

PLCC       SOIC       PLCC

Plated-Through Via Hole (PWR)    Ceramic Resistors    SOICs
(a) Type I exclusively surface-mounted components and devices on top or on top and bottom

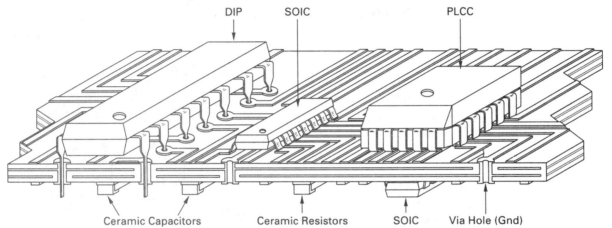

DIP       SOIC       PLCC

Ceramic Capacitors    Ceramic Resistors    SOIC    Via Hole (Gnd)
(b) Type II insertion components on top with surface-mounted components on one or both sides

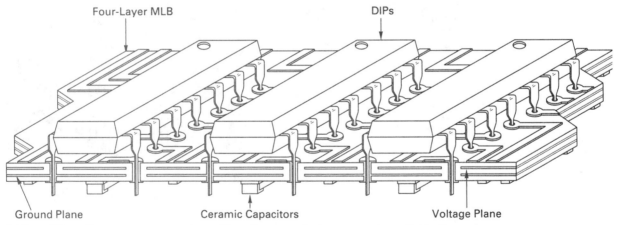

Four-Layer MLB       DIPs

Ground Plane    Ceramic Capacitors    Voltage Plane
(c) Type III insertion components on top with surface-mounted components on the bottom

**Figure 18.8**   Type I, II, and III SMT assemblies.

**Figure 18.9** Single Arm Pick/Place/ Dispense System. *Courtesy of OK International, Yonkers, NY.*

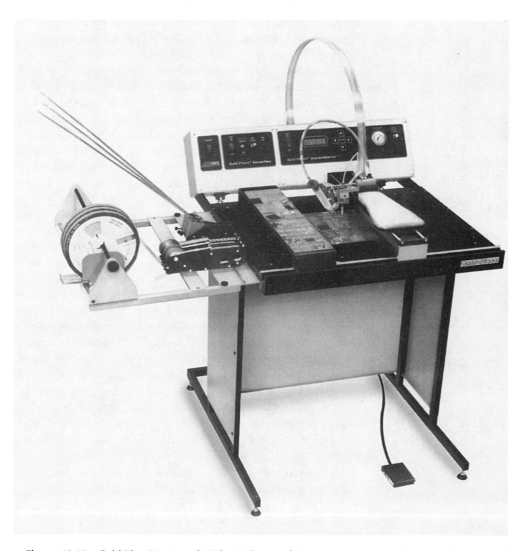

**Figure 18.10** Gold-Place™ Manual Pick & Place and Dispensing System. *Courtesy of Automated Production Systems, Inc., Huntingdon Valley, PA.*

In any process of assembly used, the component terminations or leads must register squarely on lands. Even though the surface tension of the melting solder paste tends to pull misregistered components into place, this should not be expected. If misregistration is not detected on SMCs that have been secured with adhesive, correction is difficult and time consuming.

## 18.3 SOLDER PASTE

To solder SMCs to pc boards, solder paste or solder cream is first applied to the footprints by automatic dispensing with a syringe, screening, or stenciling, after which the components are assembled. *Solder paste* is the standard term, although rosin-based materials are referred to as *creams*. Solder paste is a homogeneous combination of minute prealloyed solder particles that are suspended in a flux system. The solder particles are spherical as a result of blowing or splashing melted solder onto a spinning wheel in a controlled atmosphere. Their round shape has less tendency to clog screens, as well as having minimum surface oxidation.

The flux system is made up of a flux base and a gelling or suspension agent, which are blended together. The formulation of this flux system is a well-guarded secret in industry. The proprietary ingredients involved give the paste body, the ability to withstand soldering temperature and time without breaking down or losing efficiency (thermal resistance), chemical strength (effectiveness), and a consistency that makes it suitable for application by screening, stenciling, or syringe dispensing.

Factors that govern the selection of a solder paste for a specific application are *particle size, flux activators, metal content,* and *rheological properties* such as viscosity, tackiness, slump, and working life. Each of these factors will be described in detail.

If the screening process of flux application is to be used, the particle size should be small enough to pass through a 200-mesh (200 holes per square inch) screen. Larger particle sizes may be used if stencils or syringes are the method of application.

The flux promotes wetting action between component leads and the surface mount lands. Solder paste is available with three different types of flux: rosin flux (R), mildly activated rosin (RMA), and fully activated rosin (RA). The R flux contains no activators and may not have an aggressive enough cleaning action for acceptable solderability. The RMA fluxes are the most generally used for surface mount applications. The RA fluxes are used only when a more potent cleaner is required. They should be avoided whenever possible, however, since the characteristics of surface-mounting techniques prevent thorough flux cleaning under and around components.

The metal content of the solder paste determines the size of the solder fillets that will form about the component leads. Solder pastes have different melting temperatures. For the soldering of plastic surface-mounted packages, a lower-melting-point paste should be used compared to that used for ceramic packages. If higher shear and tensile strength are required at elevated temperatures, tin-antimony alloy solders are used. Tin–lead alloy solders are not compatible with gold-plated terminations. For this application, indium–lead alloys are recommended. Compatibility with silver-plated leads requires a tin–lead solder containing 2% silver. The most widely used solder pastes for mounting plastic components to FR-4-type pc boards are those having either 63% tin and 37% lead (eutectic) with a melting point of 360°F (183°C) or 62% tin, 36% lead, and 2% silver, which has a melting point of 355°F (179°C).

The rheological properties of the solder paste are controlled by the gelling or thickening agents and secondary solvents. The solvent dissolves the flux and controls the tackiness of the paste as a result of its evaporation rate. The degree of tackiness is an important factor in the selection of paste since it holds the SMCs in place during placement, baking, and handling before reflow soldering.

The viscosity of solder paste changes when agitated by syringe dispensing, screening, or stenciling. This is due to the fact that these pastes are *thixotropic* fluids; that is, they become more fluid when worked and more viscous when allowed to rest. The recommended viscosity of solder pastes as related to their method of application is as follows:

*Syringe dispensing:* 300,000 to 400,000 centipoise (cP)
*Screening:* 450,000 to 550,000 cP
*Stenciling:* 550,000 to 750,000 cP

These viscosity measurements may be taken with a Brookfield viscometer using a TF spindle at 5 revolutions per minute.

The term *slump* refers to the ability of the paste to spread, yet stay within the bounds of footprints after it has been deposited. The degree of slump is determined mainly by the percentage of metal in the paste. Sample test boards with land patterns are used to evaluate how far the paste spreads after it is applied. This is to ensure that it does not spread onto adjacent lands before or after component assembly.

The working life of solder paste is defined as the length of time from when it is removed from the container and goes through all of the fabrication processes—its application, SMC placement, and baking—to the solder reflow process. The degree of tackiness determines if the working life of the paste has elapsed.

When pc boards with SMT components and devices are soldered by the vapor phase or the infrared reflow process, solder paste often throws small spherical particles of solder referred to as *solder balls.* These can measure 2 to 5 mils in diameter. Since they usually end up on the insulated surface around footprints, they are easily dislodged and create a potential short-circuit problem.

Solder balls form either as small particles that float away with the melting flux before the larger mass of paste melts or as heavily oxidized particles that cannot be removed by the flux and activators. These particles are pushed aside as the molten solder melts and flows about the connection.

To test for paste acceptability in terms of solder ball formation, several 6- to 10-mil-thick paste patterns are screened onto a suitable sample of nonmetallic material, such as FR-4 pc board or ceramic—or frosted glass. The test panel is then put through a solder reflow process and the paste examined. It is acceptable if no solder balls are generated and the solder deposits are large shiny spheres having smooth surfaces. A few particles having diameters of 5 mils or less are acceptable. Anything larger, or a halo or clusters of particles, is unacceptable. Some of the methods used to eliminate the formation of solder balls are better control of the amount of paste applied and using fresh paste with the most active flux allowable.

The application of the solder paste to the lands by any method must be checked for coverage and for registration. A number of acceptable and unacceptable configurations of paste deposits on land patterns are shown in Fig. 18.11. Figures 18.11a through d are acceptable, with part (a) preferred. Those labeled as unacceptable (Figs. 18.11e through h) would result in short circuits or unreliable solder joints. Soldering of pc boards is discussed in Sec. 18.5.

Different types of bench-top SMT stencil/screen equipment from OK International and Automated Production Systems, Inc., are shown in Figs. 18.12 and 18.13, respectively. OK International's manual SMT Screen/Stencil Printer, shown in Fig. 18.12, is intended for batch, production, or prototyping applications. It uses frameless stencils and framed stencils and screens as well as commercial screens. The printer is designed for precise fine-pitch registration through X, Y, Z, and Theta adjustments and assurance of both volumetric and geometric deposit repeatability. This type of printer is available in either a maximum print size of 6" × 10" or 10" × 14" and comes with a squeegee and adjusting tools.

Automated Production Systems' bench-top Stencil/Screen Printers, shown in Fig. 18.13, are designed for low- to medium-volume production. They have X, Y, Z, and true centerpoint, Theta adjustment control, and an independent 4-point z-axis adjustment feature for precise alignment and consistent solder paste deposition for SMT. Printers are available having a maximum print area of 12" × 15" (Fig. 18.13a) and 19" × 19" (Fig. 18.13b). Both sizes of printer come with a squeegee, holder, and tubular frame. The larger

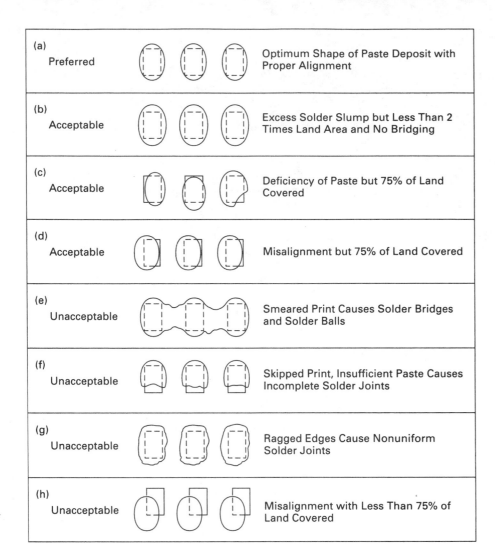

| (a) Preferred | | | | Optimum Shape of Paste Deposit with Proper Alignment |
| (b) Acceptable | | | | Excess Solder Slump but Less Than 2 Times Land Area and No Bridging |
| (c) Acceptable | | | | Deficiency of Paste but 75% of Land Covered |
| (d) Acceptable | | | | Misalignment but 75% of Land Covered |
| (e) Unacceptable | | | | Smeared Print Causes Solder Bridges and Solder Balls |
| (f) Unacceptable | | | | Skipped Print, Insufficient Paste Causes Incomplete Solder Joints |
| (g) Unacceptable | | | | Ragged Edges Cause Nonuniform Solder Joints |
| (h) Unacceptable | | | | Misalignment with Less Than 75% of Land Covered |

**Figure 18.11** Acceptable and unacceptable solder paste deposits.

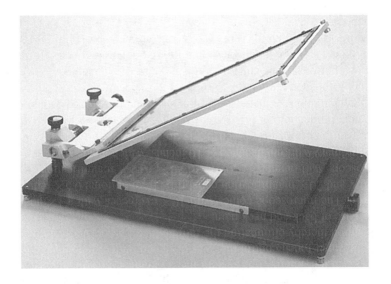

**Figure 18.12** SMT Screen/Stencil Printer. *Courtesy of OK International, Yonkers, NY.*

 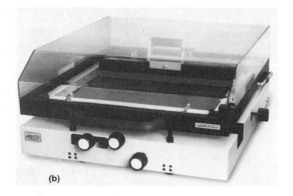

**Figure 18.13** Gold-Print™ Manual Stencil/Screen Printer: (a) Model SPR20: (b) Model SPR30. *Courtesy of Automated Production Systems, Inc., Huntingdon Valley, PA.*

printer will accommodate any tubular frame or cast frame stencil up to 20" × 20" and has a squeegee guide control to obtain precise print strokes.

## 18.4 ADHESIVES

The two most common types of adhesives used for SMCs are *epoxies* and *acrylics.* Unlike thermoplastic adhesives, which harden and soften with temperature cycles, these adhesives, once hardened, will not soften or lose their bond strength with elevated temperatures. Because epoxies and acrylics are classified as thermosetting adhesives, they will withstand the heat of the solder wave and will retain the components tightly to the pc board as it passes over the wave.

To be most effective, the adhesive selected must be properly applied and cured. In addition, the adhesive must have specific precure, cure, and postcure properties that are compatible with SMT. The adhesive must also have sufficient *green strength,* which is similar to the tackiness of solder paste, to hold the component in place before curing. It should also have the proper viscosity to fill the gap between the component and the board without spreading onto the footprints.

For most applications, one-part adhesives are preferred over the two-part (mixing) type, which requires more time to prepare and has a short working life since it begins to harden when the two parts are mixed. Even though one-part adhesives have a shorter shelf life, refrigeration extends this life.

An adhesive that cures in the shortest time at lower temperatures is preferred, provided that it possesses sufficient bond strength. Greater bond strength results from elevated temperatures with the same curing time. However, the temperature must not be so high as to adversely affect heat-sensitive components or deform the board.

The adhesive should have minimum shrinkage properties when cured to prevent excessive stress on component bodies. Curing time should not be accelerated, since voids may occur, which would decrease the effectiveness of the adhesive. If outgassing occurs during the cure process, flux may become entrapped from the solder paste, which will result in difficult cleaning problems.

The curing of adhesives is accomplished in infrared (IR) ovens for epoxies, and with a combination of UV light and heat for acrylics. The thermal-cured epoxies are popular since the IR oven can also be used for reflow soldering without requiring the expense for a UV light system. Acrylic adhesives have a faster curing time, however, and are preferred for high-volume production.

Acrylic adhesive must extend beyond the component so that the UV light will strike it and begin the chain reaction of curing. If this polymerization does not occur, heating alone will allow uncured adhesive under the component, resulting in voids and entrapped flux during the soldering process. Ultraviolet thermal systems use a 2-kilowatt lamp set at a distance of about 4 inches from the board. The adhesive is exposed for about 15 seconds, and the full cure will result in less than 2 minutes in an IR oven.

Epoxies have a curing time of about 5 minutes when the temperature is gradually increased to about 305°F (150°C). The board is then allowed to cool slowly to room temperature for another 10 minutes. A typical thermal profile curve of epoxy adhesive cure in an IR oven is shown in Fig. 18.14. (The procedure for obtaining a thermal profile is discussed in Chapter 17.)

After the reflow soldering process has been completed, the adhesive no longer serves a useful function. Its postcure properties are essential, however, in that it must not absorb moisture and must remain nonconductive and noncorrosive.

The application of adhesives is accomplished by three common methods: screening, pin transfer, and syringing. Syringing is generally preferred because, in many instances, there are components on the board before the adhesive is to be applied. In these situations, screen and pin transfer applications cannot be used. Syringing forces the adhesive through an orifice by electric, hydraulic, or pneumatic pressure. This system, although not as fast as other methods of application, controls the rate of adhesive flow very accurately to ensure uniformity of deposit.

The pin transfer method employs a grid of pins that are dipped simultaneously into the adhesive. This array of pins is then automatically positioned and lowered to the board. The adhesive has a greater affinity for the nonmetallic substrate than for the metal pins. Gravity causes a uniform amount of adhesive to be deposited from the pins.

Screening is done using the same techniques as the silk-screen process described in Chapter 10. The application of surface-mount adhesives, however, requires a stainless steel screen and the appropriate mesh number. As in the use of any adhesive, the manufacturer's instructions should be carefully observed.

The correct application of adhesive for large and small components is shown in Fig. 18.15 together with the correct dot (adhesive deposit) sizes for various land sizes. The deposit of adhesives for a large component after the application of solder paste is shown in Fig. 18.15a. Note that there are two adhesive dots on the substrate between the lands and a

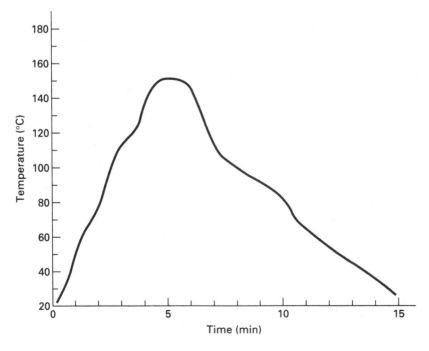

**Figure 18.14** Thermal profile of epoxy cure in an IR oven.

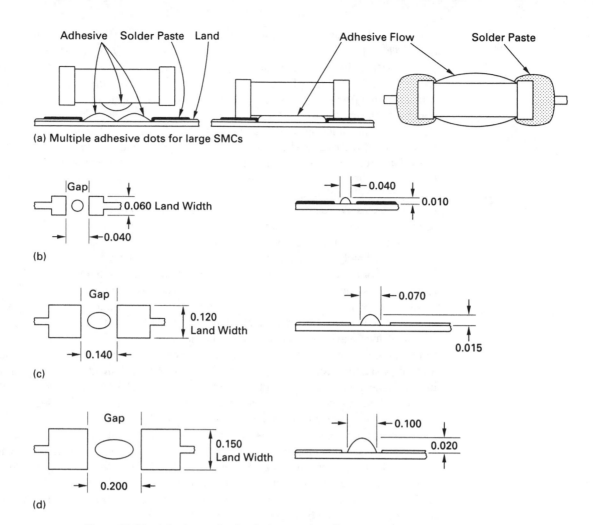

**Figure 18.15** Adhesive application for large and small components.

smaller dot on the bottom of the component. This will ensure that the space under the component is filled with adhesive. However, excess adhesive should be avoided since it will interfere with proper solder joint formation. Approximate adhesive dot sizes used for typical lands are shown in Figs. 18.15b, c, and d.

## 18.5 SOLDERING FOR SMT

The soldering process in SMT is termed *reflow soldering.* The joining of metal surfaces of the lands with the terminations of the components and devices is completed by the mass heating of the predeposited solder paste. When the paste melts, solder fillets are formed between the lands and the terminations. Automated volume production methods include *wave soldering, vapor-phase soldering,* and *infrared reflow.* Laser soldering is another automated process that is faster than hand soldering but not as fast as the other production processes. Because laser soldering is a relatively new process, it is more expensive and the technical problems involved with solder defects and component damage have not all been resolved. For these reasons, laser soldering is confined to specialized applications and is viewed as a complement to rather than a replacement of the other soldering processes.

Wave soldering for pc boards has been discussed in Sec. 17.7. Some modifications of the process are required for boards that incorporate both conventional and surface-mounted components and devices. These modifications require special features to accommodate higher board densities as well as components being mounted on the bottom side of the board. These features, shown in Fig. 18.16, are primarily *air knives* and a *dual solder wave.*

The first air knife overcomes the problem of not being able to smooth the applied flux with a brush roller, due to the surface-mounted components on the bottom side of the board. The air knife effectively removes excess flux and smooths the flux coating over lands and component terminations.

The dual soldering wave is a required feature because of the components mounted to the bottom of the board. The first wave is turbulent, which ensures sufficient distribution of solder across the board as it passes. The velocity of the solder allows it to penetrate between tightly spaced components and leads. The solder jet is pointed in the direction of the conveyer travel, which results in no solder skips. However, the applied solder is very uneven and excessive, causing bridges and icicles. The second wave is the conventional type used for through-hole components. As the board passes over this smooth wave, excess solder is removed along with any bridges and icicles.

Immediately after the board leaves the second solder wave, it passes over a hot air knife. A jet of hot air, whose temperature is above the melting temperature of solder, is directed at the soldered joints. A properly wetted solder connection will be unaffected by the jet of air since the surface tension of the solder exceeds the force of the air. Any solder that is not bonded, such as bridges, icicles, and excess amounts that were not removed by the second solder wave, is removed by the hot air knife. To prevent the blowing of solder through holes in the board and at the same time remove just the right amount of solder, the air knife's spacing from the wave and its angle to the board, in addition to the air temperature, volume, and velocity, must be accurately controlled. The hot air process also acts as a method of quality control since the superheated air stress-tests each solder fillet and uncovers any land or pad that is not properly soldered.

To maintain and monitor precise temperature control of each phase of the wave soldering process, a temperature profile of the board is required. This is done by placing one or more thermocouples at strategic locations on the board and recording the temperature variations as it passes through the preheat and wave soldering stages. The recommended temperature parameters for preheat rate, temperature differential between preheat and solder wave, time of the board in the wave, and cool-down rate are 2°C/second maximum, 100°C maximum, 3 to 4 seconds, and −5°C/second, respectively. The procedure for obtaining a temperature profile is discussed in Chapter 17. A temperature profile for a pc board as it is passed through a single solder wave is shown in Fig. 17.19.

Vapor-phase and infrared soldering processes are much different from wave soldering, where the source of solder and most of the required heat are provided by the wave. This significant difference for both processes will be described in detail.

For type I assemblies, vapor-phase, infrared, or convection soldering processes are used exclusively, while type II and III assemblies require the use of wave soldering as well. The source of heat for vapor phase comes from the condensation of vapors from a boiling liquid onto the metal terminations, solder paste, and lands. Since heat is also transferred by convection, the work is heated uniformly. Also, the board is protected from heat damage because its temperature cannot exceed that of the boiling point of the liquid. This automatic

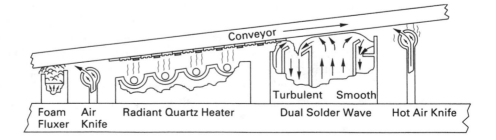

**Figure 18.16** Air knives and dual solder wave reduce solder defects of surface-mounted components.

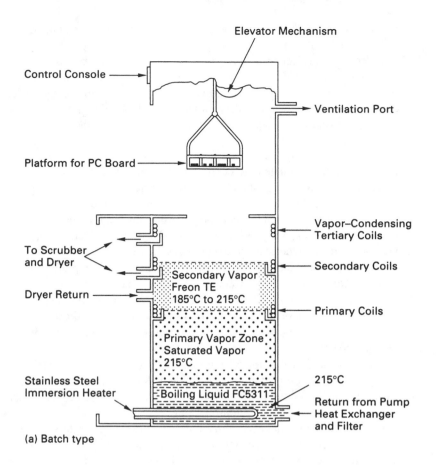

(a) Batch type

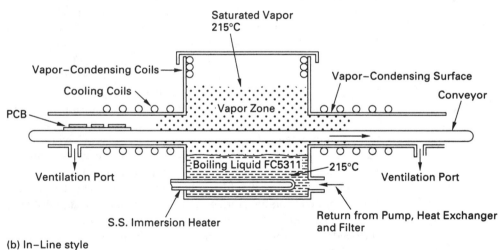

(b) In−Line style

**Figure 18.17** Batch and in-line vapor-phase soldering systems.

vapor-phase soldering system makes use of either *batch* or *in-line* equipment. These are shown in Fig. 18.17. In the batch system shown in Fig. 18.17a, the work is lowered through a secondary vapor layer, usually Freon TE, into the primary vapor, which is an inert per-fluorocarbon such as FC-5311 from the 3M Corporation. The board remains in the primary vapor zone, whose temperature is approximately 419°F (215°C), for 20 to 30 seconds. The board is then raised into the secondary vapor zone and allowed to drain off the primary fluid (30 seconds to 1 minute). The temperature of this zone is about 225°F (107°C). Because of the secondary vapor zone, the cooling coils at the top of the chamber, and the high density of both vapors, the loss of expensive perfluorocarbon is kept to a minimum. The high density also acts to reject air and moisture from entering the process.

The in-line system, shown in Fig. 18.17b, does not have a secondary vapor zone. The board is passed through a saturated vapor zone in 20 to 30 seconds. There are vapor condensing surfaces at the top of the chamber as well as at the input and output locations. With an optional vapor recovery system added to the machine, the vapor loss can approach that of the batch system.

The infrared reflow soldering process passes the boards along a conveyor between top- and bottom-mounted IR quartz iodide lamps having ceramic reflectors for higher wavelength or IR panels called area source emitters. The basic arrangement of this system is shown in Fig. 18.18. Both natural and forced-air convection are used on these systems. Test panels should be run through the process to ensure that the emitted wavelength is correct, so that nonuniform heating of the board will be avoided. Also, the optimum air convection needs to be determined to cure the solder paste properly and prevent the formation of solder balls or splatter. Infrared soldering is the generally preferred process because it is more controllable for higher yields and lower operating costs.

Some automated soldering systems provide a forced cooling stage as the final process in the soldering operation. This has been shown to improve the strength of solder joints. A benchtop SMT reflow oven and a full convection batch oven are shown in Figs. 18.18 and 18.19, respectively, with temperature profiles.

Automated Production Systems' Gold-Flow™ Benchtop SMT Reflow Oven, shown in Fig. 18.18a, is a forced air convection and IR oven. It has an easy-lift clamshell design with a 12" wide stainless steel conveyor and chamber. The heated tunnel is 24" long. The oven has viewing ports and interior lighting. This oven features computer control. The heating elements, conveyor speed, forced air, Cyclonic™ generators, and cooling fans are all programmable. It has 100 menu-profile storage and a 7-day programmable timer for automated machine start-up. By using the included real-time temperature profiler port and a thermocouple attached to the pc board, the reflow temperature profile is displayed graphically, as shown in Fig. 18.18b, as the board travels through the oven. The oven also features SPC (statistical process control) fault monitoring and reporting and battery backup. The model GF-12 oven has 3 top and 2 bottom heating zones (1 Cyclonic™), while the model GF-12A has 3 top and 3 bottom heating zones (3 Cyclonics™) as shown in Fig. 18.18c. The oven operates at 220 VAC 50/60 Hz, single phase, 25 amps.

OK International's Full Convection Batch Oven, shown in Fig. 18.19a, is designed for prototype and low-volume production in laboratory environments to satisfy precise heating requirements. The oven has an 11" × 11" glass viewing window, a 12" × 12" convection reflow area, a lock to prevent unwanted profile changes, and a 3" × 6" backlit LCD digital display that shows either temperature and times or a temperature profile such as that shown in Fig. 18.19b. It is fully grounded and ESD safe. This computer-controlled oven features closed-loop monitoring to ensure absolute repeatability of temperature profiles. Ninety-nine profiles can be stored with quick profile changes possible by design. Independent control of both front and rear convection heaters allows easy profiling of boards having varying mass. Also, both front and rear heaters can be profiled for the Preheat 1, Preheat 2, Reflow, and Cooldown "time" zones. One model of this oven also has a "nitrogen input."

For small soldering tasks, hand soldering is commonly employed. It is also used as a means of reworking or correcting soldering defects resulting from an automated process. The power of the soldering iron used on SMCs, such as ceramic resistors, capacitors, and diodes, should not exceed 22 watts. These components could easily be damaged if subjected to extreme heat. A pointed soldering tip is preferred and its temperature should not exceed 700°F (370°C). Excessive pressure against connections should be avoided to prevent delamination of lands. In addition, the tip should be in contact with the connection for only 1 to 3 seconds, depending on the type of component.

(a)

Gold-Flow™ Reflow Profile

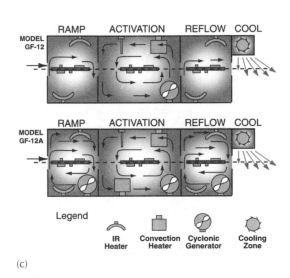

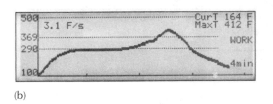

(b)

(c)

**Figure 18.18** Gold-Flow™ Benchtop SMT Reflow Oven: (a) oven with cover open: (b) Gold-Flow™ Reflow Profile; (c) Cyclonic™ System with legend. *Courtesy of Automated Production Systems, Inc., Huntingdon Valley, PA.*

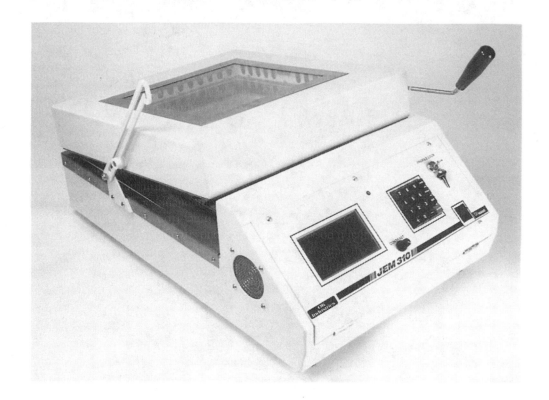

(a)

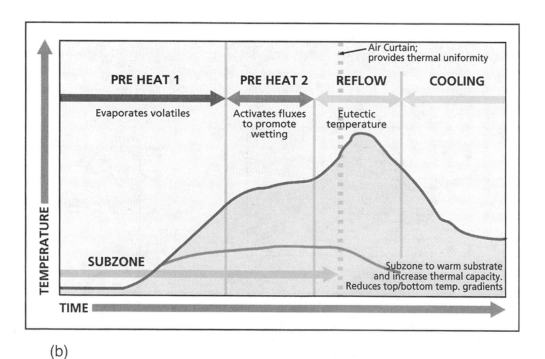

(b)

**Figure 18.19** Full Convection Batch Oven: (a) Convection oven having 12" × 12" reflow area; (b) typical temperature profile. *Courtesy of OK International, Yonkers, NY.*

Because of their small size and light weight, along with the fact that they have no leads inserted through drilled holes, SMCs tend to move off of the land when hand soldering. This problem can be overcome by placing a droplet of adhesive onto the substrate between lands. Using solder that has a diameter of 0.031 inch is ideal for transferring the adhesive to the board. Any cyanoacrylate adhesive will prove effective. These adhesives are more commonly known by their trade names, such as Crazy Glue, Instant Glue, or Super Glue. It is important that the deposit of glue be accurately placed and that the amount be controlled so that it does not spread onto the lands when the component is pressed into place. This would inhibit the formation of a proper solder connection. Because this adhesive is fast acting and very strong, care must also be taken to ensure that the component is properly placed, since it is virtually impossible to move it after the adhesive has set. Once the adhesive has cured, soldering of the connection may begin. Solder having a maximum diameter of 0.031 inch is shaped to a fine point for this delicate work. The time to solder each connection should be no more than 2 to 3 seconds for chip resistors and 1 to 2 seconds for delicate IC leads.

For easier assembly and closer inspection of printed circuit boards, an inspection system such as that made by Luxo Corporation, shown in Fig. 18.20, is employed. This "MicroLux" microscope system has a balanced positioning arm that has a 21-inch reach with finger touch control that was designed with ergonomics in mind. It can be set in any position thereby eliminating the fatigue factor associated with bending over, which can be a cause of back and neck pain.

The microscope comes with $10\times$ eyepieces with a $1\times$ and $2\times$ objective that gives $10\times$ and $20\times$ magnification capability. If an optional $0.5\times$ reducing lens is used, the microscope will range $5\times$, $10\times$, and $20\times$. At $5\times$ magnification, the microscope has an extraordinary working distance of 10 inches, which makes it ideal for inspection and rework. Built-in dual 5-watt halogen lights are adjustable, having flexible necks to deliver crisp white illumination where desired. This system is available with either a horizontal mounting clamp or weighted base and when used with illumination, complies with MIL-STD-2000, DOD-STD-2000, WS-6536E, and MIL-S-45743E.

Because of the small size of the parts and the requirement for precision craftsmanship, a magnifying lamp, such as that shown in Fig. 18.21, is also useful.

Flowcharts for type I, II, and III assembly and processes are shown in Fig. 18.22.

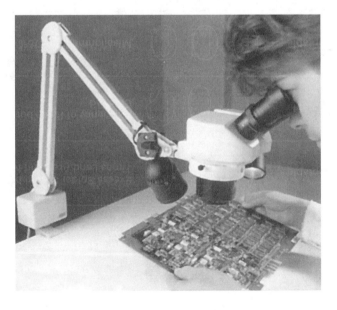

**Figure 18.20** Luxo model 18000-MicroLux Microscope system. *Courtesy of Contact East, Inc., N. Andover, MA.*

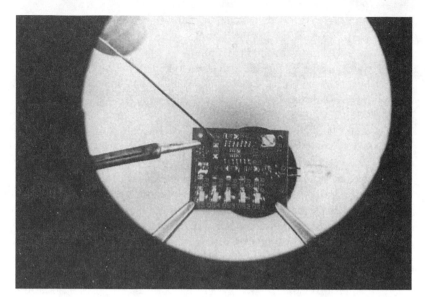

**Figure 18.21**
Soldering and rework of small surface mount technology devices are best accomplished under a magnifying lamp.

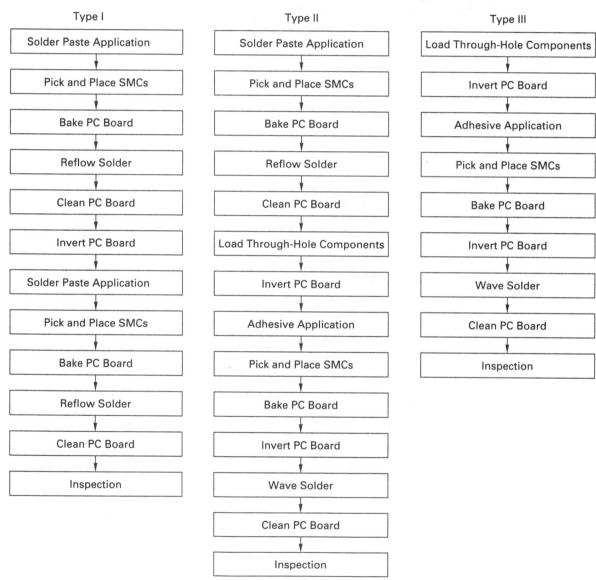

| Type I | Type II | Type III |
|---|---|---|
| Solder Paste Application | Solder Paste Application | Load Through-Hole Components |
| Pick and Place SMCs | Pick and Place SMCs | Invert PC Board |
| Bake PC Board | Bake PC Board | Adhesive Application |
| Reflow Solder | Reflow Solder | Pick and Place SMCs |
| Clean PC Board | Clean PC Board | Bake PC Board |
| Invert PC Board | Load Through-Hole Components | Invert PC Board |
| Solder Paste Application | Invert PC Board | Wave Solder |
| Pick and Place SMCs | Adhesive Application | Clean PC Board |
| Bake PC Board | Pick and Place SMCs | Inspection |
| Reflow Solder | Bake PC Board | |
| Clean PC Board | Invert PC Board | |
| Inspection | Wave Solder | |
| | Clean PC Board | |
| | Inspection | |

**Figure 18.22**   Flowcharts of process sequence for SMT type I, II, and III pc boards.

378

## 18.6 SOLDERING AND DE-SOLDERING SYSTEMS

Seldom is the yield 100% where there are no components damaged or solder joint defects after the assembly and soldering process. For SMT, rework and repair are less damaging to the components and the pc board than for conventional through-hole boards. Also, less operator skill is required for working with surface-mounted boards.

De-soldering equipment for SMT falls into two general categories: *soldering iron conductive tip configurations* and *hot air systems*. Conductive tips are much less expensive than hot air systems.

OK International's SA-400 series variable temperature (600° to 800°F) solder and de-solder station is shown in Fig. 18.23a. Static dissipative construction and closed-loop RTD (Resistance Temperature Detector) sensor feedback with zero voltage switching meet all MIL standards for ESD safety. This conductive soldering/de-soldering system features three interchangeable handpieces from which to choose. The Macro iron, the Micro iron, and the SMTweezers® are suitable for through-hole, mixed technology, and SMT applications. The Macro iron has a 90-watt element and is used for standard and high thermal demand soldering. The Micro iron has a 40-watt element and is designed for SMT work and precision soldering. The SMTweezers® have 2 × 40-watt elements and are used for SMT removal. Fig. 18.23b shows blade tips being employed. Figure 18.23c shows SMTweezers® with chip removal tips. One of a variety of Micro soldering iron tips (i.e., conical, angled conical, and double-flat) are shown in Fig. 18.23d. Figure 18.23e shows a Macro soldering iron being used for solder removal.

In using conductive tips to remove SMCs, all of the solder joints must be melted simultaneously with uniform heat. The component is then removed: first, by a slight twisting, and then by lifting off while the solder is still molten. Twisting is important to prevent damage to the leads, especially if the component is held with adhesive. As the tips heat the solder joints, the adhesive is raised to its *glass transition* temperature ($T_g$). This is the temperature at which the adhesive changes from its hardened state to a viscous condition. The twisting motion shears the adhesive. If no adhesive had been used, the twisting is still necessary, since it keeps the component flush to the surface of the board as the leads are separated from the molten solder and prevents deforming them.

Some of the common tip styles used for de-soldering SMCs are shown in Fig. 18.24. Figure 18.24a shows a tip style designed for leadless chip packages such as ceramic resistors, capacitors, and MELFs. They are also available in widths that will de-solder SOT packages. The tip style shown in Fig. 18.24b is used on SOIC packages, and the arrangement in Fig. 18.24c is designed for de-soldering flatpacks and LCCC and PLCC packages. The tip style selected must have the correct dimensions to span all of the leads of the specific package size.

When using conductive tips for de-soldering, certain precautions must be taken to avoid damage. A liquid RMA flux should first be applied to the solder joints and then the tips placed against the connections with minimum pressure. The length of time for the solder to become molten should not be more than 2 seconds. The tweezer action of the tips, along with their being wetted with molten solder, will easily allow the component to be removed from the board. It should then be placed onto a damp sponge and the lands cleaned with a suitable solvent such as isopropyl alcohol. The tips should also be wiped on the sponge to remove excess solder and flux residue.

Figure 18.25a shows PACE's PRC 1500 Complete Conductive and Convective Rework System with MBT SMT/Thru-Hole Assembly and Repair System and ThermoFlo™ Programmable Hot Air Station for BGA/SMD Assembly and Rework. The MBT 250A is a 3-channel system. It does not require shop air and operates on 115 VAC 50/60 Hz. A microprocessor-controlled power supply allows handpieces connected to each channel to be operated simultaneously at three different temperatures up to a total power of 180 watts. An auto-tip temperature offset compensation allows setting and displaying true tip temperature of each channel in °C/°F between 100°F and 900°F. A 1.2-second minimum vacuum on-time (Auto-Snap-Vac®), when the handpiece switch is depressed, prevents

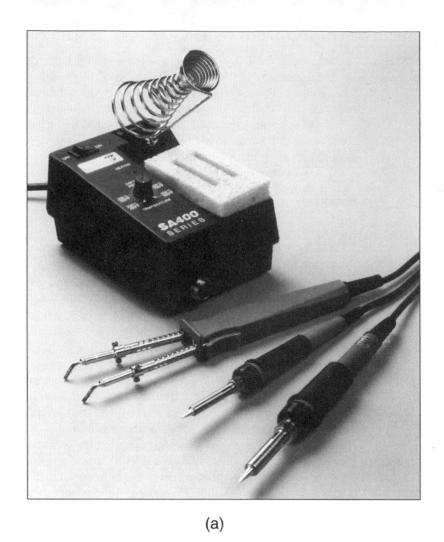

(a)

(b)       (c)       (d)       (e)

**Figure 18.23** Variable solder and de-solder station: (a) single handpiece system; (b) blade set; (c) SMTweezers®; (d) Micro soldering iron with conical tip; (e) Macro soldering iron with solder removal tip. *Courtesy of OK International, Yonkers, NY.*

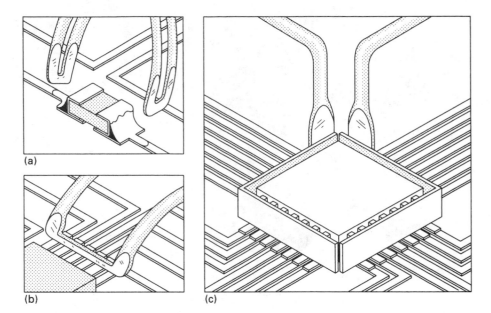

**Figure 18.24** Desoldering tips.

clogging due to a short airflow duration. The system meets MIL STD 2000 and all EOS/ESD requirements. It extends tip life and saves energy with setback and auto-off features. The setback feature reduces tip temperature to 350°F after a preselected 10–90 minutes and the auto-off eliminates the possibility of leaving the unit on overnight. This unit also has a nonvolatile memory that retains previous settings and calibrations even after the power has been removed. As shown in Fig. 18.25 (b through f), the system comes with a de-soldering piece (SX70 Sodr-X-Tractor) for de-soldering through-hole components, a handpiece (TP65 ThermoPik) having a vacuum pickup for reflow and removal of PQFPs, a heavy-duty soldering iron (SP2A) for small SMDs and heavy through-hole soldering, a high-power tweezer (TT65 ThermoTweez) for SMD removal, and an optional focused hot-gas reflow handpiece (TJ70 Mini-Thermojet).

The ThermoFlo® programmable hot-air station features a self-contained air source with a static-safe handle for removing sensitive SMDs. The built-in self-adjusting vacuum pickup adjusts to the height of components. A programmable power supply offers a selection of manual (simple operations), timed (user-specified cycle times), and program (full process control) modes of operation. This unit can store up to 80 user-defined profiles, has password lockout that prevents unauthorized changes in any mode, comes with an advanced 700-watt static safe handpiece for both small and large components, and over 25 standard nozzles. The nozzles are easily installed or replaced with a simple twisting action.

PACE's patented Mini-Wave tip operates with the SP-2A solder pen and is illustrated in Fig. 18.26a. This type of tip is designed especially for rapid, high-reliability soldering of

**Figure 18.25** Conductive and convective rework systems: (a) rework systems with handpieces.

(a)

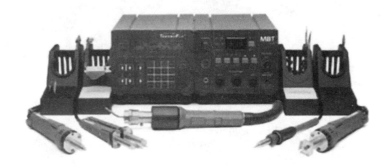

**(b)**

**Applications**

Thru hole. ThermoDrive desoldering, Flo desoldering, PCMCIA fine pitch TSOP and small PQFP removal

**(c)**

Flat pack removals

**(d)**

Micro, general purpose, heavy duty and extended reach soldering, SMD removals, regular and fine pitch Mini-Wave SMD installation

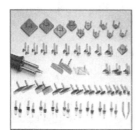

**(e)**

SMD removals: Chips, SOICs, connectors, flat packs, PLCCs, etc.

**(f)**

SMD hot air component installations

**Figure 18.25 continued** (b) SX70 Sodr-X-Tractor; (c) TP65 ThermoPik; (d) SP2A soldering iron; (e) TT65 ThermoTweez; (f) TJ70 Mini-Thermojet. *Courtesy of PACE Inc., Laurel, MD.*

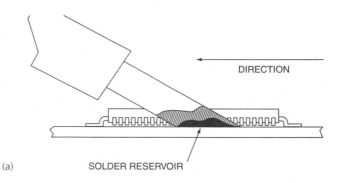

DIRECTION

(a)          SOLDER RESERVOIR

**Figure 18.26** Mini-Wave tip design: (a) solder wave reservoir; (b) standard tip; (c) fine pitch tip; (d) angled tip. *Courtesy of PACE, Inc., Laurel, MD.*

(b)

(c)

(d)

all types of surface-mounted component geometrics. The tip has a concave elliptical well that holds a miniature reservoir of solder. This unique tip design assures safety since low temperature is used and only the molten solder wave, not the tip, touches the leads and lands. Perfect solder joints are formed every time as the reservoir of solder is drawn across a row of component leads. Just the right amount of heat is delivered and the wave action deposits the right amount of solder without the formation of solder bridges. A variety of Mini-Wave tip sizes including standard, fine pitch, and angled tips, shown in Figures 18.26b, c, and d, makes the Mini-Wave process possible for fine-pitch components and high-density boards.

Figure 18.27 shows PACE's SensaTemp® PQFP (Plastic Quad Flat Pack) and BGA (Ball Grid Array) Rework Station. This unit requires shop air (80 psi), is ESD safe, and meets MIL STD 2000. This system comes with a dual thermopick handpiece for removing PQFPs and BGAs, shown in Figs. 18.26b and c. The handpiece has a self-adjusting vacuum pick, vacuum switch, and cushion grips. The SensaTemp® heat control affords high capacity, low temperature de-soldering, including heavy multilayer boards. This unit also features Auto-Snap Vac®, which is a quick-rise vacuum having an automatic minimum vacuum on-time of 1.2 seconds. It operates on 115 VAC 50/60 Hz and requires shop air from a regulated 80 psi (90 psi maximum) source.

Hot air systems are available for removal and repair of both passive and active devices. As a result of the special designs of these systems, hot air is directed onto the solder joints through a nozzle. This prevents the bodies of the components from overheating. The temperature reached during the de-soldering process will be less than that involved in either the vapor-phase or IR reflow operations. The hot air temperatures reach as high as 400°F (205°C). Two hot-air system nozzle designs are shown in Fig. 18.28. The open wall nozzle shown in Fig. 18.28a is useful for de-soldering gull wing style terminations. When fitted with a precision insert or pilot that fits over the case of the device and a vacuum tube with a suction cup, the system can also be used to remount devices. To remount, a microscope with controls is required to result in perfect lead-to-land registration. Where there is concern for overheating the solder connections of adjacent components that will not be desoldered, the vented nozzle design shown in Fig. 18.28b is employed.

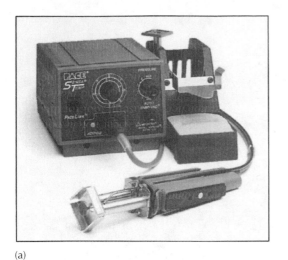
(a)

(b)

(c)

**Figure 18.27** PQFP and BGA Rework Station: (a) rework station with handpiece; (b) removing a PQFP; (c) removing a BGA. *Courtesy of PACE Inc., Laurel, MD.*

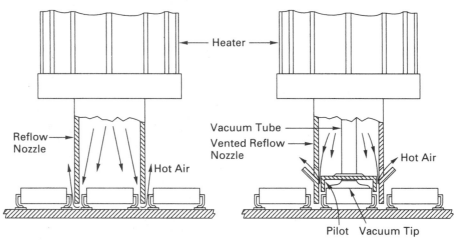

**Figure 18.28** Hot air de-soldering systems.

(a) Open wall nozzle for de-soldering

(b) Vented nozzle for de-soldering and replacement

Many systems also have the feature of heating the bottom surface of a board in preparation for de-soldering components on the top surface. The purpose of this is to reduce thermal shock to the board. The temperature used, however, should not be so excessive as to adversely affect bottom surface solder joints or components. Heating the board for 30 to 40 minutes at a temperature of 200°F (94°C) has been found to eliminate thermal shock problems, while not causing any damage to joints or components.

OK International's Hot Air Rework Station, shown in Fig. 18.29a, has a heater table with heater control. This focused convection SMT rework system with built-in vacuum pick-up has a power base and self-contained vacuum and air pumps that regulate temperature and airflow. It operates on 115VAC 50/60 Hz, and is fully grounded and ESD safe. Temperature can be controlled from 392°F to 932°F and air flow 3 to 20 l/min. Heater timing is adjustable from 0 to 180 seconds. The timing used in conjunction with the air-flow meter on the front panel affords precise repetitive process control.

The system is designed for rework of discretes, PLCC, QFPs, and SOICs without damage to the pc board or component leads. The 300-watt handpiece has power and vacuum lines

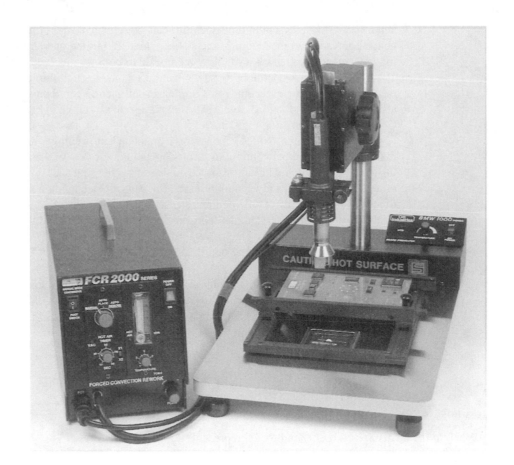

(a)

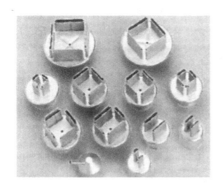

(b)

(c)

**Figure 18.29** Hot air rework station: (a) tool stand and board holder with pre-heater table; (b) hot air nozzle kit; (c) chip removal. *Courtesy of OK International, Yonkers, NY.*

and a spring-loaded vacuum tube that eliminates mechanical stress on contact with components. The vacuum tube length is adjustable, which allows the operator better viewing during placement or to retract it to allow the nozzle to be fully involved with the component. The system includes a straight nozzle, a handpiece that has a multiple nozzle holder, a 10-piece silicon rubber cup set, and tools for nozzle and suction cup removal. The manufacturer also supplies an instruction manual. The hot air nozzle kit shown in Fig. 18.29b is available. Chip removal is shown in Fig. 18.29c.

There are three modes of operation: manual, auto-remove, and auto-replace. The manual mode has either continuous or on-demand hot-air flow and on-demand vacuum using a dual foot switch. The auto-remove mode times and sequences functions of hot air, vacuum, hot air shut-off, and vacuum release. The auto-replace mode times and sequences the functions of vacuum, hot air, vacuum release, and hot air shut-off.

The tool stand/board holder features Z axis control of the handpiece, and affords free-hand operation to support precise repetitive rework operations. The board holder is adjustable and will accommodate boards up to 12" × 18". It also has Teflon feet for ease of X and Y positioning.

After assembling components and hardware to the printed circuit boards, the next phase of a project's development is to construct the sheet metal chassis elements. In the next unit, Unit 5, information will be given on sheet metal fabrication techniques. Topics include metal shearing, layout, drilling, filing, bending, and finishing.

# EXERCISES

## A.  Questions

18.1    What do the terms *SOIC, MELF,* and *PLCC* represent?

18.2    Describe the soldering defect called *tombstoning.* How is it avoided?

18.3    How are conventional and SMT components oriented on type II assemblies?

18.4    What factors govern the selection of solder paste for a specific application? Briefly discuss each factor.

18.5    What are *rheological properties?*

18.6    What are some of the methods used to prevent the formation of solder balls during reflow soldering?

18.7    What is the difference between *pot life* and *shelf life?*

18.8    What is the advantage to the use of epoxy and acrylic adhesives when compared to the thermoplastic type?

18.9    Why does a wave soldering system for type II assemblies use a dual solder wave?

18.10   Why is it recommended to heat the bottom surface of pc boards prior to de-soldering SMCs?

18.11   What adjustments are available to ensure volumetric and geometric repeatability of solder paste deposits?

18.12   What magnifications are possible when using the 0.5× reducing lens with the LUXO model 18000 MicroLux microscope?

18.13   How are real-time temperature profiles obtained when using benchtop convection and IR ovens?

18.14   To what does *ESD protected* refer?

18.15   Describe manual, auto-remove, and auto-replace modes.

## B.  True or False

Circle *T* if the statement is correct, or *F* if any part of the statement is false.

18.1    Resistors are much thinner than capacitors and require a longer land to prevent misalignment.                                                                                    **T    F**

18.2    Lands for MELFs are partially slotted to prevent these devices from rolling out of position when adhesive is not used.                                                        **T    F**

| 18.3 | Type III assemblies have SMCs exclusively on one or both sides of the board. | **T** | **F** |
|------|---|---|---|
| 18.4 | Adhesives must be used to hold components and devices to the bottom side of the board for any soldering process used. | **T** | **F** |
| 18.5 | RMA flux is the most active of the three types used in solder paste. | **T** | **F** |
| 18.6 | Solder paste to be deposited by stenciling requires a higher viscosity than that applied by screening. | **T** | **F** |
| 18.7 | When solder paste is applied, a minimum of 75% of the land must be covered. | **T** | **F** |
| 18.8 | Epoxies and acrylics are classified as thermoplastic adhesives. | **T** | **F** |
| 18.9 | Acrylic adhesives require both UV light and heat for curing. | **T** | **F** |
| 18.10 | The maximum soldering iron tip temperature for use on surface-mounted components is 750°F. | **T** | **F** |
| 18.11 | It is possible to dispense solder paste, adhesive, and potting compounds with pick and place systems. | **T** | **F** |
| 18.12 | True center point adjustment is done in the *z* axis. | **T** | **F** |
| 18.13 | A thermocouple must be attached to the pc board in order to obtain a temperature profile. | **T** | **F** |
| 18.14 | A 0.5× reducing lens is used to extend the range of a microscope. | **T** | **F** |
| 18.15 | Conductive rework units primarily use forced hot air to solder and de-solder components. | **T** | **F** |

## C. Multiple Choice

Circle the correct answer for each statement.

18.1    Manually assembled SMCs should be handled by their (*bodies, terminations*).
18.2    Plastic SMCs require solder paste having (*lower, higher*) melting temperatures.
18.3    Epoxy and acrylics are classified as (*thermoplastic, thermosetting*) adhesives.
18.4    Acrylic adhesives can be cured in about (*2 hours, 2 minutes*).
18.5    Solder joints should not remain in a solder wave for more than (*4, 10*) seconds.
18.6    Center point alignment is accomplished by (*X, Y, Z; Theta rotation*) adjustment.
18.7    Zone 2 in an SMT IR oven (*activates fluxes, reflows solder*).
18.8    A 10-inch working distance of 5× magnification is achieved by using a combination of a 0.5× (*objective, reducing*) lens with a 10× eyepiece.
18.9    Removing PLCCs is accomplished by using (*blade, tweezer*) tips.
18.10   Tip life is extended and energy is saved by returning the handpiece tip temperature to 350°F when not in use after a preset time by a(n) (*Auto-off, setback*) feature.

## D. Matching

Match each item in Column A with the most appropriate item in Column B.

| COLUMN A | COLUMN B |
|---|---|
| 1. Air knife | a. Footprint |
| 2. Land pattern | b. Syringe |
| 3. MELFs | c. Flux |
| 4. Acrylic | d. Rheological property |
| 5. Type I assemblies | e. Round bodies |
| 6. RMA | f. SMCs only |
| 7. Slump | g. Adhesive |
| 8. Solder paste | h. Wave solder |
| 9. Theta | i. Real-time |

10. Microscope      j. Quad

11. ESD      k. Tip life

12. Profile      l. Dispense

13. Macro      m. Objective

14. Pick and place      n. Rotation

15. Setback      o. Protection

Unit

# FIVE

# Chassis Fabrication and Metal Finishing Techniques

# 19

# Shearing

*Upon completion of this chapter on sheet metal shearing, the student should be able to*

- Properly and safely use the foot-operated squaring shear, the hand-operated squaring shear, and the hand notcher to accurately shear sheet metal.
- Make the necessary adjustments in these metal cutting shears to obtain precise cuts and notches.

## 19.0 INTRODUCTION

Industrial descriptions of machine tools include sheet metal *shears* in the category of *metal-forming machines.* The action of a shear is primarily to cut pieces of sheet and bar stock (or structural shapes) to desired dimensions. The construction of electronic equipment involves metal-cutting operations. Selecting and properly using the most appropriate tools for metal cutting is essential for quality workmanship. Bulk sheet stock must be cut into blanks of the desired overall dimensions for the chassis (or enclosure elements) before layout work begins.

Two basic types of shears are typically used in electronic fabrication: the *blade shear* and the *notcher.* The vertical drive for shear machines is either manual or power-operated. Some shear machines are also equipped for *reciprocating* vertical motion. Shears are capable of producing a neat, straight cut without edge imperfections.

This chapter considers in detail the alignment and practical operating information for the foot-operated and hand-operated *squaring shear* and the *hand notcher.*

*CAUTION:* Prior to performing any of the operations described in this chapter, approved safety glasses with sideshields in compliance with ANSI Z87.1-1989 must be worn. When using any of the shearing equipment, do not stand inside hazard tapes that may be bonded to the floor around the machine to avoid personal injury from a moveable foot tread or operating handle. Keep hands and fingers clear of hold-down bars and cutting blades. Abrasion-resistant gloves should be worn to protect your hands from sharp corners and edges of sheet metal when moving a large piece onto the shear. Also, the section on safety outlined at the beginning of the book should be reviewed for more specific information.

## 19.1 FOOT-OPERATED SQUARING SHEAR

The foot-operated squaring shear, shown in Fig. 19.1, is used for (1) *cutting* large pieces of sheet metal into *blanks,* (2) *trimming* stock to obtain a finish cut or reference edge, and (3) *squaring.* The basic parts of this shear, shown in Fig. 19.2, are two *cutting blades, bed, foot treadle,* and four *alignment gauges.*

Accurate and professional chassis layout begins with blank stock that is not only cut to the correct overall dimensions, but also is cut squarely. Before the 90-degree align-

**Figure 19.1** Foot-operated squaring shear. Proper positioning of hands and feet is illustrated together with the correct implementation of the back stop.

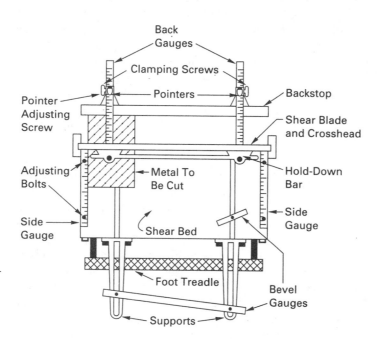

**Figure 19.2** Parts identification of foot-operated shear and correct metal positioning for shearing (top view).

ment of the shear blades with the side gauges is tested, the following inspections of the shear should be made: (1) uniform hold-down bar pressure, (2) proper shear blade clearance, and (3) even tension when the foot treadle is depressed. Each of these tests is described here.

A test piece of scrap metal of uniform thickness is used to test for uniform hold-down bar pressure. With the test piece under the hold-down bar and the foot treadle fully depressed, proper adjustment is made via two adjusting screws located on the front side of the hold-down bar. The screws are set so that the work is held firmly, with care taken so that excessive pressure does not deform the metal.

Proper adjustment of shear blade clearance between the upper and lower blades is a uniform gap, typically from three to five thousandths of an inch. Adjustment screws for varying clearance to this distance are located on the front end of the shear frame. To ensure a clean cut, there must be uniform translation of motion between the foot treadle and the upper cutting blade. This tension adjustment is made by turning the two *turnbuckles* that link the *foot treadle* and the *crosshead.*

When the foregoing tests and adjustments are completed, the 90-degree alignment of the blades and the side gauges may be tested. To determine this alignment, a *carpenter's square* is placed on the shear bed, with the longer leg flush with the lower cutting blade. The shorter leg of the square is placed against one of the side gauges. For square cuts, the side gauges must be perpendicular to the cutting edge and flush against the square. Any necessary correction can be made by loosening the two *securing bolts* holding the side gauge in place and positioning the gauge so that it rests flush against the leg of the square. When the securing bolts are tightened, the shear is properly aligned for square cuts. *Both* the gauges on the shear bed should be checked for alignment before starting any cutting operations. Once the side gauges are properly adjusted, stock placed against them for cutting will have a right angle between the edge of the stock against the gauge and the cut edge. Although either side gauge may be used, the gauge on that side of the shear where the cutting action *first* occurs is used when cutting small pieces of stock.

Many shears are equipped with *back gauges* and a *stop* for *back shearing,* as shown in Fig. 19.2. Although front shearing is more accurate when using the side gauges, the back gauges are convenient when a large quantity of stock is to be cut to the same dimensions. The back gauges are calibrated in sixteenths of an inch measured from the rear edge of the lower blade. The gauges also serve as the supporting arms of the back stop. To ensure that the back gauge pointers are properly set, a long piece of stock of a known uniform width is inserted between the back stop and the rear edge of the lower blade. The pointers should be readjusted if they do not read the correct dimensions.

The desired depth of cut is easily set by moving the stop along the back gauges to the required dimension using the clamping screws to secure the back stop in place. Metal to be cut is placed on the shear bed and moved under the upper blade until it contacts the back stop. With the metal held firmly in position, pressure is applied to the foot treadle. The hold-down bar will contact the metal first, securing it firmly against the shear bed. Additional pressure applied to the treadle engages and depresses the upper blade until the metal is cut. Removing the pressure from the treadle releases the blade and then the hold-down bar, in sequence. To aid in shearing large pieces of stock, *extension arms* are available. These arms bolt to the front edge of the shear bed and support the material.

Careful attention to all the adjustments mentioned in this section results in a neatly cut edge free of waves and burrs. The *cutting capacity* of the shears should *never be exceeded.* The cutting capacity of the type of shear shown in Fig. 19.1 is approximately 16 gauge for mild steel and 12 gauge for the softer, nonferrous metals (aluminum, brass, and copper), with a maximum cutting length of 36 inches.

Several safety precautions must be observed when using a foot-operated squaring shear. Of course, fingers must be kept clear of the hold-down bar and cutting blades at all times. In addition, the operator's feet must be positioned so that no injury occurs during depression of the foot treadle.

## 19.2 HAND-OPERATED SQUARING SHEAR

For work on smaller pieces of stock, the *hand shear* is usually preferable since it is more accurate and easier to operate than the foot shear. The hand shear consists of a *table,* a *side gauge, back gauges,* and two *cutting blades.* A *hand lever* causes two *eccentric cams* to move the *crosshead.* This imparts the vertical motion to the *upper blade.*

Use of the hand shear is basically the same as that of the foot shear. Hand shear tables vary in size from 6 inches to 36 inches and the shearing capacities are similar to those of foot shears. Sheet metal is cut by holding the stock firmly against the side gauge or back

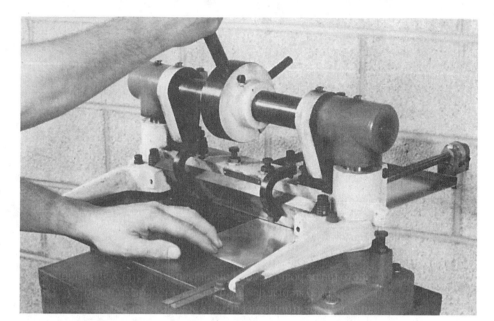

**Figure 19.3** Using the hand-operated shear to trim chassis blanks. Metal to be cut is positioned against the right-side gauge and held firmly with the left hand. The micrometer back gauge is also shown.

gauge stop with one hand and pulling the shear handle forward with the other. This operation is shown in Fig. 19.3.

The hand shear should also be checked for correct adjustment prior to use. The adjustment procedures are basically the same as those for the foot shear (Sec. 19.1). If the shear table is not wide enough to accommodate a carpenter's square, a piece of sheet metal that has been previously squared serves the purpose.

The *side gauge* on the shear table is calibrated from the *cutting edge* of the *lower* shear blade. This gauge should be checked for accuracy before any cutting operations by placing a piece of square stock of known dimension on the shear table with one edge flush with the lower blade and the other against the side gauge. The known length of stock should exactly measure the dimension shown on the gauge. If adjustment is necessary, the securing screws on the gauge are loosened and the gauge is moved until the correct calibration mark exactly aligns itself with the edge of the stock. With the side gauge properly adjusted, measurements can be taken directly from drawings and stock can be cut without the need for layout marks scribed into the work. The degree of accuracy of any cut is directly related to the accuracy to which the side gauge is adjusted. The side gauge is normally calibrated in sixteenths of an inch. If the drawing requires a closer tolerance than permitted by this calibration, lines scribed into the stock may be used to determine the correct cutting position. For an even greater degree of dimensional accuracy, *micrometer back gauges* are available with precise settings as low as 0.001 inch.

## 19.3 HAND NOTCHER

In the construction of electronic equipment, it is often necessary to shear *notches* of various sizes and shapes into a piece of stock. The *hand notcher* is a useful tool for this purpose. The basic parts of the notcher are the *operating handle, table, dies,* and *blades.* The operating handle is attached to an *eccentric cam.* A *V-ram* and a *vertical block,* to which the blades are attached, move vertically as the hand lever is operated.

The notcher is used for shearing short lengths of stock, cutting 90-degree notches, and making various angular trim cuts. The two cutting blades are mounted at right angles to each other and positioned over a pair of dies that are secured to the table. The edges of the blades and dies are at 45-degree angles to the sides of the table. Figure 19.4 shows a hand notcher and illustrates its use.

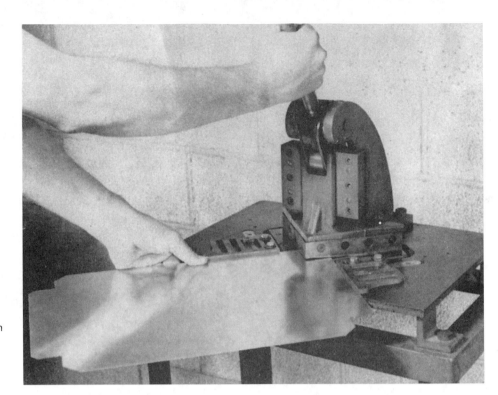

**Figure 19.4** Metal to be cut is held firmly against the notcher table and stops with one hand, while the handle is operated with the other. The stops assist in quickly positioning metal for duplicating work.

For notches having angles *less than* 90 degrees, a *bevel square* or *protractor head* provides a means of transferring drawing dimensions directly to the work. The bevel square or protractor head is set to the desired angle of cut. The base of the square or head is then placed against the bevel blade or rule. While maintaining the desired angle, the work is moved along the table scale for the required depth of cut. For notches having angles *greater than* 90 degrees, it is necessary to perform *two* cuts.

When quantity duplication of work is performed, it is tedious to use the table scales alone or in combination with a square to align the stock. Stops are available, similar to those shown in Fig. 19.4, which are easily bolted to the notcher table over both scales. Each stop is slotted for ease of adjustment.

Extreme care must be taken to keep fingers away from notcher blades. Unlike the hand- and foot-operated shears, which provide some degree of protection to the operator, the notcher is *not* equipped with a hold-down bar. The correct cutting position of the notcher blades and dies must be continuously maintained; otherwise, both will be damaged. The maximum capacity of the hand notcher is approximately the same as for the foot shear shown in Fig. 19.1 (16 gauge for mild steel and 12 gauge for the softer metals), with a cutting length of 6 inches.

## 19.4 HYDRAULIC AND MECHANICAL POWER-DRIVEN SHEARS

Hydraulic or mechanical power-driven shears have a wider range of metal capacities than those of the manually operated type. Power shear capacities range from 1/8 inch by 24 inches to 1 inch by 30 feet mild steel with cutting pressures up to 8000 tons. Power shears for even thicker and wider stock are also available for special applications.

Hydraulic shears offer some advantage over the mechanically powered motor-driven models. The hydraulic drive provides both variable speed and variable blade pressure. This allows wider adjustments to obtain optimum cutting action for any specific type of stock material.

Power shears have distinct advantages over manually operated models if quantity cutting is a major concern. These shears may be set for single or multiple shearing operations, thereby increasing cutting rates and lowering costs.

## EXERCISES

### A. Questions

19.1 What are the two basic types of shears used in sheet metal fabrication for electronic packaging? *the blade shear and the notcher.*

19.2 What are the four basic parts of a foot-operated squaring shear? *2 cutting blades, bed, foot treadle and 4 alignment gauges*

19.3 Answer Question 19.2 for a hand-operated squaring shear. *table, side gauge, back gauge, 2 cutting blades*

19.4 What is the maximum thickness, in decimal form, of aluminum and of mild steel that can be cut without exceeding the capacity of the foot- or hand-operated shears?

19.5 Explain how a properly adjusted side gauge on a hand shear can save considerable time when shearing stock to drawing specifications.

19.6 How does a hand notcher differ from a hand-operated shear?

19.7 What is used with a hand-operated shear when a greater degree of dimensional accuracy is required? *micrometer back gauge*

19.8 What are typical metal thicknesses that can be cut with hydraulic and mechanical power shears? *1/8 in by 2ft to 1in by 30ft*

19.9 Why is it important that the hold-down bar pressure of a foot-operated shear be properly adjusted? What could happen if this pressure is not sufficient? What could happen if the pressure is excessive?

19.10 What safety precautions need to be emphasized when using the foot-operated shear?

### B. True or False

Circle *T* if the statement is true, or *F* if any part of the statement is false.

19.1 Cutting 12-gauge mild steel on the squaring shear will not exceed its capacity.     T  F

19.2 Micrometer back gauges make the hand-operated shear more accurate.     (T)  F

19.3 Micrometer back gauges are available down to precise settings of 0.025 inch. *as low as 0.001 inch*     T  (F)

19.4 The hold-down bar on a hand notcher must be uniformly pressing down on the metal before it is sheared.     T  F

19.5 It is possible to cut a 125-degree notch with the hand notcher.     (T)  F

19.6 The maximum width of stock that can be cut on a foot-operated shear is 4 feet.     (T)  F

19.7 The maximum length of cut using a hand notcher is 6 inches.     (T)  F

### C. Multiple Choice

Circle the correct answer for each statement.

19.1 Cutting large pieces of sheet metal into blanks requires the use of a (*notcher,* *squaring shear*).

19.2 The cutting capacity of a foot-operated squaring shear is given in units of (*metal length, metal thickness*).

19.3 The side gauges of a squaring shear are (*parallel, perpendicular*) to the cutting edge.

19.4 Extension arms on a squaring shear are bolted to the (*front, rear*) edge. *of the shear bed & support the material.*

19.5 The hand notcher requires a minimum of (*two, three, four*) cuts to produce a sheared angle of greater than 90 degrees.

### D. Matching

Match each item in Column A to the most appropriate item in Column B.

COLUMN A

1. Cutting capacity
2. Cutting pressure
3. Side gauges
4. Hand notcher
5. Squaring shear
6. Cutting length
7. Hold-down bar

COLUMN B

a. Foot treadle
b. Clamp
c. 90-degree cuts
d. 8000 tons
e. 90 degrees to cutting edge
f. 16 gauge
g. 36 inches

### E. Problems

19.1   Test the 90-degree alignment of the side gauges to the cutting blades, side gauge 1-inch graduation distance from cutting blades, and parallel alignment of back stop to cutting blades for identical back gauge settings on a foot-operated shear.

19.2   Check the accuracy of the 45-degree angle between the cutting blades and front edge of a sheet metal notcher. Also inspect the alignment of the scale zero graduation marks and the leading ends of the cutting blades.

19.3   Fabricate a piece of 16-gauge aluminum into the configuration shown in Fig. 19.5 for a transistor heat sink. The heat-sink layout will be completed in Problem 20.2, drilling and deburring in Problem 21.2, and bending in Problem 23.2.

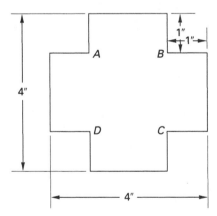

**Figure 19.5**

# 20 Chassis Layout

*Upon completion of this chapter on laying out sheet metal chassis elements, the student should be able to*

- Properly use layout dye for marking sheet metal.
- Properly use the layout table for sheet metal layout.
- Use common layout tools, such as scribes, scales, hermaphrodite calipers, combination squares, punches, and dividers to produce good-quality sheet metal layouts.
- Lay out an uneven dimension into a number of equal parts.

## 20.0 INTRODUCTION

In Chapter 19, stock was sheared into *blanks* and all the information relative to hole size, hole position, bend lines, and so on, resulting from the planning stage will now be transferred from the chassis layout drawings to the work. Techniques and necessary tools for prototype development are considered in this chapter and attention is given to methods for producing accurate layout. Selecting and using the proper tools is stressed to guide the technician in developing greater skills to minimize the inherent errors associated with this type of work.

Specific information is provided in this chapter on *layout dye, layout tables,* and procedures using the *scribe* and *scale.* Also included are discussions on the proper use of the *hermaphrodite caliper, combination square, punches,* and *dividers.*

*CAUTION:* Applying layout dyes and removing them with solvent must be done only in a well-ventilated station that is working properly. Approved face shields and goggles in compliance with ANSI Z81.1-1989 and chemical-resistant clothing and footwear and neoprene or nitrile gloves must be worn. When using layout tools, care must be taken to avoid injury from the extremely fine, sharp scribe points. Also, the section on safety outlined at the beginning of the book should be reviewed for more specific information.

## 20.1 LAYOUT DYE

Layout dye is a convenient aid in sheet metal layout work. Commercial dyes are generally *blue* or *yellow,* the selection depending on the application. Blue dye is preferable for chassis layouts since it provides more contrast with the light metal surface. Layout dyes are fast-drying and are easily removed. For this reason, care must be exercised to prevent oil or solvents from contacting the surface during the layout operations.

The surface to be dyed must be free from dirt and oil for the dye to adhere. Any common detergent will serve as a cleaning agent. A light coat of dye can easily be applied with a small nylon brush or cotton swab. The dye should be applied to the side of the metal that will *not* be visible after the metal is bent to the desired form. After approximately 10 minutes' drying time, the surface is ready for working. Light pressure applied with a pointed object on the dye results in a smooth, clean line where the dye is removed. The resulting contrast between the dark color of the dye and the exposed underlying metal facilitates rapid and accurate location and identification of layout marks for punching, drilling, cutting, and bending operations. If the applied pressure is excessive, the surface will be gouged. Enough pressure must be applied, however, to ensure the removal of a suitable amount of dye to achieve the desired contrast. Removal of the dye, after the chassis fabrication operations have been completed, requires the application of a suitable solvent, such as wood alcohol, and light rubbing with a cloth or cotton swab.

## 20.2 LAYOUT TABLE

For precision chassis layout work, a *layout table* will aid in producing good quality results. Layout tables are commonly made of cast iron and are generally referred to as *bench plates* or *surface plates*. These plates are machined on the top and sides with the edges perpendicular to the top surface and at right angles to each other. This level surface can serve as an excellent means of testing for flatness. Any rocking motion of a surface placed on the bench plate indicates a variance from true flatness.

Figure 20.1 shows a typical layout table. In scribing layout lines on a chassis blank, the following procedure is followed. The blank is placed in one of the corners of the layout table with two of its adjacent edges exactly flush with two adjacent edges of the table. *Parallel clamps* firmly secure the metal to the table. Excessive clamp pressure should be avoided to prevent damage to the work. Usually, finger-tight pressure is sufficient. It is often necessary to reposition the clamps when they obstruct the layout operations. With the metal thus properly positioned, the edges of the layout table serve as a convenient reference for measuring, tool positioning, and alignment.

**Figure 20.1**
Transferring dimensional information to the sheet metal with scales and scribe.

398

The *scribe* serves as the instrument to mark construction lines on the metal surface. The scribe, as shown in Figure 20.1, is a sharp-pointed implement that makes a fine groove in the dye covering the metal. The handle is knurled to reduce the possibility of slipping. The tip has a taper of approximately 15 degrees, resulting in an extremely fine point, to facilitate a high degree of accuracy in scribing lines. Many scribes have *tungsten carbide* tips and seldom require sharpening. Tips made from hardened tool steel need to be sharpened occasionally. Care must be exercised in this sharpening process to maintain the original taper over the entire length of the tip to avoid *rounding.* Rounding prevents the tip from getting as close as possible to the scale edge when laying out measurements.

Dimensional information from the chassis layout drawing is transferred to the blank stock with the aid of a scale. The end of the scale should never be used as a measuring reference unless it is established that the dimension from each end to the first major graduation is accurate. It is common practice to use one of the major numbered graduations as the reference. This procedure is shown in Fig. 20.1. A small *light* scratch is made on the metal with the point of the scribe to the desired measurement on the scale. To position a hole, *two* perpendicular dimensions transferred from the drawing must be constructed on the work. The point of intersection of these two lines represents the center of the hole.

To minimize errors, all measurements should be made from a reference edge of the metal. Measuring from a previously marked point could compound errors. It may be necessary, then, to add or subtract measurements from the drawings before transferring them to the metal.

When scribing lines such as bend lines along the entire length of the metal, two *location marks,* spaced as far apart as possible, are first made. The edge of the scale is then placed along these marks. The scribe, with the point along the scale edge and tilted for close contact with this edge, is drawn *lightly* across the metal. The resultant line drawn on the dye should completely include or cover the location marks to ensure accuracy. If a correction has to be made, a new line must be drawn. To avoid later confusion brought about by a double line, the incorrect line should be blotted out with dye.

## 20.4 HERMAPHRODITE CALIPERS

A simple way of producing scribed lines parallel to the edge of the metal blank is to use the *hermaphrodite caliper.* This instrument is shown in Fig. 20.2a. It consists of an *adjusting knob* and *two legs,* one having a *scribe point* and the other a *hooked end* for alignment along the metal edge. The legs are joined at the top with a *spring crown* that provides tension. To set the desired width of scribe, a scale is first placed on the layout table. The scribe tip of the caliper is then set on a convenient reference graduation. The adjusting knob is rotated until the face of the hooked end exactly aligns with the desired measurement of the scale. A scribed line is made on the dye by holding the hooked end perpendicular to the edge of the blank and resting it on the surface of the layout table.

Care must be taken to maintain the correct position when using the caliper and also to apply the proper amount of pressure. The caliper should be tilted at approximately 45 degrees toward the operator (Fig. 20.2b). While maintaining this 45-degree angle, a constant downward pressure is applied with the index finger to the scribe leg as the caliper is drawn across the work from the farthest end. Although the hermaphrodite caliper legs are rigidly positioned, lateral pressure can cause the tip distance to change as the legs tend to flex. If an error in marking is made, it can be blotted out with dye and another line drawn. An effort should be made to draw the scribe tip across the work only once. Otherwise, a double line could result, inviting error. Since alignment and pressure must be maintained simultaneously for optimum results, experience will provide the necessary "feel" for this instrument.

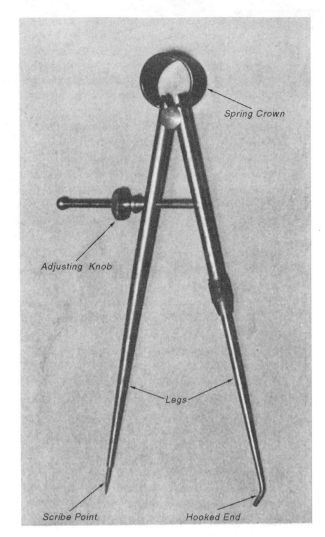

(a)

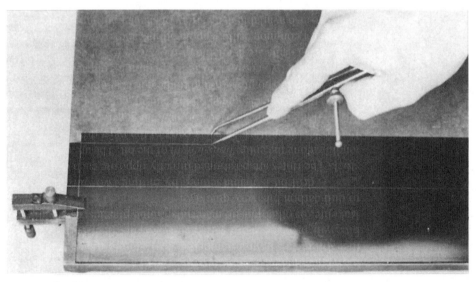

**Figure 20.2**
Hermaphrodite calipers;
(a) nomenclature;
(b) correct pressure and
position are important.

(b)

## 20.5 COMBINATION SQUARE

The combination square, shown in Fig. 20.3, is a useful and versatile tool in layout work. It consists of four parts: the *grooved blade,* a *squaring* and *miter head,* a *bevel protractor head,* and a *centering head.* The blade is a steel rule normally graduated on all four edges, providing two 1/16-inch scales, one 1/8-inch scale, and one 1/32-inch scale. This blade is grooved to accept a *binding pin* operated by a *thumbscrew* associated with each of the three removable heads. This binding pin clamps the blade in place. To mark a 90-degree line with respect to a reference edge of the work, the blade is inserted into the squaring head. With the thumbscrew loosened, sufficient length of blade is exposed to traverse the intended layout. After securing the blade, the 90-degree inside face of the squaring head is placed along a reference edge of the layout table and blank, as shown in Fig. 20.4. With the head and blade in this position, a scribed line along either edge of the blade will be perpendicular to the reference edge. A line parallel to the reference edge can be drawn by firmly holding a scribe on the work against the *end* of the blade while moving the head along the reference edge. A caution should be pointed out. As mentioned previously, the end of a scale may not be accurate for dimensioning. If checked and found to be accurate, direct positioning of the blade for scribing parallel lines can be made using the scale's graduations. If, however, the scale is not accurate at the end, a location mark to the desired dimension is first constructed. The end of the rule is then positioned to this mark and the head is slid to the reference edge and secured.

The combination square, using the squaring head, provides a rapid and convenient means of transferring dimensions for hole locations from drawings. The measurement from the reference edge to the center of the hole is first set on the blade. With the head moved to cover the approximate position of the hole center, a line is scribed against the blade end. The head is then moved to a second reference edge with the correct dimension set between blade and head and another line is scribed perpendicular to the first. The intersection of the scribed lines locates the center of the hole.

To scribe 45-degree angle lines, the 45-degree angle edge of the miter head is used. With the inside face of the head against the reference edge, the blade makes a 45-degree angle with respect to that edge (Fig. 20.5). Lines may be scribed at this angle using either edge of the blade.

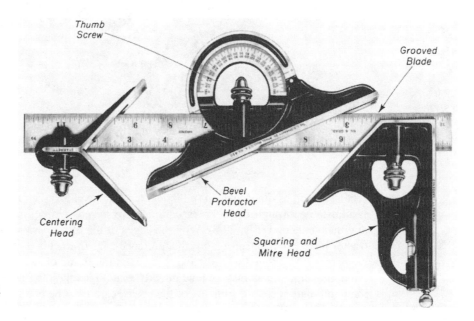

**Figure 20.3**
Combination square.
*Courtesy of the L.S. Starrett Company, Athol, MA.*

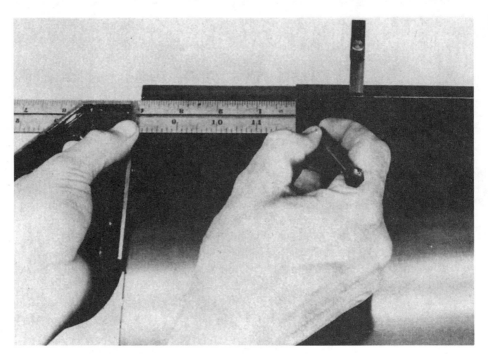

**Figure 20.4** Use of the squaring head allows rapid transfer of dimensional information.

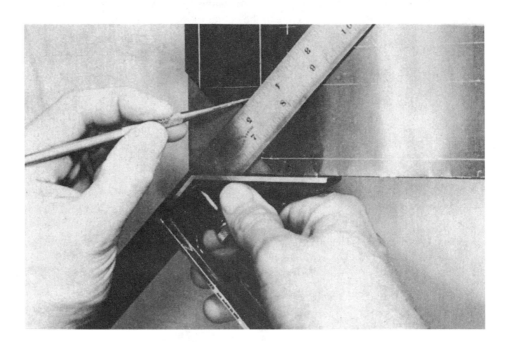

**Figure 20.5** Miter head is used to aid in scribing lines at 45-degree angles.

For angles other than 45 or 90 degrees, the bevel protractor head is used. This head is adjustable through 180 degrees, calibrated from zero degrees center scale to 90 degrees at either scale end (Fig. 20.6). With the edges of the blade parallel to the face of the protractor head when placed against the reference edge of the blank, the scale reading is zero degrees and is aligned with the index mark on the frame of the head. As the blade is rotated, the angle made may be read directly as it aligns with the index mark. Before using the combination square with any of the heads, care must be taken to assure that the blade is secured to the head by tightening the thumbscrew. Otherwise, angular and scalar errors may occur.

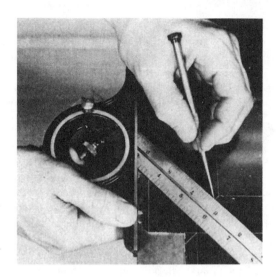

**Figure 20.6** Bevel protractor head and blade.

## 20.6 PUNCHES

To facilitate the drilling process, as discussed in the next chapter, a conical indentation must be made in the metal at the intersection of the location marks of all hole centers. These indentations are made with the use of a *hammer* and a *punch*.

Punches are classified as *prick* and *center,* depending on the angles of the point. Both punches have hardened steel points and knurled handles for ease in gripping. The prick punch has a point ground to 30 degrees and serves to introduce a small, round indentation primarily for location purposes. The center punch has a point ground to 90 degrees. Its function is to provide a larger-diameter indentation in the metal at the proper angle to accept the narrowest part of the drill bit web for alignment or centering. This indentation also prevents the drill bit from "walking" from the center position of the hole during the drilling operation.

With the punch held perpendicular to the work and the point at the center of the intersecting lines, it is *lightly* tapped with a hammer. An 8-ounce ball peen hammer is sufficient for use in punching. It is important to place the work on a hard, flat surface for complete support during the punching operation. Use of a heavy hammer or excessive striking force will damage the work as well as the punch. It is advisable to test for the correct striking force on a sample piece of stock, applying only one strike with the hammer. Any error in the location of the indentation will result in a corresponding error in the location of the hole. To correct an off-center error in punching, the tip of the center punch is placed in the indentation at an angle such that when struck the hammer will tend to drive the point *toward* the intended intersection.

*Automatic punches* are available with prick or center punch tips (Fig. 20.7). These punches eliminate the need for a hammer because they contain an adjustable spring-loaded driving mechanism in the handle that forces the tip into the work when pressure is applied to the handle. The amount of striking force required for the proper indentation is determined by the type and hardness of the metal used. The pressure is adjusted by turning the upper

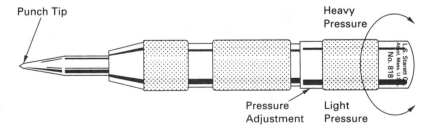

Punch Tip

Heavy Pressure

Pressure Adjustment

Light Pressure

**Figure 20.7** Automatic punch.

portion of the punch clockwise to compress the spring for increased pressure or counter-clockwise to expand the spring for reduced pressure. To use the automatic punch, it is positioned perpendicular to the work surface with the tip exactly at the point of intersection of the location marks. A steady downward pressure is applied to the handle until the spring is released, driving the punch tip into the work. As with most hand tools, practice on sample stock with the automatic punch will help in selecting the optimum setting.

## 20.7 DIVIDERS

Dividers serve several purposes in layout work. In addition to locating centers of evenly spaced holes and dividing lengths into equally divided parts, they also are used to scribe circles about hole location marks to ensure accuracy of drilling or chassis-punching operations. The common divider, as shown in Fig. 20.8, consists of *two legs* with scribe points, *spring crown,* and *adjusting knob.* The adjusting knob is used to set the distance between the divider scribe points.

To set the divider for scribing a circle, one tip is placed at a convenient graduation mark on a scale, such as the 1-inch mark. The adjusting knob is then rotated until the other tip is on the graduation mark corresponding to the radius of the circle. To position the divider for use, one tip is placed in the indentation previously made in the punching operation with the other tip contacting the work surface. While maintaining the first divider tip firmly in place, the other is swung around the pivot point inscribing an outline of the circumference of the hole in the dye. The divider should be positioned so that the spring crown is perpendicular to the work surface as the divider is rotated. To divide an uneven dimension into several equal parts, the divider is used with the technique illustrated in the solution to the following example.

*Example.* It is necessary to position five holes along a front panel, the total length of which is 14 inches. The spacing between each hole and the distance from each end of the panel must be equal.

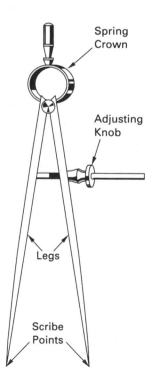

**Figure 20.8** Dividers.

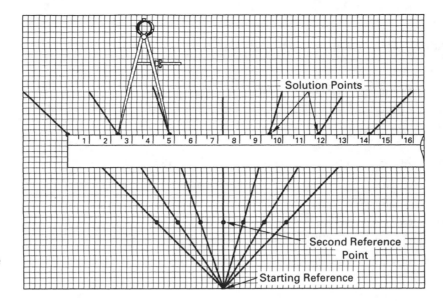

**Figure 20.9**
Construction technique for equal spacing of location marks.

*Solution.* A sheet of quadruled paper is placed lengthwise and a point is marked on a grid intercept close to the bottom and halfway between either side (Fig. 20.9). (Although 17- by 22-inch paper with four divisions to the inch is used for this solution, the size and divisions per inch are not important as long as linear graph paper is used.) At a point approximately 3 inches directly above the first reference point, another point is marked. Several points are then marked at 1-inch intervals both to the left and to the right of the second point on the same horizontal grid line. For this problem, seven points, including the center point, were made to account for the six spaces necessary. Straight lines are now drawn through each of the seven points and intersecting at the first point at the lower edge of the paper. It can now be seen that each horizontal grid line above the first reference point represents a division of six equal parts, progressively increasing with each vertical increment. To determine the correct distance between hole centers for this problem, the edge of a scale is placed over the construction line on the extreme left. With the scale aligned with the horizontal grid lines and maintaining the reference point on the scale edge, the scale is moved up until the 14-inch mark on the scale intercepts the construction line on the extreme right. (Refer to Fig. 20.9.) Dividers can now be set to one of the equal horizontal dimensions on the edge of the scale and this dimension can be transferred to the metal. A scribed center line on the work is necessary to locate the height of the position of the holes. The dimension set on the divider can now be transferred along this center line. One scribe tip is brought gently to the edge of the panel where the center line runs off the surface. Care must be exercised that this tip does not drop below the face of the panel; otherwise, an error in layout will result. With the first scribe tip held carefully in this position, the other scribe tip is brought down on the center line and a small arc is drawn on the dye. The same scribe tip is maintained at the intersection of that arc and the center line and the divider is gently rotated until the other scribe tip is brought along the center line to cut another arc. By such successive 180-degree rotations of the divider, the required number of hole centers will be accurately located.

In the following chapter we provide detailed information concerning the proper selection and use of drills, bits, and chassis punches.

# EXERCISES

## A. Questions

20.1 How should a scale be positioned on the sheet metal surface when transferring dimensions using a scribe?

20.2 How does the use of a hermaphrodite caliper reduce layout time?

20.3 What care must be taken when using a hermaphrodite caliper to avoid the introduction of layout errors?

20.4 When a hermaphrodite caliper is not available, what other method is used to rapidly locate and scribe lines parallel to a reference edge?

20.5 What tool is used to locate angles other than 45 or 90 degrees on sheet metal?

20.6 Why is a center punch used to make an indentation at the intersection of layout lines at the center of all holes to be drilled?

20.7 How are errors in center-punched locations corrected?

20.8 List three purposes that dividers serve in sheet metal layout.

20.9 Why is layout dye used in sheet metal layout work?

## B. True or False

Circle *T* if the statement is true, or *F* if any part of the statement is false.

20.1 Layout dye should be removed from the metal surface before any bending is done. **T** **F**

20.2 Sheet metal layout is done from a top view. **T** **F**

20.3 Hermaphrodite calipers should be held perpendicular to the metal surface when scribing layout marks. **T** **F**

20.4 A prick punch makes a 90-degree indentation for drill bit alignment. **T** **F**

20.5 Layout tables are commonly made of hardened wood. **T** **F**

## C. Multiple Choice

Circle the correct answer for each statement.

20.1 The preferred color of layout dye used on aluminum is (*yellow, blue*).

20.2 Layout lines made with a scribe require (*light, heavy*) pressure.

20.3 The quickest way to scribe layout lines parallel to the edge of sheet metal is with the use of a (*combination square and scribe, hermaphrodite caliper*).

20.4 The layout of angles other than 45 or 90 degrees with a combination square requires the use of a (*bevel protractor, miter*) head.

20.5 Automatic punches (*require, do not require*) use with a ball peen hammer.

## D. Matching

Match each item in Column A to the most appropriate item in Column B.

| COLUMN A | COLUMN B |
|---|---|
| 1. Hermaphrodite calipers | a. 22.5-degree layout angle |
| 2. Dividers | b. Center |
| 3. Bevel protractor head | c. Tungsten carbide bit |

4. Miter head  c
5. Punch  b
6. Layout dye  g
7. Scribe

✓ d. Scribing circles
✓ e. Combination square
✓ f. Parallel layout lines
✓ g. Yellow

## E. Problems

20.1 Using the layout techniques discussed in this chapter and quadruled graph paper, locate six evenly spaced hole locations across 11 inches.

20.2 Lay out the bend lines on the stock cut for the heat sink in Problem 19.3. The bend lines are positioned from *A* to *B, B* to *C, C* to *D,* and *D* to *A.* This type of heat sink is suitable for device case styles such as DO-4, DO-5, TO-3, TO-61, and TO-68. Determine the hole spacing required for the TO-3 style and locate and center punch this hole location pattern on the center section of the heat sink. Drilling and deburring are performed in Problem 21.2 and bending in Problem 23.2.

20.3 Shear a 3 1/2- by 8-inch piece of 16-gauge aluminum to lay out an oil pressure, battery, and temperature gauge panel shown in Fig. 20.10. The gauge hole locations are to be evenly spaced across the panel. Drilling and deburring are completed in Problem 21.1, hole cutting and bandsaw operations are performed in Problem 22.3, and bending is done in Problem 23.1.

20.4 Lay out a 2- by 5-inch piece of 14-gauge steel as shown in Fig. 20.11 for a drill gauge. Using the layout techniques discussed in this chapter, center punch five pairs of evenly spaced hole locations along the length of the gauge in addition to the 3/8-inch hole position. Drilling and deburring of the gauge are done in Problem 21.3, and radius cutting, punching, and filing are performed in Problem 22.2.

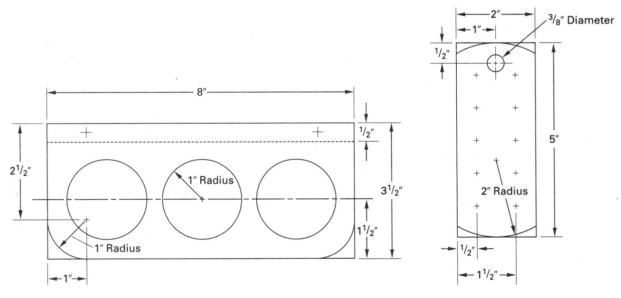

**Figure 20.10**                    **Figure 20.11**

# 21 Drilling, Reaming, and Punching

## LEARNING OBJECTIVES

*Upon completion of this chapter on the drilling, reaming, and punching of sheet metal, the student should be able to*

- Know the characteristics of twist drills.
- Understand how to use a drill gauge and a micrometer to determine the size of a drill bit.
- Use the drill press to perform drilling operations.
- Use the depth stop to limit the depth of a drilled hole.
- Change the spindle speed of a drill press.
- Use the proper lubricant for a specific drilling task.
- Properly use deburring tools.
- Properly use a hand expansion reamer.
- Properly use a hand-operated sheet metal punch.
- Understand the use of draw and turret punches.

## 21.0 INTRODUCTION

In preceding chapters we discussed the detailed techniques for preparing sheet metal for chassis construction. All the layout work, including the accurate positioning of all required holes, has been located on the metal blanks. Cutting operations can now begin. This chapter presents the basic cutting operation techniques of *drilling, reaming,* and *punching.*

In this chapter drill bit design is discussed, including *spiral flutes* for twist drills, *straight* and *taper* shanks, *standard drill sizes,* and *drill gauges.* Drill point configurations are also presented. To implement the drill bit, the *drill press,* together with its associated hardware and general operating instructions, is considered with respect to obtaining accurate hole size for common sheet metal applications. In addition, the proper lubrication for drilling specific metals is tabulated.

Deburring tools are also discussed, in addition to the techniques of reaming, including the use, advantages, and disadvantages of the *hand-expansion reamer.* Practical operational information on punches, including the *hand, draw,* and *turret* punches is presented. Techniques for using *square* and *rectangular* punches are discussed and punching and drilling are compared.

*CAUTION:* When performing any of the operations described in this chapter, approved safety glasses with sideshields and hearing protection, that is earplugs, hearing bands, or earmuffs where applicable, in compliance with ANSI Z87.1-1989 and S3.19-1974, respectively, must be worn.

Do not use drill presses when wearing loose clothing or allow long hair to be unrestrained that could become entangled in the machine. Also, clamp work to the drill press table, especially when drilling large holes. Serious injury can result if the spinning drill bit becomes lodged into the workpiece. Finally, shut off the drill press when finished or interrupted. Never leave power equipment running when unattended.

Handle drill bits only by the shank to avoid cutting fingers on the sharp cutting edges of the margin.

Be careful not to touch sharp spur cutting edges when assembling or disassembling hand-operated chassis punches to avoid cutting fingers.

When using a lever-operated turret punch, do not stand inside the hazard tapes that may be bonded to the floor around the machine to avoid personal injury from the operating handle. Keep hands and fingers clear from the space between the turrets holding the punches and dies. Also, the section on safety outlined at the beginning of the book should be reviewed for more specific information.

## 21.1 BASIC TWIST-DRILL CHARACTERISTICS

*Drill bits* are manufactured of *carbon* or *tungsten-molybdenum* steel rods. The carbon steel bits are made for drilling into soft metals such as aluminum, copper, brass, or mild-tempered steel. These bits become dull after a relatively short period of use. Tungsten-molybdenum bits, commonly referred to as *high-speed* (HS) bits, are used for drilling either soft metals or high-tempered steel, stainless steel, and other hard metals. High-speed bits will produce more satisfactory results than carbon bits when used on fiberglass, Bakelite, or other plastic materials. Some plastics, especially glass-cloth laminates, tend to dull carbon bits very quickly.

A drill bit is manufactured by milling two *spiral grooves* called *flutes* into the rod stock. The flutes are positioned directly opposite each other, beginning at the *point* and extending along the entire *body length* of the bit. The area between the flutes is called the *web*. Figure 21.1 identifies the basic parts of the twist drill.

Along the outer surface of the *land* generated when the flutes are milled, a *margin* is ground to provide clearance on the cutting edge and also to ream and maintain drill alignment. When the point is ground to the correct angle, a *chisel edge* is formed at dead center in addition to two *cutting lips* at the beginning of each flute.

Twist drills are available in *short lengths* for center-hole or starting purposes, *jobber lengths* for common drilling applications, and *extended lengths* for deep-hole drilling. Twist drills are also available in a variety of shank configurations, the most common of which are the *straight* and *tapered* shanks. These shanks are extensively used in sheet metal and panel stock drilling applications in electronic packaging. Smaller-diameter drills are made with a straight shank since they are accommodated by the conventional drill press or powered hand-drill chucks. Although more expensive for the smaller sizes (less than 1/2 inch), the taper shank has several advantages. It is more convenient to insert the bit by pressing it into a *tapered collet* on a drill press or lathe. In addition, bits can be removed and replaced much more quickly from a collet than can straight-shank bits used with a key-type chuck. The prices of comparable size drills of each type in the larger sizes are equivalent.

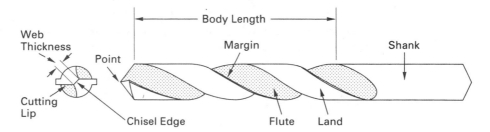

**Figure 21.1** Twist-drill nomenclature.

*Drilling, Reaming, and Punching*

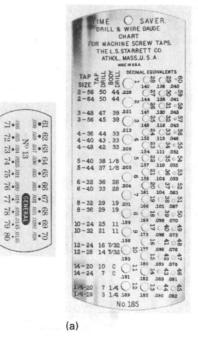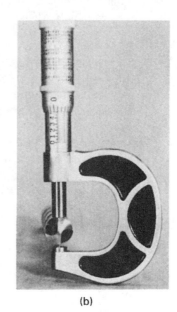

**Figure 21.2** Tools for drill size measurement. *Courtesy of General Tool Company and the L. S. Starrett Company, Athol, MA.*

(a)　　　　　　(b)

Three size systems are used for drill bits: *fractional, numerical,* and *letter.* The three lengths described previously with straight or tapered shanks are available in each system. *Fractional* bits range from 1/64 (0.0156) inch to 1/2 (0.500) inch in diameter in increments of 1/64 inch. A complete set of drills in this system would therefore include 32 bits. The *numerical* system consists of 80 bits ranging from number *80,* whose decimal equivalent size is 0.0135 inch in diameter, to a number *1* bit, 0.228 inch in diameter. *Letter* designation of bits starts with *A* size (0.234 inch in diameter) through *Z* size (0.413 inch in diameter). For a composite tabulation of the standard drill size systems, showing the fractional, number, and letter designations along with their decimal equivalents, refer to Appendix II.

The larger drill bits have their size stamped on the shank in accordance with the particular system used. For the smaller bits, *drill gauges,* such as the one shown in Fig. 21.2a, must be used to determine the size because the stamping would be too small and therefore difficult to read. Each hole in the drill gauge is designated by the size bit that will fit snugly into it. Drill sizes are checked by passing the flute portion of the bit into the gauge hole. The shank end is never used for measuring, since any burr or wear incurred in its use will result in an incorrect size determination. Drill gauges are also available for the larger bit sizes to determine diameters when the stamping on the shank has worn from extensive use. Drill bit diameters may also be measured with a micrometer. It is important, however, that this measurement be taken across the margins. This procedure is shown in Fig. 21.2b.

## 21.2 DRILL BIT POINTS

The points of twist drills have several features that must be examined in order to completely understand the cutting phenomenon. Figure 21.3a identifies the drill point nomenclature. When drilling, the bit is introduced into a center-punched hole, and the chisel edge at dead center of the point is the first cutting edge to come in contact with the work. As the rotating bit is forced against the work, the cutting lips begin to chip metal away. These chips are channeled from the work through the flutes. The flutes also allow cutting lubricants to be applied directly to the work area at the chisel edge. Once the drill bit has penetrated deeply

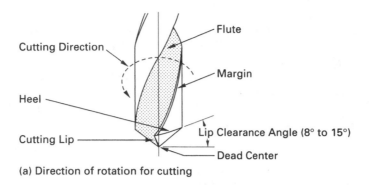

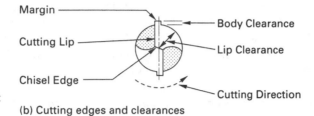

(a) Direction of rotation for cutting

**Figure 21.3** Drill point nomenclature.

(b) Cutting edges and clearances

enough into the work so that the entire length of the cutting lips are making contact, the sharp leading edge of the margin will begin to ream the inside of the hole to the final dimension. Note in Fig. 21.3b that the body clearance is achieved by the margin being *slightly raised* above the surface of the land. Lip and body clearance are essential so that the bit will not jam or bind in the work during the drilling operation. Once the point has passed through the work, the bit should be slightly raised and lowered before removing it so that the hole formed will be completely reamed.

The degree of drill point angle depends on the type of material being worked. In general, the harder the material, the flatter the point. For general-purpose applications, such as drilling aluminum and soft steels, a drill point angle of 118 degrees, a lip clearance of 12 to 15 degrees, and a standard chisel edge angle of 120 to 135 degrees is recommended. When drilling materials such as Bakelite or epoxy-glass (used extensively in printed circuit work), a steep drill point angle of 90 degrees and a lip clearance of 12 degrees is recommended.

The point configurations described above result in reasonably round holes, especially in the smaller sizes. Many times, however, chassis construction requires the drilling of larger holes (greater than 3/8 inch). Because aluminum is a soft metal, conventional drills tend to cut a hole that is out of round, especially in the larger sizes. The *sheet metal bit* (also referred to as a *brad point bit*), as shown in Fig. 21.4, reduces this problem with its *double-spurred* cutting lips. The point or chisel edge at the center of the web is used to align the bit in the center-punched hole. As the bit is forced against the work, the sharpened spurs on each side of the lands cut into the material in a circular fashion. Continued pressure on the

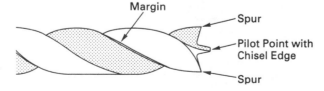

**Figure 21.4** Sheet metal drill point configuration.

*Drilling, Reaming, and Punching*

bit causes the spurs to completely cut through the metal, leaving a slug of the diameter of the hole. The margin then reams the hole to the desired dimension.

Since common sizes of bits used in printed circuit work are in the order of No. 55 to No. 67, difficulty may be encountered when attempting to rigidly secure and align these small shanks into a conventional chuck. The enlarged shank on the printed circuit drill bit overcomes this problem. However, the shorter-length bit increases the possibility of breaking. This problem is overcome by using straight-shank bits. Extreme care must be exercised when using small-diameter drills because of their delicate structure.

## 21.3 DRILL PRESS

One of the most useful power tools employed in electronic fabrication is the *drill press*. A bench-type drill press is shown in Fig. 21.5. Mounted on the vertical *column* is a horizontal *base table* to support and secure the press to the workbench. The base table may also be used as a supporting surface when unusually long pieces are to be drilled. Uppermost on the column is the *head*. Between the base table and the head, a movable *worktable* is positioned on the column and held in place by a *lever-operated clamp*. The work to be drilled is first positioned on the worktable, which is moved vertically along the column by first loosening the lever clamp. When the table is conveniently positioned and the hole in the center of the table is aligned with the *spindle* axis, the clamp is tightened, rigidly securing the table.

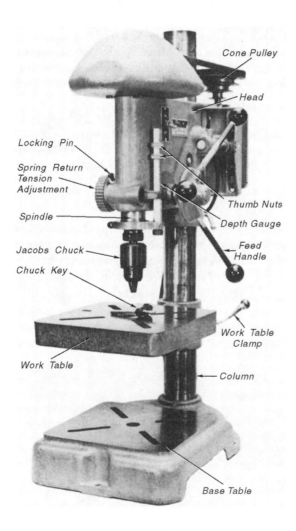

**Figure 21.5** Bench-type drill press.

Most standard drill press heads are equipped with a three-jaw *Jacobs chuck,* accommodating straight shank drill bits up to 1/2 inch. The chuck jaws are opened by a counterclockwise rotation of the *chuck key* inserted into the chuck. When the shank of the drill bit is inserted between the jaws of the chuck, clockwise rotation of the key will securely clamp the jaws onto the drill.

Chucks for both straight and tapered shanks are secured to the lower threaded end of the spindle. The drilling pressure is applied to the spindle by pulling the *feed handle* toward the operator. This action provides the *downward* motion of the normally stationary spindle and revolving chuck toward the work. The feed handle is coupled to the spindle with a *spindle return spring* that causes the spindle to return to its normal position after the downward force is removed. This upward motion should be guided with the feed handle to prevent needless jarring of the spindle and head assemblies. When the spindle does not automatically return to its normal position, *spring return tension adjustment* is necessary. The *locking screw* or *pin* on the left side of the head is loosened and the spring return tension knob is turned counterclockwise to increase tension. Since the knob is under constant spring tension, this adjustment should never be made when the drill press is on or when a drill is in the chuck. Should the knob be inadvertently released with the locking screw loose, the spindle will plunge downward (if the feed handle is not held), resulting in damage to the work, the drill, and possibly to the operator.

When a specified depth of hole is to be drilled into the work rather than a complete hole, a *depth gauge* is provided along the right side of the head. Figure 21.6 shows how the depth gauge limits penetration of the drill below the work. The depth gauge has graduations in 1/16-inch divisions and is threaded to provide for two *locking thumb nuts.* As the feed handle is turned for downward motion, the depth gauge moves through the stop, which is a machined hole in the head casting. The gauge moves with the spindle. With the thumb nuts positioned on the gauge, downward motion stops when they contact the *depth stop.* This prevents any further downward motion beyond the desired position. This position is set by feeding the spindle downward with the drill in the chuck to the desired drill point depth and then tightening the locking nuts until they contact the depth stop. Two locking nuts ensure that the vibration during drilling operations will not change the adjustment.

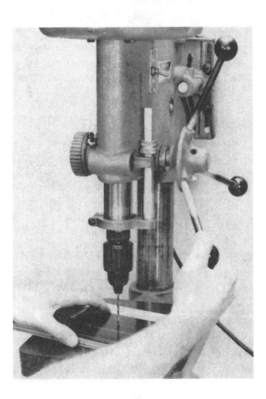

**Figure 21.6** Depth gauge set for drilling sheet metal to eliminate unnecessary spindle travel.

The spindle is rotated by a belt-driven pulley assembly powered by a motor mounted to the rear of the head. The motor base is secured to two steel rods that pass through machined holes to the rear of the head casting. The position of the motor base can be adjusted by loosening the *clamping bolts* that hold the rods in place. The belt tension between the *cone pulleys* on the motor and upper end of the spindle should not be excessive in order to avoid needless spindle and motor-bearing wear. An acceptable means of testing belt tension is to depress either side of the belt at its midpoint. Proper tension exists if approximately 1/2 inch of inward deflection is attained. More or less than this optimum deflection indicates a need for adjustment. Belt tension is increased when the motor is moved away from the head and is decreased when brought toward the head.

The spindle speed can be changed by moving the belt directly opposite to the motor and spindle pulleys. The belt must always operate in a horizontal position. When the belt position on the cone pulleys is to be changed, the motor should be brought toward the head to provide sufficient slack to avoid stretching or damaging the belt. When the largest motor pulley and smallest spindle pulley are used, the fastest spindle speed is obtained, which is approximately 3600 revolutions per minute (rpm). The slowest spindle speed of approximately 400 rpm is obtained using the smallest motor pulley and largest spindle pulley. These combinations are achieved by the motor pulleys and the spindle pulleys being inverted with respect to each other.

## 21.4 DRILLING OPERATION

All drilling operations begin by aligning and securing the stock to the worktable. *C clamps* provide good stability for drilling metal used in chassis fabrication. This arrangement is shown in Fig. 21.7. The center hole in the worktable should always be aligned with the spindle axis. In addition, when drilling sheet metal, a piece of wood of uniform thickness is used between the worktable and the stock. This allows the drill bit to completely pass through the work without contacting the worktable and provides complete support so that burring is minimized as the drill point breaks through the work (see Sec. 21.5). When drilling small parts or components, a *drill-press vise* provides the necessary support. For holes less than 1/4 inch, holding the vise manually against the table is usually sufficient. However, for larger holes, the vise should be mechanically secured. Drill-press vises are available with a slotted side or end sections that can be bolted to the table through these slots. For a further measure of security against work rotation, a *safety bar* is employed. This bar threads into the front of the vise and must rest against the *left side* of the column.

**Figure 21.7** Clamping techniques for drilling.

**414**     *Chapter Twenty-One*

**TABLE 21.1**

Drilling Recommendations for Various Materials

| Materials | Cutting Speed for High-Speed Drills* (ft/min) | Drill Point Angle (deg) | Lip Clearance Angle (deg) | Lubricants |
|---|---|---|---|---|
| Stainless steel | 40 | 136–150 | 8–12 | Sulfur and lard oil |
| Carbon tool steel | 60 | 136–150 | 8–12 | Sulfur-based oil mineral lard oil |
| Epoxy glass† | 80 | 90–110 | 12–15 | Dry |
| Cold-rolled steel | 90 | 120–130 | 12–15 | Sulfur-based oil, mineral lard oil |
| Bakelite | 100 | 60–90 | 12–15 | Dry |
| Cast iron | 120 | 90–118 | 12–15 | Dry or air jet |
| Copper | 200 | 90–110 | 12–15 | Kerosene or dry |
| Hard brass | 200 | 110–125 | 12–15 | Dry, kerosene, or lard oil |
| Bronze | 200 | 110–125 | 12–15 | Dry or mineral oil |
| Aluminum | 250 | 90–118 | 12–15 | Kerosene, lard oil, or turpentine |

*Drilling speeds for carbon steel drills are approximately one-half of those for high-speed drills.

†Use carbide tip drill for long drill tool life.

The correct drill bit size, drill point angle, and lip clearance must be compatible with the material to be drilled. Table 21.1 shows the type of drills with cutting speed, drill point angles, lip clearance angles, and proper lubricants to use for various materials. The proper drill is secured in the drill press chuck with the drill point serving as reference to align the work. Without securing the work to the table, the spindle is lowered so that the drill point nearly touches the work surface. With the drill point held stationary in this position, the work is adjusted so that the center-punched hole is aligned with the dead center of the drill bit. The power is then turned on and the drill point is allowed to gently touch the center hole and remove a small amount of material. The spindle is then returned to its normal position and the power is turned off. Without moving the work, the test cut is inspected for uniformity about the center lines of the hole. If the alignment is proper, the work is carefully secured to the table. When working with soft metal and twist drills, it is advisable to use a small bit (No. 28 to No. 32) to drill a *pilot hole* through the material to aid in alignment. Also, to ensure as round a hole as possible, a drill bit several hundredths of an inch smaller than the final dimension should be used before the final hole is drilled. This progressive *undercutting* will allow the lips rather than the chisel edge to cut and thus avoid the possibility of the drill's "walking." Of course, the use of a sheet-metal drill will eliminate the need for undercutting. However, the use of a pilot hole is recommended for either type of drill.

From Table 21.1 the average cutting speed is selected for the given material. (Note that the maximum cutting speeds for HS drills are approximately twice those for carbon-steel drills.) This information is given in units of *feet per minute* (fpm). These speeds should be considered *maximum* and should not be exceeded in order to minimize wear and not overheat the point. It is therefore important to adjust the spindle speed to the closest available speed without exceeding the maximum as prescribed in Table 21.1. This is done by the proper choice of cone pulleys on the motor and spindle. To convert average cutting speed from feet per minute to rpm, the following relationship is used:

$$\text{rpm} = \frac{12 \times \text{cutting speed (fpm)}}{\pi \times \text{diameter of the drill (inches)}} \approx \frac{4 \times \text{cutting speed}}{\text{diameter}}$$

Continuous lubrication is necessary during operation to minimize friction between the point and the work, thus reducing wear. Table 21.1 lists the recommended cutting compounds and lubricants for several common materials. Lubrication should *not* be used sparingly if the life of the drill is to be prolonged.

The following example illustrates the use of the information in Table 21.1.

*Example.* It is necessary to drill a 3/8-inch hole using twist drills through a piece of 16-gauge aluminum. The drills selected are as follows:

*Pilot size:* No. 31 (0.120 inch)

*Undercut size:* P (0.323 inch)

*Final:* 3/8 (0.375 inch)

From Table 21.1, the necessary information is obtained as follows:

| Drill Bit Size | HS Cutting Speed (ft/min) | Drill Point Angle (deg) | Lip Clearance (deg) | Lubrication | RPM | Closest RPM on Drill Press |
|---|---|---|---|---|---|---|
| No. 31 | 250 | 118 | 12–15 | Kerosene or turpentine | *7959 | 3600 |
| P | 250 | 118 | 12–15 | Kerosene or turpentine | 2956 | 2800 |
| 3/8 in. | 250 | 118 | 12–15 | Kerosene or turpentine | 2546 | 2400 |

*Note that although this calculated rpm is 7959 and that the highest available speed on the conventional drill press is 3600 rpm, this is acceptable because the maximum recommended cutting speed will not be exceeded.

## 21.5 DEBURRING

The fabrication of a chassis requires numerous *through* holes in which the drill body passes completely through the stock. In soft metals, such as aluminum, copper, and brass, the result is a rough edge generated on the reverse side of the stock. This *burr* is the result of the flex in the remaining thin metal as the trailing edges, or *heels,* of the drill push this material aside along the circumference of the hole instead of cutting it away as the chisel edge breaks through the work. The amount of burr can be minimized by using a wooden block as a support. In addition, by lowering the spindle more slowly, the remaining metal will tend to flex *less* and allow the drill to chip more of the remaining material away. This reduces the size of the resulting burr.

Since burrs can prevent flush mounting of components, in addition to marring rubber or plastic hardware, they must be removed. Two common hand deburring tools are the *countersink tool,* shown in Fig. 21.8a, and the *deburring tool,* shown in Fig. 21.8b. The standard 1/2-inch countersink is attached to a handle by a press fit for positive gripping action and control, and is constructed with three to five flutes. These flutes do not spiral as on a twist drill. The cutting lips run parallel to the flutes and converge at dead center of the point. The cutting lips have a clearance angle of approximately 15 degrees.

The deburring tool, shown in Fig. 21.8b, has three interchangeable *cutting heads* that pivot when fully seated in the handle. These heads are designed to deburr and form *chamfered edges* on holes in sheet metal, or tubing and edges on channel stock.

The countersink is commonly employed for holes less than 1/4 inch in diameter. This tool requires only moderate hand pressure to impart a circular rotation of the cutting lips against the burr. The result is a slight chamfer about the circumference of the hole. It is important to hold the countersink as perpendicular to the work as possible to ensure uniform cutting.

When harder materials need deburring, hand pressure may not be sufficient. In this application, a countersink can be used in a drill press chuck to obtain maximum pressure. The countersink should be rotated at the slowest possible spindle speed to avoid chatter, which will roughen the surface of the chamfered portion. Before the deburring operation begins on the drill press, the hole center must be aligned with the countersink point. The procedure is similar to aligning a drill point to a center-punched hole. With the drill press power turned off, the spindle is lowered until the countersink uniformly seats into the hole. The spindle is then

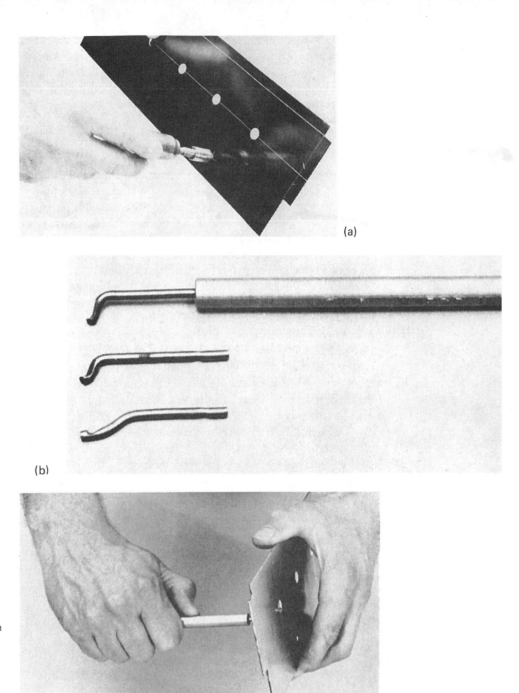

**Figure 21.8** Sheet metal deburring: (a) deburring small holes with a countersink; (b) deburring tool with interchangeable cutting heads; (c) deburring tool held at proper cutting angle.

(a)

(b)

(c)

returned to its normal position and the work is clamped securely into place. With the power on, the countersink is brought into contact with the edge of the hole to begin the deburring operation. Continuous inspection is required so that only the burr is removed. Excessive spindle pressure will cause too much material to be removed from around the edge of the hole. This will result in a sharp edge around the hole on the reverse side of the work and will weaken the material around the edge of the hole because of the excessively large chamfer. Also, if the countersink is allowed to pass too far into the hole, the inside diameter will be altered.

For holes larger than 1/4 inch, the deburring tool shown in Fig. 21.8b may be used. The proper cutting head is inserted into the work from the *reverse side* of the burr. The cutting head has a *hook-shaped blade* that is brought down against the burr. With the handle firmly gripped and held perpendicular to the work, it is rotated in a circular motion about the circumference of the hole (Fig. 21.8c). The cutting head will pivot in the handle, allowing the blade to remove the burr. It may take several turns to completely remove the burr and produce the desired chamfer. Caution must be exercised to avoid gouging or removing excessive material about the edge of the hole.

## 21.6 REAMING *khoan, khoét.*

It occasionally becomes necessary to expand the diameter of a previously drilled hole to a slightly larger size to correct an error in the selected size of a drill. Since this error is not usually detected until the assembly phase of construction, redrilling becomes cumbersome *vướng* and, in some cases, impossible. For this reason the *hand expansion reamer,* shown in Fig. *nặng nề* 21.9, is a convenient tool for enlarging holes.

Enlarging drilled holes introduces an important point in prototype chassis construction, especially if the unit being fabricated is the first of many to be produced. The undersized holes could be the result of an error on the original construction drawings. If this is the case, the drawings must be revised to show the required change in dimension of these holes so that later reference to these drawings will not result in the same error. An erroneous estimate of a hole dimension is not uncommon, but it should be corrected immediately once detected.

The hand expansion reamer consists of a *T handle* and *tapered fluted blades* running lengthwise along the body of the reamer. There are *six cutting blades* that converge at the taper point. At least six cutting blades are necessary to minimize chatter along the edge of the hole as the tool is used. Cutting is accomplished by a scraping action, and each cutting blade has a *body clearance* for relief similar to that of a countersink.

To expand a previously drilled hole, the reamer is inserted into the hole and the handle is rotated clockwise while a moderate downward pressure is applied. The result is a slightly enlarged hole tapered at the same angle as the reamer blades. When hole tolerance

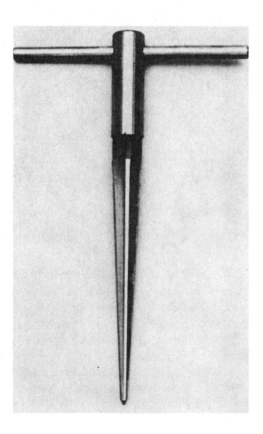

**Figure 21.9** Hand expansion reamer.

is critical, the tapered reamer should not be used to remove more than 0.030 inch because the resultant taper will weaken the surface of the metal about the hole.

The use of the tapered hand expansion reamer is a crude but effective method of enlarging holes. It also generates a burr, because of the downward pressure, which must be removed. Straight-sided reamers with either straight or spiral flutes are available for precision work. These are commonly used in conjunction with a drill press or milling machine.

## 21.7 PUNCHING

Test

Of all the methods available to the technician for fabricating holes in a chassis, the *sheet metal punch* will produce by far the neatest and most accurate results. Holes larger than 1/2 to 3 inches in diameter fabricated by punching are superior to those formed by drills. The punch forms a hole that is not only more circular but produces a negligible burr. (A slight chamfer around the edges of punched holes may still be desired to break the sharp edge in preparation for finishing.)

The common *round hand-operated punch* is shown in Fig. 21.10a, consisting of a *punch* with a threaded hole, a *die* with an unthreaded hole, and a *screw*. The punch is constructed of tempered steel with the cutting edges having two sharpened *spurs* aligned 180 degrees apart. As the screw is tightened, drawing the punch and die together, the spurs and cutting edges *shear* the metal.

Punching with a hand punch begins by drilling a guide hole to accept the screw. This guide hole is drilled in the center of the hole to be formed. The screw size is either 1/4, 3/8, or 3/4 inch in diameter, depending on the size of the punch. The guide hole thus drilled should provide a 1/16-inch clearance for the screw. Table 21.2 shows the screw size and guide hole size for punching holes from 1/2 to 3 inches.

The punch has two parallel flat sides at its base to secure it in a vise so that it will not turn. Once the correct guide hole has been drilled and the punch secured, the screw is passed through the closed end of the die. As shown in Fig. 21.10b, the sheet metal is positioned over the punch. The screw is then passed through the guide hole and threaded into the punch. The screw is tightened to clamp the punch and die against the stock, making sure that the die is centered. (If the die is not centered, the resulting hole will be off-center by a maximum of 1/32 inch, which, for most applications, is tolerable.) The head of the screw is square and should be tightened with a properly fitted open-end wrench. As the screw is further tightened, the wide edge of the die applies uniform pressure on the sheet metal about

(a)           (b)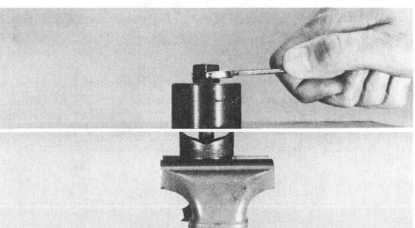

**Figure 21.10**   Hand-operated chassis punch: (a) round chassis punch; (b) chassis punch properly assembled for cutting.

*Drilling, Reaming, and Punching*

**TABLE 21.2**

Screw and Guide Hole Sizes for Chassis Punches

| Chassis Punch Size (in.) | Screw Size (in.) | Guide Hole Size (in.) |
|---|---|---|
| $\frac{1}{2}$ | | |
| $\frac{9}{16}$ | $\frac{1}{4}$ | $\frac{5}{16}$ |
| $\frac{5}{8}$ | | |
| $\frac{11}{16}$ | | |
| $\frac{3}{4}$ | | |
| $\frac{13}{16}$ | | |
| $\frac{7}{8}$ | | |
| $\frac{15}{16}$ | $\frac{3}{8}$ | $\frac{7}{16}$ |
| 1 | | |
| $1\frac{1}{16}$ | | |
| $1\frac{1}{8}$ | | |
| $1\frac{5}{32}$ | | |
| $1\frac{3}{16}$ | | |
| $1\frac{7}{32}$ | $\frac{3}{4}$ | $\frac{13}{16}$ |
| $1\frac{1}{4}$ | | |
| $1\frac{5}{16}$ | $\frac{3}{8}$ | $\frac{7}{16}$ |
| $1\frac{3}{8}$ | | |
| $1\frac{1}{2}$ | | |
| $1\frac{5}{8}$ | | |
| $1\frac{3}{4}$ | | |
| $1\frac{7}{8}$ | | |
| 2 | | |
| $2\frac{1}{8}$ | $\frac{3}{4}$ | $\frac{13}{16}$ |
| $2\frac{1}{4}$ | | |
| $2\frac{1}{2}$ | | |
| $2\frac{25}{32}$ | | |
| $2\frac{3}{4}$ | | |
| 3 | | |

the circumference of the hole being punched. As the metal is forced into the punch, the spurs start the initial cut. When the punch clears through the stock, a definite "snap" will be heard and the stock will drop down about the punch. When this occurs, the screw should not be tightened further. This would cause the cutting edges to be forced against the inside of the die resulting in unnecessary damage.

In addition to the round configuration, hand punches are also available in *square, key,* and *D* styles in various common sizes. These special-application punches consist of four parts: the *punch, die, screw,* and *nut. Keys, keyways,* or *flats* are located on portions of the dies, punches, and screws for automatic alignment and to hold the parts in the correct cutting position. These punches operate on the same principle as the round hand punches. Although the use of hand punches requires drilling a guide hole and assembling the various parts, the result obtained completely justifies the time and effort involved.

*Draw punches* and *turret punches* are preferable when many round holes of various sizes are to be punched. The draw punch assembly is secured by bolts to a work bench. This lever-operated tool has a punch and die, but the screw common to hand punches is replaced with a *draw bolt.* The draw bolt is fastened to an *off-center moment arm* of the lever. The alignment of the draw punch is identical to that just described for the hand punch in terms of positioning the metal between the punch and the die. Some draw punches have a punch with a threaded center hole, whereas others use an additional knurled thumb nut to secure the punch in position on the draw bolt. To punch the hole, the lever is drawn toward the operator, which forces the punch through the metal and into the die. Die and punch sets are interchangeable on the draw punch, providing a wide range of hole sizes.

**Figure 21.11** Lever-operated turret punch.

The most versatile of all the punches is the *turret* punch. No guide hole is necessary and a large number of die and punches are mounted on rotating drums and are readily available to the operator. A typical lever-operated turret punch is shown in Fig. 21.11. Both manual and hydraulically powered turret punches are available. The action of each is identical, with the exception of the application of power. Before any punching can be performed, a matched die and punch set of the desired size must be properly aligned. To accomplish this, the *pin locks* of both the die and punch turrets are disengaged. The two turrets are then rotated so that the selected die and its associated punch size are aligned and positioned directly below the lever arm. The locking pins are then engaged and the punch is ready for use. To operate, the metal is inserted between the punch and die. To position the metal for proper hole location, the operating lever is brought downward until the center point on the punch fits into the center-punched hole in the metal. Further downward motion of the lever engages the *cam* and drives the punch through the sheet metal into the die. The sheet metal then drops and clears the die for immediate duplication of the operation. In addition to providing rapid setup or size changes, no guide holes are required with the turret punch and no wrench or vise is necessary. The capacities of all the punches discussed in this chapter are from approximately 12-gauge aluminum to 16-gauge tempered steel.

## EXERCISES

### A. Questions

21.1    What two metals are used in the manufacture of drill bits? What is the application of each?

21.2    Describe the three systems used to designate the sizes of drill bits.

21.3    What two methods can be used to determine the size of a drill bit if it cannot be read on the shank?

21.4    List the cutting speeds and lubricants recommended for drilling (a) cold-rolled steel; (b) Bakelite; (c) copper; and (d) aluminum.

21.5    Determine the following for drilling a 1/4-inch hole through a piece of 16-gauge copper: (a) cutting speed (HS); (b) drill point angle; (c) lip clearance; (d) lubrication; (e) rpm; and (f) closest rpm on the drill press for this application.

21.6    Describe how deburring sheet metal is accomplished.

21.7    What is the purpose of a hand expansion reamer?

## B.  True or False

Circle *T* if the statement is true, or *F* if any part of the statement is false.

| | | | |
|---|---|---|---|
| 21.1 | A No. 28 drill has a larger diameter than a No. 36 drill. | **T** | **F** |
| 21.2 | Sheet metal bits should be used when drilling holes larger than ⅜ inch in soft sheet metal. | **T** | **F** |
| 21.3 | Drills designated as HS do not dull as quickly as carbide drills and are less brittle. | **T** | **F** |
| 21.4 | Work should be clamped to the drill press table to avoid injury. | **T** | **F** |
| 21.5 | Drilling speeds for HS bits are approximately one-half of those for carbon steel bits. | **T** | **F** |
| 21.6 | Aluminum can be drilled almost three times faster than cold-rolled steel. | **T** | **F** |
| 21.7 | A guide hole must be drilled in the center of each hole prior to using the lever-operated turret punch. | **T** | **F** |
| 21.8 | The hand expansion reamer is used to remove burrs in drilled holes. | **T** | **F** |

## C.  Multiple Choice

Circle the correct answer for each statement.

21.1    The spiral grooves along the length of a twist drill bit are called (*webs*, *flutes*).

21.2    Twist drills are available in short, jobber, and (*extended*, *long*) lengths.

21.3    A No. 54 drill is (*larger*, *smaller*) than a No. 58 drill.

21.4    The recommended lubricant for drilling epoxy glass printed circuit board material is (*kerosene*, *dry*).

21.5    A (*countersink*, *file*) is the most efficient tool by which to remove burrs in drilled holes.

21.6    The most efficient tool used to enlarge a previously drilled hole is a (*file*, *reamer*).

## D.  Matching

Match each item in Column A to the most appropriate item in Column B.

| COLUMN A | | COLUMN B |
|---|---|---|
| 1.  Feed rate  f | a. | Hook-shaped blade |
| 2.  HS bits  h | b. | Spurs |
| 3.  Lubricant  g | c. | Three jaws |
| 4.  Jacobs chuck  c | d. | Expansion |
| 5.  Deburring tool  a | e. | Margin |
| 6.  Chassis punch  b | f. | 60 ft/min |
| 7.  Twist drill  e | g. | Air jet |
| 8.  Reamer  d | h. | Tungsten-molybdenum |

### E. Problems

21.1 Find the clearance hole size for a No. 8 sheet metal screw from Table 25.2 and drill and deburr the mounting holes in the apron of the gauge panel of Problem 20.3. Gauge holes and radius corners will be cut in Problem 22.3.

21.2 Determine the drill sizes necessary to mount the TO-3-style case onto the heat sink of Problem 20.2. (Device lead holes should be drilled approximately 0.040 inch oversize.) All holes must be carefully deburred because flush mounting is critical. The ends of the heat sink are bent in Problem 23.2.

21.3 Tabulate the clearance and tap drill sizes for machine screw sizes 2, 4, 6, 8, and 10. Drill pairs of clearance and tap drill size holes along the length of the gauge in Problem 20.4 for the five machine screw sizes listed. Also, punch the indicated 3/8-inch hole at the end of the gauge. Deburr all holes formed. Bandsaw and filing operations are done in Problem 22.2.

# 22 Metal Cutting

## LEARNING OBJECTIVES

*Upon completion of this chapter on cutting sheet metal, the student should be able to*

- Be familiar with the characteristics of bandsaw blades.
- Use the correct blade speed for different materials to be cut.
- Relate the bandsaw blade width to the minimum radius of cut.
- Replace a blade in a vertical bandsaw.
- Make the necessary adjustments to a bandsaw blade.
- Properly use a vertical bandsaw to cut sheet metal.
- Properly use a hand nibbler to make inside cuts in sheet metal.
- Know the characteristics of files and the various types available.
- Use a file cleaner.
- Properly use files for cross-filing and draw-filing techniques.

## 22.0 INTRODUCTION

With the drilling and punching operations completed, the next step involves cutting irregularly shaped holes and slots and large round holes in the chassis elements in preparation for bending the metal into the desired forms. Information about the proper selection and use of each tool is provided.

In this chapter various metal cutting blades used in conjunction with a vertical bandsaw are examined together with the proper blade tension, guide settings, and cutting speeds. Practical cutting operations are illustrated on the bandsaw. The hand nibbler used in cutting irregularly shaped holes is described and operational information for proper use of this versatile cutting tool is supplied. In addition, the various types of files used to provide finish to the edges of cut metal are described.

*CAUTION:* When performing any of the operations described in this chapter, approved safety glasses with sideshields and hearing protection, that is, earplugs, hearing bands, or earmuffs where applicable, in compliance with ANSI Z87.1-1989 and S3.19-1974, respectively, must be worn.

Never use a file without a handle properly installed over the sharp edges of its tang to avoid hand injury. Be careful handling Swiss pattern files to avoid injury, as the ends of many come to fine sharp points.

When operating a bandsaw, keep hands and fingers clear of the blade and do not take your eyes away from the work while cutting. Never leave the bandsaw running when unattended. Shut power off when finished or interrupted. Also, the section on safety outlined at the beginning of the book should be reviewed for more specific information.

## 22.1 METAL CUTTING BLADES

The technician often encounters the problem of cutting round, extruded, or angle stock, as well as interior holes or irregular shapes in sheet metal. The shear is not suitable for these tasks. The vertically operated bandsaw with the proper metal cutting blade is ideally suited.

The shape and thickness, as well as the type of material to be cut, determine the proper blade selection. Cutting blades have three specific characteristics: *set, tooth pitch,* and *profile. Set* is the manner in which the teeth are aligned or bent away from the blade edge to provide adequate clearance between the blade and the work while cutting. Sufficient set is essential to generate adequate *kerf* (slot) so that the blade will not bind in the work during cutting. *Tooth pitch* indicates the number of teeth per inch, and *profile* is the style or shape of each tooth.

The three most common types of set are *alternate, raker,* and *wavy.* In blades with *alternate* set, every other tooth is tilted to the same side. The alternate teeth are bent in the opposite direction. This type of set is shown in Fig. 22.1a and is used primarily for cutting the softer, nonferrous metals. In the *raker* set blade, shown in Fig. 22.1b, one tooth is tilted left, the next is tilted right, and the third tooth is straight. This arrangement is continued throughout the length of the blade. The raker set blade will remove chips more easily during cutting operations than the alternate set blade and is used for cutting hard metals such as iron and steel. *Wavy* set, as shown in Fig. 22.1c, is characterized by groups of teeth tilted first left and right and then right and left along the entire blade length. Beginning with one tooth that is not offset, the first half of the group tilts increasingly until the maximum offset is reached in one direction. The second half of the group tilts decreasingly until a tooth with no offset is reached. This arrangement of outward and inward tilt is then repeated on the opposite side of the blade edge, in sequence, throughout the length of the blade. Wavy set is applicable to cutting thin sheet metal stock and tubing.

The tooth pitch must be suited to the work to result in a clean and efficient cut and to prevent damage both to the work and to the blade. For cutting thin sheet metal or tubing, blades with 32 teeth per inch are recommended. At least two teeth must contact or engage the work surface at all times. This prevents any tooth from becoming hooked on an edge or corner of the work and being stripped from the blade. The wavy set readily provides adequate kerf in thin stock. When soft or mild materials such as brass or copper or low-carbon steel are to be cut, 14 or 24 teeth per inch are recommended. When cutting large cross sections of these materials, a coarse 14-teeth-per-inch blade is best and ensures proper chip removal. When these materials are in the form of angle iron, conduit, or pipe, a finer 24-teeth-per-inch blade should be used because of the thinner cross sections. High-carbon and

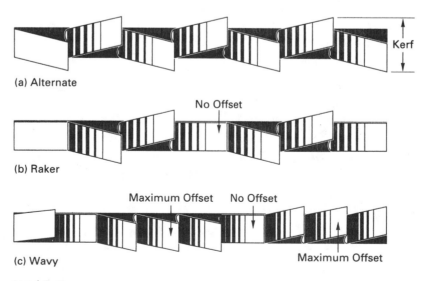

(a) Alternate

(b) Raker

(c) Wavy

**Figure 22.1** Saw blade set styles.

high-speed steels, as well as tool steel, should be cut with an 18-teeth-per-inch blade to provide adequate chip clearance.

The profile of the blade also affects the cutting operation. Figure 22.2 shows the three most common styles of tooth configuration: *standard straight face tooth, hooked tooth,* and *skipped tooth.*

The *standard straight face* design moves down over the work and scrapes the metal, removing the chips. This style of tooth is used for general-purpose work. Blades with *hooked teeth* are used almost exclusively for cutting large cross sections of steel, soft metals, and plastics. The slight *rake* on the teeth digs into the work instead of scraping it, thus removing metal more quickly than the standard straight-faced blade. For work on thin sheet metal, especially aluminum, the *skipped-tooth* blade is preferable. Fashioned with a large *gullet* (space) between each tooth, this style allows chips to form within the gullet and clears easier than the standard or hooked blade.

Both carbon-steel and high-speed steel blades are available. Carbon-steel blades do not maintain their sharpness as well as high-speed blades and should be used only for relatively soft materials. In addition, carbon-steel blades must be operated at slower speeds than high-speed blades to avoid impairing normal blade life. In general, hard materials should be cut slower than softer materials to prevent overheating and loss of temper. Most bandsaws are equipped with belt and pulley speed shift or gear reduction arrangements to provide a variation of blade surface speeds of approximately 50 to 4500 feet per minute. Table 22.1 lists the recommended speeds for various materials.

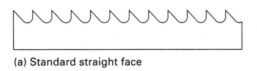

(a) Standard straight face

Rake

(b) Hooked

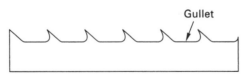

Gullet

**Figure 22.2** Saw blade profiles.

(c) Skipped

**TABLE 22.1**

Recommended Speeds for Carbon-Steel Bandsaw Blades

| Materials | Surface Speed (ft/min) |
|---|---|
| Stainless steel | 40–80 |
| Carbon tool steel | 100–150 |
| Epoxy glass | 90–200 |
| Cold-rolled steel | 125–200 |
| Bakelite | 800–1000 |
| Cast iron | 75–150 |
| Copper | 500–1000 |
| Hard brass | 200–400 |
| Bronze | 150–300 |
| Aluminum | 3000–4500 |
| Plastics | 3000–4000 |

Since the blade is often utilized to cut arcs and circles, blade width determines the minimum radius of cut. The 1/4- and 1/2-inch blades generally used in electronic fabrication will cut minimum radii of 5/8 and 2 1/2 inches, respectively. The 1/4-inch blade width is preferred for general-purpose work (Table 22.2).

**TABLE 22.2**
Bandsaw Blade Cutting Radius Guide

| Bandsaw Blade Width (in.) | Minimum Radius of Cut (in.) |
|---|---|
| 1 | $7\frac{1}{4}$ |
| $\frac{3}{4}$ | $5\frac{7}{16}$ |
| $\frac{5}{8}$ | $3\frac{3}{4}$ |
| $\frac{1}{2}$ | $2\frac{1}{2}$ |
| $\frac{3}{8}$ | $1\frac{7}{16}$ |
| $\frac{1}{4}$ | $\frac{5}{8}$ |
| $\frac{3}{16}$ | $\frac{5}{16}$ |
| $\frac{1}{8}$ | $\frac{1}{8}$ |
| $\frac{3}{32}$ | $\frac{3}{32}$ |
| $\frac{1}{16}$ | Square |

## 22.2 VERTICAL BANDSAW

A vertical bandsaw, as shown in Fig. 22.3, is a power tool found in most electronic shops. Although larger and more sophisticated models are used in production, the principles of operation and use are similar. The bandsaw consists of a *drive wheel* mounted below the *worktable* and an *idler wheel* above the table, both *V-pulley* driven. The motor is positioned within the pedestal that supports the saw. The blade is positioned around both the idler and drive wheels. The teeth are protected by the hard-rubber outer surfaces on the wheel rims that also provide friction to prevent blade slippage. To permit proper operation, the idler wheel is equipped with a *tension adjustment* and a *tracking,* or *blade centering, adjustment.*

To replace a worn or broken blade, the protective wheel covers are removed and the blade tension is reduced by rotating the spring-loaded tension adjustment knob counterclockwise. This will lower the idler wheel (Fig. 22.3). The used blade is removed and the new blade is passed into the *table slot.* (On some bandsaw tables, a large set screw is threaded into the table at the leading edge of the slot. This screw must be removed before a blade can be replaced.) The blade is then passed around the drive wheel and carefully over the idler wheel so that it is positioned between the *upper* and *lower blade guides.* These guides are located immediately above and below the worktable. With the blade centered on both wheels, the tension-adjustment knob is rotated clockwise, thus increasing the tension. Optimum tension is achieved when approximately 1/4-inch displacement is obtained when finger pressure is applied to the unsupported portion of the blade. This displacement is typical for blades whose width is from 1/4 to 3/8 inch. For blades 1/2 to 3/4 inch wide, a displacement of 3/8 inch is recommended.

It is important that work be fed into the blade from the front edge of the table when the throat of the bandsaw is to the left of the operator (Fig. 22.4). Therefore, the teeth of the blade must be facing the operator when viewed from the front edge of the bandsaw table. Also, the blade is designed so that the teeth will cut when they move in a clockwise direction. The distance in inches from the side of the cutting blade to the inside face of the column supporting the idler wheel is called the *throat clearance.* This clearance is used to designate the bandsaw sizes, which range from 16 to 60 inches. Bandsaw capacities are also given in terms of height from the table to the underside of the top blade guides when this assembly is in the uppermost position. These capacities range from 6 1/4 to 12 inches.

With the proper tension applied to the blade, the tracking is next inspected by manually rotating the idler wheel and observing if the blade deviates from its original center position on the wheels. Any deviation can be corrected by changing the tilt on the idler wheel by rotating the *tracking adjustment knob* clockwise to correct for a forward

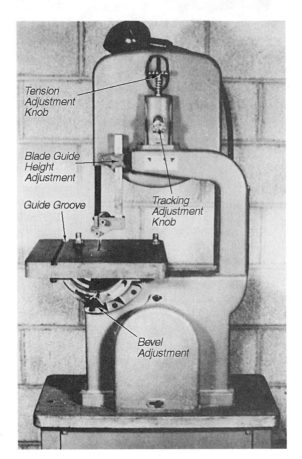

**Figure 22.3** Blade tracking and tension adjustment controls.

Tension
Adjustment
Knob

Blade Guide
Height
Adjustment

Guide Groove

Tracking
Adjustment
Knob

Bevel
Adjustment

**Figure 22.4** Proper positioning of hands when cutting sheet metal.

deviation and counterclockwise to correct for a blade that has moved back toward the rear of the wheel and table slot. Tracking is properly adjusted if the blade remains centered when manually rotated. Correct blade tracking is extremely important. Blades installed without regard for tracking could ride out from between the guides or completely off the wheels, thus damaging the work, the blade, and possibly the operator. *CAUTION: Never test the tracking by switching on the power.*

To cut sheet metal with the bandsaw, the work is placed flat against the worktable and about 1/4 inch from the blade. The *blade guard clamping screw* is loosened and the guard is lowered to within 1/8 inch of the stock. The guard is secured and the power switch is turned on. After allowing sufficient time for the blade to attain its operating speed, the stock is fed into the blade with the right hand, as shown in Fig. 22.4. Note that both hands are kept clear of the blade to prevent injury. The work should *never* be forced. If cutting requires excessive pressure, the blade should be inspected for dullness and replaced if necessary. At the end of a *through cut* (a cut that traverses the entire length of the stock), care must be taken not to close the kerf generated by the blade set. This will cause the back edge of the blade to bind and could result in damage to both the blade and the work. A through cut is made whenever possible because backing the work from the blade when the power is on could cause blade damage.

To saw a curve or an irregular shape, the layout of the cut is first scribed onto the work surface (see Sec. 20.2). Figure 22.5a is an example of a series of separate cuts necessary to remove a rectangular opening. One continuous cut is first made to remove the bulk of the rectangle. This cut is made perpendicular to the edge and close to one finished scribed line. As the first corner is approached, the smallest possible radius is cut in such a manner so that the blade is positioned close to the *inside* finish line. This process is continued to the other corner with a second radius, allowing the blade to exit along the remaining finish line. This will remove the largest possible portion of the metal from within the scribed area. To square the corners, the edge of the blade is placed along the edge of the first straight cut and the remaining curved area is fed into the blade until the 90-degree corner is encountered. The work is rotated and this procedure is repeated until all the metal is removed from the scribed area in that corner. These cuts are duplicated for the other corner (Fig. 22.5b).

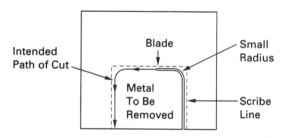

(a) Removing bulk stock

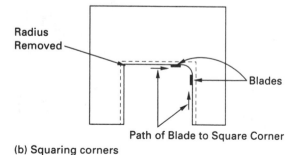

**Figure 22.5** Cutting sequence to form a rectangular opening with a bandsaw.

(b) Squaring corners

When irregularly shaped objects are to be cut on a bandsaw, special holding and clamping devices or jigs are necessary. For example, to cut the shaft of a potentiometer or switch, a *V-block* is needed for support and as a base to guide the work. In all cases, a solid support should be provided as close as possible to the cut to minimize saw and component damage owing to unnecessary vibration.

Not only do the profile, pitch, and set of a blade affect the cutting action, but the speed at which the work is fed into the blade also influences the characteristics of the resulting cut. In general, a higher rate of penetration will result when a coarse blade and a heavy feed rate are applied. For fine finish work within close tolerances, the finest pitch allowable for the particular material should be used and the feed rate should be reduced. The appropriate feed rate is best determined not only by visual inspection of the work but also by audible strain of the motor speed. Too fast a feed rate will reduce the motor speed. At all times, attention should be directed to (1) avoiding excessive feed rates that crowd or bind the blade needlessly, and (2) being constantly alert to the inherent dangers associated with the use of this power tool.

## 22.3 HAND NIBBLER

Cutting internal openings in sheet metal is a frequent requirement in electronic fabrication. The *hand-operated nibbler* is ideally suited for cutting internal holes because of its operating simplicity.

The hand nibbler, shown in Fig. 22.6, is basically a miniature hand shear consisting of two blades. The lower blade engages the underside of the work when the cutting head is

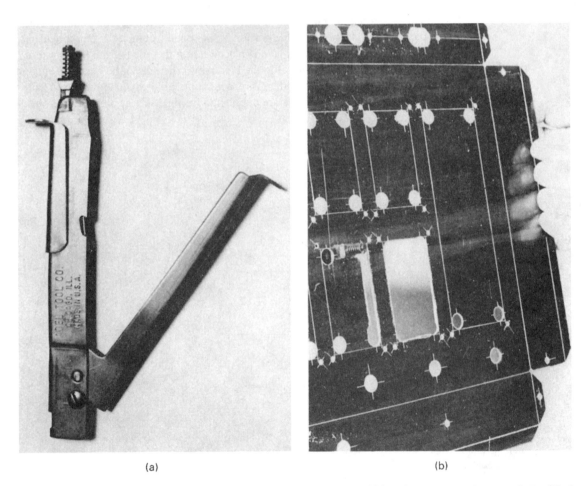

(a)  (b)

**Figure 22.6**  Hand-operated nibbler: (a) a sheet metal nibbler; (b) interior stock removal simplified through the use of a nibbler.

passed through and pressed against the edge of the access opening. The upper blade is actuated through a spring located in the handle. As the handle is squeezed, the upper blade is pulled down to contact the upper surface of the work. Further pressure on the handle causes the shearing action between the two blades. Each shearing stroke removes a small rectangular piece of metal approximately 1/16 by 3/8 inch. Thus, this tool "nibbles" a 3/8-inch-wide strip through sheet metal. Removing the applied pressure from the handles automatically raises the upper blade to its original position and the metal chip freely falls away. The nibbler requires that a hole, approximately 7/16 inch in diameter, initially be drilled into the area of metal to be removed to accommodate the cutting head through the stock. When the opening has been completed, the edges must be filed to remove the irregularities that result from the use of this tool.

## 22.4 FILES

Of all the mechanical skills necessary in electronic fabrication, perhaps the one that requires the most practice for proficiency is *filing*. To gain this proficiency, a familiarity with the file characteristics is essential so that the most suitable file for the particular application and material can be selected and used properly.

Filing is primarily a finish operation and is employed to improve the appearance of a previously made cut in a workpiece. It is also used to remove irregular and rough edges or burrs present after certain metal-cutting operations. Also critical for a quality appearance and to minimize damage to components, wiring, and possible injury to the operator is the technique of *breaking*, or *rounding*, all sharp corners and edges. This is especially true if a workpiece is to be finished by painting. Breaking ensures a better bonding of the finish over the edges of the work and prevents chipping (see Chapter 24).

Files are generally constructed of hardened high-carbon steel and consist of six basic parts: *handle, heel, face, edge, point,* and *tang* (Fig. 22.7). The sharp tang must be inserted into an appropriate size handle prior to its use to prevent injury to the operator. The wooden handles used for files have a metal *ferrule* around the end at which the tang is inserted. This prevents the wood from splitting when the tang is forced into the handle. The file is properly seated by tapping the end of the handle on a wooden block or bench after the tang has been firmly inserted by hand. This process is reversed to remove the file from the handle. By striking the edge of the ferrule next to the tang against the edge of a work bench, the file will loosen sufficiently to be withdrawn. Of course, this procedure can be avoided if a handle is provided for each file.

Files are classified with respect to *length, cut of teeth,* and *cross section.* The length of a file is measured from the heel to the point. For electronic fabrication uses, files vary in length from 3 to 12 inches. Files have four basic cuts: *single cut, double cut, curved tooth,* and *rasp cut.* The curved tooth and rasp cuts are not normally employed in electronics work. The single-cut and double-cut files are shown in Fig. 22.8. The single-cut file is characterized by parallel rows or courses of teeth that are set between 65 and 85 degrees with respect to the axis of the file. This type of file is preferable when a smooth surface finish is desired. A double-cut file has *two* courses of teeth on the same face that are set at different angles. One course is set between 40 and 45 degrees and the second course traverses the face at

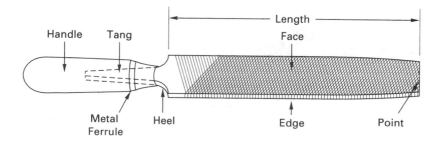

**Figure 22.7** File nomenclature.

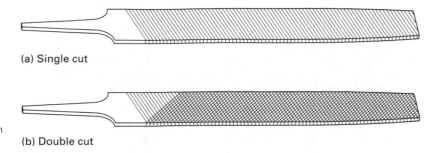

(a) Single cut

(b) Double cut

**Figure 22.8** Basic cuts of files used in electronics work.

between 70 and 80 degrees to the axis of the file. These angles are characteristic of files used for general-purpose applications. Double-cut files are also available with first and second courses at 30 degrees and 80 to 90 degrees, respectively, for finer work. Double-cut files are used primarily in *second-cut* work (i.e., in removing larger amounts of material prior to finishing with a single-cut file). The double-cut configuration produces a slicing effect that requires less effort than the single-cut file.

Tooth spacing determines the relative *coarseness* of a file. Files are categorized relative to coarseness as *rough, coarse, bastard, second cut, smooth,* and *dead smooth.* The tooth spacing is the largest with *rough cut* and becomes progressively smaller with the *dead smooth cut.* These spacings for the six categories are 20, 25, 30, 40, 50, and 100 or more teeth per inch, respectively. It is important to mention that each of the six grades are coarser for long files (10 to 12 inches) than for the equivalent grade of short files (6 inches). Consequently, length as well as grade of tooth coarseness determines the number of teeth per inch. When selecting a particular grade of file coarseness, the technician must consider the type of material to be worked. Steel and other hard metals are worked well with the less coarse grades, such as *second cut* and *smooth.* When soft metals such as aluminum, brass, or copper are to be worked, coarser grades, such as *coarse* or *bastard,* are more effective. These choices are dictated by a condition called *loading,* which is the accumulation of waste metal that clogs the file teeth. This condition builds up rapidly when using too fine a file on soft material. These bits of metal impede normal file operation and should be periodically removed with a *file cleaner.* This cleaner consists of a wooden handle to which are attached short, stiff wire bristles called the *card* and a regular bristle brush on the reverse side of the card (Fig. 22.9). To use the cleaner, the regular bristle brush is first employed to remove the loose waste material on the file face. If any material remains after briskly brushing parallel to the teeth, the cleaner handle is inverted and the card is rubbed over the file teeth. (To clean double-cut files, it is necessary to alternately brush the two courses of teeth.) To remove any remaining bits of material (called *file points*) lodged between the teeth after these two operations, a pick must be used. The pick is made of a softer metal than the file so that it will not dull the file teeth. To minimize the rate of loading, *chalk* may be rubbed into the face of a file prior to its use. The chalk will not impede the filing operation.

Many types of file shapes and cross sections are available to accommodate the numerous applications of this tool. Figure 22.10 shows most of the common types frequently used. Following is a brief description of each.

**Figure 22.9** File cleaner.

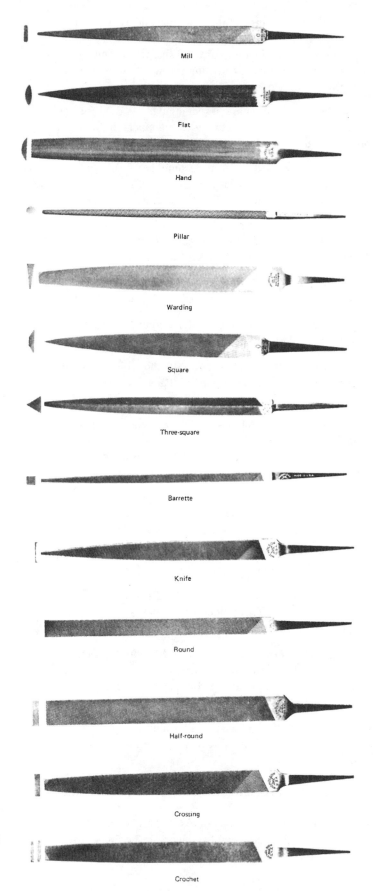

Mill

Flat

Hand

Pillar

Warding

Square

Three-square

Barrette

Knife

Round

Half-round

Crossing

Crochet

**Figure 22.10** Standard file shapes and cross sections. *Courtesy of Nicholson File Co.*

***Mill File.*** A single-cut rectangular file used primarily for lathe work and draw filing. It is available in round or smooth edges and tapers in thickness and width for one-third of its length. Other applications are tool sharpening and removing burrs and rough edges. This file is not recommended for filing flat work because of loading and the resulting marring of the work surface.

***Flat File.*** A double-cut rectangular file used for general-purpose work in finishing flat surfaces. It has single-cut edges and tapers in thickness and width toward the heel and point from the center of the face.

***Hand File.*** A double-cut rectangular file also used to finish flat surfaces. It is sometimes preferred over the flat file because of its parallel edges. One edge is *safe* (no teeth) and the other is single cut. This file is extremely useful for filing into 90-degree corners. The faces taper slightly toward the heel and point.

***Pillar File.*** A double-cut rectangular file similar to the hand file used for general-purpose work. It is slightly thicker but not as wide as the hand file.

***Warding File.*** A double-cut rectangular file much thinner than those previously described and tapers to a point. It finds wide application in filing slots and notches.

***Square File.*** A four-sided double-cut file that tapers toward the point used primarily for filing or enlarging square or rectangular holes.

***Three-Square File.*** A three-sided double-cut file that tapers toward the point. It is generally used for smoothing internal angular surfaces or corners, especially those having angles less than 90 degrees.

***Barrette File.*** A four-sided double-cut file, tapering to a point, having sharper edges than the three-square file. Only two faces cut; the other two faces are safe. This file is used for finishing in grooves, slots, and sharp angles or corners, especially where cutting must be done while an upper face is contacting a surface from which no further material is to be removed.

***Knife File.*** A file with a safe back edge, a single-cut thin edge, and two double-cut faces. It is often used in place of the Barrette file.

***Round File.*** A round double-cut file tapering toward the point and used for finishing or enlarging round holes or small radii.

***Half-Round File.*** A semicircular file usually double cut on the flat face and single cut on the curved face. This unique shape makes it one of the most versatile files for general-purpose work such as finishing curved edges or large-diameter holes as well as flat surfaces.

***Crossing File.*** An oval-shaped file with one face having a radius similar to the half-round file and the other face having a larger radius. It tapers to a point in both width and thickness and is double cut on both surfaces. This file also finds application in finishing curved edges, but affords a selection of the most appropriate face for the work radius.

***Crochet File.*** A flat file with rounded edges, double cut on all faces and edges. It tapers in both width and thickness and is used primarily for finishing slots, rounded corners, and filleted shoulders.

Most of the file cross sections just described are also available for extremely delicate work. *Swiss pattern files* find wide application in electronic packaging. The file selected is generally determined by the contour to be worked.

When filing, the metal to be worked should be firmly secured in a vise and protected from damage with *jaw protectors.* This is especially true when working with sheet metal.

These protectors are often fashioned from scraps of soft sheet metal, such as aluminum or copper. The metal is bent at 90 degrees and inserted on each jaw to either side of the work. The 90-degree bend keeps the protectors in position when the jaws are opened to release the work. To minimize vibration of the work, which causes the file to "skip" across the surface resulting in *chatter* and an irregular surface appearance, the metal should be gripped as close to the working edge as possible.

*Cross-filing* is the technique of pushing the file, under pressure, over and along the edge of the work. The file handle is held in the right hand with the thumb on the ferrule and the file point in the left hand with the base of the thumb resting on the face. The right hand provides the pushing action while the thumb controls the direction and pressure. For best results, a smooth, even *cutting stroke* must be maintained as the file is forced *away* from the operator. Improved cutting results if the file is swung at between 30 and 45 degrees to the work and level to the filing surface (Fig. 22.11). A common mistake of the beginner is to *rock* the file on the cutting stroke, thus leading to a curvature of the surface. The file must be kept level along the entire length of cut. Cutting pressure is applied on the forward stroke *only*. No pressure is applied on the return stroke.

To obtain a fine finish such as on the edges of an exposed portion of a chassis, the method of *draw-filing* is used. The handle is gripped with the right hand and the point with the left with both hands perpendicular to the file axis and both thumbs pressing against the closest file edge. This technique is shown in Fig. 22.12. With the file axis at right angles to the work surface, the file is passed *back and forth* with uniform pressure. Cutting is accomplished on both strokes. Single-cut files are considered superior to double-cut files for draw-filing.

Single-handed operation of a file is advisable only when using Swiss pattern files. Since these are so much smaller than the standard type, the file can be completely controlled with one hand. The knurled file handle should be held firmly with the index finger along the uppermost face or edge. The procedures for cutting are similar to those described for the standard-size files. Figure 22.13 shows a knife file being used to remove burrs along the tab edges of a chassis that resulted from the bandsaw cutting operations.

The proper care of files is just as important as their correct use. Files must not be allowed to come in contact with other files or tools in storage. This will cause tooth chipping and dulling. Also, the face or edge must never be struck on a surface to attempt to remove

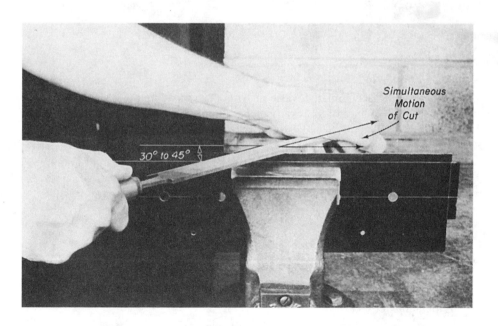

**Figure 22.11** Proper file position for cross-filing.

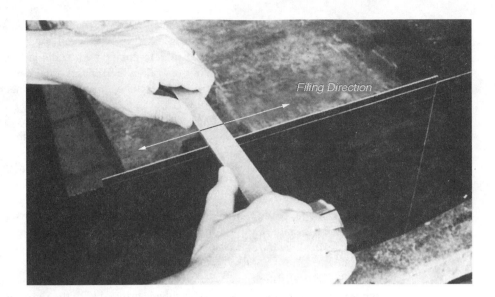

**Figure 22.12** Draw-filing technique.

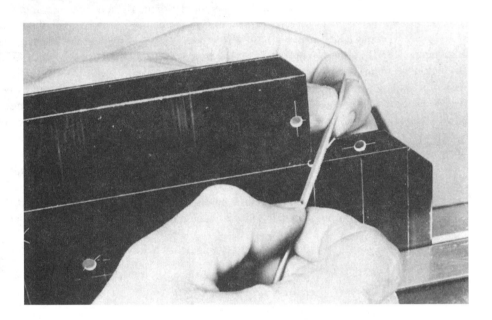

**Figure 22.13** Swiss pattern files are used where access space is limited.

accumulated waste material that is loading the file. Although the proper care and maintenance of tools is time-consuming, it is more than worthwhile in terms of longer tool life and quality workmanship.

The final process necessary for the completion of a chassis prior to applying a finish requires forming these elements into the desired configurations. Metal-forming tools and techniques are discussed in the following chapter.

## EXERCISES

### A. Questions

22.1    List and describe the three characteristics of cutting blades.

22.2    Explain the term *kerf.*

22.3    List the three most common types of saw blade sets together with their applications.

22.4    List the three saw blade profiles together with their applications.

22.5    What is the minimum radius that can be cut with a 1/4-inch-wide bandsaw blade?

22.6    List the six coarseness categories of files.

22.7    List the four basic file cuts together with their applications.

22.8    How are files cleaned?

22.9    Describe the difference between cross-filing and draw-filing, together with the application of each.

22.10   For what application are Swiss pattern files used?


## B.  True or False

Circle *T* if the statement is true, or *F* if any part of the statement is false.

22.1    Tooth pitch describes the amount of kerf on a cutting blade.                              T    **F**

22.2    The thicker the material to be cut, the fewer teeth per inch are required for proper        **T**    F
        chip removal.

22.3    Cold-rolled steel is cut at a speed approximately 20% of that for copper.                   **T**    F

22.4    The minimum radius of cut for a 1/8-inch-wide blade is 5/16 inch.                           T    **F**

22.5    The purpose of a hand nibbler is to form round holes in sheet metal.                        **T**    F

22.6    When filing into a 90-degree corner, a file having at least one safe edge is used.          T    **F**

22.7    Cutting pressure is applied on both the forward and return strokes when cross-filing.       T    **F**


## C.  Multiple Choice

Circle the correct answer for each statement.

22.1    Nonferrous metals, such as aluminum and copper, are cut primarily with a blade having a(n) (*alternate*,
        *raker*) set.

22.2    Cutting thin sheet metal is best accomplished using a blade with a (*hooked*, *skipped*) tooth profile.

22.3    The recommended speed of a bandsaw for cutting aluminum is (*300 to 450*, *3000 to 4500*) feet per
        minute.

22.4    Cutting internal holes in sheet metal is best accomplished by using a (*bandsaw*, *hand nibbler*).

22.5    A second-cut file is (*more*, *less*) coarse than a bastard file.

22.6    A file having 50 teeth per inch would be classified as (*second cut*, *smooth*).

22.7    On a barrette file, (*two*, *four*) of the faces cut.

22.8    Cutting pressure is applied to both strokes when (*cross-*, *draw-*) filing.


## D.  Matching

Match each item in Column A to the most appropriate item in Column B.

COLUMN A                          COLUMN B

1.   Hand nibbler                 a.    Cleaner

2.   Kerf                         b.    45-degree axis to work

3.   File                         c.    90-degree axis to work

4.   Throat clearance             d.    Second cut

5.   File card                    e.    Miniature shear

6.   Draw-filing                  f.    Tooth set

7.   Cross-filing                 g.    Blade profile

8.   Hooked tooth                 h.    Bandsaw

## E. Problems

22.1   Determine the maximum bandsaw blade width and speed for cutting a 2-inch radius in a piece of 16-gauge aluminum.

22.2   Using a bandsaw, cut the 2-inch radii of the drill gauge laid out and drilled in Problems 20.4 and 21.3. Break all sharp edges with a file.

22.3   Using a nibbler, cut the three holes into the gauge panel laid out and drilled in Problems 20.3 and 21.1. Cut the 1-inch radii with a bandsaw, deburr the holes, and break all sharp edges with a file. Bending is performed in Problem 23.1.

22.4   List the most appropriate type, coarseness, and cut of file for each of the filing operations required in Problems 22.2 and 22.3.

# 23

# Bending

## LEARNING OBJECTIVES

*Upon completion of this chapter on bending sheet metal, the student should be able to*

- Calculate bend allowance to overcome metal bending distortion.
- Make the necessary adjustments to a finger brake.
- Properly use the finger brake to bend sheet metal.
- Bend a chassis in the proper sequence to result in all bends being completed on the finger brake.

## 23.0 INTRODUCTION

After the cutting operations are completed, the next step in the fabrication of a chassis is to form the metal into the desired configuration. Industrial mass production techniques utilize hydraulic press brakes, which are very similar in mechanical action to the power shear (see Chapter 19). These presses can be set to automatically produce *all* the required bends in one step. The hydraulic presses are not considered in this chapter because the cost of tooling makes such automation prohibitive in prototype and small-volume work.

This chapter deals exclusively with the *hand-operated finger brake* (also called the *box* and *pan* brake), which is common to most sheet metal shops and allows the technician to produce a good-quality chassis. Information is provided on layout bend allowance, types of chassis configurations, bending sequence and brake nomenclature, adjustment, and operation. In addition, the critical limitations of most standard brakes are discussed.

*CAUTION:* While working with sheet metal, approved safety glasses in compliance with ANSI Z87.1-1989 must be worn. When operating any type of bending brake, do not stand inside hazard tapes that may be bonded to the floor around the machine to avoid personal injury from the movable wing and operating handles. Keep your fingers clear of the space between the bottom of the bending bar, or fingers, and the table as the crosshead is being lowered to clamp the work into position. Also, the section on safety outlined at the beginning of the book should be reviewed for more specific information.

## 23.1 BEND ALLOWANCE

When sheet metal is bent, there is an inherent distortion of the metal that results in a dimensional change. To maintain a degree of precision when bending, several characteristics of the metal and the equipment must be examined. The result of these examinations will be an adjustment in layout measurements termed *bend compensation,* or *bend allowance,* to overcome the distortions generated by bending. Since aluminum and mild steel are the most common metals used for chassis fabrication, the factors that affect the bending of these metals with a hand-operated finger brake will be considered.

Bending sheet metal *stretches* the outer section of the bend and *compresses* the inner section. The effect of these modifications on the bend area is to alter the position of the metal edges with respect to the bend lines established in the layout operations (see Chapter 20). Figure 23.1 shows the deviation of an edge from the original established line after bending. Note in Fig. 23.1a that an inside dimension of distance $x$ measured from edge $C$ is required after bending. This is impossible because the metal possesses a thickness $t$. After bending, the compression of the metal leaves an inside dimension of $x - B$ (Fig. 23.1b). Moreover, measured from edge $C$ is a new outside dimension of $x + A$, the result of stretching. It therefore becomes necessary to determine the desired total inside and outside finished dimensions *after bending* by computing the values $A$ and $B$ and compensating for the change in measurements due to the metal-bending characteristics. If an accurate outside dimension is required, the bend line is positioned at a distance $x - A$ from the reference edge $C$. This procedure is shown in Fig. 23.2a. When bent, the resulting outside dimension will be desired dimension $x$ (Fig. 23.2b). The *stretching* of the metal induced by the bending has been compensated for. If an *inside* dimension is critical, the bend line is positioned at a distance of $x + B$ from the reference edge $C$. This compensates for the *reduction* of the inside dimensions stemming from the compression of the metal when bent. The desired dimension $x$, as shown in Figs. 23.2c and d, is accurately obtained because of this bend allowance.

The dimensions $A$ and $B$ depend on (1) the type of metal, (2) the radius of the bend, and (3) the angle of the bend. In chassis fabrication, 90-degree bends and bend radii equal to the thickness of the metal are constructed almost exclusively. Therefore, for our purposes, the values of $A$ and $B$ will be a function of metal type only.

For 16- and 18-gauge aluminum alloys of the 1100 and 3000 series, commonly used in chassis work, a value of $A$ equal to two-thirds the thickness ($t$) of the metal ($A = 2/3\ t$) is a good approximation for most applications. It follows, then, that $B$ is equal to $1/3\ t$ ($B = 1/3\ t$).

Steel presents a high opposition to compression variations. As a result, $A$ is considered to be the *total thickness* of the metal ($A = t$), whereas $B$ can be approximated to be zero ($B = 0$). An inside bend measurement on steel will be approximately correct if laid off directly without regard to bend allowance. To compensate for an outside measurement caused by stretching, the bend line dimensions must include the thickness of the metal. Typical sheet metal gauges, together with their associated thicknesses, are shown in Appendix I to aid in evaluating bend allowance.

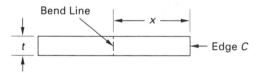

(a) Intended bend position

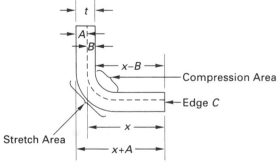

(b) Resulting bend characteristics

**Figure 23.1**
Dimensional changes
due to metal thickness.

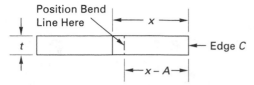

(a) Bend allowance for accurate outside
   dimension

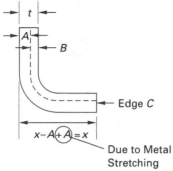

(b) Resulting bend

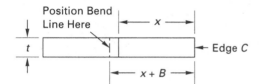

(c) Bend allowance for accurate inside
   dimension

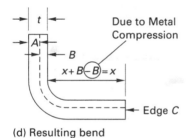

**Figure 23.2** Bend allowance methods to obtain accurate outside and inside dimensions.

(d) Resulting bend

## 23.2 BENDING BRAKE

Sheet metal forming by bending is accomplished primarily in prototype work with the aid of a *hand-operated finger brake,* as shown in Fig. 23.3. The basic parts of the brake are the *table, wing, fingers, crosshead, clamping handle, angle stop, operating handles, clamping-tension adjustment screws,* and *finger-position adjustment screws.* The metal to be bent is positioned on the brake table under the fingers and secured in place by pulling the clamping handle *forward.* This engages the crosshead, to which the fingers are attached, and applies clamping pressure to the metal. Once positioned, the metal is bent by raising the operating handles attached to the wing. This causes the wing to move to the desired angle with respect to the table. The adjustable angle stops are located on either end of the brake frame. These contact the wing when it is swung up to the desired angle and impede any further motion. When the metal is bent, the wing is returned to the downward position and the work is released by again operating the clamping handle in an *upward* motion to raise the fingers and release the clamping pressure.

*Bending*                                                                                                   **441**

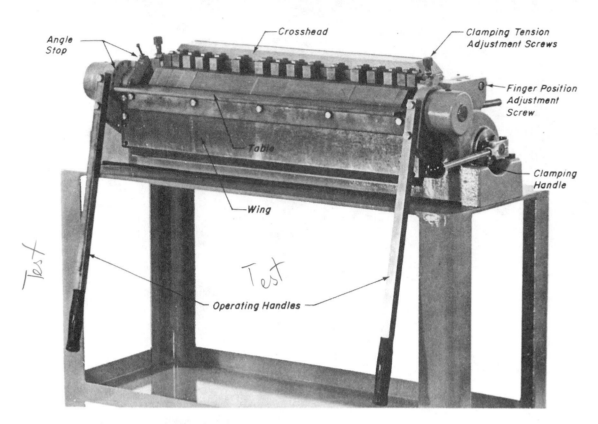

Angle Stop

Crosshead

Clamping Tension Adjustment Screws

Finger Position Adjustment Screw

Table

Wing

Clamping Handle

Operating Handles

*Test*

**Figure 23.3**  Finger brake. *Courtesy of Di-Acro.*

Before any bending operations are made on the brake, *four* adjustments must be checked and corrected, if necessary, to protect the brake and the work from damage. These adjustments, described in the following paragraphs, are (1) *finger seating,* (2) *clamping pressure,* (3) *finger clearance* (*setback distance*), and (4) *wing-stop angle.*

The fingers must first be checked to ensure that they are all properly seated and securely clamped to the crosshead. The seating is correct if all the fingers are parallel to one another with their leading edges in line and parallel to the edge of the wing. A *box-end wrench* should be the only type used when adjusting any of the hexagonal clamping or adjusting screws on the brake. Open-end adjustable wrenches should be avoided because they tend to slip and can damage the screw head.

After the fingers are properly positioned, the clamping pressure is then tested. This is done by first shearing a piece of stock (of the type used in forming the chassis) to a size equal to the brake's full lateral capacity and at least 2 to 3 inches wide. This sample piece is placed between the fingers and the table with at least 1 inch of the stock extending over the upper edge of the wing. The clamping handle is engaged and the pressure is checked by attempting to pull the metal from the brake. If the metal can be moved easily at either end, the clamping pressure must be increased. This is done by releasing the locknut securing the appropriate pressure adjustment screw. These screws are located on the frame at either end of the crosshead. The adjustment screw is rotated *clockwise* to lower the crosshead, thereby increasing the pressure. This test procedure of pulling the metal and adjusting the crosshead is continued until sufficient pressure is applied to the metal to just secure it to the table without damaging its surface. If the pressure appears to be excessive, the adjusting screws are turned *counterclockwise* to raise the crosshead. When the optimum pressure is achieved for the type of metal being worked, the locknuts are again secured. It is important that equal finger pressure be applied across the entire length of the metal to maintain positive gripping.

The finger clearance distance is tested with the same sample stock used for testing the clamping pressure. The edge of this stock is placed perpendicular to the table and touching the leading edges of the fingers, as shown in Fig. 23.4. The wing is then rotated to the 90-degree bend position. The setback clearance of the fingers should be just equal to the thickness of the metal to be worked. Any distance less than this thickness will cause the fingers to dig into the metal along the bend line as the wing is engaged. This will severely weaken the work and possibly damage the fingers. A setback distance larger than the thickness of the metal will result in a "sweeping" 90-degree bend. The radius of this type of bend may be greater than normally acceptable. If adjustment is necessary, the two locknuts securing the *finger-position adjustment screws* are released. These are located at either end and to the rear of the brake frame. *Clockwise* rotation of the screws will drive the crosshead forward, causing the fingers to move closer to the wing. *Counterclockwise* rotation will cause the fingers to be drawn away from the wing. With the fingers uniformly positioned at the desired distance across the entire length of the table, the locknuts are tightened to secure the adjusting screws.

The adjustments for setting the bending angle stops are made via the assemblies located immediately above and to either side of the wing. A rough adjustment is first made by inserting a pin into one of four holes provided on the inside face of each assembly. These rough angles are typically 30, 45, 60, and 90 degrees. Adjusting screws secured with locknuts are also associated with these assemblies. These screws allow for fine angle adjustment among the four rough adjustments obtained with the pins. The pins extend outward from the assemblies when seated in the desired hole position and prevent any further upward movement of the wing beyond the pin's angular position with respect to the brake table. It is important to remember that because of the flexing characteristics of most sheet metal, called *spring back,* the metal must be bent slightly *beyond* the desired angle to compensate for this effect. The angle adjustment settings, therefore, must be *slightly greater* than the intended angle. After initial adjustments are made, a sample piece can be bent and the resulting angle tested with a protractor head and blade (refer to Chapter 20). Final angle adjustments can then be made if necessary. The four brake adjustments outlined can be made independently and do not influence one another. However, each of the adjustments affects any resulting bend.

The work must be carefully aligned and inserted into the brake for an accurate bend. To obtain a sharp bend, the scribed bend line is positioned parallel to the leading edges of the fingers and aligned so that it *bisects* the finger clearance distance. This positioning is critical to achieve the proper dimensions resulting from the bend allowance considerations previously

**Figure 23.4** Finger setback clearance test.

discussed. When bending harder or thicker metals than 18-gauge mild steel, a larger radius is necessary to prevent fracture along the axis of the bend as the metal is formed. By adjusting the finger clearance distance to *twice* the thickness of the stock to be bent and bisecting this distance with the scribed bend line, a uniform sweeping bend will result.

The advantage of the finger brake, having individual fingers of varied widths, over the larger type brakes (*cornice* or *leaf type*), which operate with only the wing and one solid bending bar or leaf, will become apparent in the following section.

## 23.3 BENDING SEQUENCE

Any bending operation, irrespective of its simplicity, requires preplanning concerning the sequence and positioning of bends because of the limitations and restrictions of the brake. Random bends, without regard to sequence, may result in an impossible bending situation and waste of a time-consuming layout. Even with the simple U-shaped chassis shown in Fig. 23.5a, one critical limitation could result as a consequence of the dimensions of the work and the size of the brake. This configuration is to be bent from the scribed blank of Fig. 23.5b. The first bend can be either line *A* or line *B* and involves no problem as long as the width of the metal is less than the lateral capacity of the brake. In either case, however, the dimension *X* is critical. If the distance *X* is less than the distance measured from the table

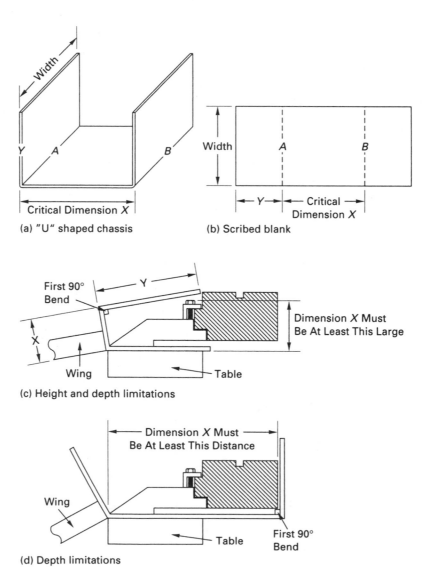

(a) "U" shaped chassis

(b) Scribed blank

(c) Height and depth limitations

(d) Depth limitations

**Figure 23.5** Bending sequence for U-shaped configuration.

to the top of the fingers, a second 90-degree bend is impossible. Attempting to form the second bend by placing the unbent side section under the fingers will result in the previously bent section contacting the top surface of the fingers before the desired bend can be completed. A bend less than 90 degrees will result, as shown in Fig. 23.5c. To bend this type of chassis, the distance X must be at least equal to the height of the finger-securing bolts. Again referring to Fig. 23.5c, note that dimension Y is also critical if finger clamp assembly or crosshead interference is to be avoided.

An alternative method of making this second 90-degree bend is to position the chassis under the fingers with the first bent section to the *rear* of the crosshead and the second bend line aligned with the finger edges (Fig. 23.5d). This technique introduces still another limitation of the brake. That is, dimension X must be greater than the distance measured from the leading edges of the fingers to the rear face of the crosshead.

Even with the simple U-type configuration, the aforementioned restrictions of the brake limit the minimum size of the chassis. These restrictions should be considered during the planning stages to produce a quality chassis easily. When a bend cannot be made on a brake or completed to the desired angle through an oversight in the bending sequence, a great deal of effort, skill, and ingenuity with wooden blocks, clamps, and soft-faced mallets will be required to obtain an acceptable bend.

To introduce further brake limitations and bending problems, a cover, as shown in Fig. 23.6a, for the U-shaped chassis will be formed. The layout of this cover, shown in Fig.

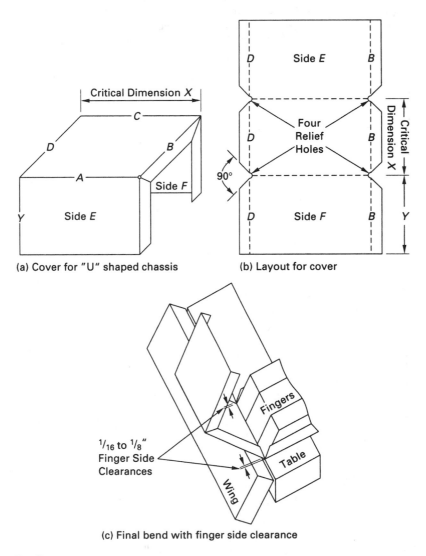

(a) Cover for "U" shaped chassis      (b) Layout for cover

**Figure 23.6** Bending sequence for U-shaped cover.

(c) Final bend with finger side clearance

*Bending*

23.6b, will require four 90-degree bends. (Two of these bends, *B* and *D,* will produce tabs.) A bending sequence that would appear to be reasonable would begin with bend *A* and continue bending clockwise or counterclockwise around the work. This sequence, however, will introduce another bending limitation. If bend *A* is formed first, the tabs associated with side *E cannot* be formed on the brake. Owing to this limitation, the following sequence must be followed to form this chassis. Bends *B* and *D* are formed first, followed by bends *A* and *C.* Figure 23.6c shows the last bend, *C,* being formed in accordance with this sequence. Since this cover has the same critical dimensions *X* and *Y* as the U-shaped chassis, similar brake limitations apply. Note also in Fig. 23.6c that just enough fingers are used to cover the bend line *A* or *C* between the previously bent tabs along lines *B* and *D.* If it is necessary to leave any length of bend line uncovered by the fingers, this distance should be limited to 1/16 to 1/8 inch between fingers or from the ends. This spacing will not cause any noticeable distortion along the bend.

It is always good practice to provide *relief holes* where stresses are incurred during bending, especially at corners where more than one bend is to be made. These relief holes, shown in Fig. 23.6b, prevent bulging or fracturing of the metal at these points of high stress. Note that four relief holes were drilled at the intersections of the bend lines. The diameter of these holes should be at least equal to the stock thickness.

The bending sequence for a box-type chassis, as shown in Fig. 23.7a, will be formed from the layout shown in Fig. 23.7b. This chassis consists of a *top,* four outside *aprons,* and four *subaprons.* The bends are numerically labeled to show the proper bending sequence to produce the desired results. All four subaprons (bends 1, 2, 3, and 4) are bent first. No special order or sequence of bending is necessary for these subaprons because their relative location and short width cause no restrictions on the brake. The four inside bends (5, 6, 7, and 8) are then made to form the aprons. Note that the first dimensional restriction of the chas-

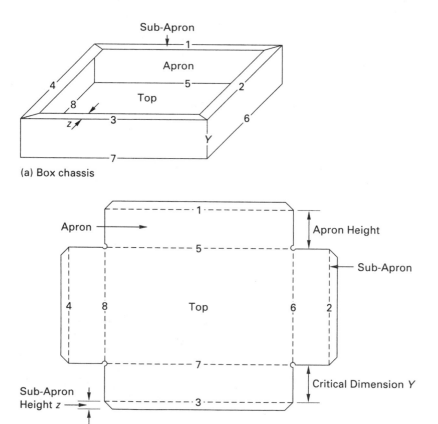

(a) Box chassis

**Figure 23.7** Bending sequence for box chassis.

(b) Bottom view of box chassis layout

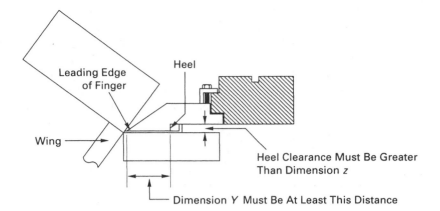

**Figure 23.8** Finger and heel clearance restrictions.

Leading Edge of Finger

Heel

Wing

Heel Clearance Must Be Greater Than Dimension *z*

Dimension *Y* Must Be At Least This Distance

sis is that the subapron height must be limited to the *heel clearance* height of the fingers because the subaprons must be placed *under* the fingers when bending the aprons (Fig. 23.8). A further restriction is the dimension *Y,* also shown in Fig. 23.8. This is the inside height of the finished apron as measured from the inside face of the subapron to the underside of the top. For the four aprons to be formed, the distance *Y* must be greater than the distance measured from the leading edge to the heel of the fingers. Figure 23.8 shows the final bend being made with all the aforementioned dimensional restrictions satisfied.

## 23.4 CORNER-TAB BOX CHASSIS

A box chassis with secured corners is the most widely used, since it is able to support heavy components. The U-type chassis and the unsecured box chassis are too weak to provide this support.

Figure 23.9 shows the unformed blank of a chassis element. This is a *corner-tab box chassis*. The bending operations begin by forming the four outside subaprons (bends 1, 2, 3, and 4). As in the unsecured box chassis, these subaprons may be bent in any order. It is

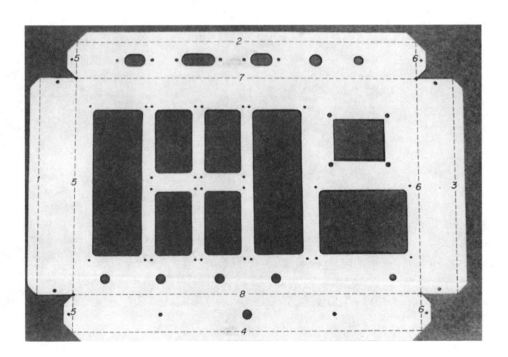

**Figure 23.9** Bending sequence for a corner tab box chassis with subaprons.

next necessary to make the inside bends for the tabs and aprons, beginning with bend 5. This will include bending *both* tabs to either side of the apron associated with bend 5 as well as the apron itself. Since the wing of most brakes is a single, straight unit, the tabs cannot be bent separately. As a result, all bends along bend line 5 must be formed at the same time. If this bend was formed in the usual manner, bends 7 or 8 would make the edges of the corner tabs contact the adjacent end apron edges, making it impossible to complete the 90-degree bend. For this reason, some means must be devised to provide the corner tabs with sufficient clearance so that they may be bent along a line that is offset behind the apron by a distance equal to the thickness of the metal. By providing this clearance, bends 7 and 8 can now be made, and the tabs will pass *behind* the end aprons, where they may be mechanically secured. To properly form line 5 to allow for this clearance, it is first necessary to reposition the fingers with the *finger-position adjustment screws* and to use *shims* to provide the proper finger offset and bend. (A shim is nothing more than a strip of sample stock used to determine the necessary setback distances.) The finger clearance is first set to a distance of *twice* the thickness of the metal. The work is next fed into the brake from *behind* the crosshead. With bend 5 aligned with the fingers, there will be enough fingers to cover *just* the length of the apron without extending beyond either relief hole onto a tab. *These* fingers are then moved *forward* from the crosshead by an amount equal to the thickness of the metal by placing a shim between the crosshead and the finger supports (Fig. 23.10). The shims used in this procedure have a width of approximately 3/4 inch. The shim length depends entirely on the width of the fingers that it is used with, but should be no longer than the length of the section of the bend line with which it is associated. (In this example, this length is bend line 5 *within the relief holes.*) It is informative to demonstrate what would happen if the bend were performed at this point. The apron section would be bent normally, whereas the tabs would have a bend radius *twice that of the apron* and *with no setback distance*. To achieve the desired setback distance, additional shims are placed between the tabs and the upper edge of the wing. These shims should *not* extend onto the brake table. Also, they should not extend laterally beyond a relief hole on either apron end. This technique is shown in Fig. 23.11. The bend resulting from this arrangement is shown in Fig. 23.12. Note that the tabs are offset the exact required amount behind the apron. Bend 6 can now be formed in the same manner as bend 5. The chassis is placed on the brake table with bend 5 (just completed) positioned to the rear of the crosshead. After bend 6 is completed, it will be necessary to remove all shims and fingers from the crosshead to retrieve the chassis.

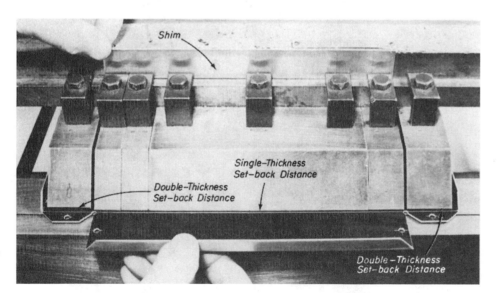

**Figure 23.10** Finger and shim arrangement to bend the short side of a corner tab box chassis.

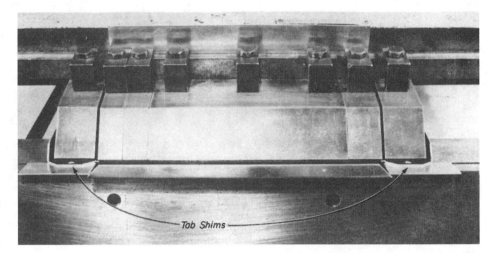

**Figure 23.11** Tab shims are used to obtain the offset single-thickness setback distance for tabs.

Tab Shims

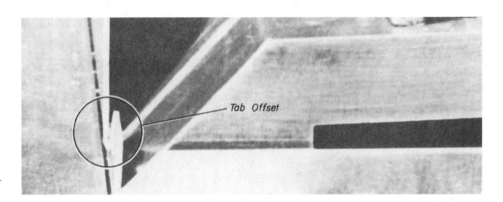

**Figure 23.12** Bend results in tab offset.

Tab Offset

To complete the bending of the chassis after bend 6 has been formed, the fingers are realigned and reset to a setback distance equal to the thickness of the metal. By using a total finger length that will accommodate the lengths of bends 7 and 8, these bends can be formed. When bend 7 or 8 is undertaken, the subaprons previously formed by bends 1 and 3 will prevent the leading edges of the outermost fingers being placed at the extreme ends of these remaining bend lines. If bend 7 is placed under the fingers from the front of the brake, a set of fingers covering the entire length of the bend can be employed to partially set the radius. A 90-degree bend from this position cannot be formed. When the apron is bent to approximately 30 degrees, the tabs will contact the faces of the fingers. This partial bend, however, is sufficient to permit the remainder of the bend to be completed by reducing the total finger width to allow for the necessary side clearance for the subaprons. Although the subaprons now prevent the fingers from covering approximately 3/8 inch at either end of the bend line, no noticeable distortion will result when the bend is completed because the radius was basically established by the first partial bend. Bend 8 is formed in the same manner to complete the chassis configuration.

The bending techniques discussed in this chapter are applied to form the chassis shown in Fig. 23.13. The final stage in the construction of a chassis is to apply a finish that will not only protect it but will also enhance its appearance. The various finishes and techniques of finishing are discussed in the following chapter.

**Figure 23.13** Chassis element bent into desired form.

## EXERCISES

### A. Questions

23.1 What is meant by *bend allowance,* and why is it an important factor in bending sheet metal?

23.2 Describe the distortions that result when bending sheet metal.

23.3 A bracket is to be bent into a U shape using 16-gauge 3000 series aluminum. Determine the inside and outside dimensions in decimals if the bend lines are 4 inches apart.

23.4 What is *setback distance?*

23.5 What is the advantage of a brake with individual fingers when compared to a cornice or leaf-type brake?

23.6 What is the purpose of relief holes?

### B. True or False

Circle *T* if the statement is true, or *F* if any part of the statement is false.

23.1 Bending sheet metal compresses the outer section of the bend and stretches the inner section.     **T**   **F**

23.2 When forming a box chassis, the subaprons are bent last.     **T**   **F**

23.3 Subaprons are bent first when forming a U-shaped cover.     **T**   **F**

23.4 The angle adjustment setting must be made slightly greater than the desired angle to compensate for the *spring-back* characteristics of sheet metal.     **T**   **F**

23.5 Gauge number 12 aluminum is thinner than gauge number 20.     **T**   **F**

### C. Multiple Choice

Circle the correct answer for each statement.

23.1 Bend allowance is required to compensate for the distortion generated during the bending operation that (*stretches, compresses*) the inner section of the bend.

23.2 On a hand-operated finger brake, the (*wing, crosshead*) holds the fingers.

23.3 Clamping pressure on a finger brake should be set using (*once, twice*) the thickness of sample stock.

23.4 Finger setback clearance for the tabs of a corner tab box chassis should be set using (*once, twice*) the thickness of sample stock.

23.5 Heel clearance limits the height of the (*apron, subapron*) of a box chassis.

23.6 The bending sequence for a standard box chassis requires that the subaprons be bent (*first, last*).

23.7 The bending sequence for a basic U-shaped chassis with tabs requires that the tabs be formed (*first, last*).

## D. Matching

Match each item in Column A to the most appropriate item in Column B.

COLUMN A

1. Finger brake    *c*
2. Setback distance    *e*
3. Relief holes
4. Angle stop
5. Bend allowance
6. Heel clearance

COLUMN B

a. Chassis corners
b. 30, 45, 60, 90 degrees
c. Box or pan
d. Finger relief
e. Finger clearance
f. Stretching

## E. Problems

23.1    Bend the apron of the gauge panel laid out, drilled, and cut in Problems 20.3, 21.1, and 22.3 to 90 degrees from the front section.

23.2    Bend the four ends of the heat sink laid out and drilled in Problems 19.3 and 20.2 to 90 degrees from the center section and all in the same direction.

23.3    Considering bend allowance, design, lay out, and construct a utility chassis similar to the one shown in Fig. 23.5a, including its cover shown in Fig. 23.6a, with final outside dimensions of 3 inches by 5 inches by 7 inches.

23.4    Considering bend allowance, design, lay out, and construct a box chassis similar to the one shown in Fig. 23.9, with 3/8-inch tabs and subaprons and final outside dimensions of 2 inches by 6 inches by 9 inches.

# 24 Metal Finishing and Labeling

## LEARNING OBJECTIVES

*Upon completion of this chapter on metal finishing and labeling, the student should be able to*

- Be familiar with the types of contaminants commonly found on aluminum and steel.
- Be familiar with the technique of spray painting onto metal surfaces.
- Use aluminum foil plates and dry transfers to label chassis and panels.
- Use a silk screen for labeling metal chassis.

## 24.0 INTRODUCTION

Most manufactured parts used in electronic packaging require a *finish* to enhance and protect their surfaces. This is especially true of metal elements whose corrosive characteristics require protection. A prerequisite for obtaining a sound finish is the proper removal of all surface contaminants. The technique of spray painting the chassis is presented, followed by information on the application of literal and numerical reference markings on chassis and panels.

*CAUTION:* The processes discussed in this chapter should be performed in adequately ventilated facilities with proper fume exhaust systems. Be sure the fume exhaust systems are turned on when using solvents to clean surfaces for finishing and applying spray finishes or silk-screen inks. A chemical hazard exists in the materials presented in this chapter. Approved face shields (goggles) in compliance with ANSI Z87.1-1989 and chemical-resistant clothing and footwear and neoprene or nitrile gloves must be worn while performing any of these operations. Also, the section on safety outlined at the beginning of the book should be reviewed for more specific information.

## 24.1 SURFACE PREPARATION AND FINISHING

Metal surfaces that are to be finished must be free of contaminants. The most common *soils* absorbed by aluminum and mild steel are (1) *water vapor,* (2) *fingerprint deposits,* (3) *oils,* (4) *lubricants,* (5) *paints,* and (6) *mill scale* (this term is used to describe the contaminants present on the sheet metal as a result of the manufacturing processes). If these soils are not removed, an inferior bond between the surface and the finish will result.

Cleaning is effectively accomplished with the use of *isopropyl alcohol.* Simply hand wiping or dipping the metal parts into the alcohol will remove most surface contaminants. After the metal has been allowed to dry, finishing and labeling may be applied.

Of all the materials available for finishing applications, the most economical and easiest to apply are *paints, enamels, varnishes,* and *lacquers.* They also offer the widest selection of colors and textures among all surface finishes. These finishes are made from *binders* or *vehicles* (i.e., oils or synthetic resins), *pigments,* and *solvents.* The composition of these finishes is as follows:

*Paint:* vehicle and a pigment

*Enamel:* paint and varnish

*Varnish:* vehicle and solvent

*Lacquer:* vehicle, solvent, and nonvolatile coloring additive

Paints and enamels are used extensively on metal surfaces common to electronic packaging. They provide a heavier and more durable coating than either lacquer or varnish and have the advantage of hiding minor surface imperfections such as scratches or nicks in metal. Paint and enamel finishes range from *lusterless* (flat) to *high gloss.* The degree of gloss is determined by the proportion of pigment to vehicle. Semigloss to flat finishes are most commonly used in electronic packaging applications. Some of the more common paint textures available are *wrinkle, crackle,* and *hammertone.* They not only hide surface imperfections but also enhance the overall appearance of the chassis. The *wrinkle*-type enamel tends to *expand* as it cures. This causes small, randomly oriented wrinkles of paint to form over the entire surface. The *crackle*-type enamel is applied over a color base. As this finish cures, it *contracts* and forms very fine random lines. The base color contrasts with the finish color, producing an extremely attractive package. The *hammertone* enamel leaves a very smooth surface when it cures, yet provides a hand-hammered appearance. This finish is available in a variety of colors.

Varnishes are used when a surface is to be protected from moisture and mildew. Lacquers are often used clear or unpigmented to protect labeling from abrasion or to help protect etched or plated surfaces from staining or tarnishing. Lacquers are water resistant but will not stand up to saltwater or prolonged exposure to sunlight. These critical environments will cause the lacquer to discolor and become brittle. Plastic sprays of the acrylic or vinyl type are readily available in aerosol cans. These sprays do not possess many of the undesirable characteristics of lacquer and varnish. Plastic sprays are available either clear or in a choice of colors.

*Aerosol-pressure spray cans* are limited to painting small surface areas because of their cost. These cans are pressurized to approximately 40 pounds per square inch with a refrigerant, which acts as the propellant for atomizing the paint. Spray painting must be done in a well-ventilated spray booth that will exhaust both the solvent and excess paint particles. In addition, atomized paints are highly combustible and should be used with caution. To aid in obtaining a uniform coating, especially over irregularly shaped surfaces, the work may be placed on a turntable. This eliminates having to handle the work to paint the reverse side. For best results, the aerosol can is held perpendicular to the work surface, which is held vertically. Spraying is begun at the top of the work with the nozzle held 6 to 10 inches from the surface. The spray is swept across the work in a straight-line motion continuing beyond the edge of the work. To ensure proper coverage, each sweep of spray should slightly overlap the last pass when moving the spray can to either the left or right. Once the surface has been completely covered, the paint must be allowed to cure. If a single coating does not provide sufficient coverage, a second coat may be applied *only after* the first coat is completely cured. For superior results, it is always good practice to apply two coats rather than one thick coat. This reduces the possibility of the paint sagging or running. For uniform paint thickness when the work is irregularly shaped, the more difficult and unexposed areas are sprayed first and then the open surfaces. Some finishes cure at normal room temperatures but others require a drying oven. Most paints can be dried in about 20 minutes at 225 to 250°F (107 to 121°C). Manufacturers' curing specifications are available in their literature. Figure 24.1 shows a panel being spray painted using an aerosol can.

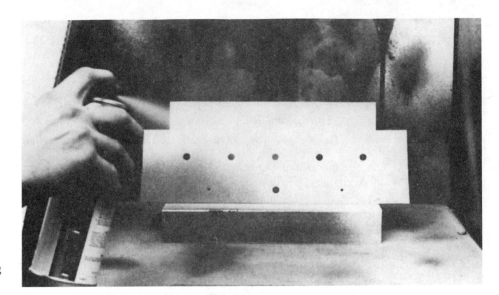

**Figure 24.1** Use of aerosol-pressure spray cans simplifies finishing small parts.

## 24.2 LABELING

The final step in completing a chassis and panel involves applying literal or numerical reference designations. The type and procedure of labeling will depend on several factors, including (1) the degree of clarity required, (2) cost, (3) time required for process, (4) durability, (5) materials and finishes involved, (6) availability of equipment and facilities, and (7) the complexity of the labeling layout pattern. The most common methods of labeling are by *decals, dry transfers,* and *silk screen.*

Pressure-sensitive decals and dry transfers, such as those shown in Fig. 24.2, provide a rapid means of labeling with crisp and sharp line characteristics. They are used extensively with the silk-screen method of labeling. Decals and transfers will adhere to practically any clean, smooth surface, such as wood, paper, cardboard, plastic, glass, or metal. To improve their durability, they should be given a protective clear coating after they have been applied.

Pressure-sensitive decals have a protective backing that must be removed before they can be applied. Because literal and numerical decals are small, it is best to use tweezers or a knife edge to remove them from the backing, which is chemically treated for easy release.

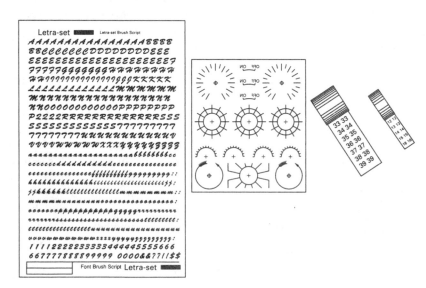

**Figure 24.2** Typical selection of pressure-sensitive decals and dry transfers.

This technique reduces the possibility of touching the adhesive with the fingers and upsetting the tack. Once the backing has been removed, the decal is brought into the desired position, aligned, and pressed firmly into place. Occasionally, it may be necessary to trim the clear support liner if overlapping occurs.

Dry transfers used in electronic packaging consist of drafting symbols, letters, words, numbers, and meter and dial markings printed on a transparent film with opaque black ink. This ink contains a pressure-sensitive heat-resistant adhesive. Before applying the transfer figure to a surface, the protective backing sheet is first removed. This sheet does not adhere to the film and therefore separates readily. The figure is brought into position with the inked side *toward* the work surface and is held firmly in place. The entire area around the figure is rubbed with a wooden burnisher, pencil, or ballpoint pen using overlapping strokes. Figure 24.3 shows dry transfers being applied to a chassis element. Since many figures or symbols are generally on the same sheet and closely spaced, care must be taken to avoid running over unwanted figures or portions of figures. By gently lifting one edge of the film, a check for complete transfer can be made. To assure perfectly transferred figures with sharp lines and no pinholes, the backing sheet should be placed back over the work surface and the figures reburnished. A wax residue resulting from the transfer process is generally formed on surfaces such as glass or metal. This residue can easily be removed with rubber cement thinner and a soft cloth without damaging the transfers. When dry transfers are used for direct labeling on a surface, a *plastic*-type protective film is necessary. These acrylic or vinyl films should be selected carefully because they attack certain materials. Vinyls are often preferred over acrylics because they are more resistant to flaking and chipping. These films are available in aerosol cans in both gloss and matte finishes.

The silk-screen process for labeling requires the preparation of a *clear polyester positive master* when identical duplication of work is required. Unlike the previously discussed methods of labeling, the silk-screen process is intended for mass production. (The silk-screen operation for producing marking masks as well as printed circuit conductor patterns is described in detail in Chapter 10.) Decals or dry transfers allow the positive artwork to be constructed quickly and eliminate time-consuming drafting procedures that would otherwise be necessary. A thin sheet of drafting *vellum,* which is slightly larger than the surface to be labeled, is first placed over the sheet metal layout drawing. The corners, mounting holes, and hole centers are carefully located by *pencil* on the vellum. Allowing for

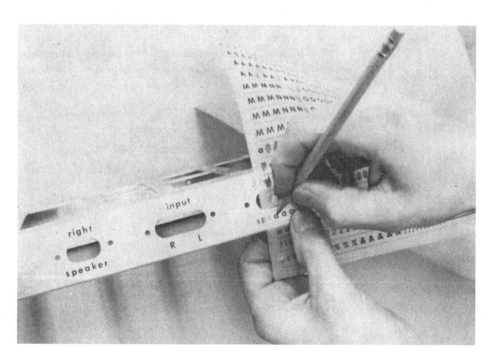

**Figure 24.3** Application of dry transfers to metal surface.

surface space to be taken up by mounting nuts, bezels, and control knobs, the required nomenclature and symbol locations together with the necessary spacing are *sketched* or *blocked* in. Crowding should be avoided and the spacing should be standardized to avoid confusion. Whenever possible, nomenclature should be consistently above or below components or controls. For a balanced appearance, all lettering should be centered on a vertical center line of a mounting hole. After all labeling positions have been located, the vellum is placed over a comparable size sheet of graph paper and taped into place on a drafting board. (Graph paper with 10 divisions to the inch is recommended.) A *polyester sheet* is then taped over the vellum. With the aid of the grid lines and the blocked-in nomenclature and lettering positions, all decals or transfers can be accurately positioned and applied onto the polyester using the procedure shown in Fig. 24.3.

For the silk-screen process, described in detail in Chapter 10, a *film negative* is contact-printed from the polyester master. The developed film is allowed to dry on a fine mesh silk screen attached to a frame. The screen is positioned over the metal panel to be labeled and the appropriate paint is spread across the screen with a squeegee. The paint will penetrate the screen and be deposited onto the panel only through those areas seen as clear. This technique is shown in Fig. 24.4a. The screen is then removed and the characters are allowed to dry. Figure 24.4b shows a panel labeled by the silk-screen method.

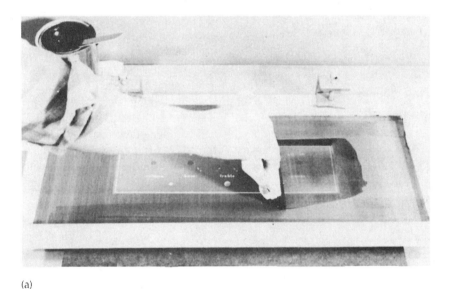

(a)

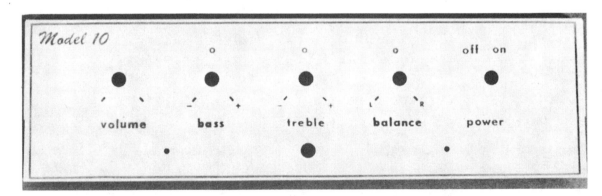

(b)

**Figure 24.4** Silk-screen method for panel labeling: (a) application of paint through the screen onto a panel; (b) screened panel.

In this chapter, only the most common types of labeling techniques have been discussed. Although many other processes are commercially available, they are not intended for prototype work. Of those labeling methods considered, decals and dry transfers are the most economical, least involved, and produce a highly acceptable appearance.

With so many variations and techniques available for finishing and labeling, the technician must consider the following factors prior to determining the most appropriate method to use: surface materials, number and complexity of markings, required sharpness of detail, equipment availability, cost, time required, and the technician's skill. An evaluation of these factors will dictate the appearance of the finished product.

The six chapters of this unit concentrated on techniques and methods for fabricating sheet metal chassis elements. Therefore, with all subassemblies for a project fabricated—printed circuit boards and sheet metal chassis elements—the technician can easily complete a project. In Unit 6, chassis hardware assembly and wiring techniques are covered to combine the various elements into a finished system.

## EXERCISES

### A. Questions

24.1 List the most common soils absorbed by aluminum and steel that must be removed prior to finishing operations.

24.2 Describe in detail the proper procedure for applying spray paint to a vertical surface.

24.3 What is the advantage of using a turntable when spray painting?

24.4 List the composition of (1) paint, (2) enamel, (3) varnish, and (4) lacquer.

24.5 List some of the factors that must be considered in determining the most appropriate finish and labeling to use.

### B. True or False

Circle *T* if the statement is true, or *F* if any part of the statement is false.

| | | | |
|---|---|---|---|
| 24.1 | The composition of enamel is paint and varnish. | **T** | **F** |
| 24.2 | Crackle-type enamel tends to expand as it cures. | **T** | **F** |
| 24.3 | Binders are oils or synthetic resins. | **T** | **F** |
| 24.4 | Reburnishing dry transfers through the backing sheet will produce pin holes. | **T** | **F** |
| 24.5 | The silk-screen process for labeling requires the preparation of a clear polyester positive master. | **T** | **F** |
| 24.6 | Vinyls are often preferred over acrylics because they are more resistant to flaking and chipping. | **T** | **F** |

### C. Multiple Choice

Circle the correct answer for each statement.

24.1 The most common method of labeling is (*decals, stenciling*).

24.2 The composition of varnish is (*vehicle and pigment, vehicle and solvent*).

24.3 Most paints can be dried in about 20 minutes at (*225 to 250°F, 275 to 300°F*).

24.4 A clear polyester (*negative, positive*) master is needed when identical duplication of work is required in the silk-screen process.

24.5 Wrinkle-type paint tends to (*contract, expand*) as it cures.

## D. Matching

Match each item in Column A to the most appropriate item in Column B.

| COLUMN A | | COLUMN B | |
|---|---|---|---|
| 1. | Soils | a. | Decal |
| 2. | Wrinkle | b. | Positive master |
| 3. | Film negative | c. | Textures |
| 4. | Polyester | d. | Silk screen |
| 5. | Pressure sensitive | e. | Water vapor |

# SIX

# Chassis Hardware
# and Wiring

# 25 Chassis Hardware and Assembly

*Upon completion of this chapter on chassis hardware and component assembly, the student should be able to*

- Be familiar with the order in which parts and hardware are assembled.
- Know the various types of machine screws, nuts, washers, and thread-forming and thread-cutting screws.
- Use the following tools for assembling parts to a chassis:
    a. Screwdrivers
    b. Nutdrivers
    c. Wrenches
    d. Pop rivet guns
    e. Pliers
- Be familiar with the various types of special fasteners and hardware used in chassis assembly.

## 25.0 INTRODUCTION

With the completion of all subassemblies (the pc boards), the assembly of hardware and components onto the main chassis element may begin. In addition to the boards, this involves the mechanical securing of jacks, connectors, fuse holders, terminal strips, switches, potentiometers, lamps, meters, power cords, and transformers.

Before beginning to assemble parts, the overall task should be examined. If there are few components and parts to be assembled and if they are small, random mounting will pose no particular problem. If, however, the assembly involves many component parts with a wide range of sizes and weights, some order will be necessary to minimize needless effort and reduce the possibility of damaging delicate components.

Small, sturdy components such as fuse holders, terminal strips, lugs, grommets, and connectors should be mounted first. Transformers, filter chokes, and other large or bulky components should be among the last to be mounted, to avoid having to work with a heavy, bulky chassis during the initial assembly phase. Fragile components such as lamps, meters, sensitive relays, and pc boards should be the last to be mounted. As working space diminishes with the addition of components, care must be exercised to avoid damaging parts with assembly tools.

Mounting electronic components and hardware to a metal chassis may appear to be a simple task. However, efficient and quality workmanship will result only if these four major factors of packaging are considered: (1) mechanical security, (2) appearance, (3) accessibility, and (4) cost. Each factor is discussed in this chapter in conjunction with specific information on common and special fastening devices. Also considered are correct tool selection and assembly techniques for securing components and hardware.

*CAUTION:* Prior to performing any of the operations described in this chapter, approved safety glasses with sideshields in compliance with ANSI Z87.1-1989 must be worn. Keep your free hand clear when using screwdrivers to avoid injury if the driver slips out of position when exerting pressure. Be sure to hold pliers and pop rivet tools properly to avoid receiving an uncomfortable pinch. Also, the section on safety outlined at the beginning of the book should be reviewed for more specific information.

## 25.1 MACHINE SCREWS

Machine screws are among the most common types of fasteners used in electronic packaging since they provide sound mechanical security and are easy to assemble and remove. Machine screws are designated by (1) *head style,* (2) *drive configuration,* (3) *diameter,* (4) *threads,* (5) *length,* (6) *material,* (7) *finish,* and (8) *fit.*

Some of the common head configurations are the *fillister, binder, pan, round, flat, oval,* and *truss* styles (see Fig. 25.1a). Although somewhat of a personal preference, the head style selected is generally determined by appearance, application, and mechanical security. Fillister, binder, and pan heads are usually preferred over round heads for appearance's sake. These head styles are mechanically superior to the round head because there is more material at the driver blade pressure points. These head styles have approximately the same-size sealing plane (surface contact area) for a particular screw size, although the fillister style has the smallest head diameter. Both the flat and oval heads require that the work surfaces be countersunk prior to assembly. The flat head is used for flush mounting and has sealing plane angles of 82 or 100 degrees. Neither the flat nor the oval heads are recommended for use with thin sheet metal, wherein countersinking might weaken the assembly. The oval head is often used with a cup-style washer for securing control panels to racks. Countersinking in this case is unnecessary since the shape of the washer provides the required seating. This arrangement tends to provide both a decorative appearance as well as excellent mechanical security. The truss head style, common to sheet metal screws, has a larger sealing plane that provides greater surface contact area with the work.

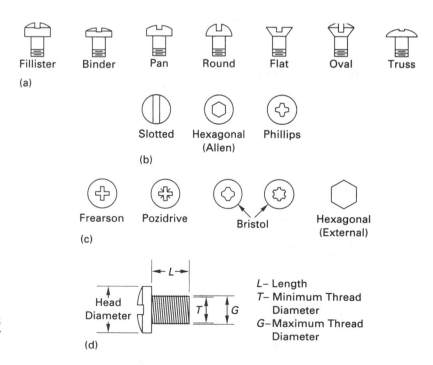

**Figure 25.1** Machine screw styles and drives: (a) machine screw head styles; (b) head-style drive configurations; (c) drive configurations requiring special driving tools; (d) machine screw thread measurements.

*Chassis Hardware and Assembly*

Drive configuration refers to the style of head for tool access. Those most common to electronic assemblies are the (1) *slotted,* (2) *Phillips,* and (3) *hexagonal socket,* or *Allen* design (see Fig. 25.1b). The slotted style is the most widely used. The Phillips head has the advantage over the slotted type in that it provides four positive-drive pressure points completely within the perimeter of the head. It is often used when the screw is visually obstructed during assembly, since it allows the screwdriver to be self-aligned. In addition, the recessed style of the Phillips head greatly reduces the possibility of the driver slipping from the screw head, which could damage the work surface. The Allen configuration allows for even more positive tool positioning and gripping and is commonly used on set screws for securing knobs to control shafts. Other less common drive configurations that may be encountered are shown in Fig. 25.1c. Each of these requires special driving tools. The various driving tools common to electronic assembly are discussed in Sec. 25.4.

The maximum thread diameter (dimension $G$ in Fig. 25.1d) varies from 0.086 to 0.216 inch for general electronic assembly; the minimum diameter, shown as dimension $T$ in Fig. 25.1d, will vary from 0.0628 to 0.1619 inch. Small sizes of machine screws are generally designated by *gauge number.* For the thread diameters referred to, the gauge numbers vary from No. 2 having a diameter of 0.086 inch, to No. 12, having a diameter of 0.216 inch. Larger thread diameters are given in fractions of an inch. Smaller-diameter screws, with gauge numbers of 1 or 0, are not common in electronic assembly. A list of some common thread gauge numbers together with their decimal equivalents is shown in Table 25.1. A more complete listing is provided in Appendix III. One way of determining the gauge number for an unknown screw size is simply to measure the thread diameter with a micrometer and compare the reading thus obtained with the corresponding gauge number of Table 25.1.

Machine screws are manufactured with *right-* and *left-*hand threads. Right-hand threads are secured by clockwise (cw) rotation and extracted by counterclockwise (ccw) rotation; left-hand threads are rotated in the opposite directions and are used only in special cases wherein vibration or rotation of the system would tend to loosen the conventional right-hand thread.

Machine screws are also designated by the number of threads per inch. The simplest method of determining this is by placing a scale against the threads and counting the number of threads in 1 inch. For a particular screw gauge or diameter, the number of threads per inch will depend on the screw thread system used. Although there are other systems of screw designations, machine screws having either *Unified Coarse* (UNC) or *Unified Fine* (UNF) designations are predominantly used for electronic assembly. Referring to Table 25.1, a UNC screw having a gauge No. 4 diameter has 40 threads per inch, whereas a UNF screw with the same gauge number has 48 threads per inch. It is apparent that the UNC system has fewer threads per inch than the UNF system. The coarseness of the screw selected will depend on its application. Generally, UNC threads are preferable for rapid assembly since thread meshing is a less exacting process. This coarser screw is also used

**TABLE 25.1**

Machine Screw Sizes Most Frequently Used in Electronics

| Gauge Number | Decimal Equivalent (in.) | Threads per Inch | |
| --- | --- | --- | --- |
| | | UNC | UNF |
| 2 | 0.0860 | 56 | 64 |
| 4 | 0.1120 | 40 | 48 |
| 6 | 0.1380 | 32 | 40 |
| 8 | 0.1640 | 32 | 36 |
| 10 | 0.1900 | 24 | 32 |
| 12 | 0.2160 | 24 | 28 |

when threading into soft material. Since a machine screw of a given size with a coarse thread has a smaller minimum diameter than the fine thread, it will mesh deeper into the material surrounding the tapped hole. This advantage greatly reduces the possibility of thread shearing when an axial load is placed on the screw. If vibrations are a consideration, fine threads are preferable. Because of their greater minimum diameter, they can be torqued more than comparably sized coarse-thread screws and will not shake loose as readily.

For all head styles except the flat and oval designs, the thread length is measured from the sealing plane to the end of the screw (Fig. 25.1d). Flat-style and oval-style screw lengths are measured from the top of the sealing plane to the end. Screw lengths between 1/8 and 5/8 inch are available in increments of 1/16 inch. Lengths from 5/8 to 1 1/4 inches are graduated in 1/8-inch increments, whereas those from 1 1/4 to 3 inches are graduated in 1/4-inch increments.

Screws for electronic assembly are made from *steel, brass, aluminum,* or insulating materials such as *nylon.* The finishes available include *black anodizing, cadmium, nickel, zinc,* or *chrome* platings. The finish selected will depend on the appearance and the necessary corrosion protection.

Screws are also designated with a specific relationship (fit) between mating threads of nuts or tapped holes and machine screws. There are four classes of fit between threaded parts: *Class 1* is an extremely loose fit with considerable "play" between the threaded parts; *Class 2,* sometimes designated as a "free fit," can be readily assembled with finger pressure; *Class 3* has more of a snug fit than Class 2 but requires only finger pressure for initial assembly; *Class 4* is characterized by the tightest available fit, requiring screwdrivers and wrenches for assembly of the threaded parts. Selecting the most appropriate class of fit will depend on such factors as (1) speed of assembly, (2) type and style of washer employed, and (3) the amount of looseness between threaded parts that can be tolerated. For most electronic applications, the Class 2 fit is standard.

Machine screws may be completely identified by the following abbreviated description:

$$\#6\text{--}40 \times 3/8 \text{ UNF---2A LH Pan Hd. Steel C.P.}$$

where

> #6 is the major thread diameter (equivalent to 0.138 inch; refer to Table 25.1).
>
> 40 is the number of threads per inch.
>
> 3/8 is the length of thread in inches measured from the sealing plane to the screw end.
>
> UNF is the type of thread (Unified Fine). If the type was UNC, this designation would normally be omitted and assumed to be understood.
>
> 2A—the 2 represents the class of fit between threaded parts and the A designates that the thread is external. A B suffix designates an internal-type thread for machine nuts.
>
> LH is the left-hand thread. If a right-hand was specified, no designation is given since it is assumed to be understood.
>
> Pan Hd. is the head style.
>
> Steel is the material.
>
> C.P. is the plating (cadmium-plated).

Clearance hole sizes for several machine screws are given in Appendix IV. The clearance hole size is determined by the maximum diameter of the screw thread. This type of hole is drilled when components and hardware are to be assembled with screws and nuts. The diameter of a clearance hole must be sufficiently larger than the thread diameter to allow complete unobstructed passage of the screw through the hole. The amount of clearance will depend on the materials to be fastened. However, the values given in Appendix IV are average sizes, which are satisfactory for most applications. A clearance hole together with its tolerance is shown in Fig. 25.2.

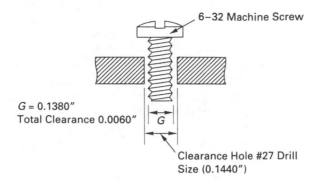

G = 0.1380"
Total Clearance 0.0060"

**Figure 25.2** Clearance hole for a 6–32 machine screw.

Clearance Hole #27 Drill Size (0.1440")

## 25.2 NUTS AND WASHERS

Nuts used with machine screws are designated by (1) *style,* (2) *chamfer,* (3) *hole diameter and threads,* (4) *fit,* (5) *material,* (6) *finish,* (7) *width,* and (8) *thickness.* Nuts common to electronic assembly are shown in Fig. 25.3a. The hexagonal style is by far the most widely used and is generally preferable to the square nut for reasons of appearance and ease of assembly. This nut may have *single chamfer, double chamfer,* or *single chamfer* and *washer face* (Fig. 25.3b). When a single chamfer nut is used, the flat side should be against the work to provide the best appearance with no exposed sharp edges. If the work surface is finished, double chamfer or washer face nuts should be used. When tightening the nut against the work surface, the chamfer or washer face will prevent any visible surface marring, such as the circular abrasions formed by the points of a single chamfer square or hexagonal nut.

The hole diameter and thread type are, naturally, consistent with the mating screw. This is also true for class of fit, material, and finish. For each size and style of nut, different widths (measured across pairs of opposite flats) and thicknesses are available. These external dimensions are shown in Fig. 25.3c. Tabulated sizes for common hexagonal nuts are given in Appendix V. Nuts with larger widths are normally used for ease of handling and rapid assembly, whereas those with smaller widths are selected if there is a possibility of interference with other hardware or if assembly clearance is limited.

The special-purpose nuts, also shown in Fig. 25.3a, have styles that dictate their application. A description and application of each follows:

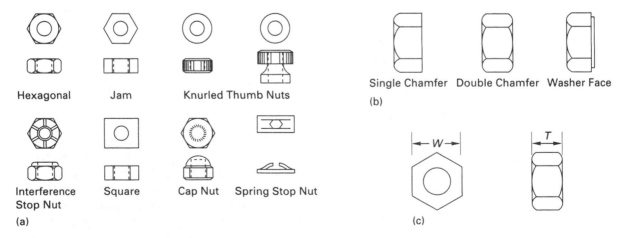

Hexagonal    Jam    Knurled Thumb Nuts

Single Chamfer   Double Chamfer   Washer Face

(b)

Interference Stop Nut    Square    Cap Nut    Spring Stop Nut

(a)

(c)

**Figure 25.3** Machine screw nuts and fasteners: (a) nuts common to electronic assembly; (b) hexagonal nut face styles; (c) width and thickness measurements for hexagonal nuts.

*Jam nut:* Basically the same as a hexagonal nut, except that it is from 1/16 to 3/8 inch thinner, depending on the hole diameter. This nut may be used alone, but it is more often employed under a regular hexagonal nut for a more secure assembly. When this nut is tightened, its flats align with those of the hexagonal nut, providing the desired locking action.

*Interference stop nut:* Used to prevent loosening due to vibration; it has an internal unthreaded plastic or fiber collar. When the nut is tightened, the screw threads are forced into the inside surface of this tapered collar. Being elastically deformed, the collar squeezes against the screw threads, providing a secure locking action and eliminating the need for a lock washer.

*Square nut:* Often used in securing channel stock in steel cabinets or racks, the flat of the nut often fits snugly or jams against the inside member when being tightened, thus requiring merely a screwdriver for assembly.

*Knurled thumb nut:* Allows for quick finger tightening. It is commonly used on terminal posts such as those on some primary cells and telephone terminal blocks.

*Cap nut:* Used primarily for purposes of appearance, it completely covers the end of the screw and its shape eliminates sharp corners or edges. Chrome-plated cap nuts add a decorative appearance to a cabinet.

*Spring stop nut* (speed nut): Pressed onto the screw with only minimal final tightening required. It is self-retaining as well as self-locking. Because it greatly reduces assembly line operations, the speed nut is often used in radio and television receivers. It may also be used to secure smooth studs or tubing.

*Washers* are classified as either *flat* or *lock type* (Fig. 25.4). Flat washers are used in contact with a screw head or the chamfer of a nut to provide more surface area to distribute the forces imposed on the screw and nut. Lock washers are used to secure square or hexagonal nuts from loosening due to vibrations or handling. For electronic assembly, the *tooth-type* lockwasher is used extensively since it provides maximum gripping action. Normally, the use of one lock washer is sufficient. However, under severe vibrations lock washers should be placed under both the nut and screw head. The tooth-type lock washer should not be reused when removed. The gripping edges of the teeth are easily dulled and the spring action becomes fatigued. These washers have from 10 to 24 combined points of contact, depending on the number of teeth. *Internal* tooth washers are preferred over the *external* type from the standpoint of appearance since they cause no visible surface marring when assembling components such as the potentiometer shown in Fig. 25.5a. External tooth washers are generally used with flat washers for mounting components having oversized clearance holes. This arrangement is shown in Fig. 25.5b.

The *spring-type,* or *split,* lock washer, also shown in Fig. 25.4, uses the temper of the steel to maintain locking action. As the washer is compressed, the spring tension exerts a force that prevents counterrotation. This type of washer is less effective than the tooth type because the locking action results only from spring tension. Flat steel washers are used in conjunction with tooth-type lock washers when securing soft materials, whereas resilient

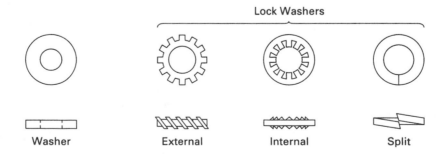

**Figure 25.4** Washer styles.

Lock Washers

Washer　　　　External　　　　Internal　　　　Split

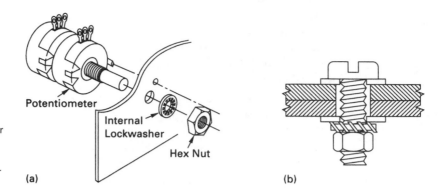

**Figure 25.5** Washer assembly techniques: (a) washer assembly for potentiometer mounting; (b) washer assembly for oversized clearance holes.

**(a)**

Potentiometer

Internal Lockwasher

Hex Nut

**(b)**

washers, such as lead or fiber, are used to provide a cushion effect when assembling brittle material.

The inside diameter of a washer must coincide with the clearance hole diameter of the screw with which it is associated. The outside diameter will depend on the application. Recommended washer sizes are given in Appendix VI.

## 25.3 THREAD-FORMING AND THREAD-CUTTING SCREWS

Thread-forming and thread-cutting screws are widely employed when the use of nuts or the tapping of machine threads is not practical. These screws do not require a prethreaded hole for fastening. Manufactured from hardened steel, they form or cut their own threads as they are turned and driven into a hole. They cannot be driven into a material harder than their own. These screws are available in slotted, hexagonal, and Phillips drives with head styles common to machine screws, in addition to the washer head style shown in Fig. 25.7. Three of the popular thread designs, shown in Fig. 25.6, are the sheet metal (thread-forming), machine screw thread-forming, and self-tapping (thread-cutting) types. The sheet metal screw, shown in Fig. 25.6a, when forced into the proper-sized predrilled hole, will form the metal around the hole into a threaded shape matching its own thread. This screw has a widely spaced thread extending to a point. The machine screw thread-forming type, shown in Fig. 25.6b, has a slightly tapered end but does not extend to a point. It has standard machine screw threads, which require greater force to drive than a sheet metal screw. This screw is used in place of the sheet metal screw when the possibility of loosening from vibration exists. The self-tapping screw, shown in Fig. 25.6c, has one or more slots in the threaded end for chip relief. As this screw is driven into a hole, it cuts threads in basically the same manner as a tap. The self-tapping screw is not used on sheet metal but is intended for fastenings to castings and soft materials, such as Bakelite, copper, aluminum, and plexiglass.

Sizes of thread-forming and thread-cutting screws are similar to those for machine screws. To result in optimum mechanical security, the size of the predrilled hole is impor-

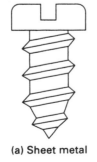

**Figure 25.6** Thread-forming and thread-cutting sheet metal screw styles.

**(a) Sheet metal**

**(b) Machine**

**(c) Self-tapping**

## TABLE 25.2
Recommended Drill Sizes for Sheet Metal Screws

| Screw Size | Thread-Forming Machine and Sheet Metal Screws | | | | | | Self-Tapping (Thread-Cutting) Screws | | | | | | |
|---|---|---|---|---|---|---|---|---|---|---|---|---|---|
| | 28 | 26 | 24 | 22 | 20 | 18 | 24 | 22 | 20 | 18 | 16 | 14 | 12 |
| *Aluminum* | | | | | | | | | | | | | |
| 4 | 45 | 44 | 42 | 42 | 42 | 40 | 45 | 44 | 44 | 44 | 44 | 43 | 43 |
| 6 | 40 | 39 | 39 | 39 | 38 | 36 | 38 | 38 | 37 | 37 | 37 | 36 | 35 |
| 8 | 34 | 34 | 33 | 32 | 32 | 31 | 33 | 32 | 32 | 31 | 30 | 29 | 28 |
| 10 | 32 | 31 | 30 | 30 | 30 | 29 | 28 | 28 | 27 | 27 | 27 | 27 | 26 |
| 12 | 29 | 28 | 27 | 26 | 25 | 24 | 21 | 21 | 20 | 20 | 20 | 19 | 17 |
| *Steel* | | | | | | | | | | | | | |
| 4 | 44 | 44 | 42 | 42 | 40 | 38 | 43 | 42 | 42 | 41 | 39 | 38 | 37 |
| 6 | 39 | 39 | 39 | 38 | 36 | 35 | 36 | 36 | 35 | 34 | 32 | 31 | 30 |
| 8 | 33 | 33 | 33 | 32 | 31 | 30 | 32 | 31 | 31 | 30 | 29 | 28 | 25 |
| 10 | 31 | 30 | 30 | 30 | 29 | 25 | 27 | 27 | 26 | 24 | 24 | 22 | 20 |
| 12 | 28 | 26 | 26 | 25 | 24 | 22 | 19 | 19 | 19 | 18 | 16 | 14 | 13 |

*Metal Gauge Numbers*

These drill sizes are intended only as a guide and may vary with application.

tant. The size of this hole will depend on the *screw size, gauge of the metal,* and *type of metal.* Generally, for the same size screw, a larger hole must be provided in harder materials. For example, if a No. 6 self-tapping screw is to be used to secure sheet metal whose thickness is 0.060 inch (16 gauge), a No. 32 drill bit is recommended for steel, whereas a No. 37 would be necessary for aluminum. Information on hole sizes for thread-forming and thread-cutting screws is provided in Table 25.2.

Thread-forming and thread-cutting screws do not provide the mechanical security obtainable with machine screws with lock washers and nuts. However, when vibration is not a factor, they overcome accessibility problems in areas where nuts cannot be installed. Access or cover plates are an example of this application.

The proper securing of two pieces of sheet metal with a sheet metal screw is shown in Fig. 25.7. The hole drilled in the metal closest to the screw head must be a clearance hole to ensure positive metal-to-metal contact when the threading is completed and the sealing plane of the screw head is flush with the metal. For sheet metal applications, screws with larger sealing planes, such as the washer-head type shown in Fig. 25.7, should be used.

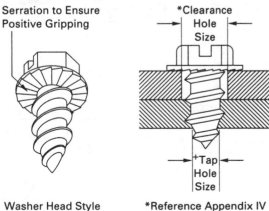

Serration to Ensure Positive Gripping

*Clearance Hole Size

Washer Head Style Sheet Metal Screw

+Tap Hole Size

*Reference Appendix IV
+Reference Table 25.2

**Figure 25.7**
Recommended method for sheet metal assembly.

Screwdrivers are the most common tools used for securing the screws discussed in Sec. 25.3. They are available in a variety of *handles, shank lengths* and *shapes,* and *driver styles.* A popular style screwdriver is shown in Fig. 25.8 together with the nomenclature of its parts. The handle is usually made of hardwood or plastic. In addition, a rubber grip may be provided for additional electrical protection, positive gripping, and comfort. For those screwdrivers without a rubber grip, large flutes are formed in the handle for improved gripping. The shank is manufactured from a tempered steel alloy specifically designed to withstand distortion caused by the excessive torque to which these tools are subjected. The handle and shank, however, will not withstand severe forces when the screwdriver is misused, such as for prying or bending. The shank may have either a round or square cross section. The square design is usually found on larger-size screwdrivers, where severe torque is expected. Shank lengths normally vary from 3 to 16 inches. For general electronic assembly, shank lengths up to 10 inches find wide application.

Driver styles are available for all screw-head drive configurations shown in Figs. 25.1b and c. The three most common types are the *standard, Phillips,* and *hexagonal* drivers. The standard driver has a tapered tip design that fits into the drive slot of a screw head. Common blade widths are 3/32, 1/8, 5/32, 3/16, 1/4, and 3/8 inch. When selecting the most suitable screwdriver to avoid damage to the screw or the work surface, the width of the blade should closely match the slot length but should not extend beyond the perimeter of the screw head.

Phillips screwdrivers are designed to fit the tapered cross-slots on screw heads made for this type of drive (Fig. 25.1b). Since more contact points are available with this design than with the slotted type, more driving force can be applied with less possibility of slippage. This style is commonly available in five number sizes: 0, 1, 2, 3, and 4. These will fit a large variety of Phillips screwhead sizes. Table 25.3 shows the screwdriver number and the screw size which each will accommodate. Those most frequently used in electronic assembly are numbers 1, 2, and 3. Phillips screwdrivers must be held in exact alignment with the drive configuration of the screw head to be effective.

The Frearson or Pozidrive (a trade name of Phillips International) styles (see Fig. 25.1c) are similar to the Phillips drive except that they allow for a much more precise fit between the driver and the screw head. This fit enables the screw to be supported unaided on the end

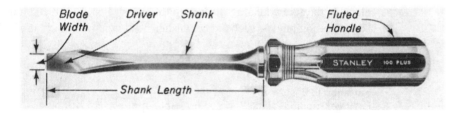

**Figure 25.8**
Screwdriver nomenclature. *Courtesy of Stanley Tools.*

**TABLE 25.3**
Phillips Screwdriver Selection

| | Machine Screws | | Sheet Metal Screws |
| --- | --- | --- | --- |
| *Number Tip Size* | *Flat or Oval* | *Round* | *Flat, Round, Oval, or Binder* |
| 0 | 0,1 | 0,1 | — |
| 1 | 2–4 | 2–4 | 2–4 |
| 2 | 5–9 | 5–10 | 5–10 |
| 3 | 10–16 | 12–16 | 12–14 |
| 4 | above 16 | above 16 | — |

of the driver, which can be a time-saving advantage in assembly work when accessibility is limited.

The *Allen* drive, and the special *Bristol* type (whose screw heads are shown in Fig. 25.1c) have a distinct advantage over the slotted style in that they minimize the possibility of the driver slipping out of the screw head and damaging the work surface. The Allen drive is encountered more often in electronics assembly than the Bristol style. A selection of Allen drives from 3/64 to 3/8 inch is suitable for most electronic applications. These sizes vary by 1/64 inch for sizes from 3/64 to 3/32 inch, by 1/32 inch for 3/32 to 1/4 inch, and by 1/16 inch for 1/4 to 3/8 inch.

For extremely small set screws, the Bristol drive is often preferable to the Allen style, since its multiple spline design prevents rounding of the driving surfaces of either the drive or the screw head through continued use. Bristol drives are available with four or six flutes.

For small and delicate work, *jewelers'* screwdriver sets may be used. The barrel of the handle is knurled with a swivel-top finger rest. This configuration allows the driver to be turned using the thumb and forefinger while applying pressure with the index finger (Fig. 25.9).

Screwdrivers for special applications are shown in Fig. 25.10 and are available in all the standard drive configurations. The "stubby" design, shown in Fig. 25.10a, is used if there is limited access to the screw head. These drivers are available in lengths from 1 to 2 inches.

*Offset* drivers, such as the one shown in Fig. 25.10b, are also used for work in restricted areas where even the stubby style would not fit. When the driver is of the blade type, alternate ends of the offset screwdriver must be used successively in order for the screw to be completely tightened by alternate partial turns. When using the offset Phillips or Allen drivers, it is unnecessary to use alternate ends. An improvement on the offset driver is the *ratchet* design shown in Fig. 25.10c. This tool is similar to a ratchet-type socket wrench. A screw is tightened or loosened by a backward-and-forward motion of the handle. A two-position lever on the drive end of the handle is set to obtain either clockwise or counter-clockwise ratchet driving action.

Screwdrivers having special screw-holding devices, such as that shown in Fig. 25.10d, retain the screw head firmly against the driver, allowing the full length of the shank to be used in gaining access to otherwise impossible assembly positions. The screw head is grasped by a spring-loaded dual-leaf device. When this assembly is forced against the driver tip, the spring compresses and the leaves spread outward. The screw head is then inserted against the driver and the leaves cup the head under its sealing plane as the spring assembly is released and forced toward the handle.

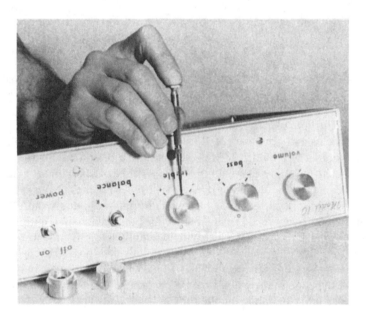

**Figure 25.9** Properly held jewelers' screwdriver.

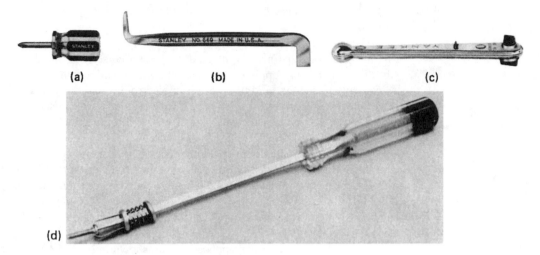

**Figure 25.10** Screwdrivers for special applications: (a) "stubby"; (b) standard offset driver; (c) offset ratchet; (d) driver with screw-holding mechanism. *Courtesy of Stanley Tools.*

The proper operating position of any screwdriver is always in alignment with the axis of the screw. This position minimizes the possibility of slippage damage and provides maximum surface contact area between the driver and the screw head.

*Nutdrivers* are used in a similar manner to screwdrivers but are designed to accommodate hexagonal-style machine nuts or screw heads. A set of nutdrivers is shown in Fig. 25.11a. They are available in sizes 6, 7, 8, 9, 10, 11, 12, 14, 16, and 18. The number represents the size in 32nds of an inch. For example, a number 6 will fit a 6/32 (3/16) hexagonal nut or screw head as measured across opposite flats. Sizes 6, 8, 10, 11, and 12 are most frequently used in electronics assembly. Sizes above 1/2 inch are useful for the larger nuts associated with potentiometers and switches. The sizes of nutdrivers are often stamped into the shank, or colored handles are provided for quick identification. A shank length of 4 inches is suitable for most applications. Nutdrivers are available with hollow shanks for threading nuts onto long machine screws that extend more than 1/4 to 3/8 inch into the driver.

A screwdriver in conjunction with a nutdriver should be used to assemble and disassemble hardware whose fasteners require considerable force. This technique is shown in Fig. 25.11b.

## 25.5 WRENCHES

Wrenches are necessary for larger nuts found on some potentiometers and switches when a suitable nutdriver is not available or when accessibility does not permit its use. The most common types of wrenches used in electronic assembly are the *box, open-end,* and *adjustable* wrenches.

The box wrench, similar to a nutdriver, requires complete access over the nut since it encircles and grips the nut at its points. This wrench is a 12-point drive as opposed to the 6 points of a nutdriver. If used to secure a hexagonal nut, all 6 of the points will be contacted, virtually eliminating the possibility of shearing or rounding the points. The box wrench is preferable over other types of wrenches because it is much less likely to gouge the work surface. The box wrench should fit snugly onto the nut. All sizes listed for the nutdriver are also available for the box wrench, in addition to larger sizes that are not normally required in electronic assembly. Lengths of 4 to 6 inches are suitable for most assembly applications. The proper use of a box wrench used to secure a potentiometer to a front panel is shown in Fig. 25.12. Masking tape may be applied to the panel where the edge of the wrench would contact the work surface around the nut to eliminate the possibility of damaging the finish.

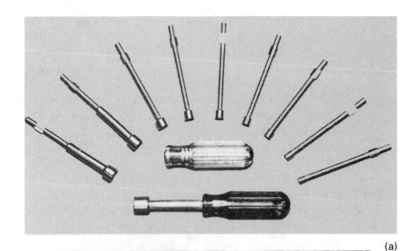

(a)

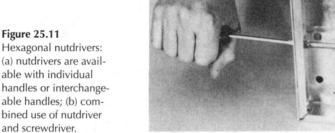

(b)

**Figure 25.11**
Hexagonal nutdrivers:
(a) nutdrivers are available with individual handles or interchangeable handles; (b) combined use of nutdriver and screwdriver.

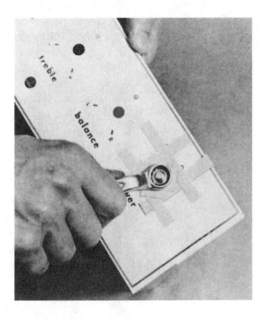

**Figure 25.12** Box wrench used in assembly.

Open-end wrenches are primarily intended for use with square nuts. The proper-size wrench can be determined when the entire length of the inside faces of the jaws fits snugly against any two opposite flats of the nut. This ensures maximum surface contact and minimizes slippage and rounding of the nut. The jaws of this wrench are generally angled at 15 degrees to the handle center line. This design makes it possible to work in confined areas where the nut can be turned only partially. Open-end wrenches for electronic assembly are available in a complete range of sizes from 3/16 to 1 inch and between 3 1/2 and

10 inches long. Although available in sizes much larger than 1 inch, these are seldom required in electronic applications.

Both the box and open-end wrenches require a complete set to accommodate all possible nut sizes that may be encountered in assembly. The adjustable wrench eliminates the need for complete sets since the distance between jaws can be changed to any size within its capacity. It does not, however, provide as secure a grip as either the box or open-end wrench. The adjustable wrench is similar to the open-end wrench with the 15-degree angle except that only one jaw is stationary. The second jaw can be moved by means of a thumb-operated *adjusting screw* located within the head of the wrench. To adjust the wrench for proper fit, the jaw is first opened sufficiently to allow both jaws to easily slip over opposite flats on the nut. The movable jaw is then closed until both jaws are gripping firmly against the flats. It is important that when positioned, the *pulling* force should be applied to the stationary jaw. If excessive force is applied to the movable jaw, the adjusting screw may be damaged or the wrench could slip, damaging the work surface or surrounding parts.

## 25.6 POP RIVETS

An alternative method of securing sheet metal or hardware by means other than machine screws and nuts or sheet metal screws is with the use of *pop rivets*. Pop riveting is rapid and requires little skill, and free access to only one side of the work is necessary. The only tool used for this process is a hand-type riveting gun.

A typical pop rivet, shown in Fig. 25.13a, consists of a *truss-style head, sealing plane, hollow shaft, stem,* and *button head.* The stem is tapered near the button head. When a tensile force is applied to the stem, the button head is drawn up into the hollow shaft. This action flares the end of the rivet against the metal, thus providing a positive fastening of the metal parts. When a pulling force of from 500 to 700 pounds is reached, the stem will break at the tapered section. The sheet metal or hardware to be fastened must be provided with clearance holes to accept the shaft of the rivet. A clearance of 3 to 5 thousandths of an inch is sufficient. The shaft is then pressed through the holes with the sealing plane flush against

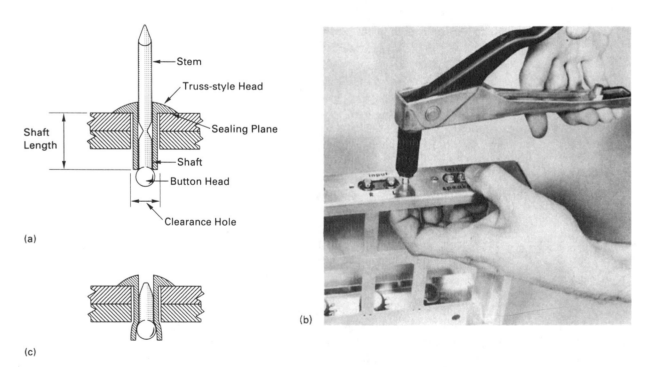

**Figure 25.13** Pop rivet and its assembly: (a) pop rivet assembly and nomenclature; (b) pop rivet gun positioned against rivet head; (c) installed pop rivet.

the metal surface. The pulling force is applied to the stem through the locking action of the pop rivet gun *nose piece* (stem collet). The nose piece is placed over the stem and pressed flush against the head of the rivet (Fig. 25.13b). As the operating lever of the gun is squeezed, the stem of the rivet is gripped, drawing the button head upward into the hollow shaft. Continued squeezing of the operating lever will create the desired flaring action. The stem is broken when sufficient pulling force is reached. The operating lever may have to be actuated more than once to complete this operation. When the lever is again released, the broken stem drops out of the nose piece. An installed pop rivet is shown in Fig. 25.13c.

Rivets are a permanent type of fastener intended for use if components, sheet metal, or hardware is not to be removed once fastened. This fact must be considered when using rivets, since they are difficult and time-consuming to remove. If it is necessary to remove a rivet, drilling through the rivet center with a drill bit size that is at least equal to the clearance hole is required.

Pop rivets of the style described are available in steel or aluminum with shaft diameters of 1/8, 5/32, and 3/16 inch and shaft lengths of 1/8 to 5/8 inch. The clearance hole drill numbers for these three diameters are Nos. 30, 20, and 11, respectively. The stems of the three available shaft diameters require their own nose piece on the pop rivet gun. These nose pieces thread into the gun assembly and are easily removed with an open-end wrench.

## 25.7 PLIERS FOR HARDWARE ASSEMBLY

The two types of pliers most commonly used in electronic assembly are the *long-nose* and the *gas* pliers (see Fig. 25.14). Both styles consist of a *pivot, jaws,* and *handles.*

Long-nose pliers consist of a pair of long, narrow tapered jaws. The jaws are smooth or may have serrations near the tips for improved gripping. These pliers are designed to aid in assembly for holding or positioning small parts in tight spaces where finger accessibility is limited. They are not intended to withstand severe gripping or twisting forces and should not be used for tightening machine nuts. Misusing the tool in this way could cause the jaws to spring out of alignment, rendering the pliers useless. Long-nose pliers holding a machine nut for assembly is shown in Fig. 25.14a.

Gas pliers have a much more rugged design than long-nose pliers and have coarse serrations over the complete inside surface of each jaw. These pliers are used primarily

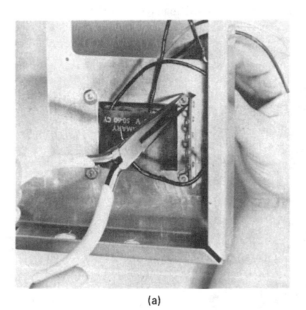
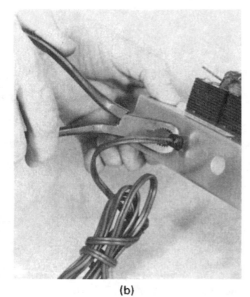

(a)                                                          (b)

**Figure 25.14**   Pliers for hardware assembly: (a) long-nose pliers aid in assembling small parts; (b) gas pliers used to install power cord strain relief.

for holding round stock, such as inserting a power cord strain relief bushing, as shown in Fig. 25.14b. Excess gripping force on soft material should be avoided because the serrated jaws will mar the surface. For this reason, gas pliers should be used with extreme care. A minimum of two layers of masking tape wrapped around the stock will help prevent the jaws from marring.

Although many other styles of pliers are available, such as the *slip-joint* and *rib-joint* types, they are designed for holding heavy stock in which marring of the surface is not a factor. They should not be used in place of any of the drivers, wrenches, or pliers discussed in this chapter. Special pliers for cutting wire are considered in Chapter 26.

## 25.8 SPECIAL FASTENERS AND HARDWARE

Many instances arise in which the conventional fastening methods do not completely satisfy the assembly requirements, for example, accessibility, orientation, frequent removal, and location. Many special fasteners and hardware that find wide application in electronic assembly are commercially available. Some of the more common of these will be discussed here.

The layout design may specify a single wire or cable to pass through a hole formed in the metal chassis element. For this requirement, a grommet should be used to prevent damage to the wire installation. A *grommet* is a rubber insulator that "cups" around the hole in the chassis and prevents wires from rubbing against the metal edges. A grommet being inserted into a chassis element is shown in Fig. 25.15a. An access hole must be provided in the chassis that is equal to dimension *A* in Fig. 25.15b. Some common size soft-rubber grommets together with the necessary size of access hole for each are listed in Table 25.4. Although dimension *A* is the grommet size designation, the selection of the proper-size grommet depends on the minimum required size of dimension *B*.

If the packaging design requires that a machine screw or other metal stud be insulated from the chassis, *fiber shoulder washers,* such as those shown in Fig. 25.16a, may be used. Two washers are necessary, as shown in Fig. 25.16b, to ensure complete electrical insulation of the stud from the surrounding metal edges of the access hole. The major diameter of the stud used determines the size of dimension *A* in Fig. 25.16c, whereas dimension *B* is the outside of the washer. Dimension *C* is the outside diameter of the shoulder for which the access hole in the metal must be formed, and dimension *D* corresponds to the shoulder height. If the height of the shoulder is more than one-half the thickness of the metal, a flat fiber washer may be necessary in place of a second shoulder washer to ensure flush mounting and total insulation. The appropriate size shoulder may be determined from Table 25.5.

**Figure 25.15**
Grommets are inserted into holes to cover sharp edges: (a) grommet installation technique; (b) important dimensions for grommet installation.

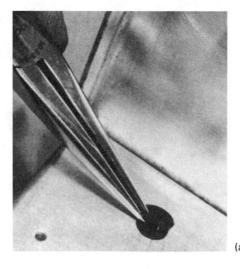

(a)    (b)

Access Hole Diameter for Sheet Metal

A

B

Inside Diameter of Grommet

### TABLE 25.4
Rubber Grommet Dimensions

| Grommet Size Designation | A (in.) | B (in.) |
|---|---|---|
| $\frac{1}{4}$ | $\frac{1}{4}$ | $\frac{1}{8}$ |
| $\frac{5}{16}$ | $\frac{5}{16}$ | $\frac{3}{16}$ |
| $\frac{3}{8}$ | $\frac{3}{8}$ | $\frac{1}{4}$ |
| $\frac{1}{2}$ | $\frac{1}{2}$ | $\frac{3}{8}$ |
| $\frac{5}{16}$ | $\frac{5}{16}$ | $\frac{3}{16}$ |
| $\frac{3}{4}$ | $\frac{3}{4}$ | $\frac{9}{16}$ |
| $\frac{9}{16}$ | $\frac{9}{16}$ | $\frac{7}{16}$ |
| $\frac{5}{8}$ | $\frac{5}{8}$ | $\frac{1}{2}$ |
| $\frac{7}{8}$ | $\frac{7}{8}$ | $\frac{3}{8}$ |
| $\frac{7}{8}$ | $\frac{7}{8}$ | $\frac{5}{8}$ |
| 1 | 1 | $\frac{11}{16}$ |

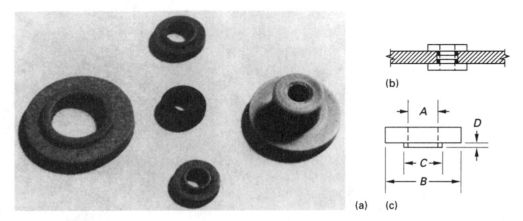

**Figure 25.16** Fiber shoulder washers: (a) common shoulder washers; (b) recommended method of assembly; (c) important shoulder washer dimensions for installation.

The use of machine screws and nuts, as used to mount pc boards into a chassis, shown in Fig. 25.19, is one acceptable mounting arrangement. However, large areas of metal must be removed from the chassis. The sheet metal work could be reduced by using *standoffs* and *circuit board supports,* which often find application in prototype as well as production assemblies. This hardware is shown in Fig. 25.17. Straight standoffs are made of plastic for applications requiring electrical insulation and are available in lengths from 1/8 to several inches. Standoffs, shown in Fig. 25.17a, are threaded at both ends for mounting and will accept a screw size range of No. 4 to No. 10. *Hinged* standoffs are available in cadmium-plated brass. Both ends of this style standoff are threaded for mounting to accept screw sizes from 4–40 to 8–32.

By drilling holes into the four corners of the pc board, straight standoffs may be used between the board and chassis to support the board away from the chassis surface. The length of standoff is determined by the tallest projection or component extending from the side of the pc board that faces the chassis. A minimum of 1/8 inch must be added to this dimension to ensure adequate clearance.

The hinged standoff is generally selected to make circuit testing or servicing easier. Two corners of the pc board are fastened to hinged standoffs and the two opposite corners are secured to standard standoffs. The height of the hardware must be selected to allow the board to rest parallel with the chassis surface with sufficient clearance. For servicing, the two screws at the corners of the board secured to the standard standoffs are unfastened. The board can then be lifted upward to an angle of 90 degrees by the pivoting action of the

**TABLE 25.5**

Fiber Shoulder Washer Dimensions

| Screw Size | I.D.<br>A | O.D.<br>B | Shoulder<br>Diameter<br>C | Height<br>D |
|---|---|---|---|---|
| 3 | 0.110 | 0.052 | 0.187 | 0.031 |
| 4 | 0.136 | 0.052 | 0.187 | 0.031 |
| 5 | 0.136 | 0.312 | 0.187 | 0.031 |
| 6 | 0.140 | 0.375 | 0.237 | 0.031 |
| 8 | 0.172 | 0.375 | 0.246 | 0.031 |
| 10 | 0.196 | 0.375 | 0.308 | 0.031 |
| 12 | 0.250 | 0.500 | 0.312 | 0.028 |
| $\frac{1}{4}$ | 0.265 | 0.500 | 0.365 | 0.031 |
| $\frac{5}{16}$ | 0.375 | 0.750 | 0.500 | 0.031 |
| $\frac{3}{8}$ | 0.385 | 0.625 | 0.500 | 0.031 |

Dimensions A, B, C, D are in inches.

(a)    (b)

**Figure 25.17** Hardware for mounting printed circuit boards: (a) threaded standoffs; (b) nylon printed circuit board support.

hinged standoffs, thereby exposing wiring and components for testing and repair. This method of mounting not only provides easy access for repair, but also can reduce the overall package size.

*Circuit board supports,* shown in Fig. 25.17b, are available in nylon and other plastic materials. They consist of two snap-lock arrangements, one at each end of the support. For mounting, one end of the support snaps securely into a 0.187-inch-diameter hole drilled into the chassis. These supports provide another convenient method of mounting pc boards to chassis. The board is secured to the upper end of the support by a spring-type locking action. Support heights range from 3/16 to 7/8 inch. Circuit board supports can be mounted on all four corners of small boards. More may be used on larger boards for greater support. They have the advantage of allowing boards to be removed and quickly replaced, and eliminate the need for removing large areas of metal from the chassis. These circuit board supports can be used in conjunction with connectors mounted for electrical interconnections between the board and chassis, which adds to the versatility of this method of mounting.

*Printed circuit board guides,* shown in Fig. 25.18, are grooved runners made of plastic and designed for use with edge connectors. These guides ensure alignment between the edge of a pc board and a connector and also provide additional support for the board. Generally fastened to the chassis with machine screws and nuts, pc board guides are also available with an adhesive backing for rapid attachment. The grooves in these guides are designed to accept all standard board thicknesses.

The final method of fastening to be considered is the use of adhesives, particularly *epoxy resins*. High-strength adhesives find wide application in electronic packaging since they provide a number of advantages over other methods of bonding. Some of these advantages are elimination of machine operations, provision of electrical insulation and low thermal resistance between parts joined, and uniform stress distribution over the entire bonded area. Although the use of adhesives requires thorough preliminary cleaning of parts and often heat and pressure during the curing period, the advantages of this method of bonding far outweigh the time necessary for its proper use.

Epoxy resins must generally be mixed with a *catalyst* (hardener) prior to use. The parts are measured by volume or weight. Manufacturers' mixing instructions must be consulted for the correct amounts for a specific epoxy since there is a wide variation among the many different adhesive types. The *working time* (time between mixing and initial hardening) must also be predetermined for the specific epoxy to avoid mixing more than can be used during the prescribed time.

For proper bonding, the surfaces must be thoroughly cleaned. Fine emery cloth to remove oxides may be used followed by degreasing with a solvent such as toluene or trichlorethylene. After the epoxy is properly mixed, a thin coating is applied to each of the mating surfaces and pressed together to squeeze out air bubbles and excess adhesive. Improved bonding to aluminum results if the surface is anodized prior to bonding. Depending on the epoxy used, curing time will vary from 5 minutes to 24 hours at room temperature. These periods can be reduced if the bonded surfaces are baked in an oven, but care must be exercised not to exceed the temperature range of the epoxy. In addition, baking is not recommended when bonding devices to heat sinks because of the possibility of damaging components.

Epoxy is used for bonding when the stresses encountered will be only tensile or shear. Since the shelf life of many adhesives is often less than one year, for sound bonds the expiration date must not be exceeded. In addition, adhesives should be stored in a cool area to ensure longest possible shelf life.

Hardware and pc boards are shown assembled to a chassis and front panel element in Fig. 25.19 using materials and techniques discussed in this section.

After assembly is completed, wiring of the boards and chassis-mounted hardware is necessary to complete a system. In the next chapter we consider wire and wiring assembly techniques.

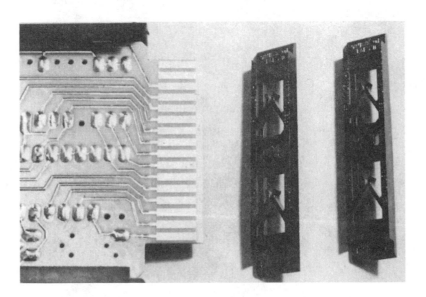

**Figure 25.18** Printed circuit grooved guides.

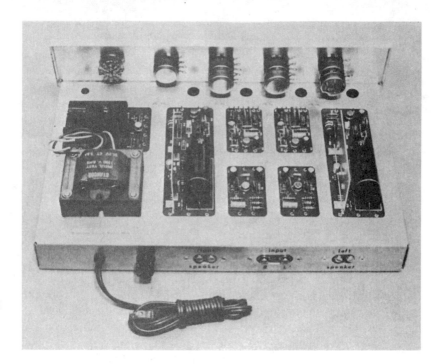

**Figure 25.19**
Completed mechanical assembly of all parts to a chassis and front panel element.

## EXERCISES

### A.  Questions

25.1    List the four major factors in electronic packaging that are required for quality workmanship.

25.2    List the eight features that distinguish machine screws.

25.3    What is the advantage of using Phillips head screws over slotted head screws?

25.4    Define the term *fit* as it pertains to machine screws.

25.5    What is the major difference between thread-forming and thread-cutting screws?

25.6    For fastening two pieces of 18-gauge aluminum with a No. 6 sheet metal screw, determine the required clearance hole size and tap hole size. Refer to Appendix IV and Fig. 25.7.

25.7    What is one disadvantage of using pop rivets rather than screws and nuts?

25.8    What hole diameter is required to install a grommet having an inside diameter of ³⁄₁₆ inch? Refer to Fig. 25.15b and Table 25.4.

### B.  True or False

Circle *T* if the statement is true, or *F* if any part of the statement is false.

| | | | |
|---|---|---|---|
| 25.1 | Flat-head machine screws are recommended for use on thin sheet metal. | **T** | **F** |
| 25.2 | The Allen drive configuration has a more positive tool positioning and gripping action than the Phillips head. | **T** | **F** |
| 25.3 | The UNC system has fewer threads per inch than the UNF system. | **T** | **F** |
| 25.4 | For most electronic assembly applications, the Class 3 machine threads are standard. | **T** | **F** |
| 25.5 | The tooth-type lock washer should not be reused when removed. | **T** | **F** |
| 25.6 | Split lock washers are more effective than any other type of lock washer. | **T** | **F** |
| 25.7 | Self-tapping screws are intended to be used for sheet metal fastening applications. | **T** | **F** |
| 25.8 | Pop rivets are used where a permanent type of fastener is required. | **T** | **F** |

25.9    The diameter of a hole for the installation of a grommet is the same dimension as the          **T    F**
        grommet size designation.

25.10   Hinged standoffs are used to improve access to wiring and components.                           **T    F**


## C.  Multiple Choice

Circle the correct answer for each statement.

25.1    Some common machine screw head styles are binder, round, and *(oval, slotted)*.

25.2    For equivalent screw sizes, *(UNC, UNF)* screws have more threads per inch.

25.3    Machine screws with a *(Class 1, Class 2, Class 3)* screw thread are considered as having a snug fit.

25.4    Pop rivets are a *(temporary, permanent)* type of fastener.

25.5    Shoulder washers are usually made from *(metal, fiber)* in order to provide electrical *(isolation, conduction)*.


## D.  Matching

Match each item in Column A to the most appropriate item in Column B.

         COLUMN A                          COLUMN B

1.   Class 2                      a.   Wire protector
2.   Hexagonal drive             b.   Washer
3.   Shoulder washer             c.   Fine thread
4.   Lock                        d.   Allen
5.   UNF                         e.   Insulation
6.   Split-joint                 f.   Free fit
7.   Grommet                     g.   Pliers


## E.  Problems

25.1    Using an appropriate epoxy, bond the two halves of an ink eraser to an aluminum handle such as that shown
        in Fig. 25.20. The handle is fabricated from 16-gauge aluminum having blank dimensions of 3/4 inch by
        10 inches and bent around a 5/8-inch-diameter dowel.

25.2    Select an appropriate epoxy and seal the four corners of the soldering iron tip cleaning sponge tray con-
        structed in Problem 17.1.

**Figure 25.20**

25.3    The meter for the tachometer printed circuit designed and fabricated in Problems 6.3, 10.3, 16.3, and 17.3
        may be mounted onto a panel of the same design as that shown in Fig. 20.10. The pc board is to be mounted
        with proper length spacers and mounting screws to the rear of the meter, which in turn is secured to the
        panel. Wiring for the installation of the tachometer is done in Problem 26.5.

# 26 Wire and Interconnection Techniques

## LEARNING OBJECTIVES

*Upon completion of this chapter on hookup wire and interconnection techniques, the student should be able to*

- Know the composition, configurations, and size designations of common hookup wire.
- Calculate the resistance of a given length of wire.
- Measure the diameter of wire.
- Be familiar with the various types of wire insulation material and their characteristics.
- Select the proper type of wire insulation for a specific application.
- Know the common types of special wire configurations and their applications.
- Be familiar with the wire color-coding system.
- Use mechanical strippers to remove wire insulation.
- Prepare wire for soldering.
- Form proper wire connections on a variety of terminals.
- Use crimp-type terminals.
- Use manually operated wire-wrapping and de-wrapping tools.
- Understand the use of semiautomatic and fully automatic wire-wrapping machines.
- Use proper bussing techniques on wire-wrapped terminals.

## 26.0 INTRODUCTION

One of the least emphasized and perhaps the most important aspect of packaging is the selection of the most suitable hookup wire for a specific assembly. From the standpoint of reliability, this selection must be based on such factors as *conductor material and plating, conductor size, type of insulation, flexibility, environmental factors,* and *current* and *voltage limits,* as well as *cost.* This chapter provides guidelines to aid in evaluating and selecting wire. Coupled with the problem of selecting are the methods for *interfacing interconnections.* Since there is a variety of standard interconnections (in addition to many special types), this chapter also includes pertinent information for evaluating the suitability of various connection methods.

Specific information also is provided here on common conductor materials and platings and American Wire Gauge (AWG) classifications of size, along with a comparison of solid wire with various types of stranded wire. Typical types of insulation are examined with respect to their application, including color-coding methods. Additional information is provided on current, voltage, and mechanical and environmental criteria for a total understanding of wire selection from the many types commercially available. Special wire types are also discussed.

Methods of interconnections are introduced, and the tools used for wire preparation and mechanical assembly, such as *strippers* and *diagonal cutters,* are described. *Wrap-around* and *feed-through* connections as they apply to swaged solder-type terminals, eyelet-type terminals, and solder lugs are discussed and expanded to include special hardware connections, such as plugs and jacks. Solderless wiring techniques, such as solderless or crimp-type connectors and wire-wrap connections, are set forth. Finally, information on cable ties, shrink tubing, and cable clamps is presented.

*CAUTION:* Prior to performing any of the operations described in this chapter, approved safety glasses with sideshields in compliance with ANSI Z87.1-1989 must be worn. Care should be exercised to prevent pieces of insulation or ends of severed wires from becoming projectiles when using wire strippers and diagonal cutters.

Place a hot soldering iron into its holder when it is not in use. Keep hands and fingers away from the heating element and soldering iron tip. Be especially watchful that the heating element or tip does not come into contact with the power cord. When in use, solder pots contain molten tin/lead. Do not hold or pour liquids over an operating solder pot.

The soldering processes discussed should be performed only in adequately ventilated facilities and with applicable equipment having proper fume exhaust systems. See Appendix XX regarding flux fumes, health issues related to flux fumes, and fume extraction.

When preparing shielded wire, point the tip of the scribe away from you when making an access hole through which the inner insulated wire can be pulled.

It is important to grip only the cushioned handles of a crimping tool to prevent the possibility of injury from the sharp cutting edges of the wire stripper portion of the tool, which is located between the handles and the crimping jaws.

Be careful not to touch the end of a heat gun or its deflector, since they become extremely hot during use. Also, the section on safety outlined at the beginning of the book should be reviewed for more specific information.

## 26.1 CONDUCTOR MATERIALS AND WIRE CONFIGURATIONS

Most of the wire used for electronic interconnections is fabricated from *annealed* (softened) copper. Copper used for this purpose is termed *Electrolytic Tough Pitch* (E.T.P.). Although other materials, such as aluminum, steel, and silver, are employed in wire construction, they are not common to electronics fabrication. Alloys of copper have been developed, however, that display superior characteristics when compared to pure copper. Although pure copper possesses excellent electrical conductivity characteristics as well as being malleable and ductile, it has the pronounced disadvantage of being highly susceptible to fractures under vibration and flexing conditions. Modern high-temperature annealing techniques, however, have made it possible to develop high-strength alloys with improved fatigue life. Some of these common high-strength alloys are *cadmium–copper, chromium–copper,* and *cadmium–chromium–copper.* One disadvantage in using these alloyed wires is that they are available only with coatings of silver or nickel, which are both expensive and difficult to solder. Tin-coated high-strength alloys are, unfortunately, not readily available.

Copper wire is usually coated, because pure copper oxidizes quickly when exposed to the atmosphere. *Tin* is the most common coating, although *silver* and *nickel* coatings are used in special applications. Tin-coated wire, recommended for applications wherein environmental temperatures will not exceed 150°C (302°F), is the least expensive of the three coatings mentioned. In addition, tin coating improves solderability. Silver coating, recommended for use in temperatures ranging from approximately 150 to 200°C (302 to 395°F) greatly improves solderability but tends to corrode at elevated temperatures. Silver is the most expensive coating of those mentioned. Nickel coating is capable of oxidation protection to temperatures of up to 300°C (575°F) but requires activated fluxes and high-temperature soldering techniques.

Tin-coated wire is by far the most commonly used in the electronics industry. Tin is applied to the bare wire by either *dipping* or *electroplating.* Dipping is the least expensive method and is used if uniform plating thickness is not critical. Electroplated tin should be specified if close tolerance automatic stripping is to be employed, since uniform plating is achievable by this method.

Wire used in the electronics industry can be divided into two groups: *solid* and *stranded.* Generally, solid wire that is comparable to a specific size of stranded wire is less expensive to manufacture and subsequently less expensive to purchase. Solid wire, however, does not possess the flexibility characteristics or the fatigue life of stranded wire. Whereas solid wire tends to fracture even under mild flexing, stranded wire remains highly flexible. Because much of the electronic equipment produced today is exposed to some form of vibration under normal use or has wiring flexibility requirements during assembly, stranded wire is the type most often specified.

Solid wire finds extensive application in the fabrication of leads for small components, such as resistors, capacitors, and solid-state devices. Hookup applications of solid wire are generally limited to conditions wherein flexibility is not a criterion, or lead runs are to be short, rigid terminations such as the direct connection of two closely spaced swaged turret-type terminals on a pc board. Solid, uninsulated tinned copper wire used as a short jumper or as a common circuit point is termed *bus* wire.

When longer lengths of bus wire are needed, it is usually recommended that the wire be straightened prior to its use. The best way to straighten wire is to secure one end of it in a vise and the other end in the chuck of a hand drill (eggbeater). All that is usually required to straighten the wire is 8 to 10 complete turns of the drill in one direction followed by an equal number of turns in the opposite direction while applying continuous tension to the wire.

Stranded wire is classified as either *bunch* or *concentric.* Bunch-type stranded wire is constructed of several small-diameter solid wires "bunched" together without regard to symmetry. Although the cost of bunch wire is low and it is extremely flexible, it does not have a consistent diameter. Because of this disadvantage, it is not recommended, since modern insulation stripping tools may nick or break individual strands, which will weaken the wire as well as reduce current capacity.

Concentric-type stranded wire overcomes the disadvantages of bunch wire in that its consistent circular cross section makes it suitable to manual as well as automatic stripping techniques. *Counterdirectional* concentric stranded wire is a multilayer wire with alternate layers of strands rotating or twisting in an opposite direction to the next succeeding lower layer. *Unidirectional* concentric stranded wire is also a multilayer wire, but each layer of strands twists in the same direction but with a different degree of *pitch* (angle at which strands cross the axis of the overall wire) as they are twisted. Unidirectional stranded wire is more flexible than the counterdirectional type, but its diameter is not as precisely uniform.

Solid and stranded wire each have their specific size designation systems. Solid wire is designated in terms of its cross-sectional area and diameter by *gauge number.* The most commonly used designation is the *American Wire Gauge* (AWG) standard system. Appendix VII lists the AWG numbers of solid bare copper wire with its equivalent cross-sectional area and diameter for gauge Nos. 00 to 50.

To better understand the information provided in wire tables, certain relationships need to be examined. In the electrical field, characteristics of different wire gauges and lengths are, for convenience, compared to a standard so that different wire sizes may be readily compared. The accepted standard is a piece of solid annealed copper wire with a diameter of 0.001 inch (1 mil) and a length of 1 foot whose resistance is 10.37 ohms at 20°C. (This standard is often referred to as the *mil-foot.*) Since most electrical conductors have round cross sections, a system of circular measure has been established to reduce the tedious task of comparing conductor characteristics. The parameters involved in such a comparison are resistance, length, and cross-sectional area. These relationships are expressed mathematically as

$$\frac{\text{resistance wire I}}{\text{resistance wire II}} = \frac{\text{length wire I}}{\text{length wire II}} \times \frac{\text{area wire II}}{\text{area wire I}} \qquad (26.1)$$

The expression for determining the area of a circle is $A = \frac{\pi}{4} (diameter)^2$. When the area of two conductors is compared, as above, $\frac{\pi}{4}$ is a constant for both wire I and wire II. Therefore, Equation (26.1) may be simplified as follows:

$$\frac{\text{resistance wire I}}{\text{resistance wire II}} = \frac{\text{length wire I}}{\text{length wire II}} \times \frac{(\text{diameter})^2 \text{ wire II}}{(\text{diameter})^2 \text{ wire I}} \tag{26.2}$$

The term $(diameter)^2$ represents the cross-sectional area of a round wire. This term is designated as the *area in circular mils* (CM), where the area now is equal to the square of the diameter ($A = d^2$). All wire tables refer to diameters in mils and to cross-sectional areas in circular mils.

To determine the resistance of copper wire, the following relationship is used:

$$\text{resistance} = \frac{10.37 \times \text{length (feet)}}{\text{diameter}^2 \text{ (CM)}} \tag{26.3}$$

If, for example, it is necessary to find the resistance of 50 feet of AWG No. 18 wire, Appendix VII shows that the area of this wire is 1624 CM. Substituting these known values into Equation (26.3) yields

$$R = \frac{(10.37)(50)}{1624} = 0.32 \text{ ohm} \tag{26.4}$$

The wire diameter in mils can be converted to inches merely by moving the decimal point three places to the left (multiply by 0.001). For example, AWG No. 18 wire has a diameter of 40.30 mils, which is equal to 0.0403 inch.

Because one of the most important criteria for selecting wire is its current-carrying capacity, wire tables provide the cross-sectional areas of each wire gauge expressed in CM. The fourth column in Appendix VII lists the maximum recommended current capacity in amperes for common wires used in electronics. These values were obtained by the "rule of thumb" relationship, that is, each 500-CM cross section will safely accommodate 1 ampere of current. For example, AWG No. 10 wire, having a cross section of approximately 10,000 CM, can handle 20 amperes. AWG No. 13 wire, which is 3 gauge numbers higher, reduces the cross-sectional area by one-half to approximately 5000 CM and therefore is capable of handling one-half the current, or 10 amperes. Again, AWG No. 23 (10 gauge numbers higher than No. 13), with its cross section of approximately 500 CM, is capable of handling 1 ampere. From these relationships, the following guidelines can be used: For each *decrease* or *increase* of three gauge numbers, the CM area doubles or halves, respectively, as does its current-carrying capacity. Correspondingly, a gauge number decrease or increase by a factor of 10 will increase or decrease the cross section by 10 or $\frac{1}{10}$, respectively, as will its current capacity. (Refer to Appendix VII.) Common gauges of hookup wire used in the electronics industry range from AWG No. 10 to No. 26, and their current-carrying capacities, from approximately 20 amperes to 1/2 ampere.

The AWG standard numbering system was developed principally for bare, solid-copper wire. This system is also used to designate stranded wire together with an additional number to specify the number of strands composing the wire. For example, in the designation 7/34, the first number specifies that there is a total of 7 strands of wire and the second number indicates that each strand is AWG No. 34. To determine the equivalent AWG solid-wire size to a specific stranded wire, the cross-sectional area of each strand is multiplied by the number of strands in the overall wire. For example, to determine the equivalent solid-wire size of No. 7/34 stranded wire, the cross-sectional area of No. 34 is found to be 40 CM (from Appendix VII). The 7 strands constitute a total cross section of $7 \times 40$, or 280 CM. Appendix VII reveals that its closest equivalent is AWG No. 26 solid wire. Therefore, both No. 7/34 and AWG No. 26 have a maximum recommended current-carrying capacity of approximately 0.5 ampere. Table 26.1 lists some of the common stranded-wire types with their equivalent AWG numbers.

To determine the AWG wire size of either solid or stranded wire, a *wire gauge,* such as that shown in Fig. 26.1a, may be used. Each slot in the gauge is accompanied by the corresponding AWG number size and a relief hole having a diameter greater than the slot width. The purpose of the relief hole is to allow for proper measurement "feel" and also for damage-free removal of the wire from the gauge after measurement. The proper method of positioning the gauge and wire to be measured is shown in Fig. 26.1b. Note from the figure that when measuring stranded wire, the insulation is not completely removed as it is for solid wire. Rather, a section of insulation is separated to expose the conductor. This section should not initially be cut closer than 1 inch from the end of the wire to prevent the individual strands from separating or flattening as the wire is placed in the gauge. When

**TABLE 26.1**

Stranded-Wire Configurations to AWG Numbers

| Equivalent AWG Number | Stranding |
| --- | --- |
| 10 | 37/26 |
| 12 | 19/25 |
| 14 | 19/27 |
| 14 | 37/29 |
| 16 | 19/29 |
| 16 | 27/30 |
| 18 | 7/26 |
| 18 | 19/30 |
| 18 | 27/32 |
| 20 | 7/28 |
| 20 | 19/32 |
| 22 | 7/30 |
| 22 | 19/34 |
| 24 | 7/32 |
| 24 | 19/36 |
| 26 | 7/34 |
| 26 | 19/38 |

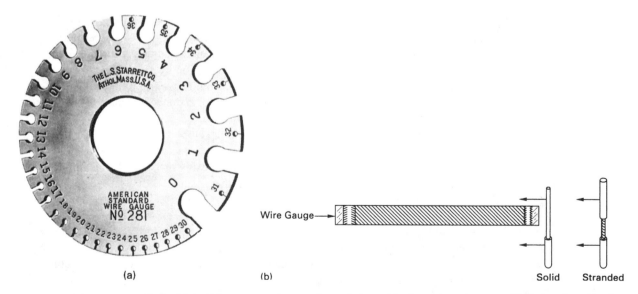

(a)  (b)  Solid  Stranded

**Figure 26.1** Wire gauge and measuring technique: (a) wire gauge to measure AWG sizes from No. 0 to No. 36, *courtesy of the L.S. Starrett Company, Athol, MA;* (b) perpendicular gauge and wire positioning for accurate measurement in gauge slot.

removing the section of insulation, the underlying wires must not be twisted, for the diameter of the conductor would be altered and an erroneous reading would result.

When measuring stranded wire, the wire gauge reading will be slightly larger than the corresponding AWG number for solid wire because of the slight increase in diameter caused by the spaces between strands. Therefore, when measuring stranded wire, the gauge slot into which the wire fits will be one AWG number *less* than the actual size and will also yield an *odd*-numbered AWG value. For example, a stranded AWG No. 20 wire will properly fit into the gauge slot marked AWG No. 19. Therefore, this rule of thumb indicates that the correct size of this wire is AWG No. 20. (It may be well to emphasize at this point that *odd* AWG number sizes are not commonly stocked and, for this reason, should not be specified.) An optional method of measuring stranded wire is to measure *one strand* in the gauge, count the number of strands in the total conductor, and refer to Table 26.1 to determine the equivalent AWG number.

To provide additional help in selecting wire, we use the following example to determine the most suitable type of wire plating, type (stranded versus solid), and size.

> *Example*
>
> *Power amplifier requirements:*
>
> *Current:*   1.6 amperes dc maximum
>
> *Operating temperature range:*   70°C (see Sec. 26.2)
>
> *Flexibility requirements:*   pc board is rigidly supported—no vibration problems—minimum flexibility required for assembly
>
> *Solution:*   Since flexibility is not a prime factor, solid wire is selected because it is the least expensive. The temperature requirements are minimal, thus allowing the use of tin coating that will further reduce costs and ensure availability. The size of the wire will now be selected based on the current requirements of the circuit. Most of the interconnections are signal lines in which current demands are very low. The only wires that will be expected to handle 1.6 amperes are those originating from the power supply. Appendix VII indicates that AWG No. 20 wire will handle 2 amperes. All signal wires could be a great deal smaller than this; but for the sake of uniformity and simplicity, a realistic choice would be AWG No. 20 wire for *all* interconnections. Although the cost of wire is proportional to size, the savings realized by the choice of a smaller wire are negligible.

## 26.2 INSULATION MATERIAL

Hookup wire used in making interconnections in electronic equipment must be provided with an insulation to prevent short circuits and to protect the conductor. The only exception to this rule involves the use of bus wire, which, because of rigidity and short length, poses no insulation problems.

The selection of the most suitable insulation for a specific application is based on three criteria: *electrical, mechanical,* and *environmental.* Each of these criteria is discussed.

Common wire insulations and insulation coatings are divided into three main categories: *rubbers, vinyls,* and *special high-temperature,* or *special-application, materials.* A comparative tabulation of various insulating materials is given in Table 26.2. This table compares the following mechanical and physical properties:

1. Voltage breakdown
2. Insulation resistance
3. Low dissipation factor
4. Abrasion resistance
5. Flexibility
6. Cut-through resistance

**TABLE 26.2**
Comparative Ratings for Wire Insulation

| Family | Material | Common Designation | Voltage Breakdown | Insulation Resistance | Low Dissipation Factor | Abrasion Resistance | Flexibility | Cut-Through Resistance | Water Absorption Resistance | Flame Resistance | Acid Resistance | Useful Temperature Range (°C) |
|---|---|---|---|---|---|---|---|---|---|---|---|---|
| | Silicone rubber | Silicone | Exc. | Exc. | Good | Fair/poor | Exc. | Poor | Fair | Fair | Fair | −60 to 200 |
| Rubber | Polychloroprene rubber | Neoprene | Fair | Poor | Fair | Exc. | Exc. | Good | Good | Good | Good | −30 to 80 |
| | Butyl rubber | Butyl | Good | Good | Good | Good/exc. | Exc. | Fair | Good | Poor | Good/exc. | −40 to 85 |
| | Polyvinyl chloride | PVC | Good | Good | Fair | Good | Good | Good | Good | Good/exc. | Good/exc. | −55 to 105 |
| Polyolefin (vinyl) | Polyethylene | PE | Exc. | Exc. | Good | Fair/good | Fair | Good | Exc. | Poor | Good/exc. | −60 to 80 |
| | Polypropylene | Poly | Good/exc. | Exc. | Good | Good/exc. | Poor | Good | Exc. | Fair/Poor | Exc. | −20 to 125 |
| | Polytetrafluoroethylene | Teflon* (TFE) | Good | Good | Exc. | Exc. | Good | Fair | Exc. | Exc. | Exc. | −70 to 250 |
| | Monochlorotrifluoroethylene | Kel-F† | Good | Exc. | Fair | Good/exc. | Good | Good | Exc. | Exc. | Exc. | −80 to 200 |
| Fluorocarbon | Polyvinylidine fluoride | Kynar‡ | Exc. | Fair | Good | Exc. | Good | Exc. | Good | Good | Good | −65 to 125 |
| | Polyurethane | Urethane | Good | Poor | Fair | Exc. | Exc. | Good | Poor | Poor | Fair | −55 to 80 |
| | Polyamide | Nylon | Poor | Poor | Fair | Exc. | Poor | Good | Fair/poor | Good/Fair | Fair | −40 to 120 |
| | Polyester film | Mylar* | Exc. | Exc. | Poor | Good | Fair | Exc. | Good | Fair/poor | Good | −65 to 120 |

Ratings are in decreasing order: excellent (exc.) good, fair, poor

*Trademark of DuPont.

†Trademark of Minnesota Mining & Manufacturing Company.

‡Trademark of Pennsalt Chemical Corp.

7. Water absorption resistance
8. Flame resistance
9. Acid resistance
10. Useful temperature range

The *voltage breakdown* and *insulation resistance* are the two factors that are of key importance in high-voltage applications. Insulations designated *excellent* are the only ones that should be considered for this purpose. As seen from Table 26.2, silicone, polyethylene (PE), and Mylar possess high corona resistance and are commonly used for high-voltage cables. Neoprene, urethane, and nylon insulations would not be suitable for this application because of their poor rating. The amount of vibration that a wire is exposed to is directly related to its insulation voltage breakdown rating. Therefore, this factor needs to be considered in the choice of insulation.

Signal attenuation at high frequencies is a function of the insulation's *dissipation factor.* Generally, insulations with a low dissipation factor are preferable. Teflon has the lowest dissipation factor among those listed in Table 26.2.

Mechanical considerations, such as *abrasion resistance, flexibility,* and *cut-through resistance,* represent a compromise between cost versus installation ease. Abrasion and cut-through resistance are generally directly related—the higher the value of one, the higher the value of the other. Of the insulations listed in Table 26.2, Kynar is the toughest, having excellent abrasion and cut-through resistance characteristics. The insulation flexibility becomes an extremely important consideration, especially in high-density packaging, for ease of installation. The *rubber* family (silicon, neoprene, and butyl) exhibits the highest flexibility and the *vinyls* (PE and poly) the lowest.

*Water-absorption resistance* is an environmental consideration. Rubber and urethane materials are more susceptible to moisture than the vinyl materials. Teflon materials are the least susceptible to water absorption.

*Flame-resistant* materials find application in electronic equipment in which explosions or flammable conditions may be encountered. Teflon and Kel-F are the materials specified for this application, since they do not support combustion. All the other insulation materials listed in Table 26.2 will support combustion in varying degrees.

Insulations that are highly *acid resistant,* such as butyl rubber, polypropylene, and Teflon, find wide application in the space industry. Urethane, nylon, and silicone rubber are inferior in this property.

The useful temperature range for PVC, nylon, and PE is narrow, whereas for Teflon and Kel-F it is wide. In considering this property, it is important to use good judgment in terms of cost. For example, it would not be advisable to choose Teflon (upper limit 250°C) based on its superior temperature characteristics if the temperature range of PE (upper limit of 80°C), as well as its other characteristics, will satisfy the design requirements, because Teflon is considerably more expensive. From the wide range of materials available, it is evident that the selection of the most appropriate wire insulation for a specific application is a compromise between specifications and cost.

To illustrate the use of Table 26.2, the following problem is considered:

*Circuit:* stereo power amplifier
*Electrical specifications*
    Voltage 36 volts dc maximum
    Current 1.6 amperes dc maximum
    Frequency 20 to 20 kilohertz (kHz)
*Mechanical specifications*
    Stationary
    No expected high-level vibrations
    Low-density packaging
    All leads pass through grommets

*Environmental specifications*

   Temperature range—+50 to 90°F (+10 to 32°C)

   Humidity—40% typical

These specifications appear to present no critical circuit requirements insofar as wire insulation selection is concerned. Since voltage and current requirements are low, any insulation with a *voltage breakdown* and *insulation resistance* comparison rating of *good* and *excellent* is acceptable. This stipulation eliminates neoprene, Kynar, urethane, and nylon from further consideration.

With a frequency range specified at 20 hertz to 20 kHz, any insulation with a dissipation factor rating of *good* or *excellent* will be suitable. Low-dissipation-loss factors become critical only at higher frequencies.

Because all wires will be passed through rubber grommets, the possibility of abrasion is nonexistent and the quality of *abrasion resistance* need not be considered in this evaluation even if a *fair* or *poor* rating is assigned a specific material. Therefore, for this consideration, any of the wire insulations listed are suitable.

*Flexibility* also is not critical because of the low-density packaging of the circuit and the resultant uncomplicated wiring. Therefore, no insulation material need be rejected because of its flexibility rating. To reduce assembly time, especially if large-scale production is considered, an insulation with a comparison rating for *cut-through resistance* of *good* or *excellent* should be considered. This eliminates silicone, butyl, and Teflon from further consideration. Since the use of this type of power amplifier is usually in the home or other location wherein temperature and humidity are controlled at levels that exclude critical environmental problems, any insulation with a *water-absorption resistance* rating of *good* or *excellent* is acceptable. *Flame* and *acid resistance* is, of course, of no consideration in the intended environment. Therefore, no insulation, even with a *fair* rating for either of these categories, is omitted from consideration.

Finally, the expected ambient temperature of the unit is approximately +10 to 32°C, with a factor of 2 used in derating on each end of the temperature range to allow for extreme conditions (the inside case temperature will increase under full power output). The range of derated temperatures now considered becomes 5 to 64°C. Table 26.2 shows that all the insulations listed have *useful temperature ranges* that are well within these specifications. As a result, no insulation material will be eliminated because of temperature limitations.

An evaluation chart is provided in Table 26.3 to show more clearly the procedure for eliminating insulation materials that are below minimum ratings. The *Analysis Results* column of Table 26.3 indicates that either polyethylene (PE) or polypropylene (poly) plastic vinyl insulation is suitable for the example problem. The final selection between these two materials now becomes a matter of cost and availability.

For purposes of identification, hookup wire insulation is provided with a color code. This provides rapid identification of wires, which is an indispensable aid when working with cables and harnesses. Color-coding is extremely useful in both assembly and in troubleshooting. Military and industrial standards have been established for color-coding insulation. These standards are listed in Table 26.4. Similar to the resistance color code adopted from EIA (Electronic Industries Association) standards, the solid color of the insulation is designated by numbers 0 through 9 and represents the first digit of the color-code number. With 10 basic colors, 10 numerical possibilities are available to identify individual wires. For complex wiring involving more than 10 leads that must be identified, insulation is also available with a broad tracer band that is a different color from the body color and spirals the length of the wire. These tracer colors now extend the available different color combinations to 100. When more than 100 wires are involved, as in the case of complex harnesses, insulation is available with a *second* tracer that is narrower than the first. Both tracers are closely spaced, running parallel with each other, and spiral the entire length of the wire. The use of the second tracer extends the numerical coding capabilities to 910 possible combinations. To illustrate the numbering system, a wire with a *red*

**TABLE 26.3**

Sample Evaluation Chart for the Selection of Wire Insulation

| Material | Voltage Breakdown | Insulation Resistance | Low Dissipation Factor | Abrasion Resistance | Flexibility | Cut-Through Resistance | Water Absorption Resistance | Flame Resistance | Acid Resistance | Useful Temperature Range | Analysis Results* |
|---|---|---|---|---|---|---|---|---|---|---|---|
| Silicone | ✓ | ✓ | ✓ | ✓ | ✓ | † | † | ✓ | ✓ | ✓ | 2 |
| Neoprene | † | † | † | ✓ | ✓ | ✓ | ✓ | ✓ | ✓ | ✓ | 3 |
| Butyl | ✓ | ✓ | ✓ | ✓ | ✓ | † | ✓ | ✓ | ✓ | ✓ | 1 |
| PVC | ✓ | ✓ | † | ✓ | ✓ | ✓ | ✓ | ✓ | ✓ | ✓ | |
| PE | ✓ | ✓ | ✓ | ✓ | ✓ | ✓ | ✓ | ✓ | ✓ | ✓ | 0—suitable |
| Poly | ✓ | ✓ | ✓ | ✓ | ✓ | † | ✓ | ✓ | ✓ | ✓ | 0—suitable |
| Teflon (TFE) | ✓ | ✓ | † | ✓ | † | ✓ | ✓ | ✓ | ✓ | ✓ | 1 |
| Kel-F | ✓ | † | † | ✓ | ✓ | ✓ | ✓ | ✓ | ✓ | ✓ | 1 |
| Kynar | ✓ | † | † | ✓ | ✓ | ✓ | † | ✓ | ✓ | ✓ | 1 |
| Urethane | ✓ | † | † | ✓ | ✓ | ✓ | ✓ | ✓ | ✓ | ✓ | 3 |
| Nylon | † | ✓ | † | ✓ | ✓ | ✓ | ✓ | ✓ | ✓ | ✓ | 3 |
| Mylar | ✓ | | | ✓ | | | | | | | 1 |

*Total number of insulation characteristics that do not meet minimum required ratings

✓Insulation characteristic meets minimum required rating.

†Insulation characteristic below example problem minimum required rating.

**TABLE 26.4**

Color-Coding Standards for Wire Insulation

| | | Numeric Wire Code | |
|---|---|---|---|
| Color Code | | Numerical Possibilities | Description |
| 0 | Black | 10 | Solid color insulation |
| 1 | Brown | 90 | Solid color with one |
| 2 | Red | | broad trace |
| 3 | Orange | 810 | Solid color with one |
| 4 | Yellow | | broad trace and |
| 5 | Green | | one narrow trace |
| 6 | Blue | | |
| 7 | Violet | 910 (total numeric combinations) | |
| 8 | Gray | | |
| 9 | White | | |

**TABLE 26.5**

Wire Color Identification Related to Circuit Function

| Functional Color Coding Chassis Wiring | | | Color Code for Power Transformers | | |
|---|---|---|---|---|---|
| Color | Code Number | Function | Winding | | Color Code |
| Black | 0 | Grounds, grounded elements, and returns | Primary { Untapped | | Black |
| Brown | 1 | Heaters, or filaments not connected to ground | Tapped | | Common black, tap black/yellow |
| Red | 2 | Power supply, $+V_{cc}$ or $B+$ | High-voltage plate | | Red |
| Orange | 3 | Screen grid | High-voltage center tap | | Red/yellow |
| Yellow | 4 | Transistor emitters or cathodes | Rectifier filament | | Yellow |
| Green | 5 | Transistor bases and control grids | Rectifier filament center-tap | | Yellow/blue |
| Blue | 6 | Transistor collectors, anodes for semiconductor elements and tube plates | Filament (No. 1) | | Green |
| | | | Filament (No. 1) center tap | | Green/yellow |
| Violet | 7 | Power supply, $-V_{cc}$ or $B-$ | Filament (No. 2) | | Brown |
| Gray | 8 | Ac power lines | Filament (No. 2) center tap | | Brown/yellow |
| White | 9 | Miscellaneous, above-ground returns | Filament (No. 3) | | Gray |
| | | | Filament (No. 3) center tap | | Gray/yellow |

body color, broad *green* tracer, and narrow *orange* tracer (red/green/orange) can be classified by the three-digit number 253.

Standards also have been established that do not employ a numbering system but rather designate wire color with respect to circuit function. Two such standards are the Government Standard MIL-STD-122 color code for chassis wiring and the EIA color code for power transformers. These classifications are shown in Table 26.5.

## 26.3 SPECIAL WIRE CONFIGURATIONS

There are several readily available wire configurations that are used for special applications. *Shielded wire,* shown in Fig. 26.2a, is a variation of hookup wire previously discussed. It consists of an insulated length of hookup wire enclosed in a conductive envelope in the form of a braided-wire shield. The shield is formed from fine strands of tinned, annealed copper wire. The strands are woven into a braid that provides approximately 85% coverage of the underlying insulated conductor. Shielded wire is also available with an outer insulation covering to prevent shorts when running the wire between termination points. Another variation of shielded wire has *two* inner conductors twisted around each other. This twist improves unwanted signal cancellation.

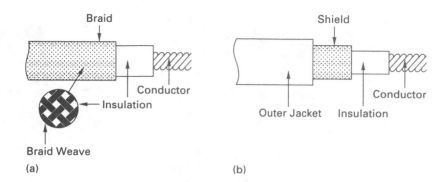

**Figure 26.2**
(a) Shielded wire;
(b) coaxial cable.

Shielded wire is employed in audio circuits at the input stage where the signal level is low and undesired noise (typically 60-hertz hum), or feedback from the output stage of a high-gain amplifier, could produce unwanted signals. The shield is either connected to the chassis at the circuit input or it may be connected to the chassis at both ends to provide the most effective method of minimizing undesired signal pickup.

*Coaxial cable* (coax), shown in Fig. 26.2b, is used exclusively in radio frequency (rf) circuits. It is similar in construction to shielded wire, with the outer insulating jacket isolating the shield from ground. The major difference between coaxial cable and shielded wire is in their electrical characteristics. Coax is designed specifically to transmit rf energy, from one point to another, with minimum loss (attenuation). Insulation material and thickness is controlled with extreme accuracy during manufacture to produce cables possessing 50, 75, or 95 ohms characteristic impedance for proper matching. Shielded wire, on the other hand, is intended for low-frequency applications in which its impedance is not critical.

Losses from skin effect at high frequencies must be minimized when using coaxial cable. *Skin effect* is the term used to describe the type of loss or attenuation of the signal caused by the current traveling along the surface (skin) of the conductor and partly into the adjacent insulation. Because of its outstanding insulating characteristics at high frequencies, nylon is the type of insulation most often used in the manufacture of coaxial cables.

Another special wire configuration is *Litzendraht* (Litz) wire, which is used in the manufacture of tuning coils. At short-wave frequencies, this type of wire has low skin effect, owing to its configuration. Litz wire is a stranded conductor containing 25 or more extremely fine wire strands. Each strand is independently insulated with cotton insulation and wound from the center to the outer surface of the overall conductor. This technique of winding tends to equalize the skin-effect problem.

## 26.4 TOOLS FOR WIRE PREPARATION AND ASSEMBLY

To form a sound electrical connection between the hookup wire used for interconnections and the terminal to which it is to be soldered, *all* insulation must be completely removed from the end of the wire to expose the conductor. In addition, it may become necessary to remove contaminants, such as dirt, finger oils, or oxidation, from the bare wire. (Oxidation will occur if the bare conductor has been exposed to the atmosphere for a prolonged time before soldering.) It is vitally important when preparing to make a solder connection that all surfaces to be soldered (alloyed) are absolutely free of all contaminants.

Mechanical wire strippers, such as the two styles shown in Fig. 26.3, are among the most common tools used to remove wire insulation. Each has its advantages and disadvantages with regard to resulting quality and ease of operation. The *step-adjustable* stripper is shown in Fig. 26.3a. All that is required to set the jaws is to rotate the *selector step wheel* until the desired gauge number is oriented with the *stop pin*. Its obvious disadvantage is that it is capable of stripping only eight gauge wire sizes (AWG No. 12 to AWG No. 26). To minimize the possibility of conductor damage when using this stripper, it should be held perpendicular to the wire when pulling off the insulation.

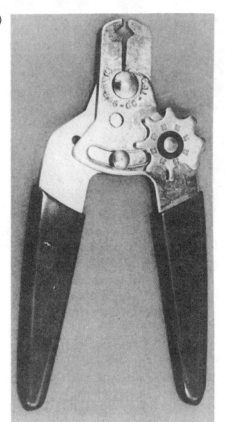

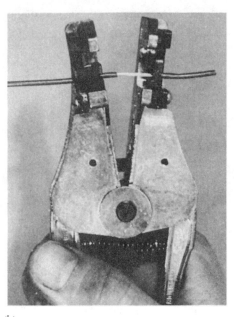

(b)

**Figure 26.3** Mechanical wire strippers: (a) step adjustable; (b) insulation removal with automatic wire stripper.

The most popular mechanical stripper is the *automatic* style shown in Fig. 26.3b. This stripper has six marked cutting positions on the jaws to accommodate various gauge sizes. The wire to be stripped is placed into the appropriate cutting jaw position and between the gripping jaws. With the wire held in position, the handles are gently squeezed together. This causes the movable upper cutting jaws to first press the insulation down into the lower stationary cutting jaws. Additional squeezing of the handles causes the insulation to be severed about the conductor while the upper movable gripping jaw firmly grasps the wire with the lower stationary gripping jaw. Further squeezing of the handle causes both sets of jaws, while engaged, to separate, thereby removing the insulation from the end of the wire (Fig. 26.3b). As pressure on the handles is removed, both sets of jaws disengage, thus releasing the stripped wire. These strippers are also available with a stop that, when set to the desired length of strip, will consistently remove the same length of insulation from wires.

*Thermal* wire strippers, although not popular in prototype construction, are another means of removing the insulation from wires. They consist of a heating element and two pivoted handle grips having electrodes that serve as the stripping jaws. The wire to be stripped is positioned between the electrodes, which are then clamped together around the wire. Squeezing the insulated handles activates a switch that causes the electrodes to heat the insulation to its melting point. With the electrodes energized, the wire is rotated so that the insulation melts uniformly. When the insulation has melted, the pressure on the handles is released, the wire is removed, and the insulation is pulled off of the conductor. Thermal strippers will not nick or cut conductors. They are more expensive than mechanical strippers, however, and may also cause irritating fumes as the insulation is melted.

Whichever method of wire stripping is employed, more wire than is necessary to make the connection should be stripped. The reasons for this are that (1) in the case of stranded wire where tinning (discussed later in this section) of the strands is required, solder buildup usually occurs at the conductor end and can be removed simply by cutting; and (2) cutting and stripping each wire to its exact required dimension will not provide the necessary "play" that the technician may need to adjust wire positions during wire operations.

As mentioned previously, it may be necessary to further clean the exposed wire prior to soldering. This can be quickly accomplished with the use of the lead cleaner shown in Fig. 25.20. Designed primarily for use on solid wire (including component leads), this eraser-style cleaner will remove contaminants, such as dirt and oxides, from the surface of the wire.

To strip enamel or any type of film insulation from solid wire, an *abrasive* lead cleaner may be used. It is made of braided wire formed about a piece of aluminum. Details for constructing this type of cleaner are given in Problem 26.6.

The final step prior to interconnection is the *tinning* of stranded wire. This process is necessary to allow for the formation around a terminal in a manner that will result in as neat a joint as obtainable with the use of solid wire. Approximately $\frac{1}{2}$ inch of insulation is first removed from the wire and the individual strands are firmly twisted together in a direction to return to their original twist orientation that is disturbed by the stripping action. This twisting also returns the wire to its original diameter, which is important to avoid difficulties in making connections to small eyelet terminals or the center prong on phono plugs.

The exposed and twisted wire is then dipped into a liquid flux such as Kester 1544 or 1589 flux. Allowing the flux to contact the insulation should be avoided so that cleaning up after the tinning process will be faster.

The tinning of stranded wire can be effectively accomplished using one of two methods. The first is with the use of a soldering iron, as shown in Fig. 26.4a. The insulated portion of the wire is gently held between the jaws of a bench vise with the fluxed conductor extended. The hot soldering iron is brought into contact with the bottom of the wire and the solder is applied to the top directly onto the strands. Heat from the iron will cause the solder to melt and be drawn through the strands since melting solder follows the direction of heat flow. Drawing the iron and the solder along the length of the wire from the end to the insulation and back again will ensure solid, uniform bonding of the individual strands.

The second method of tinning is with the use of a *solder pot,* shown in Fig. 26.4b. The pot contains a pool of molten solder maintained at a temperature between 600 and 650°F. To begin the tinning process, the dross formation on the surface of the molten solder is skimmed away with a small wooden spatula, such as a tongue depressor. The fluxed wire is then slowly placed into the solder so that the insulation is just above the surface.

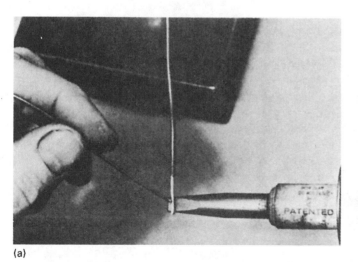

(a)

(b)

**Figure 26.4** Tinning stranded wire: (a) using a soldering iron; (b) dipping the wire into liquid flux and then into molten solder.

This slow entry allows the individual strands to arrive at a sufficient temperature so that proper wetting action and bonding result. Withdrawing the wire is again done slowly so that the surface tension of the solder will draw off any excess solder on the wire. (*CAUTION:* Safety glasses should be worn when working with molten solder, to avoid injury.)

Effective tinning results when sufficient solder has wetted all conductor surfaces forming a solid bond between them. Excessive solder should be avoided. The contour of each strand should be plainly visible.

After the wire has been tinned, the flux residue should be removed. This is done by placing the wire on a paper towel and scrubbing the end with a stiff acid brush using iso-propyl alcohol. The wire is now ready to be wrapped around any type of connection without concern for any strands becoming separated from the group.

After the tinning operation, the excess wire length along with the accumulation of solder that usually builds up at the wire end can be removed with a pair of *diagonal cutters.* Selecting the most suitable type of diagonal cutters (dykes) depends primarily on the size of the wire being used as well as accessibility during assembly. Because of the variety of cutting situations encountered, many styles of diagonal cutters are available. A selection of common styles used for electronic wiring is shown in Fig. 26.5. The size of these diagonal cutters ranges from 6 inches long for heavy-gauge wire (AWG No. 14 maximum) to 4 inches long for use with smaller wire such as AWG No. 24 and smaller.

The cutting edges of the diagonal cutters can be classified as *regular, semiflush,* or *full-flush.* Regular cutters, used for general-purpose applications, leave a "point" at the end of the wire after cutting. Semiflush cutters leave a less pronounced point. Full-flush cutters, intended for use only on annealed copper wire, leave no point or chamfer but rather a smooth, flat end perpendicular to the wire axis. Since the regular and semiflush cutters distort the end of a conductor, this could create a problem when attempting to insert the wire into a close tolerance hole such as those associated with small eyelets and holes in pc board terminal pads. In addition, the sharp points these cutters leave are undesirable when protruding from terminal connections. For these reasons, the full-flush cutter is recommended.

The jaws of most diagonal cutters produce a slight "shock" as they come together when cutting a wire. When used on hookup wire, this presents no problem. However, sensitive semiconductor devices could be damaged by this shock. For this reason, shear-type

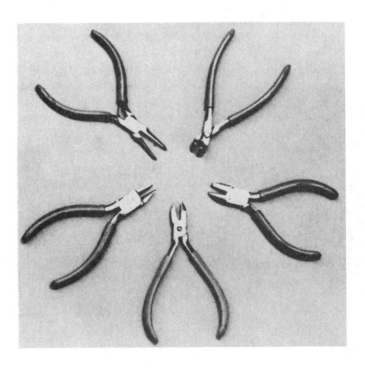

**Figure 26.5** Various styles of wire-cutting tools.

cutters are recommended for delicate work. The jaws of these cutters pass each other similar to a pair of scissors, thereby minimizing the cutting shock.

Diagonal cutters are available in *rounded, tapered,* and *offset* noses to fulfill the requirements of a variety of cutting needs. The most appropriate cutting jaws, shapes, and edge types are selected after the work to be performed is evaluated. When heavier-gauge wires are to be cut and close cutting tolerances and accessibility are not critical, less delicate cutters may be employed. Intricate wiring using finer-gauge wire will require more delicate cutters.

Rubber jaw inserts are available on some styles of cutters, which grip the end of the wire to be removed. These inserts prevent the cut end from flying out of the jaw when the wire is cut. (*CAUTION:* There is a serious safety hazard involved in using diagonal cutters not equipped with rubber inserts. It is always a good practice to point the conductor end *downward* to prevent the flying end from causing injury. Cutters with plastic grips are preferable to cutters with no grips because they not only provide more positive gripping, but also reduce hand fatigue.)

## 26.5 WIRE CONNECTIONS

One of the most common types of terminal used to make interconnections between pc boards and chassis-mounted components is the *swaged turret terminal* discussed in Chapter 16. To obtain a good electrical connection, it is vitally important that first a solid mechanical connection be made. The wire to be soldered to a turret-type terminal should be formed as tightly as possible around the turret without scraping or nicking the conductor surface or deforming the wire contour. With the insulation positioned no more than 1/8 inch away from the terminal, the conductor is first tightly formed 180 degrees around the turret with long-nose pliers (Fig. 26.6a). The excess length of wire is then cut at point *A,* which leaves a sufficient length to continue forming a complete 360-degree *wraparound* connection. This

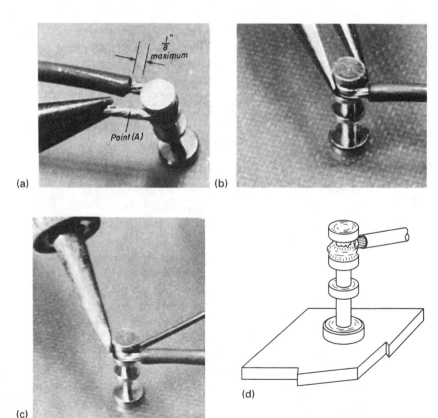

**Figure 26.6** Wiring to a turret-type terminal: (a) initial 180° wrap; (b) forming the complete 360° wrap; (c) correct iron tip and solder position; (d) properly soldered terminal.

wrap being formed with long-nose pliers is shown in Fig. 26.6b. Although 180-degree wraps are often used in electronic wiring, the 360-degree wrap provides the most secure mechanical connection. Long-nose pliers with smooth jaws should be used when making mechanical connections to any terminal configuration to avoid nicking and deforming the conductor surface.

Soldering to a turret terminal is performed using the soldering techniques discussed in Chapter 17 for soldering leads to pc board terminal pads. A small amount of solder is first applied to the soldering iron tip (solder bridge). The tip is next brought into contact with the wire and shoulder of the terminal, as shown in Fig. 26.6c. Positioning the tip directly opposite from the insulation will minimize heat damage to plastic and rubber insulation. Solder is applied immediately when the connection has reached the proper temperature (i.e., when the solder of the bridge or the solder used in tinning stranded wire begins to flow). Only a *small* amount of solder is necessary to make a sound electrical connection. If properly soldered, the contour of both the wire and the terminal will be plainly visible. A properly wired and soldered turret terminal connection is shown in Fig. 26.6d.

Wraparound connections also find wide application when wiring to any eyelet-type terminal, such as those associated with *terminal strips, switches, potentiometers,* and *pc board connectors.* Wiring to these terminals is shown in Fig. 26.7. Wire forming about an eyelet-type terminal lug begins by positioning the insulation no more than 1/8 inch from the eyelet opening. The wire wrap is then formed in a similar manner to the 360-degree wrap for turret terminals, but since the cross section of the eyelet is a very thin rectangular shape, the wrap is more difficult to form. It is recommended that the first two 90-degree contours, as shown in Fig. 26.7a, be preformed before hooking the conductor through the eyelet. Once positioned into the eyelet, the excess lead length is cut at point *A* and the final 90-degree bend is made to complete the wrap (Fig. 26.7b). It is not good practice to make more than three wraparound connections to a single eyelet or to solder them until all have been mechanically formed.

When mechanical strength requirements are minimal, wires can be soldered more easily into small-diameter holes, such as those found on solder lugs, by means of a *feedthrough* connection, shown in Fig. 26.8. For this type of connection, the diameter of the wire should be approximately equal to that of the hole in the lug. As with other types of connections, the insulation is positioned at a maximum of 1/8 inch from the terminal, with the prepared wire end passing through the lug hole. The excess wire is trimmed to within approximately 1/16 inch from the terminal. Because the diameter of the wire approximates that of the hole, soldering will properly alloy both circumferences. The resulting connection is mechanically satisfactory and electrically sound, although feed-through connections should be avoided in equipment that will be subjected to severe shock and vibration.

Wires may be *dressed* between connections by either *point-to-point* or *square-corner* wiring techniques. Point-to-point wiring, as the name implies, is a wiring technique used primarily in high-frequency circuits or any economical wiring application. Wiring interconnections are made with regard to the shortest routing path possible to minimize lead length, which is critical at high frequencies. This wiring technique is quicker and uses less

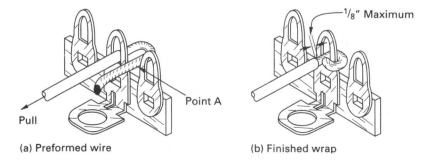

**Figure 26.7** Wire forming for eyelet terminals.

(a) Preformed wire

Pull

Point A

1/8" Maximum

(b) Finished wrap

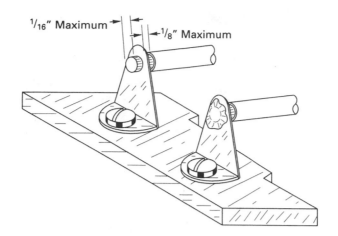

**Figure 26.8** Feed-through connection.

wire than the other methods discussed here. Point-to-point wire dress usually does not add to the neatness of the package. However, with careful wiring positioning and neat soldering connections, the resulting wiring will be reliable and will display quality workmanship.

Square-corner wiring is a wire dress technique in which all wires installed in the package run either parallel or perpendicular to one another. When any wire path direction changes, the change is made with a 90-degree bend. Although not used in high-frequency work, this technique imparts a very neat appearance to the package. The only disadvantage to square-corner wiring is the additional amount of wire and time necessary to complete a package.

## 26.6 SOLDERING TO DOD-STD-2000-3 STANDARDS

If soldering of terminal connections is required to be done to comply with DOD-STD-2000-3 standards, Figs. 26.9, 26.10, 26.11, and 26.12 should be referenced. In the case of turret terminals, the leads are formed for single- and multiple-lead connections as shown in Fig. 26.9a. For this standard, the prepared lead is wrapped about a post section between 180 and 270° leaving

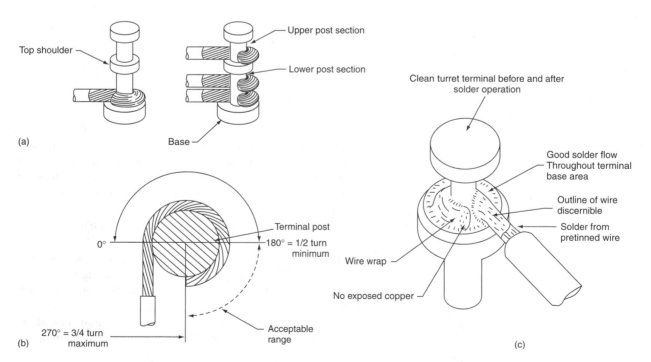

**Figure 26.9** Turret terminal wire wrap methods and solder connection.

*Wire and Interconnection Techniques*

**497**

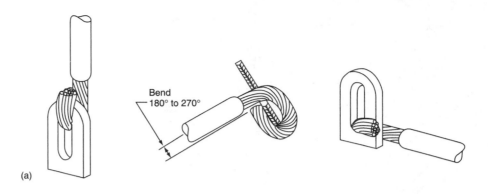

(a)

Bend
180° to 270°

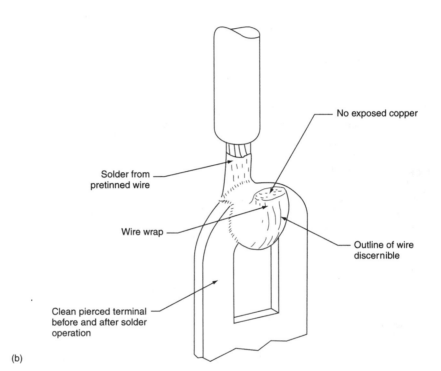

No exposed copper

Solder from
pretinned wire

Wire wrap

Outline of wire
discernible

Clean pierced terminal
before and after solder
operation

(b)

**Figure 26.10**  Pierced or perforated terminal wire wrap and solder connection.

an insulation gap between the post and the wire insulation, as shown in Fig. 26.9b, of no more than two lead diameters prior to soldering. By using the proper positioning of the soldering iron tip and solder on opposite sides of the post, as shown in Figure 26.6c, the proper wetting and concave filleting that leaves the contour of this type of connection still visible will result. The required degree of wetting and filleting is shown in Fig. 26.9c.

When soldering pierced or perforated (eyelet) terminal connections to DOD-STD-2000-3 standards, the mechanical wrap should be between 180 and 270°, as shown in Figs. 26.7 and 26.10a, with an insulation gap between the terminal and the wire insulation of no more than two lead diameters prior to soldering. The proper positioning of the soldering iron tip and the solder is on the opposite sides of the mechanically formed connection, similar to that for the soldering shown previously for a turret terminal, remembering to position the tip of the iron farthest away as possible from the wire insulation to avoid melting the insulation. Figure 26.10b shows the desired wetting and concave filleting that leaves the contour of this type of connection still visible.

Another type of terminal often encountered on connectors is the solder cup. Figure 26.11a shows a D-Subminiature connector that has cup terminals.

When the wires are prepared for installation to this type of terminal, they must be cut long enough to reach the bottom of the cup and have an insulation gap between the top of the cup and the wire insulation of no more than two lead diameters. This is also the case for attaching multiple wires to one solder cup as shown in Fig. 26.11b. Before attempting to make a solder connection that complies with DOD-STD-2000-3 standards, the cup must be heated and solder introduced into the cup, allowing it to fill one-half to two-thirds full. The tip of the iron must first be sponge wiped to remove any solder that could be transferred to

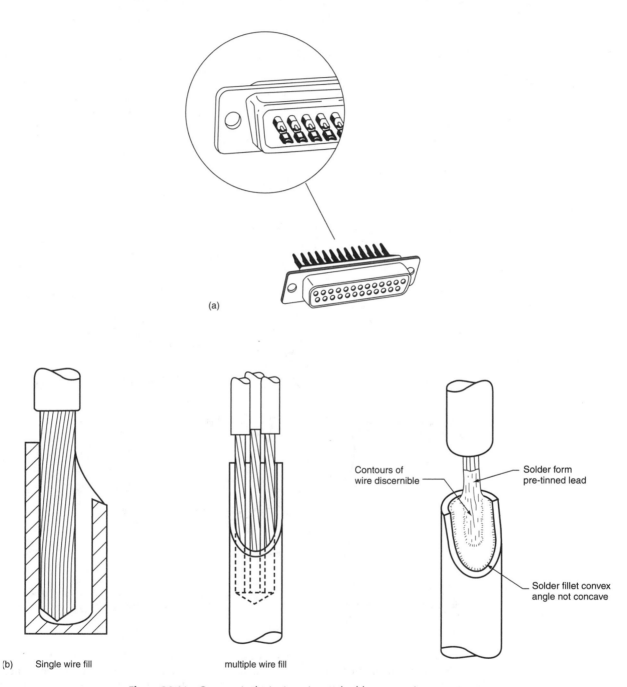

(a)

Contours of
wire discernible

Solder form
pre-tinned lead

Solder fillet convex
angle not concave

(b)    Single wire fill

multiple wire fill

**Figure 26.11**  Cup terminal wire insertion and solder connection.

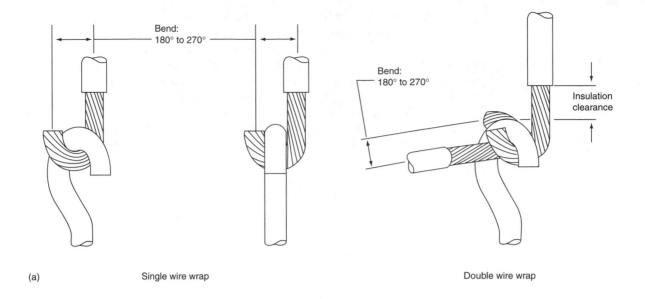

(a)

Single wire wrap

Double wire wrap

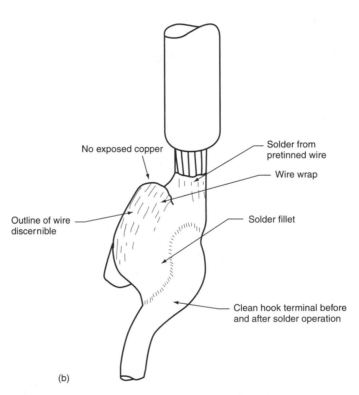

(b)

**Figure 26.12** Hook terminal wire connection methods and solder connection.

the outer surface of the cup in this part of the procedure. The solder deposited in this step is necessary to ensure that no air or flux can become trapped at the bottom of the cup when the wire connection is soldered. The proper position of the soldering iron tip, when soldering this type of connection, is against the rear surface of the cup. The tip is again sponge wiped to avoid depositing solder onto the outside of the cup. It is then pressed against the back of the cup near the top and slid down toward its base while maintaining contact between the tip of the iron and the back of the cup. Also, no solder is applied to the cup and lead when the cup is reheated. The lead is simply pushed into the cup that has been partially filled with solder. As the solder is re-melted, the lead is pushed into the cup until it bottoms. The solder should flow up the lead to a point slightly higher than the top of the cup while forming slightly convex instead of concave fillets on each side of the wire and the inside surface of the cup. After the soldering has been completed, there should be no solder void inside the cup. Figure 26.11b shows the desired wetting and filleting that leaves the contour of this type of connection visible.

Some components encountered have terminals that are formed into the shape of tiny hooks and are referred to as hook terminals. Figure 26.12a shows the method of making the 180 to 270° mechanical wrap of one or more leads to a hook terminal. The insulation clearance (gap) between the hook and the wire insulation should not be more than two lead diameters prior to soldering. When soldering one or more wires onto the hook, the rule of placing the tip of the iron and the solder on opposite sides of a connection are followed, being careful not to melt the insulation. Figure 26.12b shows the proper wetting and concave filleting that leaves the contour of the connection visible after the proper amount of heat and solder have been applied.

## 26.7 SPECIAL WIRE ASSEMBLIES AND TECHNIQUES

A common yet difficult wiring connection involves the preparation and installation of shielded wire to phono plugs and jacks. The function of the braid of the wire is to cover and thereby isolate the inner (signal) conductor. There should be minimum exposure of the signal wire insulation after the shield and signal wire connections are made to the connectors. If the shielded wire has an outer insulating jacket, approximately 1 1/2 inches of the jacket should be removed. The shield is then pushed back from the end so as to "loosen it." A pick (scriber) is then carefully worked into the braid without damaging any individual strands to form a hole large enough through which the insulated signal wire can be pulled. This technique is shown in Fig. 26.13a. The pick is then passed into the hole between the braid and the signal wire insulation so that it protrudes from the other side. Doubling the shielded wire

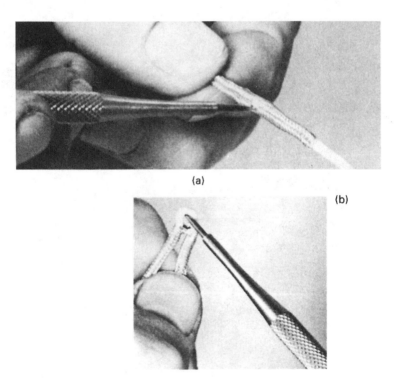

(a)

(b)

**Figure 26.13**
Technique for preparing shielded wire for soldering: (a) scribe used to provide access hole in shield; (b) removing signal lead from shield.

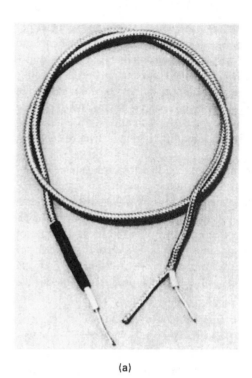

(a)

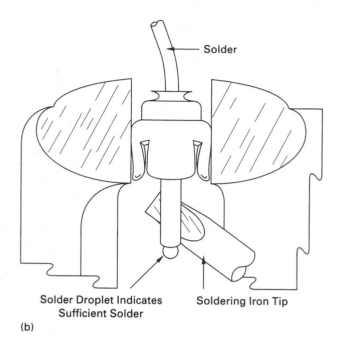

Solder

Solder Droplet Indicates
Sufficient Solder

Soldering Iron Tip

(b)

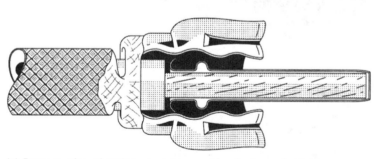

(c) Cutaway view showing correct assembly of wire and shield

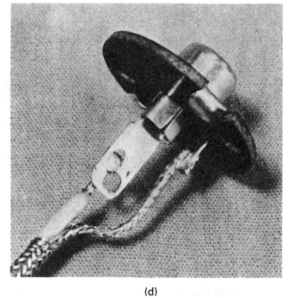

(d)

**Figure 26.14** Preparation and soldering of shielded wire to plugs and jacks: (a) shielded wire prepared for soldering; (b) preparation of prong with solder; (c) proper position of iron tip and solder; (d) correctly prepared jack.

back at the breakout point and lifting the pick will remove the signal wire from the hole in the braid. This technique is shown in Fig. 26.13b. The end of the braid is then tinned and will serve as the ground lead.

The prepared shield and signal wire ready for soldering to a connector are shown in Fig. 26.14a. Soldering the signal wire to a plug requires careful preparation. The hollow center prong of the plug is first filled with solder. This step can best be done by gently gripping the outer shell of the plug in a bench vise. The correct soldering iron tip position and application of solder is shown in Fig. 26.14b. When solder begins to appear at the tip of the prong, the iron is removed. Solder must not be allowed to flow in the outer surface of the prong or it may prove to be impossible to plug it into a jack. To install the wires onto the plug, at least 3/4 inch of signal wire is stripped and tinned, leaving only 3/16 inch of insulation between the stripped signal conductor and the braid. The braid is also tinned. The center prong of the plug is reheated and the signal wire is inserted into the top of the plug until the braid is even with the top of the ferrule. Some of the signal conductor may protrude from the end of the prong. This is removed flush with diagonal cutters. The braid is next wrapped 360 degrees around the ferrule and then soldered, using the technique shown in Fig. 26.14c. Proper installation of shielded wire to a phono jack is shown in Fig. 26.14d.

## 26.8 SOLDERLESS WIRING TECHNIQUES

*Solderless-* or *crimp*-type terminals provide a strong metal-to-metal (terminal-to-wire) bond that is suitable for applications using either stranded or solid wire. These terminals are particularly desirable in some production assemblies because they eliminate soldering. Two common configurations of solderless-type (crimp) connectors are the *ring tongue* and *fork tongue* styles shown in Fig. 26.15.

Selecting the most suitable style depends on the application. If rapid connection and disconnection are not required but a higher degree of mechanical security is necessary, the ring tongue should be used. If, however, rapid connecting and disconnecting requirements are specified with no critical mechanical requirements, the fork tongue style is preferable.

The *barrels* of crimp terminals are available in uninsulated or insulated forms. The insulator is usually polyvinyl chloride (PVC) and offers a degree of strain relief for the wire if strong vibration is expected. These insulators are commonly available in colors so that the range of AWG sizes that each barrel will accept can be quickly identified (see Table 26.6).

Solderless terminals in common use in electronics have barrels that are capable of accepting wire sizes from AWG No. 10 to AWG No. 22. A No. 10 terminal will accept both AWG Nos. 10 and 12 wire; a No. 14 terminal will accept AWG Nos. 14 and 16 wire; and a No. 18 terminal will accept AWG Nos. 18, 20, and 22 wire. The terminals are available to accept screw sizes from No. 4 to 3/8 inch.

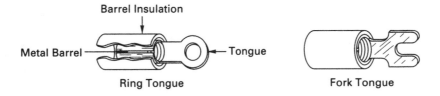

**Figure 26.15** Crimp terminals.

**TABLE 26.6**
Color Code for Insulated Crimp Connector Barrel Sizes

| Terminal Number | Barrel Insulation Color | Accepts AWG Sizes |
|---|---|---|
| No. 18 | Red | 22 to 18 |
| No. 14 | Blue | 16 to 14 |
| No. 10 | Yellow | 12 to 10 |

*Wire and Interconnection Techniques*

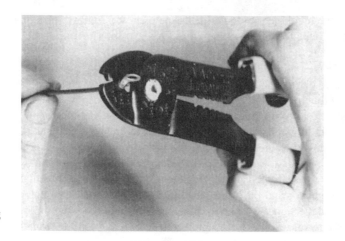

**Figure 26.16**
Technique for installing crimp terminals.

Conductors are assembled into solderless terminals with a *crimping tool* such as the one shown in Fig. 26.16. The wire insulation is stripped to a length equal to the metal barrel length. The conductor is then inserted into the terminal until the insulation seats against the end of the metal barrel. The barrel is then placed into the appropriate jaw crimping position of the tool. The split side of the barrel should be placed against the smooth portion of the jaw opposite the crimping tooth. When the handles are firmly squeezed together, as shown in Fig. 26.16, the barrel will be crimped against the conductor and will make a sound mechanical and electrical connection.

## 26.9 WIRE-WRAP CONNECTIONS

An alternative method of providing randomly located metal studs to which external connections can be made *without* the necessity of soldering wires is through the use of *wire-wrapping* techniques. Wire-wrap terminals are square or rectangular pins having at least two sharp corners that are mechanically secured to the pc board by staking (press-fitted) or soldering. Lead connections are made to these pins with mechanical or pneumatic wire-wrapping tools using solid wire. The resulting connection is both mechanically secure and electrically sound.

Two types of wire-wrap pins are shown in Fig. 26.17, each having its own configuration and mounting arrangement. The most basic wire-wrap pin is shown in Fig. 26.17a. It consists simply of a 0.045-inch-square wire from 3/4 to 2 inches long with both ends

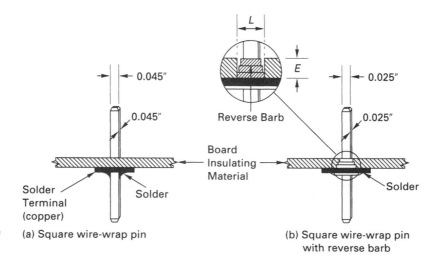

**Figure 26.17** Styles of wire-wrap terminals.

(a) Square wire-wrap pin

(b) Square wire-wrap pin with reverse barb

tapered for ease of initial insertion into the pc board. No special equipment is required for their installation since the pin is simply pressed into a predrilled hole. The recommended drill hole has a diameter of 0.050 inch with a 0.002-inch tolerance. Although commercially available vibrating tables and templates are employed to position the pins into the holes, the cost of such equipment is prohibitive in small volume or prototype work. These pins can be inserted easily by hand. The tapered end of the pin is first inserted into the drill hole of the board using finger pressure. This procedure is sufficient to temporarily hold the pin in a staking position. Similar to the staking of solder-type terminals, an arbor press is used with an anvil and flat-faced setting tool (see Sec. 16.1). The guide hole in the anvil should be large enough to accept the square configuration of the pin with sufficient clearance. The arbor press then is used to force the pin through the pc board, allowing it to extend to any desired height on the opposite side of the board. Although held in position only by the contact between the corners of the pin and the inside edge of the hole, it is sufficiently gripped in place until it can be soldered to the terminal pad. Connecting wires are then wrapped along the length of the pin extending above either side of the board.

Another wire-wrap terminal that provides uniform staking height is shown in Fig. 26.17b. It consists of double-ended square stock with either 0.025-inch or 0.045-inch flats made of brass and plated with gold alloy. Dimension $E$ in Fig. 26.17b is generally selected to be equal to the thickness of the pc board insulating material. To insert this pin, a hole is drilled that is approximately 0.002-inch smaller than the maximum diameter of the *reverse barb*. For the 0.025-inch square pin, dimension $L$ is 0.052 inch (hole size $\approx$ 0.050 inch), and for the 0.045-inch square pin it is 0.096 inch (hole size $\approx$ 0.094 inch). The pin is positioned in such a manner as to allow its crown to contact the foil side (terminal pad) of the board. The pin is then secured with an anvil and setting tool in an arbor press. The reverse barb prevents the pin from turning or backing out of the insulating material. Soldering between the crown and terminal pad will make a reliable electrical connection.

Mechanical and electrical connection for wire-wrap terminals is achieved by wrapping, under tension, solid wire around a post terminal that has at least two sharp edges. These edges, as well as the wire itself, are deformed when tightly wrapped. The average pressure between the wire and the sharp contact points of the post is in the order of 30,000 psi, which is more than sufficient to produce a sound mechanical and electrical connection. Two properly formed types of wire-wrap connections are shown in Fig. 26.18.

Wire-wrap connections are formed by manual operation and automatic or semiautomatic wire-wrapping machines. Manually operated wire-wrap tools, such as the one shown

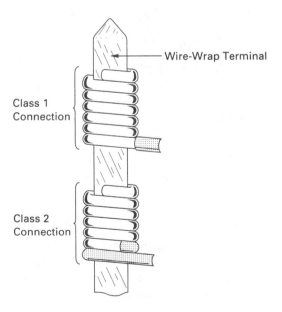

**Figure 26.18** Two styles of wire wrapping.

in Fig. 26.19a, are available to quickly form wire-wrap connections. Pneumatic wire-wrap tools are also available, but are usually employed with semiautomatic machines in mass-production applications.

To use the tool shown in Fig. 26.19a, solid wire of the desired gauge is first stripped and then inserted into the outermost hole at the end of the wire-wrapping tip (Fig. 26.19b).

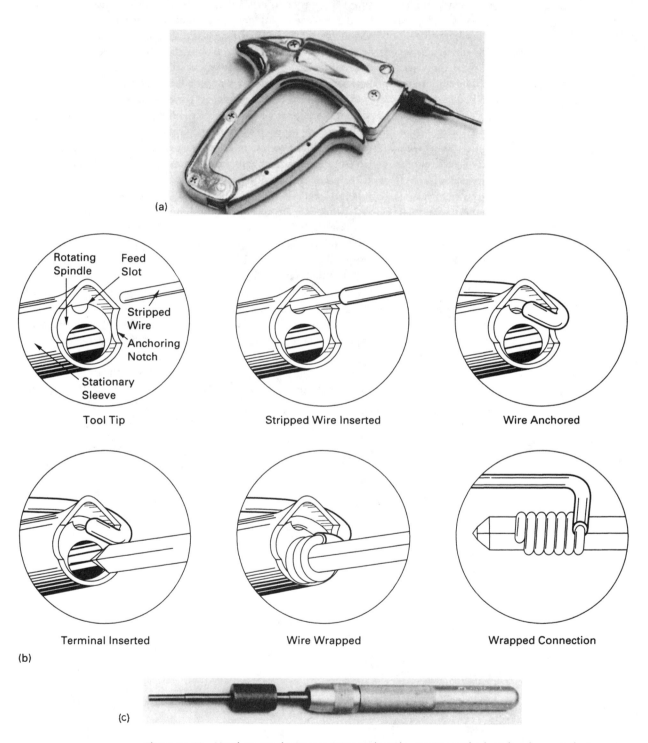

**Figure 26.19** Hand operated wire-wrapping tools and wrapping method: (a) hand operated wire-wrap tool; (b) wire-wrap connection process; (c) de-wrapping tool.

The tip is then positioned over the terminal and the trigger is quickly squeezed. The nose of the gun spins the wire around the terminal to form the desired connection.

The number of turns of wire wrapped is determined by the specifications provided in Table 26.7. The amount of insulation that must be stripped from the end of the wire can be quickly determined by trial and error.

Should it be necessary to unwrap a connection that is not readily accessible (on high-density packaging, wrapped terminals may be within 0.1 inch away from each other), *de-wrapping* tools such as the one shown in Fig. 26.19c are available. To remove a wrapped connection, the center hole in the nose is positioned over the terminal. Downward pressure of the tool against the wire will quickly unwrap the wire from the post. The tool must be kept vertical during the entire operation to avoid bending the terminal post.

When the number of terminal connections to be made becomes impossible to manually wire wrap at a cost-effective rate and without excessive errors, semiautomatic and fully automatic wire-wrapping machines are employed. One application of dense wiring includes that of large backplanes having thousands of wiring terminals on one side and heavily populated on the other side with printed circuit board or cable connectors. Another example is on panels requiring the discrete wiring of vast numbers of integrated circuit sockets.

Semiautomatic wire wrapping is performed on a machine having a vertical worktable to which the panel to be wired is clamped. Data-controlled carriages move the table in both $x$ and $y$ directions. The table movement is such that each new pin to be wired positions itself in front of the operator. This feature aids in reducing operator fatigue. A terminal pointer with a tool guide (mounted wrapping tool) traverses the $x$ axis and stops after each table movement in front of the next terminal to be wrapped. The operator feeds the specified size of wire into the tool guide and, with slight pressure, pushes the tool onto the pin to make each connection. Rates of between 450 and 600 wraps per hour are typical with semiautomatic machines.

When higher wiring densities and speeds exceed the capabilities of semiautomatic systems, a fully automatic wire-wrapping machine, such as that shown in Fig. 26.20a, is employed. This machine has an operator's console that includes the following: manual operating switches and data input, fault monitoring system, cycle counter, cycle interrupt, emergency stop, repeat wire operation, and sequence number readout.

The panel to be wired is clamped vertically inside the machine. Unlike the semiautomatic machines in which the board moves only in the $x$ and $y$ directions, the automated systems also provide for 90-degree board rotation. Two sets of programmable wire-wrapping tools and dressing fingers are mounted on two vertical carriages that are also data controlled. The dressing fingers form the specified contour of the wire as it is routed above selected channels between closely spaced terminal locations. After the wire-pattern excursion is completed between the two terminals, the machine automatically strips and cuts the wire to length. The tool bits then lower the wire ends over the two terminals to be connected and

## TABLE 26.7

Recommended Wiring Specifications for 0.025- and 0.045-Inch-Square Wire-Wrap Terminals

| AWG Number | Number of Turns | |
| --- | --- | --- |
| | Type 1 Connection | Type 2* Connection |
| 18 | 4 | 5 |
| 20 | | |
| 22 | 5 | 6 |
| 24 | | |
| 26 | 6 | 7 |
| 28 | | |
| 30 | 7 | 8 |

*Type 2 indicates one turn of insulated conductor contacting the terminal.

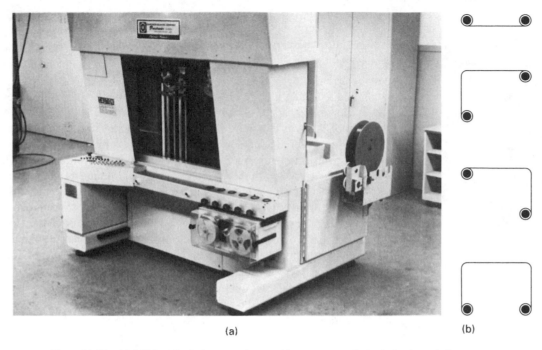

(a)　　　　　　　　　　　　　　　　　　　　　　　(b)

**Figure 26.20**　(a) 14FV vertical wire-wrapping machine, *courtesy of EPE Technology;* (b) basic wire patterns.

simultaneously form the wire wraps. One wrapping bit rotates clockwise while the other end of the wire is wrapped in a counterclockwise direction. The basic wire patterns formed are shown in Fig. 26.20b. The symbols at the bend locations represent the use of a dressing finger to form the pattern. The data also include the $z$ levels to obtain the specified height of the wire wrap on the terminal. This ensures proper spacing between multiple wraps on a single terminal. Because wire wrapped in opposite directions may result on the same terminal, the spacing between these wraps is critical to prevent a wire being installed from disturbing one previously formed.

The automatic wire-wrapping machine shown in Fig. 26.20a will accommodate a panel size of up to 26 by 26 inches with a wrap area of 22 by 22 inches. It is made to wrap AWG Nos. 26 and 30 wire at a rate as high as 1100 to 1200 wraps per hour.

Several terminals properly wrapped with solid wire are shown in Fig. 26.21. The type of wire normally used is copper-alloy solid wire with a tin, silver, or gold plating. Sizes of wire for this application range from AWG No. 18 to No. 26. The size of the wire will determine the exact number of turns to make a properly wrapped connection. This information for two types of interconnections is given in Table 26.7. The type 1 interconnection requires that the wire be wrapped around the terminal with its insulation approximately 1/16 inch from the first turn; type 2 interconnections require that at least one turn of insulation also be formed around the terminal. This second classification utilizes the flexibility of the wire insulation to reduce stress encountered during vibrations of the wire at the point where it enters into the first wrap on the terminal. In addition, quality workmanship dictates that no more than three independent lead wraps be made on a single terminal pin and that (1) turns from one independent wrap do not overlap any portion of another wrap, (2) adjacent turns on the same wrap touch but do not overlap, (3) the sum of all gaps should not exceed one wire diameter, excluding the first and last turns (spiral and open wrap), (4) there are not insufficient turns, and (5) the wire end does not protrude from the terminal (end tail). The appearance of unacceptable wire wraps is shown in Fig. 26.22a.

It is often the case that several terminals must be connected together or *bussed.* To avoid later servicing problems, bussed pairs of terminals to be connected should be pro-

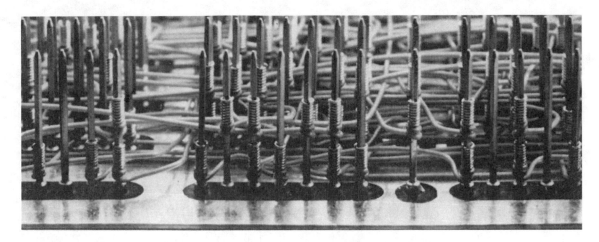

**Figure 26.21** Wire-wrap assembly. *Courtesy of Augat, Inc.*

vided with level 1 wraps first, followed by level 2 wraps to complete the string. Alternately wrapping terminals from level 2 to level 1 is unacceptable. If a wire that has its end wrapped at different levels must be relocated, the entire string may have to be removed to replace only one wire (see Fig. 26.22b).

Proper locations of both wire wraps and routing is essential to optimum circuit performance, especially in high-speed circuitry. If a signal wire in this type of design must be relocated because of stray inductance or capacitance problems, it is important that proper wire-wrap levels are adhered to throughout the design. Also, to reduce the number of parallel wires, advantage of the machine's data-controlled pallet-rotation ability should be taken. The software can be used to spread out the wires and result in a marked reduction of density buildup, in addition to a reduction in the number of parallel wires.

## 26.10 CABLE TIES

*Cable ties* are used when two or more wires are bundled together to form a harness or cable. They consist of a nylon strap with a locking eyelet. Small, closely spaced grooves are on the inside surface of the strap perpendicular to the length of the strap. These grooves will be in contact with the wires when assembled. The cable tie is first looped around the section to be bundled. The tapered end of the strap is then fed through the eyelet and pulled tightly. The strap is held securely by the locking eyelet, which engages the grooves and prevents the strap from slipping. Excess strap length is cut off with diagonal cutters close to the eyelet. The proper procedure for bundling leads with cable ties is shown in Fig. 26.23.

Special hand-operated cable tie assembly tools are available to facilitate the use of these bundling aids. One such tool is shown in Fig. 26.24. It consists of a gripping and cutting *head,* a *tension control,* and an *actuating trigger.* The tension control, roughly gauged from *loose* to *tight,* is set for the desired tension. The specific setting is determined by "feel," using the trial-and-error method. With the cable tie looped around the wires, the tapered end of the strap is fed through the eyelet and into the gripping head of the assembly tool. Each time the trigger is actuated, the head grips the strap and pulls it for a short distance through the eyelet. It may be necessary to actuate the trigger several times to reach the amount of tension that will firmly grip the leads. When the preset tension setting is reached, the cutting head shears the strap close to the eyelet. The use of this tool allows for rapid and uniformly controlled cable tie assembly.

Cable ties are positioned at regularly spaced intervals along the harness and at each breakout point. At breakout points, a single cable tie is looped in a *crossed* fashion around the breakout (see Fig. 26.25).

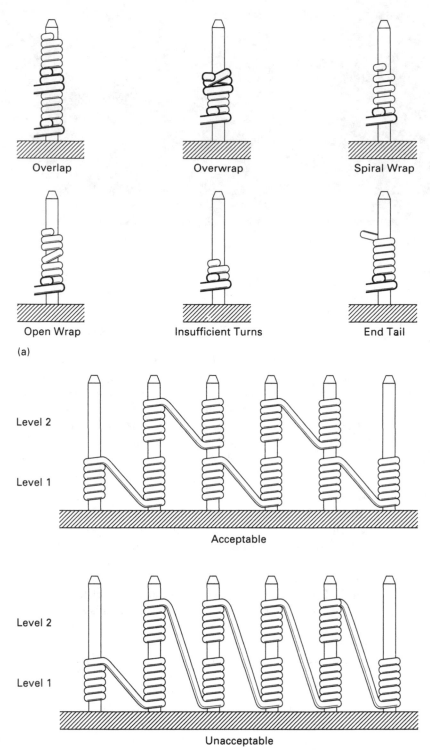

**Figure 26.22** Wire-wraps and bussing technique: (a) unacceptable wire wraps; (b) bussing.

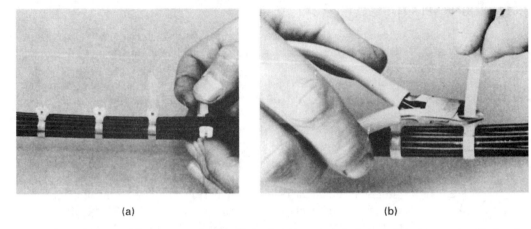

(a)                                    (b)

**Figure 26.23**  Cable ties are used to bundle leads: (a) insertion of tab through locking eyelet; (b) trimming excess tab length.

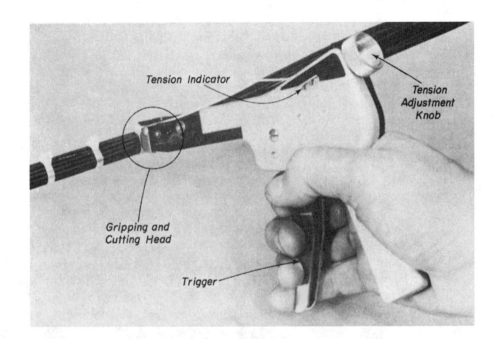

Tension Indicator

Tension
Adjustment
Knob

Gripping and
Cutting Head

Trigger

**Figure 26.24**  Hand-operated cable tie assembly tool.

**Figure 26.25**
Technique for securing breakout points.

Cable ties for harness construction is the most popular method used. Another technique of harnessing employs plastic and elastomer materials classified as *heat-shrinkable tubing.* This tubing is available in at least 12 different materials, each having its unique properties.

Manufacturers of shrinkable tubing employ special mechanical and thermal means to expand the molecules of the tubing and "fix" them in a strained state. When heated, the strains established during the manufacturing process are relieved, thus returning the tubing diameter to its original size. In addition, the wall thickness, which is reduced during manufacture, returns to its original thickness when heated. *Shrink temperatures* for some of the common tubing materials range from 175 to over 600°F (79 to 316°C).

In its expanded state, shrinkable tubing is easily slipped over the leads to be bundled after being cut to the required length (Fig. 26.26a). Shrink temperature is achieved with the use of a portable electric *heat gun,* shown in Fig. 26.26b, which concentrates heated air completely around the tubing, thus shrinking it and forming a tight fit. This process requires only one or two practice trials to determine the optimum length of time for the application of heat to shrink the tubing to the desired size.

Shrinkable tubing exhibits several desirable characteristics that are not attainable with cable tie methods. Shrinkable tubing provides a dust-free, waterproof, and abrasive- and oil-resistant covering over the harness as well as the ends of the connectors. Short

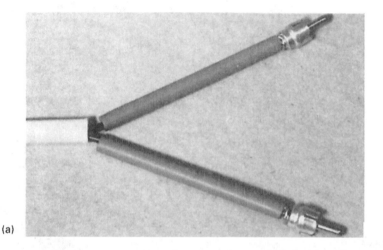

(a)

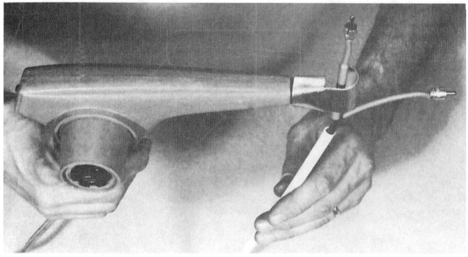

(b)

**Figure 26.26**
Installation of shrinkable tubing: (a) initial placement of tubing sections; (b) air deflector allows for uniform heating.

lengths of shrinkable tubing may also be used on a cable-tied harness where abrasion at a particular point could be a problem. The one disadvantage of this tubing is that it can be used only where accessibility will permit.

## 26.12 CABLE CLAMPS

When the harness has been completed, it is initially installed into the system. If connectors have been preassembled onto the harness, they are first mounted. All leads requiring soldered connections are then assembled onto the subassemblies. The orientation of the installed harness must be such as to protect the wire insulation from abrasion. In addition, the harness must not contact high-temperature components.

The harness must finally be fastened rigidly to the chassis to prevent movement or undue strain on terminal connections. Commercially available *cable clamps,* such as those shown in Fig. 26.27, are used for this purpose. These clamps are constructed of nylon or Teflon and are provided with either a clearance hole for typical machine screw sizes or with an adhesive backing for securing the clamp to the chassis. They are available in sizes to accept harness diameters from 1/16 inch to several inches and provide an excellent means of rigidly securing a harness to chassis elements.

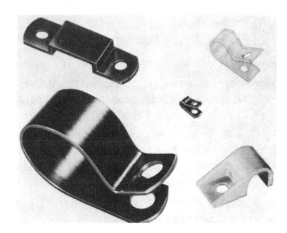

**Figure 26.27** Typical cable clamps. *Courtesy of Weckesser Company, Inc.*

## EXERCISES

### A. Questions

26.1    What is the reason that copper wire is tin-coated?

26.2    What is the advantage of stranded wire over solid wire?

26.3    Define the term *bus wire*.

26.4    Why is concentric-type wire preferred over bunch-type wire?

26.5    What does the acronym *AWG* represent?

26.6    Explain the terms *mil-foot* and *circular mil*.

26.7    Determine the resistance of 3 feet of AWG No. 22 copper wire.

26.8    What is the relationship between cross-sectional area and the AWG number of copper wire?

26.9    What is the relationship between current-carrying capacity and AWG number?

26.10   What three criteria determine the most suitable wire insulation for a specific application?

26.11   Place the following insulation materials in order of poor, fair, good, and excellent for insulation resistance: Teflon (TFE), nylon, Mylar, and Kynar.

26.12 Describe the body and tracer colors of hookup wire in a harness assembly that is assigned the code number 528.

26.13 What is Government Standard MIL-STD-122?

26.14 What is the primary difference between coaxial cable and shielded wire?

26.15 How is standard wire prepared prior to making a connection to a terminal?

26.16 What are the number and barrel color of a crimp terminal that will accept AWG wire numbers 22 to 18?

26.17 Define *insulation gap.*

26.18 How are cup terminals prepared to receive wire connections?

26.19 What do the DOD-STD-2000-3 standards specify for turret terminal connections?

26.20 How are the soldering iron tip and solder positioned to obtain a properly soldered wrapped terminal connection?

## B. True or False

Circle *T* if the statement is true, or *F* if any part of the statement is false.

26.1 Unidirectional stranded wire is more flexible than the counterdirectional type, but its diameter is not as precisely uniform.  **T  F**

26.2 In the AWG system, as the numbers increase, the corresponding wire diameters decrease.  **T  F**

26.3 Wire resistance increases by a factor of 4 when the diameter is doubled.  **T  F**

26.4 The term *circular mil* refers to the diameter of wire.  **T  F**

26.5 When measuring the size of stranded wire with a wire gauge, the slot that it snugly fits into will be one AWG number more than the gauge reading due to the space between strands.  **T  F**

26.6 Shielded wire is made to be used on low-frequency applications where its impedance is not critical.  **T  F**

26.7 When tinning stranded wire with a soldering iron, the wire is positioned between the solder and the iron's tip.  **T  F**

26.8 When forming wraparound connections, the wire insulation should touch the terminal to prevent the possibility of shorts between adjacent terminals.  **T  F**

26.9 A crimp connector having a red barrel will accept a larger-diameter wire than will a connector with a yellow barrel.  **T  F**

26.10 Only one wire may be attached to a cup terminal.  **T  F**

## C. Multiple Choice

Circle the correct answer to each statement.

26.1 Wire having an AWG number of 22 is (*smaller, larger*) in diameter than one having an AWG number of 16.

26.2 AWG No. 22 wire has a (*smaller, larger*) current-carrying capacity than AWG No. 14 wire.

26.3 The most common material used for coating copper wire is (*tin, nickel*).

26.4 The AWG wire numbering system was developed primarily to designate the size of (*stranded, solid*) wire.

26.5 (*Solid, stranded*) wire is used to make electrical connections to wire-wrap terminals.

26.6 Solder cup fillets should be slightly (*concave, convex*).

26.7 To comply with DOD-STD-2000-3, connections made onto turret terminals should be wrapped (*180 to 270°, 360°*).

## D. Matching

Match each item in Column A to the most appropriate item in Column B.

| COLUMN A | COLUMN B |
|---|---|
| 1. Diagonal cutters | a. Insulation |
| 2. AWG | b. Cross-sectional area  *(circular mils = CM)* |
| 3. Wire-wrap | c. Full-flush |
| 4. Stranded wire | d. Rf circuitry  *(similar – shielded wire)* |
| 5. Kel-F | e. American Wire Gauge |
| 6. Coaxial cable | f. Solderless |
| 7. CM | g. Concentric |
| 8. Wetting | h. Shoulder |
| 9. Cup | i. Contour |
| 10. Turret terminal | j. Fill |

## E. Problems

26.1 Construct a pair of test leads such as those shown in Fig. 26.28. The test leads are to be made from 36-inch lengths of plastic insulated No. 18 AWG stranded wire. Strip 1/2 inch of insulation from all four lead ends, and twist and tin the stranded conductors. Slip the rubber insulated grips onto the leads before soldering the alligator clips to the lead ends. After soldering the clips, slide the rubber insulators into position.

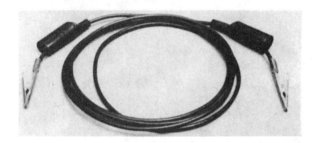

**Figure 26.28**

26.2 Construct a set of interconnecting leads for a stereo system by first centering a 2-foot length of shrinkable tubing about two 3-foot lengths of insulated shielded conductor cables. Slide 2-inch lengths of shrinkable tubing over all four ends of the cables. Remove sufficient outer insulation from all cable ends to prepare the copper braid and the inner insulated conductors for soldered connections to RCA phono plugs. Once the phono plugs have been soldered, slip the shrinkable tubing onto the plugs until the end of the tubing is even with the leading edges of the outer portion of the connector. Finally, shrink the tubing about the cables and plugs.

26.3 Construct four 6-foot speaker leads from plastic insulated No. 16 AWG stranded wire. Slide 1 1/2-inch lengths of shrinkable tubing over all lead ends. Strip and tin all lead ends to a length of 1/2 inch. Position the uninsulated barrels of solderless fork tongue connectors onto all prepared conductor ends and crimp. Slide the shrinkable tubing over the entire barrel length and shrink into place.

26.4 Using the technique discussed in this chapter for straightening bus wire, prepare a 32 1/2-inch length of tinned No. 14 AWG bus wire to construct a UHF antenna. Form the wire into a loop with a 7 3/4-inch diameter. Adjust the loop so that both loop ends overlap equally. Crimp solderless spade lugs onto both loop ends and form two 90-degree bends on each end for attachment to terminal lugs on the rear of a television set. The first 90-degree bends are to be made 4 inches from the end of the spade lugs and perpendicular to the circumference of the loop. The remaining 90-degree bend is to be located 2 inches from the ends of the lugs for the most convenient attachment.

26.5 Install interconnection wires to the tachometer designed, fabricated, and assembled in Problems 6.3, 10.3, 16.3, 17.3, and 25.3. All leads will be plastic-insulated No. 18 AWG stranded wire. The power lead insulation color is to be red, with lead length sufficient for connection to the battery side of the fuse block. The ground lead insulation color should be black. This lead is to be terminated with a solderless spade lug and attached to any convenient screw on the car chassis. Select green for the insulation color for the remaining lead to be connected to the distributor side of the coil. A solderless ring tongue connector of sufficient size for the coil connection should be crimped onto the end of the green lead.

26.6 Fabricate a lead cleaner such as that shown in Fig. 26.29. Use a 5/8- by 10-inch strip of 16-gauge aluminum, two 5-inch lengths of 5/8-inch-wide copper braid, and two 1-inch lengths of shrinkable tubing.

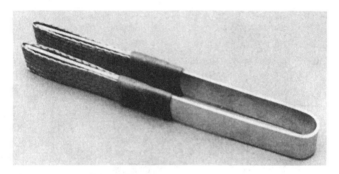

**Figure 26.29**

# SEVEN

## Student Projects

# 27

# Electronic Devices for Student Projects

## 27.0 INTRODUCTION

A review of the next two chapters of this unit will show that a somewhat large and diverse selection of electronic circuits and semiconductor devices and components have been assembled in the student projects, which are the main elements of this unit. Thus, this chapter has been added to provide both the theory of operation as well as the practical background knowledge necessary to more completely understand and troubleshoot the student projects that are to come. While broad in overall range and coverage, the material presented here is specific to the devices chosen for student projects that follow and is of sufficient detail as to be useful. Although not absolutely necessary, some previous background in electronic devices would be helpful. A working knowledge of both DC and AC circuit analyses is essential if one is to take full advantage of this material.

## 27.1 SEMICONDUCTOR DIODES

**The Circuit Symbol**    The typical semiconductor diode, shown symbolically in Fig. 27.1, is a two-terminal device that allows current to flow easily in one direction but prevents any substantial current from flowing in the opposite direction. From Fig. 27.1, note that an arrowhead symbol is used to identify the *anode* terminal, while a short perpendicular bar represents the *cathode* terminal of a diode. The arrowhead, in addition to locating the anode terminal, also identifies the direction of conventional current flow through the diode in the forward direction. Also relevant here, by way of introduction, is the fact that the anode of these types of devices is made from *p-type* semiconductor material and the cathode consists of *n-type* semiconductor material. *P*-type semiconductor material is composed predominately of positively charged locations (holes), while *n*-type semiconductor material is chiefly made from negatively charged particles (electrons). Finally, it should be noted that today's semiconductor diodes are primarily made from a base material of silicon. While germanium devices are also available, they seem to currently have very limited applications and since none are used in any of our student projects, they will not be considered in this chapter.

**The 1N914 Silicon Diode Characteristics**    Figure 27.2 shows the composite characteristics of a 1N914 silicon diode including both its forward- and reverse-biased characteristics. This graph is a Cartesian system plot of the current through the device (plotted along

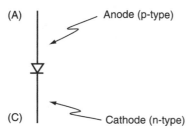

**Figure 27.1** The symbol for a semiconductor diode.

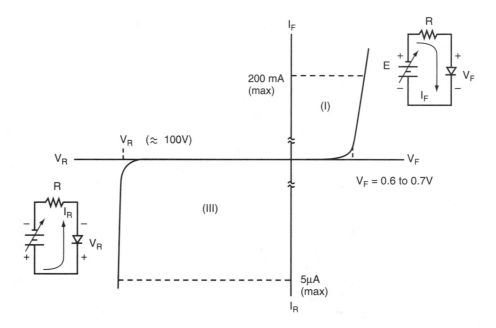

**Figure 27.2** Both the forward (I) and reverse (III) characteristics of a small-signal 1N914 semiconductor diode.

the *y-axis*) vs. the voltage across the device (plotted along the *x-axis*). The graphical representation of a device is a convenient way to display its characteristics and is not only used in this section to introduce some typical semiconductor diodes but will be used again when we discuss semiconductor transistors.

The first quadrant (I) of Fig. 27.2 shows the forward-biased or forward *I–V characteristics* of a small-signal 1N914 silicon diode. Note that the series circuit used to forward-bias a diode applies a positive potential or voltage to the anode terminal and a negative potential is applied to the cathode terminal. Resistor R is added to limit circuit current. When the applied voltage E is below the forward *turn-on* voltage—0.6 to 0.7 V for a silicon diode like the 1N914—the current flow into the anode and out the cathode of the device is very small and for all practical purposes we can think of it as being zero amperes in magnitude. However, when the applied voltage is above the forward *turn-on* voltage $V_F$, the forward current flow $I_F$ through the diode increases rapidly, limited only by the circuit resistance. For the 1N914 diode, forward current is limited by the power dissipation capability of the package to a maximum of approximately 200 mA.

The third quadrant (III) of Fig. 27.2 shows the reverse-biased or reversed *I–V characteristics* of the same 1N914 silicon diode. To reverse-bias a semiconductor diode one applies a positive potential to the cathode and a negative potential to the anode terminal. The resistor R is again added to limit circuit current. When the applied voltage (E) is increased from 0 V, the current flow $I_R$ in the reverse direction—from cathode through the device to the anode terminal—is extremely small. Consisting of a temperature-dependent leakage current component that is generally taken to be nanoamperes ($10^{-9}$ A) in value, this reverse current for all practical purposes can be considered to be zero amperes.

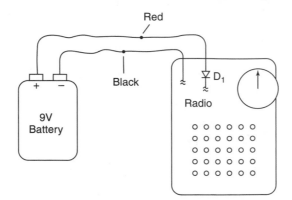

**Figure 27.3** Diode $D_1$ protects sensitive circuitry from possible reverse potentials.

However, when the applied voltage (E) is forced above the *minimum reverse breakdown voltage* $V_R$—typically taken to be 100 V for the 1N914 diode, the current flow through the diode increases rapidly, limited only by the circuit resistance R. For the 1N914 diode, reverse current is again limited by the power dissipation capability of the package to a maximum of approximately 5 μA. A word of caution here is warranted: This type of diode should not be operated above the minimum reverse breakdown voltage $V_R$.

Applications for silicon diodes are numerous. They include digital logic circuitry, protection of sensitive components and devices from reverse potential, as well as measuring temperatures that range from $-25°C$ to $150°C$, just to name a few. A simple but typical protection application is shown in Fig. 27.3. Note that diode $D_1$ is located in series with the radio circuitry, allowing power to be applied to the radio and current to flow—if the 9-V battery terminals are connected correctly—but preventing a current to flow if the battery is connected incorrectly to the radio.

**The 1N4001 Rectifying Diode**   The 1N4001 rectifying diode, similar to the 1N914 silicon diode, is one of a larger family of devices whose reverse breakdown voltage $V_R$ is specified by the last numeral of the component designating number. For example, the 1N4001 has a $V_R = 50$ V, the 1N4002 has a $V_R = 100$ V, the 1N4003 has a $V_R = 200$ V, and so on. Commonly housed in a DO-41 package—where DO stands for d̲iode o̲utline— this family of diodes has the following forward and reverse current specifications:

$$I_F = 1 \text{ A @ } 75°C$$

$$I_R = 10 \text{ μA @ } 25°C$$

With an I–V characteristic plot similar to that for the 1N914 diode shown in Fig. 27.2, this diode is used primarily when the application requires rectification. A typical full-wave application—the building block for most power supply designs—is shown in Fig. 27.4. Circuit operation is as follows: During the positive half-cycle of the AC input wave, diodes $D_1$ and $D_2$ are forward biased and diodes $D_3$ and $D_4$ are reversed biased. Starting at the positive input terminal, point *x*, conventional current flows through $D_1$, the load resistor $R_L$, and then through $D_2$ before returning to the negative input terminal, point *y*. The voltage across the load is shown with a solid line as a positive pulsating wave. During the negative half-cycle of the AC input wave, diodes $D_3$ and $D_4$ are forward biased and diodes $D_1$ and $D_2$ are reversed biased. Starting this time at the new positive input terminal, point *y*, conventional current flows through $D_3$, the load resistor $R_L$ (in the same direction as previously shown), and then through $D_4$ before returning to the nega-

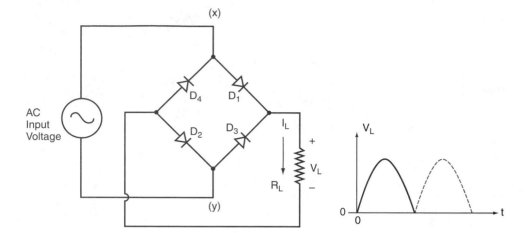

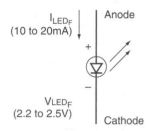

**Figure 27.4** A full-wave bridge rectifying circuit converts an AC input voltage to a pulsating DC voltage.

**Figure 27.5** A typical light-emitting diode requires 10 to 20 mA of forward current.

tive input terminal, point *x*. For this half cycle of the input the voltage across the load is shown with a dashed line. Therefore, although the input wave is an AC signal, the output voltage across the load resistor is a positive pulsating DC wave.

**Light Emitting Diodes** Light emitting diodes, or LEDs, are special types of semiconductor diodes and are shown symbolically in Fig. 27.5. A two-terminal device consisting of both an anode and a cathode, this diode is forward biased when the current through it is from anode to cathode. Made chiefly from gallium arsenide phosphide (GaAsP) or gallium phosphite (GaP), these diodes emit visible light (electroluminescence) when their *p-n* junction is forward biased. Constructed to emit only one particular color, LEDs are commonly available as red, yellow, or green emitters—although other colors are now being made.

The typical standard-sized LED requires an average forward current $I_{LEDF}$ of from 10 to 20 mA and at that current will have a forward terminal voltage $V_{LEDF}$ of approximately 2.2 to 2.5 V. It should be noted here that the specifications above are for standard-sized LEDs housed separately as a single-emitter. Physically larger devices (jumbos) will often require more current. LED forward terminal voltage $V_{LEDF}$ varies somewhat from the nominal values stated here depending on the color of the emitted light.

Figure 27.6 illustrates a common LED application. When the output of a microprocessor output control line PB0 is at a logic HIGH state (5 V) the LED is on and when the output control line is at a logic LOW state (0 V) the LED is off. The transistor $Q_1$ is needed to interface the low-power level of output line PB0 and will be considered in Section 27.3. Light emitting diodes are also available in several monolithic diode array configurations such as the popular seven-segment display discussed in the following section.

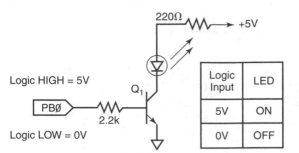

**Figure 27.6** A typical LED application is to visually identify the state of a digital logic line.

Logic HIGH = 5V

Logic LOW = 0V

PBØ

2.2k

$Q_1$

| Logic Input | LED |
|---|---|
| 5V | ON |
| 0V | OFF |

## 27.2 DIGITAL DISPLAYS

**A Seven-Segment Display**  The seven-segment LED array is a very popular display configuration. Shown in Fig. 27.7a, the MAN6760E configuration consists of eight LEDs, the anodes of each being connected together—or common to one another—and accessible externally at either pin 3 or pin 8 of the package. The cathode of the decimal point LED is attached to external pin 5 and the remaining segment LEDs, identified with the letters *a-b-c-d-e-f-g,* have their individual cathodes connected to separate external pins, with pin 7 for the cathode of segment *a,* pin 6 for the cathode of segment *b,* and so on.

Testing this display can be done statically as shown in Fig. 27.7b for segment *e.* The positive terminal of a 5-VDC power supply is applied to one of the common anode (CA) pins (3 or 8) and pin 1 is attached to the negative terminal of the supply through a 220 Ω resistor, limiting the forward current through segment *e* to approximately 15 mA. If working properly, segment *e* should light. By separately probing each segment pin in this way, we can check to see the functionality of each segment.

Normally the individual cathode pins of this type of seven-segment display are connected to separate driver IC pins, which are logically selected depending on which segments are to be lit to display a particular numerical value. Table 27.1 shows which segments must be lit—that is, which cathodes must be grounded through a current limiting resistor—to display each possible numeral from 0 to 9.

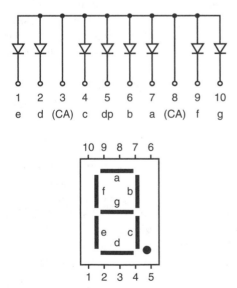

| 1 | 2 | 3 | 4 | 5 | 6 | 7 | 8 | 9 | 10 |
|---|---|---|---|---|---|---|---|---|---|
| e | d | (CA) | c | dp | b | a | (CA) | f | g |

10 9 8 7 6

1 2 3 4 5

(b) Static testing an LED array

**Figure 27.7** Seven-segment LED displays are popular digital indicators.

(a) MAN6760 E common anode (CA) LED configuration

**TABLE 27.1**

Numerical display and segments lit

| Number | Segments |
| --- | --- |
| 0 | a, b, c, d, e, f |
| 1 | b, c |
| 2 | a, b, g, e, d |
| 3 | a, b, c, d, g |
| 4 | b, c, d, f, g |
| 5 | a, c, d, f, g |
| 6 | a, c, d, e, f, g |
| 7 | a, b, c |
| 8 | a, b, c, d, e, f, g |
| 9 | a, b, c, d, f, g |

The seven-segment display illustrated here is referred to as a common anode (CA) configuration, but common cathodes (CC) are also available, as well as multi-segment displays. While red LEDs are common for these types of readouts, it is also possible to find other colors such as green and yellow.

One of the main advantages of using an LED-based readout is that they are highly visible in most ambient light conditions. However, their main disadvantage is the amount of power that they require. Take, for example, the display of the numeral 8 including the decimal point. Table 27.1 shows that a total of seven individual LED segments must be lit to display the numeral 8 plus one more for the dp. If each LED segment requires an average forward current of say 15 mA, then the total display current $I_T$ is

$$I_T = 15 \text{ mA/segment (8 segments)} = 120 \text{ mA} = 0.12 \text{ A}.$$

While acceptable for equipment powered from an Edison supply, this type of display is certainly not a reasonable choice for battery-powered or portable applications.

**A Liquid Crystal Display**   When power consumption is of concern, as with most battery applications, the liquid crystal display (LCD) is often the most appropriate choice. Requiring only microwatts of power—or at worst low milliwatts depending on the configuration—and having a life expectancy (useful life) in excess of 10,000 hours, field-effect LCDs are categorized as being either transmissive or reflective in function. Shown diagrammatically in Fig. 27.8 is a reflective-type LCD. While field-effect LCDs are available in a fairly large choice of colors, they are considerably less rugged than the seven-segment LED display discussed previously and are limited to a display height of approximately 2 inches.

The FE-LCD shown in Fig. 27.8 is constructed with a front glass plate and a rear glass plate that is coated with conductive/reflective material—the segments pattern. Liquid-crystal material is encapsulated between these glass plates and their edges sealed. Finally, light-polarizing films are applied to the outer layers of the glass and the polarizing film on the rear of the display is outfitted with a reflective material, usually silver or gold foil.

The reflective-type FE-LCD operates as follows: Light entering horizontally from the rear of the display passes through the horizontally polarized film to the reflector where it reflects off this rear reflector and is thrown back into the liquid crystal material. The light is then forced to pass through the vertical polarizing film at the viewing side. If there is no voltage applied between the imbedded segments and the common electrode, the display is uniformly lit. When a voltage is applied between a segment and the common electrode, the vertically incident light will encounter the rear horizontal polarizing filter and be reflected back. Thus the segment will appear dark from the viewer's side.

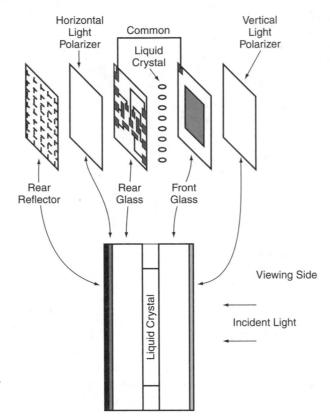

**Figure 27.8** Reflective-type field-effect liquid crystal display.

Labels in figure:
Horizontal Light Polarizer
Common
Liquid Crystal
Vertical Light Polarizer
Rear Reflector
Rear Glass
Front Glass
Viewing Side
Incident Light
Liquid Crystal

Unlike the LED arrays discussed in the previous section, LCDs require an AC segment drive voltage, typically 5 V rms, at a frequency that ranges anywhere from 40 to 100 Hz.

## 27.3 SEMICONDUCTOR TRANSISTORS

**General Theory of Operation**   In general, today's semiconductor transistors are manufactured from silicon and are classified as being either bipolar junction (BJT) or field-effect (FET) transistors. Since the latter type devices are not used in any of the student projects to follow this chapter, we will restrict our coverage here on the BJT. Furthermore, the BJT is not encountered in any student project requiring linear amplification of a signal and therefore we will not consider the BJT as an amplifier—in any substantial way—but further limit our discussion chiefly to transistors used only as electronic switches. These restrictions are warranted since the area of semiconductor transistors is very broad and it is not uncommon for entire textbooks to be devoted to this single topic.

The BJT, a three-terminal electronic device, is fabricated with alternating layers of *n*-type and *p*-type semiconductor material. Two types are possible: the NPN and the PNP bipolar junction transistors (see Fig. 27.9a). It should be noted here that no attempt has been made to draw the three transistor regions to scale. The emitter (E)—a region with multiple edges like the fingers of one's hand—provides a ready supply of charged carriers that are uniformly injected into the very thin—of the order of $10^{-6}$ inches in width—base (B) region and retrieved at the physically large surface area of the collector (C). The base region controls the amount of emitted carriers retrieved by the collector. Figure 27.9b shows the symbolic representation of both device types. Note that the arrowhead always identifies the

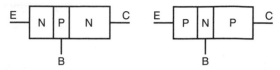

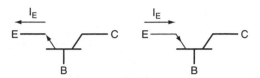

(a) The layer structure of a BJT transistor

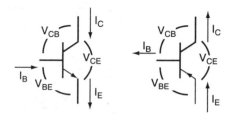

(c) Terminal currents follow the relationship:
$I_E = I_C + I_B$
Terminal voltages follow the relationship:
$V_{CE} = V_{BE} + V_{CB}$

(b) Symbolic representation of both an NPN and a PNP transistor

**Figure 27.9**   The structure, symbol, and current flows of BJTs.

emitter terminal of a BJT, and points in the direction of conventional current flow for the emitter current $I_E$.

The relationship between the three external DC terminal currents can be studied by referring to Fig. 27.9c. In both types of transistor, the emitter current $I_E$ is always the largest in magnitude of the three terminal currents. Furthermore $I_E$ is made up of the collector current $I_C$ and the base current $I_B$. This relationship can be described algebraically as

$$I_E = I_C + I_B \qquad (27.1)$$

Note that for an NPN transistor, both $I_C$ and $I_B$ enter their respective terminals and $I_E$ leaves the emitter terminal. Conversely, for the PNP transistor, $I_E$ enters the emitter terminal and both $I_C$ and $I_B$ leave their respective terminals. Finally, when the transistor is operated as a linear amplifier, the emitter current is approximately equal to the collector current, shown as

$$I_E \approx I_C \qquad (27.2)$$

and the base current is very small. The relationship between the collector current and base current is extremely important and is termed beta ($\beta$)—the current amplification factor—and is given by the relationship

$$\beta = I_C/I_B \qquad (27.3)$$

The value of beta can range from a low of approximately 30 to upwards of 500 for the typical small signal-transistor discussed in the next section. Rewriting Equation 27.3, one can think of the collector current $I_C$ as being equal to beta times the base current $I_B$:

$$I_C = \beta\, I_B \qquad (27.4)$$

Figure 27.9c also illustrates the three terminal voltages of interest for a semiconductor transistor. These are the voltages measured between the collector-to-emitter $V_{CE}$, the voltage between the base-to-emitter $V_{BE}$, and the voltage measured between the collector-to-base $V_{CB}$ terminals. When operated as a linear amplifier, the following relationship between these three voltages applies:

$$V_{CE} = V_{BE} + V_{CB} \qquad (27.5)$$

Note that the largest terminal voltage, $V_{CE}$, is equal to the sum of both $V_{BE}$ and $V_{CB}$.

Basic transistor action for both NPN and PNP transistors is shown in Figs. 27.10a and 27.10b, respectively. Battery voltage $V_{BB}$ forward biases the base-to-emitter junction ($V_{BE}$) of the transistor, causing an emitter current $I_E$ to flow. Since the base region is very thin,

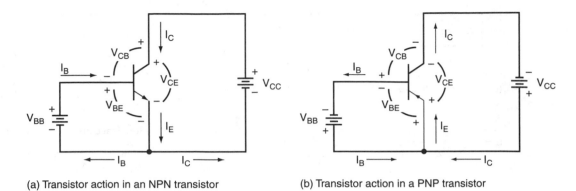

(a) Transistor action in an NPN transistor          (b) Transistor action in a PNP transistor

**Figure 27.10**   An illustration of basic current flow through a BJT transistor.

most of the current that defuses across the forward-biased base junction is accelerated by the $V_{CC}$ supply that reverse-biases the collector junction ($V_{CE}$) and becomes the collector current $I_C$. $I_B$ is often quite small and amounts to that portion of the emitter current that does not reach the collector.

**The 2N3904 and 2N3906 Transistors**   Two very popular low-power small-signal devices often chosen for applications requiring the switched control lamps, relays, and other typical loads are the 2N3904 and its complement, the 2N3906 transistor. Typically housed in a TO-92 plastic case, these devices are characterized in Table 27.2 as to type, maximum collector-to-emitter voltage, maximum collector current, and typical values of beta or current amplification factor.

**Using a BJT as a Switch**   When used as a switch the base of the transistor can be thought of as controlling the amount of current flowing through the collector-to-emitter, that is, the collector current $I_C$. As can be seen from Equation 27.4, if the base current is very small or approximately zero, the collector current is also small or zero. We refer to this switch condition as *cutoff* or simply that the transistor is OFF—conducting no collector current. However, when the base is supplied with a current, the collector current increases at a proportionally constant rate—assuming beta does not change—eventually reaching a point where any further increase in $I_B$ does not increase the collector current at a proportionally constant rate. Under these conditions Equation 27.4 no longer applies and we refer to this switch condition as *saturation* or identify the transistor as ON—conducting a collector current.

Figure 27.11a illustrates a typical switching application using the 2N3904 NPN transistor. When the transistor switch is in saturation (ON) the LED will be lit and when the transistor is at cutoff (OFF) the lamp will not be lit. Switch control is accomplished by either providing a forward-biased potential to the base-to-emitter ($V_{BE}$) of the transistor to turn it ON, or provide a reverse-biased potential to the base-to-emitter to turn it OFF. Resistor $R_B$ is

**TABLE 27.2**

Two common transistors used as electronic switches

| Number | Type | $V_{CE(MAX)}$ | $I_{C(MAX)}$ | β |
|--------|------|---------------|--------------|---|
| 2N3904 | NPN | 40 V | 100 mA | 30 |
| 2N3906 | PNP | 40 V | 100 mA | 30 |

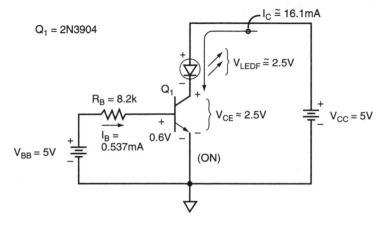

(a) Transistor in saturation (ON) and LED lit

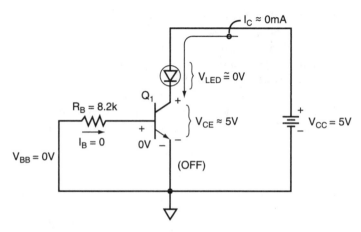

**Figure 27.11** Using a 2N3904 NPN transistor as a switch.

(b) Transistor $Q_1$ in cutoff (OFF) and LED is not lit

chosen to limit the base current and subsequently fix the collector current. The polarity of the DC supply, $V_{BB}$, applied in Fig. 27.11a forward-biases the base-to-emitter junction of the transistor forcing $V_{BE}$ to equal approximately 0.6 V—the turn-on voltage of a silicon *p-n* junction. The current $I_B$ is limited to

$$I_B = \frac{5\,V - 0.6\,V}{8.2\,k} = 0.537\,mA$$

Given that $\beta \approx 30$ and the LED current is equal to the collector current, $I_C$ can be approximated from Equation 27.4 to be equal to

$$I_C = \beta\,I_B = (30)\,0.537\,mA = 16.1\,mA$$

If the voltage across the terminals of the LED is taken to be 2.5 V, then $V_{CE}$ is 2.5 V. A similar test circuit for the 2N3904 NPN configuration in cutoff is shown in Fig. 27.11b. Note here that since the base current $I_B = 0$ mA, the collector current $I_C = 0$ mA. The LED is not lit and all the $V_{CC}$ supply voltage is forced across the collector-to-emitter $V_{CE}$ terminals of the transistor. Note that this voltage is well below the 40 V maximum value from Table 27.2. Similar circuits, using PNP transistors, can be configured as electronic switches when positively grounded supplies are encountered.

**General Theory** Another device, designed primarily for use as an ON–OFF switch, is the silicon-controlled rectifier, or simply SCR. A multilayer device fabricated with four alternating layers of *p*-type and *n*-type silicon material, the SCR is intended to conduct current in only one direction. In effect, it is a switch designed to control the flow of DC control currents. Figure 27.12a shows the simplified structural model of an SCR, while Fig. 27.12b is its accepted symbolic representation. Three external terminals are associated with the symbol: the anode (A), the cathode (C), and the gate (G). The arrowhead originating on the anode indicates the direction of conventional current flow—internal to the SCR—from anode-to-cathode $I_{AC}$. An SCR can be operated in either its ON or OFF states—only these two states are possible. An SCR that is ON allows a current to flow from the anode-to-cathode, thus acting like a mechanical switch whose terminals are closed. When an SCR is OFF, the anode-to-cathode acts like a mechanical switch whose terminals are open, and no current $I_{AC}$ is allowed to flow.

One can force an SCR into its ON state by applying a small voltage between the gate-to-cathode terminals ($V_{GC}$) of the SCR, causing a small gate-trigger current ($I_{GT}$) to flow into the gate. Unlike the transistor switches discussed in the previous sections, once this gate-trigger voltage ($V_{GC} = V_{GT}$) fires the SCR into conduction, it has no further control over the operating condition of the SCR. In fact, the $V_{GC}$ voltage can be removed from the SCR and the device will remain in its ON state. Once on, an SCR can be turned off only by reducing the anode-to-cathode $I_{AC}$ current below its *minimum holding current* $I_H$. Figure 27.13 illustrates the forward and reverse characteristics of a typical silicon-controlled rectifier, along with conceptual test circuitry to generate both the forward (ON) and reverse (OFF) characteristics.

**The C106B1 SCR as a Switch** The C106B1 SCR, housed in a TO-220 plastic package, is characterized with the following specification: $V_{GT} = 0.5$ V typical, $I_{GT} = 500$ μA maximum, $I_H = 1$ mA, $V_{AC(MAX)} = 2.2$ V typical, and $I_{AC(MAX)} = 4$ A. Figure 27.14 shows a conventional method of statically testing this SCR. Note that the load, $L_1$, selected is a 24 V DC, 1 A incandescent lamp. To test an SCR, build Fig. 27.14—and with switch SW-1 closed and a voltmeter across the gate-to-cathode terminals—slowly in-

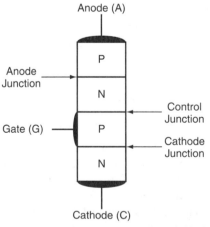

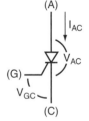

(b) Symbol for an SCR showing the direction of current flow from anode to cathode $I_{AC}$

**Figure 27.12** A silicon-controlled rectifier (SCR).

(a) A 4 layer, 3 junction semiconductor device used as a switch to control DC current

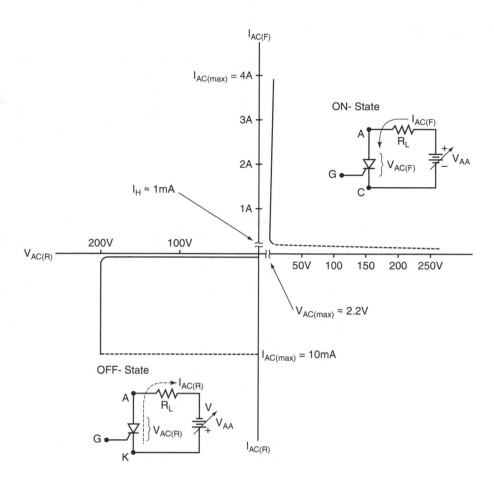

**Figure 27.13** The forward and reverse characteristics of a silicon-controlled rectifier.

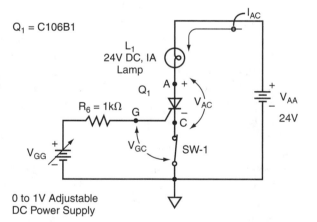

**Figure 27.14** A conventional test circuit to statically test the characteristics of an SCR.

crease the $V_{GG}$ supply until the lamp lights. $V_{GT}$, the voltage necessary to just fire the SCR ON (and into conduction), for the C106B1 should be between 0.3 to 0.8 V DC. Once the SCR is ON remove the voltmeter and measure the voltage between anode-to-cathode. For moderate-to-low load currents (i.e., less than 1 A DC), $V_{AC}$ should be approximately 1 V DC but always less than the $V_{AC(MAX)} = 2.2$ V typical specification. Finally shut off the SCR by opening SW-1 and reducing the anode-to-cathode current $I_{AC}$ below the holding current $I_H$ of 1 mA.

Operational amplifiers, or simply *op amps,* are ordinarily fabricated as integrated circuits and conveniently housed for use in several popular-styled packages. Their internal circuitry, extremely complex in nature, is designed with either BJT or FET devices, while some hybrid units are currently available that employ both technologies. We begin our study of this important device by looking first at a general-purpose device followed by some practical limitations and include information on both dual- and single-supply units. A survey of the student projects to follow shows that op amp use is limited to either comparator applications or as linear amplifiers and adder circuits. We therefore restrict our treatment of this important analog device to these two broad areas.

**General Theory**   A general-purpose op amp is a solid-state semiconductor device that is designed to have a very large voltage gain capability (from 50,000 to 5,000,000) and an extremely large input impedance specification (1 MΩ to 200 GΩ) on both differential input terminals. The input voltage that is amplified by this very large gain is the differential (or difference) voltage ($E_d$) applied between both input terminals, not the voltage applied at either input with respect to circuit common. The typical op amp can accept either AC or DC input signals and produce an output voltage, related to the input signals, that is presented basically as an ideal voltage source; that is, the op amp's output resistance is very small—essentially zero. Available as either single- or dual-supply devices, the operational amplifier—as the name implies—can be configured to solve a variety of problems or *operations.* We first look at the op amp as a comparator followed by several amplifier configurations and conclude with a typical summing amplifier or adder circuit.

**The Op Amp as a Comparator**   One application of an op amp is to monitor and use the polarity of voltage at the output terminal to indicate the magnitude and polarity of voltages applied to the two input terminals. This application gets its name from the fact that we are *comparing* the voltage on the noninverting input terminal ($+$) with the voltage on the inverting input terminal ($-$). Figure 27.15 shows the use of one of the four op amps, an LM324, as a comparator. Note that a $\pm 15$ V dual-supply is applied to the power supply terminals (pins 4 and 11) and a load resistor $R_L$ is connected between the output terminal (pin 1) and power supply common. Also note that the voltage applied to the ($+$) input at pin 3 (with respect to common) is labeled $V_1$ and the voltage applied to the ($-$) input at pin 2 (with respect to common) is labeled $V_2$. We can determine the magnitude and polarity of the output voltage by first determining the magnitude and polarity of differential input voltage $E_d$ from Equation 27.6.

$$E_d = V_{(+)} - V_{(-)} \qquad (27.6)$$

If $E_d$ is determined to be positive from Equation 27.6, as is illustrated in Fig. 27.15a, the polarity of $V_O$ measured with respect to common will also be positive. If $E_d$ is negative, as is illustrated in Fig. 27.15b, then the polarity of $V_O$ will be negative with respect to common. Therefore, by monitoring the polarity of output voltage one can determine the voltage conditions at the input.

- If the polarity of $V_O$ is positive, then the voltage on the ($+$) input is more positive than the voltage on the ($-$) input.
- If the polarity of $V_O$ is negative, then the voltage on the ($-$) input is more positive than the voltage on the ($-$) input.

In the example shown in Fig. 27.15a, $V_1 = V_{(+)}$, $V_2 = V_{(-)}$ and $E_d = 3$ V $- 2$ V $= +1$ V. Thus, we know that the voltage on the ($+$) input is more positive than the voltage on the ($-$) input. In the example shown in Fig. 27.15b, $E_d = -3$ V $- 5$ V $= -8$ V. We know that the voltage on the ($-$) input is more positive than the voltage on the ($+$) input.

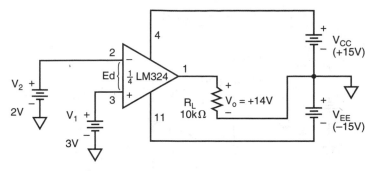

(a) The positive $V_O$ indicates that the voltage $V_1$ is more positive than the voltage $V_2$

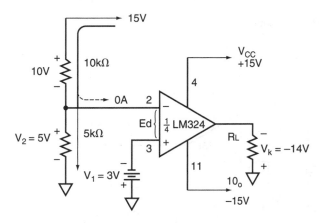

**Figure 27.15** A general-purpose dual-supply op amp used as a comparator.

(b) The negative $V_O$ indicates that the voltage $V_2$ is more positive than the voltage $V_1$

In practical terms, the magnitude of the output voltage for both applications shown will be approximately 1 V away from the power supply's "rails," or we say that the output is in *saturation.* Therefore, for the examples being shown in Fig. 27.15, the expected output voltage is $\pm V_O = \pm V_{SAT} = \pm 14$ V for the $\pm 15$ V dual-supply powering the circuitry.

Figure 27.16 illustrates the use of a single-supply op amp design using the LM358. The LM358, in which two general-purpose devices are housed, can also be used effectively in comparator applications. In this illustration the positive terminal of a 5 V DC supply is connected to pin 8 and the negative terminal of the power supply is connected to pin 4.

The output terminal, pin 1, is connected to a series limiting-resistor R and the anode of an LED. When the input voltage on the (+) input is more positive than the input voltage on the (−) input, as shown in Fig. 27.16a, the output $V_O$ is positive at approximately + 4 V and a forward current, limited by R, flows through the LED. The LED is lit, indicating this input condition ($V_{PROBE} > 2.0$ V). When the input voltage on the (−) input is more positive than the input voltage on the (+) input, as shown in Figure 27.16b, the output $V_O$ is positive but approximately 0 V. No current flows through the LED and the LED is not lit. Again the LED, this time being off, indicates the input condition that $V_{PROBE} < 2.0$ V.

**The Op Amp as an Amplifier** Many other op amp applications, other than comparators, are possible. One very popular and useful function is to use an op amp as the building block for a host of linear amplification applications. Three common configurations are shown in Fig. 27.17.

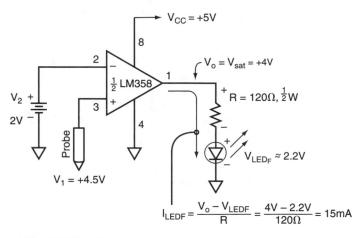

$$I_{LEDF} = \frac{V_o - V_{LEDF}}{R} = \frac{4V - 2.2V}{120\Omega} = 15mA$$

(a) The lit LED indicates that the probe voltage is above the 2.0V reference $V_2$

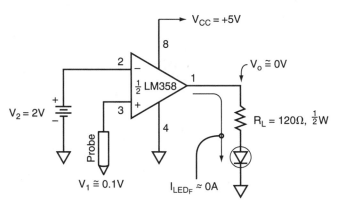

**Figure 27.16** Single-supply op amp comparator applications.

(b) The LED is off. No current flows through the LED when the (−) input voltage is more positive than the (+) input terminal voltage, $V_2$

Note that in each amplifier configuration shown, negative feedback is included around the op amp. Negative feedback is present when an electrical connection, which can cause a DC current to flow, exists between the output terminal and the inverting or (−) input terminal of the op amp. While the precision, dual-supply, OP-07A op amp is used to illustrate the amplifier applications in Fig. 27.17, it should be noted that virtually any general-purpose op amp could be substituted.

The amplifier shown in Fig. 27.17a is classified as an *inverting amplifier* and its circuit gain or amplification factor, $A_{CL}$, is given by Equation 27.7.

$$A_{CL} = -\frac{R_f}{R_i} \tag{27.7}$$

Gain is set to a first-order approximation, independent of op amp characteristics, by the ratio of the feedback resistor $R_f$ and the input resistor $R_i$. A gain of 20 is computed for the resistor values shown in Fig. 27.17a. As the name implies, this inverting amplifier configuration not only amplifies the input signal so that the output voltage is larger—in this case, larger by 20 times—it also has a polarity at its output terminal opposite that

$$V_O = A_{CL} V_i \tag{27.8}$$

of the input if the input signal is DC. If the input signal is AC, the output voltage will be 180° out of phase with the input.

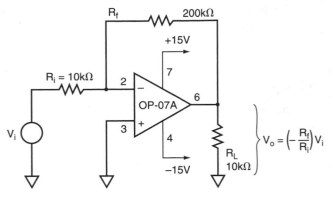

(a) An inverting amplifier with a voltage gain set by the values of $R_f$ and $R_i$ and an output polarity opposite the input

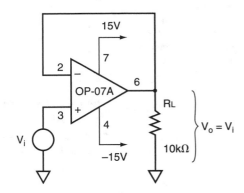

(c) The buffer, isolation, or voltage follower has no phase shift and a gain of 1

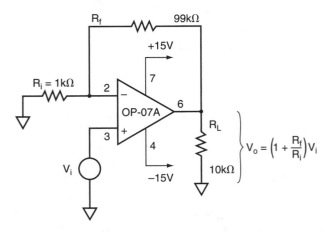

(b) A non-inverting amplifier will cause the output voltage, $V_0$, to be an amplified version of $V_i$ without polarity reversal or phase shift

**Figure 27.17**   Three common amplifier configurations using general-purpose op amps.

The amplifier shown in Fig. 27.17b is classified as a *noninverting amplifier* and its circuit gain or amplification factor, $A_{CL}$, is given by Equation 27.9.

$$A_{CL} = 1 + \frac{R_f}{R_i} \qquad (27.9)$$

Again the ratio of the feedback resistor $R_f$ and the input resistor $R_i$ set gain. A gain of 100 is computed for the resistor values shown in Fig. 27.17b. As the name implies, this non-inverting amplifier configuration does not invert the input polarity at its output terminal. The polarity of the input and output terminals are the same. One additional advantage of this configuration is the very high input impedance it presents to the input source $V_i$. This means that for all practical purposes the input source need supply virtually none, or an extremely small—microamperes or nanoamperes—current, at most.

Finally the *buffer, isolation,* or *voltage follower* amplifier shown in Fig. 27.17c is a special case of the noninverting amplifier shown previously. Characterized with a voltage gain $A_{CL} = 1$, it faithfully outputs a voltage that is an exact duplication of the input. A common application of the buffer amplifier is to transform the impedance of the input voltage source to one with an impedance of essentially zero, the output impedance of an op amp.

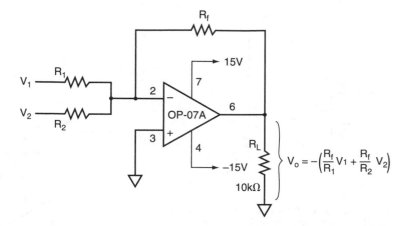

**Figure 27.18** This inverting summing amplifier can be used to signal-condition sensors. If $R_f = R_1 = R_2$, then the performance equation reduces to $V_O = -(V_1 + V_2)$.

$$V_o = -\left(\frac{R_f}{R_1}V_1 + \frac{R_f}{R_2}V_2\right)$$

**The OP-07A as an Adder** Our final op amp circuit is classified in the literature as an inverting adder or inverting summing amplifier. Used in the student project section of this book, the circuit shown in Fig. 27.18 has two input channels, that is, two inputs: $V_1$ and $V_2$—and its output voltage can be described algebraically with the following equation:

$$V_O = -\left(\frac{R_f}{R_1}V_1 + \frac{R_f}{R_2}V_2\right) \tag{27.10}$$

Circuit operation is as follows: Input voltage $V_1$ is amplified by channel gain $R_f/R_1$ and added to input voltage $V_2$, which is first amplified by channel gain $R_f/R_2$. Since this is an inverting configuration—all inputs applied indirectly to the $(-)$ op amp input—the final output voltage is inverted as predicted by the minus sign in Equation 27.10.

One useful application of this circuit configuration is to signal-condition the voltage output of a sensor to provide the appropriate gain and, if necessary, to add or eliminate an offset voltage. View the input voltage $V_1$ being applied to one input of the adder shown in Fig. 27.18 as a sensor's output voltage—changing in value as the physical parameter being measured changes—and $V_2$ as a fixed reference voltage. Then Equation 27.10 can be written as the very common expression of a straight line:

$$y = mX + b$$

where $y$ equals the output voltage $V_O$, $m$ is the channel gain $R_f/R_1$, and $b$ is the offset voltage set by $-(R_f/R_2)V_2$. In effect this circuit can be used to create the mathematical expression of a straight line.

If all three resistors of Fig. 27.18 are made equal, say all are set to 10 k$\Omega$, then Equation 27.10 reduces to

$$V_O = -(V_1 + V_2) \tag{27.11}$$

which is a simple summing amplifier that produces an output voltage $V_O$ that is the algebraic inverted sum of two inputs, $V_1$ and $V_2$.

## 27.6 DIGITAL LOGIC

**The 74H07 Hex Buffer with Open Collector** When a digital logic device or gate is termed a *buffer* or *driver*, it is designed to accept at its input the standard digital logic level and output a corresponding logic level at a higher than normal current and voltage. The algebraic expression used to describe this operation is given as

$$Y = A \tag{27.12}$$

where $Y$ represents the output voltage level and $A$ corresponds to the input voltage level. The 74H07 buffer—a version of the standard 7400 series logic that is also designed for high-speed

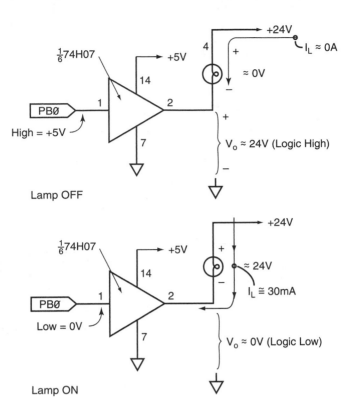

**Figure 27.19** A digital buffer like the 74H07 can easily make interfacing a microcontroller to the external world.

applications—is just such a device. Packaged typically as six separate buffers in one 14-pin DIP, each individual gate output is provided as an open collector. This type of output configuration increases the normal load current capacity $I_{OL}$ from approximately 1.6 mA to a maximum of 40 mA, and the normal load or output voltage capacity $V_{OH}$ from typically 5 V to a maximum of 30 V. Figure 27.19 illustrates a typical 74H07 interface between a single-output port line of a 68HC11 microcontroller ($\mu$C) and a test lamp that is rated at 24 V and 30 mA.

When the $\mu$C-output port line is at a logic HIGH—approximately 5 V for this illustration—the buffer output is pulled up to approximately 24 V and the test lamp is OFF, load current is zero. Note that the $V_{OH}$ rating (30 V) of the gate is not exceeded. However, when the $\mu$C-output port line is at a logic LOW—approximately 0 V in this application—the voltage across the lamp is 24 V, the full secondary supply voltage. Note that the lamp current of 30 mA supplied by the secondary supply does not exceed the buffer's $I_{OL}$ level of 40 mA.

Finally, it should be noted that open-collector outputs could be wired together safely, thus increasing the $I_{OL}$ specification. For example, two buffers wired in parallel increase the $I_{OL}$ level or load current capacity from 40 mA to approximately 80 mA (see Fig. 29.22a).

**The CA4030 Exclusive-OR Gate**  The CA4030 digital integrated circuit, housed typically in a 14-pin DIP, is configured with four (quad) exclusive-OR gates. Each exclusive-OR, or EX-OR, gate has two input terminals, normally labeled $A$ and $B$, and one output terminal identified as $Y$. Circuit operation is capsulated in the following *truth table* and the output of the EX-OR function can be described algebraically as

$$Y = A \oplus B \tag{27.13}$$

The $\oplus$ symbol is interpreted as the exclusive-OR function and Equation 27.13 is read as output $Y$ equals input $A$ exclusive-ORed with input $B$. One should note from the truth table below that this digital gate outputs a logic level HIGH (5 V) whenever the inputs are at opposite logic levels. The output of an EX-OR gate is at a logic level LOW (0 V) when both inputs are at the same logic level.

*Electronic Devices for Student Projects*     **535**

| A | B | Y |
|---|---|---|
| 0 V | 0 V | 0 V |
| 0 V | 5 V | 5 V |
| 5 V | 0 V | 5 V |
| 5 V | 5 V | 0 V |

One interesting application of this gate—the control of the location of the decimal point (dp) for a liquid crystal display (LCD)—is presented in several student projects to follow and is illustrated here in Fig. 27.20. When the CONTROL input to the EX-OR gate is

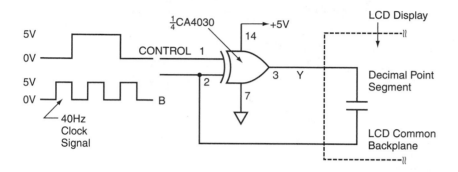

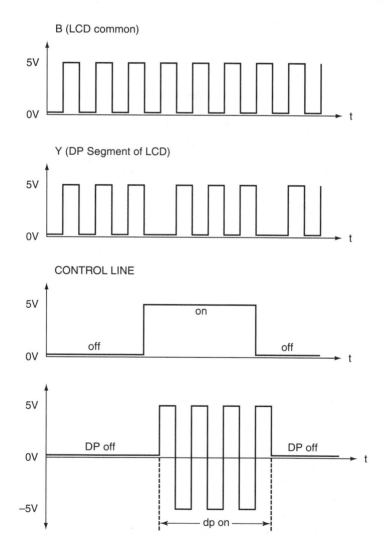

**Figure 27.20** The EX-OR gate used to select a decimal point (dp) on an LCD.

LOW (0 V), the output $Y$ will be exactly equal to the square-wave input $B$. Since the signal applied to decimal point segment and the back plane (common) of the LCD are the same (the differential voltage is 0 V), the decimal point is OFF. When the CONTROL input terminal to the EX-OR gate is at a logic HIGH (5 V), the output $Y$ will be the exact opposite of the square-wave shown. Since the signal applied to the dp segment and the back plane of the LCD are opposite (180° out of phase), the differential voltage is ($\pm$ 5 V) the same and the decimal point is ON.

## 27.7 HIGH-POWER VOLTAGE REGULATOR ICs

A modern AC-to-DC converter, or simply DC power supply, is designed today with three basic building blocks: (1) the transformer, (2) a rectifier and filter circuit, and (3) a voltage regulator. The first building block or transformer, generally wired in a step-down configuration, isolates the electric company's (Edison) earth or ground connection from the remaining circuitry, as well as reduces the amplitude of the applied AC voltage. A rectifier section follows—either half-wave or full-wave in design—which converts the transformer's AC output voltage into a pulsating DC voltage. A filter circuit, often a capacitor, follows the rectifier and results in a "smoother" output—approximately a constant DC voltage with some output ripple (AC) voltage—whose actual value depends on load current. Finally, a voltage regulator circuit is added to complete the typical DC power supply. This circuitry is designed to hold the output voltage constant under varying load conditions (i.e., regulate the output at a constant voltage level) and reduce the AC ripple superimposed on the DC signal to a minimum value. We discuss the voltage regulator section of a DC power supply by considering two types of linear regulators, the fixed and adjustable elements.

The 7805 three-terminal fixed-output and the LM317T three-terminal adjustable-output voltage regulators will be discussed in the next sections and are just two of a group of popular linear ICs that are used extensively when a low-current (less than 1 A) low-cost (less than \$1) easy-to-design voltage regulator solution is mandated.

**The 7805 Three-Terminal Fixed-Output Voltage Regulator** The 78xx series of linear IC voltage regulator is a three-terminal device that outputs a positive voltage at a fixed value designated by the value assigned by xx. For example, the 7805 device has a fixed +5 V DC output where the 7812 regulator has an output of +12 V DC. The 79xx complement series generates a negative DC output voltage with the 7915 regulator, for example, outputting a fixed −15 V DC.

Figure 27.21a illustrates a typical configuration for the 7805, a three-terminal linear voltage regulator IC. The output of the "raw" supply—transformer, rectifier, and filter—is applied between the input (IN) and common (COM) terminal and the load (modeled here as $R_L$) is connected between the output (OUT) terminal and COM.

Figure 27.21b is a plot of voltage vs. load-varying current $I_L$ and is used to discuss circuit operation. Under no-load conditions (i.e., $I_L = 0$ A) the output of the "raw" supply, and therefore the input to the 7805 regulator, is 16.7 V DC for the components shown. The ripple voltage is 0 V at no-load current. The regulated output voltage developed across the load is $V_L = 5.00$ V. At full-load conditions (i.e., $I_L = 500$ mA) the output of the "raw" supply has dropped to approximately 12 V DC and the ripple voltage has increased to approximately 5 V(peak-to-peak). Under full-load conditions the regulated output voltage has fallen by only 50 mV to $V_L = 4.950$ V. Peak-to-peak ripple is approximately 2 mV at a full-load current of 500 mA. Clearly the 7805 regulator has controlled the output voltage across the load to 5.0 V DC, while simultaneously substantially reducing the ripple voltage of the "raw" supply.

Power dissipation and the need for device heat sinking is an important consideration when dealing with high-power voltage regulators. Therefore, we next check the power dissipated by the 7805 at both no-load and full-load conditions to develop a further awareness of circuit operation. From the manufacturer's data sheet, a typical bias current of 4.2 mA

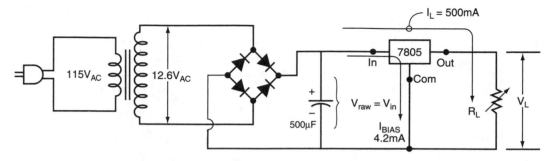

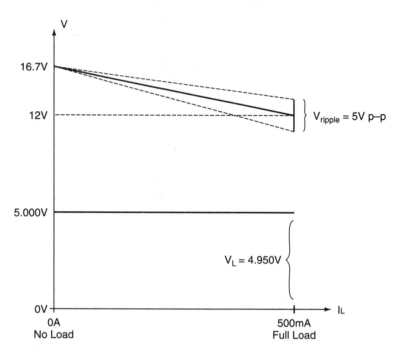

(a) A typical AC-to-DC converter designed to regulate the output of a 0.5A supply to 0.5V

(b) A plot of voltage vs load current for the circuit shown

**Figure 27.21** A typical application for the 7805 positive voltage regulator.

(see Fig. 27.21a) can be expected to flow through the 7805 IC with a no-load voltage across it of 16.6 V DC. The power wasted in the regulator under no-load conditions is therefore $P_D = I_{BIAS} V_L = (4.2 \text{ mA})(16.6 \text{ V}) \approx 70 \text{ mW}$. Under full-load conditions the load current $I_L$, which flows through the regulator from IN to OUT terminals, is 500 mA and the average voltage across the regulator, also from IN to OUT terminals, is 12 V $-$ 4.950 V $=$ 7.05 V DC. The power wasted in the regulator under full-load conditions is therefore $P_D = I_L V_{(IN\_OUT)} = (500 \text{ mA})(7.05 \text{ V}) \approx 3.5 \text{ W}$. A review of the 7805 data sheets reveals that a TO-220 package will dissipate a maximum of 2 W. Thus, heat-sinking will be required for this design. For completeness, it was found that mounting the package tap to a 3- by 5-inch piece of anodized aluminum was adequate at removing the heat generated under full-load conditions.

**The LM317 Three-Terminal Adjustable-Output Voltage Regulator**  The LM317 three-terminal adjustable-output voltage regulator, shown in Fig. 27.22a, is similar in many re-

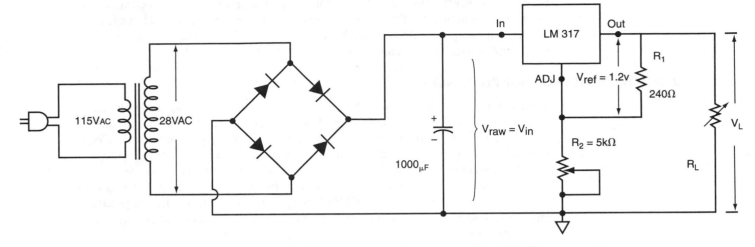

(a) The LM317 3-terminal regulator design controls the load voltage $V_L$ from a low of 1.2V to a high of 26.2V

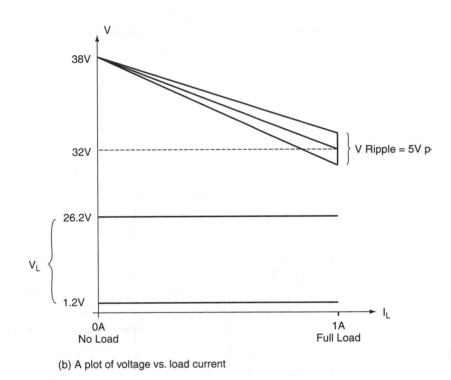

(b) A plot of voltage vs. load current

**Figure 27.22** The LM317 adjustable voltage regulator.

spects to the 7805 voltage regulator with the additional feature of having an adjustable output voltage. For the component values shown, this circuit is designed to regulate the output voltage from a minimum of 1.2 V DC to a maximum of 26.2 V DC. Load current for this design is set to a maximum of 1 A when the TO-3 package is properly heat-sinked.

The adjustability of load voltage $V_L$ is controlled by both fixed resistor $R_1$ and potentiometer $R_2$ value settings. When powered, the voltage between the OUT and ADJ terminals of the LM317 is set to a constant reference voltage of $V_{REF} = 1.2$ V DC. With $R_1$ chosen to be a 240 $\Omega$ resistor, the current through $R_1$, and therefore through $R_2$, is fixed at $I_{REF} = V_{REF}/R_1 = 1.2$ V/240 $\Omega = 5$ mA. Note that the regulated voltage across the load resistor is set by the voltage across $R_1$—fixed at 1.2 V DC—plus the adjusted voltage across the potentiometer $R_2$. Therefore, if the potentiometer is adjusted to $0\ \Omega$ the load voltage will be at a minimum of 1.2 V DC. When $R_2$ is set to a maximum value of 5 k$\Omega$, the regulated voltage at the output is $1.2\ \text{V} + (5\ \text{mA})(5\ \text{k}\Omega) = 1.2\ \text{V} + 25\ \text{V} = 26.2\ \text{V DC}$. Figure 27.22b is a plot of voltage vs. load current for the circuit shown in Fig. 27.22a.

## 27.8 LOW-POWER VOLTAGE REFERENCE ICs

The electronics needed to convert the electrical output of a sensor into a useful measurement value for display often requires the addition or subtraction of a precise and stable DC voltage. This is especially true and necessary for portable battery-powered electronic equipment where the terminal voltage of a battery naturally varies due to both usage and age and therefore cannot be used directly to provide the needed stable voltage reference.

Two precision references useful in solving the above problem are covered in this section. The first reference IC, the REF-02, is a voltage source that outputs a precise 5 V and the second integrated circuit, the REF-200, is a current source that produces a constant 100 $\mu$A DC current.

**The REF-02 5 V Reference IC**   The REF-02 is one of a family of popular three-terminal, low-cost, low-power voltage reference ICs commonly packaged in an 8-pin mini-DIP, or dual-in-line, integrated circuit package. Shown conceptually in Fig. 27.23, this IC consists

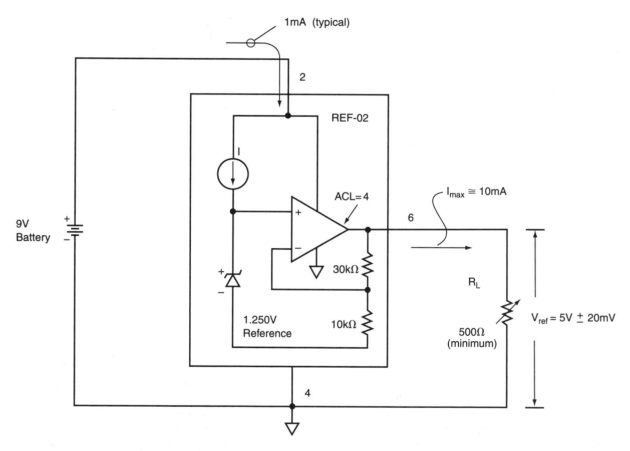

**Figure 27.23**   The internal circuitry of REF-02 voltage reference IC consists of a current source (I), a 1.250 V reference, and a noninverting amplifier (ACL-4) to set the output between pin 6 and pin 4 precisely at 5.00 V.

of a DC current source driving a precision voltage reference—modeled here as a simple zener diode—and an operational amplifier for output voltage level adjustment. The internal DC current source provides an unchanging and fixed drive to the band-gap reference device setting an internal voltage of 1.250 V. This precise internal voltage is then amplified by a non-inverting operational amplifier—designed with a closed-loop amplification factor ($A_{CL}$) of 4—to produce an output of precisely 5.000 V at pin 6.

By adjusting the amplification factor to say, $A_{CL} = 2$, the output can be set to 2.500 V (REF-03). For $A_{CL} = 8$, the output voltage is set to 10.000 V (REF-10). All three of these ICs form a family of low-power voltage references for applications requiring precision voltages to be established and maintained.

Figure 27.23 shows DC power applied to pin 2 and ground applied to pin 4. For the REF-02, this supply voltage can range from approximately 7 to 30 V. For a typical 9-V battery, the supply current required by this device is only about 1 mA. Designed to be a low-power device, the REF-02 can source a maximum of 10 mA at its output—the typical safe output current level for an op-amp—thus preventing a load resistor less than 500 Ω from being connected between pin 6 and ground. Figure 27.23 can be used to test the performance of the output voltage of a REF-02. With no-load connected ($R_L$ = infinite ohms) to full-load ($R_L$ = 500 Ω), the output will be consistently within ±20 mV to the 5.000 V specified by the manufacturer.

**A 100 μA Precision Current Reference—REF200**    A unique but practical device designed to deliver a constant DC current is the REF200. Figure 27.24 conceptually shows the internal architecture of this device. Like the REF-02 voltage source considered previously, this dual-current source is also available in an 8-pin mini-DIP and is configurable to deliver precise currents of 50 μA, 100 μA, 200 μA, 300 μA, or 400 μA by simply wiring the appropriate pins. Two common configurations, one to yield a 100 μA current and a second to produce a 200 μA current source, are shown in Figs. 27.25a and 27.25b, respectively, for two typical applications. It is the manufacturer's recommendation that when working with this device all unused pins be connected to ground or the positive supply voltage. Further, it is recommended that pin 6, the substrate for this device, be connected to the most positive constant potential of the circuit.

In addition to the two typical sensor excitation applications shown in Fig. 27.25, the REF200 is also well-suited to a variety of other uses where a highly-precise, constant DC, low-current level is required.

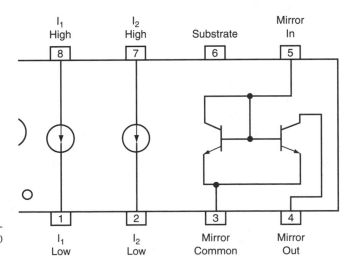

**Figure 27.24** The internal design of an REF200 DC current source.

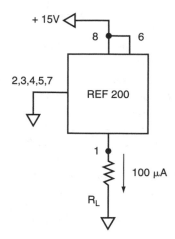
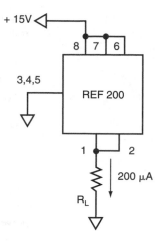

**Figure 27.25** Two practical REF200 circuit configurations.

## 27.9 NE555 TIMER

The ubiquitous 555 timer/oscillator IC has proven to be a remarkably versatile integrated circuit that since its inception has found wide applications. One of the first devices to combine both analog and digital technologies into a single hybrid integrated circuit, this component is predominantly configured for use as either a monostable or astable multivibrator. Both applications will be considered in the sections to follow.

**General Theory of Operation** We begin our discussion of the 555 IC by looking at its eight external terminals and a conceptual representation of its internal circuitry (see Fig. 27.26). Positive DC power, $V_{CC}$, which can range from $+5$ V to $+18$ V, is applied to pin 8 with the common supply terminal connected to pin 1. This supply voltage applies power to all the internal circuitry of the 555 device including three 5-k$\Omega$ resistors connected in series between $V_{CC}$ and common. As seen in Fig. 27.26, this series resistor network sets the $(-)$ input of comparator A to an *upper-threshold voltage* determined by the relationship $V_{UT} = 2/3 \ V_{CC}$. The $(+)$ input of comparator A is connected to pin 6, which is identified as the external *threshold* pin. The three-resistor divider network also sets the $(+)$ input of comparator B to a *lower-threshold voltage* determined by the relationship $V_{LT} = 1/3 \ V_{CC}$. The $(-)$ input of comparator B is connected to pin 2, which is identified as the external *trigger* pin.

While capable of exerting some influence on both trigger levels, $V_{UT}$ and $V_{LT}$, the external control voltage terminal, pin 5, is usually connected to a power supply common through a 0.01 $\mu$F bypass capacitor. This capacitor acts to reduce power supply noise and attempts to maintain more constant threshold voltages.

The *reset* terminal at pin 7 can be used to unconditionally override normal IC functions. While useful in some applications, this reset feature—not often used in this mode—can be disabled by connecting it directly to the positive power terminal $V_{CC}$ at pin 8.

All 555 timer/oscillator ICs are equipped with a *discharge* terminal connected externally to pin 7. As can be seen from Fig. 27.26, this external pin is connected internally to the collector of an NPN transistor that acts as one contact of a switch that is either ON or OFF. The other switch terminal is circuit common. Controlled by the internal logic of the 555, this discharge transistor is ON (a switch whose terminals are closed) when the output terminal at pin 3 is in its LOW state. The internal discharge transistor is OFF (a switch whose terminals are open) when the output terminal is in its HIGH state. As the terminal name implies, the discharge terminal is used primarily to *discharge* the timing capacitor and will be discussed more fully in the two applications to follow this introductory section.

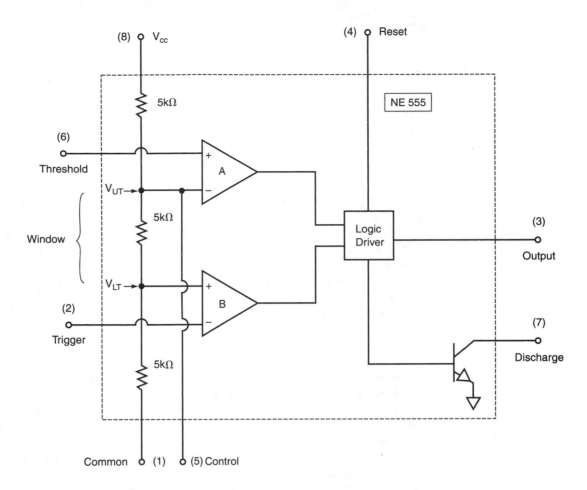

**Figure 27.26** The internal structure of an NE555 timer/oscillator IC.

A single *output* terminal, pin 3, is made available to the user. The condition of the output terminal is controlled by internal output logic circuitry, which is in turn controlled by the external conditions on the two input terminals at pins 2 and 6. When the output is in its LOW state at a voltage level of approximately 0.1 V with respect to common, it is capable of sinking (or accepting) an output current of approximately 200 mA. When pin 3 is in its HIGH state, at a voltage level of approximately $V_{CC}$ − 0.5 V with respect to common, it is capable of sourcing (or delivering) a continuous output current of 20 mA and a pulsed current of as much as 200 mA.

As stated above, the conditions on the trigger (pin 2) and threshold (pin 6) terminals control the state of the output terminal. Table 27.3 helps us understand the interrelationship between input and output terminals as well as the discharge pin.

Looking at the first entry in Table 27.3, we note that the trigger terminal is below $V_{LT}$ and the threshold terminal is also below $V_{UT}$. For this input condition the output is in its HIGH state at a voltage level of approximately 0.5 V below the power supply rail ($V_{CC}$ − 0.5 V) and the discharge terminal transistor OFF or, thinking of the internal transistor as a mechanical switch, its terminals are open.

Looking next at the last entry in Table 27.3, we note that the trigger terminal is above $V_{LT}$ and the threshold terminal is also above $V_{UT}$. For this input condition the output is in its LOW state at a voltage level of approximately 0.1 V and the discharge terminal transistor ON or, again thinking of the internal transistor as a mechanical switch, its terminals are closed. Note here that these first two input conditions discussed cause the output terminal to *unconditionally* go to one of two possible states.

**TABLE 27.3**

All possible operating states of the 555 timer

| Trigger (pin 2) | Threshold (pin 6) | Output (pin 3) | Discharge (pin 7) |
|---|---|---|---|
| $<V_{LT}$ | $<V_{UT}$ | HIGH | OFF |
| $<V_{LT}$ | $>V_{UT}$ | — | — |
| $>V_{LT}$ | $<V_{UT}$ | No change | No change |
| $>V_{LT}$ | $>V_{UT}$ | LOW | ON |

We next look at the third entry in Table 27.3 and note that for this input condition—where the trigger terminal is above $V_{LT}$ and the threshold terminal is below $V_{UT}$—the output is in whatever its previous state was and the condition of the discharge terminal does not change. Thus, when the inputs are within the *window* of voltages established by both $V_{UT}$ and $V_{UT}$, we can think of this condition as the *memory,* or *remember,* state for the output.

Finally, note that the second entry in Table 26.3 details an input condition where both inputs are outside the window established by the threshold and trigger voltage references and at opposite extremes. No entries are assigned to this condition, which we will never encounter because pins 6 and 2 will always be connected.

**Using the NE555 as a Time Delay Power-On**  Figure 27.27 uses the NE555 IC in a time-delayed power-on application. This circuit solves the problem of generating an output voltage of $V_O = V_{CC} - 0.5\ V = 12\ V - 0.5\ V = 11.5\ V$ at pin 3 to Circuit A approximately 5 seconds after power is initially applied to Circuit B. Circuit operation is as follows. When switch SW-1 is first closed, the voltage across capacitor C is 0 V and therefore the voltage across R is 12 V. Both inputs (pins 2 and 6) are above $V_{UT}$ and the output terminal is in its LOW state at a voltage of approximately 0.1 V. Using Equation 27.14 we compute the time for the capacitor to discharge by an amount equal to 2/3 $V_{CC}$ and the voltage across the resistor to just drop below $V_{LT}$.

$$T = 1.1\ RC \tag{27.14}$$

The output terminal will go HIGH to approximately 11.5 V. For the values shown in Fig. 27.27, the time delay is $T = 1.1\ (1\ M\Omega)(4.7\ \mu F) = 5.2$ seconds.

**The NE555 as an Oscillator**  In Fig. 27.28a the NE555 IC is wired as an astable multivibrator or square wave oscillator. When power is first applied to this circuit, the voltage across the timing capacitor C is at 0 V, forcing both pins 2 and 6 below the lower threshold voltage $V_{LT}$ and causing the output high to 4.5 V and the discharge transistor at pin 7 to be OFF. Current flowing through both $R_A$ and $R_B$ charges the capacitor C towards $V_{UT}$ and causes the NE555 output to remain HIGH for a period of time set by

$$T_{HIGH} = 0.695(R_A + R_B)C \tag{27.15}$$

When the voltage across the capacitor reaches—actually just slightly exceeds $V_{UT}$—the output terminal goes low to 0.1 V and the discharge terminal at pin 7 acts as an ON switch whose terminals are closed. The capacitor is forced to discharge through resistor $R_B$ and the discharge pin towards $V_{LT}$. The time that the output is LOW is set by the following relationship:

$$T_{LOW} = 0.695(R_B)C \tag{27.16}$$

When the voltage across the capacitor reaches—actually just goes below $V_{LT}$—the output terminal again goes HIGH to 4.5 V and the cycle of operation repeats. Figure 27.28b shows

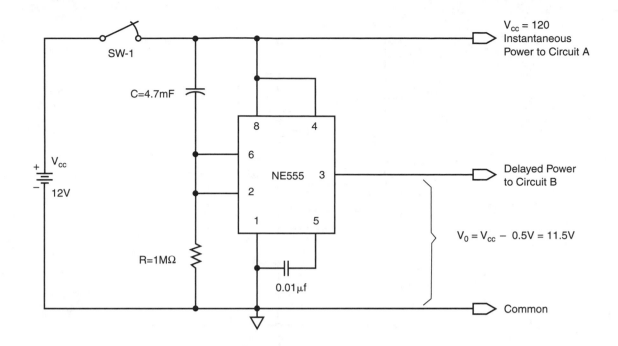

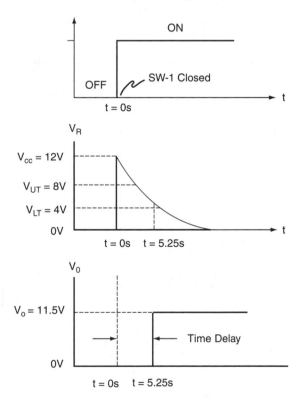

**Figure 27.27** A 555 IC used to delay power to circuit B by 5 seconds.

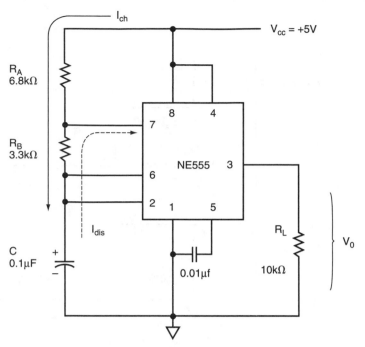

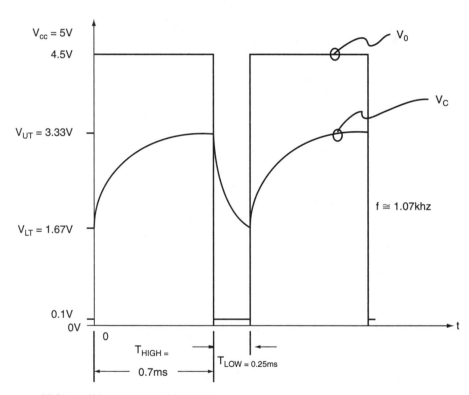

(a) Basic circuit configuration for a square wave oscillator

(b) Plots of Vo vs. time and Vc vs. time

**Figure 27.28** The NE555 used as an adjustable multivibrator.

the pertinent waveforms. The period of time for one complete charge-discharge cycle of the timing capacitor is expressed as

$$T = T_{HIGH} + T_{LOW} \qquad (27.17)$$

And therefore the frequency of oscillation can be calculated by the following expression:

$$f = 1.44/(R_A + 2R_B)C \qquad (27.18)$$

For the values shown in Fig. 27.28a the high time is 0.7 ms, the low time is 0.23 ms for a total period T = 0.93 ms or a frequency of square wave oscillation of approximately f = 1.07 kHz.

## 27.10 SENSORS

Temperature is an important parameter to be determined for many data acquisition and process control systems. In fact, process control engineers often tout temperature as being the most commonly measured physical parameter. In this section we look at two sensors, or transducers, to measure temperature. The thermistor, a nonlinear sensor, is considered first. The AD590—a complete integrated circuit that exhibits a linear change in output current as a function of change in temperature—is then discussed. Both sensors are popular devices used to measure temperature.

**The UUA41J1 Thermistor** The UUA41J1 negative temperature coefficient (NTC) thermistor is a two-terminal sensor whose terminal resistance decreases exponentially with increases in temperature. A ceramic semiconductor, this thermistor is fabricated by a process called *sintering,* which is the formation of a coherent nonporous substance by heating without melting. Most NTC thermistors are manufactured from the sintering of metal oxides such as magnesium, nickel, cobalt, copper, and iron with two leads attached as part of the fabrication process.

Thermistors are frequently used for surface-temperature measurements and are also employed to measure the surrounding temperature of fluids—both liquids and gaseous. In addition, thermistors are utilized extensively in medical instrumentation where measurement repeatability is often a more important design factor than the actual measured value. Thermistors are highly repeatable. These sensors also find application in household appliances—microwave ovens, freezers, cordless power tools, and camcorders to name a few, as well as automobiles, telecommunications, and the aerospace industries.

The UUA41J1 temperature detector is a glass-coated bead-type sensor typical of a family of devices fabricated as small as 0.095 inch in diameter with 0.006-inch diameter leads. Because of their small physical size, these sensors have very fast response times—less than 1 second—but require some care in implementation. The response time of a thermistor, referred to as its *time constant,* is the time in seconds needed to reach 63% of the difference between its initial temperature value and that of the new temperature environment.

Regarding proper implementation, the excitation voltage used to power a thermistor must be controlled carefully to maintain a power dissipation level below say 1 mW, or output measurement errors due to internal self-heating will result. In this regard a thermistor's *dissipation constant* is defined as the milliwatt power needed to raise the temperature of the thermistor 1°C above its surrounding temperature. For the UUA41J1 the dissipation constant is 1 mW. Therefore it is generally recommended that the UUA41J1 thermistor be operated at a current not to exceed 100 μA.

Internally, the exponential decrease in terminal resistance due to an increase in temperature is caused by an increase in the number of electrons capable of supporting a current. These conducting electrons are thermally generated, resulting from the increased thermal energy of the surface or environment being detected. Figure 27.29a shows an exponentially decaying plot of terminal resistance ($R_T$) vs. temperature (T) for the UUA41J1 thermistor and

Fig. 27.29b is the manufacturer's tabulation of $R_T$ vs. T. This tabulation is useful and can be interpreted as follows: At 0°C, for example, note that the resistance of the UUA41J1 is 32,650 Ω, while at room temperature—taken to be 25°C—the resistance has decreased to precisely 10,000 Ω, falling to only 678.3 Ω at 100°C. Thus, by simply measuring the terminal resistance of a thermistor with a good-quality digital multimeter (DMM), one can convert a measured resistance reading into a precise temperature using the manufacturer's data.

**An AD590 Current Transmitter**   Whenever temperature must be remotely measured—several hundred feet or more from the monitoring equipment—the selection of a sensor that outputs a changing voltage (or resistance) that is proportional to a changing temperature is not recommended. Voltage output sensors are sensitive to both voltage drops along long wires and also to induced noise. Resistance output sensors are sensitive to lead-wire resistance. But a high impedance current-output transducer renders circuit performance insensitive to both voltage drops and noise. To eliminate measurement errors due to long lead

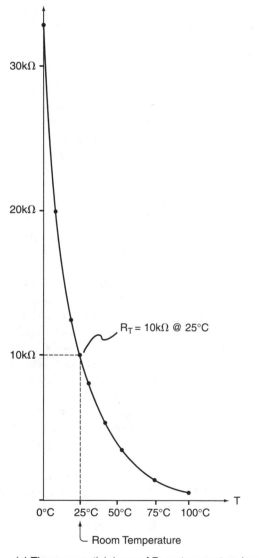

(a) The exponential decay of $R_T$ as temperature increases

| T | $R_T$ | T | $R_T$ |
|---|-------|---|-------|
| 0 | 32,650 | 51 | 3,467.0 |
| 1 | 31,030 | 52 | 3,340.0 |
| 2 | 29,500 | 53 | 3,217.0 |
| 3 | 28,050 | 54 | 3,099.0 |
| 4 | 26,690 | 55 | 2,986.0 |
| 5 | 25,390 | 56 | 2,878.0 |
| 6 | 24,170 | 57 | 2,774.0 |
| 7 | 23,010 | 58 | 2,675.0 |
| 8 | 21,920 | 59 | 2,579.0 |
| 9 | 20,880 | 60 | 2,488.0 |
| 10 | 19,900 | 61 | 2,400.0 |
| 11 | 18,970 | 62 | 2,316.0 |
| 12 | 18,090 | 63 | 2,235.0 |
| 13 | 17,250 | 64 | 2,157.0 |
| 14 | 16,460 | 65 | 2,083.0 |
| 15 | 15,710 | 66 | 2,011.0 |
| 16 | 15,000 | 67 | 1,942.0 |
| 17 | 14,320 | 68 | 1,876.0 |
| 18 | 13,680 | 69 | 1,813.0 |
| 19 | 13,070 | 70 | 1,752.0 |
| 20 | 12,490 | 71 | 1,693.0 |
| 21 | 11,940 | 72 | 1,636.0 |
| 22 | 11,420 | 73 | 1,582.0 |
| 23 | 10,920 | 74 | 1,530.0 |
| 24 | 10,450 | 75 | 1,479.0 |
| 25 | 10,000 | 76 | 1,431.0 |
| 26 | 9,573.0 | 77 | 1,384.0 |
| 27 | 9,167.0 | 78 | 1,340.0 |
| 28 | 8,777.0 | 79 | 1,297.0 |
| 29 | 8,407.0 | 80 | 1,255.0 |
| 30 | 8,057.0 | 81 | 1,215.0 |
| 31 | 7,723.0 | 82 | 1,177.0 |
| 32 | 7,403.0 | 83 | 1,140.0 |
| 33 | 7,097.0 | 84 | 1,104.0 |
| 34 | 6,807.0 | 85 | 1,070.0 |
| 35 | 6,530.0 | 86 | 1,036.0 |
| 36 | 6,267.0 | 87 | 1,004.0 |
| 37 | 6,017.0 | 88 | 973.70 |
| 38 | 5,777.0 | 89 | 944.00 |
| 39 | 5,547.0 | 90 | 915.30 |
| 40 | 5,327.0 | 91 | 887.70 |
| 41 | 5,117.0 | 92 | 861.00 |
| 42 | 4,917.0 | 93 | 835.30 |
| 43 | 4,727.0 | 94 | 810.30 |
| 44 | 4,543.0 | 95 | 786.70 |
| 45 | 4,370.0 | 96 | 763.30 |
| 46 | 4,200.0 | 97 | 741.00 |
| 47 | 4,040.0 | 98 | 719.30 |
| 48 | 3,890.0 | 99 | 698.70 |
| 49 | 3,743.0 | 100 | 678.30 |
| 50 | 3,603.0 | | |

(b) Tabulated data for $R_T$ vs. T

**Figure 27.29**   The thermal characteristics of a thermistor.

length, we choose the AD590 IC temperature transducer or current transmitter. Manufactured by Analog Devices, Inc., the AD590, shown symbolically as a current source in Fig. 27.30a, is a sensor that outputs a current $I_T$ that is directly proportional to absolute temperature (PTAT). The still-air thermal characteristics for the AD590 device, capable of measuring temperatures that range from $-55°C$ to $150°C$, are shown in Fig. 27.30b.

According to the manufacturer this temperature sensor will output a $1 \mu A$ increase in current $I_T$ for each $1°K$ (Kelvin) or $°C$ (Centigrade) increase in surrounding temperature. Therefore, as a Kelvin thermometer, the AD590 can be described by the following performance equation:

$$I_T = (1 \ \mu A/°K) \ T_K \tag{27.19}$$

The performance of the AD590 can also be described for both Centigrade and Fahrenheit temperature scales.

$$I_T = (1 \ \mu A/°C)T_C + 273.2 \ \mu A \tag{27.20}$$

$$I_T = (0.555 \ \mu A/°F)T_F + 255.4 \ \mu A \tag{27.21}$$

The test circuit shown in Fig. 27.30c can be used to test both the sensor and the theory of its operation. If we build Fig. 27.30c we can experimentally determine the output current by first measuring the air temperature surrounding the AD590 with a good quality laboratory thermometer, then connect a digital multimeter or DMM—configured to read a DC current in units of $\mu A$—in series with the sensor. The current measurement can be

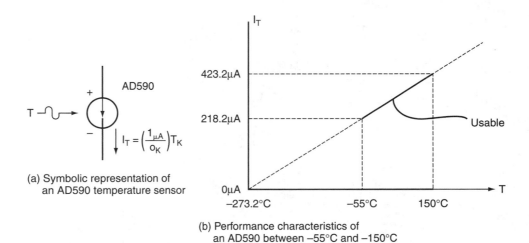

(a) Symbolic representation of an AD590 temperature sensor

(b) Performance characteristics of an AD590 between $-55°C$ and $-150°C$

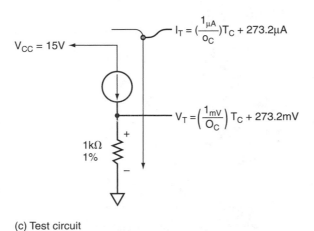

**Figure 27.30** The AD590 current transmitter is a popular choice when measuring temperature remotely.

(c) Test circuit

predicted from Equation 27.20. For example, if the surrounding air temperature being measured is known to be 25°C then the expected output current $I_T$ is calculated to be

$$I_T = (1 \ \mu A/°C)25°C + 273.2 \ \mu A = 298.2 \ \mu A$$

Note that the 1 kΩ, 1% calibrating resistor is really a current-to-voltage converter and the voltage across $R_T$ is given by

$$\begin{aligned}
V_T = R_T I_T &= R_T \left[ (1 \ \mu A/°C)T_C + 273.2 \ \mu A \right] \\
&= 1 \ k\Omega \left[ (1 \ \mu A/°C)T_C + 273.2 \ \mu A \right] \\
&= (1 \ mV/°C)T_C + 273.2 \ mV
\end{aligned} \qquad (27.22)$$

For the same test at 25°C, Equation 27.22 predicts an output voltage of 298.2 mV across $R_T$.

**Semiconductor Pressure Sensors**   Pressure measurements have become commonplace due in part to the availability of low-cost reliable semiconductor-based transducers. The conceptual side-view or cross-section of a typical semiconductor-type pressure sensor is presented in Fig. 27.31a. The structure begins with a silicon diaphragm where the bottom portion has been micro-machined chemically to form a seismic cavity. Then four piezo-resistive strain gage resistors are diffused into the top surface of this diaphragm and interconnected electrically to form a conventional Wheatstone bridge configuration. Finally, the diaphragm is bonded to a Pyrex support.

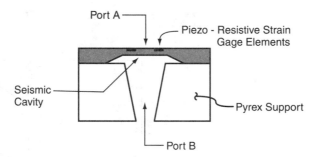

(a) The side view of a semiconductor-type differential pressure sensor

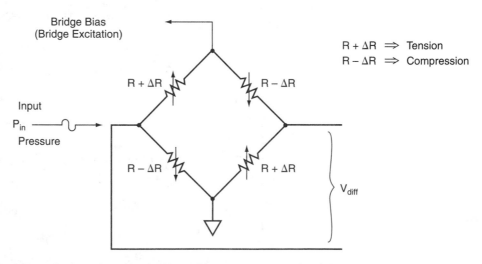

(b) 4 strain-gage elements wired in a Wheatstone bridge configuration

**Figure 27.31** A basic semiconductor pressure sensor.

Figure 27.31b shows an energized pressure sensor. When an input pressure $P_{IN}$ is applied to the sensor, the diaphragm deflects, creating a differential output voltage ($V_{DIFF}$) at the differential output of the Wheatstone bridge circuitry that is directly proportional to input pressure. This output voltage input-pressure relationship is extremely linear since opposite arms of the bridge are simultaneously in *tension* (piezo-resistive elements increase by an amount $+\Delta R$) or *compression* (piezo-resistive elements decrease by an amount $-\Delta R$).

**Types of Pressure References**   All pressure readings are made as differential measurements. That is, the measured pressure value is made with respect to a reference pressure. Figure 27.32 illustrates each of the three possible types of pressure measurement.

*Absolute pressure sensors* measure an input pressure with respect to a zero pressure (see Fig. 27.32a). One port (port B) is sealed to create a near perfect vacuum. Port A becomes the active or working port of the sensor. Barometric or atmospheric pressure measurements are made with this type of sensor. Normal atmospheric pressure at sea level is taken to be 14.7 pounds per square inch absolute (14.7 psia).

*Gage pressure sensors* measure input pressures with respect to a normal atmospheric pressure reference (see Fig. 27.32b). Port B is the active port for gage pressure and port A is exposed to the atmosphere as the reference pressure. One of the more common types of gage pressure is automobile tire pressure. Gage pressure is given in units of psig (pounds per square inch gauge).

*Differential pressure sensors* measure the difference in pressure between two independently variable pressures (see Figure 27.32c). Differential pressure measurements are the general form of the more specific gage pressure measurements. Differential pressure is given in units of psid (pounds per square inch differential).

**SCX C Series Semiconductor Pressure Sensors**   SenSym of Sunnyvale, California, produces the SCX C series of semiconductor pressure sensors, illustrated in Fig. 27.32. The sensors in this series can measure all three types of pressure and are available in ranges from 0 to 1 psi (SCX01) to 0 to 100 psi (SCX100). An SCX C series sensor is activated by applying DC power between pins 2 and 4. While this sensor is able to handle a maximum $+30$ V of excitation, 12 V to 15 V is typically available and therefore used for bridge excitation. For absolute pressure measurements, the differential output voltage ($V_{DIFF}$) across the sensor will cause pin 5 to be more positive than pin 3. Therefore this sensor's output can be described as

$$V_{DIFF} = (V_5 - V_3)\, P_{IN} \tag{27.23}$$

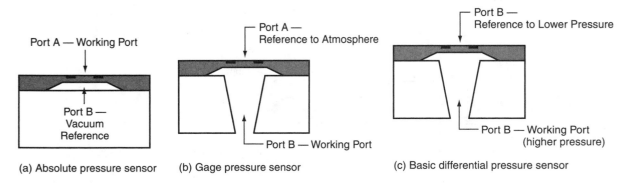

(a) Absolute pressure sensor   (b) Gage pressure sensor   (c) Basic differential pressure sensor

**Figure 27.32**   The three types of pressure sensors.

**Measuring with the SCX15 ANC Absolute Pressure Sensor**   The data sheet for the SCX15 ANC absolute pressure sensor specifies sensitivity Sd of 6 mV/psia of absolute pressure when biased from a 12 V DC source. To ratiometrically correct this data sheet sensitivity Sd for a new bridge supply voltage, we use the following equation:

$$Sc = \frac{V_{Bias}}{12\ V}\ Sd \tag{27.24}$$

For a bridge bias applied between pins 2 and 4 of an SCX15 ANC absolute pressure sensor powered with a 5 V supply, the corrected sensitivity would be Sc = (5 V/12 V)6 mV/psia = 2.5 mV/psia. Therefore this transducer's performance equation can be written as a modification of Equation 27.24.

$$V_{DIFF} = (2.5\ mV/psia)\ P_{IN} \tag{27.25}$$

## 27.11 PIEZOELECTRIC DEVICES

**An Audible Buzzer**   One defines *piezoelectricity* as a mechanical strain that induces an electric field or vice versa: as an electric field that develops a mechanical strain. It is this latter definition of the *piezoelectric* phenomenon that is exploited to create low-cost highly efficient audible devices, or annunciators, which are finding increased use in such common household products as smoke and gas detectors, as well as burglar alarms, to name just a few.

Fabrication of a typical piezoelectric-effect audible element begins with a very thin, circular brass disk to which another (smaller) circular disk—this time consisting of a piezoelectric material like ceramic—has been bonded (see Fig. 27.33). A number of ceramics—notably zirconata titanate or PZT—exhibit a piezoelectric effect when polarized by an electric field and are one choice for this class of low-cost annunciators. Attaching electrodes to both the brass disk and the ceramic material completes fabrication of the audible element.

Electrically, the piezoelectric-effect audible element—in effect a capacitor—becomes part of the tuned (LC) circuit of an oscillator. The electric field of the oscillator is applied across the piezoelectric disk, developing a voltage first with one polarity and then with an opposite polarity. One polarity attempts to reduce the diameter of the ceramic disk, which is restrained on one of its surfaces by the brass disk. The result of this oscillator polarity is to cause the piezoelectric capacitor to become slightly convex in shape. When the oscillator polarity is reversed the diameter of the ceramic disk attempts to increase. Again restrained on one surface by the brass disk, the result of this polarity is to cause the piezoelectric capacitor to become slightly concave in shape. Mounted properly in a suitable housing, the result of this material flexing—at or near resonant frequency—is an audible sound.

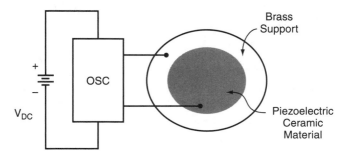

**Figure 27.33**  Basic structure of an audible buzzer.

Using the piezoelectric-effect phenomenon and the fabrication techniques outlined above, audible annunciators can be manufactured to develop an output sound level of 85 dB or greater at a distance of up to 15 feet.

## 27.12 INTEGRATED CIRCUIT CONVERTERS

**The 7660 Switched-Capacitor Inverter**   The *charge pump*, a special class of DC-to-DC converter that has recently been finding more applications in portable electronic devices such as cellular phones and handheld data-logging and measurement equipment, converts DC power efficiently without the use of inductors. The term *charge pump* refers to a type of DC-to-DC converter that uses capacitors instead of inductors and transformers to transfer energy. Offered by several manufacturers with various improved features, the basic 7660 building block will be discussed here.

Figure 27.34 illustrates the internal architecture of these devices, which are also referred to as a switched-capacitor converter. Constructed internally with four BJT or FET switches and shown here as their mechanical equivalents—an inverter and an internal digital clock—this IC requires two externally connected capacitors—tantalum or polystyrene preferred—to complete its utility. The capacitor $C_1$ connected between external pins 2 and 4 is referred to as the *charge-storage* capacitor and $C_2$ connected between external pins 3 and 5 is termed the *charge-pump,* or *reservoir,* capacitor. A typical application, to convert $+ 9$ V to $- 9$ V, is discussed next.

Note that the positive terminal of a 9 V DC supply is wired to pin 8 of this IC—housed in an 8-pin mini-DIP—and that the power-supply common is wired to pin 3. The internal oscillator, normally designed to run at an internal clock frequency of 10 kHz, drives four internal switches ON or OFF. When the internal 10 kHz clock is HIGH, the *charge phase* begins. The terminals of switches SW-1 and SW-3 are closed and the terminals of SW-2 and SW-4—because of the inverter logic—are open. Current, limited by the resistance of the ON switches, flows to charge the capacitor $C_1$ to 9 V DC with the polarity shown. When the clock is LOW, the *pump phase* begins. The terminals of switches SW-1 and SW-3 are open and the terminals of SW-2 and SW-4 are closed. The charge developed on the capacitor $C_1$ is redistributed and some is transferred to $C_2$ with the *inverted-with-respect-to-common* polarity shown. When power is first applied to this IC and one charge-pump cycle is completed, the voltage across $C_2$ will be 4.5 V DC or 50% of the applied voltage of 9 V DC. After a second charge-pump cycle is completed, the voltage across $C_2$ will rise to 75% of the applied 9 V DC and so on until the voltage across $C_2$ will equal the voltage applied after approximately five complete charge-pump cycles. Thus the output voltage across $C_2$ reaches its full input value in approximately 0.5 ms.

A few words of caution are warranted here when using this technology. First the internal regulator at pin 6, the LV or low-voltage pin, should be connected to pin 3 (common) for inputs below 6 V DC and left unconnected for inputs greater than 6 V DC. Finally the 7660 core, a first generation charge pump, does not have a regulated output, and therefore is only useful for loads that do not exceed about 10 to 20 mA of current.

**An ICL7106 Analog-to-Digital Converter**   Most portable handheld electronic equipment—frequency-counters, component test and measurement functions, or even complete digital multimeters (DMMs)—use as their core for measurement and display an analog-to-digital converter (ADC) designed to interface directly with low-cost liquid crystal displays (LCDs). Developed using large-scale integration or LSI technology, chips like the 7106 ADC with driver circuit have become a popular basic building block for the portable electronics industry. The following discusses briefly how the 7106 core works and some of its features.

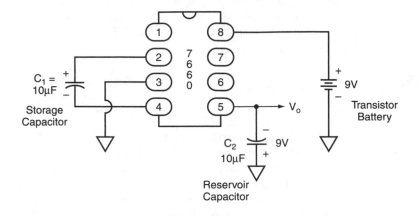

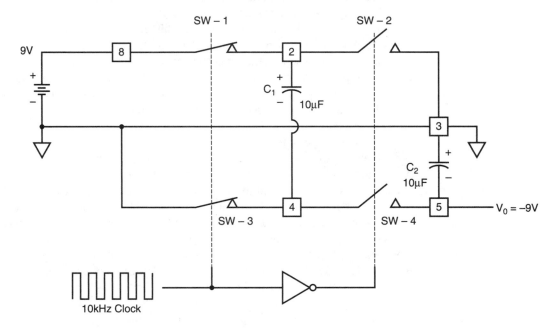

**Figure 27.34**  The 7660 charge pump configured to create −9V DC from a +9V transistor battery.

The 7106 converter is shown as a block diagram in Fig. 27.35a. Fig. 27.35b is a graphical representation of dual-slope integration, the method by which this and other similar ICs convert an analog voltage into a digital representation. When an unknown voltage $V_{IN}$ is applied to the input, a measurement process is initiated by charging a storage capacitor $C_{int}$ for a fixed period of time, referred to as the *integration phase,* or integration time. Integration time is set to 83.33 ms, or precisely 1000 cycles of an internal counter running at 12,000 cycles per second (see Fig. 27.36). During the second portion of the measurement, referred to as the *reference phase,* the stored charge is removed from the storage capacitor (discharged to 0 V) at a fixed rate with a known reference voltage, $V_{ref}$. An internal counter measures discharge time, which can range from 0 ms to a maximum of 166.67 ms, or from 0 to 2000 cycles of the internal counter. The count determined after the reference phase is completed is directly proportional to voltage at the input being measured and a scaled version of this count is processed by the internal driver circuitry, which then is presented to the LCD as the digital representation of this analog signal. After the reference phase an *auto zero* phase completes one voltage measurement of the converter. This phase is characterized as an "internal housekeeping" phase—preparing the circuitry of the converter for the

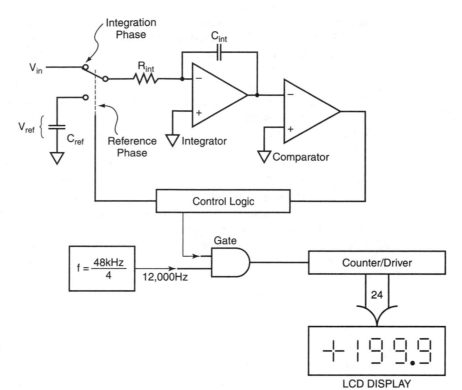

**Figure 27.35** A structural block diagram of the internal architecture of a 7106 dual-slope, analog-to-digital converter.

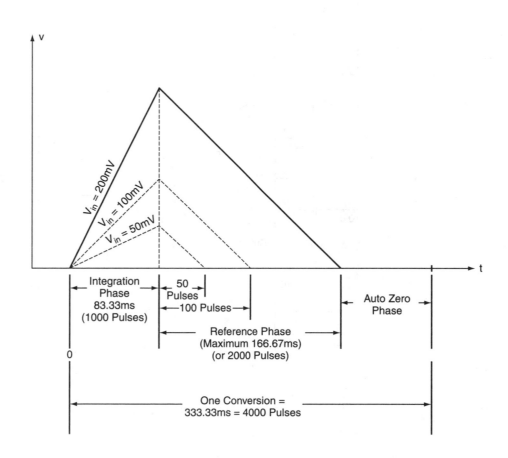

**Figure 27.36** A graphical representation of a dual-slope, integrating analog-to-digital converter.

next measurement. The conversion time to complete just one voltage measurement equals 333.33 ms (1/3 second), or a total of 4000 internal clock cycles.

The 2000-count maximum possible with the 7106 IC core during its reference phase translates into a full-scale voltage resolution of 199.9 mV. This is referred to as 3 1/2 digits of resolution and in theory this unit can display or resolve a voltage change as small as 0.1 mV. This is, however, not actually possible because of quantization errors inherent with this type of ADC architecture. Input scaling can also be added externally to the basic converter allowing the ADC to be configured to measure inputs up to 1.999 V (2-V range) or even 19.99 V (20-V range).

We conclude our discussion of the 7106 ADC by looking at its external pin connections shown in Fig. 27.37, and discussing, in general functional terms, the circuitry associated with each. The unknown input voltage $V_{IN}$ is applied to pins 31 and 30 (INPUT HI and INPUT LO, respectively), allowing indirect access to the internal integrator. Note that for most applications the *analog common* at pin 32 is wired directly to pin 30 and pin 35 (REF LO). The external 1-k$\Omega$ potentiometer and 24-k$\Omega$ fixed resistor connected to pins 35 and 36 form a circuit that is used to set the precise reference across $C_{ref}$. The potentiometer is adjusted until the

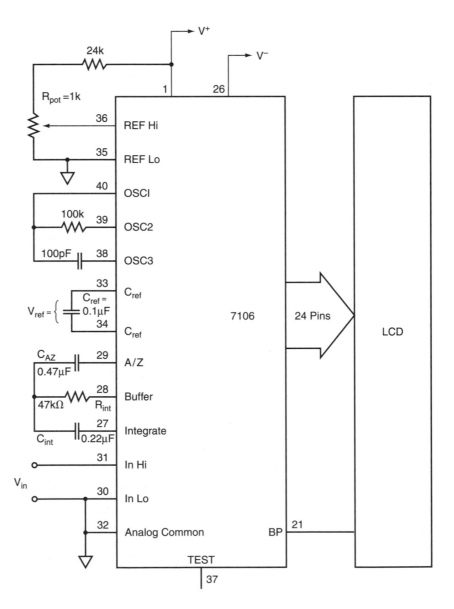

**Figure 27.37** The pin configuration of a standard 7106 ADC.

desired $V_{ref}$ is recorded at pin 36. $V_{ref}$ must be set equal to 100 mV of a full-scale design of 200 mV. Internally this voltage is applied across the 0.1 $\mu$F reference capacitor $C_{ref}$ attached to pins 33 and 34. The internal oscillator is set at 40 kHz, and then divided down internally to 12,000 Hz, by the 100 k$\Omega$–100 pF combination wired to pins 38, 39, and 40.

Finally, a 0.22 $\mu$F integrating capacitor ($C_{int}$) at pin 27 and a 47-k$\Omega$ integrating resistor ($R_{int}$) at pin 28 form the external components needed to configure the internal op amp integrator. The auto zero housekeeping feature of this chip requires an externally connected 0.47 $\mu$F ($C_{AZ}$) connected to pin 29 and the $C_{int}$ and $R_{int}$ combination.

The LCD is wired pin-for-pin to appropriate converter pins. A total of 24 pins in all—two are needed to drive the 1/2 digit, three groups of seven segments each for the three full digits, and one pin for the decimal point segment. In addition pin 37 (TEST) and pin 21 (BACK PLANE) make up the final LCD wiring.

**The CA3162E/CA3161E IC Chip-Set**  The final pair of devices to be considered in this chapter are the CA3162E analog-to-digital converter and the CA3161E BCD-to-seven segment (LED) decoder/driver ICs. This chip set, similar in functional operation to the 7106, is one possible solution when the output display is to be a large power consumption LED readout instead of a low power LCD.

Refer to Fig. 29.7 and note that the CA3162E is a dual-slope integrating ADC with external integrating capacitor and both gain and zero adjustments. This ADC accepts at its input an analog voltage and converts this signal to a BCD output at pins 1, 2, 15, and 16. The CA3161E IC interfaces with these pins of the ADC to decode the BCD output of the ADC into the needed seven-segment drivers. In conjunction with the digit drivers—pins 3, 4, and 5 of the CA3162E IC—the CA3161E IC processes the analog input voltage into a digital representation.

# 28 Introductory-Level Student Projects

## 28.0 INTRODUCTION

The projects presented in this chapter are for the beginning student who wants to practice the packaging design, planning, and construction skills discussed throughout this book. From the information contained in each of the six units of this book, the authors have created an introductory-level project to highlight the more important material covered in those chapters. Even so, the projects presented in this chapter are wide-ranging, including not only simple drawing exercises, but also practice (drill) exercises, and even a few exercises that will yield a completely fabricated project.

Each presentation begins with *project objectives* and a *project overview*. Project objectives identify what should be learned from completing the project, and in the project overview section, general information about the project is provided and cautions, if any, are identified. This project overview concludes with a *materials* or *parts list,* if applicable. Finally, each project is presented with a detailed *procedure* to help guide the student through the design, practice drill, or fabrication of the project.

When shown, the finished designs illustrated in this chapter are to be considered typical solutions to the project. This is especially true of the conceptual design projects. A word of caution, however, is necessary before we continue. The projects presented in this chapter are intended for the beginner and, while in most cases students are not restricted to the designs shown, changes should be made with caution and only after consulting with your instructor.

Finally, a word on safety is in order here. Before working with any tools, equipment, or materials presented in these projects, you should review the technical information relating to the project as well as read and understand the relevant safety practices and information provided throughout this book and from other sources. In addition, ask your instructor for help before undertaking anything that is unfamiliar to you, especially if it seems unsafe.

The following introductory-level unit projects are presented in this chapter:

- Packaging a Laboratory Power Supply
- Negative-Acting One-Shot Multivibrator PCB Design
- Fabricating a Single-Sided PCB for the TTL Logic Probe Project
- Soldering to a Practice Printed Circuit Board
- Sheet Metal Tap and Drill Index
- Soldering to a Practice Printed Wire Board

558

## 28.1 PACKAGING A LABORATORY POWER SUPPLY

### Project Objectives

*Upon completion of this project, the packaging design of a laboratory power supply, the student should be able to*

- Understand how to use the selection criteria to choose which components and hardware should be assigned to the front-and-side panel element of a power supply design.

- Understand how to use the selection criteria to choose which components and hardware should be assigned to the subchassis elements of a power supply design.

- Know how to use the positioning criteria to optimize the placement of the components and hardware chosen to be assigned to the front-and-side panel element.

- Know how to use the positioning criteria to optimize the placement of the components and hardware chosen to be assigned to the subchassis element.

- Develop a preliminary sketch of both the front and side panel as well as the subchassis.

**Project Overview**   The planning and design considerations studied in Unit 1 will be revisited here with this first introductory-level project on packaging a laboratory-style power supply. In fact the circuit schematic and accompanying parts list, shown in Fig. 28.1, has an unregulated power supply section (transformer, bridge rectifier, and filter) that is very similar to the circuit studied in Chapter 1. However, the regulator section for this project, unlike the fixed-voltage dc-to-dc converter developed in Sections 1.10 and 1.11, uses a linear IC to generate an adjustable output voltage. In addition, a panel meter and switching circuit has been included to complete this project. This new section will display both terminal voltage and load current depending on the position of selector switch $SW_2$.

Circuit operation can be described briefly as follows: When $SW_1$ is closed, 115 Vac is applied directly across the primary winding of transformer $T_1$, lighting the pilot lamp $L_1$, and powering all remaining circuitry. At the secondary winding of transformer $T_1$, the voltage is stepped down to approximately 25.2 Vac under full-load conditions of 2 A and the four bridge diodes ($D_1$ to $D_4$) convert this ac voltage into pulsating dc. The output voltage of the diode bridge is filtered by parallel capacitors $C_1$ and $C_2$ resulting in an approximate dc voltage at the input to the linear regulator $IC_1$. Adjustment of $R_2$ controls the regulated terminal voltage at the positive ($+$) and negative ($-$) output jacks. This output voltage will range from approximately 1.2 to 21 V depending on the setting of $R_2$. Finally, when switch $SW_2$ is in the *V* position, the basic meter movement $M_1$ is in series with $R_{mult}$ to read terminal voltage. When $SW_2$ is in the *I* position, meter movement $M_1$ is in parallel with $R_{sh}$ to read (approximate) load current. Note that the values for both $R_{mult}$ and $R_{sh}$ given in the parts list are for a basic meter movement with an internal resistance of 1 kΩ.

As with the power supply package developed in Chapter 3, this design project will incorporate the same three basic chassis elements: front-and-side, subchassis, and cover. While it is expected that much of the package planning done on this project will result in a design similar to that of the subchassis shown in Fig. 3.4 and front-and-side element with subchassis and cover illustrated in Fig. 3.5, there will necessarily be some differences.

This packaging design project is divided into three parts as follows: (1) to identify and list all components and hardware for each of the three chassis elements, (2) to develop a freehand engineering sketch for the subchassis, and (3) to create an engineering sketch of the main chassis elements similar to Fig. 3.5.

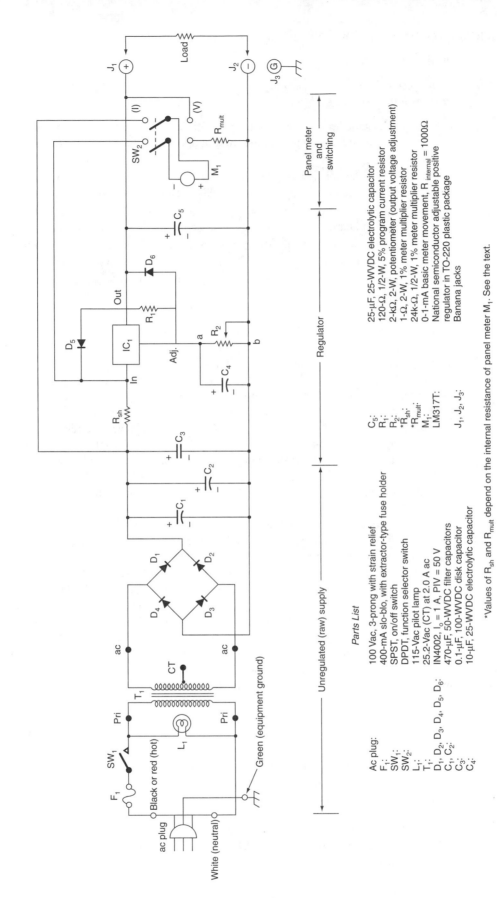

## Parts List

| | |
|---|---|
| Ac plug: | 100 Vac, 3-prong with strain relief |
| $F_1$: | 400-mA slo-blo, with extractor-type fuse holder |
| $SW_1$: | SPST, on/off switch |
| $SW_2$: | DPDT, function selector switch |
| $L_1$: | 115-Vac pilot lamp |
| $T_1$: | 25.2-Vac (CT) at 2.0 A ac |
| $D_1, D_2, D_3, D_4, D_5, D_6$: | IN4002, $I_o$ = 1 A, PIV = 50 V |
| $C_1, C_2$: | 470-μF, 50-WVDC filter capacitors |
| $C_3$: | 0.1-μF, 100-WVDC disk capacitor |
| $C_4$: | 10-μF, 25-WVDC electrolytic capacitor |
| $C_5$: | 25-μF, 25-WVDC electrolytic capacitor |
| $R_1$: | 120-Ω, 1/2-W, 5% program current resistor |
| $R_2$: | 2-kΩ, 2-W, potentiometer (output voltage adjustment) |
| *$R_{sh}$: | 1-Ω, 2-W, 1% meter multiplier resistor |
| *$R_{mult}$: | 24k-Ω, 1/2-W, 1% meter multiplier resistor |
| $M_1$: | 0-1-mA basic meter movement, $R_{internal}$ = 1000Ω |
| LM317T: | National semiconductor adjustable positive regulator in TO-220 plastic package |
| $J_1, J_2, J_3$: | Banana jacks |

*Values of $R_{sh}$ and $R_{mult}$ depend on the internal resistance of panel meter $M_1$. See the text.

**Figure 28.1**  Laboratory-style linear power supply, adjustable from 1.2 Vdc to 21 Vdc.

560

*Parts List*

The parts list for this design project is included with the circuit schematic shown in Fig. 28.1.

## Procedure

1. Review the factors affecting package design presented in Chapter 3.
2. Next list the components and hardware that you select for mounting to the subchassis. Further subdivide this list to identify which components are PCB mounted and which are chassis mounted.
3. List the components and hardware that you select for mounting to the front-and-side panel element. Further subdivide this list to identify which components are front-panel mounted and which are side-panel mounted.
4. Review the factors affecting component and hardware placement presented in Chapter 3.
5. Using 10-squares-to-the-inch grid paper, develop a freehand 1:1 scale engineering sketch detailing the hardware and component placement for your subchassis.
6. Again on 10-squares-to-the-inch grid paper, develop a freehand 1:1 scale engineering sketch detailing the hardware and component placement for your front-and-side panel and cover elements. Use Fig. 3.5 as a model and estimate that the meter requires a 1 1/2-inch diameter mounting hole and is housed in a 2 × 2-inch plastic case.

## 28.2 NEGATIVE-ACTING ONE-SHOT MULTIVIBRATOR PCB DESIGN

### Project Objectives

*Upon completion of this project, the component layout design of a negative-acting one-shot multivibrator circuit, the student should be able to*

- Interpret the results of the Packaging Feasibility Study for Printed Circuits presented in Appendix XVIII and use these results to select a PCB design type.
- Position components, parts, and devices according to imposed board size limitations and external connection restrictions.
- Place and reposition components, parts, and devices in such a way as to minimize board area wasted space while maintaining a uniform distribution and balance of components.
- Electrically interconnect components accurately from the schematic and according to other specified restrictions.
- Use a CAD software package to complete a component layout drawing for this project.

**Project Overview**   Designing a pc board component layout drawing for the negative-acting one-shot multivibrator circuit shown in Fig. 28.2 will require the use of much of the material contained in Unit Two—Designing Printed Circuit Boards with CAD Techniques. However, before beginning with this design we give a brief description of the functionality of the circuit.

Negative signals are applied to the input (IN) terminal at point 1, where their amplitude is adjusted by potentiometer R101 and then applied to the inverting input (pin 4) of U1, an IC amplifier. U1 both amplifies and inverts the signals and applies them to the

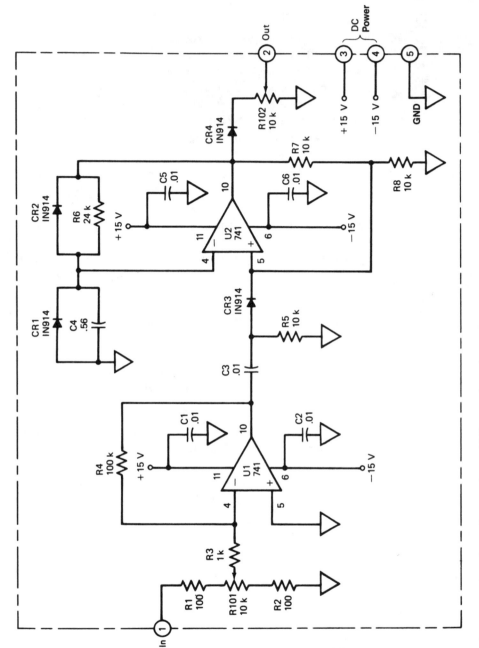

**Figure 28.2** Negative-acting one-shot multivibrator schematic.

562

noninverting input terminal (pin 5) of U2. The signals are shaped into pulses by the circuit associated with U2 and only their positive portion is passed to the output by the steering diode CR4. The portion of the signal that appears across R102 also exists at the output (OUT) terminal at point 2. Note that output signal amplitude is controlled by adjustment of potentiometer R102.

To minimize circuit confusion, some shorthand methods of schematic drawing have been used in Fig. 28.2. Note that pin 11 of both U1 and U2 are to be electrically connected to $+15$ Vdc and that both are connected to point 3, an external printed circuit board connection. In like manner, pin 6 of both U1 and U2 are to be electrically connected to $-15$ Vdc and both pins are connected to another external board connection at point 4. This method of schematic shorthand is extended to include all common, or ground, connections, which are symbolically represented by triangles. With ten separate electrical connections symbolically represented by triangles, it is implied that they all must be electrically connected together and then connected to the system common, an external board connection located at point 5.

The component layout drawing for the multivibrator circuit shown in Figure 28.2 must be designed within the $3.5 \times 5$-inch board outline shown in Fig. 28.3. All parts are to be placed in such a way as to minimize any wasted board space, yet maintain uniform component distribution and balance. The layout is restricted to the unhatched area and no component or conductor is to be placed within a minimum of 0.20 inch from any board edge. The minimum spacing between electrical conductors (paths and pads) is to be 0.025 inch. Finally, the position and orientation of all electrical connections and mounting holes are to be placed in accordance with Fig. 28.3. Pads for external connections are to be 0.1 inch in diameter.

The common, or ground, connections are to be divided into three separate groups with an individual conductor path provided for each group to the external connection point at point 5. This is to be done as follows: (1) the common points for all of the decoupling capacitors (C1, C2, C5, and C6) along with the common points for C4 and R5 will be connected together with a separate return path to the external common point; (2) all common points associated with input circuitry (bottom end of R2 and pin 5 of U1) will be connected and returned separately to the external common point; and (3) all common points associated with the output circuitry (bottom ends of R8 and R102) will be connected and returned separately to the external common point.

A review of the material presented in Appendix XVIII on the Packaging Feasibility Study for Printed Circuits reveals that the $3.5 \times 5$-inch board specified for this project has approximately 10.9 in$^2$ of usable area. With a total of 75 connection points (holes) for the multivibrator circuit the packaging density was calculated to be about 6.9 holes / in$^2$, which is identified as a moderate density for single-sided designs. Therefore, it is expected that this layout can readily be accomplished using single-sided design techniques.

*Parts List*

    U1, U2—741 Op-amp, TO-116 style package

    R1, R2—100 $\Omega$, 1/2 W, 10%

    R3—1k $\Omega$, 1/2 W, 10%

    R4—100 k$\Omega$, 1/2 W, 10%

    R5, R7, R8—10 k$\Omega$, 1/2 W, 10%

    R6—24 k$\Omega$, 1/2 W, 10%

    R101, R102—10 k$\Omega$, trim pots

    C1, C2, C3, C5, C6—0.01 $\mu$F disk capacitors

    C4—0.56 $\mu$F disk capacitor

    CR1, CR2, CR3, CR4—1N914 diodes, DO-35 style case

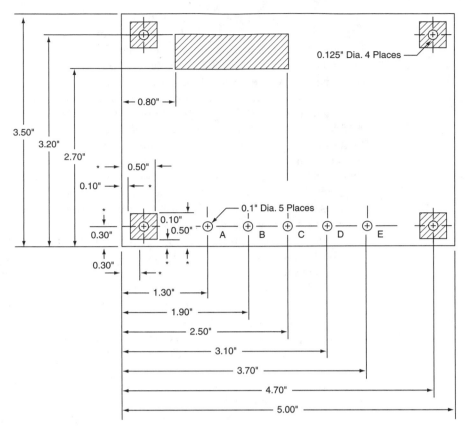

0.125" Dia. 4 Places

0.1" Dia. 5 Places

A   B   C   D   E

* Dimension at each corner of board

**Figure 28.3**
Dimensional information for the multivibrator PCB layout including placement of external connections.

A – INPUT (IN)
B – GROUND (GND)
C – (–15 Vdc)
D – OUTPUT (OUT)
E – (+ 15 Vdc)

Electrical orientation of external connections

## Procedure

1. For the single-sided component layout drawing of the negative-acting one-shot multivibrator circuit, select a 0.100-inch grid system and using a 2:1 scale, draw the board outline (7 × 10-in.) on your grid.

2. Add appropriate corner brackets to delineate the board edges. Then, using the dimensional information provided in Fig. 28.3, locate the unusable areas around the mounting holes and identify these areas with hatch marks. In a similar manner, identify the unusable area in the upper left-hand corner of the board, again identifying this area with hatch marks.

3. Complete the initial component layout by including the five external connections. Label them *A, B, C, D,* and *E* as shown on Fig. 28.3.

4. Of the three layout *viewpoints* discussed in Chapter 6, you should rely most heavily on the *schematic viewpoint* for this project. In that context, begin the parts layout by positioning U1 approximately centered over external connection *B* and midway within the board border. Position U2 approximately between external

connections *D* and *E* and centered on U1. Position decoupling capacitors C1 and C2 above and below U1, and capacitors C5 and C6 above and below U2.

5. Next position CR1, CR2, C4, and R6 horizontally in the upper right-hand corner of the layout to the right of the unusable area.

6. Finally position the remaining components vertically to the left and right of U1 and U2 and between both ICs. Note that the potentiometers R101 and R102 should be positioned so that their screw adjustment is facing the external connections.

7. Arrange and reposition the components until you are satisfied with your layout. Try to achieve a balance of parts, evenly distributed in the usable area, without wasting board space or crowding components.

8. Begin routing the signal path conductors first. Signal path conductors are all interconnections that are not power ($\pm$ 15 V) or common (ground). Include the routing to both external connections *A* and *D* (i.e., IN and OUT).

9. Complete the conductor routing by adding power and ground connections and wiring to the external connections labeled *B, C,* and *E.*

10. Complete your layout by checking that all connections are made correctly.

11. Once you are satisfied that your layout is complete and correct, use a CAD software package to produce a component layout drawing of your design.

## 28.3 FABRICATING A SINGLE-SIDED PRINTED CIRCUIT BOARD FOR THE TTL LOGIC PROBE PROJECT

### Project Objectives

*Upon completion of this project, the fabrication of a PCB for the TTL logic probe project, the student should be able to*

- Handle and use presensitized photo resist PCB stock for single-sided fabrication.
- Use a commonly available sun lamp and the negative phototool of the TTL logic probe design given in this project to expose and then develop the imaged photo resist.
- Etch the imaged PCB using ferric chloride etchant in a simple Pyrex-type baking dish.
- Clean the copper conductor paths, shear the PCB to size, and drill all needed holes for eventual component assembly.
- Remove the processed photo resist from the etched board.

**Project Overview**   If you refer ahead to Section 29.2 you will see a relatively simple, but very useful instrument for today's electronic technician: a TTL logic probe that is redrawn here in Fig. 28.4. Designed around the still popular Transistor-Transistor-Logic family of digital logic devices, this self-contained instrument is easy to build and can be an effective tool to help understand and troubleshoot digital circuitry. Those interested in how this circuit functions should consult the information under Circuit Operation in Section 29.2 before continuing.

Since our primary objective in this introductory-level project is to employ the simplest and least expensive method to fabricate a single-sided PCB, we have chosen this basic but useful circuit as the vehicle. If you are interested in constructing the complete project you can begin here by fabricating the printed circuit board. Then you can finish the project (Section 29.2) by assembling and soldering parts to this printed circuit board, wiring the probe point and power leads, and fabricating and assembling the probe housing.

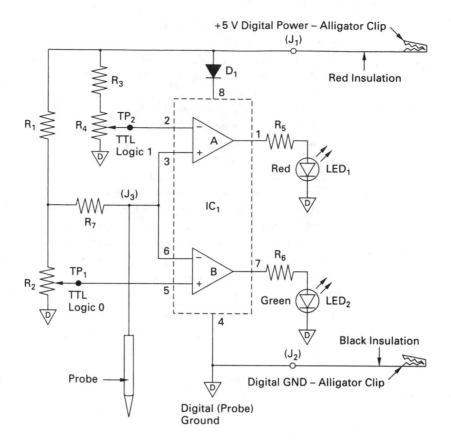

**Figure 28.4** TTL logic probe schematic. A parts list is included with Fig. 29.4.

The least expensive and least involved method of producing a one-of-a-kind single-sided printed circuit board is outlined in Chapter 10 and is called the print-and-etch technique. Using copper stock presensitized with light-sensitive photo resist, a few proprietary chemicals, and some commonly available items such as Pyrex cooking dishes and commercial sun lamps, the fabrication of a PCB can be accomplished in just a few hours. The *procedure* section of this project provides a useful and sequential guide to help fabricate a single-sided printed circuit board using the print-and-etch method. However, it is strongly recommended that before beginning this project, you review the material presented in Chapter 10 and not depend only on this procedural guide. The successful fabrication of even a simple single-sided PCB requires many steps, each requiring attention to detail that is available only in Chapter 10. Finally, if more advanced methods of board fabrication are available to you, consult your instructor for the relevant information necessary to use them.

*Material List*

> $3 \times 6 \times 1/16$-inch thick, 1-ounce copper, FR-4 type, single-sided copper-laminate PCB stock. Presensitized stock should be chosen if available.

*Parts List*

> For those interested in completing this entire project, a parts list accompanies the circuit schematic shown in Fig. 29.4.

## Procedure

> 1. Prepare a 1:1 negative photo tool from the 1:1 positive conductor pattern (foil view) shown in Fig. 28.5b.

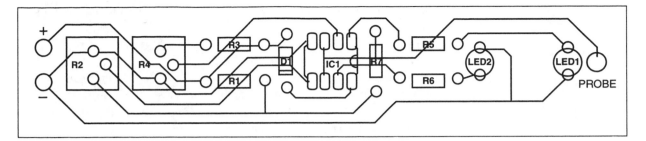

(a)

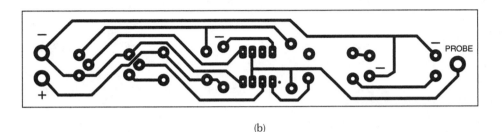

(b)

**Figure 28.5** Logic probe artworks: (a) 2:1 scale component layout generated using the TTL logic probe circuit shown in Figure 28.4; (b) 1:1 scale positive artwork of the foil side of the printed circuit board.

2. Working under safe-light conditions, remove the presensitized PCB stock from its light-tight plastic bag and place it sensitized-copper side up on a table (*Note:* If presensitized stock is not available you will have to apply liquid photo resist onto cleaned and prepared PCB stock by one of the methods detailed in Chapter 10.)

3. Orient the negative photo tool, foil view up, onto the sensitized copper, and place a small sheet of heat-resistant clear glass above the assembly.

4. Expose the light-sensitive photo resist through the negative photo tool for about three to four minutes with a 275-watt sun lamp placed about 12 inches directly above the glass.

5. Develop the exposed board by immersing, copper side up, in a Pyrex-type dish of developer suitable for the photo resist being used. Developing time is critical and should not exceed 2 minutes.

6. Use chemical resistant tongs to grasp the board by its edges. Remove it from the developing tray, apply inspection dye (either dipping or spraying), and flush for 30 seconds under cold tap water.

7. Under normal lighting, dry the board in a preheated oven for 10 minutes at 250°F. Remove the board from the oven and allow it to cool to room temperature. Inspect the conductor pattern for defects, repairing any that are found.

8. A simple single-sided board of this type can be etched in a Pyrex-type tray filled with ferric chloride ($FeCl_3$) etchant. Place the board, foil side up, in the tray. Gently rocking the tray will help speed the etching process. Using "fresh" etchant at room temperature and in this manner, it should take about 30 minutes to etch a 1-ounce copper pc board.

9. After etching, rinse the board in cold tap water for 30 seconds and pat dry with paper towels.

10. Next, using the corner brackets as a guide, shear the PCB to its finished size (see Figure 28.5a). Complete by drilling all component lead access holes and three holes for the turret terminals—one for the probe at one end of the board and two for the power leads labeled + and − at the other end of the board.

11. Finally, remove the dyed photo resist with an appropriate chemical stripper. Rinse the board in cold tap water and dry.

If you plan to complete this project, you must remove the surface oxide present on the copper conductor pattern by scrubbing with a 3M Scotch-Brite 96 pad before component assembly and soldering. In addition, it is suggested that you gain experience by completing the introductory-level project on Soldering to a Practice Printed Circuit Board (Section 28.4) before doing any component assembly and soldering to this project's pc board.

## 28.4 SOLDERING TO A PRACTICE PRINTED CIRCUIT BOARD

### Project Objectives

*Upon completion of this project on component assembly and soldering to a practice printed circuit board, the student should be able to*

- Properly clean the printed circuit board copper prior to component assembly and soldering.
- Correctly assemble components and hardware.
- Properly solder all component leads and hardware.
- Properly clean the PCB.

**Project Overview**   The "art" of soldering is a skill that is acquired only with practice—sometimes a lot of practice! The acquisition of this skill takes patience but the results of a properly soldered project are well worth the effort. To that end, the printed circuit board illustrated here as a project is not designed to be functional but instead to provide a platform to practice soldering. Therefore, it follows that any single-sided board configurations, other than the one being shown here, could be substituted without any loss of utility. If a substitute is to be made, however, choose one that has a variety of device package styles, hardware needs, and a number of components of varying lead configurations and body types.

If possible, gain experience by soldering to circuits boards that have been previously solder plated (tinned) and then repeat the process by soldering to boards with a pure copper surface. Remember, pure copper oxidizes quickly, and proper cleaning is necessary if you are to expect any type of acceptable results. While not discussed specifically in this project, it would be of some good to also practice de-soldering, another "art" that is not easily acquired and takes patience. Remember, in de-soldering the objective is often to remove the part without damaging it, and *always* without damaging the printed circuit board.

A final word about soldering before we begin. A correctly soldered joint is *not* possible without a good soldering iron, most particularly with a clean soldering iron tip that is properly prepared and correctly sized for the job. It has often been our experience that students attempt to "get away" with an inferior iron, perhaps through a lack of understanding and to save some money. Unfortunately this is one area where the uninformed approach is doomed from the start. You must invest in the proper tool if you have any hope of doing acceptable soldering.

*Parts List*

(1) Single-sided, practice printed circuit board

(1) D1—DO-41 or DO-35 style diode case

(1) Q1—TO-92 style transistor case

(1) Q1—TO-5 style transistor case

(1) U1—TO-116 style IC package

(1) U1—TO-99 style IC package

(4) 1/4 W resistors, value unimportant

(4) Ceramic-type disk capacitors, value unimportant

(4) Single-ended turret terminals

## Procedure

1. Using the information contained in Fig. 28.6, drill the correct size holes to assemble all the hardware and components identified in the parts list.

2. If the practice printed circuit board you are working with is solder plated, no initial cleaning should be necessary and you can proceed to step 4.

3. However, if your practice board is copper clad with no other finish applied, it is desirable to initially remove any surface oxidation by buffing the copper with a 3M Scotch-Brite 96 pad. Once cleaned in this fashion you should avoid touching the copper surface with your hands.

4. Using either *hand* or *pressure staking* techniques (see Chapter 16), secure the single-ended turret terminals to the practice pc board noting that the post end is facing the component side of the board.

5. Assemble the TO-116 and TO-99 IC packages using a dead-head lead termination. Remember it may be necessary to staircase the leads of the TO-99 style package to orient and insert the leads properly into the PCB holes.

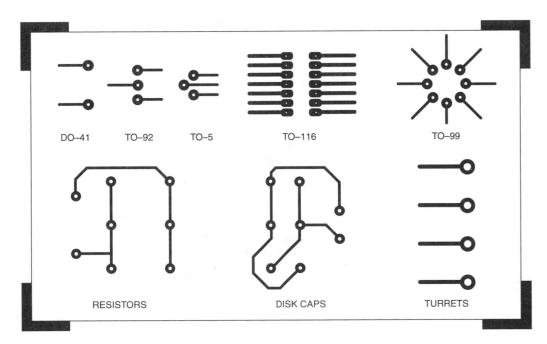

**Figure 28.6**  1:1 scale positive artwork of the foil view of the practice printed circuit board.

6. Assemble all the remaining components shown in Figure 28.6 using a fully clinched lead termination. Pay particular attention to the ceramic disk capacitors. Care should be taken to prevent them from rotating during lead clinching.

7. Once all the hardware and components are assembled, secure the PCB into a *circuit board holder* in preparation to solder. Next, using a fine artist's brush or toothpick, very selectively apply liquid (rosin) flux to only the pads to be soldered.

8. Solder the leads of all components to the board using the four-step procedure highlighted in Fig. 17.10. Inspect your work. If necessary rework any joints that appear unsatisfactory by using solder wick to remove excess solder or apply additional solder if there is a deficiency.

9. To solder the shank of the turret terminals to the PCB, repeat step 8.

10. Finally, use a stiff-bristle brush, isopropyl alcohol, or one of the aerosol cleaners discussed in Section 17.5 to clean the flux residue from your pc board.

## 28.5 SHEET METAL TAP AND DRILL INDEX

### Project Objectives

*Upon completion of this project, a sheet metal tap and drill index, the student should be able to*

- Become acquainted with sheet metal layout drawings.
- Apply the correct procedures for sheet metal layout.
- Develop the ability to work accurately and according to specifications.
- Become familiar with using some of the more common tools used in sheet metal layout and fabrication including the bending procedure.

**Project Overview**   The sheet metal tap and drill index, or stand, shown in Fig. 28.7, will involve using much of the information discussed in the six chapters of Unit Five. Consisting of two sheet metal assemblies (the *top plate* and a *base and side plate*) and a *Bakelite bottom block,* this tap and drill stand will accommodate the drills (both tap and body) as well as the taps for four common tap sizes from 4-40 to 10-32. This project involves sheet metal

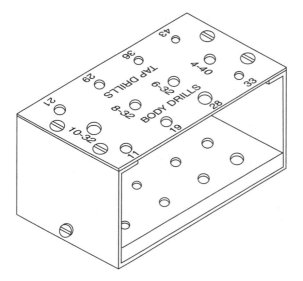

**Figure 28.7** This sheet metal tap and drill index accommodates tap and body drills and taps for sizes 4-40, 6-32, 8-32, and 10-32 machine and sheet metal screws.

layout work, metal shearing, and drilling, as well as sheet metal bending. In addition, some information on machine screws (Unit Six) will be required for its final assembly.

Before any sheet metal work can be performed, the layout details must be transferred to the work, and in this project maximum transfer accuracy is desired for all the assemblies to fit properly. The layout must be carefully scribed onto the metal. The scribe lines should be made on the *bottom* side of the work using minimum scribe pressure. Remember, deep scratches or gouges can weaken the metal. This is especially true along bend lines. For most layout work, fine scribe lines will give sufficient contrast on metal to facilitate cutting and bending. However, use of layout dye is recommended for obtaining a higher degree of contrast without scratching into the surface of the metal.

Once the layout has been completed, there is a basic sequence that should be followed to keep the sheet metal work as uncomplicated as possible. The basic steps in this sequence are (1) to perform as many cutting operations (shearing and drilling) as possible while the stock is in the flat plate form, (2) to bend the workpiece into the desired shape, (3) to finish the surface, when possible, before assembly, and (4) to apply appropriate identification markings. Thought should be given not only to the best order of operations but also to the best technique for each step.

A correct sequence must also be developed for any bending operations. An error at this point would be costly both in time lost and material wasted. There are generally certain bends that must be made first. Although the bending sequence can be worked out through a strictly mental process before the actual bending operation, it is preferably done on paper. This is especially true of more complicated chassis configurations. The bending sequence for the base-and-side plate of this project is given in the procedure.

Bend allowance must be taken into consideration when both the inside and outside dimensions of a chassis are critical. Remember, when a bend is formed the metal will stretch along the outside of the bend and compress along the inside. When working with aluminum stock, it can be expected that each inside measurement, that is, the distance from one scribe line to another, will reduce by approximately one-third the thickness of the stock after it is bent. Each outside dimension will be increased by approximately two-thirds the thickness of the stock.

*Material List*

(1) 2 × 3-inch, 16 gauge, H-1100 aluminum
(1) 2 × 8-inch, 16 gauge, H-1100 aluminum
(1) 2 × 3 1/2 × 1/4-inch Bakelite
(2) 4-40 × 1/4-inch machine screws
(4) 6-30 × 1/4-inch machine screws
(4) 6-32 nuts
Pressure sensitive dry transfers

**Procedure**

1. Transfer the layout design shown in Figs. 28.8a and 28.8b to the 16-gauge, H-1100 aluminum stock. In the case of the top plate, the layout marks should be scribed into the top surface. They should be made small enough so that the drill will remove them.

2. Determine the drill size required for the center row of holes to accommodate (clear) the taps in the top plate for the layout shown in Fig. 28.7 and then drill all the holes in this plate and the base-and-side plate. Deburr all holes.

3. Shear both the top and the base-and-side plates to exact size. Gently file all corners and edges. A slight chamfer is required. Do not round corners and edges.

4. Refer to the bending sequence shown in Fig. 28.9. Note that you should bend the two outside tabs [(1) and (2)] of the base-and-side plate before completing the

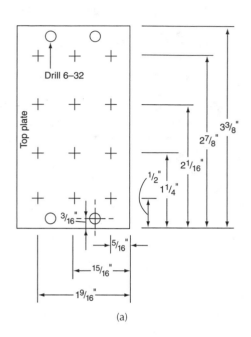

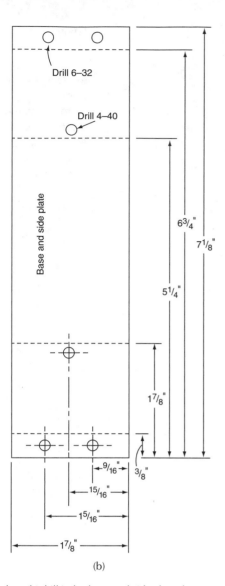

**Figure 28.8** Sheet metal layout drawings: (a) drill index top plate; (b) drill index base-and-side plate element.

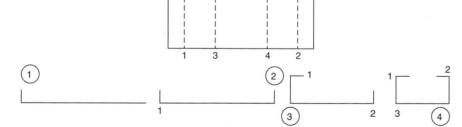

**Figure 28.9** Bending sequence for the base-and-side plate element.

90-degree bends at (3) and (4). Bend the base-and-side plate into the desired configuration as shown in Figs. 28.7 and 28.10.

5. Fit the Bakelite block into the U-shaped aluminum base-and-side plate by sanding the edges smooth and filing the corners slightly. Take care to neatly fit the Bakelite to the base-and-side plate.

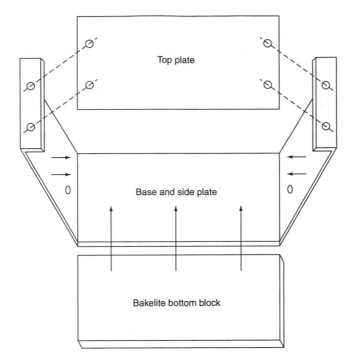

**Figure 28.10**
Assembly drawing for the sheet metal tap and drill index.

6. Using the previously drilled clearance holes as a guide, and with the Bakelite bottom block centered in the base-and-side plate, locate the two holes to be drilled and tapped into the ends of the Bakelite.

7. Remove the Bakelite block from the U-shaped piece of aluminum that forms the base and sides of the tap and drill index, center the drilled aluminum top plate over the Bakelite block, and locate all holes to be drilled.

8. Drill and tap the two holes in the end of the Bakelite block. The total depth of these holes should not exceed 1/4 inch.

9. Drill the holes in the top of the Bakelite block. The depth of these holes should not exceed 1/8 inch.

10. Lightly buff all aluminum pieces with 000 steel wool and use pressure-sensitive dry transfers to label the top plate as shown in Figure 28.7.

11. Finally, assemble the Bakelite bottom block to the base-and-side piece with two 4-40 × 1/4-inch machine screws. Assemble the top plate with four 6-32 × 1/4-inch machine screws and nuts.

## 28.6 SOLDERING TO A PRACTICE WIRE BOARD

### Project Objectives

*Upon completion of this project, soldering to a practice wire board, the student should be able to*

- Properly strip insulation from both solid and stranded hook-up wire.
- Correctly reform (dress) and tin stranded wire.
- Properly form wire to a variety of terminal and soldering lug configurations.
- Solder wire properly.
- De-solder wire from terminals without damaging insulation, terminals, or adjacent wiring.
- Inspect and properly clean soldered connections.

**Project Overview** In Section 28.4 a practice printed circuit board project was proposed and used as a tool to begin the process of acquiring some of the skills necessary to solder properly. At that time we said that sometimes a lot of practice is required and that becoming a skillful technician who can solder properly takes patience. To that end we present here another soldering project. This one is intended to give you practice in stripping insulation from standard hook-up wire, tinning stranded wire, lead forming and dress, and soldering wire to a practice wire board.

Again the wire board illustrated here as a project is not intended to be functional but is intended only to provide a platform to practice soldering. As previously mentioned in Section 28.4, any wire board configurations could be substituted here without any loss of utility. If a substitution is to be made, however, choose one that has a variety of solder terminals and lug types for several wire sizes to give experience in as many aspects of soldering hook-up wire as possible.

Unlike the practice printed circuit board project on soldering where you were not asked specifically to de-solder parts, here we will require you to practice de-soldering techniques. De-soldering is another skill, or art, that is not easily acquired and takes both practice and patience. Again, remember, in de-soldering the objective is to remove the wire without causing any damage to its insulation or to any closely related parts or wires.

Again a comment about the need for a quality soldering iron before we begin. A correctly soldered joint is *not* possible without a good soldering iron and a clean soldering iron tip that is properly prepared and correctly sized for the job. You must begin with a proper tool if you have any hope of making a reliable solder connection.

Finally it should be noted that the procedure given below will guide you through the process of soldering to DOD-STD-2000-3.

*Material List*

> (2) 8-lug terminal strips
> (2) solder lugs
> (2) single-ended turret terminals
> (1) 6 × 6-inch sheet of aluminum
> (6) 4-40 × 1/4-inch machine screws
> (6) 4-40 × 1/4-inch machine nuts
> (2) 1/4-watt resistors, value unimportant
> Solid wire—18 AWG
> Stranded wire—22 AWG

## Procedure

1. Transfer the layout shown in Fig. 28.11 to the aluminum stock and drill the appropriate size clearance holes.
2. Assemble the hardware, securing the two turret terminals first by *hand* or *pressure staking*. Mount the two solder lugs with machine screws and nuts; then bend the lug 90 degrees. Assemble the terminal strips as shown, again using machine screws and nuts to secure them to the aluminum plate.
3. Using Fig. 28.11 as an orientation guide, and assembling to lugs (d) and (e), form the leads of both 1/4-watt resistors into a 180- to 270-degree *wrap* connection. Be sure that the insulation gap does *not* exceed one lead diameter.
4. Terminals (a), (b), and (c) will use 18-AWG solid hook-up wire.
5. Begin with terminal (c). Strip approximately 1/2 inch of insulation from one end. Inspect the wire for nicks. If you nicked the wire, cut off the defect and try again. Estimate the wire length for the connection, noting that you will also have to strip

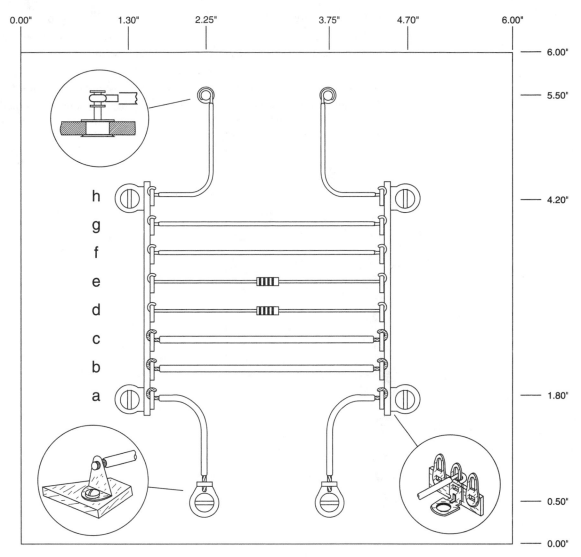

**Figure 28.11** Practice wire board with ordinate dimensioning.

another 1/2 inch from the other end. Cut the wire to size and strip the remaining insulation. Form a 180- to 270-degree *wrap* connection at both (c) lugs. Again, be sure that the insulation gap does *not* exceed one lead diameter.

6. Repeat step 5 for terminal (b).

7. For terminal (a) use a 180- to 270-degree *wrap* connection on the terminal strip lugs and a *dead-head* connection on the solder lugs. Be sure that the insulation gap does *not* exceed one lead diameter.

8. Terminals (f), (g), and (h) will use 22-AWG stranded hook-up wire.

9. Beginning with terminal (f), strip approximately 1/2 inch of insulation from one end of the wire. Again inspect the wire for nicks. If you nicked any of the strands of wire, cut off the defect and try again. Firmly twist the strands together and *tin* the end using your soldering iron and the techniques illustrated in Fig. 26.4a. Estimate the wire length for the connection, noting that you will also have to strip another 1/2 inch from the other end. Cut the wire to size and strip the remaining insulation. Tin this end before forming a 180- to 270-degree *wrap* connection at both (f) lugs.

10. Repeat step 9 for terminal (g). Check for correct insulation gap.

11. For terminal (h) use a 180- to 270-degree *wrap* connection on the terminal strip lugs and the turret terminals. Check the insulation gap.

12. Finally, use a stiff-bristle brush, isopropyl alcohol, or one of the aerosol cleaners discussed in Chapter 17 to clean the flux residue from the solder joints.

13. Using solder wick, practice de-soldering the dead-head solder connections at terminal (a). Remember, take care *not* to melt or in any way damage the wire insulation.

# 29 Advanced-Level Student Projects

## 29.0 INTRODUCTION

The projects presented in this chapter are for the advanced student who has acquired a good understanding of the skills discussed in this book and is willing to take on a challenge. While the projects were selected to illustrate varying degrees of circuit and packaging complexity, all require the student to plan carefully and execute care throughout the fabrication process. We would also like to mention that the cost of constructing these projects was one of our selection considerations but our primary emphasis was placed on the application of current packaging technology as well as the projects' potential interest and usefulness to the student.

The authors are aware of the necessity to substitute parts and devices because of availability considerations. Many devices and components are not critical, and where indicated, equivalencies may be substituted. For this reason, and also to familiarize the student further with the selected circuit, it is strongly recommended that the circuit be breadboarded and tested before fabrication. This approach also provides an opportunity for the advanced student to modify a project.

The presentation of each project begins with a general discussion of the circuit operation to acquaint the student with its function. This is followed by a circuit schematic and a complete parts list. Finally, construction hints are provided to avoid packaging difficulties.

The finished packages shown pictorially are to be considered as typical. The student, of course, is not restricted to the design shown and should feel free to employ ingenuity in fabricating a package that will reflect his or her own requirements and desires.

The following projects are presented in this chapter:

- Darkroom Timer
- TTL Logic Probe
- 1.2- to 20-volt Laboratory Power Supply with Digital Readout
- Portable Fahrenheit/Celsius Thermometer with LCD Readout
- Portable Sportsman's Barometer with LCD Readout
- Automobile Security Alarm System
- Microcontroller-Based Refrigeration Alarm System

As an aid in construction, pin configurations for all of the devices used in these projects are provided in Fig. 29.1.

## Diodes

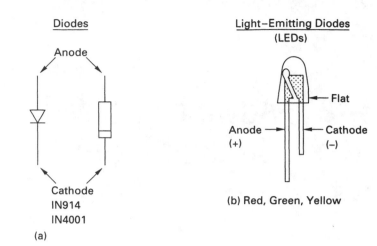

Anode

Cathode
IN914
IN4001

(a)

## Light–Emitting Diodes (LEDs)

Flat

Anode (+)  Cathode (−)

(b) Red, Green, Yellow

## Seven–Segment Display

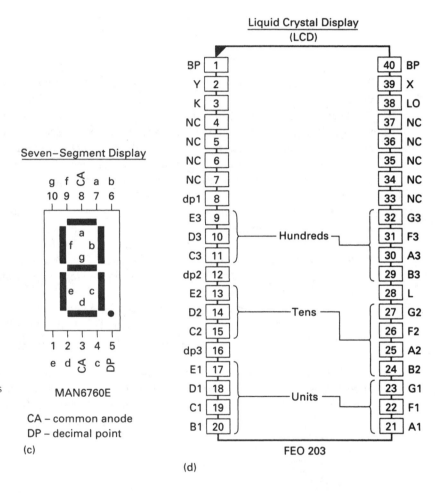

g  f  CA  a  b
10 9  8   7  6

a
f   b
g
e   c
d

1  2  3  4  5
e  d  CA  c  DP

MAN6760E

CA – common anode
DP – decimal point

(c)

## Liquid Crystal Display (LCD)

| | | | | |
|---|---|---|---|---|
| BP | 1 | | 40 | BP |
| Y | 2 | | 39 | X |
| K | 3 | | 38 | LO |
| NC | 4 | | 37 | NC |
| NC | 5 | | 36 | NC |
| NC | 6 | | 35 | NC |
| NC | 7 | | 34 | NC |
| dp1 | 8 | | 33 | NC |
| E3 | 9 | | 32 | G3 |
| D3 | 10 | Hundreds | 31 | F3 |
| C3 | 11 | | 30 | A3 |
| dp2 | 12 | | 29 | B3 |
| E2 | 13 | | 28 | L |
| D2 | 14 | Tens | 27 | G2 |
| C2 | 15 | | 26 | F2 |
| dp3 | 16 | | 25 | A2 |
| E1 | 17 | | 24 | B2 |
| D1 | 18 | Units | 23 | G1 |
| C1 | 19 | | 22 | F1 |
| B1 | 20 | | 21 | A1 |

FEO 203

(d)

**Figure 29.1** Device configurations for projects: (a) diodes; (b) light-emitting diodes (LEDs): red, green, yellow; (c) seven-segment display, CA—common anode, DP—decimal point; (d) liquid crystal display (LCD);

## Transistors

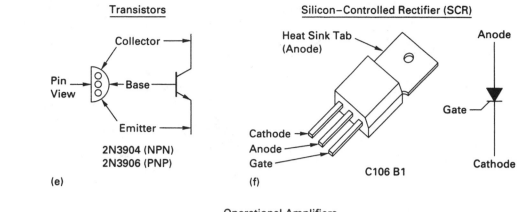

**Collector**

**Pin View** — **Base**

**Emitter**

2N3904 (NPN)
2N3906 (PNP)

(e)

## Silicon–Controlled Rectifier (SCR)

Heat Sink Tab (Anode)

Cathode
Anode
Gate

C106 B1

Anode

Gate

Cathode

(f)

## Operational Amplifiers

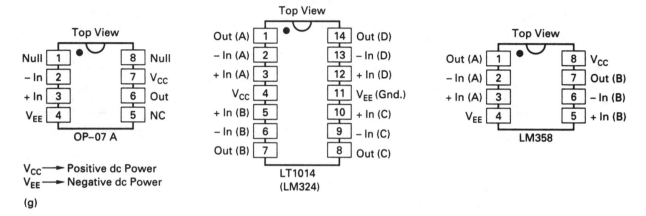

**Top View**

| | | | |
|---|---|---|---|
| Null | 1 | 8 | Null |
| – In | 2 | 7 | $V_{CC}$ |
| + In | 3 | 6 | Out |
| $V_{EE}$ | 4 | 5 | NC |

OP–07 A

$V_{CC}$ → Positive dc Power
$V_{EE}$ → Negative dc Power

(g)

**Top View**

| | | | |
|---|---|---|---|
| Out (A) | 1 | 14 | Out (D) |
| – In (A) | 2 | 13 | – In (D) |
| + In (A) | 3 | 12 | + In (D) |
| $V_{CC}$ | 4 | 11 | $V_{EE}$ (Gnd.) |
| + In (B) | 5 | 10 | + In (C) |
| – In (B) | 6 | 9 | – In (C) |
| Out (B) | 7 | 8 | Out (C) |

LT1014
(LM324)

**Top View**

| | | | |
|---|---|---|---|
| Out (A) | 1 | 8 | $V_{CC}$ |
| – In (A) | 2 | 7 | Out (B) |
| + In (A) | 3 | 6 | – In (B) |
| $V_{EE}$ | 4 | 5 | + In (B) |

LM358

## Digital Logic

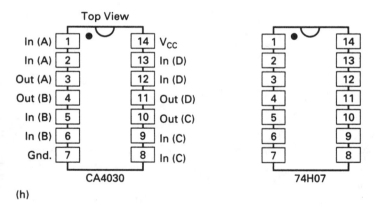

**Top View**

| | | | |
|---|---|---|---|
| In (A) | 1 | 14 | $V_{CC}$ |
| In (A) | 2 | 13 | In (D) |
| Out (A) | 3 | 12 | In (D) |
| Out (B) | 4 | 11 | Out (D) |
| In (B) | 5 | 10 | Out (C) |
| In (B) | 6 | 9 | In (C) |
| Gnd. | 7 | 8 | In (C) |

CA4030

| | | | |
|---|---|---|---|
| 1 | | | 14 |
| 2 | | | 13 |
| 3 | | | 12 |
| 4 | | | 11 |
| 5 | | | 10 |
| 6 | | | 9 |
| 7 | | | 8 |

74H07

(h)

**Figure 29.1** (Cont.) (e) transistors; (f) silicon-controlled rectifier (SCR); (g) operational amplifiers; (h) digital logic;

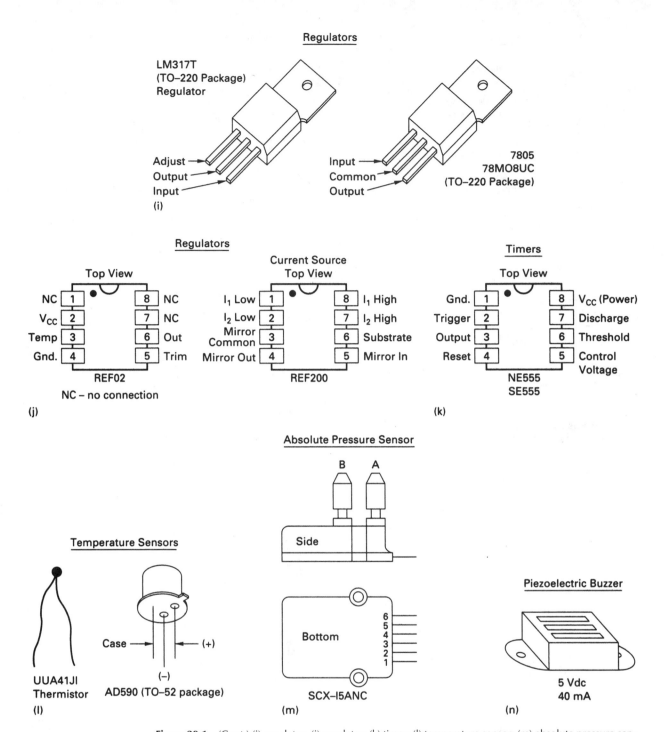

**Figure 29.1** (Cont.) (i) regulator; (j) regulator; (k) timer; (l) temperature sensor; (m) absolute pressure sensor; (n) transducer;

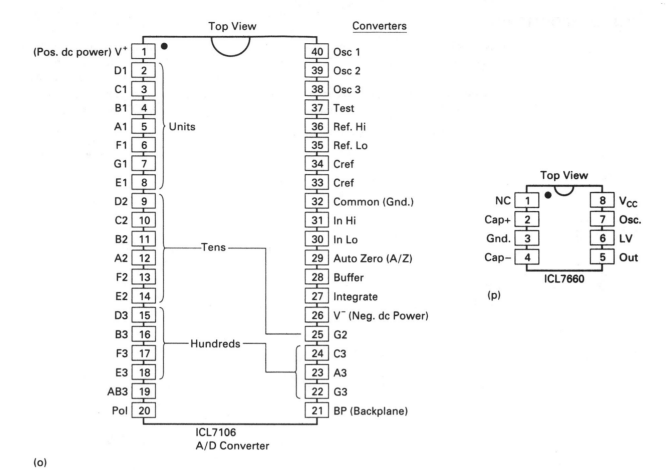

**Top View**                                   **Converters**

| (Pos. dc power) V⁺ | 1 | | | 40 | Osc 1 |
| D1 | 2 | | | 39 | Osc 2 |
| C1 | 3 | | | 38 | Osc 3 |
| B1 | 4 | | | 37 | Test |
| A1 | 5 | Units | | 36 | Ref. Hi |
| F1 | 6 | | | 35 | Ref. Lo |
| G1 | 7 | | | 34 | Cref |
| E1 | 8 | | | 33 | Cref |
| D2 | 9 | | | 32 | Common (Gnd.) |
| C2 | 10 | | | 31 | In Hi |
| B2 | 11 | | | 30 | In Lo |
| A2 | 12 | Tens | | 29 | Auto Zero (A/Z) |
| F2 | 13 | | | 28 | Buffer |
| E2 | 14 | | | 27 | Integrate |
| D3 | 15 | | | 26 | V⁻ (Neg. dc Power) |
| B3 | 16 | | | 25 | G2 |
| F3 | 17 | Hundreds | | 24 | C3 |
| E3 | 18 | | | 23 | A3 |
| AB3 | 19 | | | 22 | G3 |
| Pol | 20 | | | 21 | BP (Backplane) |

**ICL7106**
**A/D Converter**

(o)

**Top View**

| NC | 1 | | 8 | V_CC |
| Cap+ | 2 | | 7 | Osc. |
| Gnd. | 3 | | 6 | LV |
| Cap− | 4 | | 5 | Out |

**ICL7660**

(p)

**Top View**

| BCD Output ($2^1$) | 1 | | 16 | ($2^3$) BCD Output |
| BCD Output ($2^0$) | 2 | | 15 | ($2^2$) BCD Output |
| NSD | 3 | | 14 | V_CC |
| MSD | 4 | | 13 | Gain Adj. |
| LSD | 5 | | 12 | Cint |
| Hold/Bypass | 6 | | 11 | In Hi |
| Gnd. | 7 | | 10 | In Lo |
| Zero Adj. | 8 | | 9 | Zero Adj. |

Multiplex outputs* { NSD, MSD, LSD }

**CA3162E**

*MSD – most significant digit
 NSD – next significant digit
 LSD – least significant digit

(q)

**Top View**

| BCD Input ($2^1$) | 1 | | 16 | V_CC |
| BCD Input ($2^2$) | 2 | | 15 | f |
| Gnd. | 3 | | 14 | g |
| NC | 4 | | 13 | a |
| NC | 5 | | 12 | b |
| BCD Input ($2^3$) | 6 | | 11 | c |
| BCD Input ($2^0$) | 7 | | 10 | d |
| Gnd. | 8 | | 9 | e |

Segment Driver Outputs

**CA3161E**

NC-No Connection

**Figure 29.1**   (Cont.) (o-q) converters.

## 29.1 DARKROOM TIMER

This darkroom timer circuit employs the popular 555 timer IC, which is readily available from electronic parts suppliers. The timer features four *fixed time-delay* intervals, a *timing cycle in progress* indicator, a *timed-out* indicator, and separate *start* and *reset* controls. Although this timer circuit was designed to aid in photographic film developing, it is sufficiently flexible to be adapted to many purposes where a portable timer is necessary.

**Circuit Operation** Power is applied to the timer circuit shown in Fig. 29.2 by closing switch SW$_1$. The voltage at pin 3 of IC$_1$ is at approximately 0 volts. This forces the *timing cy-*

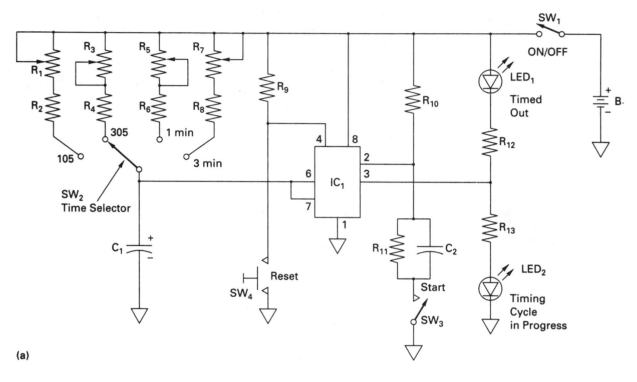

(a)

Parts List

IC$_1$ – NE555 timer
LED$_1$ – green LED
LED$_2$ – yellow LED
R$_1$, R$_3$ – 1-MΩ, trim potentiometer
R$_2$ – 220-kΩ, $\frac{1}{4}$-W, 5% resistor
R$_4$ – 1-MΩ, $\frac{1}{4}$-W, 5% resistor
R$_5$, R$_7$ – 1-MΩ, trim potentiometer
R$_6$ – 1.8-MΩ, $\frac{1}{4}$-W, 5% resistor
R$_8$ – 6.8-MΩ, $\frac{1}{4}$-W, 5% resistor
R$_9$, R$_{10}$ – 22-kΩ, $\frac{1}{4}$-W, 5% resistors
R$_{11}$ – 100-kΩ, $\frac{1}{4}$-W, 5% resistor
R$_{12}$, R$_{13}$ – 680-Ω, $\frac{1}{2}$-W, 10% resistors
*C$_1$ – 22-$\mu$F, 16-WVDC tantalum electrolytic capacitor
C$_2$ – 0.01-$\mu$F, 50-WVDC ceramic capacitor
SW$_1$ – SPST
SW$_2$ – four-position wafer switch
SW$_3$, SW$_4$ – momentary contact switches
B$_1$ – 9-V transistor battery and connector
Assorted hardware and chassis
*See text.

(b)

**Figure 29.2** Circuit schematic for the darkroom timer.

*cle in progress* indicator $LED_2$ off and the *timed-out* indicator $LED_1$ to be on. One of four time delays can be selected with the time-selector switch $SW_2$. The time delay is set by $C_1$ and the two-resistor combinations functioning with $SW_2$. The timer shown in Fig. 29.2 is designed for delays of 10 seconds, 30 seconds, 1 minute, and 3 minutes. Potentiometers $R_1$, $R_3$, $R_5$, and $R_7$ are used to adjust the individual time delays. A description on setting these potentiometers for the desired delays is given in the circuit calibration section.

Momentarily depressing the start switch $SW_3$ forces pin 2 of $IC_1$ to about 0 volts and begins the selected time delay. Pin 3 goes to a positive voltage, lighting $LED_2$ and shutting off $LED_1$. At the completion of the delay time, pin 3 returns to about 0 volts, turning on $LED_1$ (*timed-out* indicator) and shutting off $LED_2$ (*timing cycle in progress*).

Normally, $R_9$ holds pin 4 of $IC_1$ high at about 9 volts. If, however, the *reset* switch $SW_4$ is depressed during a timing cycle, pin 3 of $IC_1$ will go immediately to about 0 volts, which activates $LED_1$ and cancels the cycle. A new delay time cycle can then be initiated with the start switch $SW_3$. It is important that the timing capacitor $C_1$ have a good-quality tantalum or polystyrene dielectric.

**Circuit Calibration**  The time delays for the circuit of Fig. 29.2 are calculated using the relationship $T \cong 1.1RC_1$, where $T$ is in seconds, $C_1$ is in farads, and $R$ is the two-resistor combination $(R_1 + R_2)$ or $(R_3 + R_4)$ or $(R_5 + R_6)$ or $(R_7 + R_8)$ in ohms, determined by the position of the time selector switch $SW_2$.

The setting of a 10-second time delay will be used to demonstrate the calibration of the circuit. Resistor $R_1$ is first adjusted to one end of its range. The circuit is cycled and the delay timed. $R_1$ is then adjusted to the other end of its range and again the cycle is timed. These two readings will provide the range of the time delays possible for the two-resistor combination of $R_1$ and $R_2$. To obtain an exact 10-second delay time, $R_1$ is adjusted within the range and the time cycle recorded. This fine tuning is repeated until the exact 10-second delay results. To set and fine tune for the other three time-delay periods, the calibration procedure described is repeated for each position of the selector switch $SW_2$.

**Construction Hints**  There are no special construction problems with this project. As shown in Fig. 29.3, a suggested package is a small, commonly available chassis. Note

**Figure 29.3**  Suggested package for the darkroom timer.

that LED$_1$ and LED$_2$ are positioned to be prominently visible. Since the *start* and *reset* switches will require extensive use, SW$_3$ is positioned on the top and SW$_4$ on the front face of the package. The power switch SW$_1$ is placed along the side.

## 29.2 TTL LOGIC PROBE

Digital circuitry plays a major role in today's electronics industry. While new families of devices are introduced to keep up with the demands of changing technology, TTL (transistor-transistor-logic) devices maintain their dominance due in part to their ready availability and simplicity of use. A technician working with and troubleshooting digital circuits using these devices finds a time-saving instrument in a *TTL logic probe.* The probe described in this project uses readily available parts, is easy to build, and provides effective results. A single-sided printed circuit board for this project is detailed in Sec. 28.3.

**Circuit Operation** The TTL logic probe circuit, shown in Fig. 29.4, is designed around a LM358 low-power dual op amp capable of single-supply operation. Power for the probe is applied when the red alligator clip, connected to J$_1$, is attached to the +5-volt power supply bus of the digital circuit to be tested. The probe ground is made when the black alligator clip, connected to J$_2$, is attached to the digital circuit's ground bus. The purpose of diode D$_1$ is to prevent damage to the probe circuitry if these power supply connections are inadvertently reversed.

With power applied, the probe is calibrated as follows: The positive lead of a dc voltmeter is connected to test point 1 (TP$_1$) and the negative lead to J$_2$. Potentiometer R$_2$ is adjusted until a reading of exactly 0.8 volt appears at TP$_1$. This is the maximum allowable voltage level for a TTL logic 0. The positive lead of the voltmeter is next connected to test point 2 (TP$_2$), and potentiometer R$_4$ is adjusted for a voltage reading of 2.0 volts, which is the minimum allowable TTL logic 1 level.

The circuit can be tested by first contacting the probe to a known logic 0 (or to digital ground). The output of op amp B at pin 7 will go about +3.0 volts and the green light-emitting diode (LED$_2$) will light to indicate a logic 0 state. The red diode (LED$_1$) will remain off. The probe is next contacted to a known logic 1 (or to the digital power supply's +5-volt bus). This will cause the output of op amp A to go to about +3.0 volts, lighting LED$_1$, which indicates a logic 1 level. Resistors R$_5$ and R$_6$ limit the LED "on" current to about 10 mA to conserve power.

The purpose of R$_7$ is to maintain the probe voltage above 0.8 volt and below 2.0 volts, which is called the "forbidden" area for TTL logic. Thus, when the probe contacts neither the logic 0 nor logic 1 state, both of the LEDs will be off.

**Construction Hints** Construction of the TTL logic probe requires no special considerations. Fabrication of a pc board is recommended but is not essential. The probe package shown in Fig. 29.5 was constructed using a modified probe case. The sharp metal probe is isolated from the case and wired to terminal J$_3$ on the pc board with a short length of hookup wire.

Power (red) and ground (black) leads are AWG No. 24 stranded wire fitted with miniature alligator clips and extend into the probe case. All probe circuitry shown with the digital ground symbol $\bigtriangledown$ must be electrically connected together and to the black external ground lead J$_2$.

A final consideration in the packaging of the probe is in the composition of the case. If it is metal, care must be exercised to ensure that none of the circuitry touches it electrically. Any type of round hollow material, such as a short length of PCV piping or a hard plastic tube, may be used as the probe case.

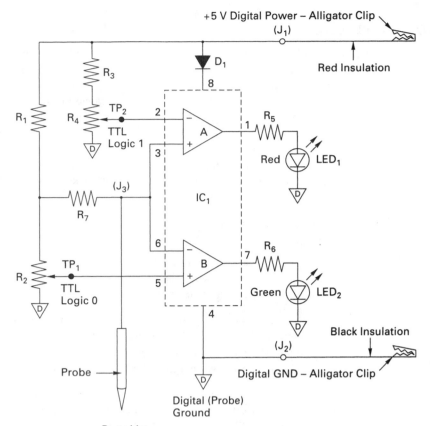

Parts List

IC$_1$ – LM358 low-power dual op amp
R$_1$ – 4.7-k$\Omega$, $^1/_4$-W, 10% resistor
R$_2$ – 2.5-k$\Omega$, 10-turn trim pot
R$_3$ – 2.7-k$\Omega$, $^1/_4$-W, 10% resistor
R$_4$ – 5-k$\Omega$, 10-turn trim pot
R$_5$, R$_6$ – 120-$\Omega$, $^1/_2$-W, 10% resistor
R$_7$ – 3.9-M$\Omega$, $^1/_4$-W, 10% resistor
LED$_1$ – light-emitting diode (RED)
LED$_2$ – light-emitting diode (GREEN)
D$_1$ – IN4001 diode (I$_0$ = 1A, PIV = 50 V)
        Two alligator clips and assorted hardware
Case – see text

**Figure 29.4**  TTL logic probe schematic.

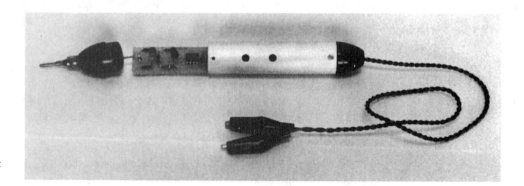

**Figure 29.5**  TTL logic probe package.

One of the most useful projects the experimenter can construct is a laboratory-type dc power supply. Applications of this type of circuit are almost unlimited. However, for a power supply to be most useful, it should have the following features: (1) adjustable output-voltage capability over a wide range, (2) high-load-current capability, (3) excellent voltage regulation at all load currents, (4) current limiting with short-circuit protection to prevent damage to the supply when accidental overloads or shorts are applied to the output terminals, and (5) metering to monitor output voltage.

The *variable-voltage*-regulated power supply circuit, shown in Fig. 29.6, will meet all of the foregoing requirements with the following specifications:

*Output voltage:* adjustable from 1.2 to 20 volts dc

*Output current:* 0 to 1 ampere

*Voltage regulation:* 0.1%

*Including:*    Load-current limiting
                Thermal-overload protection
                Safe-area-device protection

The requirements listed for a basic laboratory-type power supply are easily met with the use of an integrated regulator chip having only three external terminals: input (in), output (out), and adjustment (adj), shown in Fig. 29.6. The resulting circuit is (1) low in overall parts count, (2) inexpensive, and (3) simpler to package than a comparable-type discrete power supply.

**Circuit Operation**  For purposes of analysis, the power supply circuit schematic diagram, shown in Fig. 29.6, can be divided into two basic sections: an unregulated power supply and a voltage regulator section. The unregulated power supply consists of transformer $T_1$, diodes $D_1$ to $D_4$ in a full-wave bridge configuration, and the filter capacitor $C_1$.

The basic regulator section consists of the LM317T regulator chip and support components $R_2$, $R_3$, $R_4$, $C_2$, and $C_3$. The LM317T regulator chip requires that the output of the unregulated supply, across filter capacitor $C_1$, be limited to a maximum of 40 volts dc. This specification is easily met by $T_1$, since its theoretical peak output voltage under no-load current conditions is approximately 34.5 volts dc. Under full-load current conditions of 1 ampere, the dc output voltage of the unregulated supply is approximately 27 volts dc.

All of the electronics necessary to produce an adjustable, positive regulated power supply are contained in one three-terminal integrated circuit package, available in several standard package styles. The TO-220 plastic case was selected for this project. An internal reference voltage of 1.2 volts (constant) is established between the output (out) and the adjustment (adj) terminals of the LM317T regulator. This reference voltage is applied directly across the 240-ohm resistor ($R_2$) establishing a constant reference current of 5 mA flowing through $R_2$. This reference current also flows through the potentiometer ($R_3$) and parallel padding resistor $R_4$. An increase in the resistance of $R_3$ results in larger output voltages while a decrease in $R_3$ results in lower voltages. The minimum dc output voltage can never be lower than the reference voltage of 1.2 volts.

An ac ripple voltage is present on the input terminal of the regulator chip. At full-load currents of 1 ampere, this ripple can be as high as 5 volts peak to peak. On the output terminal of the regulator, there is essentially no ripple voltage (5 mV or less). The ripple voltage that is present at the input terminal is absorbed by the regulator.

Capacitor $C_2$ is a 0.1-$\mu$F disk required if the lead length from the unregulated supply terminal is excessive (2 inches or longer). It is, however, good practice to always include this capacitor. Capacitor $C_3$ (10-$\mu$F tantalum) is connected from the output terminal of the regulator to common to improve the transient response of the regulator.

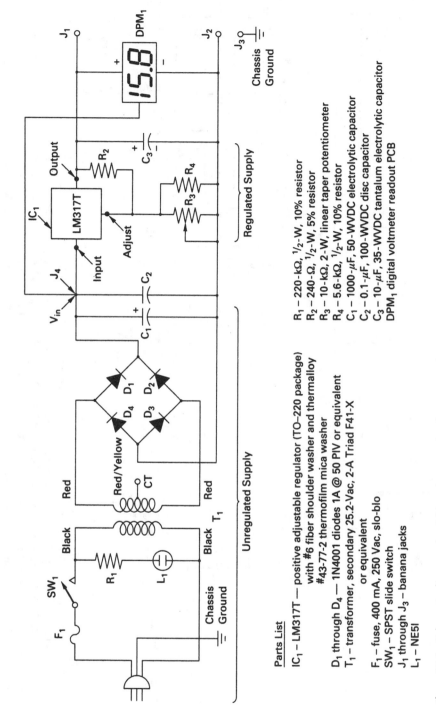

**Parts List**

IC$_1$ – LM317T — positive adjustable regulator (TO–220 package) with #6 fiber shoulder washer and thermally #43-77-2 thermofilm mica washer

D$_1$ through D$_4$ — 1N4001 diodes 1A @ 50 PIV or equivalent

T$_1$ – transformer, secondary 25.2-Vac, 2-A Triad F41-X or equivalent

F$_1$ – fuse, 400 mA, 250 Vac, slo-blo

SW$_1$ – SPST slide switch

J$_1$ through J$_3$ – banana jacks

L$_1$ – NE5l

R$_1$ – 220-k$\Omega$, $^1\!/_2$-W, 10% resistor
R$_2$ – 240-$\Omega$, $^1\!/_2$-W, 5% resistor
R$_3$ – 10-k$\Omega$, 2-W, linear taper potentiometer
R$_4$ – 5.6-k$\Omega$, $^1\!/_2$-W, 10% resistor
C$_1$ – 1000-$\mu$F, 50-WVDC electrolytic capacitor
C$_2$ – 0.1-$\mu$F, 100-WVDC disc capacitor
C$_3$ – 10-$\mu$F, 35-WVDC tantalum electrolytic capacitor
DPM$_1$ digital voltmeter readout PCB

**Figure 29.6**   Schematic diagram of variable-voltage-regulated power supply.

587

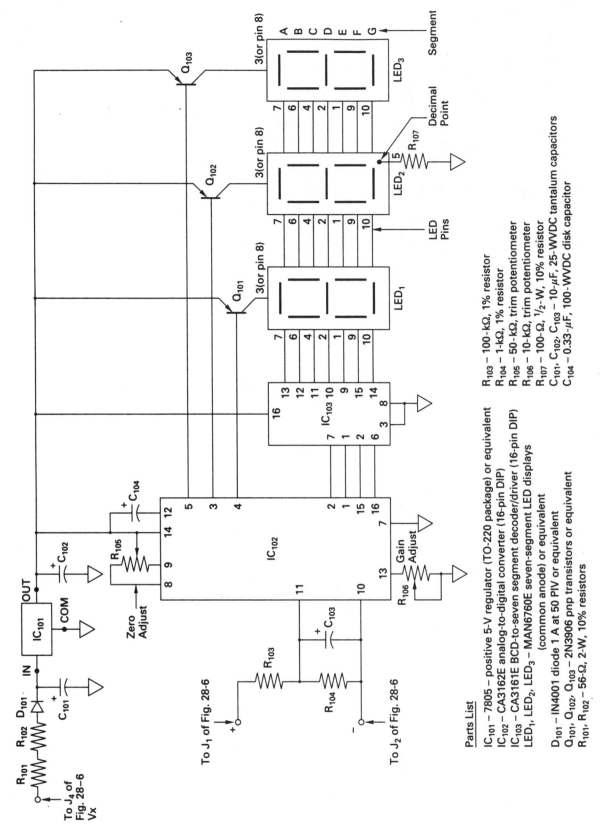

**Parts List**

IC101 – 7805 – positive 5-V regulator (TO-220 package) or equivalent
IC102 – CA3162E analog-to-digital converter (16-pin DIP)
IC103 – CA3161E BCD-to-seven segment decoder/driver (16-pin DIP)
LED1, LED2, LED3 – MAN6760E seven-segment LED displays
    (common anode) or equivalent

D101 – IN4001 diode 1 A at 50 PIV or equivalent
Q101, Q102, Q103 – 2N3906 pnp transistors or equivalent
R101, R102 – 56-Ω, 2-W, 10% resistors

R103 – 100-kΩ, 1% resistor
R104 – 1-kΩ, 1% resistor
R105 – 50-kΩ, trim potentiometer
R106 – 10-kΩ, trim potentiometer
R107 – 100-Ω, 1/2-W, 10% resistor
C101, C102, C103 – 10-μF, 25-WVDC tantalum capacitors
C104 – 0.33-μF, 100-WVDC disk capacitor

**Figure 29.7**  Schematic diagram of the digital panel meter readout for the regulated power supply project.

588

The dc output voltage appearing between terminals $J_1$ and $J_2$ is monitored with a digital panel meter (DPM), which is described here in detail.

**Digital Panel Meter** Recent advances in the development of integrated circuit technology have made it possible to replace older, less reliable analog panel meters with digital readout at lower costs and moderate increases in circuit complexity. To illustrate this, the power supply described in this project will be provided with a digital panel meter instead of the traditional analog meter. This DPM is designed around two 16-pin dual-in-line packaged ICs and support components as shown in Fig. 29.7. The core of this DPM system is the RCA CA3162E integrating analog-to-digital (A/D) converter and its companion, the CA3161E BCD-to-seven-segment decoder/driver IC.

Dc power for the DPM is provided from the unregulated output section at $J_4$ of Fig. 29.6. This voltage ($V_{in}$) is then converted by $IC_{101}$ (Fig. 29.7) to +5.0 volts.

The output voltage of the power supply to be monitored is between $J_1$ and $J_2$ of Fig. 29.6. This voltage is connected between the + and − input of the DPM (see Fig. 29.7). The voltage is divided by $R_{103}$ and $R_{104}$ and feeds into the A/D convertor ($IC_{102}$), which converts the analog voltage into a binary-coded decimal (BCD) equivalent that appears at pins 2, 1, 15, and 16. In turn, $IC_{103}$ takes this BCD code and converts it into a seven-line output that represents a decimal number. These seven-segment outputs are multiplexed to $LED_1$, $LED_2$, and $LED_3$. The digits are selected by pins 3, 4, and 5 of $IC_{102}$, which provide base drive to turn on $Q_{101}$, $Q_{102}$, or $Q_{103}$. These transistors then feed current to the anodes of the appropriate LED readouts.

The DPM assembled on a double-sided pc board is shown in Fig. 29.8. The DPM requires calibration prior to being integrated into the power supply project. This is done with the calibration setup shown in Fig. 29.9. A power supply of approximately 30 volts is connected between the $V_{in}$ and − terminals of the DPM. To zero the meter, the + and − inputs are shorted and $R_{105}$ is adjusted for an output reading of 00.0. To calibrate the meter, the

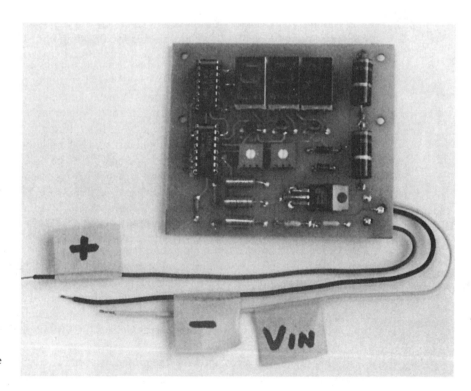

**Figure 29.8** Printed circuit board assembly of the digital panel meter (DPM) section of the power supply project.

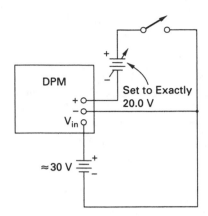

**Figure 29.9**
Calibration setup for the digital panel meter.

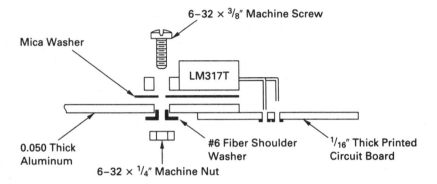

**Figure 29.10**
Assembly diagram of LM317T regulator package to chassis.

short is removed and a precise voltage, for example, 20 volts, is applied between the + and − inputs. Potentiometer $R_{106}$ is then adjusted for a reading of exactly 20.0 on the DPM.

**Construction Hints** For optimum performance, the power supply circuit should be packaged in a well-ventilated enclosure. The LM317T regulator package will require heat sinking to achieve the required specifications. A metal chassis element can be used as a suitable heat sink provided that it consists of at least 25 square inches of 1/16-inch-thick aluminum. The regulator chip must be electrically insulated from the chassis. This is done with the use of a silicon-greased mica washer and insulating shoulder washer (Fig. 29.10). Note that the leads are bent at 90 degrees onto a pc board. The pc board is used to assemble all the small fragile circuit components. A three-pronged plug is essential for electrical grounding of the chassis. The basic power supply package is shown in Fig. 29.11.

## 29.4 PORTABLE FAHRENHEIT/CELSIUS THERMOMETER WITH LCD READOUT

The accurate measurement of temperature requires a sensor that is linear and its readings repeatable. To satisfy both of these requirements, this electronic thermometer project is designed around the popular AD590 linear integrated circuit from Analog Devices, Inc. This device is a practical current source producing an output current that is directly proportional to temperature in degrees Kelvin (°K). Because temperature readings are more commonly read in degrees Fahrenheit (°F) and degrees Celsius (°C), additional electronic circuitry is required for conversion.

The thermometer designed for this project is portable, accurate, and measures temperature in the ranges of 0 to 199.9 degrees Fahrenheit and 0 to 100 degrees Celsius. Its electronics can be divided into three main subdivisions: the power supply circuit, the temperature sensor with its associated signal-conditioning circuitry (SCC), and an integrating-

**Figure 29.11** Basic package configuration for the power supply project.

type analog-to-digital converter (ADC) with its liquid crystal display (LCD) readout. Each of those circuits is described in detail.

**Universal Power Supply** A common design specification for portable hand-held equipment is that all of the electronics be powered from a single energy source such as a 9-volt transistor battery. While standard CMOS digital logic can operate at this voltage level, much of the current analog devices, such as operational amplifiers, require split (dual) voltage sources. The power supply circuit shown in Fig. 29.12 provides a regulated ±5-volt source from a 9-volt battery. When $SW_1$ (READ) is closed, $IC_2$ converts the +9-volt battery level into a precise +5.000 volts reference at pin 6 of $IC_2$. This reference at $J_1$ is thus made independent of reductions in battery voltage due to age and use. A negative nonregulated output of −9 volts is then developed with $IC_1$ and two noncritical 10-μF tantalum capacitors, $C_1$ and $C_2$. $IC_3$ regulates this output to −5 volts at $J_2$.

**Signal-Conditioning Circuit** The power supply circuit shown in Fig. 29.12 provides a precise +5-volt reference for both the AD590 temperature sensor and its associated bridge circuitry. This is shown in Fig. 29.13. Resistor $R_{101}$ converts the sensor's temperature-dependent current into a temperature-dependent voltage that is applied to the + input of the instrumentation amplifier (IA). The − input of the IA is returned to the null (zero offset) circuitry of the bridge reference through the scale selector switch $SW_2$. When $SW_2$ is *down*, the offset voltage at $TP_3$ is applied to the IA to zero the Fahrenheit scale. When $SW_2$ is *up*, the voltage at $TP_3$ is returned to the − input of the IA to zero the Celsius scale. The three-op amp IA, consisting of $IC_{102}$ A, B, and C and its associated circuitry, performs two major functions. First, it provides a single-ended output voltage to drive the input of the ADC voltmeter and second, it applies the correct amplification so that the output voltage of $TP_4$ is scaled to (10 mV/°F)$T_F$ or (10 mV/°C)$T_C$, depending on the position of $SW_2$.

**Circuit Calibration—Fahrenheit Scale** With power supply switch $SW_1$ closed and $SW_2$ down (°F position), a digital voltmeter is connected between $TP_1$ and circuit ground. At room temperature of approximately 77.5°F, the voltage across $R_{101}$ should be about 293 mV. By pinching the AD590 sensor between the index finger and thumb, the body temperature will increase this voltage at $TP_1$ by about 7 mV for a reading in the vicinity of 300 mV. With this basic test completed, the calibration procedure continues. The *offset adjustment* is made by connecting a digital multimeter (DMM) to $TP_2$ and adjusting $R_{103}$ for a

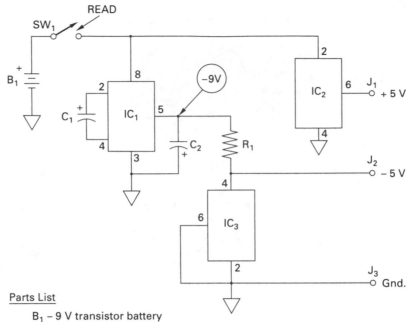

**Figure 29.12** Power supply electronics to develop ±5 volts for the thermometer project.

Parts List

    $B_1$ – 9 V transistor battery
    $SW_1$ – SPST pushbutton switch
    $IC_1$ – ICL7660 voltage converter
   $IC_2$, $IC_3$ – 5-V precision voltage reference ICs REF-02
    $C_1$, $C_2$ – 10-$\mu$F, 25-WVDC tantalum capacitors
    $R_1$ – 2.2-k$\Omega$, $^1/_4$-W, 10% resistor

voltage reading of precisely 255.4 mV. *Span adjustment* (or gain adjustment) requires that the DMM be connected to read the voltage of $TP_4$. The room temperature is then measured, in °F, with a high-quality laboratory thermometer, and the output voltage reading at $TP_4$ corresponding with this temperature is calculated using the relationship $V_o = (10$ mV/°F$)T_F$. For example, if the room temperature measures 77.5°F, then $V_o = (10$ mV/°F$)$ 77.5°F = 775 mV. For this example, $R_{109}$ would be adjusted for a reading at $TP_4$ of 775 mV. With a fixed decimal point positioned between the 7 and the 5, the digital voltmeter would provide a direct reading in °F (i.e., 77.5).

**Circuit Calibration—Celsius Scale** The *offset adjustment* is made by first placing $SW_2$ in the *up* position for the °C scale. A DMM is then connected to $TP_3$, and $R_{106}$ is adjusted for a voltage reading of exactly 273.2 mV. (No span adjustment is required on the Celsius scale.) The DMM is moved to $TP_4$ and, again, room temperature is precisely measured in °C. The corresponding voltage reading of $TP_4$ for the temperature is calculated with the equation $V_o = (10$ mV/°C$)T_C$. For example, if the room temperature is 24.5°C, then $V_o = (10$ mV/°C$)$ 24.5°C = 245 mV, which corresponds to 24.5°C.

**Circuit Description—3½-Digital Readout System** The readout system for the thermometer project is built around the familiar ICL7106 dual-slope integrating-type A/D converter and a 0.6-inch high-character LCD (see Fig. 29.14). Both the ADC and the LCD are housed in 40-pin dual-in-line packages. The supporting components have been selected to optimize circuit performance for a 0- to 199.9-mV full-scale input range and a corresponding output display.

Output signals from the SCC and $TP_4$ of Fig. 29.13 are applied to the A/D converter through a low-pass filter ($R_{201}$ and $C_{201}$) having a cutoff frequency of approximately 15.9 Hz. The input signals that are passed are processed by the dual-slope method of conversion employed by the 7106 IC. This conversion method samples input voltage and converts it into a current that charges capacitor $C_{202}$ for a fixed interval. At the end of this charging

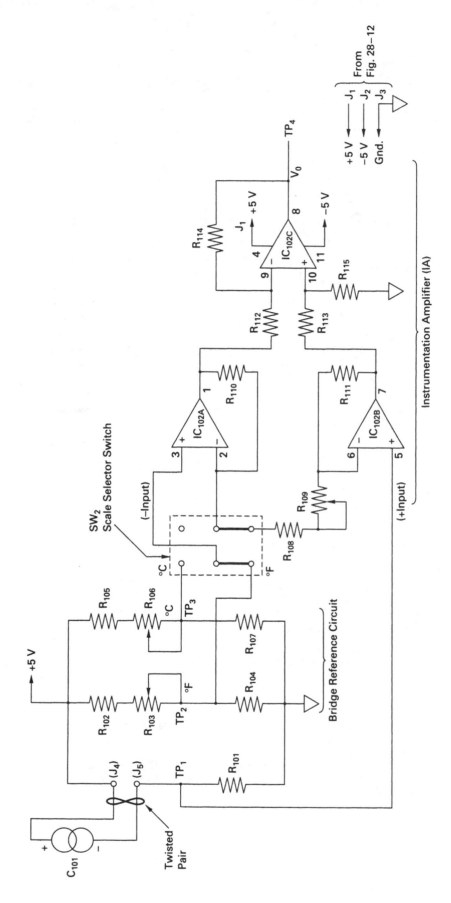

**Parts List**

IC$_{101}$ – AD590 two-terminal temperature sensor from Analog Devices
*IC$_{102}$ – LT1014 quad precision op amp from Linear Technology
R$_{101}$ – 1-kΩ, $^1/_8$-W, 1% resistor
R$_{102}$, R$_{105}$ – 20-kΩ, $^1/_8$-W, 1% resistors

R$_{103}$, R$_{106}$, R$_{109}$ – 25-kΩ, 10-turn potentiometers
R$_{104}$, R$_{107}$ – 2.49-kΩ, $^1/_8$-W, 1% resistor
R$_{108}$ – 12-kΩ, $^1/_8$-W, 5% resistor
R$_{110}$, R$_{111}$, R$_{112}$, R$_{113}$, R$_{114}$, R$_{115}$ – 10-kΩ, $^1/_8$-W, 1% resistors
SW$_2$ – DPDT miniature toggle switch

* The LM324 quad op amp from National Semiconductor is a substitute

**Figure 29.13** Sensor and signal conditioning circuitry (SCC) for the thermometer project.

593

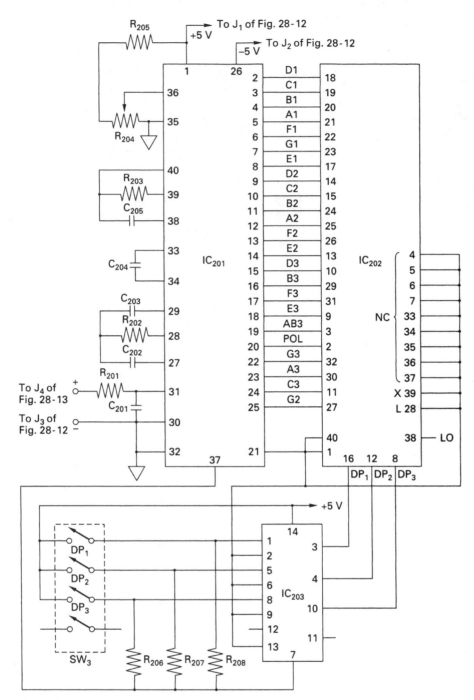

## Parts List

IC$_{201}$ – ICL7106 integrating-type A/D converter
IC$_{202}$ – FE0203 liquid crystal display (LCD)
IC$_{203}$ – CA4030 quad exclusive OR gate
R$_{201}$, R$_{206}$, R$_{207}$, R$_{208}$ – 1-MΩ, $\frac{1}{8}$-W, 10% resistors
R$_{202}$ – 47-kΩ, $\frac{1}{8}$-W, 10% resistor
R$_{203}$ – 100-kΩ, $\frac{1}{8}$-W, 10% resistor
R$_{204}$ – 1-kΩ, 10-turn potentiometer

R$_{205}$ – 24- kΩ, $\frac{1}{8}$-W, 5% resistor
C$_{201}$ – 0.01-μF, 50-WVDC disk capacitor
*C$_{202}$ – 0.22-μF capacitor
*C$_{203}$ – 0.47-μF capacitor
*C$_{204}$ – 0.1-μF, capacitor
C$_{205}$ – 100-pF, 50-WVDC disk capacitor
SW$_3$ – SPST, PC mount, switch block

*See text.

**Figure 29.14** Circuit schematic of the ICL7106 A/D converter and LCD readout.

time interval, the input signal is replaced by a precise reference voltage that is developed across $C_{204}$ and established externally by adjusting $R_{204}$. This reference voltage allows $C_{202}$ to be discharged at a constant rate until its charge is completely removed. The time for this discharge is therefore directly proportional to the value of the input voltage level and is measured with an internal counter operating from an internal clock.

To test that the internal clock is operating correctly, an oscilloscope (or counter) is connected between pins 37 and 32 (GND) of $IC_{201}$. With a $10\times$ probe, a clock frequency of 48 kHz $\pm$ 20% is an acceptable reading.

All digital decoding and drive necessary to interface to a low-power LCD is provided on the 7106 IC. The decimal-point position for the LCD is selected by depressing one of the three switches labeled *dp* on $SW_3$. For this project, $dp_3$ should be the choice so that the output of the SCC will be interpreted directly into units of °F or °C.

As a final note, it is recommended that capacitors $C_{202}$, $C_{203}$, and $C_{204}$ be manufactured with a polypropylene or Mylar dielectric.

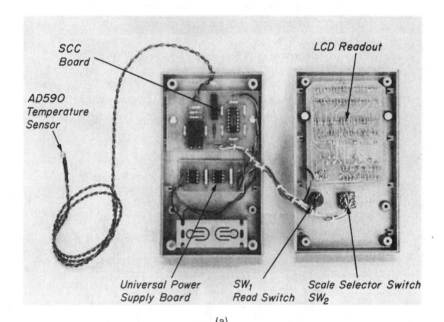

(a)

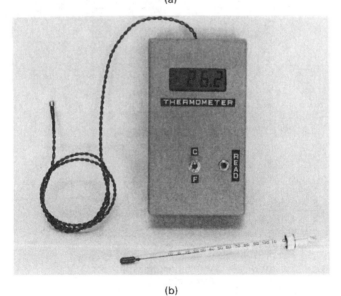

**Figure 29.15** Portable Fahrenheit/Celsius thermometer with liquid crystal display: (a) major pcb assemblies; (b) completed project.

(b)

**Final Circuit Calibration** To complete the final calibration of the digital readout system, a DMM is connected between pins 36 and 32 (GND). Potentiometer $R_{204}$ is adjusted until an exact reading of 100.0 mV is obtained. The system is then ready to accept the input signals from the SCC circuit.

**Construction Hints** The electronic thermometer shown in Fig. 29.15 is housed in a readily available plastic instrument box. All of the signal-conditioning circuitry was assembled on a 2- by 3-inch single-sided pc board that is piggybacked with a 3- by 5-inch double-sided board on which is packaged the ADC and LCD display. The universal power supply circuit is assembled on a 1- by 2-inch pc board. The AD590 temperature sensor, soldered to a 3-foot length of twisted pair AWG No. 24 wire, is external to the chassis, as shown in Fig. 29.15a.

To complete the package, a 1- by 2-inch cutout was formed in the plastic case to expose the LCD. A small DPDT switch ($SW_2$, the scale selector switch) is mounted on the front of the case and labeled. Last, the pushbutton SPST switch ($SW_1$) is also mounted on the case front. Depressing this button applies power and displays the temperature. Upon releasing the button, power is removed in order to conserve battery life. These are shown on the completed thermometer package in Fig. 29.15b.

## 29.5 PORTABLE SPORTSMAN'S BAROMETER WITH LCD READOUT

The sportsman's barometer project, shown in Fig. 29.16, is built around a solid-state absolute pressure sensor (type SCX15ANC from Sensym Corporation). The rugged construction of this sensor provides high reliability at moderate cost.

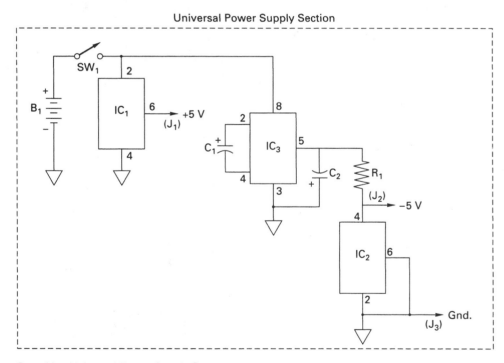

**Figure 29.16**
Complete circuit schematic and parts list for the sportsman's barometer project:
(a) universal power supply section;

**Parts List: Universal Power Supply Section**
$B_1$ – 9-V transistor battery
$SW_1$ – SPST switch
$IC_3$ – ICL7660 voltage converter
$IC_1$, $IC_2$ – 5-V precision voltage reference ICs REF-02
$C_1$, $C_2$ – 10-$\mu$F, 25-WVDC tantalum capacitors
$R_1$ – 2.2-k$\Omega$, $^1/_4$-W, 10% resistor

(a)

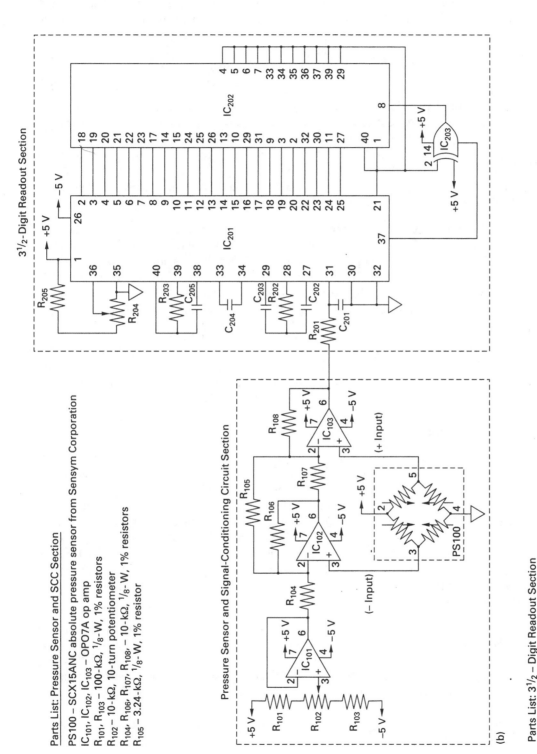

Parts List: Pressure Sensor and SCC Section

PS100 – SCX15ANC absolute pressure sensor from Sensym Corporation

IC₁₀₁, IC₁₀₂, IC₁₀₃ – OPO7A op amp
R₁₀₁, R₁₀₃ – 100-kΩ, 1/8-W, 1% resistors
R₁₀₂ – 10-kΩ, 10-turn potentiometer
R₁₀₄, R₁₀₆, R₁₀₇, R₁₀₈, – 10-kΩ, 1/8-W, 1% resistors
R₁₀₅ – 3.24-kΩ, 1/8-W, 1% resistor

Parts List: 3¹/₂ – Digit Readout Section

IC₂₀₁ – ICL7106 integrating-type A/D converter
IC₂₀₂ – FEO203 liquid crystal display (LCD)
IC₂₀₃ – CA4030 quad exclusive OR gate
R₂₀₁ – 1-MΩ, 1/8-W, 10% resistor
R₂₀₂ – 470-kΩ, 1/8-W, 10% resistor
R₂₀₃ – 100-kΩ, 1/8-W, 10% resistor
R₂₀₄ – 25-kΩ, 10-turn potentiometer

R₂₀₅ – 24- kΩ, 1/8-W, 5% resistor
C₂₀₁ – 0.01-µF, 50- WVDC disk capacitor
*C₂₀₂ – 0.22-µF capacitor
*C₂₀₃ – 0.047-µF capacitor
*C₂₀₄ – 0.1-µF capacitor
C₂₀₅ – 100-pF, 50-WVDC disk capacitor

*Use polypropylene or Mylar capacitors for best results

**Figure 29.16** (Cont.) (b) pressure sensor and signal conditioning circuit section.

597

A portable barometer has significant advantages to those who spend a great deal of time outdoors hiking, skiing, fishing, and so on, where changing weather conditions may alter one's plans suddenly. A drop in barometric pressure signifies that stormy weather is approaching, while a rising pressure indicates fair-weather conditions ahead. A steady pressure denotes unchanging weather patterns.

This portable barometer outputs absolute pressure digitally in units of inches of mercury (in. Hg) in a range from 28.0 to 32.0 in. Hg. As an aside, fishermen have found that the best fishing occurs when sea-level pressure is between 29.6 and 30.4 in. Hg and rising.

Note in Fig. 29.16 that both the universal power supply section and the 3 1/2-digit readout sections are similar to those described in the section on the thermometer project. They are shown here for purposes of completeness, but their operation will not be repeated.

**Circuit Operation—Pressure Sensor and Signal-Conditioning Circuit** Pressure sensor PS100 outputs a differential voltage between pins 5 and 3 that is directly proportional to absolute pressure. This voltage is applied to the instrumentation amplifier consisting of $IC_{102}$, $IC_{103}$, and resistors $R_{104}$, $R_{105}$, $R_{106}$, $R_{107}$, and $R_{108}$. $R_{105}$ is a precision 1% 3.24-k$\Omega$ resistor, which sets the IA's gain and scales the output at pin 6 of $IC_{103}$ into units of in. Hg. The output voltage of $IC_{103}$ is applied to the readout section for the digital display. Dc power for the three op amps as well as for the sensor is obtained from the universal power supply.

**Circuit Calibration** Prior to calibrating the sensor SCC, the reference voltage for the digital readout section should be adjusted just as it was in the thermometer project. A DMM is connected between pins 36 and 32 (GND) and $R_{204}$ is adjusted for an exact dc voltage reading of 1.00 volt. The readout section is now ready to accept input signals. To calibrate the barometer, it is first necessary to obtain the barometric pressure by calling the local airport or weather station. Closing $SW_1$ to apply power to the system, $R_{102}$ is adjusted until the digital readout displays the correct pressure.

**Construction Hints** The completed sportsman's barometer project is shown in Fig. 29.17. Note that it is packaged in the same type of plastic case as the previous project on the thermometer. For this reason, all of the construction hints presented in that section also apply to this project.

## 29.6 AUTOMOBILE SECURITY ALARM SYSTEM

Although no alarm system is foolproof, alarms do serve as a deterrent to automobile thefts, especially those thefts committed by the amateur. The most positive types of automobile alarm systems are those that employ mechanical door, hood, and trunk switches. Although this type of system is more difficult to install than the voltage- or current-sensing type of alarm systems, it is less susceptible to false triggering due to climate changes, extreme variations in temperature, or outside electrical interference. The most effective alarm systems are those that are completely concealed inside the automobile, thus giving potential thieves a false sense of security. The automobile security alarm system and parts list, shown in Figs. 29.18 and 29.19, is a reliable, relatively inexpensive system that will provide a high degree of security for an automobile.

**Circuit Operation** An automobile security alarm system is usually installed in the trunk of the automobile. All the required electrical connections are then made from this location.

Switch $SW_1$ is a DPDT slide or toggle switch that serves two functions: It "arms" or "disarms" the alarm system. In the "armed" position, the switch acts to connect the distributor side of the ignition points to ground, thus preventing the car from being started. This arm/disarm switch is usually mounted under the dashboard of the car, hidden from view but accessible to the driver.

The alarm system includes (1) an exit delay, (2) an entrance delay, (3) alarm and recycle, and (4) an "instant on" protection feature for both the hood and trunk. To activate the

**Figure 29.17**
Sportsman's barometer.

system, the arm/disarm switch $SW_1$ is moved from the "off" or "disarmed" position to the "armed" position. After the switch has been activated, the operator has a preset exit delay time of approximately 17 seconds to leave the car and lock the doors. (If closing the doors takes longer than 17 seconds, the horn will sound.) The exit delay time interval is set by the timing capacitor $C_3$ and resistor $R_2$. At the end of the exit delay cycle, the output of $IC_2$, at pin 3, goes from approximately 0 volts to 6 volts, applying power to the reset terminals of $IC_3$ and $IC_4$ at pin 4. The system is now armed and ready to detect an intruder entering any protected door. If desired, $LED_1$ can be connected between jacks $J_2$ and $J_3$ as shown in Fig. 29.18. $LED_1$ will light at the end of the 17-second exit delay interval, indicating that the system is armed. This light can be mounted on the dashboard of the car.

Once the circuit is armed, the door switch control circuitry, consisting of $R_4$, $R_5$, $Q_1$, $D_3$, $SCR_1$, $R_6$, $D_4$, $D_5$, $Q_4$, and $Q_2$, is also powered. As long as all door switches are open (door switches are open when car doors are closed since the door position depresses the switch plunger), transistor $Q_4$ remains on and acts as a short circuit across $C_7$, preventing $IC_3$ from operating. An intruder opening a door causes the switch plunger to release and closes the switch, which causes the collector of $Q_1$ to go high, firing the gate of $SCR_1$. The base of $Q_4$ will go to approximately 0 volts, causing its collector-to-emitter to act as an open circuit across $C_7$. The timing capacitor $C_7$ (associated with $IC_3$) begins to charge. The intruder now has 11 seconds to find the arm/disarm switch and return it to the "off" position. If the switch is not thrown within the preset entrance delay time of 11 seconds, the charge developed across $C_7$ (through resistor $R_9$) will cause the output of $IC_3$ (pin 3) to go high, thus triggering $IC_4$. With $IC_4$ triggered, its pin 3 goes high, turning on diode $D_8$, which places 6 volts across the coil of the relay, Ry. The relay switches from its normally open condition to cause

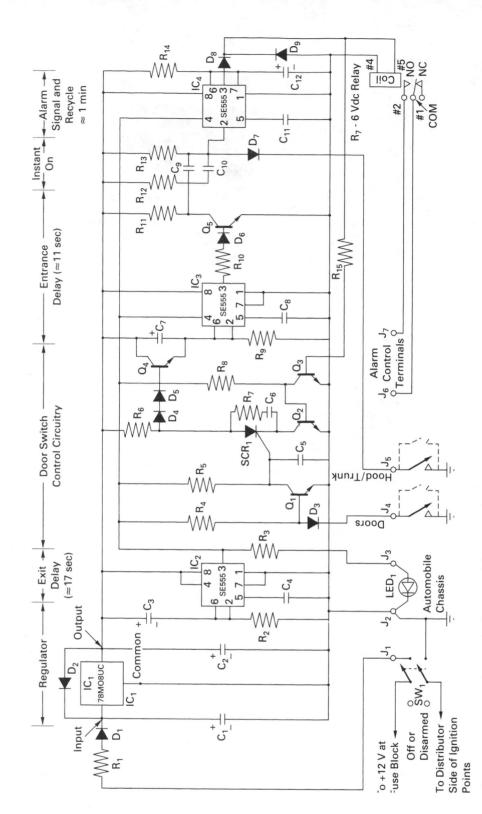

**Figure 29.18** Circuit schematic for the automobile security alarm system.

$IC_1$—78MO8UC positive 8 Vdc IC regulator
$IC_2$, $IC_3$, $IC_4$—SE555 timer
Ry—6 Vdc relay (Guardian series 1345 pc mount SPDT—NO)
$SCR_1$—C106B1
$Q_1$, $Q_2$, $Q_3$, $Q_4$, $Q_5$—2N3904 NPN transistor
$D_1$, $D_3$, $D_7$, $D_8$, $D_9$—1N4001 diode ($I_0$ = 1A PIV = 50 V)
$D_2$, $D_4$, $D_5$, $D_6$—1N914 diodes ($I_0$ = 10 mA PIV = 75 V)
$C_1$, $C_2$, $C_3$, $C_7$, $C_{12}$—10 μF @ 15 WVDC tantalum capacitors
$C_5$, $C_6$, $C_9$—0.1 μF disk capacitors
$C_4$, $C_8$, $C_{11}$—0.01 μF disk capacitors
$C_{10}$—0.001 μF disk capacitor
$R_1$—15 Ω, 1/2 W, 10%
$R_2$—1.5 MΩ, 1/2 W, 10%
$R_3$—220 Ω, 1/2 W, 10%
$R_4$—100 kΩ, 1/2 W, 10%
$R_5$, $R_{10}$—10 kΩ, 1/2 W, 10%
$R_6$—180 Ω, 1/2 W, 10%
$R_7$—10 Ω, 1/2 W, 10%
$R_8$, $R_{11}$, $R_{12}$, $R_{15}$—1 KΩ, 1/2 W, 10%
$R_9$—1.0 MΩ, 1/2 W, 10%
$R_{13}$—22 kΩ, 1/2 W, 10%
$R_{14}$—4.7 MΩ, 1/2 W, 10%
$SW_1$—DPDT switch
$LED_1$—Light emitting diode
Seven terminal barrier strip and assorted hardware

**Figure 29.19** Parts list for the automobile alarm project of Fig. 29.18.

a closed condition at points $J_6$ and $J_7$. As can be seen in Fig. 29.20a, b, or c, when $J_6$ and $J_7$ are shorted together by the switching action of relay Ry, they complete the alarm circuit. This causes the alarm siren or automobile horn to sound for a duration of approximately 1 minute, this time interval being set by $R_{14}$ and $C_{12}$. After the alarm sounds for 1 minute, the system will recycle. If, for example, during the 1-minute sound interval, the intruder is frightened off and closes the door initially opened, the system will recycle. Even though the door has been returned to its closed position, the sound will complete its 1-minute cycle, after which the system is set and ready to detect another door intrusion. If, on the other hand, the intruder leaves without closing the door initially opened, the alarm will sound for 1 minute, shut down for an 11-second interval, and then sound again for 1 minute. This 1 minute on/11 seconds off cycle will continue until the opened door is finally closed.

Since it is not advisable to have any delay on either the hood or trunk switches, the "instant-on" feature accomplished by $C_{10}$, $R_{12}$, and $D_7$ is included. Opening either the hood or trunk after the system has been armed will instantly sound the alarm for the same preset 1-minute interval as established by $R_{14}$ and $C_{12}$.

Diode $D_1$ has been added to prevent damage to any of the components in the system should the correct polarity be reversed when wiring the system into the automobile. In addition, an on-board voltage regulator circuit has been added. This circuit consists of $IC_1$, $C_1$, $C_2$, and $D_2$, and its purpose is to minimize any voltage spikes from the car's electrical system being transferred to the IC timers, causing damage or false triggering.

**Construction Hints** Because of the complexity of the circuit, it is strongly recommended that all components be assembled onto a pc board. The pc board should be mounted onto a chassis and protected by a sturdy enclosure. This unit is best secured in a safe position in the trunk.

Jack $J_2$ should be wired to the automobile chassis frame close to where the circuit enclosure is mounted. At least AWG No. 18 wire should be used and the connection tested to ensure that a good electrical contact has been made.

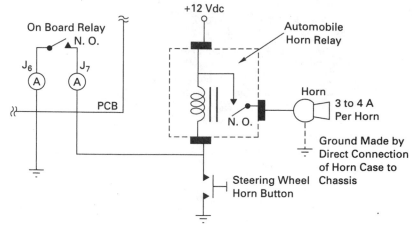

(a) One-Wire horn system

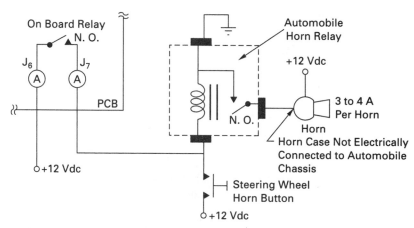

(b) Two-Wire horn system

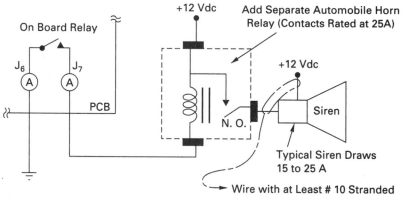

**Figure 29.20** Horn and siren wiring diagrams.

(c) Optional siren hook-up

It is generally unnecessary to buy switches for the door, trunk, or hood, since these are often standard equipment on many cars. All existing switches should be checked for perfect working order. If a switch must be installed at any of the locations, the plunger type is most suitable for this application. A newly installed switch must be wired in parallel with either those using exit and entrance delay (door switches) or instant-on (trunk or hood). For example, hood and trunk switches should be wired in parallel with each other but not in parallel with the door switches (Fig. 29.18). Since diode $D_3$ isolates the alarm circuit from the

12-volt source and all door switches are in parallel, only a single wire connection is necessary between $J_4$ and the terminal connection on one door switch. In like manner, only one wire is necessary between $J_5$ and either the trunk or hood switch.

**Caution** Never attempt to switch any automobile horn directly with the pc on-board relay. These horns draw in excess of 3 amperes each and should be controlled only from the car's horn relay. Depending on the type of automobile horn system, wiring according to Fig. 29.20a or b is recommended. As seen in Fig. 29.18, $J_6$ and $J_7$ have been left uncommitted for controlled switching of any particular type of horn circuit. This allows for convenient wiring connections to be made between $J_6$ and $J_7$ and any of the circuits shown in Fig. 29.20a, b, or c. If the car has no horn relay or if a siren is desired, an automobile horn relay must be installed. Sirens for alarm systems for use on a 12-volt dc source can draw in excess of 20 amperes. A horn relay capable of handling at least this much current must be selected. The siren, $+12$ volts, and ground connections should be made with a minimum of AWG No. 10 wire.

The finished security alarm package is shown in Fig. 29.21.

## 28.7 MICROCONTROLLER-BASED REFRIGERATION ALARM SYSTEM

This project illustrates an industrial trend in the area of data acquisition (DA) and process control, that is, the development of small, computer-based systems to first measure physical parameters, such as temperature, barometric pressure, and weight, and then input the measured analog signals into either a microprocessor ($\mu$P) or microcontroller ($\mu$C) for further data processing and/or control. Within the processor, the analog signal is immediately converted into a digital equivalent code that is stored in memory. It is this digital code that represents the physical measurement and is a necessary first step to create any computer-based system.

Microcontroller-based data acquisition and process control systems are commonplace today. They are found in such diverse fields as automotive testing, biomedical engineering, and the meteorological sciences, to name just a few.

The software needed to run a microcontroller-based process control system is called an *application program* and most application programs begin with software code, called a *driver,* that performs the actual measurement and stores the data in memory. Therefore, the driver acts as the software interface between input hardware (sensor and analog interface circuitry) and the code to be written to control the output devices and solve the engineering problem.

Application programs can be written to take the stored digital code and display it on a video monitor, send it to a printer, mathematically manipulate it to perform scaling, compare it against a previously input set-point value to make a control decision, such as turning an output device *on* or *off,* or any number of other operations limited only by the engineering problem to be solved and the programmer's imagination. It is the application software, running on the processor, controlling and manipulating the stored data, that provides the tremendous versatility within a microcontroller-based system. Often, changes to software are all that are needed to completely redefine a control system's entire function.

**Statement of the Problem** The object is to steadily monitor the internal temperature of a standard home refrigerator that is typically set below 38°F (3.3°C) for normal operation. When working correctly, a green status light must be *on* continuously. If a fault occurs and the refrigerator's temperature rises above the normal set-point temperature, the green status light should begin blinking and an audible alarm should be activated to alert the user to a potential problem—the possibility of food spoilage—if the difficulty is not immediately corrected.

**A Basic Overview of the Solution** As with any measurement and control problem, this project begins with the selection of an appropriate sensor to measure a physical quantity, in this case the normal refrigerator storage temperature of 38°F (3.3°C). A thermistor

(a)

**Figure 29.21** Completed alarm project: (a) alarm electronics mounted on a double-sided pcb; (b) chassis for alarm.

(b)

or *therm*ally sensitive res*istor* is a negative *temp*erature *co*efficient (TEMPCO) device and a good choice to measure temperature in this range and then convert it into an analog voltage. This voltage is then buffered and applied directly to one input of the on-chip analog-to-digital converter (ADC) of the microcontroller.

Two separate microcontroller output control lines (TTL compatible) will be needed to complete the project. One line will drive the green status indicator (LED) circuitry and the second will control a low-voltage, low-current, solid-state piezoelectric buzzer. In addition, the entire system will be powered by a single 5-volt dc voltage regulator.

Finally, an assembly-language application program is included to both acquire the temperature information and then solve the monitor/alarm problem and complete the project. The application program for this project compares the measured input data (in hexadecimal form) with the hexadecimal equivalent of 38°F (3.3°C). When the input voltage drops below the set-point value (temperature exceeds 38°F), the LED must begin to flash at a rate of approximately 1/2 hertz and the alarm should sound at the same rate.

Of course many other applications that solve useful engineering problems can be developed for the data acquisition and process control hardware presented in this project. The uses are virtually unlimited.

**Circuit Operation** Figure 29.22 shows the complete block diagram and circuit schematic for this refrigerator temperature monitor and alarm project, which was designed around Motorola's low-cost M68HC11 evaluation board (EVBU). The system hardware, external to the EVBU board, includes (1) a linear voltage regulator for system power, (2) sensor and interface circuitry to monitor temperature, and (3) status lamp and audio alarm. What follows is a detailed description of each hardware element group.

**Single 5-Volt dc Power Supply Circuitry** Dc power is supplied to all functions from a single three-terminal linear regulator, $IC_1$. Diode $D_1$ is added to prevent damage if power is applied incorrectly and the output of the $+5$-volt dc regulator is wired directly to connector $P_2$ on the EVBU printed circuit board. $R_1$ diverts a small amount of power (HEAT) from the IC regulator package.

**Sensor and Analog Interface Circuit** Figure 29.22b is the complete schematic needed to interface the output signal of a UUA41J1 thermistor ($T_1$) into one ADC input of the microcontroller. This circuit consists of a stable 100-$\mu$A current source ($IC_2$) thermistor sensor and a single-supply operational amplifier ($IC_3$) configured as a voltage-follower with a gain of $+1$.

$IC_2$ delivers a precise 100 $\mu$A to the grounded sensor, thermistor $T_1$. This drive current, $I$, is constant in value and independent of power supply variations. $IC_3$ buffers thermistor output voltage $V_T$, and presents a single-ended voltage ($V_o$) to the ADC input within its 5-volt range. To check operation of the sensor and analog interface circuit shown in Fig. 29.22b, apply dc power and connect a good-quality digital multimeter to pin 1 of $IC_3$ to measure $V_o$. Temporarily replace the thermistor with a decade-resistor box set to 27,600 ohms (thermistor's resistance at 38°F) and measure $V_o$. The voltage $V_o$ should equal $V_T = (I)R_s = (100 \ \mu\text{A}) (27.6 \ \text{K}) = 2.76$ volts. Replace the decade box with the thermistor when proper circuit operation is verified.

$R_4$ acts as a pull-up resistor, applying a 5-volt reference to the ADC's $V_{RH}$ pin (pin 52 on connector $P_4$). Capacitor $C_3$ is connected between the ADC's $V_{RH}$ and $V_{RL}$ pins and provides filtering. See the ADC wiring detail shown in Fig. 29.22b.

Access to the 8-bit ADC (via Port E) on the M68HC11 microcontroller chip is made available to the user at connector $P_4$ of the EVBU board. See the pin assignments for $P_4$ given in Fig. 29.23a. While there are 8 input channels on the M68HC11 chip, we will use only one, $PE_1$ (pin 45 on connector $P_4$), for this project.

**Green Status LED and Audio Alarm.** A medium-size green light-emitting diode ($LED_1$), used as a status indicator, is shown in Fig. 29.22a with accompanying drive circuitry. Output control line $PB_0$, accessible at pin 42 of connector $P_4$, is wired to the base of $Q_1$ through a 4.7-k$\Omega$ current-limiting resistor, and the LED's anode is connected to $Q_1$'s emitter. Under software control, when $PB_0$ is driven high (Logic 1), $Q_1$ conducts and the LED lights. A Logic 0 on $PB_0$ will cause the LED to be off.

In a similar fashion, port B's line $PB_4$ is used to control the audio alarm $BZ_1$. However, for this circuitry, the alarm will sound when $PB_4$ goes low to a Logic 0 and will be off when $PB_4$ goes high to a Logic 1. To supply 40 mA of drive current for $BZ_1$, two 7407 buffer/driver gates, $IC_{4A}$ and $IC_{4B}$, are connected in parallel as shown in Fig. 29.22a.

**Communicating with the M68HC11.** Note that Fig. 29.22 also shows the EVBU board connected to a personal computer's serial input through a ribbon cable connected to connector $P_2$. The pin assignments for connector $P_2$ are shown in

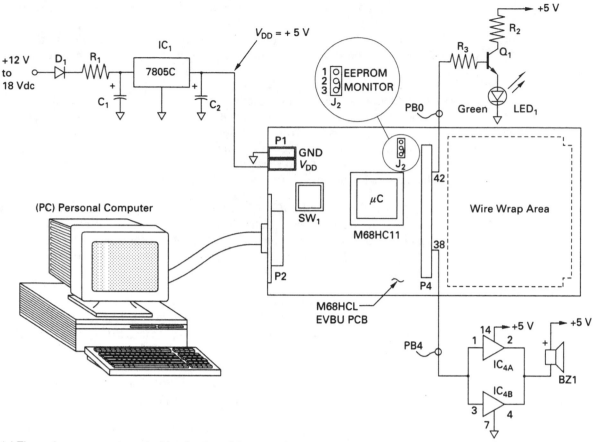

(a) The main components and wiring for the refrigerator alarm

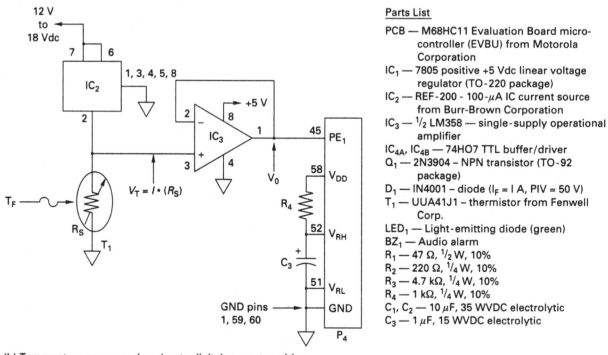

**Parts List**

PCB — M68HC11 Evaluation Board microcontroller (EVBU) from Motorola
Corporation

$IC_1$ — 7805 positive +5 Vdc linear voltage
regulator (TO-220 package)

$IC_2$ — REF-200 - 100-$\mu$A IC current source
from Burr-Brown Corporation

$IC_3$ — $\frac{1}{2}$ LM358 — single-supply operational
amplifier

$IC_{4A}$, $IC_{4B}$ — 74HO7 TTL buffer/driver

$Q_1$ — 2N3904 – NPN transistor (TO-92
package)

$D_1$ — IN4001 – diode ($I_F$ = I A, PIV = 50 V)

$T_1$ — UUA41J1 – thermistor from Fenwell
Corp.

$LED_1$ — Light-emitting diode (green)

$BZ_1$ — Audio alarm

$R_1$ — 47 $\Omega$, $\frac{1}{2}$ W, 10%

$R_2$ — 220 $\Omega$, $\frac{1}{4}$ W, 10%

$R_3$ — 4.7 k$\Omega$, $\frac{1}{4}$ W, 10%

$R_4$ — 1 k$\Omega$, $\frac{1}{4}$ W, 10%

$C_1$, $C_2$ — 10 $\mu$F, 35 WVDC electrolytic

$C_3$ — 1 $\mu$F, 15 WVDC electrolytic

(b) Temperature sensor and analog-to-digital converter wiring

**Figure 29.22** Microcontroller-based refrigerator alarm system.

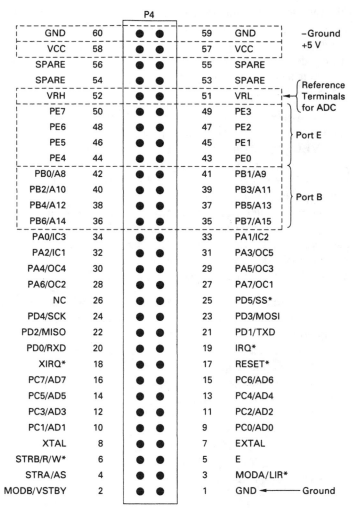

**P4**

| | | | | |
|---|---|---|---|---|
| GND | 60 | ● ● | 59 | GND |
| VCC | 58 | ● ● | 57 | VCC |
| SPARE | 56 | ● ● | 55 | SPARE |
| SPARE | 54 | ● ● | 53 | SPARE |
| VRH | 52 | ● ● | 51 | VRL |
| PE7 | 50 | ● ● | 49 | PE3 |
| PE6 | 48 | ● ● | 47 | PE2 |
| PE5 | 46 | ● ● | 45 | PE1 |
| PE4 | 44 | ● ● | 43 | PE0 |
| PB0/A8 | 42 | ● ● | 41 | PB1/A9 |
| PB2/A10 | 40 | ● ● | 39 | PB3/A11 |
| PB4/A12 | 38 | ● ● | 37 | PB5/A13 |
| PB6/A14 | 36 | ● ● | 35 | PB7/A15 |
| PA0/IC3 | 34 | ● ● | 33 | PA1/IC2 |
| PA2/IC1 | 32 | ● ● | 31 | PA3/OC5 |
| PA4/OC4 | 30 | ● ● | 29 | PA5/OC3 |
| PA6/OC2 | 28 | ● ● | 27 | PA7/OC1 |
| NC | 26 | ● ● | 25 | PD5/SS* |
| PD4/SCK | 24 | ● ● | 23 | PD3/MOSI |
| PD2/MISO | 22 | ● ● | 21 | PD1/TXD |
| PD0/RXD | 20 | ● ● | 19 | IRQ* |
| XIRQ* | 18 | ● ● | 17 | RESET* |
| PC7/AD7 | 16 | ● ● | 15 | PC6/AD6 |
| PC5/AD5 | 14 | ● ● | 13 | PC4/AD4 |
| PC3/AD3 | 12 | ● ● | 11 | PC2/AD2 |
| PC1/AD1 | 10 | ● ● | 9 | PC0/AD0 |
| XTAL | 8 | ● ● | 7 | EXTAL |
| STRB/R/W* | 6 | ● ● | 5 | E |
| STRA/AS | 4 | ● ● | 3 | MODA/LIR* |
| MODB/VSTBY | 2 | ● ● | 1 | GND |

–Ground
+5 V

Reference Terminals for ADC

Port E

Port B

GND ◄——— Ground

(a) Pin assignments for I/O connector, P4 on the EVBU PCB (top view)

**P2**

| | | | | |
|---|---|---|---|---|
| GND | 1 | ○   ○ | | |
| TXD | 2 | ○   ○ | 14 | NC |
| RXD | 3 | ○   ○ | 15 | NC |
| NC | 4 | ○   ○ | 16 | NC |
| CTS | 5 | ○   ○ | 17 | NC |
| DSR | 6 | ○   ○ | 18 | NC |
| SIG-GND | 7 | ○   ○ | 19 | NC |
| DCD | 8 | ○   ○ | 20 | DTR |
| NC | 9 | ○   ○ | 21 | NC |
| NC | 10 | ○   ○ | 22 | NC |
| NC | 11 | ○   ○ | 23 | NC |
| NC | 12 | ○   ○ | 24 | NC |
| NC | 13 | ○ | 25 | NC |

**Figure 29.23**
Transducers are interfaced to the ADC of the M68HC11 microcontroller at port E located on the $P_4$ connector. Port B output lines are also accessible at the $P_4$ connector. Communication between the μC and the PC is serial, using connector $P_2$.

(b) Pin designation for the P2 communications terminal on the EVBU board

Fig. 29.23b. For communication between the EVBU and the PC, the authors used a software communication program called KERMIT. It should be noted here also that once the applications software has been assembled into the EEPROM area of the microcontroller, the PC can be disconnected from the EVBU. This refrigerator monitor and alarm system is a stand-alone project requiring the PC only for software assembly, not for actual project usage.

**Construction Hints** Assembly of all the analog interface and output drive circuitry, shown in Fig. 29.22, is best accomplished on the small 3-inch by 3-inch wire-wrap area on the EVBU printed circuit board. Use the wire-wrap techniques discussed in Sec. 26.9 for component assembly and wiring. Using wire-wrap sockets for $LED_1$, driver $Q_1$, current source $IC_2$, buffer $IC_3$, and alarm buffer/driver $IC_4$ will greatly facilitate troubleshooting and repair. Power supply connections complete the wiring of the analog interface circuit to the microcontroller evaluation board. Once wired, the M68HC11 pcb is mounted in a commercially purchased case using standoffs, and small stick-on rubber feet complete the package. Figure 29.24 illustrates the complete system assembly. Note that no provision was made to make connector $P_2$ accessible from the case. The $P_2$ connector is needed only for software assembly, not for system utility.

**The Applications Program** The applications program needed to complete this project must first condition the ADC to accept data, then acquire data from the sensor and analog interface circuit, store the measured value in a memory location, and then perform a comparison to determine if the measured temperature is below 38°F (normal operation) or above the set point temperature. If the measured temperature is above 38°F, the LED must begin flashing and the audio alarm must sound. Figure 29.25 shows the flowchart to accomplish these tasks, and the actual assembly-language code is given in Fig. 29.26. In a later section, specific information will be given on how to assemble the applications program code into the M68HC11 microcontroller's EEPROM area of memory.

Starting at EEPROM memory location $B600, the first two lines of code send the correct control word to the μC's *options* register, instructing it on power up to activate the on-chip ADC. Then lines 3 and 4 turn on $LED_1$, indicating normal operation.

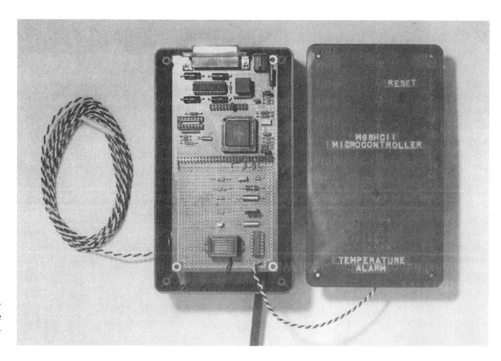

**Figure 29.24** The completed package for the microcontroller refrigeration alarm system.

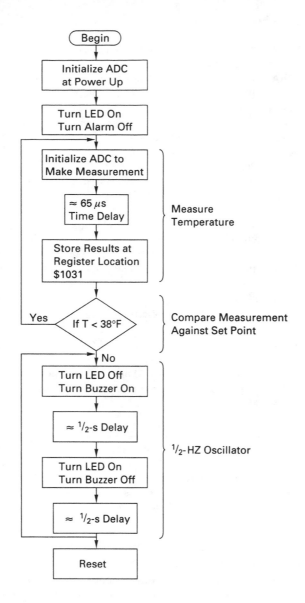

**Figure 29.25**
Flowchart details the applications program for the refrigerator project.

The flowchart in Fig. 29.25 shows the *driver* software beginning at line 5. The driver must first properly initialize the ADC control register and then provide a necessary 65-microsecond time delay to perform the analog-to-digital conversion. Note that the M68HC11 microcontroller's one-line assembler interprets all numbers as hexadecimal. Note also that we use the $ symbol here in the text to denote hexadecimal numbers, but the one-line assembler needs no such prefix. Figure 29.26 shows the code properly written without the $ prefix.

Lines 5 and 6 of the program immediately load the control word $01 (0000 0001) into accumulator A and then store it in the ADC control/status register at location $1030. The $01 control word sets the ADC for a single measurement at port PE$_1$. Next, lines 7, 8, and 9 establish a time delay of about 65 microseconds, giving the ADC sufficient time to make the conversion and then transfer the data to an internal data storage register at location $1031.

The hex equivalent of 38°F ($80) is next loaded in accumulator A and then temporarily stored at location $40 where it is compared with the measured value. The compare and branch instructions (lines 13 and 14) return the operation back to location $B60A for another measurement if the previous measurement was below 38°F or to line 15 if the temperature was above 38°F.

| | | | | |
|---|---|---|---|---|
| B600 | LDAA | #93 | \ Initialize ADC at power up | ;line 1 |
| | STAA | 1039 | / | ;line 2 |
| | LDAA | #01 | \ Set PB0 line high and PB4 low and | ;line 3 |
| | STAA | 1004 | / turn LED on, and alarm off | ;line 4 |
| B60A | LDAA | #01 | \ | ;line 5 |
| | STAA | 1030 | Measure refrigerator | ;line 6 |
| | LDAB | #1A | temperature | ;line 7 |
| B611 | DECB | | | ;line 8 |
| | BNE | B611 | / | ;line 9 |
| | LDAA | #C0 | \ Store 38 degrees (HEX $C0) | ;line 10 |
| | STAA | 40 | / in RAM memory | ;line 11 |
| | LDAA | 1031 | \ | ;line 12 |
| | CMPA | 40 | Compare set point against | ;line 13 |
| | BLS | B60A | / measurement | ;line 14 |
| B61F | LDAA | #01 | \ Set PB0 low and PB4 high | ;line 15 |
| | STAA | 1004 | / turn LED off and buzzer on | ;line 16 |
| | LDAA | #FF | \ | ;line 17 |
| | LDAB | #FF | | ;line 18 |
| B628 | DECA | | 1/2-Hertz oscillator | ;line 19 |
| | BNE | B628 | | ;line 20 |
| | DECB | | | ;line 21 |
| | BNE | B628 | / | ;line 22 |
| | LDAA | #00 | \ Set PB0 high and PB4 low | ;line 23 |
| | STAA | 1004 | / turn LED off and buzzer on | ;line 24 |
| | LDAA | #FF | \ | ;line 25 |
| | LDAB | #FF | | ;line 26 |
| B637 | DECA | | 1/2-Hertz oscillator | ;line 27 |
| | BNE | B637 | | ;line 28 |
| | DECB | | | ;line 29 |
| | BNE | B637 | / | ;line 30 |
| | BRA | B61F | Repeat osc. until reset | ;line 31 |
| | SWI | | Software interrupt | ;line 32 |

**Figure 29.26** The assembly language code needed to monitor refrigerator temperature and sound an alarm if temperature exceeds 38°F.

Oscillation of the LED and audio alarm follows. Lines 15 to 22 of this applications program switch off the LED and turn on the buzzer for approximately 1/2 second. Lines 23 to 30 switch the LED on and the buzzer off for approximately 1/2 second.

Line 31 completes the oscillation loop by returning the program to address $B61F to repeat the oscillation cycle.

**Loading the Applications Program.** To assemble an applications program into the microcontroller, connect your PC's serial communications port to connector $P_2$ of the EVBU, and use a communications software package like KERMIT. Then with power off, be sure that jumper $J_2$ is connected to the MONITOR position (pins 2 and 3 wired together) as shown in Fig. 29.22a.

We will assume that KERMIT is loaded in a subdirectory called MS-Kermit on your personal computer, and that you wish to load the application program into the M68HC11's EEPROM memory. Change to the KERMIT directory by typing the underscored information below:

C:\>KERMIT<CR>

Then at the KERMIT prompt, type

MS-Kermit>SET PORT 1<CR>

This instructs the software as to which serial communication port on your computer you have connected to $P_2$ of the EVBU. Port 1 is shown but port 2 should be substituted if it applies to your hardware. Then set the baud rate in software to match the pin selector rate on the EVBU. Again at the KERMIT prompt, type

MS-Kermit>SET BAUD 9600&lt;CR>

A baud rate of 9600 is shown, but rates of 4800 and 2400 can also be used, depending on which pins are strapped together on the EVBU. Finally, to connect to the M68HC11, type

C:>\Kermit>C&lt;CR>

Then pressing the Reset button on the EVBU will display the BUFFALO message:

BUFFALO 2.5 (ext) -<br/>Bit User Fast Friendly Aid to Logical Operation

Next key in a carriage return &lt;CR> to display the BUFFALO prompt (>). The EVBU is supplied with monitor/debugger firmware (in EPROM memory) called BUFFALO, which also includes a one-line assembler/disassembler. These features allow the user to program the μC. The **ASM** command allows the user to access the one-line assembler and enter assembly-language instructions, like the applications program shown in Fig. 29.26. To load the software, use the ASM command below followed by the starting address location you wish for the program:

>ASM B600&lt;CR>

You will see the specified address, followed by the current instruction or information at that location, and then the BUFFALO prompt on the next line. As an example,

B600 SBCA, $AA,X<br/>>

might be the current information at $B600. Refer again to Fig. 29.26 and load the instructions, following each with a carriage return. Note that the descriptive information and line numbers are not part of the assembly language program, but are presented as documentation only. For example, the first line of the driver program is loaded at location $B600, after the prompt as

>LDAA #93&lt;CR>

After the program is assembled, you are ready to exit the one-line assembler by holding down the **CTRL** key and simultaneously pressing the **A** key. The BUFFALO prompt is once again displayed.

**System Operation.** With power off, reconnect jumper $J_2$ to the EEPROM position (pins 1 and 2 wired together). When power is once again applied, the system will immediately jump to the starting EEPROM memory location at $B600 and begin measuring temperature.

## APPENDIX I

Comparison of Aluminum and Steel Sheet Metal Gauges

| Gauge Number | Aluminum (American or Brown & Sharpe) | Steel (U.S. Standard) |
|---|---|---|
| 0000000 | — | 0.500 |
| 000000 | 0.5800 | 0.46875 |
| 00000 | 0.5165 | 0.4375 |
| 0000 | 0.4600 | 0.40625 |
| 000 | 0.4096 | 0.375 |
| 00 | 0.3648 | 0.34375 |
| 0 | 0.3249 | 0.3125 |
| 1 | 0.2893 | 0.28125 |
| 2 | 0.2576 | 0.265625 |
| 3 | 0.2294 | 0.25 |
| 4 | 0.2043 | 0.234375 |
| 5 | 0.1819 | 0.21875 |
| 6 | 0.1620 | 0.203125 |
| 7 | 0.1443 | 0.1875 |
| 8 | 0.1285 | 0.171875 |
| 9 | 0.1144 | 0.15625 |
| 10 | 0.1019 | 0.140625 |
| 11 | 0.09074 | 0.125 |
| 12 | 0.08081 | 0.109375 |
| 13 | 0.07196 | 0.09375 |
| 14 | 0.06408 | 0.078125 |
| 15 | 0.05707 | 0.0703125 |
| 16 | 0.05082 | 0.0625 |
| 17 | 0.04526 | 0.05625 |
| 18 | 0.04030 | 0.05 |
| 19 | 0.03589 | 0.04375 |
| 20 | 0.03196 | 0.0375 |
| 21 | 0.02846 | 0.034375 |
| 22 | 0.02535 | 0.03125 |
| 23 | 0.02257 | 0.028125 |
| 24 | 0.02010 | 0.025 |
| 25 | 0.01790 | 0.021875 |
| 26 | 0.01594 | 0.01875 |
| 27 | 0.01420 | 0.0171875 |
| 28 | 0.01264 | 0.015625 |
| 29 | 0.01126 | 0.0140625 |
| 30 | 0.01003 | 0.0125 |
| 31 | 0.008928 | 0.0109375 |
| 32 | 0.007950 | 0.01015625 |
| 33 | 0.007080 | 0.009375 |
| 34 | 0.006305 | 0.00859375 |
| 35 | 0.005615 | 0.0078125 |
| 36 | 0.005000 | 0.00703125 |
| 37 | 0.004453 | 0.006640625 |
| 38 | 0.003965 | 0.00625 |
| 39 | 0.003531 | — |
| 40 | 0.003145 | — |

All decimal values in inches.

# APPENDIX II

Twist Drill Sizes and Decimal Equivalents of Number (Wire Gauge), Letter, and Fraction Size Drills

| Drill Size | Decimal Equivalent (in.) | Drill Size | Decimal Equivalent (in.) | Drill Size | Decimal Equivalent (in.) |
|---|---|---|---|---|---|
| 80 | 0.0135 | 40 | 0.0980 | 2 | 0.2210 |
| 79 | 0.0145 | 39 | 0.0995 | 1 | 0.2280 |
| $\frac{1}{64}$ in. | 0.0156 | 38 | 0.1015 | A | 0.2340 |
| 78 | 0.0160 | 37 | 0.1040 | $\frac{15}{64}$ in. | 0.2344 |
| 77 | 0.0180 | 36 | 0.1065 | B | 0.2380 |
| 76 | 0.0200 | $\frac{7}{64}$ in. | 0.1094 | C | 0.2420 |
| 75 | 0.0210 | 35 | 0.1100 | D | 0.2460 |
| 74 | 0.0225 | 34 | 0.1110 | E | 0.2500 |
| 73 | 0.0240 | 33 | 0.1130 | $\frac{1}{4}$ in. | 0.2500 |
| 72 | 0.0250 | 32 | 0.1160 | F | 0.2570 |
| 71 | 0.0260 | 31 | 0.1200 | G | 0.2610 |
| 70 | 0.0280 | $\frac{1}{8}$ in. | 0.1250 | $\frac{17}{64}$ in. | 0.2656 |
| 69 | 0.0292 | 30 | 0.1285 | H | 0.2660 |
| 68 | 0.0310 | 29 | 0.1360 | I | 0.2720 |
| $\frac{1}{32}$ in. | 0.0312 | 28 | 0.1405 | J | 0.2770 |
| 67 | 0.0320 | $\frac{9}{64}$ in. | 0.1406 | K | 0.2810 |
| 66 | 0.0330 | 27 | 0.1440 | $\frac{9}{32}$ in. | 0.2812 |
| 65 | 0.0350 | 26 | 0.1470 | L | 0.2900 |
| 64 | 0.0360 | 25 | 0.1495 | M | 0.2950 |
| 63 | 0.0370 | 24 | 0.1520 | $\frac{19}{64}$ in. | 0.2969 |
| 62 | 0.0380 | 23 | 0.1540 | N | 0.3020 |
| 61 | 0.0390 | $\frac{5}{32}$ in. | 0.1562 | $\frac{5}{16}$ in. | 0.3125 |
| 60 | 0.0400 | 22 | 0.1570 | O | 0.3160 |
| 59 | 0.0410 | 21 | 0.1590 | P | 0.3230 |
| 58 | 0.0420 | 20 | 0.1610 | $\frac{21}{64}$ in. | 0.3281 |
| 57 | 0.0430 | 19 | 0.1660 | Q | 0.3320 |
| 56 | 0.0465 | 18 | 0.1695 | R | 0.3390 |
| $\frac{3}{64}$ in. | 0.0469 | $\frac{11}{64}$ in. | 0.1719 | $\frac{11}{32}$ in. | 0.3438 |
| 55 | 0.0520 | 17 | 0.1730 | S | 0.3480 |
| 54 | 0.0550 | 16 | 0.1770 | T | 0.3580 |
| 53 | 0.0595 | 15 | 0.1800 | $\frac{23}{64}$ in. | 0.3594 |
| $\frac{1}{16}$ in. | 0.0625 | 14 | 0.1820 | U | 0.3680 |
| 52 | 0.0635 | 13 | 0.1850 | $\frac{3}{8}$ in. | 0.3750 |
| 51 | 0.0670 | $\frac{3}{16}$ in. | 0.1875 | V | 0.3770 |
| 50 | 0.0700 | 12 | 0.1890 | W | 0.3860 |
| 49 | 0.0730 | 11 | 0.1910 | $\frac{25}{64}$ in. | 0.3906 |
| 48 | 0.0760 | 10 | 0.1935 | X | 0.3970 |
| $\frac{5}{64}$ in. | 0.0781 | 9 | 0.1960 | Y | 0.4040 |
| 47 | 0.0785 | 8 | 0.1990 | $\frac{13}{32}$ in. | 0.4062 |
| 46 | 0.0810 | 7 | 0.2010 | Z | 0.4130 |
| 45 | 0.0820 | $\frac{13}{64}$ in. | 0.2031 | $\frac{27}{64}$ in. | 0.4219 |
| 44 | 0.0860 | 6 | 0.2040 | $\frac{7}{16}$ in. | 0.4375 |
| 43 | 0.0890 | 5 | 0.2055 | $\frac{29}{64}$ in. | 0.4531 |
| 42 | 0.0935 | 4 | 0.2090 | $\frac{15}{32}$ in. | 0.4687 |
| $\frac{3}{32}$ in. | 0.0937 | 3 | 0.2130 | $\frac{31}{64}$ in. | 0.4844 |
| 41 | 0.0960 | $\frac{7}{32}$ in. | 0.2187 | $\frac{1}{2}$ in. | 0.5000 |

# APPENDIX III

## Machine Screw Sizes and Decimal Equivalents

*unify national route*

| Gauge No. / Fraction | Maximum Diameter* (G) | Minimum Diameter* (T) | Threads per Inch | |
|---|---|---|---|---|
| | | | UNC | UNF |
| 0 | 0.0600 | 0.0438 | — | 80 |
| 1 | 0.0730 | 0.0527 | 64 | |
| | | 0.0550 | | 72 |
| 2† | 0.0860 | 0.0628 | 56 | |
| | | 0.0657 | | 64 |
| 3 | 0.0990 | 0.0719 | 48 | |
| | | 0.0758 | | 56 |
| 4† | 0.1120 | 0.0795 | 40 | |
| | | 0.0849 | | 48 |
| 5 $\frac{1}{8}$ | 0.1250 | 0.0925 | 40 | |
| | | 0.0955 | | 44 |
| 6† | 0.1380 | 0.0974 | 32 | |
| | | 0.1055 | | 40 |
| 8† | 0.1640 | 0.1234 | 32 | |
| | | 0.1279 | | 36 |
| 10† | 0.1900 | 0.1359 | 24 | |
| | | 0.1494 | | 32 |
| 12† | 0.2160 | 0.1619 | 24 | |
| | | 0.1696 | | 28 |
| $\frac{1}{4}$ | 0.2500 | 0.1850 | 20 | |
| | | 0.2036 | | 28 |
| $\frac{5}{16}$ | 0.3125 | 0.2403 | 18 | |
| | | 0.2584 | | 24 |
| $\frac{3}{8}$ | 0.3750 | 0.2938 | 16 | |
| | | 0.3209 | | 24 |
| $\frac{7}{16}$ | 0.4375 | 0.3447 | 14 | |
| | | 0.3726 | | 20 |
| $\frac{1}{2}$ | 0.5000 | 0.4001 | 13 | |
| | | 0.4351 | | 20 |
| $\frac{9}{16}$ | 0.5625 | 0.4542 | 12 | |
| | | 0.4903 | | 18 |
| $\frac{5}{8}$ | 0.6250 | 0.5069 | 11 | |
| | | 0.5528 | | 18 |
| $\frac{3}{4}$ | 0.7500 | 0.6201 | 10 | |
| | | 0.6688 | | 16 |
| $\frac{7}{8}$ | 0.8750 | 0.7307 | 9 | |
| | | 0.7822 | | 14 |
| 1 in. | 1.0000 | 0.8376 | 8 | |
| | | 0.9072 | | 14 |

*Diameters in inches.

†Sizes most frequently used in electronics.

# APPENDIX IV

## Clearance Hole Sizes and Hole Tolerance Requirements for Machine Screws

| Screw Size | Clearance Drill | *Decimal Equivalent of Drill (in.) | *Hole Tolerance (in.) |
|---|---|---|---|
| 0–80 | 51 | 0.0670 | 0.0070 |
| 1–64 | $\frac{5}{64}$ | 0.0781 | 0.0081 |
| 1–72 | | | |
| 2–56 | 42 | 0.0935 | 0.0075 |
| 2–64 | | | |
| 3–48 | 37 | 0.1040 | 0.0050 |
| 3–56 | | | |
| 4–40 | 31 | 0.1200 | 0.0080 |
| 4–48 | | | |
| 5–40 | 29 | 0.1360 | 0.0110 |
| 5–44 | | | |
| 6–32 | 27 | 0.1440 | 0.0060 |
| 6–40 | | | |
| 8–32 | 17 | 0.1730 | 0.0090 |
| 8–36 | | | |
| 10–24 | 8 | 0.1990 | 0.0090 |
| 10–32 | | | |
| 12–24 | 1 | 0.2280 | 0.0120 |
| 12–28 | | | |
| $\frac{1}{4}$–20 | G | 0.2610 | 0.0110 |
| $\frac{1}{4}$–28 | | | |
| $\frac{5}{16}$–18 | $\frac{21}{64}$ | 0.3281 | 0.0156 |
| $\frac{5}{16}$–24 | | | |
| $\frac{3}{8}$–16 | $\frac{25}{64}$ | 0.3906 | 0.0156 |
| $\frac{3}{8}$–24 | | | |
| $\frac{7}{16}$–14 | $\frac{29}{64}$ | 0.4531 | 0.0156 |
| $\frac{7}{16}$–20 | | | |
| $\frac{1}{2}$–13 | $\frac{33}{64}$ | 0.5156 | 0.0156 |
| $\frac{1}{2}$–20 | | | |
| $\frac{9}{16}$–12 | $\frac{37}{64}$ | 0.5781 | 0.0156 |
| $\frac{9}{16}$–18 | | | |
| $\frac{5}{8}$–11 | $\frac{41}{64}$ | 0.6406 | 0.0156 |
| $\frac{5}{8}$–18 | | | |
| $\frac{3}{4}$–10 | $\frac{25}{32}$ | 0.7812 | 0.0312 |
| $\frac{3}{4}$–16 | | | |
| $\frac{7}{8}$–9 | $\frac{29}{32}$ | 0.9062 | 0.0312 |
| $\frac{7}{8}$–14 | | | |

*All dimensions in inches.

# APPENDIX V
## Dimensional Information for Hexagonal Machine Screw Nuts

| Machine Screw Nut Size | Width Across Flats (in.) (W) | Thickness (in.) (T) |
| --- | --- | --- |
| 0–80 | $\frac{5}{32}$ | $\frac{3}{64}$ |
| 1–64 | | |
| 1–72 | $\frac{5}{32}$ | $\frac{3}{64}$ |
| 2–56 | $\frac{5}{32}$ | $\frac{1}{16}$ |
| 2–56 | | |
| 2–64 | $\frac{3}{16}$ | $\frac{1}{16}$ |
| 3–48 | | |
| 3–56 | $\frac{3}{16}$ | $\frac{1}{16}$ |
| 4–40 | $\frac{3}{16}$ | $\frac{1}{16}$ |
| 4–40 | | |
| 4–48 | $\frac{1}{4}$ | $\frac{3}{32}$ |
| 5–40 | $\frac{1}{4}$ | $\frac{3}{32}$ |
| 5–40 | | |
| 5–44 | $\frac{5}{16}$ | $\frac{7}{64}$ |
| 6–32 | $\frac{1}{4}$ | $\frac{3}{32}$ |
| 6–32 | | |
| 6–40 | $\frac{5}{16}$ | $\frac{7}{64}$ |
| | $\frac{1}{4}$ | $\frac{3}{32}$ |
| 8–32 | $\frac{5}{16}$ | $\frac{7}{64}$ |
| | $\frac{11}{32}$ | $\frac{1}{8}$ |
| 8–36 | $\frac{11}{32}$ | $\frac{1}{8}$ |
| 10–32 | $\frac{5}{16}$ | $\frac{1}{8}$ |
| | $\frac{11}{32}$ | $\frac{1}{8}$ |
| | $\frac{3}{8}$ | $\frac{1}{8}$ |
| 10–24 | $\frac{3}{8}$ | $\frac{1}{8}$ |
| 12–24 | $\frac{7}{16}$ | $\frac{5}{32}$ |
| $\frac{1}{4}$–20 | | |
| $\frac{1}{4}$–28 | $\frac{7}{16}$ | $\frac{3}{16}$ |
| $\frac{5}{16}$–18 | $\frac{9}{16}$ | $\frac{7}{32}$ |
| $\frac{5}{16}$–24 | | |
| $\frac{3}{8}$–16 | | |
| $\frac{3}{8}$–24 | $\frac{5}{8}$ | $\frac{1}{4}$ |

# APPENDIX VI
## Dimensional Information for Metal Washers

| Screw Size | Flat | | | Lock Washers Internal | | | Lock Washers External | | | Split | |
|---|---|---|---|---|---|---|---|---|---|---|---|
| | O.D. | I.D. | Thick. | O.D. | I.D. | Thick. | O.D. | I.D. | Thick. | W | T |
| 0 | 5/32 | 1/16 | 0.020 | — | — | — | — | — | — | 0.022 | 0.022 |
| 1 | 3/16 | 5/64 | 0.016 | — | — | — | — | — | — | 0.022 | 0.022 |
| | | | | | | | | | | 0.030 | 0.015 |
| | | | | | | | | | | 0.030 | 0.020 |
| | 7/32 | 3/32 | 0.018 | | | | | | | | |
| 2 | 1/4 | 3/32 | 0.016 | 0.200 | 0.095 | 0.015 | 0.275 | 0.095 | 0.019 | 0.030 | 0.015 |
| | 1/4 | 3/32 | 0.031 | | | | | | | 0.035 | 0.020 |
| | 7/32 | 7/64 | 0.022 | | | | | | | | |
| 3 | 1/4 | 7/64 | 0.016 | 0.225 | 0.109 | 0.015 | — | — | — | 0.035 | 0.020 |
| | 1/4 | 7/64 | 0.031 | | | | | | | 0.040 | 0.025 |
| | 1/4 | 1/8 | 0.017 | | | | | | | | |
| | 1/4 | 1/8 | 0.028 | | | | | | | 0.035 | 0.020 |
| 4 | 5/16 | 1/8 | 0.016 | 0.270 | 0.123 | 0.019 | 0.260 | 0.123 | 0.019 | 0.040 | 0.025 |
| | 5/16 | 1/8 | 0.025 | | | | | | | | |
| | 5/16 | 1/8 | 0.031 | | | | | | | 0.047 | 0.031 |
| | 3/8 | 1/8 | 0.036 | | | | | | | | |
| 5 | 9/32 | 9/64 | 0.025 | 0.280 | 0.136 | 0.019 | 0.280 | 0.136 | 0.019 | 0.040 | 0.025 |
| | | | | | | | | | | 0.047 | 0.031 |
| | 5/16 | 5/32 | 0.027 | | | | | | | 0.040 | 0.025 |
| | 11/32 | 11/64 | 0.015 | | | | | | | | |
| 6 | 3/8 | 5/32 | 0.016 | 0.295 | 0.150 | 0.021 | 0.320 | 0.150 | 0.022 | 0.047 | 0.031 |
| | 3/8 | 5/32 | 0.031 | | | | | | | | |
| | 3/8 | 5/32 | 0.036 | | | | | | | 0.055 | 0.040 |
| | 7/16 | 5/32 | 0.036 | | | | | | | | |
| | 3/8 | 11/64 | 0.016 | | | | | | | 0.047 | 0.031 |
| | 3/8 | 11/64 | 0.031 | | | | | | | | |
| 8 | 3/8 | 3/16 | 0.036 | 0.340 | 0.176 | 0.023 | 0.381 | 0.176 | 0.023 | 0.055 | 0.040 |
| | 1/2 | 3/16 | 0.036 | | | | | | | 0.062 | 0.047 |

| Screw Size | Flat O.D. | Flat I.D. | Flat Thick. | Internal O.D. | Internal I.D. | Internal Thick. | External O.D. | External I.D. | External Thick. | Split W | Split T |
|---|---|---|---|---|---|---|---|---|---|---|---|
| 10 | $\frac{3}{8}$ | $\frac{13}{64}$ | 0.047 | | | | | | | | |
| | $\frac{13}{32}$ | $\frac{13}{64}$ | 0.036 | | | | | | | 0.055 | 0.040 |
| | $\frac{7}{16}$ | $\frac{13}{64}$ | 0.031 | 0.381 | 0.204 | 0.025 | 0.410 | 0.204 | 0.025 | 0.062 | 0.047 |
| | $\frac{7}{16}$ | $\frac{13}{64}$ | 0.062 | | | | | | | | |
| | $\frac{7}{16}$ | $\frac{7}{32}$ | 0.036 | | | | | | | | |
| | $\frac{1}{2}$ | $\frac{7}{32}$ | 0.062 | | | | | | | 0.070 | 0.056 |
| | $\frac{9}{16}$ | $\frac{13}{64}$ | 0.036 | | | | | | | | |
| | $\frac{9}{16}$ | $\frac{1}{4}$ | 0.051 | | | | | | | | |
| 12 | $\frac{1}{2}$ | $\frac{15}{64}$ | 0.040 | 0.410 | 0.231 | 0.025 | 0.475 | 0.231 | 0.025 | 0.062 | 0.047 |
| | | | | | | | | | | 0.070 | 0.056 |
| $\frac{1}{4}$ | $\frac{1}{2}$ | $\frac{17}{64}$ | 0.031 | | | | | | | 0.107 | 0.047 |
| | $\frac{1}{2}$ | $\frac{17}{64}$ | 0.062 | | | | | | | | |
| | $\frac{1}{2}$ | $\frac{9}{32}$ | 0.056 | | | | | | | 0.109 | 0.062 |
| | $\frac{5}{8}$ | $\frac{9}{32}$ | 0.051 | 0.478 | 0.267 | 0.028 | 0.510 | 0.267 | 0.028 | | |
| | $\frac{5}{8}$ | $\frac{9}{32}$ | 0.062 | | | | | | | 0.110 | 0.077 |
| | $\frac{11}{16}$ | $\frac{17}{64}$ | 0.050 | | | | | | | | |
| | $\frac{3}{4}$ | $\frac{9}{32}$ | 0.056 | | | | | | | | |
| | $\frac{3}{4}$ | $\frac{5}{16}$ | 0.051 | | | | | | | | |
| | $\frac{9}{16}$ | $\frac{21}{64}$ | 0.031 | | | | | | | 0.117 | 0.056 |
| | $\frac{9}{16}$ | $\frac{21}{64}$ | 0.062 | | | | | | | | |
| | $\frac{11}{16}$ | $\frac{11}{32}$ | 0.051 | | | | | | | | |
| $\frac{5}{16}$ | $\frac{11}{16}$ | $\frac{11}{32}$ | 0.062 | 0.610 | 0.332 | 0.034 | 0.610 | 0.332 | 0.033 | 0.125 | 0.078 |
| | $\frac{5}{8}$ | $\frac{11}{32}$ | 0.056 | | | | | | | | |
| | $\frac{3}{4}$ | $\frac{11}{32}$ | 0.050 | | | | | | | 0.130 | 0.097 |
| | $\frac{7}{8}$ | $\frac{3}{8}$ | 0.064 | | | | | | | | |
| $\frac{3}{8}$ | $\frac{5}{8}$ | $\frac{25}{64}$ | 0.031 | | | | | | | 0.136 | 0.070 |
| | $\frac{5}{8}$ | $\frac{25}{64}$ | 0.062 | | | | | | | | |
| | $\frac{3}{4}$ | $\frac{13}{32}$ | 0.056 | | | | | | | | |
| | $\frac{13}{16}$ | $\frac{13}{32}$ | 0.062 | 0.692 | 0.398 | 0.040 | 0.694 | 0.398 | 0.040 | | |
| | $\frac{13}{16}$ | $\frac{13}{32}$ | 0.051 | | | | | | | | |
| | 1 | $\frac{7}{16}$ | 0.064 | | | | | | | 0.141 | 0.094 |

All dimensions in inches.

# APPENDIX VII

## Wire Table for Uninsulated Round Copper Conductors

| AWG* Number | Diameter (mils) (approximate) | Area (circular mils) | Maximum Recommended Current Capacity† (amperes) |
|---|---|---|---|
| 00 | 365 | 133,080 | |
| 0 | 325 | 105,560 | |
| 1 | 289 | 83,690 | |
| 2 | 257 | 66,360 | |
| 3 | 229 | 52,620 | |
| 4 | 204 | 41,740 | |
| 5 | 182 | 33,090 | |
| 6 | 162 | 26,240 | |
| 7 | 144 | 20,820 | |
| 8 | 128 | 16,510 | 33.0 |
| 9 | 114 | 13,090 | 26.0 |
| 10 | 102 | 10,380 | 20.5 |
| 11 | 90.7 | 8,226 | 16.5 |
| 12 | 80.8 | 6,529 | 12.0 |
| 13 | 72.0 | 5,184 | 10.0 |
| 14 | 64.1 | 4,109 | 8.0 |
| 15 | 57.1 | 3,260 | 6.5 |
| 16 | 50.8 | 2,580 | 5.0 |
| 17 | 45.3 | 2,052 | 4.0 |
| 18 | 40.3 | 1,624 | 3.0 |
| 19 | 35.9 | 1,289 | 2.5 |
| 20 | 32.0 | 1,024 | 2.0 |
| 21 | 28.5 | 812 | 1.5 |
| 22 | 25.3 | 640 | 1.0 |
| 23 | 22.6 | 511 | 1.0 |
| 24 | 20.1 | 404 | 0.80 |
| 25 | 17.9 | 320 | 0.60 |
| 26 | 15.9 | 253 | 0.50 |
| 27 | 14.2 | 202 | 0.40 |
| 28 | 12.6 | 159 | 0.30 |
| 29 | 11.3 | 128 | 0.25 |
| 30 | 10.0 | 100 | 0.20 |
| 31 | 8.9 | 79 | |
| 32 | 8.0 | 64 | |
| 33 | 7.1 | 50 | |
| 34 | 6.3 | 40 | |
| 35 | 5.6 | 31 | |
| 36 | 5.0 | 25 | |
| 37 | 4.5 | 20 | |
| 38 | 4.0 | 16 | |
| 39 | 3.5 | 12 | |
| 40 | 3.1 | 10 | |
| 41 | 2.8 | 8 | |
| 42 | 2.5 | 6 | |
| 43 | 2.2 | 5 | |
| 44 | 2.0 | 4 | |
| 45 | 1.8 | 3 | |
| 46 | 1.6 | 2.5 | |
| 47 | 1.4 | 2 | |
| 48 | 1.2 | 1.5 | |
| 49 | 1.1 | 1.2 | |
| 50 | 1.0 | 1.0 | |

*American Wire Gauge Standard (AWG).

†Current capacity based on 500 circular mils per ampere.

Proper exposure time for a specific resist is determined empirically for each exposure unit with the use of a *step table* such as the Stouffer 21-step exposure chart shown in Fig. A. This step table consists of a sheet of plastic with 21 areas (windows), each having increasing optical density. Step 1 is essentially clear and totally transparent, whereas step 21 is the most opaque. For any given exposure time and light source, the amount of light striking the resist surface is a function of the density of the window on the step table. Use of this table to determine the correct exposure requires a knowledge of the resist manufacturer's recommendations. Dynachem recommends the following for Laminar AX resist. After exposure and developing, the results of the resist exposed under a 21-step exposure table should be a "solid No. 4" with high-gloss green resist existing totally under that window. Under step 5, some resist will have been attacked during developing and the remaining resist under that window will be a dull-green color. Under step 6 and above to step 21, only clear copper will be visible.

Too long an exposure time will result in "high-gloss" resist occurring under a *higher* step number than 4 as a result of enough light having penetrated the more-opaque windows of the step table. This exposure time will result in resist brittleness. Too short an exposure time will result in high-gloss resist occurring under a *smaller* step number than 4 because of the opaqueness of the smaller step numbers being sufficient to block enough light from striking the resist. Incomplete resist polymerization results. This may appear, after developing, as a dull resist image or simply an incomplete image of the conductor pattern.

The use of a 21-step exposure table to determine the correct exposure time for a particular exposure unit will require that six small pc boards be cleaned, dried, preheated, laminated, and allowed to normalize to room temperature. (These processes are detailed in Chapter 12.) Each board is then exposed with the Stouffer exposure chart as the phototool for a predetermined amount of time with this time designated on the board. The first board should be exposed for 10 seconds, the second for 20 seconds, the third for 30 seconds, and so on, until all boards are exposed. After all boards have been normalized to room temperature for 10 minutes, they should be developed, rinsed, neutralized in a 10% sulfuric acid bath, rinsed, and dried. (Again, refer to Chapter 12.) Six properly processed samples are shown in Fig. B. The time required to obtain a "solid No. 4" on the step table is the proper exposure time for the specific resist and exposure employed.

**Figure A** Stouffer 21-step exposure chart.

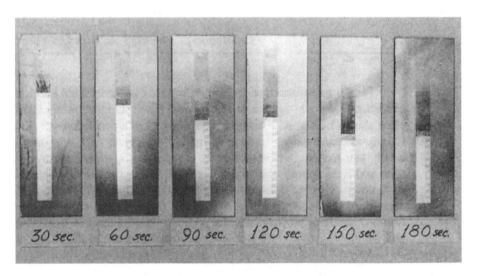

**Figure B** Exposure time to obtain a solid No. 4 is approximately 30 seconds.

As discussed in Chapters 11 and 13, plating solutions are extremely susceptible to organic and metallic contaminations. It is therefore essential that all contaminations be removed from all objects that will contact the plating solution. New plating tanks must be thoroughly leached before being used for the first time. The contaminants that must be removed from these tanks are cements, plastic welding residues, mold-release residues, machine oils, airborne particles, and simple dirts and greases. A new tank may look clean but is probably the single largest source of contamination that will result in plating failures.

Leaching involves first exposing the tank to an extremely strong alkaline solution followed by a relatively mild acid solution. The alkaline solution is used to remove the contaminations, and the acid solution neutralizes the alkaline residue.

A 1000-watt quartz heater with temperature control is first positioned in the tank with all hardware removed from the tank. The tank is filled with tap water to within 1/2 inch of the top rim. The following are then added to the tank:

Trisodium phosphate ($Na_3PO_4$), 6 ounces per gallon

Sodium hydroxide (NaOH)–caustic soda, 3 ounces per gallon

Plastic gloves should be worn when handling these strongly alkaline compounds. The chemicals are stirred into the solution with a plastic paddle and the temperature is raised to 130°F (54°C). The tank is subjected to this temperature for 8 hours, after which the heater is shut off, allowed to cool, and removed from the tank. The alkaline solution is then discarded. The tank is then *twice* filled with cold tap water and emptied.

To neutralize any remaining alkalinity, the tank is then filled with a 10% by volume solution of fluoboric acid. The solution is slowly stirred with a clean plastic paddle. This solution is allowed to remain in the tank for 24 hours at room temperature before being discarded. Again the tank is twice filled with cold tap water and emptied. The use of deionized water as a final rinse is recommended, but cold tap water will suffice if deionized water is unavailable.

The tank is now properly leached and ready for the plating solution.

## APPENDIX X

Conversions: Degrees Fahrenheit into Degrees Celsius

| °F* → | °C | °F* → | °C | °F* → | °C | °F* → | °C | Weights and Measures |
|---|---|---|---|---|---|---|---|---|
| 0 | −17.8 | 140 | 60.0 | 280 | 137.8 | 420 | 215.6 | 1 pound = 453.6 grams |
| 5 | −15.0 | 145 | 62.7 | 285 | 140.6 | 425 | 218.3 | 1 ounce = 28.35 grams |
| 10 | −12.2 | 150 | 65.6 | 290 | 143.3 | 430 | 221.1 | 1 gallon = 3.785 liters |
| 15 | − 9.4 | 155 | 68.3 | 295 | 146.1 | 435 | 223.9 | 1 quart = 946 milliliters |
| 20 | − 6.7 | 160 | 71.1 | 300 | 148.9 | 440 | 226.7 | 1 ounce = 29.57 milliliters |
| 25 | − 3.9 | 165 | 73.9 | 305 | 151.7 | 445 | 229.4 | 1 inch = 25.4 millimeters |
| 30 | − 1.1 | 170 | 76.7 | 310 | 154.4 | 450 | 232.2 | 1 foot = 304.8 millimeters |
| 35 | 1.7 | 175 | 79.4 | 315 | 157.2 | 455 | 235.0 | 1 square inch = 645.2 square millimeters |
| 40 | 4.4 | 180 | 82.2 | 320 | 160.0 | 460 | 237.8 | |
| 45 | 7.2 | 185 | 85.0 | 325 | 162.8 | 465 | 240.6 | |
| 50 | 10.0 | 190 | 87.8 | 330 | 165.6 | 470 | 243.3 | |
| 55 | 12.8 | 195 | 90.6 | 335 | 168.3 | 475 | 246.1 | |
| 60 | 15.6 | 200 | 93.3 | 340 | 171.1 | 480 | 248.9 | |
| 65 | 18.3 | 205 | 96.1 | 345 | 173.9 | 485 | 251.7 | |
| 70 | 21.1 | 210 | 98.9 | 350 | 176.7 | 490 | 254.4 | |
| 75 | 23.9 | 215 | 101.7 | 355 | 179.4 | 495 | 257.2 | |
| 80 | 26.7 | 220 | 104.4 | 360 | 182.2 | 500 | 260.0 | |
| 85 | 29.4 | 225 | 107.2 | 365 | 185.0 | 505 | 262.8 | |
| 90 | 32.2 | 230 | 110.0 | 370 | 187.8 | 510 | 265.6 | |
| 95 | 35.0 | 235 | 112.8 | 375 | 190.6 | 515 | 268.3 | |
| 100 | 37.8 | 240 | 115.6 | 380 | 193.3 | 520 | 271.1 | |
| 105 | 40.6 | 245 | 118.3 | 385 | 196.1 | 525 | 273.9 | |
| 110 | 43.3 | 250 | 121.1 | 390 | 198.9 | 530 | 276.7 | |
| 115 | 46.1 | 255 | 123.9 | 395 | 201.7 | 535 | 279.4 | |
| 120 | 48.9 | 260 | 126.7 | 400 | 204.4 | 540 | 282.2 | |
| 125 | 51.7 | 265 | 129.4 | 405 | 207.2 | 545 | 285.0 | |
| 130 | 54.4 | 270 | 132.2 | 410 | 210.0 | 550 | 287.8 | |
| 135 | 57.2 | 275 | 135.0 | 415 | 212.8 | 555 | 290.6 | |

*For each °F between those listed add 0.56°C.

$°C = \frac{5}{9}(°F - 32)$     $°F = \frac{9}{5}°C + 32$

   Celsius          Fahrenheit

*Example:* 74°F = 21.1°C + 4(0.56°C)

          74°F = 23.3°C

*(continued)*

| Decimals to Millimeters | | | | Fractions to Decimals to Millimeters | | | | | |
|---|---|---|---|---|---|---|---|---|---|
| Decimal | mm | Decimal | mm | Fraction | Decimal | mm | Fraction | Decimal | mm |
| 0.001 | 0.0254 | 0.500 | 12.7000 | $\frac{1}{64}$ | 0.0156 | 0.3969 | $\frac{33}{64}$ | 0.5156 | 13.0969 |
| 0.002 | 0.0508 | 0.510 | 12.9540 | $\frac{1}{32}$ | 0.0312 | 0.7938 | $\frac{17}{32}$ | 0.5312 | 13.4938 |
| 0.003 | 0.0762 | 0.520 | 13.2080 | $\frac{3}{64}$ | 0.0469 | 1.1906 | $\frac{35}{64}$ | 0.5469 | 13.8906 |
| 0.004 | 0.1016 | 0.530 | 13.4620 | | | | | | |
| 0.005 | 0.1270 | 0.540 | 13.7160 | | | | | | |
| 0.006 | 0.1524 | 0.550 | 13.9700 | $\frac{1}{16}$ | 0.0625 | 1.5875 | $\frac{9}{16}$ | 0.5625 | 14.2875 |
| 0.007 | 0.1778 | 0.560 | 14.2240 | | | | | | |
| 0.008 | 0.2032 | 0.570 | 14.4780 | | | | | | |
| 0.009 | 0.2286 | 0.580 | 14.7320 | $\frac{5}{64}$ | 0.0781 | 1.9844 | $\frac{37}{64}$ | 0.5781 | 14.6844 |
| | | 0.590 | 14.9860 | $\frac{3}{32}$ | 0.0938 | 2.3812 | $\frac{19}{32}$ | 0.5938 | 15.0812 |
| 0.010 | 0.2540 | | | | | | | | |
| 0.020 | 0.5080 | | | $\frac{7}{64}$ | 0.1094 | 2.7781 | $\frac{39}{64}$ | 0.6094 | 15.4781 |
| 0.030 | 0.7620 | | | | | | | | |
| 0.040 | 1.0160 | 0.600 | 15.2400 | $\frac{1}{8}$ | 0.1250 | 3.1750 | $\frac{5}{8}$ | 0.6250 | 15.8750 |
| 0.050 | 1.2700 | 0.610 | 15.4940 | | | | | | |
| 0.060 | 1.5240 | 0.620 | 15.7480 | | | | | | |
| 0.070 | 1.7780 | 0.630 | 16.0020 | $\frac{9}{64}$ | 0.1406 | 3.5719 | $\frac{41}{64}$ | 0.6406 | 16.2719 |
| 0.080 | 2.0320 | 0.640 | 16.2560 | $\frac{5}{32}$ | 0.1562 | 3.9688 | $\frac{21}{32}$ | 0.6562 | 16.6688 |
| 0.090 | 2.2860 | 0.650 | 16.5100 | | | | | | |
| | | 0.660 | 16.7640 | $\frac{11}{64}$ | 0.1719 | 4.3656 | $\frac{43}{64}$ | 0.6719 | 17.0656 |
| 0.100 | 2.5400 | 0.670 | 17.0180 | | | | | | |
| 0.110 | 2.7940 | 0.680 | 17.2720 | $\frac{3}{16}$ | 0.1875 | 4.7625 | $\frac{11}{16}$ | 0.6875 | 17.4625 |
| 0.120 | 3.0480 | 0.690 | 17.5260 | | | | | | |
| 0.130 | 3.3020 | | | | | | | | |
| 0.140 | 3.5560 | | | | | | | | |
| 0.150 | 3.8100 | | | $\frac{13}{64}$ | 0.2031 | 5.1594 | $\frac{45}{64}$ | 0.7031 | 17.8594 |
| 0.160 | 4.0640 | 0.700 | 17.7800 | $\frac{7}{32}$ | 0.2188 | 5.5562 | $\frac{23}{32}$ | 0.7188 | 18.2562 |
| 0.170 | 4.3180 | 0.710 | 18.0340 | $\frac{15}{64}$ | 0.2344 | 5.9531 | $\frac{47}{64}$ | 0.7344 | 18.6531 |
| 0.180 | 4.5720 | 0.720 | 18.2880 | | | | | | |
| 0.190 | 4.8260 | 0.730 | 18.5420 | | | | | | |
| | | 0.740 | 18.7960 | $\frac{1}{4}$ | 0.2500 | 6.3500 | $\frac{3}{4}$ | 0.7500 | 19.0500 |
| 0.200 | 5.0800 | 0.750 | 19.0500 | | | | | | |
| 0.210 | 5.3340 | 0.760 | 19.3040 | | | | | | |
| 0.220 | 5.5880 | 0.770 | 19.5580 | $\frac{17}{64}$ | 0.2656 | 6.7469 | $\frac{49}{64}$ | 0.7656 | 19.4469 |
| 0.230 | 5.8420 | 0.780 | 19.8120 | $\frac{9}{32}$ | 0.2812 | 7.1438 | $\frac{25}{32}$ | 0.7812 | 19.8438 |
| 0.240 | 6.0960 | 0.790 | 20.0660 | $\frac{19}{64}$ | 0.2969 | 7.5406 | $\frac{51}{64}$ | 0.7969 | 20.2406 |
| 0.250 | 6.3500 | | | | | | | | |

(*continued*)

**Decimals to Millimeters**

| Decimal | mm | Decimal | mm |
|---|---|---|---|
| 0.260 | 6.6040 | | |
| 0.270 | 6.8580 | | |
| 0.280 | 7.1120 | 0.800 | 20.3200 |
| 0.290 | 7.3660 | 0.810 | 20.5740 |
| | | 0.820 | 20.8280 |
| 0.300 | 7.6200 | 0.830 | 21.0820 |
| 0.310 | 7.8740 | 0.840 | 21.3360 |
| 0.320 | 8.1280 | 0.850 | 21.5900 |
| 0.330 | 8.3820 | 0.860 | 21.8440 |
| 0.340 | 8.6360 | 0.870 | 22.0980 |
| 0.350 | 8.8900 | 0.880 | 22.3520 |
| 0.360 | 9.1440 | 0.890 | 22.6060 |
| 0.370 | 9.3980 | | |
| 0.380 | 9.6520 | | |
| 0.390 | 9.9060 | | |
| | | 0.900 | 22.8600 |
| 0.400 | 10.1600 | 0.910 | 23.1140 |
| 0.410 | 10.4140 | 0.920 | 23.3680 |
| 0.420 | 10.6680 | 0.930 | 23.6220 |
| 0.430 | 10.9220 | 0.940 | 23.8760 |
| 0.440 | 11.1760 | 0.950 | 24.1300 |
| 0.450 | 11.4300 | 0.960 | 24.3840 |
| 0.460 | 11.6840 | 0.970 | 24.6380 |
| 0.470 | 11.9380 | 0.980 | 24.8920 |
| 0.480 | 12.1920 | 0.990 | 25.1460 |
| 0.490 | 12.4460 | 1.000 | 25.4000 |

**Fractions to Decimals to Millimeters**

| Fraction | Decimal | mm | Fraction | Decimal | mm |
|---|---|---|---|---|---|
| $\frac{5}{16}$ | 0.3125 | 7.9375 | $\frac{13}{16}$ | 0.8125 | 20.6375 |
| $\frac{21}{64}$ | 0.3281 | 8.3344 | $\frac{53}{64}$ | 0.8281 | 21.0344 |
| $\frac{11}{32}$ | 0.3438 | 8.7312 | $\frac{27}{32}$ | 0.8438 | 21.4312 |
| $\frac{23}{64}$ | 0.3594 | 9.1281 | $\frac{55}{64}$ | 0.8594 | 21.8281 |
| $\frac{3}{8}$ | 0.3750 | 9.5250 | $\frac{7}{8}$ | 0.8750 | 22.2250 |
| $\frac{25}{64}$ | 0.3906 | 9.9219 | $\frac{57}{64}$ | 0.8906 | 22.6219 |
| $\frac{13}{32}$ | 0.4062 | 10.3188 | $\frac{29}{32}$ | 0.9062 | 23.0188 |
| $\frac{27}{64}$ | 0.4219 | 10.7156 | $\frac{59}{64}$ | 0.9219 | 23.4156 |
| $\frac{7}{16}$ | 0.4375 | 11.1125 | $\frac{15}{16}$ | 0.9375 | 23.8125 |
| $\frac{29}{64}$ | 0.4531 | 11.5094 | $\frac{61}{64}$ | 0.9531 | 24.2094 |
| $\frac{15}{32}$ | 0.4688 | 11.9062 | $\frac{31}{32}$ | 0.9688 | 24.6062 |
| $\frac{31}{64}$ | 0.4844 | 12.3031 | $\frac{63}{64}$ | 0.9844 | 25.0031 |
| $\frac{1}{2}$ | 0.5000 | 12.7000 | $1$ | 1.0000 | 25.4000 |

*(continued)*

| | | | | Millimeters to Decimals | | | | | |
|------|---------|------|---------|------|---------|------|---------|------|---------|
| mm | Decimal | mm | Decimal | mm | Decimal | mm | Decimal | mm | Decimal |
| 0.01 | 0.00039 | 0.41 | 0.01614 | 0.81 | 0.03189 | 21 | 0.82677 | 61 | 2.40157 |
| 0.02 | 0.00079 | 0.42 | 0.01654 | 0.82 | 0.03228 | 22 | 0.86614 | 62 | 2.44094 |
| 0.03 | 0.00118 | 0.43 | 0.01693 | 0.83 | 0.03268 | 23 | 0.90551 | 63 | 2.48031 |
| 0.04 | 0.00157 | 0.44 | 0.01732 | 0.84 | 0.03307 | 24 | 0.94488 | 64 | 2.51969 |
| 0.05 | 0.00197 | 0.45 | 0.01772 | 0.85 | 0.03346 | 25 | 0.98425 | 65 | 2.55906 |
| 0.06 | 0.00236 | 0.46 | 0.01811 | 0.86 | 0.03386 | 26 | 1.02362 | 66 | 2.59843 |
| 0.07 | 0.00276 | 0.47 | 0.01850 | 0.87 | 0.03425 | 27 | 1.06299 | 67 | 2.63780 |
| 0.08 | 0.00315 | 0.48 | 0.01890 | 0.88 | 0.03465 | 28 | 1.10236 | 68 | 2.67717 |
| 0.09 | 0.00354 | 0.49 | 0.01929 | 0.89 | 0.03504 | 29 | 1.14173 | 69 | 2.71654 |
| 0.10 | 0.00394 | 0.50 | 0.01969 | 0.90 | 0.03543 | 30 | 1.18110 | 70 | 2.75591 |
| 0.11 | 0.00433 | 0.51 | 0.02008 | 0.91 | 0.03583 | 31 | 1.22047 | 71 | 2.79528 |
| 0.12 | 0.00472 | 0.52 | 0.02047 | 0.92 | 0.03622 | 32 | 1.25984 | 72 | 2.83465 |
| 0.13 | 0.00512 | 0.53 | 0.02087 | 0.93 | 0.03661 | 33 | 1.29921 | 73 | 2.87402 |
| 0.14 | 0.00551 | 0.54 | 0.02126 | 0.94 | 0.03701 | 34 | 1.33858 | 74 | 2.91339 |
| 0.15 | 0.00591 | 0.55 | 0.02165 | 0.95 | 0.03740 | 35 | 1.37795 | 75 | 2.95276 |
| 0.16 | 0.00630 | 0.56 | 0.02205 | 0.96 | 0.03780 | 36 | 1.41732 | 76 | 2.99213 |
| 0.17 | 0.00669 | 0.57 | 0.02244 | 0.97 | 0.03819 | 37 | 1.45669 | 77 | 3.03150 |
| 0.18 | 0.00709 | 0.58 | 0.02283 | 0.98 | 0.03858 | 38 | 1.49606 | 78 | 3.07087 |
| 0.19 | 0.00748 | 0.59 | 0.02323 | 0.99 | 0.03898 | 39 | 1.53543 | 79 | 3.11024 |
| 0.20 | 0.00787 | 0.60 | 0.02362 | 1.00 | 0.03937 | 40 | 1.57480 | 80 | 3.14961 |
| 0.21 | 0.00827 | 0.61 | 0.02402 | 1 | 0.03937 | 41 | 1.61417 | 81 | 3.18898 |
| 0.22 | 0.00866 | 0.62 | 0.02441 | 2 | 0.07874 | 42 | 1.65354 | 82 | 3.22835 |
| 0.23 | 0.00906 | 0.63 | 0.02480 | 3 | 0.11811 | 43 | 1.69291 | 83 | 3.26772 |
| 0.24 | 0.00945 | 0.64 | 0.02520 | 4 | 0.15748 | 44 | 1.73228 | 84 | 3.30709 |
| 0.25 | 0.00984 | 0.65 | 0.02559 | 5 | 0.19685 | 45 | 1.77165 | 85 | 3.34646 |
| 0.26 | 0.01024 | 0.66 | 0.02598 | 6 | 0.23622 | 46 | 1.81102 | 86 | 3.38583 |
| 0.27 | 0.01063 | 0.67 | 0.02638 | 7 | 0.27559 | 47 | 1.85039 | 87 | 3.42520 |
| 0.28 | 0.01102 | 0.68 | 0.02677 | 8 | 0.31496 | 48 | 1.88976 | 88 | 3.46457 |
| 0.29 | 0.01142 | 0.69 | 0.02717 | 9 | 0.35433 | 49 | 1.92913 | 89 | 3.50394 |
| 0.30 | 0.01181 | 0.70 | 0.02756 | 10 | 0.39370 | 50 | 1.96850 | 90 | 3.54331 |
| 0.31 | 0.01220 | 0.71 | 0.02795 | 11 | 0.43307 | 51 | 2.00787 | 91 | 3.58268 |
| 0.32 | 0.01260 | 0.72 | 0.02835 | 12 | 0.47244 | 52 | 2.04724 | 92 | 3.62205 |
| 0.33 | 0.01299 | 0.73 | 0.02874 | 13 | 0.51181 | 53 | 2.08661 | 93 | 3.66142 |
| 0.34 | 0.01339 | 0.74 | 0.02913 | 14 | 0.55118 | 54 | 2.12598 | 94 | 3.70079 |
| 0.35 | 0.01378 | 0.75 | 0.02953 | 15 | 0.59055 | 55 | 2.16535 | 95 | 3.74016 |
| 0.36 | 0.01417 | 0.76 | 0.02992 | 16 | 0.62992 | 56 | 2.20472 | 96 | 3.77953 |
| 0.37 | 0.01457 | 0.77 | 0.03032 | 17 | 0.66929 | 57 | 2.24409 | 97 | 3.81890 |
| 0.38 | 0.01496 | 0.78 | 0.03071 | 18 | 0.70866 | 58 | 2.28346 | 98 | 3.85827 |
| 0.39 | 0.01535 | 0.79 | 0.03110 | 19 | 0.74803 | 59 | 2.32283 | 99 | 3.89764 |
| 0.40 | 0.01575 | 0.80 | 0.03150 | 20 | 0.78740 | 60 | 2.36220 | 100 | 3.93701 |

## APPENDIX XI
### Resistance Color Code

| Band Colors | 1st Band (A) | 2nd Band (B) | 3rd Band (C) | 4th Band (D) |
|---|---|---|---|---|
| Black ⟶ | 0 | 0 | 0 | |
| Brown ⟶ | 1 | 1 | 10 | |
| Red ⟶ | 2 | 2 | 100 | |
| Orange ⟶ | 3 | 3 | 1,000 | |
| Yellow ⟶ | 4 | 4 | 10,000 | |
| Green ⟶ | 5 | 5 | 100,000 | |
| Blue ⟶ | 6 | 6 | 1,000,000 | |
| Violet ⟶ | 7 | 7 | — | |
| Gray ⟶ | 8 | 8 | — | |
| White ⟶ | 9 | 9 | — | |
| Gold ⟶ | — | — | 0.1 | ± 5% |
| Silver ⟶ | — | — | 0.01 | ± 10% |
| No color ⟶ | — | — | — | ± 20% |

Arrows point to possible numerical values for each BAND COLOR depending upon position

| Example: | R | = | (A) Yellow | before | (B) Violet | × | (C) Orange | ± | (D)% Silver |
|---|---|---|---|---|---|---|---|---|---|
| | R | = | 4 | | 7 | × | 1,000 | ± | 10% |

Engineering notation
1,000 = K
10,000 = 10 K
100,000 = 100 K
1,000,000 = M

R = 47,000 Ω ± 10% or 47 KΩ ± 10%

**Material Safety Data Sheet**
May be used to comply with
OSHA's Hazard Communication Standard,
29 CFR 1910.1200. Standard must be
consulted for specific requirements.

**U.S. Department of Labor**
Occupational Safety and Health Administration
(Non-Mandatory Form)
Form Approved
OMB No. 1218-0072

IDENTITY (*As Used on Label and List*)

*Note: Blank spaces are not permitted. If any item is not applicable, or no information is available, the space must be marked to indicate that.*

### Section I

| Manufacturer's Name | Emergency Telephone Number |
|---|---|
| Address (*Number, Street, City, State, and Zip Code*) | Telephone Number for Information |
| | Date Prepared |
| | Signature of Preparer (*optional*) |

### Section II – Hazardous Ingredients/Identity Information

| Hazardous Components (Specific Chemical Identity; Common Name(s)) | OSHA PEL | ACGIH TLV | Other Limits Recommended | % (*optional*) |
|---|---|---|---|---|
| | | | | |
| | | | | |
| | | | | |
| | | | | |
| | | | | |
| | | | | |
| | | | | |
| | | | | |
| | | | | |
| | | | | |

### Section III – Physical/Chemical Characteristics

| Boiling Point | | Specific Gravity (H$_2$O = 1) | |
|---|---|---|---|
| Vapor Pressure (mm Hg.) | | Melting Point | |
| Vapor Density (AIR = 1) | | Evaporation Rate (Butyl Acetate = 1) | |

Solubility in Water

Appearance and Odor

### Section IV – Fire and Explosion Hazard Data

| Flash Point (Method Used) | Flammable Limits | LEL | UEL |
|---|---|---|---|

Extinguishing Media

Special Fire Fighting Procedures

Unusual Fire and Explosion Hazards

(Reproduce locally)

OSHA 174, Sept. 1985

### Section V – Reactivity Data

| Stability | Unstable | | Conditions to Avoid |
|---|---|---|---|
| | Stable | | |

Incompatibility (*Materials to Avoid*)

Hazardous Decomposition or By-products

| Hazardous Polymerization | May Occur | | Conditions to Avoid |
|---|---|---|---|
| | Will Not Occur | | |

### Section VI – Health Hazard Data

| Route(s) of Entry: | Inhalation? | Skin? | Ingestion? |
|---|---|---|---|

Health Hazards (*Acute and Chronic*)

| Carcinogenicity | NTP? | IARC Monographs? | OSHA Regulated? |
|---|---|---|---|

Signs and Symptoms of Exposure

Medical Conditions
Generally Aggravated by Exposure

Emergency and First Aid Procedures

### Section VII – Precautions for Safe Handling and Use

Steps to Be Taken in Case Material is Released or Spilled

Waste Disposal Method

Precautions to Be Taken in Handling and Storing

Other Precautions

### Section VIII – Control Measures

Respiratory Protection (*Specify Type*)

| Ventilation | Local Exhaust | | Special |
|---|---|---|---|
| | Mechanical (*General*) | | Other |
| Protective Gloves | | Eye Protection | |

Other Protective Clothing or Equipment

Work/Hygienic Practices

* U.S.G.P.O. 1986-491-529/45775

Organizations

American National Standards Institute (ANSI)
11 W. 42nd Street, New York, NY 10036
Phone (212) 642-4900
(http://www.ansi.org)

Chemical Abstracts Service (CAS)
Division of American Chemical Society
Box 3012, Columbus, OH 43210
Substance Identification
(Identify chemical substance by CAS number and name)
Phone 1-800-753-4227
(http://www.cas.org)

Environmental Protection Agency (EPA)
401 M Street, SW, Washington, DC 20460
Phone (202) 260-2090
(http://www.epa.gov)

National Fire Protection Association (NFPA)
1 Batterymarch Park, Quincy, MA 02269
Phone (617) 770-3000
(http://www.nfpa.org)

National Institute of Occupational Safety and Health (NIOSH)
4676 Columbia Parkway, Cincinnati, OH 45226
Phone 1-800-356-1674
(http://www.cdc.gov/niosh/homepage.html)

Occupational Safety and Health Administration
(OSHA), Health Standards
200 Constitution Avenue, Washington, DC 20210
Phone (202) 219-7075
(http://www.osha.gov)

Superintendent of Documents
Government Printing Office
Washington, DC 20402
Phone (202) 512-1800
(http://www.access.gpo.gov/su_docs)

Distributors

Contact East, Inc.
335 Willow Street
North Andover, MA 01845-5595
Phone (978) 682-9844
(http://www.contacteast.com)

Global Occupational Safety
22 Harbor Park Drive, Dept. 7147
Port Washington, NY 11050
Phone 1-800-433-4848
(http://www.800-8-global.com)

Lab Safety Supply, Inc.
PO Box 1368, Janesville, WI 53547
Phone (608) 754-2345
Safety TechLine 1-800-356-2501
Catalog          1-800-356-0783
(http://www.labsafety.com)

Using an Excel Spreadsheet to Convert Thermocouple Voltage into Temperature °C

Converting the nonlinear output voltage of a thermocouple into a human-engineered linear temperature scale is made easy with any number of today's spreadsheet software packages. For example, and by way of illustration, the *Trendline* feature in the *Chartwizard* function of Microsoft Excel 5.0, can be used to generate a second-order polynomial equation that will accept a reference-junction corrected thermocouple voltage at its input ($x$ in Excel or $A$ in HP VEE notation) and output ($y$) the corresponding equivalent temperature in either Celsius or Fahrenheit units. The sequential steps necessary to generate a voltage-to-temperature conversion equation from a set of data points are covered and illustrated in this appendix.

First, open a new Excel Spreadsheet and load the data, that is, a series of $x, y$ points. Using a Type J thermocouple table (available from a manufacturer such as Omega Engineering, Inc., see Appendix XV) load thermoelectric voltages, $x$-axis values in column 1, and their corresponding temperatures, $y$-axis values, in column 2. Note from the data presented in this appendix that when the Type J thermocouple outputs 0.001277 V (1.27 mV) it is sensing 25°C, and when its output is 0.023225 V (23.225 mV) it is sensing 425°C.

Next highlight the data in columns 1 and 2, call up *Chartwizard* and perform the following five steps:

**Step 1** Verify that the proper data has been highlighted and then point and click the left mouse button on *NEXT* ($\rightarrow$ NEXT)

**Step 2** Select the *XY Scatter* chart type. $\rightarrow$ NEXT

**Step 3** Select chart type *3.* $\rightarrow$ NEXT

**Step 4** Verify. $\rightarrow$ NEXT

**Step 5** Complete the following chart entries. (See the example chart in this appendix.)

(a) *Chart Title:* VOLTAGE-TO-TEMPERATURE CONVERSION EQ.

(b) *Category [X]:* Thermocouple voltage in volts

(c) *Value [Y]:* Temperature in degrees C

(d) Click the left mouse button on *FINISH*

To generate a voltage-to-temperature conversion equation, double-click the left mouse button anywhere inside the chart border. A blue shaded outline will appear around the chart border. Continue by clicking *once* on any of the data points. Some of the data points will be displayed in yellow. Then select *Insert* from the toolbar and highlight and select *Trendline.* Finally, to generate an equation that describes the data, highlight *Polynomial* and then *Order 2.* Complete this sequence of steps by clicking on the tab labeled *Options,* select *[] Display Equation on Chart,* and click *OK.*

As an example, assume that the output voltage of a Type J thermocouple is (a) 0.005268 V and (b) 0.021846 V. Use the second-order polynomial shown to convert this voltage into its equivalent temperature.

(a) at 100°C; $y = -12,153(0.005268)^2 + 18,500(0.005268) + 1.865 = 98.98°C$

(b) at 400°C; $y = -12,153(0.021846)^2 + 18,500(0.021846) + 1.865 = 400.22°C$

| | |
|---|---|
| 0 | 0 |
| 0.001277 | 25 |
| 0.002585 | 50 |
| 0.003917 | 75 |
| 0.005268 | 100 |
| 0.006633 | 125 |
| 0.008008 | 150 |
| 0.00939 | 175 |
| 0.010777 | 200 |
| 0.012165 | 225 |
| 0.013553 | 250 |
| 0.01494 | 275 |
| 0.016325 | 300 |
| 0.017708 | 325 |
| 0.019089 | 350 |
| 0.020467 | 375 |
| 0.021846 | 400 |
| 0.023225 | 425 |
| 0.024607 | 450 |
| 0.025994 | 475 |
| 0.027388 | 500 |

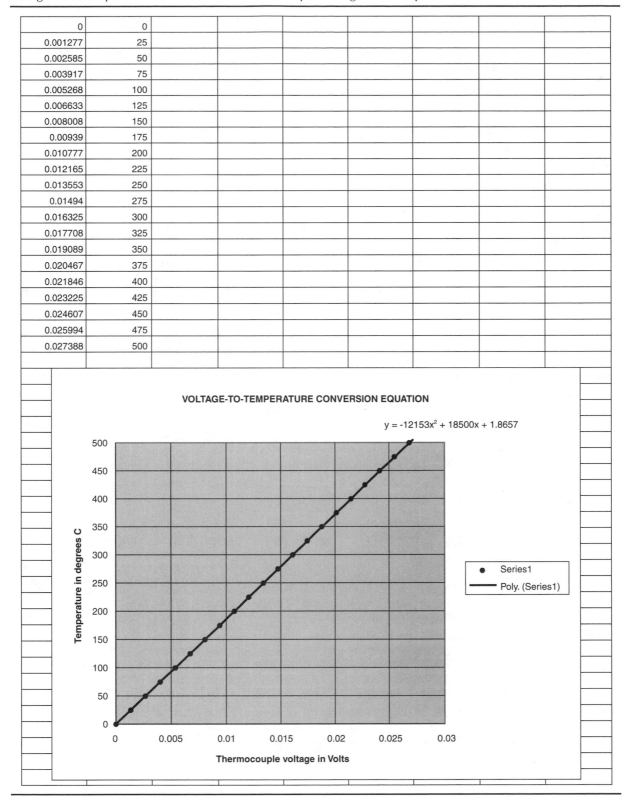

**VOLTAGE-TO-TEMPERATURE CONVERSION EQUATION**

$$y = -12153x^2 + 18500x + 1.8657$$

## J Type Thermocouple Reference Table

| DEG C | 0 | 1 | 2 | 3 | 4 | 5 | 6 | 7 | 8 | 9 | 10 | DEG C |
|---|---|---|---|---|---|---|---|---|---|---|---|---|
| 0 | 0.000 | 0.050 | 0.101 | 0.151 | 0.202 | 0.253 | 0.303 | 0.354 | 0.405 | 0.456 | 0.507 | 0 |
| 10 | 0.507 | 0.558 | 0.609 | 0.660 | 0.711 | 0.762 | 0.813 | 0.865 | 0.916 | 0.967 | 1.019 | 10 |
| 20 | 1.019 | 1.070 | 1.122 | 1.174 | 1.225 | 1.277 | 1.329 | 1.381 | 1.432 | 1.484 | 1.536 | 20 |
| 30 | 1.536 | 1.588 | 1.640 | 1.693 | 1.745 | 1.797 | 1.849 | 1.901 | 1.954 | 2.006 | 2.058 | 30 |
| 40 | 2.058 | 2.111 | 2.163 | 2.216 | 2.268 | 2.321 | 2.374 | 2.426 | 2.479 | 2.532 | 2.585 | 40 |
| 50 | 2.585 | 2.638 | 2.691 | 2.743 | 2.796 | 2.849 | 2.902 | 2.956 | 3.009 | 3.062 | 3.115 | 50 |
| 60 | 3.115 | 3.168 | 3.221 | 3.275 | 3.328 | 3.381 | 3.435 | 3.488 | 3.542 | 3.595 | 3.649 | 60 |
| 70 | 3.649 | 3.702 | 3.756 | 3.809 | 3.863 | 3.917 | 3.971 | 4.024 | 4.078 | 4.132 | 4.186 | 70 |
| 80 | 4.186 | 4.239 | 4.293 | 4.347 | 4.401 | 4.455 | 4.509 | 4.563 | 4.617 | 4.671 | 4.725 | 80 |
| 90 | 4.725 | 4.780 | 4.834 | 4.888 | 4.942 | 4.996 | 5.050 | 5.105 | 5.159 | 5.213 | 5.268 | 90 |
| 100 | 5.268 | 5.322 | 5.376 | 5.431 | 5.485 | 5.540 | 5.594 | 5.649 | 5.703 | 5.758 | 5.812 | 100 |
| 110 | 5.812 | 5.867 | 5.921 | 5.976 | 6.031 | 6.085 | 6.140 | 6.195 | 6.249 | 6.304 | 6.359 | 110 |
| 120 | 6.359 | 6.414 | 6.468 | 6.523 | 6.578 | 6.633 | 6.688 | 6.742 | 6.797 | 6.852 | 6.907 | 120 |
| 130 | 6.907 | 6.962 | 7.017 | 7.072 | 7.127 | 7.182 | 7.237 | 7.292 | 7.347 | 7.402 | 7.457 | 130 |
| 140 | 7.457 | 7.512 | 7.567 | 7.622 | 7.677 | 7.732 | 7.787 | 7.843 | 7.898 | 7.953 | 8.008 | 140 |
| 150 | 8.008 | 8.063 | 8.118 | 8.174 | 8.229 | 8.284 | 8.339 | 8.394 | 8.450 | 8.505 | 8.560 | 150 |
| 160 | 8.560 | 8.616 | 8.671 | 8.726 | 8.781 | 8.837 | 8.892 | 8.947 | 9.003 | 9.058 | 9.113 | 160 |
| 170 | 9.113 | 9.169 | 9.224 | 9.279 | 9.335 | 9.390 | 9.446 | 9.501 | 9.556 | 9.612 | 9.667 | 170 |
| 180 | 9.667 | 9.723 | 9.778 | 9.834 | 9.889 | 9.944 | 10.000 | 10.055 | 10.111 | 10.166 | 10.222 | 180 |
| 190 | 10.222 | 10.277 | 10.333 | 10.388 | 10.444 | 10.499 | 10.555 | 10.610 | 10.666 | 10.721 | 10.777 | 190 |
| 200 | 10.777 | 10.832 | 10.888 | 10.943 | 10.999 | 11.054 | 11.110 | 11.165 | 11.221 | 11.276 | 11.332 | 200 |
| 210 | 11.332 | 11.387 | 11.443 | 11.498 | 11.554 | 11.609 | 11.665 | 11.720 | 11.776 | 11.831 | 11.887 | 210 |
| 220 | 11.887 | 11.943 | 11.998 | 12.054 | 12.109 | 12.165 | 12.220 | 12.276 | 12.331 | 12.387 | 12.442 | 220 |
| 230 | 12.442 | 12.498 | 12.553 | 12.609 | 12.664 | 12.720 | 12.776 | 12.831 | 12.887 | 12.942 | 12.998 | 230 |
| 240 | 12.998 | 13.053 | 13.109 | 13.164 | 13.220 | 13.275 | 13.331 | 13.386 | 13.442 | 13.497 | 13.553 | 240 |
| 250 | 13.553 | 13.608 | 13.664 | 13.719 | 13.775 | 13.830 | 13.886 | 13.941 | 13.997 | 14.052 | 14.108 | 250 |
| 260 | 14.108 | 14.163 | 14.219 | 14.274 | 14.330 | 14.385 | 14.441 | 14.496 | 14.552 | 14.607 | 14.663 | 260 |
| 270 | 14.663 | 14.718 | 14.774 | 14.829 | 14.885 | 14.940 | 14.995 | 15.051 | 15.106 | 15.162 | 15.217 | 270 |
| 280 | 15.217 | 15.273 | 15.328 | 15.383 | 15.439 | 15.494 | 15.550 | 15.605 | 15.661 | 15.716 | 15.771 | 280 |
| 290 | 15.771 | 15.827 | 15.882 | 15.938 | 15.993 | 16.048 | 16.104 | 16.159 | 16.214 | 16.270 | 16.325 | 290 |
| 300 | 16.325 | 16.380 | 16.436 | 16.491 | 16.547 | 16.602 | 16.657 | 16.713 | 16.768 | 16.823 | 16.879 | 300 |
| 310 | 16.879 | 16.934 | 16.989 | 17.044 | 17.100 | 17.135 | 17.210 | 17.266 | 17.321 | 17.376 | 17.432 | 310 |
| 320 | 17.432 | 17.487 | 17.542 | 17.597 | 17.653 | 17.708 | 17.763 | 17.818 | 17.874 | 17.929 | 17.984 | 320 |
| 330 | 17.984 | 18.039 | 18.095 | 18.150 | 18.205 | 18.260 | 18.316 | 18.371 | 18.426 | 18.481 | 18.537 | 330 |
| 340 | 18.537 | 18.592 | 18.647 | 18.702 | 18.757 | 18.813 | 18.868 | 18.923 | 18.978 | 19.033 | 19.089 | 340 |
| 350 | 19.089 | 19.144 | 19.199 | 19.254 | 19.309 | 19.364 | 19.420 | 19.475 | 19.530 | 19.585 | 19.640 | 350 |
| 360 | 19.640 | 19.695 | 19.751 | 19.806 | 19.861 | 19.916 | 19.971 | 20.026 | 20.081 | 20.137 | 20.192 | 360 |
| 370 | 20.192 | 20.247 | 20.302 | 20.357 | 20.412 | 20.467 | 20.523 | 20.578 | 20.633 | 20.688 | 20.743 | 370 |
| 380 | 20.743 | 20.798 | 20.853 | 20.909 | 20.964 | 21.019 | 21.074 | 21.129 | 21.184 | 21.239 | 21.295 | 380 |
| 390 | 21.295 | 21.350 | 21.405 | 21.460 | 21.515 | 21.570 | 21.625 | 21.680 | 21.736 | 21.791 | 21.846 | 390 |
| 400 | 21.846 | 21.901 | 21.956 | 22.011 | 22.066 | 22.122 | 22.177 | 22.232 | 22.287 | 22.342 | 22.397 | 400 |
| 410 | 22.397 | 22.453 | 22.508 | 22.563 | 22.618 | 22.673 | 22.728 | 22.784 | 22.839 | 22.894 | 22.949 | 410 |
| 420 | 22.949 | 23.004 | 23.060 | 23.115 | 23.170 | 23.225 | 23.280 | 23.336 | 23.391 | 23.446 | 23.501 | 420 |
| 430 | 23.501 | 23.556 | 23.612 | 23.667 | 23.722 | 23.777 | 23.833 | 23.888 | 23.943 | 23.999 | 24.054 | 430 |
| 440 | 24.054 | 24.109 | 24.164 | 24.220 | 24.275 | 24.330 | 24.386 | 24.441 | 24.496 | 24.552 | 24.607 | 440 |
| 450 | 24.607 | 24.662 | 24.718 | 24.773 | 24.829 | 24.884 | 24.939 | 24.995 | 25.050 | 25.106 | 25.161 | 450 |
| 460 | 25.161 | 25.217 | 25.272 | 25.327 | 25.383 | 25.438 | 25.494 | 25.549 | 25.605 | 25.661 | 25.716 | 460 |
| 470 | 25.716 | 25.772 | 25.827 | 25.883 | 25.938 | 25.994 | 26.050 | 26.105 | 26.161 | 26.216 | 26.272 | 470 |
| 480 | 26.272 | 26.328 | 26.383 | 26.439 | 26.495 | 26.551 | 26.606 | 26.662 | 26.718 | 26.774 | 26.829 | 480 |
| 490 | 26.829 | 26.885 | 26.941 | 26.997 | 27.053 | 27.109 | 27.165 | 27.220 | 27.276 | 27.332 | 27.388 | 490 |

## J Type Thermocouple Reference Table *(Continued)*

| DEG C | 0 | 1 | 2 | 3 | 4 | 5 | 6 | 7 | 8 | 9 | 10 | DEG C |
|---|---|---|---|---|---|---|---|---|---|---|---|---|
| 500 | 27.388 | 27.444 | 27.500 | 27.556 | 27.612 | 27.668 | 27.724 | 27.780 | 27.836 | 27.893 | 27.949 | 500 |
| 510 | 27.949 | 28.005 | 28.061 | 28.117 | 28.173 | 28.230 | 28.286 | 28.342 | 28.398 | 28.455 | 28.511 | 510 |
| 520 | 28.511 | 28.567 | 28.624 | 28.680 | 28.736 | 28.793 | 28.849 | 28.906 | 28.962 | 29.019 | 29.075 | 520 |
| 530 | 29.075 | 29.132 | 29.188 | 29.245 | 29.301 | 29.358 | 29.415 | 29.471 | 29.528 | 29.585 | 29.642 | 530 |
| 540 | 29.642 | 29.698 | 29.755 | 29.812 | 29.869 | 29.926 | 29.983 | 30.039 | 30.096 | 30.153 | 30.210 | 540 |
| 550 | 30.210 | 30.267 | 30.324 | 30.381 | 30.439 | 30.496 | 30.553 | 30.610 | 30.667 | 30.724 | 30.782 | 550 |
| 560 | 30.782 | 30.839 | 30.896 | 30.954 | 31.011 | 31.068 | 31.126 | 31.183 | 31.241 | 31.298 | 31.356 | 560 |
| 570 | 31.356 | 31.413 | 31.471 | 31.528 | 31.586 | 31.644 | 31.702 | 31.759 | 31.817 | 31.875 | 31.933 | 570 |
| 580 | 31.933 | 31.991 | 32.048 | 32.106 | 32.164 | 32.222 | 32.280 | 32.338 | 32.396 | 32.455 | 32.513 | 580 |
| 590 | 32.513 | 32.571 | 32.629 | 32.687 | 32.746 | 32.804 | 32.862 | 32.921 | 32.979 | 33.038 | 33.096 | 590 |
| 600 | 33.096 | 33.155 | 33.213 | 33.272 | 33.330 | 33.389 | 33.448 | 33.506 | 33.565 | 33.624 | 33.683 | 600 |
| 610 | 33.683 | 33.742 | 33.800 | 33.859 | 33.918 | 33.977 | 34.036 | 34.095 | 34.155 | 34.214 | 34.273 | 610 |
| 620 | 34.273 | 34.332 | 34.391 | 34.451 | 34.510 | 34.569 | 34.629 | 34.688 | 34.748 | 34.807 | 34.867 | 620 |
| 630 | 34.867 | 34.926 | 34.986 | 35.046 | 35.105 | 35.165 | 35.225 | 35.285 | 35.344 | 35.404 | 35.464 | 630 |
| 640 | 35.464 | 35.524 | 35.584 | 35.644 | 35.704 | 35.764 | 35.825 | 35.885 | 35.945 | 36.005 | 36.066 | 640 |
| 650 | 36.066 | 36.126 | 36.186 | 36.247 | 36.307 | 36.368 | 36.428 | 36.489 | 36.549 | 36.610 | 36.671 | 650 |
| 660 | 36.671 | 36.732 | 36.792 | 36.853 | 36.914 | 36.975 | 37.036 | 37.097 | 37.158 | 37.219 | 37.280 | 660 |
| 670 | 37.280 | 37.341 | 37.402 | 37.463 | 37.525 | 37.586 | 37.647 | 37.709 | 37.770 | 37.831 | 37.893 | 670 |
| 680 | 37.893 | 37.954 | 38.016 | 38.078 | 38.139 | 38.201 | 38.262 | 38.324 | 38.386 | 38.448 | 38.510 | 680 |
| 690 | 38.510 | 38.572 | 38.633 | 38.695 | 38.757 | 38.819 | 38.882 | 38.944 | 39.006 | 39.068 | 39.130 | 690 |
| 700 | 39.130 | 39.192 | 39.255 | 39.317 | 39.379 | 39.442 | 39.504 | 39.567 | 39.629 | 39.692 | 39.754 | 700 |
| 710 | 39.754 | 39.817 | 39.880 | 39.942 | 40.005 | 40.068 | 40.131 | 40.193 | 40.256 | 40.319 | 40.382 | 710 |
| 720 | 40.382 | 40.445 | 40.508 | 40.571 | 40.634 | 40.697 | 40.760 | 40.823 | 40.886 | 40.950 | 41.013 | 720 |
| 730 | 41.013 | 41.076 | 41.139 | 41.203 | 41.266 | 41.329 | 41.393 | 41.456 | 41.520 | 41.583 | 41.647 | 730 |
| 740 | 41.647 | 41.710 | 41.774 | 41.837 | 41.901 | 41.965 | 42.028 | 42.092 | 42.156 | 42.219 | 42.283 | 740 |
| 750 | 42.283 | 42.347 | 42.411 | 42.475 | 42.538 | 42.602 | 42.666 | 42.730 | 42.794 | 42.858 | 42.922 | 750 |
| 760 | 42.922 | | | | | | | | | | | 760 |
| DEG C | 0 | 1 | 2 | 3 | 4 | 5 | 6 | 7 | 8 | 9 | 10 | DEG C |

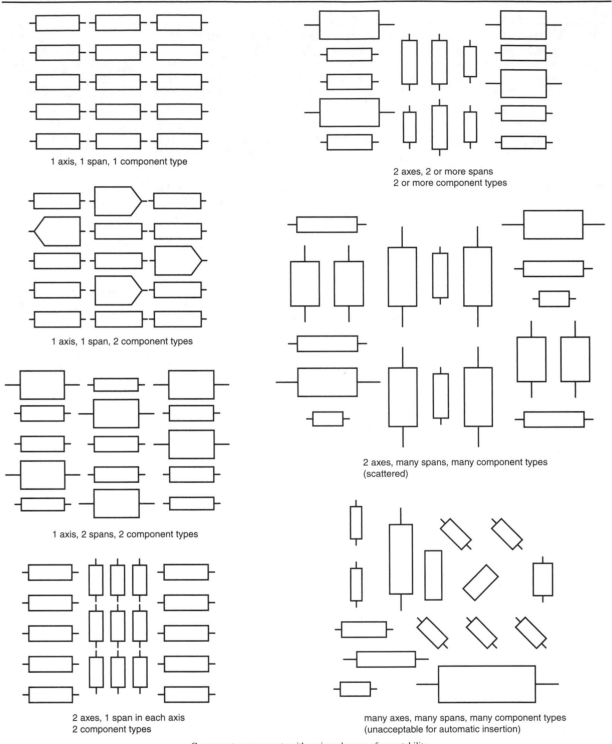

1 axis, 1 span, 1 component type

1 axis, 1 span, 2 component types

1 axis, 2 spans, 2 component types

2 axes, 1 span in each axis
2 component types

2 axes, 2 or more spans
2 or more component types

2 axes, many spans, many component types
(scattered)

many axes, many spans, many component types
(unacceptable for automatic insertion)

Component arrangements with various degrees of acceptability

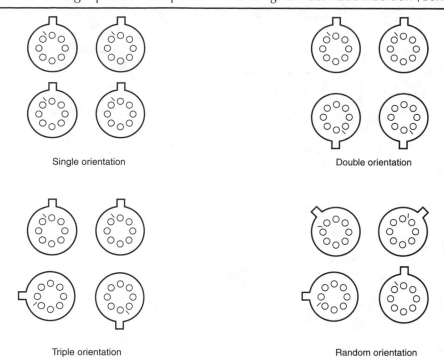

Single orientation

Double orientation

Triple orientation

Random orientation

Four different orientations for keyed radial lead ICs

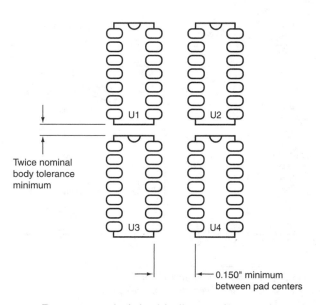

Twice nominal body tolerance minimum

0.150" minimum between pad centers

Recommended dual-in-line package orientation

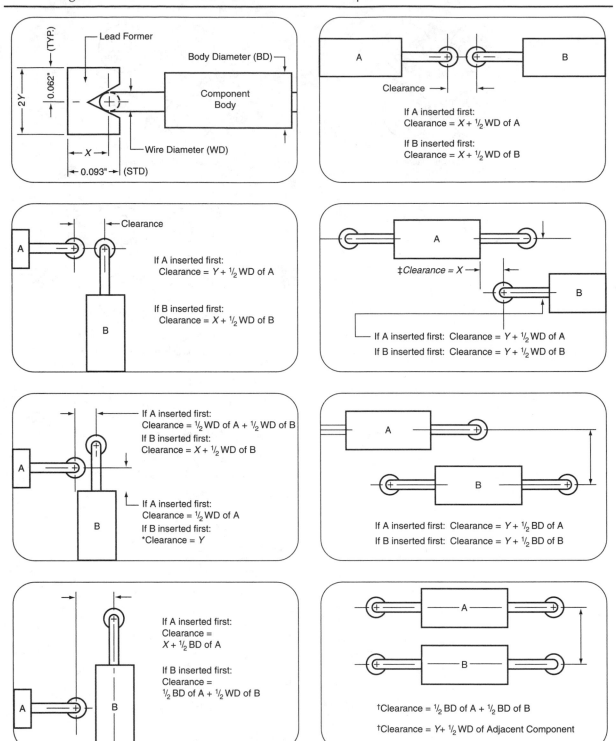

*Check *Y* dimension for possible interference with body length of component B.
†Use whichever clearance value is larger.
‡Check *X* dimension for possible interference with body length of component A.

Since high-density packaging requires the placement of more components and circuitry into less space, the task of reducing package size results in more complicated boards and increased layout difficulties. Prior to initiating a layout sketch, the PCB designer may be confronted with problems concerning the size of the board and whether it will be a single-sided, double-sided, or multilayer design. Several possibilities exist. If the size of the finished board is predetermined, the designer must consider which type of layout is best suited for the amount of circuitry that is to be packaged in that given space. On the other hand, if the overall board size *and* the type of layout is specified, the designer must determine if the task of layout is plausible. This determination is the result of a *packaging feasibility study* that involves both *interconnection density* and *package density.*

Interconnection density refers to the number of conductor paths that can be placed on a board layer per given area. For example, it is apparent that it is possible to place more than 10 mil conductors with 10 mil spacings in a 1 in$^2$ area than it is to place 50 mil paths with 50 mil spacings in that same 1 in$^2$ area. Interconnection density is typically specified in conductor width and spacing. To maximize the interconnection density, the designer must minimize conductor path widths and spacings and pad diameters.

Package density refers to the number of components mounted onto a given area or the number of lead access holes per given area. For example, it is a great deal easier to design a layout for a board having one resistor (2 holes) per square inch than it is to design a layout having five resistors (10 holes) per square inch. To increase package density, the designer must increase the number of components or holes per square inch of board space.

It can be seen that interconnection density and packaging density go hand in hand. If the designer initially decides that the design will incorporate minimum conductor path widths and spaces (maximum interconnection density), then the determination of how many components (or holes) can be packaged per square inch must be made. This will establish whether a single-sided design will suffice or if it will be necessary to provide a double-sided or even a multilayer design.

One method of *estimating* packaging feasibility is with the lead access *holes per square inch* (holes/in$^2$) technique. The designer first determines the usable surface of the board in square inches by multiplying the length by the width. From this value, all areas of the board that will not be used for component mounting and conductor routing are subtracted. These unused spaces include spaces around mounting holes and hardware, the spacing around the edge of the board, and cutouts and forbidden areas inside the board edges. The result of subtracting these unusable areas is the *total usable layout area* in square inches.

The next step in this feasibility analysis is to count all of the circuit interconnection points with the use of the circuit schematic. This is done as follows:

1. Count all two-terminal devices (resistors, diodes, capacitors, etc.) and multiply this count by two.
2. Count all three terminal devices (transistors, UJTs, etc.) and multiply by three.
3. Count all 14-pin IC packages and multiply by 14, all 16-pin packages and multiply by 16, and so on.
4. Count all input/output points, that is, external connections including power and ground.

The total of the value obtained is the number of interconnection points (holes) to be made on the board. To determine the package density in units of holes per square inch, simply divide the total interconnection points by the usable board area. This is shown in the following equation:

$$Package\ Density\ (holes/in^2) = \frac{total\ number\ of\ holes}{usable\ board\ area\ (in^2)}$$

To serve as an example of the calculation of package density, we will use the single-sided design of a one-shot multivibrator circuit shown below in Fig. 1 and the following design problem.

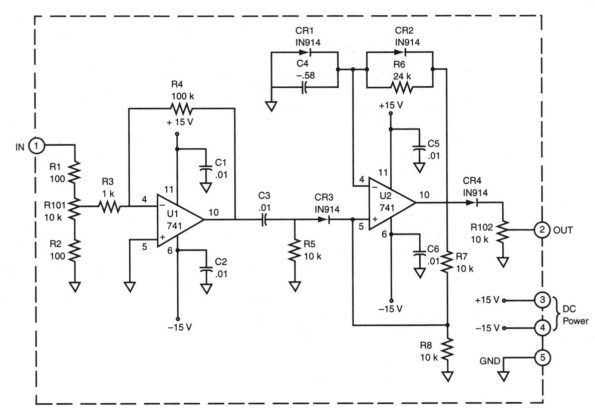

All resistors are 1/2 watt ± 10% unless otherwise specified.
All capacitors are in microfarads unless otherwise specified.

Parts List

U1, U2 – 741 Op-Amp, TO–116 style case
R1, R2 – 100Ω, 1/2 W, 10%
R3 – 1kΩ, 1/2 W, 10%
R4, – 100kΩ, 1/2W, 10%
R5, R7, R8 – 10kΩ, 1/2W, 10%
R6 – 24kΩ, 1/2W, 10%
R101, R102 – 10kΩ, trim pots
C1, C2, C3, C5, C6 – 0.01 µF disk capacitors
C4 – 0.56 µF disk capacitors
CR1, CR2, CR3, CR4, IN914 diodes, DO–35 style case

**Figure 1**   Negative-acting one-shot multivibrator.

## DESIGN PROBLEM

Design a single-sided component layout drawing for the circuit shown in Fig. 1. The PCB stock is to be FR-4 with a thickness of 0.059 inch clad on one side to include all necessary labels and codes. The layout is to be restricted to the unhatched area in the detail drawing shown in Fig. 2. All parts are to be placed in such a way as to minimize any wasted board space, yet maintain uniform distribution and balance. No component body or conductor path is to be placed within a minimum of 0.20 inch from any board edge. The minimum spacing between all electrical conductors (paths and pads) is to be 0.025 inch. The position and orientation of all external connections and mounting holes are to be placed in accordance with Fig. 2. Pads for external connections are to be made to accept 0.045 $in^2$ wire wrap pins.

The ground connections are to be divided into three separate groups with an individual path provided for each group to the external ground connection point. This is to be done as follows: (1) the ground points for all of the decoupling capacitors (C1, C2, C5, and C6) will be connected together with a separate return path to the external ground point; (2) all ground points associated with input circuitry (bottom end of R2 and pin 5 of U1) will be connected and returned to the ground point; and (3) all ground points associated with the output circuitry (bottom ends of R8 and R102) will be connected and returned to the ground point.

The dc power supply terminals are to be capable of handling a maximum current of 100 mA.

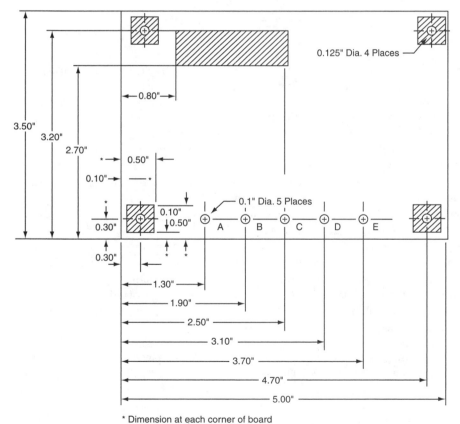

A – INPUT (IN)
B – GROUND (GND)
C – (–15 Vdc)
D – OUTPUT (OUT)
E – (+ 15 Vdc)

Electrical orientation
of external connections

0.125" Dia. 4 Places

0.1" Dia. 5 Places

* Dimension at each corner of board

**Figure 2**   Dimensional and electrical specifications for multivibrator board.

To begin the feasibility study for this design problem, refer to the design problem statement and Fig. 2. The overall dimensions of the board are 3.50 × 5.00 inches. Since the specifications require that no component or conductor path be placed closer than 0.20 inch to any board edge, the actual overall dimensions for our calculations become 3.10 × 4.60 inches. There is a total of five areas on the drawing where no components are to be mounted. These are four areas around the mounting hole locations (each 0.600 × 0.600 in.) and the forbidden area (0.500 × 1.70 in.). The total usable board area can be calculated as follows:

$$Usable\ Area = 3.10 \times 4.60 - [4(0.600)(0.600) + (0.500)(1.70)] \approx 11.9\ in^2$$

In order to calculate the total number of interconnection points, refer to the circuit schematic shown in Fig. 1. We count that there are eight resistors, two trim pots, six capacitors, four diodes, two 14-pin DIP ICs, and five external connections. The total count is as follows:

| | |
|---|---|
| 8 resistors × 2 leads | = 16 |
| 2 potentiometers × 3 leads | = 6 |
| 6 capacitors × 2 leads | = 12 |
| 4 diodes × 2 leads | = 8 |
| 2 DIPs × 14 leads | = 28 |
| 5 external connections | = 5 |
| Total | 75 holes |

The package density can now be calculated as follows:

$$Package\ density = \frac{75\ holes}{11.9\ in^2} \approx 6.3\ holes/in^2$$

To better understand the significance of this value, see Fig. 3. Figure 3a is a plot of holes/in² versus recommended type of design, and Fig. 3b is a graph of package density versus board type or degree of layout difficulty versus packaging density in holes/in. Figure 3a is used as follows. After calculating the package density (holes/in²), it is located on the graph. As shown on this graph, an estimate of holes/in² below 10 can be considered suitable for a single-sided design. A package density of between 10 and 20 holes/in² usually requires a double-sided design. Densities of 20 to 30 holes/in² require a multilayer design.

Figure 3b shows that as package density increases, the degree of layout difficulty also increases as the holes/in² count approaches the upper limit of each board type. For example, a double-sided design with a package density of 11 holes/in² is normally much easier to lay out than would be a double-sided design with a much higher density of 19 holes/in².

The graphs of Figs. 3a and b should be used only as a rough approximation for determining board type and difficulty of design since other factors also influence the nature of design. Some of these factors are the type of electrical circuit (analog/digital or digital) and special electrical and mechanical restrictions in the specifications. In general, however, these graphs can be used to determine (1) the type of design (single-sided, double-sided, or multilayer), (2) the feasibility of that level of design, and (3) the approximate degree of layout difficulty that can be expected.

Note in Fig. 3b that a horizontal projection, shown as a dashed line, extends across the graph from the midpoint of the layout difficulty scale and intersects a vertical dashed line at the midpoint of each of the packaging density groupings. Readings of package density of these vertical projections are 5 holes/in² for single-sided boards, 15 holes/in² for double-sided boards, and 25 holes/in² for multilayer designs. These projection points represent the ideal level of difficulty for each board type, that is, designs that can be laid out in a reasonable amount of time. Above this level, significantly more time would be required for completing a design for each board type.

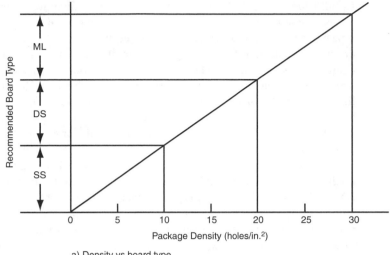

a) Density vs board type

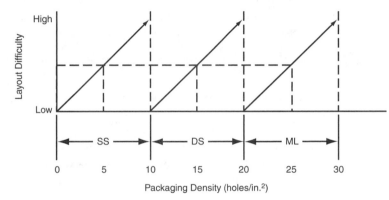

b) Layout difficulty vs density

**Figure 3** Selection of the most appropriate type of layout as a function of package density and level of design difficulty.

Recall from Appendix XVIII that a procedure for evaluating the levels of package density and packaging difficulty was presented. That evaluation aided in the determination of what type of printed circuit board design should be employed, that is, single-sided, double-sided, or multilayer design. That packaging feasibility study also showed that for a given usable board area, as the number of holes/in$^2$ increases, so does the density of the parts. As the density increased, the level of layout difficulty became greater. Rough guidelines were established, which recommended that single-sided boards have a maximum density of 10 holes/in$^2$, double-sided boards be considered for packaging densities of 10 to 20 holes/in$^2$, with 15 holes/in$^2$ considered a typical value. Multilayer boards would be recommended for densities above 20 and toward 30 holes/in$^2$.

A similar packaging feasibility study is applied to digital designs. However, the method of determination and the units will be different from those presented in Appendix XVIII. Because digital ICs are typically enclosed in dual-in-line packages (DIPs), the packaging density is defined in units of number of *DIPs per square inch* (DIPs/in$^2$). Since DIP packages vary in size from 14-pin to 64-pin, it becomes necessary to develop the concept of an *equivalent DIP package,* which will represent this range of sizes. Typically, the 16-pin DIP is considered one equivalent dual-in-line package (EDIP). The number of EDIPs in a design is determined by dividing the total number of lead holes (to include the number of fingers used for edge connections) by the number of holes for one EDIP (16):

$$EDIPs = \frac{number\ of\ holes}{16}$$

The packaging density is now given as

$$Packaging\ Density = \frac{number\ of\ EDIPs}{in^2\ of\ usable\ board\ area} = EDIPs/in^2$$

For the application of the two equations shown above, we will consider the following packaging problem.

## DESIGN PROBLEM

On a 4.5 × 4-inch double-sided printed circuit board having 15 in$^2$ of usable layout area, the following are to be mounted: ten 16-pin DIPs, two 24-pin DIPs, and six decoupling capacitors. The design will be fabricated on a board having 36 fingers (18 per side), 15 of which will be used for external connection points.

We will begin the determination of packaging density by finding the number of holes that will be required in the design, including the count of fingers used for making external connections.

| | | |
|---|---|---|
| 10 × 16-pin ICs | = | 160 |
| 2 × 24-pin ICs | = | 48 |
| 6 × 2 lead capacitors | = | 12 |
| 15 external connections | = | 15 |
| Total | | 235 holes |

The number of EDIPs in the design is:

$$EDIPs = \frac{235\ holes}{16} \approx 14.7$$

Finally, the packaging density is determined as follows:

$$Packaging\ Density = \frac{14.7\ EDIPs}{15\ in^2} = 0.98 \quad or \quad \cong 1\ EDIP/in^2$$

This example illustrates a typical packaging density for a double-sided digital design. The 1.0- $EDIP/in^2$ density is considered a valid compromise for dense packaging if 10-mil conductor path widths and spacings, 50-mil via hole pad diameters, and minimum pad diameters for plated-through-component lead holes are used in the design.

At a packaging density level of 1 $EDIP/in^2$, it is estimated that an experienced designer can produce a layout at the rate of one EDIP every 1 1/2 hours. For our example, the design problem with 14.7 EDIPs would take approximately 22 hours to complete the layout. Of course, as packaging density decreases below 1 $EDIP/in^2$, a double-sided layout becomes less difficult and requires less time to complete. Double-sided designs may be employed for densities up to approximately 1.5 $EDIPs/in^2$. This is regarded as the point at which consideration needs to be given to the use of multilayer boards. Clearly, the time required to complete these layouts will be markedly increased.

One advantage of digital circuit layouts that is rarely seen in analog circuits is that there is a high degree of built-in symmetry of parts. This is because the vast majority of the parts used are ICs that have the same size and shape, and the designer is able to produce much denser layouts on double-sided plated-through-hole printed circuit boards.

Health and safety in the workplace continues to become of greater importance. There are growing demands in all businesses to achieve higher standards of quality in effective performance and efficiency in production. This is especially true in industry to keep a competitive edge. Investments in the technologies of ergonomics, production, repair, and quality control must be complemented by a concerted effort to provide people with safe working conditions in which the environment in no way hinders workers from operating efficiently and producing the highest quality workmanship possible. There is a global effort in academia, government, and industry to make both the environment and the workplace not only cleaner but safer and healthier. It is important for both employers and employees to be abreast of health issues in the workplace and knowledgeable of sources of safety information such as those listed in Appendix XIII. Occupational health and safety laws are becoming more strictly enforced to protect workers. The micro-electronics industry especially is taking the initiative to protect employees to reduce work-related illness, which is both ethical and prudent. One of the dangers in the electronics industry is the exposure to solder flux used in soldering procedures associated with assembly operations. In this appendix, information in regard to the composition of solder flux, health issues related to exposure to solder flux fumes, and fume extraction techniques is presented.

Soldering is a thermal bonding process where two metal surfaces are joined together. In this process, as a result of the solder melting over the metal surfaces and solidifying in the form of an alloy, an electrical connection of high integrity is made between the metal surfaces. The solder used in the soldering process is made of a tin–lead alloy that melts between 361 and 750°F (183 and 399°C). To ensure proper alloying to occur in the soldering process, the metal surfaces must be free of oxides. Solder flux is used to remove oxides. The solder flux is applied either externally by means of a separate applicator, or in combination within the solder wire as a core of flux. It is important to understand that as the flux is heated to remove the oxides, airborne pollutants are produced that pose a potential health hazard to workers.

Most fluxes used in the electronics industry for cleaning prior to soldering have been generally based on a translucent amber-colored substance known as colophony, or rosin. It is obtained as a by-product after turpentine has been distilled from the oleoresin and canal resin of pine trees and is commonly used in both manual and automated soldering procedures. The composition of rosin is approximately 90% resin acid, which is mostly abietic acid with about 10% neutral materials such as stilbene derivatives and a variety of hydrocarbons. A number of fluxes also contain an organic activator that enhances the cleaning activity. When flux becomes heated in the soldering process, certain gaseous products such as formaldehyde, hydrochloric acid, benzene, chlorophenol, isopropyl alcohol, phenol styrene, and toluene and airborne particulate will be generated. Particle sizes are classified into three categories of risk as follows. First, particle sizes of 10 microns in diameter or more are trapped in the mouth or nose (inhalable fraction). Second, particle sizes of less than 10 microns in diameter can reach the lung airways (thoracic fraction). Third, particles less than 3.5 microns can become deposited in the lower respiratory tract (respirable fraction). Figure 1 identifies the three sections of the respiratory system where different size irritants are deposited.

A chart comparing particulate sizes is shown in Fig. 2. It can be seen that the 0.01- to 1.0-micron size particles produced when soldering are in the particle inhalation range of tobacco smoke, which allows them to reach the lower respiratory system if inhaled. Particles of this size are generally retained for a long time in this gas-exchange section of the lungs, which can lead to pulmonary health problems.

There are water-soluble fluxes, referred to as no clean fluxes, that have little or no rosin but require the presence of a powerful chemical activator. Although this type of flux has gained some popularity, the fumes can still be dangerous because of the common presence of a flux activator such as phosphorous hexate. Therefore, this type of soldering flux

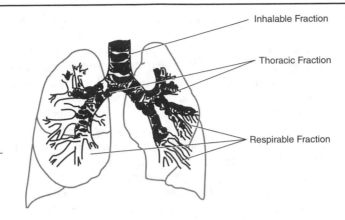

**Figure 1** Sites of different particulate-size respiratory irritants. *Courtesy of PACE, Inc., Laurel, MD.*

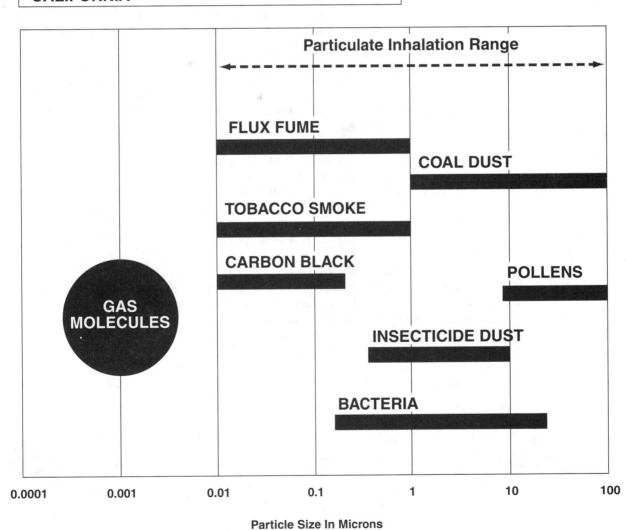

**Figure 2** Comparison of particulate sizes. *Courtesy of PACE, Inc., Laurel, MD.*

can be a source of a strong respiratory irritant since it produces acid fumes when heated above 253°F (123°C).

In regard to health issues, material safety data sheets (MSDS) are required by law to be made available to users of solder and fluxes from manufacturers of these products. MSDS identify which health hazards are associated with using solder and flux. There are many potential health hazards due to the use of these products. The following information describes health hazards which colophony or rosin flux fumes are known to cause: chest pain, chemical sensitization, eye and nose irritation, headaches, dizziness, chronic bronchitis, occupational asthma (a debilitating disease), and contact dermatitis.

The inhalation of pollutants from solder and flux can easily go undiagnosed because the initial symptoms such as chest pain, shortness of breath, bronchitis, eye and nose irritation, and headaches are much the same as those of having a cold or the flu. Since acid fumes such as hydrochloric acid are produced when flux is heated, eye and nose irritation are often experienced after soldering procedures have been performed. Allergic sensitization and associated symptoms can develop over a period of just a few months to, in some cases, many years of exposure. Once becoming sensitized, wheezing and shortness of breath can occur each day there is exposure to the fumes, and these symptoms can continue well after one is no longer being exposed. A reason for the misdiagnosis of occupational asthma is the late reaction that is generally experienced many hours after exposure. It is also possible to experience airborne contact dermatitis from rosin due to the strong skin sensitizers that are produced, such as hydrazine, when it is heated. In order to avoid the consequences of exposure that exceeds specified levels of these pollutants and the health related costs, industry implements engineering controls to prevent exposure to these health hazards and to be in compliance with the Occupational Safety and Health Administration (OSHA).

Factory air-conditioning or general ventilation systems fail to prevent solder fumes from entering breathing zones, that is, air space close to work stations. Changing the type of flux used or requiring the use of personal respirators for protection are options of engineering control. However, neither method is satisfactory. The most common engineering control method used to solve the problem of exposure to fumes containing airborne pollutants is to intercept them between the source and the worker. This is accomplished by installing some type of local exhaust ventilation system as illustrated in Fig. 3.

The process of fume extraction, to be effective, requires the four elements shown in Fig. 4.

The first element is a collection device that captures fumes and is followed by a medium, such as ductwork or tubing, for containing and transporting the fumes. The remaining two elements are a central filtration unit that removes the fume pollutants and returns the cleaned air back into the breathing zone and a vacuum pump that draws the fumes from the collection device into the central filtration unit.

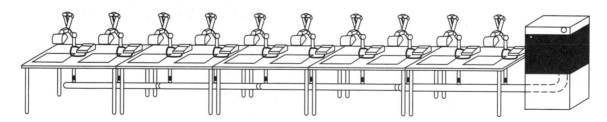

**Figure 3**   Local exhaust ventilation system. *Courtesy of PACE, Inc., Laurel, MD.*

Composition of Solder Flux, Health Issues of Concern, and Fume Extraction *(Continued)*

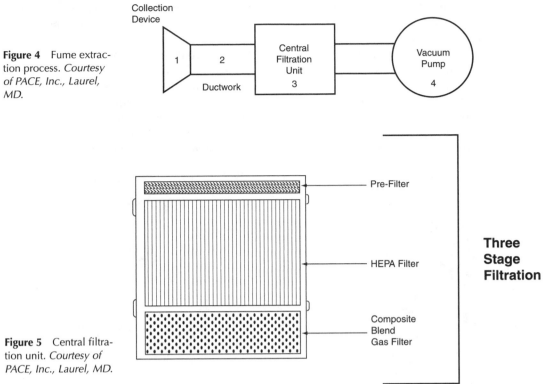

**Figure 4** Fume extraction process. *Courtesy of PACE, Inc., Laurel, MD.*

**Figure 5** Central filtration unit. *Courtesy of PACE, Inc., Laurel, MD.*

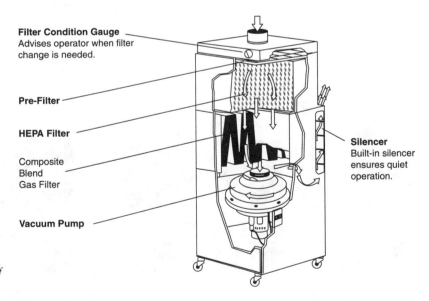

**Figure 6** Central filtration system. *Courtesy of PACE, Inc., Laurel, MD.*

 The central filtration system provides three stages of filtration as illustrated in Fig. 5. The prefilter removes large particles from the air that would clog the second stage, which is a high efficiency particulate filter. This filter removes 0.01- to 0.3-micron particles. The last stage is a Composite Blend Gas filter, which Chemisorbs absorbs gaseous pollutants in the air stream before it is vented back into the working environment. Figure 6 is an example of how the central filtration unit and vacuum pump are normally packaged in one enclosure with additional features such as a filter condition gauge to indicate the status of the filter and silencer, which affords more quiet operation.

It should be noted that the vacuum pump selected for a central filtration system must not only meet the American Council of Government Industrial Hygienists (ACGIH) requirements to provide the proper airflow to meet minimum duct design velocity for the fumes but have a preferred rating of 20,000 hours or more of continuous operation. Also, composite blend gas filters are preferred over carbon impregnated mats that have low absorbent capacity or troughs filled with loose carbon. Loose carbon can provide insufficient filtration of gaseous pollutants and emit carbon dust into the environment.

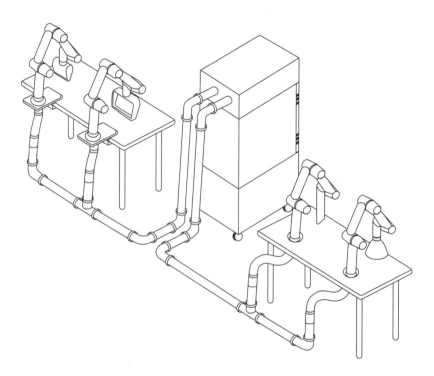

(a)

The Arm-Evac Articulated Arm has a wide range of quick-fit accessories to meet a variety of extraction applications.

Two Flex Arms with Bench Mounting Kits to Multi Arm-Evac II

Two-Section Articulated Arm to Multi Arm-Evac II

(b)                                                                                     (c)

**Figure 7**  Arm extraction system: (a) arm extraction installation; (b) collection devices; and (c) adjustable extraction arms. *Courtesy of PACE, Inc., Laurel, MD.*

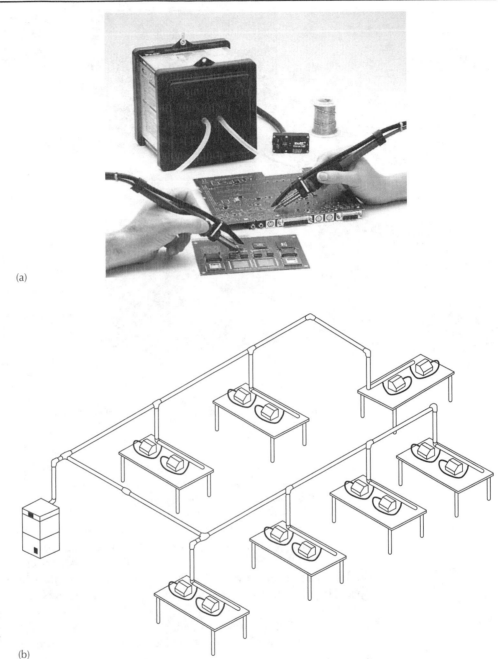

**Figure 8** Solder tip fume extraction: (a) dual-tip Venturi-Evac system (b) tip extraction installation. *Courtesy of PACE, Inc., Laurel, MD.*

Arm extraction systems, as illustrated in Fig. 7a, are designed to remove large volumes of air from breathing zones by employing adjustable extraction arms. Figures (b) and (c) show a variety of static safe collection devices such as canopies, hoods, nozzles, and suction tubes and adjustable extraction arms that are available for use in a variety of flux fume removal applications requiring arm extraction.

Figure 8a shows a method of solder tip extraction. In this method a suction tube is attached by clips to the soldering iron element and connected to a dual-tip Venturi-Evac unit.

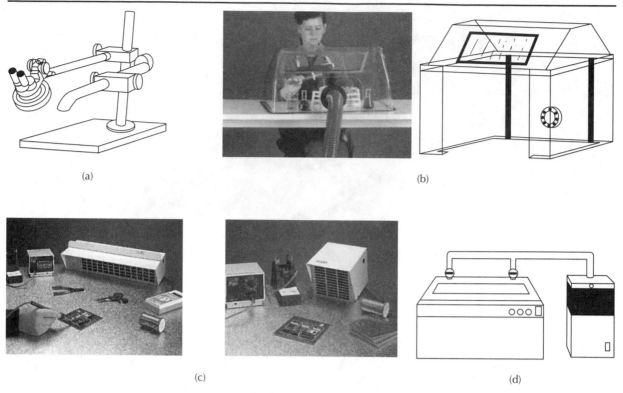

(a)

(b)

(c)

(d)

**Figure 9**  Specialized fume collection and extraction systems: (a) microscope attachment; (b) fume cabinets; (c) benchtop exhausters; and (d) production equipment extraction. *Courtesy of PACE, Inc., Laurel, MD.*

Flux fumes produced at the soldering iron tip are immediately extracted without affecting tip temperature. Figure 8b illustrates a solder tip extraction installation employing a central filtration unit that has a high static vacuum pump.

Figure 9 shows a number of specialized fume collectors and extraction systems. A microscope attachment, shown in Fig. 9a, is used in the case of detailed repair and rework. Fume cabinets, such as those shown in Fig. 9b, are used to remove fumes over a larger working area. Large fume cabinets are available for manual operations, such as employing solder pots, which provide efficient fume extraction while still giving proper working space. If portability is desired and only a small area requires fume extraction, a benchtop exhauster, such as that shown in Fig. 9c, may be considered. However, they are not as efficient as central filtration systems. For automated production equipment, such as infrared reflow ovens and wave soldering machines, both fume extraction and dust removal are major concerns. Therefore, high flow extraction systems and high suction dust removal systems having large capacity filtration are employed. Figure 9d illustrates production equipment fume extraction.

Many of the advantages of flux fume extraction by local exhaust systems where there is central filtration of the fumes are quite clear. The removal of pollutants results in a cleaner and healthier environment for the solderer, who would be most closely affected. Capturing fumes at the source and using a central ventilation unit prevents pollutants from being introduced into other breathing areas in the workplace. Therefore secondary flux fumes that could affect nearby workers are eliminated. Efficiency, increased productivity, and higher quality workmanship result when the annoyance of fume irritants is removed, especially when close detailed work is involved. There are also advantages related to operational costs. Due to the design of extraction system collectors, tubing and ductwork, and central filtration units and vacuum pumps, they can be dismantled without difficulty and reconfigured and reinstalled in other locations. Finally, not only are housekeeping costs reduced with this type of filtration, but by not having to vent the filtered air outside results in energy saving as well as costly environmental permits and pollution control devices.

(Source: Adapted from *Fume Extraction Handbook P/N 5050-0371,* Rev. A. Courtesy of PACE, Inc., Laurel, MD)

| Software | |
|---|---|
| | IsoPro–Isolation software, product code OA-ISO-PRO, is used to prepare board files for milling on the Quick Circuit. |
| | $D_2G$–Is a stand-alone DXF to Gerber conversion program, product code OA–$D_2G$. |

| Quick Circuit Files | |
|---|---|
| QCAMW.ZIP | QuickCAM 3.2 for Windows. Latest version of Windows Quick Circuit driver. Supports Windows 3.1x, Windows 95/98, and Windows NT. |
| QCAM.ZIP | QCAM 1.08 for DOS. This is the final version of the DOS Quick Circuit driver for the newer serial controllers. |
| CAM4_02.ZIP | CAM 4.02 for DOS. This is the final version of the DOS driver for the older 6000 model controllers which have the ISA bus indexer card. |

| IsoPro Files | |
|---|---|
| IsoPro.zip | IsoPro Version 1.3 (Build 70). 3,668,264 bytes. Requires Windows 95/98/NT/2000. You must have an Isolation Key dated 5/14/97 or later. To install, unzip to any sub-directory and run SETUP.EXE. |
| IsoPro_73.exe Change History Readme.txt | IsoPro **beta** Version 1.3 (Build 73). 3,726,238 bytes. Requires Windows 95/98/NT/2000. You must have an Isolation Key dated 5/14/97 or later. To install, unzip to any sub-directory and run SETUP.EXE. |
| Demo.zip | IsoPro Demo Version 1.21 (Build 63). 2,109,329 bytes. Requires Windows 95/98/NT/2000. No Key needed. To install, unzip to any sub-directory and run SETUP.EXE. The IsoPro Demo does not isolate or export. |
| D2G.zip | D2G Version 1.20 (Build 62). 2, 458,870 bytes. Requires Windows 95/98/NT/2000. You must have a D2G License Code. To install, unzip to any sub-directory and run SETUP.EXE. D2G is a stand-alone DXF to Gerber converter. |
| Net-Key.zip | IsoPro Network Key Server. Requires a special Network version of the Isolation key. To install, unzip to any sub-directory and follow the instructions in the README.TXT file. |
| find_w32.zip | Finder is a utility that will search for any of Rainbow's standalone Sentinel Keys on your PC. |

| Isolator Files | |
|---|---|
| ISOW3_3.ZIP | Isolator for Windows version 3.3. Support Windows 3.11. |
| ISOW323.ZIP | Isolator for Windows version 3.2.3. Support Windows 3.11. |
| WIN32S.ZIP | Win32s version 1.3. This is required to run Isolator under Windows 3.11. |
| APERTURE.ZIP | Aperture Converter Disk. |
| OLDISOLA.PDF | Use this PDF if you use CAM 4.x with Iso 3.x |
| DISPENSE.PDF | Isolator PDF for use with dispensing. |
| ISOD321.ZIP | Isolator for DOS 3.2.1. Supports MS-DOS. |
| ISO2_1.ZIP | Isolator 2.1 for DOS. Supports MS-DOS. |

| Quick Plate Files | |
|---|---|
| QPWIZ.ZIP | Plating Wizard. Supports Windows 9x, 2000, and NT. |

Courtesy of T-Tech, Inc. Atlanta, GA www.t-tech.com

## APPENDIX XXII

Quick Circuit Mechanical and Electrical Specifications, Computer Requirements and Additional Capabilities

| | *Quick Circuit Specifications* | | | |
| --- | --- | --- | --- | --- |
| | *QC Model 5000* | | *QC Model 7000* | |
| *Model* | *Standard* | *HS* | *Standard* | *HS* |
| Spindle Speed (software controlled) | 8,000 to 23,000 RPM | 5,000 to 60,000 RPM | 8,000 to 23,000 RPM | 5,000 to 60,000 RPM |
| Machine Weight | 19 kg (42 lbs.) | 29 kg (64 lbs.) | 41 kg (90 lbs.) | 51 kg (112 lbs.) |
| Smallest Drill Hole Size | 0.330 mm (0.013") | 0.200 mm (0.008") | 0.330 mm (0.013") | 0.200 mm (0.008") |
| Milling Speed | 1.524 m (60") /min | 2 m (80") / min | 1.524 m (60") /min | 2 m (80") / min |
| Board Work Area | 254 × 279 mm (10" × 11") | | 330 × 480 mm (13" × 19") | |
| Machine Dimensions | 390 W × 430 L × 255 H mm (15.3" × 17" × 9.75") | | 480 W × 710 L × 260 H mm (19" × 28" × 10") | |
| Resolution | 0.006 mm (0.00025") | | | |
| Repeatability | 0.006 mm (0.00025") | | | |
| Minimum Trace Width | 0.100 mm (0.004") | | | |
| Minimum Engraving Width | 0.100 mm (0.004") | | | |
| Contour Route Speed | 380 mm (15") per minute, Adjustable | | | |
| Tool Size | 3.175 mm (0.125") shank diameter, 38.1 mm (1.5") length | | | |
| Tool Change | Semi-automatic | | | |
| Power Supply | 100-240 Volt, 50-60 Hz, Selectable | | | |
| Computer Requirements | IBM Compatible 386 with Math Coprocessor or Higher | | | |
| Supported File Formats | Gerber, Excellon, DXF, and HPGL files | | | |
| Other Features | Includes Isolator CAD/CAM software capable of importing and editing circuit board files | | | |
| Additional Capabilities | Clients with Multiple Quick Circuit Systems Include: | | | |

- RF/Microwave Prototypes
- Surface Mount Prototypes
- Depaneling Production Circuit Boards
- Gerber Viewing and Editing

- AT&T Bell Labs
- Delco
- George Washington University
- Hewlett Packard
- Honeywell
- IBM

- Lockheed
- Motorola
- MIT
- NASA
- U.S. Army, Navy, and Air Force

Courtesy of T-Tech, Inc. Atlanta, GA www.t-tech.com

ProtoDrill Mechanical and Electrical Specifications and Computer Requirements

| *ProtoDrill Specifications* | | |
|---|---|---|
| Work Area | 610mm × 610mm | 24″ × 24″ |
| Maximum Panel Size | 610mm × 610mm | 24″ × 24″ |
| Maximum Work Height | 50.8mm | 2.0″ |
| Machine Weight | 205 kg | 450 lb. |
| Machine Dimensions | 120cm × 147cm × 92cm | 47″ × 58″ × 36″ |
| Positional Accuracy | ± 0.025mm (25 micron) | ± 0.0010″ |
| Drilling Accuracy | ± 0.05mm (50 micron) | ± 0.0020″ |
| Positional Repeatability | ± 0.025mm (25 micron) | ± 0.0010″ |
| Positional Resolution | 0.025mm (25 micron) | 0.0010″ |
| Smallest Drill Hole Size | 0.02mm | 0.008″ |
| XY Traverse Speed | 7.6 m/min | 300 IPM |
| Z Feed Speed | 7.6 m/min | 300 IPM |
| Spindle Speed | 5,000–50,000 RPM | |
| Spindle Torque | 25 N-cm | 35 oz-in |
| Hit Rate | 100 holes/min (based on 0.0625″ material thickness and 0.032″ diameter holes on 0.1″ centers in FR4) | |
| Tool Change | 18 position Automatic Tool Change (ATC) | |
| Tool Size (diameter;length) | standard 1/8″ shank; 1 1/2″ long | |
| Power Supply | 100-240 VAC, 15 Amps, 50–60 Hz | |
| Supported File Formats | Gerber, Excellon, DXF, and HPGL files | |
| Requirements | 100-240 VAC, 15 Amps, 50–60 Hz; Compressed air at 75 psi and 5 CFM, filtered and dried. Calibration of machine at customer site required. | |
| Min. Computer Requirements | Pentium class computer with 16MB RAM running Windows 95/98/NT. | |

Courtesy of T-Tech, Inc. Atlanta, GA www.t-tech.com

Carbide Drill Bits, Milling Tools and Contour Routers Used with the ProtoDrill and Quick Circuit Systems

## Carbide Drill Bits

T-Tech's Carbide Drill Bits (Fig. 1) provide accurate and consistent circuit board drilling. T-Tech offers a wide range of sizes to match varying hole diameter needs. Made from sub-micrograin carbide, these top quality drill bits are extremely durable. Four facet point geometry, matched with high micro flute and O.D. finishing, allows for unsurpassable drilling.

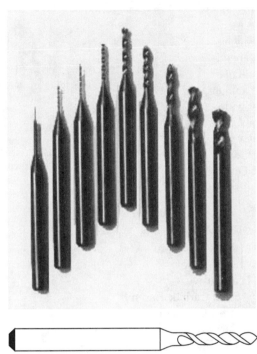

(a)

(b)

**Figure 2**

Courtesy of T-Tech, Inc. Atlanta, GA www.t-tech.com

Carbide Drill Bits, Milling Tools and Contour Routers Used with the ProtoDrill and Quick Circuit Systems *(Continued)*

## Carbide Endmills

T-Tech's Endmills (Fig. 2) are designed for effective maximum copper removal. They feature a premium surface finish for exceptional performance. They are made from sub-micrograin carbide and designed for extended tool life. Endmills are recommended for use in all RF and microwave applications because they offer a dependable milling width, leave a perpendicular edge wall and remove a minimum amount of dielectric. Endmills are used for wide path milling and rubout applications. Stub length endmills (Fig. 3) offer greater rigidity and are recommended for the majority of endmill operations on the Quick Circuit. Three and four flute endmills are available for some sizes.

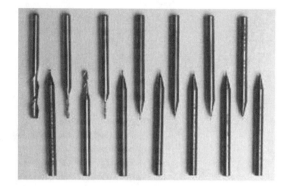

**Figure 2**

**Features:**
- Solid Carbide Endmill
- Sub-Micrograin Carbide
- Two Flute
- 1-1/2″ Overall Length
- 1/8″ Shank
- Right Hand Spiral
- Right Hand Cut

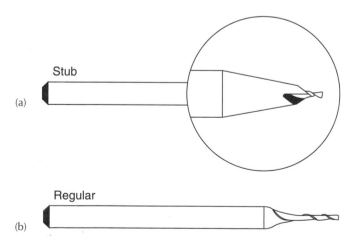

Stub

(a)

**Figure 3**          Regular

(b)

### Carbide Milling Tools

T-Tech offers three different Carbide Milling Tool styles. These milling tools are V-shaped in design (Fig. 4); thus, the deeper the depth of cut, the wider the mill path. This design allows one tool to mill a multitude of widths. All of T-Tech's Milling Tools are made from sub-micrograin carbide for extended tool life.

**Figure 4**

### T1 Milling Tool

The missile-shaped T1 milling tools are the highest precision of T-Tech's milling tools. They are recommended for milling 0.008″ to 0.012″ mill paths. The shank diameter missiles from 1/8″ near the base to 1/16″ near the 60 degree angle point. T-Tech recommends the T-1 milling tools for the majority of Quick Circuit applications.

### T2 Milling Tool

The T2 milling tools are T-Tech's original milling tools with a 90 degree angle point and an approximate 1/16″ diameter near the tip. The two-fluted style of these tools works best for milling paths between 0.010″ and 0.012″. These tools are useful for milling wider paths.

### T3 Milling Tool

T-Tech's T3 Milling Tools are 1/8″ shank to the tip. The 60 degree point works well for mill paths 0.008″ to 0.012″. These tools offer the greatest rigidity. The T3 can also be used for scoring operations.

### T4 Milling Tool

The missile shaped, T4 Ultra Fine Mill is one of T-Tech's new ultra fine line Milling Tool. The T4 is recommended for milling 0.004″ to 0.006″ mill paths, although some experienced users have successfully done mill paths finer than 0.004″. The shank diameter missiles from 1/8″ near the base to 1/32″ near the 90 degree angle point. The T4 allows the user to make the finest mill paths although only recommended for 1/2 ounce copper (0.0007″ thick/.0175mm copper).

### T8 Milling Tool

The missile shaped, T8 Ultra Fine Mill is T-Tech's newest high precision Milling Tool. The T8 is recommended for milling 0.006″ to 0.010″ mill paths although mill paths as small as 0.004″ can be made. The shank diameter missiles from 1/8″ near the base to 1/32″ near the 90 degree angle ping. The T8 offers the industry a flexible and long lasting fine pitch cutting tool. The T8 is recommended primarily for 1/2 ounce copper (0.0007″ thick/.0175mm copper) although 1 ounce copper (0.0014″ thick/.035mm copper) can be used. T-Tech recommends this new tool for the majority of Quick Circuit applications.

**All Milling Tools Feature:**

- Solid Carbide Milling Tool
- Sub-Micrograin Carbide
- 1.425″ Overall Length
- 1/8″ Shank

## Carbide Contour Routers

T-Tech's Carbide Routers (Fig. 5) create burr-free paths through a number of different materials. Offered in two different styles, all T-Tech Routers exhibit increased tool life due to a quality flute design and the use of sub-micrograin carbide. T-Tech provides drill point Routers as a standard, but fishtail Routers are available upon request.

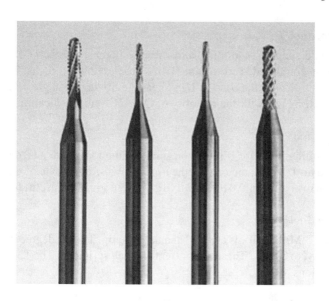

**Figure 5**

### Fiberglass Routers (R1 Style)

T-Tech's Fiberglass Routers are designed especially for glass epoxy and other materials that are phenolic and abrasive. They are engineered for high feed rate applications.

### Chipbreakers Routers (R2 Style)

T-Tech's Chipbreaker Routers are designed to provide a clean cut and a superior finish. T-Tech recommends these Routers for most applications.

**Features:**

- Solid Carbide Router
- Sub-Micrograin Carbide 1/8″ Shank
- Cut Type: Coarse/Medium
- 130 Deg Drill Point, Fishtail

Correct Spindle Speed and Feed Rate for Carbide Tooling Described in Appendix XXIV

| | Feed Rate (inches per minute) | | Spindle Speed (RPM) | |
|---|---|---|---|---|
| END MILLS | 0.5 oz Copper | 1.0 oz Copper | High speed spindle | 24000 Rpm Spindle |
| EM-0050 (0.005″ / 0.13mm)* | 2 | 1 | 60000 | 24000 |
| EM-0060 (0.006″ / 0.15mm)* | 2 | 1 | 60000 | 24000 |
| EM-0070 (0.007″ / 0.18mm)* | 3 | 2 | 60000 | 24000 |
| EM-0080(0.008″ / 0.20mm) | 5 | 3 | 60000 | 24000 |
| EM-0090 (0.009″ / 0.23mm) | 10 | 5 | 60000 | 24000 |
| EM-0100 (0.010″ / 0.25mm) | 15 | 10 | 60000 | 24000 |
| EM-0110 (0.011″ / 0.28mm) | 20 | 15 | 60000 | 24000 |
| EM-0150 (0.015″ / 0.38mm) | 30 | 20 | 54000 | 19000 |
| EM-0200 (0.020″ / 0.50mm) | 45 | 35 | 38000 | 15000 |
| EM-0310 (0.31″ / 0.80mm) | 60 | 60 | 25000 | 9900 |
| EM-0500 (0.050″ / 1.3mm) | 60 | 60 | 15000 | 6100 |
| EM-0625(0.0625″ / 1.6mm) | 60 | 60 | 12000 | 5000 |
| EM-1250(0.125″ / 3.175mm) | 60 | 60 | 9000 | 3500 |

*May have better results by scoring first at 0.003″ to 0.004″ with MILL-T8

| POINTED MILLING TOOLS | | | | |
|---|---|---|---|---|
| MILL-T1 | | | | |
| 0.008″ (0.20mm) to 0.010″ (0.25mm) | 45 | 35 | 60000 | 24000 |
| 0.011″ (0.28mm) to 0.013″ (0.33mm) | 55 | 45 | 60000 | 24000 |
| MILL-T2 | | | | |
| 0.008″ (0.20mm) to 0.010″ (0.25mm) | 45 | 35 | 60000 | 24000 |
| 0.011″ (0.28mm) to 0.013″ (0.33mm) | 55 | 45 | 60000 | 24000 |
| MILL-T3 | | | | |
| 0.008″ (0.20mm) to 0.010″ (0.25mm) | 45 | 35 | 60000 | 24000 |
| 0.011″ (0.28mm) to 0.013″ (0.33mm) | 55 | 45 | 60000 | 24000 |
| MILL-T4 | | | | |
| 0.004″ (0.10mm) to 0.006″ (0.15mm) | 5 | 2 | 60000 | 24000 |
| MILL-T8 | | | | |
| 0.004″ (0.10mm) to 0.006″ (0.15mm) | 45 | 30 | 60000 | 24000 |
| 0.007″ (0.18mm) to 0.011″ (0.28mm) | 60 | 35 | 60000 | 24000 |

| ROUTERS | | | | |
|---|---|---|---|---|
| CR-0310 | 5 | 5 | 48000 | 24000 |
| CR-0620 | 15 | 15 | 36000 | 24000 |

| DRILLS | Entry Material Recommended |
|---|---|
| 0.013 (0.3mm) to 0.031″ (0.79mm) | yes |
| 0.032″ (0.81mm) to 0.125″ (3.175mm) | no |

Note: The "traverse rate" can be set at 60 inches per minute for all routines.

Courtesy of T-Tech, Inc. Atlanta, GA www.t-tech.com

Alchemitron
*www.alchemitron.com*

Allied Electronics
*www.allied.avnet.com*

Analog Devices
*www.analog.com*

AutoDesk
*www.autodesk.com*

Automatic Production Systems
(APS)
*www.apsgold.com*

Belden Wire and Cable Company
*www.belden.com*

Burr Brown
*www.burr-brown.com*

Contact East
*www.contacteast.com*

Erem
*www.coopertools.com*

Etchomatic
*www.etchomatic.com*

Fenwal Electronics
*www.fenwal.com*

Future Active
*www.future-active.com*

Hewlett Packard
*www.hp.com*

Hexacon
*www.hexacon.com*

Kester
*www.kester.com*

Labtech
*www.labtech.com*

Linear Technologies
*www.linear.com*

Luxo
*www.hubmaterial.com*

M. E. Baker
*www.mebaker.com*

Motorola
*www.motorola.com*

National Semiconductor
*www.nationalsemiconductor.com*

Newark Electronics
*www.newark.com*

Nicholson
*www.coopertools.com*

NuArc
*www.nuarc.com*

O.K. International
*www.okinternational.com*

Pace
*www.paceusa.com*

Sensym
*www.sensym.com*

T-Tech
*www.t-tech.com*

Techni-Tool
*www.techni-tool.com*

Texas Instrument
*www.ti.com*

Thomas Regional Directory
Company
*www.thomasregional.com*

Ulano
*www.ulano.com*

Weller
*www.coopertools.com*

Wire Wrap
*www.coopertools.com*

Xcelite
*www.coopertools.com*

# Bibliography

Allied Chemical Corporation Technical Data. Allied Chemical, Morristown, NJ.

*AutoCAD-Release 12 Handbook.* Autodesk, Inc., Sausalito, CA.

**Baer, C. J., and J. R. Ottaway.** *Electrical and Electronics Drafting,* 4th ed. McGraw-Hill Book Company, New York, 1980.

*Chemical Hazard Communication,* OSHA #3084. U.S. Department of Labor, Occupational Safety and Health Administration.

*Chemical Labels: Your Signposts to Safety,* #34G, 8/91 Edition. Business and Legal Reports Inc. Madison, CT.

*Contact East 1997 General Catalog, 1997 Fall Edition.* Contact East, Inc. North Andover, MA.

**Coombs, C. F.** *Printed Circuit Handbook,* 2d ed. McGraw-Hill Book Company, New York, 1979.

**Coughlin, R. F., and F. F. Driscoll.** *Operational Amplifiers and Linear Integrated Circuits,* 5th ed. Prentice Hall, Upper Saddle River, NJ, 1998.

Department of Defense DOD STD 2000-3 *Criteria for High Quality High Reliability Soldering Technology.*

**Driscoll, F. F.** *Introduction to 6800/68000 Microprocessors.* Breton Publishers, North Scituate, MA, 1987.

**Driscoll, F. F.** *Microprocessor–Microcomputer Technology.* Breton Publishers, North Scituate, MA, 1983.

**Driscoll, F. F., R. F. Coughlin, and R. S. Villanucci.** *Data Acquisition and Process Control with the M68HC11 Microcontroller,* 2d ed. Prentice Hall, Upper Saddle River, NJ, 2000.

Dynachem Technical Data. Dynachem Corporation, Santa Fe Springs, CA.

*EAGLE 2.6 Manual.* CadSoft Computer, Inc., Delray Beach, FL, 1992.

*EIA Standards.* Electronics Industries Association, Washington, DC.

    EIA-PDP-100 *Mechanical Outlines for Registered Standard Electronic Parts.*

    EIA-JEP-95 *JEDEC Registered and Standard Outlines for Semiconductor Devices.*

    EIA-46 *Test Procedure for Resistance to Soldering for Surface Mount Devices.*

    EIA-49 *Solderability Test Procedures.*

**Fuller, J. E.** *Using AutoCAD® Release 12.* Delmar Publishers, Inc., Albany, NY, 1993.

*Hazard Communication Guidelines for Compliance,* OSHA #3111. U.S. Department of Labor, Occupational Safety and Health Administration.

*IPC Standards.* The Institute for Interconnecting and Packaging Electronic Circuits, Lincolnwood, IL.

    IPC-A-600 *Acceptability of Printed Wiring Boards.*

    IPC-CM-770 *Guidelines for Printed Circuit Component Mounting.*

    IPC-D-310 *Artwork Generation and Measurement Techniques.*

IPC-M-675 *Microsectioning Handbook.*

IPC-R-700 *Modification and Repair for Printed Boards and Assemblies.*

IPC-S-804 *Solderability Test Methods for Printed Wiring Boards.*

IPC-S-805 *Solderability Test for Component Leads and Terminations.*

IPC-S-815 *General Requirements for Soldering Electronic Interconnections.*

IPC-SF-818 *General Requirements for Electronic Soldering Fluxes.*

IPC-SM-780 *Electronic Component Packaging and Interconnection With Emphasis on Surface Mounting.*

IPC-SM-782 *Surface Mount Land Patterns.*

IPC-SM-817 *General Requirements for Surface Mount Adhesives.*

IPC-SM-819 *General Requirements and Test Methods for Electronic-Grade Solder Paste.*

*Kahn Companies Technical Data.* Kahn Companies, Wethersfield, CT.

*Lab Safety Supply, Inc. 1997 General Catalog.* Lab Safety Supply, Inc. Janesville, WI.

*LabTech Acquire Handbook.* Laboratory Technologies Corporation, Wilmington, MA, 1989.

*Lea Ronal Technical Data.* Lea Ronal, Freeport, NY.

**Lindsey, D.** *The Design and Drafting of Printed Circuits,* 2d ed. Bishop Graphics, Inc., Westlake Village, CA, 1982.

*Making Magnification Work for You,* 1997. Luxo Corporation, Port Chester, NY.

**Manko, H. H.** *Soldering Handbook for Printed Circuits and Surface Mounting.* Van Nostrand Reinhold, New York, 1986.

**Military Standard MIL-STD-275.** *Printed Wiring for Electronic Equipment.*

*MSDS Pocket Dictionary,* Edited by Joseph O. Accrocco. Genium Publishing Corporation, Schenectady, NY, 1991.

*MSDS - Your Guide to Chemical Safety,* #200 039 00 9/91 Edition. Business and Legal Reports Inc., Madison, CT.

*nuArc Technical Data.* nuArc Company, Inc., Chicago.

*PACE® Fume Extraction Handbook,* P/N 5050-0371, Rev. A, 1994. PACE, Inc., Laurel, MD.

**Prasad, R. P.** *Surface Mount Technology Principles and Practice.* Van Nostrand Reinhold, New York, 1989.

PSpice Circuit Analysis, Cadence Corporation.

*Safety & Compliance Directory 1997 Edition.* Lab Safety Supply, Inc., Janesville, WI.

*SCR Manual,* 4th ed. General Electric Company, Auburn, NY.

*Shipley Technical Data.* Shipley Company, Inc., Newton, MA.

*6B Series User's Manual.* Analog Devices Corporation, Norwood, MA, 1989.

*Systems for Development, Production, and Repair of Electronic Assemblies, Part #5400-0077 Rev. 2/97.* PACE, Inc., Laurel, MD.

**Villanucci, R. S.** Adding one breakpoint linearizes thermistor, *EDN,* 7 January 1993.

**Villanucci, R. S.** Using computer-assisted techniques to design and analyze nonlinear analog circuits, *ASEE Journal of Engineering and Technology,* Spring 1993.

# Index